HANDBOOK OF MATHEMATICAL FLUID DYNAMICS

VOLUME III

HANDBOOK OF MATHEMATICAL FLUID DYNAMICS

Volume III

Edited by

S. FRIEDLANDER
University of Illinois-Chicago, Chicago, Illinois, USA

D. SERRE
Ecole Normale Supérieure de Lyon, Lyon, France

2004

ELSEVIER
NORTH
HOLLAND

Amsterdam • Boston • Heidelberg • London • New York • Oxford • Paris •
San Diego • San Francisco • Singapore • Sydney • Tokyo

ELSEVIER B.V.
Sara Burgerhartstraat 25
P.O. Box 211, 1000 AE Amsterdam
The Netherlands

ELSEVIER Inc.
525 B Street, Suite 1900
San Diego, CA 92101-4495
USA

ELSEVIER Ltd
The Boulevard, Langford Lane
Kidlington, Oxford OX5 1GB
UK

ELSEVIER Ltd
84 Theobalds Road
London WC1X 8RR
UK

First edition 2004

Library of Congress Cataloging in Publication Data
A catalog record is available from the Library of Congress.

British Library Cataloguing in Publication Data
A catalogue record is available from the British Library.

ISBN: 0 444 51556 9

♾ The paper used in this publication meets the requirements of ANSI/NISO Z39.48-1992 (Permanence of Paper).
Printed in The Netherlands

Contents of the Handbooks

Volume I

Volume II

Preface

This volume comprises the third volume in the current 3 volume series of articles covering many aspects of mathematical fluid dynamics. In the three volumes we have endeavored to illustrate the breadth and depth of a subject that has its roots stretching back several centuries. Yet it remains a vital source of open questions that are challenging from both mathematical and physical points of view. We have concentrated on mathematical issue arising from fluid models, partly to keep the length under control but mainly because physical and numerical aspects of fluid dynamics are well developed in other comprehensive collections.

Volume 1 is more or less specialized to compressible fluids and begins with kinetic models that are the very source of almost all other fluid models under various limiting regimes. Volume 2 contains a wide range of material with the majority of the articles addressing issues related to incompressible fluids and specific physical problems. The tables of contents of Volumes 1 and 2 can be found on p. v of Volume 3.

Volume 3 begins with the fundamental and challenging issues of how and why a discrete system of a huge number of particles can be replaced by a fluid continuum. The Euler equations for an inviscid fluid are in some sense the most basic analytical description of this continuum. Even in two dimensions there are important current developments for the Euler equations; one such concerns the shape and motion of vortex patches which are discussed in the second article. The remaining articles in Volume 3 concentrate on the Navier–Stokes equations for viscous fluids. An area of mathematics that is becoming increasingly important in fluid dynamics is harmonic analysis. Techniques and results obtained for the Navier–Stokes equations using harmonic analysis are described in the third article. The fourth and sixth articles discuss mathematical issues that are directly motivated by physical and geophysical phenomena, namely boundary layers and fluid dynamics relevant to the oceans and atmosphere. The monograph length fifth article returns to the topic of compressible fluids and gives a comprehensive treatment of the stability of large amplitude shock waves.

We repeat our heartfelt thanks to all the authors who have worked so hard to write the excellent articles in the three volumes of the Handbook of Mathematical Fluid Dynamics. We also thank the many referees who have generously contributed much time to help ensure the high quality of the articles. Once again we thank the editors and staff of Elsevier for the quality of the production of these substantial volumes. We hope that the fascinating variety of mathematics that arises from the study of fluid motion will continue to inspire, motivate and challenge scientists in the future as it has in the past.

Chicago and Lyon, January, 2004
Susan Friedlander and Denis Serre
susan@math.uic.edu and denis.serre@umpa.ens-lyon.fr

List of Contributors

Cannone, M., *Université de Marne-la-Vallée, Marne-la-Vallée, France* (Ch. 3)
Chemin, J.-Y., *Centre de Mathématiques de l'École Polytechnique, Palaiseau, France* (Ch. 2)
Esposito, R., *Centro Linceo Interdisciplinare 'Beniamino Segre', Roma and Università di L'Aquila, L'Aquila, Italy* (Ch. 1)
Grenier, E., *U.M.P.A., E.N.S. Lyon, Lyon, France* (Ch. 4)
Jenssen, H.K., *North Carolina State University, Raleigh, NC 27695, USA* (Ch. 5, App. C)
Lyng, G., *University of Michigan, Ann Arbor, MI 48109-1109, USA* (Ch. 5, App. C)
Pulvirenti, M., *Università di Roma 'La Sapienza', Roma, Italy* (Ch. 1)
Temam, R., *Indiana University, Bloomington, IN, USA, and Université Paris-Sud, Orsay, France* (Ch. 6)
Ziane, M., *University of Southern California, Los Angeles, CA, USA* (Ch. 6)
Zumbrun, K., *Indiana University, Bloomington, IN, USA* (Ch. 5)

Contents

CHAPTER 1

From Particles to Fluids

Raffaele Esposito

Centro Linceo Interdisciplinare 'Beniamino Segre' Via della Lungara 10, 00165 Roma, and Dipartimento di Matematica pura ed applicata, Università di L'Aquila, Coppito, 67100 L'Aquila, Italy
E-mail: esposito@univaq.it

Mario Pulvirenti

Dipartimento di Matematica Università di Roma 'La Sapienza', P. le A. Moro, 5, 00185 Roma, Italy
E-mail: pulvirenti@mat.uniroma1.it

Contents

HANDBOOK OF MATHEMATICAL FLUID DYNAMICS, VOLUME III
Edited by S.J. Friedlander and D. Serre

1. Introduction

The famous German mathematician David Hilbert, in his well-known speech at the International Congress of Mathematicians in Paris in 1900 [56], addressed the audience towards 23 problems which should be the basis of the mathematical research in the forthcoming century. Among those, the sixth problem is rather a future field of investigation and concerns the role of Mathematics in Physics. It reads:

> *MATHEMATICAL TREATMENT OF THE AXIOMS OF PHYSICS*
>
> The investigations on the foundation of geometry suggest ... to treat in the same manner, by means of axioms, those physical sciences in which mathematics play an important part; in the first rank are the theory of probabilities and mechanics.

At a first sight it seems that the problem is to have a complete formalization of that part of Mathematics necessary for the treatment of Physics and nowadays we know how this could be meaningless. Of course Hilbert means something different as he explains later:

> As to the axioms of the Theory of probabilities, it seems to me desirable that their logical investigation should be accompanied by a rigorous and satisfactory development of the method of mean values in Mathematical Physics, and in particular in the kinetic theory of gases. Important investigations by physicists on the foundations of mechanics are at hand ... Thus Boltzmann's work on the principles of mechanics suggests the problem of developing mathematically the limiting processes, there merely indicated, which lead from the atomistic view to the laws of the motion of continua.

It is now clear that Hilbert's hope is that future investigations, logically well founded and mathematically rigorous, will clarify the transition from the fundamental model of the particle dynamics described by the Newton laws (atomistic view) to the macroscopic pictures which are usually used for fluids or rarefied gases.

More generally, in the sixth problem the role of the Mathematics in investigating how different mathematical models of the real world are connected is precisely outlined:

> ... the mathematician has the duty to test exactly in each instance whether the new axioms are compatible with the previous ones. The physicist, as his theories develop, often finds himself forced by the results of the experiments to make new hypotheses, while he depends, with respect to the compatibility of the new hypotheses with the old axioms, solely upon these experiments or upon certain physical intuition, a practice which in the rigorously logical building up of a theory is not admissible.

Therefore to clarify the scopes and methodologies of a physicist and a mathematician in some sense Hilbert establishes the role of the modern Mathematical Physics in which the concept of the mathematical model, that is a noncontradictory system of axioms, is fundamental.

As we mentioned before, the problems explicitly outlined by Hilbert are about the rigorous connection between the microscopic Physics and the Kinetic Theory of dilute gases and, more generally the fluid description of the macroscopic matter.

In the present chapter we want to review the most important steps forward developed in the last century as regards these problems which, we anticipate, are far to be completely solved.

One must say that the Relativistic and Quantum revolutions in the physical sciences certainly presented new realities and models to investigate from the point of view of Mathematical Physics in the first three decades of the century, so that it is not surprising that a systematic investigation of those problems started after the Second War only. We first mention the work of Morrey in the fifties (see [72]). He tried to show what are the steps necessary to prove that the Euler equations for a compressible fluid can be derived by the Newton laws under a suitable scaling. This work, even though not rigorous (actually the problem is still unsolved) has the merit to show, in a logically clear way, what is the link between the microscopic and macroscopic description of the fluids. This kind of arguments are presented in Section 2.

More or less in the same period Grad (see [52]) also postulated in what kind of limiting situation the description given by the Boltzmann equation is expected to hold. He gave also a nice way to write down the BBGKY hierarchy of equations for hard spheres, which turned out to be an important technical tool. Further investigations in this direction are due to Cercignani (see [25]). Finally in 1975, following also the work of Kac (see [60]) who clarified the probabilistic meaning of the Boltzmann equation, Lanford showed rigorously how to derive the Boltzmann equation from the fundamental laws of Mechanics (see [65]). In particular, it was proved that there is no contradiction between the reversible character of the Newton laws and the irreversibility delivered by the Boltzmann dynamics. The limit of Lanford's result relies on the fact that he was able to prove it for a sufficiently short time only. Generalizations to arbitrary times are now available but for very special situations. We present Lanford's argument in Section 3.

It is probably true that the short time limitation is due to the fact that, although we have a result due to DiPerna and Lions (see [37]) establishing the global existence of solutions to the Cauchy problem for the Boltzmann equation, we have not yet a sufficiently good control of the Boltzmann dynamics to approach the global validity for arbitrary times. We address the reader to Villani's contribution (see [87]) as regards the existence problem for the Boltzmann equation.

Another important point in connecting the various regimes of a rarefied gas is the investigation of its fluid behavior. In this field we have many recent progresses. The problem of deriving the Euler equations (for a perfect gas) starting from the Boltzmann equation, although technically difficult, is certainly much easier than the derivation of the Euler equation (for real gases) starting from the particle dynamics. This investigation starts with Hilbert himself who developed an important technical tool, the so-called Hilbert expansion, to approach the problem (see [57]). After the work of Grad (see [53,54]), Nishida and Caflisch (see [74] and [18], respectively) derived the Euler equation for a short time (before the occurrence of shocks). We review the relevant arguments in Section 4. The most relevant open problem in this framework is, in our opinion, to understand how to describe the shocks from a kinetic view point.

Also the incompressible Euler and Navier–Stokes equations are derivable from the Boltzmann dynamics, by means of a parabolic space–time scaling, in the low Mach numbers regime, as we show in Section 5. The problems related to the derivation of the incompressible Navier–Stokes equations from particle dynamics are outlined at the end of Section 5. We address the recent review paper [11], where many arguments presented here are discussed in a similar spirit.

Our contribution ends with a didactic part presented in Section 6. The whole program of deriving kinetic and hydrodynamic equations from the particle dynamics, can be carried out for linear systems. The basic dynamics is that of a single particle moving in a (possibly random) distribution of scatterers. For this system one can prove (under suitable scaling limits) a rigorous validation of a linear Boltzmann equation and, from this, of a diffusion equation. It is also possible to show the diffusive behavior of the particle as proved by Bunimovich and Sinai (see [17]) for a periodic distribution of obstacles. This result is somehow fundamental: It shows that the heat equation (or the Brownian motion) is rigorously derivable from a mechanical system. Unfortunately this result is technically hard because it makes use of the ergodic properties of Sinai's billiard, so that we limit ourselves to describe it qualitatively.

2. Euler equations from particle systems

The fundamental model we will consider is a system of N identical point particles of mass $m > 0$ moving in $\mathbb{R}^d$, $d \geqslant 1$, roughly modeling the molecules of a d-dimensional fluid. Although they are in their nature quantum objects, to our purposes their quantum properties can be disregarded and therefore we assume they obey the rules of the Classical Mechanics. In particular, this means that the configuration of the system is specified by the coordinates q_i and velocities v_i, $i = 1, \ldots, N$, with respect to a fixed inertial system. The evolution of the system is given by the Newton equations:

$$\begin{aligned} &\frac{\mathrm{d}q_i}{\mathrm{d}\tau}(\tau) = v_i(\tau), \\ &\frac{\mathrm{d}v_i}{\mathrm{d}\tau}(t) = -\sum_{i \neq j} \nabla\phi(q_i - q_j), \end{aligned} \tag{2.1}$$

where τ is the time, $\phi(r)$, $r \in \mathbb{R}^d$, is the potential describing the interaction between the particles, which we assume bounded, smooth, depending only on $|r|$ and, for simplicity, of compact support. Many of these assumptions can be suitably relaxed, but we do not discuss here the possible generalizations. Finally, $\nabla\phi$ denotes the gradient of ϕ with respect to r. In the above equations we have assumed unit mass for each particle, because this is a parameter which will be kept fixed in all the discussion. The variables q and τ are measured in microscopic units. In view of the fact that we are going to look at the system in the macroscopic space–time scale, it is convenient to introduce a parameter $\varepsilon > 0$ representing the ratio between a microscopic space scale (say the range of the interaction) and a macroscopic one (say the diameter of the region where the fluid is confined). Then we introduce the "macroscopic coordinates" $x_i = \varepsilon q_i$, $i = 1, \ldots, N$, and the "macroscopic time" $t = \varepsilon\tau$. In terms of the macroscopic variables the Newton equations become

$$\begin{aligned} &\frac{\mathrm{d}x_i}{\mathrm{d}t}(t) = v_i(t), \\ &\frac{\mathrm{d}v_i}{\mathrm{d}t}(t) = -\varepsilon^{-1} \sum_{i \neq j} \nabla\phi\left(\frac{x_i - x_j}{\varepsilon}\right). \end{aligned} \tag{2.2}$$

Note that the two systems are strictly equivalent, since they differ just for a change of variable. The real assumption leading to macroscopic equations will be that the system is initially in a state where the space variations of appropriate observables are on the macroscopic scale. This will be made precise below. The variables x_i are assumed in a finite cubic region $\Lambda \subset \mathbb{R}^d$ with periodic boundary conditions, making Λ a d-dimensional torus. Therefore the phase space of the system (2.2) is

$$\Gamma_N = \left(\Lambda \times \mathbb{R}^d\right)^N.$$

The role of the number N is crucial. In a typical fluid it is of the order of the Avogadro number ($\approx 10^{23}$) and hence in our discussion we will be interested in the asymptotic behavior when $N \to \infty$. The particle density of the system is given by

$$n = \frac{N}{|\Lambda|_{\text{micr}}} = \frac{\varepsilon^d N}{|\Lambda|}.$$

Here $|\Lambda|_{\text{micr}}$ is the volume of the microscopic region $\varepsilon^{-1}\Lambda$ (which shrinks to Λ after rescaling) given by $\varepsilon^{-d}|\Lambda|$. In order to keep the density finite, we assume

$$N \approx \varepsilon^{-d}.$$

The system under investigation is Hamiltonian; thus, due to the invariance under space and time translations, the total momentum and energy are conserved. Together with the total mass, which is obviously conserved in our setup, this gives us a specific set of conserved quantities which play a special role in the hydrodynamical considerations. In view of this, for any smooth function f on Λ we consider the quantities:

$$\begin{aligned}
I_f^0(t) &= \frac{1}{N}\sum_{i=1}^N f\big(x_i(t)\big),\\
I_f^\alpha(t) &= \frac{1}{N}\sum_{i=1}^N v_i^\alpha(t) f\big(x_i(t)\big), \qquad \alpha = 1, \dots, d,\\
I_f^{d+1}(t) &= \frac{1}{N}\sum_{i=1}^N \left[\frac{1}{2}v_i(t)^2 + \frac{1}{2}\sum_{j\neq i}\phi\big(\varepsilon^{-1}\big|x_i(t) - x_j(t)\big|\big)\right] f\big(x_i(t)\big),
\end{aligned} \tag{2.3}$$

where v_i^α denotes the αth component of v_i.

If f is the characteristic function of some space region $A \subset \Lambda$, then the $I_f^\alpha(t)$ represent the mass, momentum and energy of the particles in the region A at time t (the interaction between A and its complement is included in I_f^{d+1}). It is more convenient to have f smooth, but the interpretation of the I_f^α's is essentially the same.

If one introduces the distributions

$$\xi^0(x,t) = \frac{1}{N}\sum_{i=1}^{N}\delta\big(x - x_i(t)\big),$$
$$\xi^\alpha(x,t) = \frac{1}{N}\sum_{i=1}^{N}\delta\big(x - x_i(t)\big)v_i^\alpha(t), \quad \alpha = 1,\dots,d, \tag{2.4}$$
$$\xi^{d+1}(x,t) = \frac{1}{N}\sum_{i=1}^{N}\delta\big(x - x_i(t)\big)\left[\frac{1}{2}(v_i t)^2 + \frac{1}{2}\sum_{j\neq i}\phi\big(\varepsilon^{-1}\big|x_i(t) - x_j(t)\big|\big)\right],$$

to be interpreted as empirical mass momentum and energy densities, then the I_f^α's can be written as

$$I_f^\alpha(t) = \int_\Lambda \mathrm{d}x\, \xi^\alpha(x,t) f(x), \quad \alpha = 0,\dots,d+1.$$

Since we are interested in the time dependence of the I_f^α's, we differentiate them with respect to t and use (2.2) to compute the right-hand side. It is straightforward to check that

$$\begin{aligned}\frac{\mathrm{d}}{\mathrm{d}t}I_f^\alpha(t) &= \int_\Lambda \mathrm{d}x \sum_{\beta=1}^{d}\zeta_{\alpha,\beta}(x,t)\,\partial_\beta f(x) + \mathrm{O}_\alpha(\varepsilon)\\ &= \int_\Lambda \mathrm{d}x\, \zeta_\alpha(x,t)\cdot\nabla f(x) + \mathrm{O}_\alpha(\varepsilon)\end{aligned} \tag{2.5}$$

for some fields ζ_α defined below, where ∂_β denotes ∂_{x_β} and $\mathrm{O}_\alpha(\varepsilon)$, depending on f, go to 0 as ε goes to 0, provided that $N\varepsilon^d$ stays bounded uniformly in ε. Here we are implicitly assuming that the dynamic does not create large fluctuations in the density. This is one of the problems one has to face when dealing with the existence of the dynamics for systems of infinitely many particles and has been solved with suitable assumptions on the potential. We do not discuss this here and refer the interested readers to the specialized papers [22,38,65], among the others.

The empirical fields $\zeta_{\alpha,\beta}$, $\alpha = 0,\dots,d+1$, $\beta = 1,\dots,3$, called "empirical currents" of mass, momentum and energy, are:

$$\zeta_{0,\beta}(x,t) = \frac{1}{N}\sum_{i=1}^{N}\delta\big(x_i(t) - x\big)v_i^\beta(t),$$
$$\zeta_{\alpha,\beta}(x,t) = \frac{1}{N}\sum_{i=1}^{N}\delta\big(x_i(t) - x\big) \times\left[v_i^\alpha(t)v_i^\beta(t) + \frac{1}{2}\sum_{j=1}^{N}\Psi_{\alpha,\beta}\big(\varepsilon^{-1}\big(x_i(t) - x_j(t)\big)\big)\right],$$

$$\zeta_{d+1,\beta}(x,t) = \frac{1}{N}\sum_{i=1}^{N}\delta\big(x_i(t)-x\big) \tag{2.6}$$

$$\times\left[v_i^{\beta}(t)\left[\frac{1}{2}\big(v_i(t)\big)^2+\frac{1}{2}\sum_{j\neq i}\phi\big(\varepsilon^{-1}\big|x_i(t)-x_j(t)\big|\big)\right]\right.$$

$$\left.+\frac{1}{2}\sum_{j=1}^{N}\sum_{\gamma=1}^{d}\Psi_{\gamma,\beta}\big(\varepsilon^{-1}\big(x_i(t)-x_j(t)\big)\big)\frac{1}{2}\big[v_i^{\gamma}(t)+v_j^{\gamma}(t)\big]\right],$$

with

$$\Psi_{\alpha,\beta}(z) = -z_\beta(\partial_\alpha\phi)(z). \tag{2.7}$$

The first equation of (2.6) is immediate and actually $O_0(\varepsilon)$ vanishes. We now show how to obtain the second one, dropping the time dependence for sake of shortness. For $\alpha = 1,\dots,d$, we have:

$$\begin{aligned}\frac{\mathrm{d}}{\mathrm{d}t}I_f^{\alpha} &= \frac{1}{N}\sum_{i=1}^{N}\sum_{\beta=1}^{d}(\partial_\beta f)(x_i)v_i^{\alpha}v_i^{\beta}\\ &\quad-\frac{1}{N}\sum_{i=1}^{N}f(x_i)\varepsilon^{-1}\sum_{j\neq i}(\partial_\alpha\phi)\big(\varepsilon^{-1}(x_i-x_j)\big).\end{aligned} \tag{2.8}$$

Because of the symmetry of the potential, the second term can be written as

$$\begin{aligned}&-\frac{1}{N}\sum_{i=1}^{N}f(x_i)\varepsilon^{-1}\sum_{j\neq i}(\partial_\alpha\phi)\big(\varepsilon^{-1}(x_i-x_j)\big)\\ &\quad=-\frac{1}{N}\sum_{i,j=1}^{N}\frac{1}{2}\big(f(x_i)-f(x_j)\big)\varepsilon^{-1}(\partial_\alpha\phi)\big(\varepsilon^{-1}(x_i-x_j)\big).\end{aligned}$$

Since the potential has finite range, say R, only the terms such that $|x_i-x_j|<\varepsilon R$ survive in the last expression. This means that the particles interacting with the ith particle are in a microscopically finite region and hence they are a finite number because of our previous assumption ensuring small fluctuations of the density. Moreover, we have

$$\varepsilon^{-1}\big(f(x_i)-f(x_j)\big) = \varepsilon^{-1}\sum_{\beta=1}^{d}(\partial_\beta f)(x_i)(x_i-x_j)+O\big(\varepsilon R^2\big),$$

so that we obtain the second equation (2.6). The third one is obtained in the same way.

We want to stress that the relations (2.5) are nothing but identities obtained with the only assumptions that N is not larger than $O(\varepsilon^{-d})$, that the fluctuations of the density are controlled and that ϕ is a short range potential. The last assumption may indeed be relaxed to some sufficiently fast decay condition, but we are not looking here to the maximal generality.

Although (2.5) and (2.6) appear as the weak forms of a system of conservation laws, at this point the currents involved can only be computed after solving the Newton equations. The main obstacle to obtaining hydrodynamical equations (which are formally quite similar to (2.5)) is the difficulty of computing the currents in terms of the densities. This is achieved by the assumption of "local equilibrium" which we are going to discuss below. We anticipate that, unfortunately, the state of the art does not provide sufficiently good techniques to prove such an assumption for deterministic models.

In order to introduce the *local* equilibrium assumption, let us first remind briefly a few facts about systems in *global* equilibrium, which we discuss rather informally. We refer to [81] for more accurate statements and proofs.

We begin by considering an isolated system in a cubic box Q. We consider below large values of N (but not larger than ε^{-d}, for small ε). Let $\Sigma_{N,e,u,Q}$ be the hypersurface in $\Gamma_N = (Q \times \mathbb{R}^d)^N$ of all the phase points with given energy

$$\frac{1}{2}\sum_{i=1}^{N} v_i^2 + \sum_{i<j=1}^{N} \phi\big(|q_i - q_j|\big) = eN \tag{2.9}$$

and with fixed velocity of the center of mass

$$\frac{1}{N}\sum_{i=1}^{N} v_i = u. \tag{2.10}$$

Of course, by the Galilean invariance, for thermodynamic purposes only we could assume u to vanish with a proper choice of the inertial frame. However, in view of the hydrodynamical context we leave the u dependence explicit.

The hypersurface $\Sigma_{N,e,u,Q}$, plays a major role because, as a consequence of the conservation laws, it is invariant with respect to the evolution induced by the Newton equations (2.1).

The relevant states of our system are the probability distributions on the phase space Γ_N concentrated on $\Sigma_{N,e,u,Q}$ and absolutely continuous with respect to the surface measure induced on $\Sigma_{N,e,u,Q}$ by the Liouville measure.

According to Boltzmann, the equilibrium state is given by the uniform distribution on $\Sigma_{N,e,u,Q}$ (*microcanonical ensemble*). Under suitable assumptions on the potential ϕ, it makes sense to look at the asymptotes as N and $|Q|$ go to ∞ provided that

$$\frac{N}{|Q|} = n, \tag{2.11}$$

the particle density, is kept fixed (*thermodynamic limit*). The limiting state, which exists in some suitable sense, depends on the parameters n, u and e and describes the behavior of isolated thermodynamic system with particle density, n, energy per particle, e, and center of mass velocity, u.

If instead of an isolated system one considers a system in thermal contact with a heat reservoir at temperature $T > 0$, then the equilibrium probability distribution of the system is no more concentrated on a particular $\Sigma_{N,e,u,Q}$, but rather, each energy shell has its statistical weight (*canonical ensemble*). In fact in this case the equilibrium distribution is given by the Boltzmann–Gibbs distribution, whose density $P_{N,Q,T,u}$ with respect to the Liouville measure

$$\frac{dq_1\,dv_1\cdots dq_N\,dv_N}{N!}$$

is given by

$$P^{\mathrm{c}}_{N,Q,T,u}(q_1,v_1,\ldots,q_N,v_N) = \frac{1}{Z^{\mathrm{c}}(N,Q,T,u)}\exp\left[-\frac{1}{2T}\sum_{i=1}^{N}(v_i-u)^2+\frac{1}{T}\sum_{i<j}\phi\big(|q_i-q_j|\big)\right], \tag{2.12}$$

with $Z^{\mathrm{c}}(N,Q,T,u)$ the appropriate normalization factor

$$Z^{\mathrm{c}}(N,Q,T,u) = \int_{(Q\times\mathbb{R}^d)^N}\frac{dq_1\,dv_1\cdots dq_N\,dv_N}{N!} \times\exp\left[-\frac{1}{2T}\sum_{i=1}^{N}(v_i-u)^2+\frac{1}{T}\sum_{i<j}\phi\big(|q_i-q_j|\big)\right]. \tag{2.13}$$

The limit as N and $|Q|$ go to ∞ with the condition (2.11) fulfilled does exist, depends on the parameters n, u and T and describes a thermodynamic system in thermal contact with a heat reservoir at temperature T. Moreover, with the appropriate relation between T, e and n, the two states are equivalent (*equivalence of the ensembles*). Note that u does not play any role for the thermodynamic systems, as expected, because of the Galilean invariance.

Finally, if the number of particles N in the region Q is not fixed, but is itself distributed according some statistical weight, then the equilibrium state of the thermodynamic system is given by the *grand-canonical ensemble*. It is defined as a probability measure $P^{\mathrm{gc}}_{Q,z,T,u}$ on

$$\bigcup_{N\geqslant 0}\Gamma_N,$$

such that $P^{\mathrm{gc}}_{Q,z,T,u}(\Gamma_0)=1$ while the probability density of N particles is

$$P^{\mathrm{gc}}_{Q,z,T,u}(N;q_1,v_1,\ldots,q_N,v_N) = \frac{z^N}{Z^{\mathrm{gc}}(Q,z,T,u)}\exp\left[-\frac{1}{2T}\sum_{i=1}^{N}(v_i-u)^2+\frac{1}{T}\sum_{i<j}\phi\big(|q_i-q_j|\big)\right], \tag{2.14}$$

with

$$Z^{\mathrm{gc}}(Q,z,T,u)$$
$$=\sum_{N=0}^{\infty} z^N \int_{(Q\times\mathbb{R}^d)^N} \frac{dq_1\, dv_1 \cdots dq_N\, dv_N}{N!}$$
$$\times \exp\left[-\frac{1}{2T}\sum_{i=1}^{N}(v_i-u)^2+\frac{1}{T}\sum_{i<j}\phi(|q_i-q_j|)\right]. \tag{2.15}$$

Here $z>0$ is the *activity*. In the thermodynamic limit $|Q|\to\infty$, the thermodynamic system in the grand-canonical ensemble is parametrized by z, T and u.

The particle density n is, in this case, the average number of particles divided by the volume. Its value in the limit $|Q|\to\infty$ depends on z and T. Moreover, with a suitable relation between z, n and T, the grand-canonical state is equivalent to the canonical state at the same T and particle density n. Note that the correspondence between z and n may not be one-to-one in some region of the space of the parameters because of the phenomenon of phase transitions. We disregard this complication and confine ourselves to an open region in the space of the parameters where the map is one-to-one. Therefore, fixed the temperature T and the density n, the activity

$$z=z(n,T)$$

is uniquely defined. Therefore we can think the grand-canonical state parametrized by n, T, u as well.

We conclude the excursus on the equilibrium Statistical Mechanics by introducing a thermodynamic quantity which plays an important role also in hydrodynamics, the "pressure", defined as

$$\widetilde{P}(n,T)=\lim_{|Q|\to\infty}|Q|^{-1}\log Z^{\mathrm{gc}}\big(Q,z(n,T),T,u\big) \tag{2.16}$$

for a reasonable choice of the sequence of regions invading the space. The dependence of $\widetilde{P}$ on n and T is called in Thermodynamics the *equation of state* of the gas.

The hydrodynamical behavior of a particle system corresponds in general to nontranslational invariant states and hence the above considerations cannot be immediately applied. Therefore, let us partition our macroscopic volume Λ into macroscopic regions $\Delta\subset\Lambda$, whose diameter δ is sufficiently small compared with the diameter of Λ and let x_Δ be the centers. The size of Δ is, in microscopic units, of order $\delta\varepsilon^{-1}$, hence large enough, for ε small, to consider the state of the particles in Δ close to an equilibrium state (provided one neglects the interaction of the particles outside Δ with those in Δ). Moreover, since δ is small, the macroscopic parameters do not change appreciably in Δ. Therefore the system in Δ is essentially a thermodynamic system described by the grand-canonical ensemble with parameters z_Δ, T_Δ and u_Δ. If we start with a state of the whole system which is essentially in a thermodynamic equilibrium in each region Δ, the system will stay in such a state, as far as the interactions between different macroscopic regions can be disregarded. In fact on times much shorter than $\varepsilon^{-1}t$ the big majority of the particles in Δ cannot leave Δ and do not interact with those outside, so the previous considerations are sufficiently accurate. However, in the long run the effects of the neglected interactions may become relevant: In the hydrodynamical context we are interested in times of order $\varepsilon^{-1}t$ and on this time scale many particles in Δ can reach neighboring regions and interact with particles outside Δ, so that the initial state is no more stationary. But the equilibrium states are distinguished from all the other states because of their strong stability properties, so it is reasonable to conjecture that, rather than destroying the structure of the equilibrium in each region Δ, the perturbations from the outside just modify the values of the equilibrium parameters.

To make the above considerations more precise, let us define *local equilibrium state with parameters* $z(\cdot,t)$, $T(\cdot,t)$, $u(\cdot,t)$ and the probability distribution $P_{z(\cdot,t),T(\cdot,t),u(\cdot,t)}$ which restriction to Γ_N for any $N \geqslant 1$, is given by

$$\begin{aligned} &P^N_{z(\cdot,t),T(\cdot,t),u(\cdot,t)}(x_1,v_1,\dots,x_N,v_N) \\ &\quad = Z^{-1}\prod_{i=1}^N \exp\biggl[\mu(x_i,t) - \frac{1}{2T(x_i,t)}\biggl[\bigl(v_i - u(x_i,t)\bigr)^2 \\ &\qquad + \sum_{j\neq i}\phi\bigl(\varepsilon^{-1}|x_i - x_j|\bigr)\biggr]\biggr], \end{aligned} \tag{2.17}$$

where $\mu(x,t) = \log z(x,t)$ is the chemical potential and Z is the normalization factor.

We explicitly underline that the local equilibrium distribution, but for the case of constant z, T and u, is not an exact solution of the *Liouville equation*

$$\partial_t f_N + \sum_{i=1}^N v_i \cdot \partial_{x_i} f_N - \varepsilon^{-1}\sum_{i=1}^N\sum_{j\neq i}(\nabla\phi)\bigl(\varepsilon^{-1}(x_i - x_j)\bigr)\cdot\partial_{v_i} f_N = 0, \tag{2.18}$$

which rules the evolution of an initial measure according to the Newton law (2.2). On the other hand, by the previous arguments, it is reasonable to expect that, when ε is very small, if the initial distribution is a local equilibrium, then the distribution at later times t is *close* (up to corrections vanishing with ε) to a local equilibrium with suitably chosen functions $z(\cdot,t)$, $T(\cdot,t)$, $u(\cdot,t)$.

More precisely, let us denote by $\langle\cdot\rangle_t$ and $\langle\cdot\rangle_t^{\mathrm{loc}}$ the averages with respect to the solution of the Liouville equations and the local equilibrium. We will only consider a restricted class of observables on the phase space including at least the functions ξ^α and $\zeta_{\alpha,\beta}$ defined in (2.4) and (2.6) and possibly not much larger.

In the sequel it is crucial the following local equilibrium assumption:

ASSUMPTION. There is a choice of the functions $z(\cdot,t)$, $T(\cdot,t)$, $u(\cdot,t)$ such that for any observable Φ in the above class,

$$\lim_{\varepsilon\to 0}\bigl|\langle\Phi\rangle_t - \langle\Phi\rangle_t^{\mathrm{loc}}\bigr| = 0. \tag{2.19}$$

Let us return now to the identities (2.5). As $\varepsilon \to 0$ all the quantities involved depend on a larger and larger number of variables. Therefore we can expect that a law of large numbers holds for such quantities, so that, as $\varepsilon \to 0$, they converge in probability to their averages with respect to the probability distribution solution of the Liouville equation. The local equilibrium assumption implies that, in computing such averages we can replace the true nonequilibrium distribution with the local equilibrium distribution. Hence all the quantities involved in (2.5) can be computed in terms of the functions $z(\cdot, t)$, $T(\cdot, t)$, $u(\cdot, t)$ thus transforming the identities (2.5) into equations for the hydrodynamical fields z, T and u.

The computation is straightforward and the results are:

$$
\begin{aligned}
I_f^0(t) &\to \int \rho(x,t) f(x)\,\mathrm{d}x, \\
I_f^\alpha(t) &\to \int \rho(x,t) u_\alpha(x,t) f(x)\,\mathrm{d}x, \quad \alpha = 1,\dots,d, \\
I_f^{d+1}(t) &\to \int \rho(x,t)\left[\frac{1}{2}\big(u(x,t)\big)^2 + e(x,t)\right] f(x)\,\mathrm{d}x,
\end{aligned}
\tag{2.20}
$$

where $\rho(x,t)$ is the density corresponding to the activity $z(x,t)$ (remember that we assumed to be in the one phase region); e is the *specific internal energy* of a thermodynamic system at temperature T and density ρ and can be computed explicitly (at least in principle):

$$
\begin{aligned}
&\int \rho(x,t) e(x,t) f(x)\,\mathrm{d}x \\
&\quad = \lim_{\varepsilon \to 0} \left\langle \frac{1}{N} \sum_{i=1}^{N} \left[\frac{1}{2} \tilde{v}_i^2 + \frac{1}{2} \sum_{j \neq i} \phi\big(\varepsilon^{-1}|x_i - x_j|\big) \right] \right\rangle_t^{\mathrm{loc}},
\end{aligned}
\tag{2.21}
$$

where we adopt the short notation

$$
\tilde{v}_i = v_i - u(x_i, t)
$$

and use the fact that the local equilibrium average of any odd polynomial in $\tilde{v}_i$ vanishes.

By above remark, we also get

$$
\begin{aligned}
&\int \zeta_{\alpha,\beta}(x,t) f(x)\,\mathrm{d}x \\
&\quad \to \int \mathrm{d}x\, f(x)\big[\rho(x,t) u_\alpha(x,t) u_\beta(x,t) + \Pi_{\alpha,\beta}(x,t)\big], \\
&\int \zeta_{d+1,\beta}(x,t) f(x)\,\mathrm{d}x \\
&\quad \to \int \mathrm{d}x\, f(x)\left\{\rho(x,t) u_\beta(x,t)\left[\frac{1}{2}\big(u(x,t)\big)^2 + e(x,t)\right] + h_\beta(x,t)\right\}.
\end{aligned}
\tag{2.22}
$$

Here $\Pi_{\alpha,\beta}$ and h_β are the components of the *stress tensor* and *energy flux*, defined by the positions

$$\int \mathrm{d}x\, f(x)\Pi_{\alpha,\beta}(x,t)$$
$$= \lim_{\varepsilon\to 0}\left\langle \frac{1}{N}\sum_{i=1}^{N} f(x_i)\left[\tilde{v}_i^\alpha \tilde{v}_i^\beta + \sum_{j\neq i}\Psi_{\alpha,\beta}\big(\varepsilon^{-1}(x_i - x_j)\big)\right]\right\rangle_t^{\mathrm{loc}}, \tag{2.23}$$

$$\int \mathrm{d}x\, f(x)h_\beta(x,t)$$
$$= \lim_{\varepsilon\to 0}\left\langle \frac{1}{N}\sum_{i=1}^{N} f(x_i)\left\{\sum_{\alpha=1}^{d} u^\alpha(x_i,t)\tilde{v}_i^\alpha \tilde{v}_i^\beta\right.\right.$$
$$\left.\left. + \sum_{j\neq i}\sum_{\gamma=1}^{d}\frac{1}{2}\Psi_{\gamma,\beta}\big(\varepsilon^{-1}(x_i - x_j)\big)\big[v_i^\gamma + v_j^\gamma\big]\right\}\right\rangle_t^{\mathrm{loc}}. \tag{2.24}$$

Using again that $\langle \tilde{v}_i\rangle_t^{\mathrm{loc}} = 0$, and the smoothness of u it is immediate to check that

$$h_\beta = \sum_{\alpha=1}^{d} \Pi_{\alpha,\beta} u_\alpha. \tag{2.25}$$

Finally, we note that, by the isotropy of the potential, for $\alpha \neq \beta$ it results:

$$\big\langle \Psi_{\alpha,\beta}\big(\varepsilon^{-1}(x_i - x_j)\big)\big\rangle_t^{\mathrm{loc}} = 0, \tag{2.26}$$

while

$$\big\langle \Psi_{\alpha,\alpha}\big(\varepsilon^{-1}(x_i - x_j)\big)\big\rangle_t^{\mathrm{loc}}$$

does not depend on α. Therefore we define the "hydrodynamic pressure" by

$$\int \mathrm{d}x\, f(x)P(x,t)$$
$$= -\lim_{\varepsilon\to 0}\left\langle \frac{1}{\mathrm{d}N}\sum_{i=1}^{N} f(x_i)\left[\tilde{v}_i^2 + \sum_{j\neq i}\sum_{\alpha=1}^{d}\Psi_{\alpha,\alpha}\big(\varepsilon^{-1}(x_i - x_j)\big)\right]\right\rangle_t^{\mathrm{loc}}, \tag{2.27}$$

so that the stress tensor reduces to

$$\Pi_{\alpha,\beta} = -P\delta_{\alpha,\beta}. \tag{2.28}$$

It is a classical result of the Equilibrium Statistical Mechanics, the *Virial* theorem (see, for example, [84], p. 14), that the hydrodynamic pressure P coincides with the thermodynamic pressure given by (2.16):

$$P(x,t) = \widetilde{P}\big(\rho(x,t), T(x,t)\big). \tag{2.29}$$

This relation, together with the expression of the internal energy as a function of ρ and T provides the state equations for the fluid. When such relations are substituted in the conservation laws (2.5), we obtain the weak form of the Euler equation for a compressible fluid. If the *hydrodynamical fields* ρ, T and u are sufficiently regular, an integration by parts provides the usual differential form of the Euler equations in form of conservation laws:

$$\begin{aligned}
&\partial_t \rho + \sum_{\alpha=1}^{d} \partial_{x_\alpha}[\rho u_\alpha] = 0,\\
&\partial_t[\rho u_\beta] + \sum_{\alpha=1}^{d} \partial_{x_\alpha}[\rho u_\alpha u_\beta + P\delta_{\alpha,\beta}] = 0,\\
&\partial_t\left[\rho\left(\frac{1}{2}u^2 + e\right)\right] + \sum_{\alpha=1}^{d} \partial_{x_\alpha}\left[u_\alpha\left[\rho\left(\frac{1}{2}u^2 + e\right) + P\right]\right] = 0.
\end{aligned} \tag{2.30}$$

Remarks and bibliographical notes

We have already mentioned that above derivation of the Euler equations, which goes back basically to Morrey [72], is not rigorous in many respects, the main difficulty being the local equilibrium assumption. Its formulation is delicate and the one used here is, to some extent, vague. A precise meaning of the notion of local equilibrium requires indeed a more careful discussion of the interaction between each cell Δ and the surrounding cells, particularly when the parameters are in the phase transition region. We refer to [33] for details.

It is worth to mention the rigorous proof of the hydrodynamical limit has been given in a very simplified model, the hard rods model in one space dimension [14], where the microscopic solution is essentially explicit.

Other interesting results have been obtained for deterministic models including somewhat artificial mechanisms forcing the system to a local equilibrium. This is the case of the result in [75], where a particle system with a suitable long range (Kac) interaction potential is considered. The system is assumed initially in a special kind of local equilibrium, corresponding to a "cold" gas, in the sense that its initial velocity distribution has vanishing variance, and for any $i = 1, \dots, N$, the velocities v_i are given by $u(x_i, 0)$ where $u(\cdot, 0)$ is some assigned smooth initial velocity profile. The property is preserved, in time in the limit $N \to \infty$, and the velocity profile at time t, $u(\cdot, t)$ and the density profile $\rho(\cdot, t)$ solve the continuity equation and the Euler momentum equation with pressure law $P = \rho^2/2$. Another example is given in [73] where the author studies a one-dimensional system of particles interacting with a short range repulsive force and including a velocity dependent friction force among the particles, strong enough to force the velocities of nearby particles to be very close. The resulting equations differ from the Euler equations for a density dependent second-order term.

It should be remarked that such results, although related to the topic under discussion, are of rather different nature in the fact that the limiting equations are not obtained via a pure space–time scaling, the microscopic dynamics itself depending on the scaling parameter. In particular, this explains how it is possible, in the result in [73], to obtain a limiting equation for the momentum which is not invariant under a pure space–time scaling, as we will discuss in Section 5.

For more realistic models the state of the art is very poor, the only result being due to Olla, Varadhan and Yau [76]. First they give a new statement of the local equilibrium, which is more effective in a rigorous context, in terms of the relative entropy.

If μ and ν are two probability measures the relative entropy $H(\mu|\nu)$ is defined as

$$H(\mu|\nu) = \int \mathrm{d}\nu \frac{\mathrm{d}\mu}{\mathrm{d}\nu} \log \frac{\mathrm{d}\mu}{\mathrm{d}\nu},$$

where $\frac{\mathrm{d}\mu}{\mathrm{d}\nu}$ is the Radon–Nikodym derivative of μ with respect to ν. If μ is not absolutely continuous with respect to ν, we set $H(\mu|\nu)$ to $+\infty$.

Let us fix the number of particles N and let S_N be the relative entropy of the non-equilibrium solution to the Liouville equation with respect to the local equilibrium (2.17), divided by N.

If μ and ν are both probability measures absolutely continuous with respect to a reference measure m_0, with densities f and g, then the relative entropy dominates the $L^1(m_0)$-norm of the difference $f - g$. Therefore a possible formulation of the statement of local equilibrium can be given by requiring that

$$S_N \to 0, \quad N \to \infty. \tag{2.31}$$

It is not difficult to realize that given (2.31) at any time, it is possible to derive the convergence of the particle dynamics to the Euler equations in the hydrodynamical limit.

The question is how to prove (2.31). The basic idea is to get a bound for the relative entropy at time t in terms of the initial relative entropy, so that, assuming (2.31) initially, one obtains the propagation of the local equilibrium in time.

The main difficulty in obtaining such a bound is in the lack of good ergodic properties of the Newton system. Since the microscopic dynamics is intrinsically unstable, sensible dependence on the initial conditions is expected so that the behavior of the individual particles is essentially erratic and in some sense equivalent to the behavior of a stochastic system.

Although the picture is very reasonable, no proof of such an equivalence is available at the present time. The clever idea in [76] is to add an artificial stochasticity to the Newton dynamics, weak enough not to appear in the macroscopic equations, but still sufficient to get the ergodic properties necessary to the proof. Even with such a change of the model the result is strongly non trivial and additional assumptions are required in order to get a validity theorem for the Euler equations.

The first assumption is a modification of the particle kinetic energy, namely the classical kinetic energy is replaced by the relativistic one, so that the energy flow can be controlled with respect to the local equilibrium measure.

The second assumption, typical of Yau's entropy method [89] employed in [76], is that the Euler equations, with the assumed initial conditions, have a unique smooth solution in some time interval. The derivation of the Euler equations is so restricted to time intervals where the solution does not develop shocks or even weaker singularities. No validity result of the Euler equations has been proved, as far as we know, when the limiting equations develop a shock.

A remarkable exception is due to Rezakhanlou [80], which has been able to get the Euler limit even on the shocks, in the case of the Simple Exclusion Process (SEP). SEP is a model representing a caricature of a realistic particle system. Roughly, the process can be described as follows: Particles positions are restricted to the sites of a lattice $\mathbb{Z}^d$. Moreover at most one particle can stay in each site (exclusion rule). The particles move by making jumps at random exponential times (independent in each site), from one site x to a neighboring site y, chosen at random with probability p_{y-x}, provided the site y is empty, otherwise stay at x. Although the model is very simple, its formal hydrodynamical limit in the Euler time scale is the inviscid Burgers equation,

$$\partial_t \rho + \nabla_x \cdot \big[F\rho(1-\rho)\big] = 0, \tag{2.32}$$

with the drift vector F depending on the jump probability, measuring the bias to jump in one direction rather than in the opposite one. The case of vanishing F corresponds to a jump probability symmetric under reflections and is known as the *Symmetric Simple Exclusion Process*; its Euler limit is trivial. As it is well known, for F not null, the Burgers equation may develop shocks and in general one can only state existence and uniqueness of weak entropic solution to (2.32). Informally speaking, the result in [80] states that under the Euler scaling, the empirical density of the simple exclusion process converges weakly in probability to the unique entropic solution of the Burgers equation.

3. The Boltzmann equation for particle systems

3.1. *The heuristic derivation of the Boltzmann equation*

The Boltzmann equation is the main topic of the present section. This is an evolution equation for the probability density of a particle of a rarefied gas and its scope is to describe the thermodynamic behavior of a suitable class of macroscopic systems possibly not in thermal equilibrium. Here we want to illustrate the delicate passage from the Hamiltonian particle dynamics to the Boltzmann equation itself.

In 1872 Ludwig Boltzmann, starting from the mathematical model of elastic balls and using mechanical and statistical considerations, established an evolution equation to describe the behavior of a rarefied gas, in a situation in which even the microscopic structure of the matter was unclear. The reaction of the scientific community to Boltzmann's discovery showed that even great scientists (see the notes at the end of this section) did not understand the deep meaning of Boltzmann's arguments which constitute the basis of the modern kinetic theory.

The importance of this equation is two-fold. From one side it provides (as well as the hydrodynamical equations) a reduced description of the microscopic world. On the other, it is also an important tool for the applications, especially for dilute fluids when the hydrodynamical equations fail to hold.

The starting point of the Boltzmann analysis (but first Maxwell went in the same direction) is to renounce to study the behavior of a gas in terms of the detailed motion of the molecules which constitute it, because of their huge number. It is preferable to investigate a function $f(x, v)$ which is the probability density of a given particle, where x and v denote position and velocity of such a particle. Actually $f(x, v)\,\mathrm{d}x\,\mathrm{d}v$ is often confused with the fraction of molecules falling in the cell of the phase space of size $\mathrm{d}x\,\mathrm{d}v$ around x, v. The two concepts are not exactly the same, they are asymptotically equivalent (when the number of particles is diverging) if a law of large numbers holds. We shall return on this point later on; for the moment, in a purely heuristic context, we ignore it.

Thus we want to find an evolution equation for the quantity $f(x, v)$. The molecular system we are initially considering is the hard sphere model which we are going to illustrate.

We have N identical particles of diameter r in the whole space $\mathbb{R}^3$ and we denote by $Z_N = (X_N, V_N) = (x_1, v_1, \ldots, x_N, v_N)$ a state of the system, where x_i and v_i indicate the position and the velocity of the particle i. The particles cannot overlap (that is the centers of two particles cannot be at distance smaller than the diameter r) so that the phase space of the system is $\Gamma_r = \Lambda_r \times \mathbb{R}^{3N}$, where Λ_r is the configuration space

$$\Lambda_r = \bigl\{X_N \bigm| |x_i - x_j| > r,\ i \neq j\bigr\}. \tag{3.1}$$

The particles are moving freely up to the first contact instant, that is the first time in which two particles arrive at distance r. Then the two particles, say i and j, satisfy the relation $(v_i - v_j) \cdot (x_j - x_i) \geqslant 0$. The pair interacts performing an elastic collision. This means that they change instantaneously their velocities, according to the following rule:

$$\begin{aligned} v_i' &= v_i - n\bigl[n \cdot (v_i - v_j)\bigr], \\ v_j' &= v_j + n\bigl[n \cdot (v_i - v_j)\bigr], \end{aligned} \tag{3.2}$$

where n is the unit vector directed along $x_j - x_i$ and v_i' and v_j' are the outgoing velocities after the collision; see Figure 1.

Equations (3.2) follow easily by the conservation of the energy, linear and angular momentum. From (3.2) we easily get that, if V and V' denote the relative velocities before and after the collision, then

$$V' = V - 2n(n \cdot V). \tag{3.3}$$

After the first collision the system goes on by iterating the procedure.

REMARK. It is not clear that the dynamics for such a system is well defined globally in time. For instance situations leading to triple collisions create problems. A triple collision

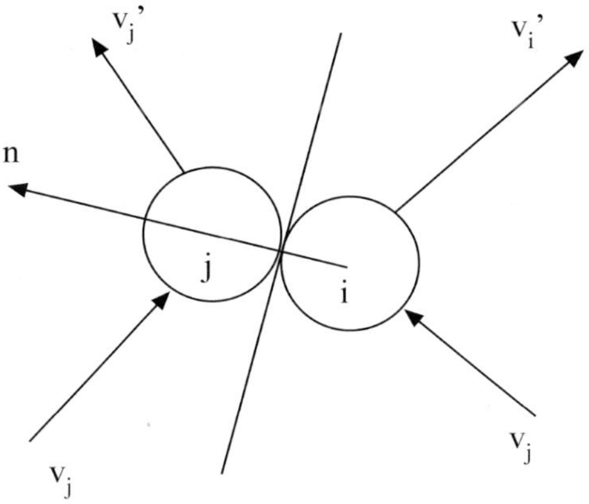

Fig. 1.

is a situation in which we have a cluster of three (or more) particles at contact. In this case we cannot extend the dynamics for further times because the result is ambiguous. Indeed the outgoing velocities of two consecutive collisions depend on the order in which these collisions take place. However, this is not a major problem because the initial phase points yielding triple collisions belong to a manifold of co-dimension one so that its Lebesgue measure is negligible. Therefore the best we can do is to define the dynamics for almost all initial conditions. However another more serious pathology can a priori occur, namely a particle could perform an infinite number of collisions in a finite time after which we do not know how to extend the dynamics anymore. Although it is not known whether this feature can really occur, we can prove that the initial conditions leading to this pathology have null Lebesgue measure. We do not give the proof of this fact because not relevant for the future analysis. We address the reader to [1] and [29] for a discussion on this point.

Once defined the hard-sphere dynamics, we write down the evolution law for f which is

$$(\partial_t + v \cdot \nabla_x) f = \text{Coll}, \tag{3.4}$$

where Coll denotes the variation of f due to the collisions.

We have

$$\text{Coll} = G - L, \tag{3.5}$$

where L and G (loss and gain term respectively) are the negative and positive contribution to the variation of f due to the collisions. It is clear that, in absence of collisions, the time derivative along the characteristics $x + vt$ (i.e., $(\partial_t + v \cdot \nabla_x) f$) vanishes. In presence of collisions the f varies along the trajectories. More precisely, $L \, \mathrm{d}x \, \mathrm{d}v \, \mathrm{d}t$ is the probability of our test particle to disappear from the cell $\mathrm{d}x \, \mathrm{d}v$ of the phase space because of a collision in the time interval $(t, t + \mathrm{d}t)$ and $G \, \mathrm{d}x \, \mathrm{d}v \, \mathrm{d}t$ is the probability to appear in the same time interval for the same reason.

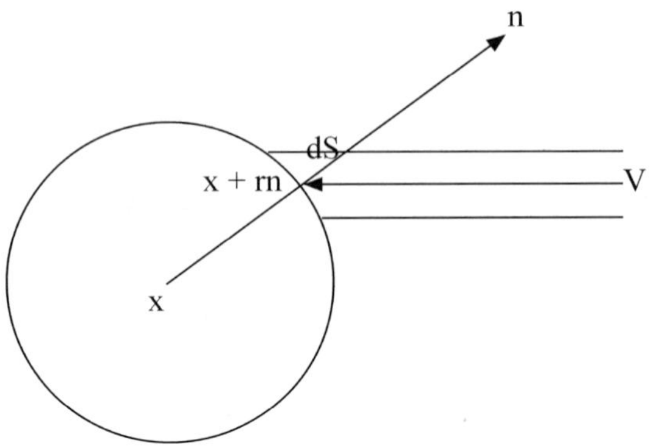

Fig. 2.

To compute these probability we consider binary collisions only.

Let us consider the sphere of center x with radius r (see Figure 2) and a point $x + rn$ over the surface, where n denotes the generic unit vector. Consider also the cylinder with base area $\mathrm{d}S = r^2\,\mathrm{d}n$ and height $|V|\,\mathrm{d}t$ along the direction of the relative velocity V given by $v_2 - v$.

Then a given particle (say particle 2) with velocity v_2, can contribute to L because it can collide with our test particle in the time $\mathrm{d}t$, provided it is localized in the cylinder and $V \cdot n \leqslant 0$. Therefore the contribution to L due to particle 2 is the probability of finding such a particle in the cylinder (conditioned to the presence of the first particle in x). This quantity is $f_2(x, v, x + nr, v_2)|(v_2 - v) \cdot n| r^2\,\mathrm{d}n\,\mathrm{d}v_2\,\mathrm{d}t$, where f_2 is the joint distribution of two particles. Integrating in $\mathrm{d}n$ and $\mathrm{d}v_2$ we obtain that the total contribution to L due to any predetermined particle is

$$r^2 \int \mathrm{d}v_2 \int_{S_-} \mathrm{d}n\, f_2(x, v, x + nr, v_2)\big|(v_2 - v) \cdot n\big|, \tag{3.6}$$

where S_- is the unit hemisphere $\{v_2 \mid (v_2 - v) \cdot n < 0\}$. Finally we obtain the total contribution multiplying by the total number of particles:

$$L = (N - 1)r^2 \int \mathrm{d}v_2 \int_{S_-} \mathrm{d}n\, f_2(x, v, x + nr, v_2)\big|(v_2 - v) \cdot n\big|. \tag{3.7}$$

The gain term can be derived analogously by considering that we are looking at particles which have velocities v and v_2 after the collisions so that we have to integrate over the hemisphere $S_+ = \{v_2 \mid (v_2 - v) \cdot n > 0\}$:

$$G = (N - 1)r^2 \int \mathrm{d}v_2 \int_{S_+} \mathrm{d}n\, f_2(x, v, x + nr, v_2)\big|(v_2 - v) \cdot n\big|. \tag{3.8}$$

Summing G and $-L$ we get

$$\text{Coll} = (N-1)r^2 \int \mathrm{d}v_2 \int \mathrm{d}n\, f_2(x, v, x+nr, v_2)(v_2 - v) \cdot n, \tag{3.9}$$

which, as we shall see in a moment, is not a very useful expression.

Anyway we have not yet derived an equation for f because its time derivative is expressed in term of another object namely f_2. An evolution equation for f_2 involve f_3, the joint distribution of three particles and so on up to arrive to the total particle number N. Here the Boltzmann main assumption enters, namely that two given particles are uncorrelated if the gas is rarefied:

$$f_2(x, v, x_2, v_2) = f(x, v) f(x_2, v_2). \tag{3.10}$$

Condition (3.10), called propagation of chaos (or *Stosszahlansatz* in German) seems contradictory at a first sight: If two particles collide, correlations are created. Even though we assume (3.10) at some time, if the test particle collides with the particle 2 (3.10) cannot be satisfied anymore after the collision.

Before discussing the propagation of chaos hypothesis, we first analyze the size of the collision operator. We remark that, in practical situations for a rarefied gas, the combination $Nr^3 \approx 10^{-4}\ \text{cm}^3$ (that is the volume occupied by the particles) is very small, while Nr^2 is of order 1. This implies that G is also of order 1. Therefore, since we are dealing with a very large number of particles, we are tempted to perform the limit $N \to \infty$ and $r \to 0$ in such a way that $r^2 = \mathrm{O}(N^{-1})$. As a consequence the probability that two tagged particles collide (which is of the order of the surface of a ball, that is $\mathrm{O}(r^2)$) is negligible. However, the probability that a given particle performs a collision with any one of the remaining $N-1$ particles (which is $\mathrm{O}(Nr^2)$ and hence of order 1), is not negligible. Condition (3.10) is referring to two pre-selected particles (say particles 1 and 2) so that it is not unreasonable to conceive that it holds in the limiting situation in which we are working.

However, we cannot insert (3.10) in (3.9) because this latter equation refers both to instants before and after the collision and, if we know that a collision took place, we certainly cannot invoke (3.10). Hence it is more convenient to assume (3.10) in the loss term and work over the gain term to keep advantage of the factorization property which will be assumed only before the collision.

Coming back to (3.8) for the outgoing pair velocities v, v_2 (satisfying the condition $(v_2 - v) \cdot n > 0$) we make use of the continuity property

$$f_2(x, v, x+nr, v_2) = f_2\bigl(x, v', x+nr, v_2'\bigr), \tag{3.11}$$

where the pair v', v_2' is pre-collisional. On f_2 expressed before the collision we can reasonably apply condition (3.10) obtaining

$$\begin{aligned} G - L = {} & (N-1)r^2 \\ & \times \int \mathrm{d}v_2 \int_{S_+} \mathrm{d}n \\ & \times (v - v_2)n\bigl[f\bigl(x, v'\bigr) f\bigl(x - nr, v_2'\bigr) - f(x, v) f(x+nr, v_2)\bigr], \end{aligned} \tag{3.12}$$

after the replacement of n with $-n$ in the gain term. This transform the pair v', v_2' from a pre-collisional to a post-collisional pair.

Finally, in the limit $N \to \infty$, $r \to 0$, $Nr^2 = \lambda^{-1}$, we find

$$(\partial_t + v \cdot \nabla_x) f = Q(f, f), \tag{3.13}$$

where Q, called the collision operator, has the form

$$\lambda^{-1} \int \mathrm{d}v_2 \int_{S_+} \mathrm{d}n \, (v - v_2) n \big[f(x, v') f(x, v_2') - f(x, v) f(x, v_2) \big]. \tag{3.14}$$

The parameter λ, called *mean free path*, represents, roughly speaking, the typical length a particle can cover without undergoing any collision.

Equation (3.13) is the Boltzmann equation for hard spheres. Such an equation has a statistical nature and it is not equivalent to the Hamiltonian dynamics from which it has been derived, as we shall discuss in more detail in the next section.

The heuristic arguments we have developed so far can be extended to different potentials than that of the hard-sphere systems. Consider, for instance, a system of point particles interacting by means of a positive and short range potential. Again we are in a situation in which the number N of particles is very large and the range of the potential is $r = \mathrm{O}(N^{-1/2})$. Then the two-body collision (we are ignoring many-body collision terms because, in the limit we have in mind, they are negligible) is almost explicitly solvable.

The two-body problem can be reduced to a central collision passing to the variables

$$V = v_1 - v_2, \qquad W = \frac{v_1 + v_2}{2}, \tag{3.15}$$

that are the relative velocity and the velocity of the center of mass. We assume to work in an inertial frame for which W is null. The relative velocity satisfies the Newton equation (with the reduced mass $\frac{1}{2}$) associated to the same potential. Referring to Figure 3, if V is the velocity before the collision, the outgoing velocity V' (that is the velocity after the collision, when the particle exits from the sphere of radius r in which the potential differs from zero) is given by

$$V' = V - 2\omega(V \cdot \omega) \tag{3.16}$$

and ω is the angle which bisects the two directions of V and V'. The angle θ in Figure 3 is called the scattering angle. Of course the collision happens in the plane generated by the initial velocity V and the center. Hence V' is determined by ω (or equivalently by the impact parameter b or the scattering angle θ). Passing to the two-body problem we can therefore determine the outgoing relative velocity and, by the momentum conservation we finally arrive to the expression for the outgoing velocities v_1', v_2':

$$\begin{aligned} v_1' &= v_1 - \omega\big[(v_1 - v_2) \cdot \omega\big], \\ v_2' &= v_2 + \omega\big[(v_1 - v_2) \cdot \omega\big]. \end{aligned} \tag{3.17}$$

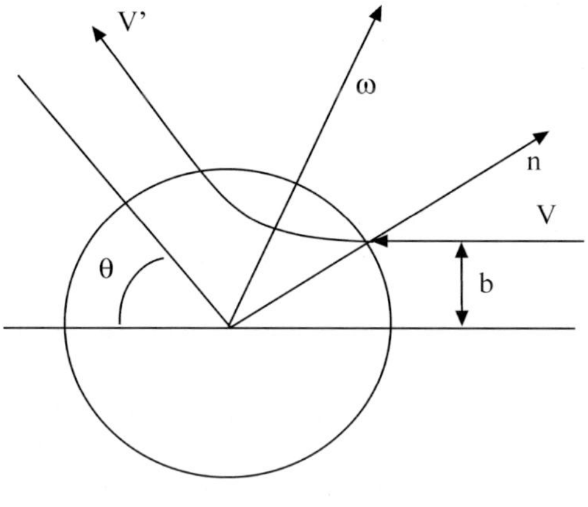

Fig. 3.

Note that the unit vector n, that is parallel to the line joining the two point particles when they are going to interact and ω, that is parallel to the line joining the two point particles when they are at the minimal distance, are not the same but for the hard-sphere interaction. We find more convenient to express the collision in terms of ω and therefore we want to express $V \cdot n$ in terms of ω. We observe now that $\frac{|V \cdot n|}{|V|}\,\mathrm{d}n$ is the fraction of particles scattered in the solid angle corresponding to $\mathrm{d}\omega(n)$, which is, by definition, the cross section $\sigma(|V|,\omega)\,\mathrm{d}\omega$. Setting

$$B\big(|V|,\omega\big) = |V|\sigma\big(|V|,\omega\big), \tag{3.18}$$

we can write the Boltzmann equation associated to our potential as Equation (3.13) with

$$Q(f,f)(x,v) = \lambda^{-1}\int \mathrm{d}v_2 \int_{S_+} \mathrm{d}\omega\, B\big(|V|,\omega\big) \\ \times \big[f\big(x,v'\big)f\big(x,v_2'\big) - f(x,v)f(x,v_2)\big]. \tag{3.19}$$

This concludes our heuristic preliminary analysis of the Boltzmann equation. We certainly know that the above arguments are delicate and require a more rigorous and deeper analysis. If we want that the Boltzmann equation is not a phenomenological model, derived by assumptions *ad hoc* and justified by its practical relevance, but rather a consequence of a mechanical model, we must derive it rigorously. In particular the propagation of chaos should be not a hypothesis but the statement of a theorem.

NOTES. To have a vague idea of the framework in which Boltzmann's work was developed, one has to mention the most significant contributions before him. After the ancient Greeks (400 BC) who postulated the matter to be constituted by atoms, a long time passed before Hermann (1716) suggested the equivalence between the heat and the motion of molecules. The first computations trying to link macroscopic quantities, like the pressure, to microscopic quantities, like the molecular kinetic energy, are due to Euler and Daniel Bernoulli in the first half of the 18th century. The first systematic attempt to explain the thermodynamics in kinetic terms is due to Waterstone (1843) who derived the correct law of pressure in terms of the mean square velocity. Unfortunately such results were not appreciated and were published posthumous.

The next remarkable step forward is due to Maxwell in 1867 [70] who introduced the equilibrium Maxwellian distribution for the thermal equilibrium of point particles. Moreover, he derived evolution equations for the moments of the velocity.

The last step was performed by Boltzmann in 1872. He derived his equation for the time evolution of the probability density of one particle of a rarefied gas. The Boltzmann equation provoked a debate involving Loschmidt, Zermelo and Poincaré who outlined inconsistencies between the irreversibility of the equation and the reversible character of the Hamiltonian dynamics. Boltzmann argued the statistical nature of his equation and his answer to the irreversibility paradox was that “most” of the configurations behave as expected by the thermodynamic laws. However, he did not have the probabilistic tools for formulating in a precise way the statements of which he had a precise intuition.

Finally Grad (1949) stated clearly the limit $N \to \infty$, $r \to 0$, $Nr^2 \to$ const, where N is the number of particles and r is the diameter of the molecules, in which the Boltzmann equation is expected to hold. This limit is usually called the Boltzmann–Grad limit (B–G limit in the sequel).

The problem of a rigorous derivation of the Boltzmann equation was an open and challenging problem for a long time. In 1975 Lanford [65] showed that, although for a very short time, the Boltzmann equation can be derived starting from the mechanical model of the hard-sphere system. The proof has a deep content but is extremely simple from a technical view point as we shall see in the next section.

We address the reader to [28] for an account of the history of the kinetic theory and for the life and personality of Boltzmann as a man and a scientist.

3.2. *Deriving the Boltzmann equation rigorously*

Our starting point is a system of N identical point particles of unitary mass, interacting by means of a two-body potential ϕ. We assume that the interaction potential ϕ is isotropic, smooth and with range one, namely $\phi(r)$ vanishes if $|r| > 1$.

Denoting by $(q_1, v_1, \dots, q_N, v_N)$ a state of the system, where q_i and v_i indicate the position and the velocity of the particle i, the dynamical flow is obtained by solving the Newton equations

$$\frac{\mathrm{d}^2}{\mathrm{d}\tau^2} q_i(\tau) = \sum_{j \neq i} F\big(q_i(\tau) - q_j(\tau)\big), \tag{3.20}$$

where

$$F(q_i - q_j) = F_{i,j} = -\nabla\phi\big(|q_i - q_j|\big) \tag{3.21}$$

is the force acting on the particle i, due to the particle j.

At this stage our starting point is exactly the same as that in Section 2.

The time evolution of a probability distribution f^N, defined on the phase space of the system, is given by the Liouville equation, which reads as

$$(\partial_\tau + \mathcal{L}_N) f^N = 0, \tag{3.22}$$

where the Liouville operator $\mathcal{L}_N$ is

$$\mathcal{L}_N = \mathcal{L}_N^0 + \mathcal{L}_N^{\mathrm{I}}, \tag{3.23}$$

with

$$\mathcal{L}_N^0 = \sum_{i=1}^{N} v_i \cdot \nabla_{q_i} \tag{3.24}$$

and

$$\mathcal{L}_N^{\mathrm{I}} = \sum_{i=1}^{N} \sum_{i \neq j} F_{i,j} \cdot \nabla_{v_i}. \tag{3.25}$$

Through the present analysis we will consider that the probability distribution f^N is initially (and hence at any positive time) symmetric in the exchange of the particles.

We are interested in the behavior of the system in the limit $N \to \infty$. Of course the probability distribution f^N is not expected to converge to something, so that we introduce the marginal distributions for which the convergence problem makes at least sense. We set

$$\begin{aligned} &\tilde{f}_j^N(q_1, v_1, \ldots, q_j, v_j, t) \\ &\quad = \int \mathrm{d}q_{j+1}\,\mathrm{d}v_{j+1} \cdots \mathrm{d}q_N\,\mathrm{d}v_N\, f^N(q_1, v_1, \ldots, q_N, v_N, t), \end{aligned} \tag{3.26}$$

that is the probability density of the first j particles (or, due to the symmetry, of any other group of j fixed particles). We find an evolution equation for the set of distribution functions $\{\tilde{f}_j^N\}_{j=1}^N$ by integrating (3.22) over the variables with indices $j+1, \ldots, N$. We obtain the well-known BBGKY hierarchy of equations

$$\partial_\tau \tilde{f}_j^N + \mathcal{L}_j \tilde{f}_j^N = (N-j) C_{j,j+1} \tilde{f}_{j+1}^N, \quad j = 1, \ldots, N-1, \tag{3.27}$$

where

$$\begin{aligned} &C_{j,j+1} \tilde{f}_{j+1}^N(q_1, v_1, \ldots, q_j, v_j) \\ &\quad = \sum_{k=1}^{j} \int F(q_k - q_{j+1}) \cdot \nabla_{v_i} \tilde{f}_{j+1}^N(q_1, v_1, \ldots, q_{j+1}, v_{j+1})\,\mathrm{d}q_{j+1}\,\mathrm{d}v_{j+1} \end{aligned} \tag{3.28}$$

and

$$\partial_\tau \tilde{f}_j^N + \mathcal{L}_j \tilde{f}_j^N = 0 \quad \text{for } j = N. \tag{3.29}$$

The name of the hierarchy is due to the names (Born, Bogoliubov, Kirkwood and Yvon) of the scientists who introduced it.

Although we shall not make a direct use of the above hierarchy of equations, we introduce it to understand which kind of limit we are going to perform in order to outline the behavior of a rarefied gas from our mathematical model, being mandatory, since we are dealing with large systems, that $N \to \infty$.

According to what we have seen in the previous section we pass to macroscopic variables defining

$$x = \varepsilon q, \qquad t = \varepsilon \tau,$$

where ε is a very small parameter to be sent to zero.

When the macroscopic variables are O(1), the times which we are considering from a microscopic point of view are very large. In this scale of times a tagged particle (say particle 1) must deliver a finite number of collisions. As consequence the density $N\varepsilon^3 \to 0$ must vanish. To be more precise we introduce (normalized) distribution functions of the macroscopic variables

$$g_j^N(x_1, v_1, \ldots, x_j, v_j) = \varepsilon^{-3j} \tilde{f}_j^N\left(\varepsilon^{-1} x_i, v_1, \ldots, \varepsilon^{-1} x_j, v_j\right) \tag{3.30}$$

and, in terms of the new variables, the BBGKY hierarchy becomes:

$$\begin{aligned}\left(\partial_t + \sum_{i=1}^{j} v_i \cdot \nabla_{x_i}\right) g_j^N + \sum_{i=1}^{j} \sum_{k \neq i} \varepsilon^{-1} F\left(\frac{x_i - x_k}{\varepsilon}\right) \cdot \nabla_{v_i} g_j^N \\ = \frac{N-j}{\varepsilon} \int \mathrm{d}x_{j+1} \int \mathrm{d}v_{j+1}\, F\left(\frac{x_i - x_{j+1}}{\varepsilon}\right) \cdot \nabla_{v_{j+1}} g_{j+1}^N.\end{aligned} \tag{3.31}$$

For j fixed, the interaction term in the left-hand side of (3.31) is negligible in the $L^1(x_1, v_1, \ldots, x_j, v_j)$ sense. Moreover, the integral in the right-hand side is $\mathrm{O}(\varepsilon^3)$. Therefore the interaction part of the group of the first j particles with the rest of the system is O(1) whenever N is $\mathrm{O}(\varepsilon^{-2})$. This is the scaling which we are considering, namely

$$N \to \infty, \; \varepsilon \to 0, \quad N\varepsilon^2 = \lambda^{-1} > 0, \tag{3.32}$$

for a system of N particles interacting via a smooth, short range potential of range ε.

Note that this is exactly the scaling we have postulated in the previous section, on the basis of heuristic arguments. Equation (3.32) is usually called a low density limit or, in macroscopic variables, the Boltzmann–Grad limit.

We write down once more the Liouville equation (3.22) in macroscopic variables with the agreement that the range of the potential is ε and that N and ε are related by the scaling law (3.32).

Instead of considering the usual BBGKY hierarchy however, we derive another more convenient equation for a sequence of quantities slightly differing from the marginal distributions, but asymptotically equivalent to them. We define

$$\begin{aligned} f_j^N(x_1, v_1, \dots, x_j, v_j, t) &= \int \mathrm{d}x_{j+1}\,\mathrm{d}v_{j+1}\cdots \mathrm{d}x_N\,\mathrm{d}v_N \\ &\quad \times \prod_{i=1}^{j}\prod_{k=j+1}^{N} \chi\bigl(\{|x_i - x_k| > \varepsilon\}\bigr) f^N(x_1, v_1, \dots, x_N, v_N, t), \end{aligned} \tag{3.33}$$

where $\chi(A)$ is the characteristic function of the set A. It is clear that the functions f_j, for any j, are asymptotically equivalent (in the limit $N \to \infty$) to the functions $\tilde{f}_j$, so that it will be the same to work with this new sequence.

Moreover, if $\prod_{i=1}^{j}\prod_{k=j+1}^{N} \chi(|x_i - x_k| > \varepsilon)$ equals to 1, since the range of the interaction is ε, the interaction between the group of the first j particles and the rest of the system is vanishing so that the Liouville equation becomes

$$\partial_t f^N + \mathcal{L}_j^{\mathrm{I}} f^N + \mathcal{L}_{N-j}^{\mathrm{I}} f^N + \sum_{i=1}^{N} v_i \cdot \nabla_{x_i} f^N = 0, \tag{3.34}$$

where $\mathcal{L}_{N-j}^{\mathrm{I}}$ is the interaction part of the Liouville operator relative to the last $N - j$ particles. Integrating (3.34) with respect to $\mathrm{d}x_{j+1}\,\mathrm{d}v_{j+1}\cdots \mathrm{d}x_N\,\mathrm{d}v_N$ we obtain

$$\begin{aligned} \bigl(\partial_t f_j^N + \mathcal{L}_j^{\mathrm{I}}\bigr) f_j^N(x_1, v_1, \dots, x_j, v_j) &= -\int_{S_{N-j}(x_1,\dots,x_j)} \mathrm{d}x_{j+1}\,\mathrm{d}v_{j+1}\cdots \mathrm{d}x_N\,\mathrm{d}v_N \\ &\quad \times \sum_{i=1}^{N} v_i \cdot \nabla_{x_i} f^N(x_1, v_1, \dots, x_j, v_j, x_{j+1}, v_{j+1}, \dots, x_N, v_N). \end{aligned} \tag{3.35}$$

Here $S_{N-j}(x_1, \dots, x_j) \subset \mathbb{R}^{3(N-j)}$ denotes the set of all configurations of the last $N - j$ particles not interacting with the first group, that is

$$S_{N-j}(x_1, \dots, x_j) = \bigcap_{i=1}^{j}\bigcap_{k=j+1}^{N} \bigl\{(x_{j+1}, \dots, x_N) \bigm| |x_k - x_i| > \varepsilon\bigr\}. \tag{3.36}$$

We also used that

$$\int \mathrm{d}v_{j+1}\cdots \mathrm{d}v_N\, \mathcal{L}_{N-j}^{\mathrm{I}} f^N = 0. \tag{3.37}$$

Note that the outward normal to $\partial S_{N-j}(x_1,\ldots,x_j)$ is defined almost everywhere. We now split the sum

$$\sum_{i=1}^{N} v_i\cdot\nabla_{x_i} f^N = \left(\sum_{i=j+1}^{N}+\sum_{i=1}^{j}\right) v_i\cdot\nabla_{x_i} f^N \tag{3.38}$$

in (3.35).

Introducing the shorthand notation $X_j=(x_1,\ldots,x_j)$, $V_j=(v_1,\ldots,v_j)$, $X_{N-j}=(x_{j+1},\ldots,x_{j+N})$ and $V_{N-j}=(v_{j+1},\ldots,v_N)$, the first sum is handled by the divergence theorem. We get

$$\begin{aligned} &-\sum_{k=j+1}^{N}\int_{S(X_j)}\mathrm{d}x_{j+1}\int_{S(X_j)}\mathrm{d}x_{j+2}\cdots\int_{\partial S(X_j)}\mathrm{d}\sigma(x_k)\cdots\int_{S(X_j)}\mathrm{d}x_N \\ &\qquad\times\int \mathrm{d}V_{N-j}(v_k\cdot n_{i,k}) f^N(X_j,V_j,X_{N-j},V_{N-j}), \end{aligned}\tag{3.39}$$

where

$$S(X_j)=\bigcap_{i=1}^{j}\{x\mid |x-x_i|>\varepsilon\}. \tag{3.40}$$

Note that $\partial S(X_j)$ is the disjoint union of pieces of spherical surfaces:

$$\partial S(X_j)=\bigcup_{i=1}^{j}\sigma_i(X_j), \tag{3.41}$$

where

$$\sigma_i(X_j)=\bigl\{x\in\partial S(X_j)\bigm| |x-x_i|=\varepsilon\bigr\}. \tag{3.42}$$

Therefore if x_k belongs to $\partial S(X_j)$, there exists $i\in\{1,\ldots,j\}$, the index of a particle of the first group, such that $|x_i-x_k|=\varepsilon$. Note that there is only one of such an index for almost all x_k with respect to the surface measure denoted by $\sigma(\mathrm{d}x_k)$. We also set

$$n_{i,k}=\frac{x_i-x_k}{|x_i-x_k|}. \tag{3.43}$$

Using the symmetry of f^N we have $N-j$ identical integrals for which (3.39) becomes

$$-(N-j)\sum_{i=1}^{j}\int_{\sigma_i(X_j)}\mathrm{d}\sigma(x_{j+1})\,\mathrm{d}v_{j+1}\,(v_{j+1}\cdot n_{i,j+1}) f_{j+1}^N(X_j,x_{j+1},V_j,v_{j+1}). \tag{3.44}$$

To treat the second sum we consider

$$\begin{aligned}(V_j \cdot \nabla_{X_j}) f_j^N(X_j, V_j) &= (V_j \cdot \nabla_{X_j}) \int_{S_{N-j}(X_j)} \mathrm{d}X_{N-j}\,\mathrm{d}V_{N-j}\, f^N(X_j, V_j, X_{N-j}, V_{N-j}) \\ &= \frac{\mathrm{d}}{\mathrm{d}t} \int_{S_{N-j}(X_j+V_jt)} \mathrm{d}X_{N-j}\,\mathrm{d}V_{N-j} f^N(X_j + V_jt, V_j, X_{N-j}, V_{N-j})\Big|_{t=0}.\end{aligned} \tag{3.45}$$

We compute the right-hand side of (3.45) and find

$$\begin{aligned}(3.45) = &\int_{S_{N-j}(X_j)} \mathrm{d}X_{N-j}\,\mathrm{d}V_{N-j}(V_j \cdot \nabla_{X_j}) f^N(X_j, V_j, X_{N-j}, V_{N-j}) \\ &+ \sum_{k=1}^{j} \int_{\sigma_k(X_j)} \mathrm{d}\sigma(x_{j+1})\,\mathrm{d}v_{j+1}(v_k \cdot n_{k,j+1}) f_{j+1}^N(X_{j+1}, V_{j+1}).\end{aligned} \tag{3.46}$$

In conclusion we arrive to the following hierarchy of equations

$$\partial_t f_j^N + \mathcal{L}_j^\varepsilon f_j^N = \varepsilon^2 (N-j) C_{j,j+1}^\varepsilon f_{j+1}^N, \quad j < N, \tag{3.47}$$

where

$$\begin{aligned}C_{j,j+1}^\varepsilon f_{j+1}^N(x_1, v_1, \ldots, x_j, v_j) &= \frac{1}{\varepsilon^2} \sum_{i=1}^{j} \int_{\sigma_i(X_j)} \mathrm{d}\sigma(x_{j+1}) \int \mathrm{d}v_{j+1}\, n_{k,j+1} \cdot (v_k - v_{j+1}) \\ &\quad \times f_{j+1}^N(x_1, v_1, \ldots, x_{j+1}, v_{j+1}).\end{aligned} \tag{3.48}$$

Here we have denoted $\mathcal{L}_j^\varepsilon$ the Liouville operator relative to the dynamics of the j-particle system. For j coinciding with N we have nothing else than the Liouville equation. The suffix ε in the definition of $C_{j,j+1}^\varepsilon$ is just to remember the interaction range and we extracted an ε^2 for reasons which will be clear in a moment.

We note that, if the first j particles are at mutual distance larger than ε (what is always true, but for a small measure set) the collision operator $C_{j,j+1}^\varepsilon$ assumes the more expressive form

$$\begin{aligned}C_{j,j+1}^\varepsilon f_{j+1}^N(x_1, v_1, \ldots, x_j, v_j) &= \sum_{k=1}^{j} \int \mathrm{d}n \int \mathrm{d}v_{j+1}\, n \cdot (v_k - v_{j+1}) \\ &\quad \times f_{j+1}^N(x_1, v_1, \ldots, x_j, v_j, \ldots, x_j - \varepsilon n, v_{j+1}),\end{aligned} \tag{3.49}$$

where n is the unit vector. Therefore we see that $\varepsilon^2(N-j)C^{\varepsilon}_{j,j+1}$ has size j but is uniformly bounded in N in the limit we are considering.

Equation (3.47), which is the starting point of our analysis, is completely equivalent to the Liouville equation or to the BBGKY hierarchy. Up to now we did nothing else than to manipulate the basic equations to make easier our future job.

We can express the solution of (3.47) by means of a perturbative expansion:

$$f_j^N(t) = \sum_{m\geqslant 0} \varepsilon^{2m}(N-j)(N-j-1)\cdots(N-j-m+1) \times \int_0^t \mathrm{d}t_1 \int_0^{t_1} \mathrm{d}t_2 \cdots \int_0^{t_{m-1}} \mathrm{d}t_m\, U^{\varepsilon}(t-t_1)C^{\varepsilon}_{j,j+1}\cdots U^{\varepsilon}(t_{m-1}-t_m) \times C^{\varepsilon}_{j+m-1,j+m}U^{\varepsilon}(t_m)f^N_{0,m+j} \tag{3.50}$$

where $f^N_{0,j}$ are defined by (3.33) in terms of the initial probability distribution f_0^N and

$$U^{\varepsilon}(t)g_j(X_j,V_j) = \mathrm{e}^{-\mathcal{L}^{\varepsilon}_j t}g_j(X_j,V_j) = g_j\big(\Phi_j^{-t}(X_j,V_j)\big), \tag{3.51}$$

where $\Phi_j^t(X_j,V_j)$ denotes the j-particle Hamiltonian flow. We have written $\sum_{m\geqslant 0}$ as a series, but this is just a sum ending when $m+j$ reaches the value N. We now want to analyze the asymptotic behavior of the generic term in the above expansion. Namely we want to perform the limit $\varepsilon \to 0$ being j and m fixed.

Consider the string

$$U^{\varepsilon}(t-t_1)C_{j,j+1}\cdots U^{\varepsilon}(t_{m-1}-t_m)C_{j+m-1,j+m}U^{\varepsilon}(t_m)f^{\varepsilon}_{0,m+j} \tag{3.52}$$

appearing in the right-hand side of (3.50). It consists of an integration of the function $f_{0,j+m}$ over a portion of the $(j+m)$-particle phase space. By definition, we can write it as

$$\int \mathrm{d}x_{j+1}\int \mathrm{d}v_{j+1}\cdots\int \mathrm{d}x_{j+m}\int \mathrm{d}v_{j+m}\, n_1\cdot(u_1-v_{j+1})\cdots n_m\cdot(u_m-v_{j+m}) \times f^{\varepsilon}_{0,m+j}(Y_{j+m},W_{j+m}), \tag{3.53}$$

where (Y_{j+m},W_{j+m}) is an initial phase point constructed in the following way. We start by the initial datum (X_j,V_j) and evolve it back in time up to the time $t-t_1$ and then we join a new particle, (x_{j+1},v_{j+1}) with x_{j+1} exactly at distance ε from one of the particles of this new configuration and at distance larger than ε from the others. We evolve back in time (with the $j+1$ flow) up to the time t_2-t_1 and then we join a new particle with the same rule and so on. In formulas:

$$\begin{aligned}(Y_{j+m},W_{j+m}) &= \phi^{-(t_{m-1}-t_m)}\Big(\cdots\big(\phi^{-(t_1-t_2)}\big(\phi^{-(t-t_1)}(X_j,V_j)\cup(x_{j+1},v_{j+1})\big)\\ &\quad \cup\cdots\cup(x_{j+m},v_{j+m})\big)\Big).\end{aligned} \tag{3.54}$$

The velocities $u_1, \ldots, u_m$ are the velocities of the particles partners of the new particle at each step.

If ε is very small the above expression simplifies considerably by using (3.49). To see this let us consider the easiest term, namely

$$\begin{aligned} &\varepsilon^2(N-1)U^\varepsilon(t-t_1)C_{1,2}U^\varepsilon(t_1)f_2(x_1,v_1) \\ &\quad = \varepsilon^2(N-1)\int \mathrm{d}v_2 \int \mathrm{d}n\,(v_1-v_2) \\ &\qquad \times n f_2\big(\phi_2^{-t_1}\big(x_1-v_1(t-t_1), v_1, x_1-\varepsilon n, v_2\big)\big). \qquad (3.55)\end{aligned}$$

Repeating the argument, we are going back with particle 1 for the time $t-t_1$, we join a new particle (say 2), with velocity v_2 at distance ε, solve the two-body problem for the time $-t_1$. Finally we integrate with respect to n_1, v_2 and t_1. There are two possibilities which is convenient to separate. Either $(v_1-v_2)\cdot n$ is positive (remind that n is $\frac{y-x_2}{|y-x_2|}$ where y is given by $x_1-v_1(t-t_1)$) and then the velocities (v_1, v_2) is a post-collisional pair. This means that the back collision takes place. Or $(v_1-v_2)\cdot n$ is negative. In this case the velocities are pre-collisional and the argument of f_2 is

$$x_1-v_1t,\ v_1,\ x_1-\varepsilon n-v_2t_1,\ v_2.$$

In the first case we have to solve the two-body problem. However, we observe that the time in which the pair of particles are really interacting is $\mathrm{O}(\varepsilon)$ so that the argument of $f_{0,2}$ is very close to

$$x_1-v_1(t-t_1)-v_1't_1,\ v_1',\ x_1-v_2't_1,\ v_2',$$

where v_1', v_2' are the pre-collisional pair associated to v_1, v_2 according to (3.17). They are, as in the previous section, expressed in terms of ω so that, assuming continuity of the initial distributions, we arrive to

$$\begin{aligned} &\varepsilon^2(N-1)U^\varepsilon(t-t_1)C_{1,2}U^\varepsilon(t_1)f_2(x_1,v_1) \\ &\quad \to \lambda^{-1}\int \mathrm{d}v_2 \int \mathrm{d}\omega\, B\big(|v_1-v_2|;\omega\big) \\ &\qquad \times f_{0,2}\big(x_1-v_1(t-t_1)-v_1't_1, v_1', x_1-v_2't_1, v_2'\big) \\ &\qquad - f_{0,2}(x_1-v_1t, v_1, x_1-v_2t_1, v_2). \qquad (3.56)\end{aligned}$$

The argument can be extended, using (3.54), with some care, to the general case.

It is convenient to represent the generic term of order m, in the series expansion (3.50) by means of a diagram which is a collection of j binary tree of the form shown in Figure 4, which expresses at what particle of the previous configuration the new particle has to be attached. It is clear that, given a diagram as in Figure 4 and the sequences $t_1, \ldots, t_m$, $v_{j+1}, \ldots, v_{j+m}, \omega_1, \ldots, \omega_m$, we can construct the initial configuration (Y_{j+m}, W_{j+m}) by

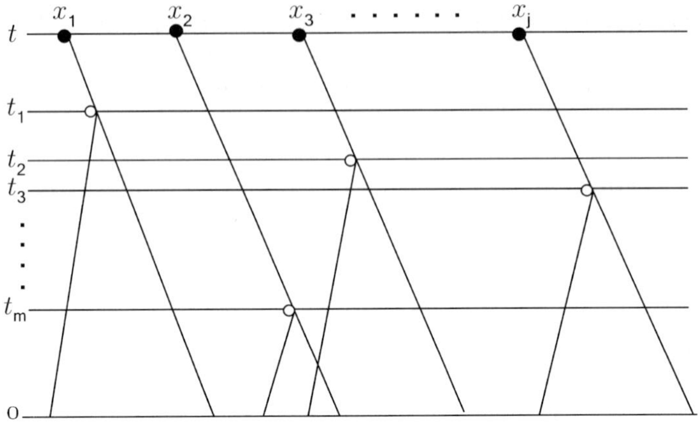

Fig. 4.

solving the equation of the motion. Moreover, for almost all (X_j, V_j) and outside a small measure set in $\mathrm{d}t_1 \cdots \mathrm{d}t_m\, \mathrm{d}v_{j+1} \cdots \mathrm{d}v_{j+m}\, \mathrm{d}\omega_1 \cdots \mathrm{d}\omega_m$, if ε is small, the motion is a collection of distinct two-body interactions which are practically instantaneous.

We now observe that the formal limit

$$\begin{aligned} &(N-j)\varepsilon^2 C^{\varepsilon}_{j,j+1} f^N_{j+1}(x_1, v_1, \dots, x_j, v_j) \\ &\quad \to C_{j,j+1} f^N_{j+1}(x_1, v_1, \dots, x_j, v_j) \end{aligned} \tag{3.57}$$

defines the operator

$$\begin{aligned} &C_{j,j+1} f_{j+1}(x_1, v_1, \dots, x_j, v_j) \\ &\quad = \lambda^{-1} \sum_{k=1}^{j} \int \mathrm{d}v_{j+1} \int_{S^+} \mathrm{d}\omega\, B\big(|v_k - v_{j+1}|, \omega\big) \\ &\qquad\qquad \times \big[f_{j+1}\big(x_1, v_1, \dots, x_k, v'_k, \dots, x_k, v'_{j+1}\big) \\ &\qquad\qquad\quad - f_{j+1}(x_1, v_1, \dots, x_k, v_k, \dots, x_k, v_{j+1})\big]. \end{aligned} \tag{3.58}$$

We have also a formal limit hierarchy which is

$$\partial_t f_j + \sum_{i=1}^{j} v_i \cdot \nabla_i f_j = C_{j,j+1} f_{j+1}, \tag{3.59}$$

because $\mathcal{L}^{\varepsilon}_j \to -\sum_{i=1}^{j} v_i \cdot \nabla_i$ whenever $\varepsilon \to 0$, since j is fixed.

Notice that (3.59) are an unbounded set of equations while (3.47) consist in a finite set of equations. Indeed, in spite of their formal similarity, (3.59) and (3.47) are very different, the former being associated to a time reversible Hamiltonian system and the latter to an irreversible (stochastic) system. We shall remark later this important aspects. For the moment we outline a remarkable property of the hierarchy (3.59). Suppose that initially the sequence of distribution functions $\{f_{0,j}\}_{j=1}^{\infty}$ factorizes, namely

$$f_{0,j}(x_1, v_1, \ldots, x_k, v_k) = \prod_{k=1}^{j} f_0(x_k, v_k). \tag{3.60}$$

Then a solution of (3.59) can be produced by putting

$$f_j(x_1, v_1, \ldots, x_k, v_k; t) = \prod_{k=1}^{j} f(x_k, v_k; t), \tag{3.61}$$

where $f(x_k, v_k; t)$ coinciding with $f_1(x_k, v_k; t)$ solves

$$(\partial_t + v \cdot \nabla) f = Q(f, f), \tag{3.62}$$

where

$$Q(f, f)(x, v) = \lambda^{-1} \int \mathrm{d}v_1 \int_{S^+} B\big(|v - v_1|, \omega\big) \times \big[f\big(x, v'\big) f\big(x, v_1'\big) - f(x, v) f(x, v_1)\big] \tag{3.63}$$

is the Boltzmann collision operator. For this reason (3.59) is called the Boltzmann hierarchy.

The factorization property (3.61) (propagation of chaos) states that if initially position and momentum of a given particle are distributed independently of the positions and momenta of all the others, such property is maintained during the time evolution. This is not true for the real particle dynamics which creates correlations. Therefore the propagation of chaos can be verified only in a limiting situation and it will be consequence of the convergence result we are going to illustrate. The physical meaning of this property is now transparent. If we look at the two-particle distribution $f_2(x_1, v_1, x_2, v_2, t)$ at time t, this is a sum of contributions of diagrams obtained by the perturbative expansion (3.50). If $f_2(x_1, v_1, x_2, v_2, 0)$ factorizes this is essentially the product of the sum of the diagrams arising in the expansions of $f_1(x_1, v_1, t)$ and $f_1(x_2, v_2, t)$ unless the particles of the two diagrams interact. But this is a negligible event ($\mathrm{O}(\varepsilon^2)$ for each pair of particles) so that the factorization holds in the limit.

We now want to show that the solution of the hierarchy (3.47) given by the expansion (3.50) converges to the solutions of the Boltzmann hierarchy (3.59) given by the analogous expansion

$$f_j(t) = \sum_{m \geqslant 0} \lambda^{-m} \int_0^t \mathrm{d}t_1 \int_0^{t_1} \mathrm{d}t_2 \cdots \int_0^{t_{m-1}} \mathrm{d}t_m\, U(t - t_1) C_{j,j+1} \cdots U(t_{m-1} - t_m) \times C_{j+m-1,j+m} U(t_m) f_{0,m+j}^N, \tag{3.64}$$

where

$$U(t)g_j(X_j - V_j t, V_j). \tag{3.65}$$

We have discussed the term by term convergence a.e. only in the case when j and m equal to 1, however as we said, it holds in general (see [29] for details). Therefore it is enough to show that the sum (3.50) and the series (3.64) are bounded by a converging positive series whose terms are independent of N. To show this we do an unnecessary but simplifying hypotheses, namely that we consider only collision with bounded relative velocity. With this assumption we first establish the obvious estimates

$$\left\| C^{\varepsilon}_{j,j+1} f_{j+1} \right\|_{L^\infty} \leqslant Cj \| f_{j+1} \|_{L^\infty}. \tag{3.66}$$

Using (3.66), the identity

$$\left\| U^{\varepsilon} f_j \right\|_{L^\infty} = \| f_j \|_{L^\infty}, \tag{3.67}$$

the bound on the initial datum

$$\left\| f^{\varepsilon}_{0,j} \right\|_{L^\infty} \leqslant C \| f_0 \|^j_{L^\infty} \tag{3.68}$$

and the formula

$$\int_0^t \mathrm{d}t_1 \cdots \int_0^{t_{m-1}} \mathrm{d}t_m = \frac{t^m}{m!}, \tag{3.69}$$

we control the mth term in the series expansion (3.50) by

$$\lambda^{-m} \| f_0 \|^{m+j}_{L^\infty} C^m j(j+1) \cdots (j+m-1) \frac{t^m}{m!}. \tag{3.70}$$

Since

$$\frac{j(j+1) \cdots (j+m-1)}{m!} \leqslant 2^j 2^{m-1}, \tag{3.71}$$

we have estimated the series expansion (3.50) by a convergent series with positive terms not depending on N provided that t is small. The same can be done for the series (3.64) so that the term by term convergence implies the convergence of the solutions.

Finally, we remark once more that if the initial datum for the Boltzmann hierarchy factorizes, i.e.,

$$f_{0,j}(x_1, v_1, \ldots, x_j, v_j) = \prod_{k=1}^{j} f_0(x_k, v_k), \tag{3.72}$$

then the unique solution of the Boltzmann hierarchy also factorizes

$$f_j(x_1, v_1, \ldots, x_j, v_j; t) = \prod_{k=1}^{j} f(x_k, v_k, t), \tag{3.73}$$

where $f(x_k, v_k, t)$ is given by $f_1(x_k, v_k, t)$. Indeed we know that the right-hand side of (3.73) is also a solution to the hierarchy and the uniqueness of the solutions is ensured by the control on the series expansion. Moreover, $f(x, v, t)$ is also a local (mild) solution of the Boltzmann equation. This allows us to establish the following theorem:

THEOREM 3.1. *Let f_0 be a probability density on $\mathbb{R}^3 \times \mathbb{R}^3$ such that $f_0 \in C \cap L^\infty$. Consider the initial datum for the BBGKY hierarchy given by (3.72) and $f_j^\varepsilon(X_j, V_j, t)$ the solution of this hierarchy. Then, in the B–G limit and if the time t is sufficiently small, $f_j^\varepsilon(X_j, V_j, t)$ converges to $\prod_{k=1}^{j} f(x_k, v_k, t)$ a.e. where $f(t)$ is the unique (mild) solution of the Boltzmann equation.*

REMARK 1. We have formulated the result for cutoff cross section. The general case can be handled with a minor effort assuming a Gaussian decay on the velocities for the initial datum.

REMARK 2. The time for which the result holds is proportional to λ which is the mean free path, actually it is a fraction of the mean free time. This is of course very unsatisfactory.

REMARK 3. The validity result we have discussed is much more delicate than it can appear at first sight. Indeed the same result does not hold in the case of a system of diamonds in the plane. The collision rules for a pair collision between two diamonds is described in Figure 5.

The kinetic equation of the model is

$$\begin{aligned}
&(\partial_t + v \cdot \nabla_x) f(x, v, t) \\
&\quad = \lambda^{-1}\big[f\big(x, v^\perp\big) f\big(x, -v^\perp\big) - f(x, v) f(x, -v)\big],
\end{aligned} \tag{3.74}$$

where $v^\perp$ is obtained by rotating v of $\pi/2$. Equation (3.74) describes the two-dimensional Broadwell model. Notice that the second collision in Figure 5 does not play any role in the limit, due to the fact that the particles are identical. Equation (3.74) can be derived formally as we did for the Boltzmann equation in the previous section. However this heuristic derivation is false. Indeed it can be proved that the term by term convergence fails as follows by constructing a counterexample showing that the term with $m = 3$ in the analog of the expansion (3.50) does not converge to its heuristic limit. This remarkable counterexample is due to Uchiyama [85] (see also [29]).

This shows how dangerous can be to believe in the formal arguments in this kind of problems.

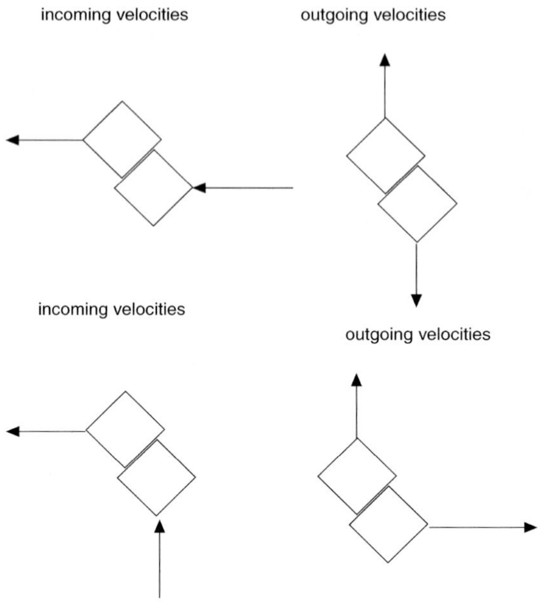

Fig. 5.

REMARK 4. The convergence result we have formulated is really a law of large number. If we consider the "empirical distribution" for our particle system, namely

$$\mu_N(\cdot,\cdot,t)=\frac{1}{N}\sum_{j=1}^{N}\delta_{x_j(t),v_j(t)}, \tag{3.75}$$

we ask how the above one-particle measure differs from $f(x,v,t)$, the solution to the Boltzmann equation. We first note that the measure (3.75) is, for all $t>0$, a measure valued random variable, the randomness arising from the distribution of X_N, V_N at time zero. We are assuming that positions and velocities of the particles are independently and identically distributed according to f_0, the initial datum for the Boltzmann equation. As $N\to\infty$ the fluctuations disappear. Indeed

$$\mathbb{E}_N\left(\left|\int f(x,v,t)\varphi(x,v)-\int\mu_N(\mathrm{d}x,\mathrm{d}v,t)\varphi(x,v)\right|^m\right)\to 0 \tag{3.76}$$

as $N\to\infty$, where φ is a smooth test function, and m is an arbitrary integer. $\mathbb{E}_N$ denotes the expectation with respect to $f_0^{\otimes N}$. Equation (3.76) is an easy consequence of the convergence of the marginals.

NOTES. As we said the local validity result for the Boltzmann equation for hard-spheres is due to Lanford (see [65]). There exists also a global result for the particular case of a rare cloud of gas in the vacuum. This was obtained by Illner and Pulvirenti in [59]. Here we have chosen to present a slight modification of Lanford's arguments for smooth potentials following the lines of an old and unpublished thesis by King. We address the reader for a more extended discussion on these validity results to [29]. There it is also explained why irreversibility can emerge from deterministic reversible systems and the probabilistic meaning of these results is widely illustrated.

If the collision kernel in the Boltzmann equation is suitably delocalized by a mollifying approximate δ-function h, for instance the quadratic part in the loss term is transformed according to the rule

$$f(x, v) f(x, v_1) \to \int \mathrm{d}y\, f(x, v) f(y, v_1) h(x - y), \tag{3.77}$$

then the collision operator is Lipschitz continuous in L^1 and the analytical treatment of the Boltzmann equation becomes much easier. The Boltzmann equation with this mollification is called the Povzner equation (by the name of the Russian mathematician who first studied this modified equation).

One can ask if this equation has a particle interpretation. Indeed a stochastic particle model can be constructed (see [26,71,77,78] and references quoted therein) yielding, under a suitable scaling limit, the Povzner equation. See also [64] for the study of the hydrodynamic limit for such a system.

The interest of these studies is not purely speculative. Indeed one of the most popular numerical codes for the Boltzmann equation, the Bird scheme, is based on this particle approximation (see [88] and also [29] for a general discussion and references quoted therein).

We finally mention an important and poorly investigated problem. In most applications the Boltzmann equation is seen in a stationary regime because, quite often, a rarefied gas, after an inessential time-dependent transient, reaches a stationary nonequilibrium state, exhibiting physically relevant transport phenomena. It would be nice to have a control of the Hamiltonian particle approximation of these stationary states. Of course nothing is known for various reasons. First, the stationary solutions to the Boltzmann equation are not well understood especially as regards their uniqueness properties (see [2,3] and references quoted therein for the state of art of the problem). Second, even having a good control globally in time of the Boltzmann–Grad limit (which is however far from being achieved), this convergence is not expected to be uniform in time, so that this potential result would not give any real insight into the stationary problem which requires new ideas and techniques. The only result in this direction concerns the derivability of the stationary Povzner equation (under smallness assumptions) in terms of stochastic particle systems (see [24] for details).

4. Hydrodynamics of the Boltzmann equation: The Euler limit

The derivation of hydrodynamical equations from the Boltzmann equation is a problem as old as the equation itself and in fact it goes back to Maxwell and Hilbert who developed the main formal tool that is the famous Hilbert expansion we shall discuss later on. The main ideas involved in this derivation are similar to those introduced before for the particle dynamics: Conservation laws and local equilibrium. The essential difference is that the local equilibrium assumption in the framework of the Boltzmann equation becomes, under suitable assumptions, a theorem. The major limitation here is in the fact that the Boltzmann equation is valid for rarefied gases only, hence the hydrodynamical equations derived from it are those for an ideal gases.

Preliminary to the discussion of the hydrodynamic limit are a few properties of the collision kernel, namely the conservation of mass, momentum and energy and the entropy dissipation. We give just a brief summary of them for sake of completeness and refer to Villani's article [87] for a more detailed discussion of such topics.

We are considering a collision operator for the Boltzmann equation given by (3.63) with unitary mean free path and a given cross section B defined in $\mathbb{R}_+ \times S^2$. If the molecular interaction potential $U(r)$ is given by the power law Cr^{1-n}, the function $B(q,\omega)$ can be written as

$$B(q,\omega) = q^\gamma b(\omega). \tag{4.1}$$

Here γ denotes the ratio $\frac{n-5}{n-1}$. The *hard potentials* are those for which γ belongs to the interval $[0,1]$, $\gamma = 1$ being the hard spheres case and $\gamma = 0$ the so-called Maxwell molecules, corresponding to a molecular interaction potential with $n = 5$. The potentials with $n < 5$ are referred as *soft potentials*. The function b is in general singular, due to the grazing collisions. By the *Grad angular cutoff* assumption [53] it is replaced by a bounded smooth function on S^2. In the sequel we only consider hard potentials and cross sections satisfying the Grad angular cutoff condition, hence B is given by (4.1) with $\gamma \in [0,1]$ and b bounded and smooth.

All the properties we are interested in are easily derived from the following lemma:

LEMMA 4.1. *Assume h, f, g be smooth real valued functions on $\mathbb{R}^3$ such that the integrals below make sense. Then*

$$\begin{aligned}
&\int_{\mathbb{R}^3} dv\, h(v) Q[f,g](v) \\
&\quad = \frac{1}{8}\int_{\mathbb{R}^3\times\mathbb{R}^3\times S^2} dv\, dv_*\, d\omega\, B(|v-v_*|,\omega)\big[h(v)+h(v_*)-h(v')-h(v'_*)\big] \\
&\qquad \times \big[f(v')g(v'_*)+f(v'_*)g(v')-f(v)g(v_*)-f(v_*)g(v)\big],
\end{aligned} \tag{4.2}$$

where $Q[f,g]$ is the symmetrized collision operator, defined as

$$\begin{aligned}
&Q[f,g](v) \\
&\quad = \frac{1}{2}\int_{\mathbb{R}^3}\int_{S_+} dv_*\, d\omega\, B(|v-v_*|,\omega) \\
&\qquad \times \big[f(v')g(v'_*)+f(v'_*)g(v')-f(v)g(v_*)-f(v_*)g(v)\big]
\end{aligned} \tag{4.3}$$

and

$$v' = v - [(v - v_*) \cdot \omega]\omega, \qquad v'_* = v_* + [(v - v_*) \cdot \omega]\omega. \tag{4.4}$$

The proof of the above identity follows immediately from changes of variables and the remark that the Jacobian of the map $(v, v_*) \to (v', v'_*)$ is equal to 1, because of the Liouville theorem.

We say that h is a *collision invariant* if h is such that

$$h(v) + h(v_*) - h(v') - h(v'_*) = 0 \tag{4.5}$$

for any $(v, v_*) \in \mathbb{R}^6$. Clearly, (4.5) is verified if $h = 1$. Moreover, from the conservation of momentum in a collision, it follows that (4.5) is verified also by $h(v) = v_\alpha$, $\alpha = 1, \ldots, 3$, v_α being the αth component of the vector v. Finally (4.5) is verified by $h(v) = v^2/2$, by the conservation of the kinetic energy during an elastic collision. It can be proved (see, for example, [27,29]) under very general assumptions on the class of functions h, that any collision invariant is a linear combination of the above invariants:

LEMMA 4.2. *The measurable function $h : \mathbb{R}^3 \to \mathbb{R}$ is a collision invariant if and only if*

$$h(v) = \sum_{\alpha=0}^{4} c_\alpha \chi_\alpha(v), \tag{4.6}$$

where

$$\chi_0(v) = 1, \qquad \chi_\alpha(v) = v_\alpha, \quad \alpha = 1, \ldots, 3, \qquad \chi_4(v) = \frac{1}{2}|v|^2. \tag{4.7}$$

For any positive function $f : \mathbb{R}^3 \to \mathbb{R}$ consider the quantity

$$D[f] = \int_{\mathbb{R}^3} \mathrm{d}v \log f(v) Q[f, f](v).$$

By Lemma 4.1, it can be written as

$$D[f] = \frac{1}{4} \int_{\mathbb{R}^3} \int_{\mathbb{R}^3} \int_{S_+} \mathrm{d}v \, \mathrm{d}v_* \, \mathrm{d}\omega \, B(|v - v_*|, \omega) \times [f(v') f(v'_*) - f(v) f(v_*)] \log \frac{f(v) f(v_*)}{f(v') f(v'_*)}. \tag{4.8}$$

Since the log is increasing, it follows that $D[f]$ is nonpositive and vanishes if and only if $\log f$ is almost everywhere a collision invariant. Hence we have proved the lemma:

LEMMA 4.3. *For any $f \geqslant 0$,*

$$D[f] \leqslant 0 \tag{4.9}$$

and

$$D[f]=0 \quad \textit{if and only if} \quad \log f = \sum_{\alpha=0}^{4} c_\alpha \chi_\alpha(v) \quad \textit{a.e.} \tag{4.10}$$

Above lemmas imply the conservation laws for mass, momentum and energy and the H-theorem.

In this section we consider the Boltzmann equation as our microscopic starting point. So, as we have already done in the previous sections, we use the notation q for the microscopic position of the particles (measured in units of mean free path) and τ for the microscopic time (measured in units of mean free time). The Boltzmann equation, in the absence of external forces is then written as

$$(\partial_\tau + v \cdot \nabla_q) f = Q[f, f]. \tag{4.11}$$

Multiplying both sides by χ_α and integrating with respect to the velocity, by Lemma 4.1 we obtain the identities

$$\partial_\tau \xi_f^\alpha + \nabla_q \cdot \zeta_f^\alpha = 0, \quad \alpha = 0, \ldots, 4, \tag{4.12}$$

where

$$\begin{aligned} \xi_f^0 &= \rho_f = \int_{\mathbb{R}^3} \mathrm{d}v\, f(\cdot, \cdot v), \\ \xi_f^\alpha &= \rho_f u_f^\alpha = \int_{\mathbb{R}^3} \mathrm{d}v\, v_\alpha f(\cdot, \cdot v), \quad \alpha = 1, \ldots, 3, \\ \xi_f^4 &= \frac{1}{2}\rho_f u_f^2 + \rho_f e_f, \quad \rho_f e_f = \int_{\mathbb{R}^3} \mathrm{d}v\, \frac{1}{2}(v-u)^2 f(\cdot, \cdot v), \end{aligned} \tag{4.13}$$

and ρ_f, u_f and e_f are interpreted as mass density, mean velocity and specific internal energy associated to the probability density f. The currents ζ_f^α are given, for $\beta = 1, \ldots, 3$, by

$$\begin{aligned} \zeta_{f,\beta}^0 &= \rho_f u_f^\beta, \\ \zeta_{f,\beta}^\alpha &= \int_{\mathbb{R}^3} \mathrm{d}v\, v_\alpha v_\beta f(\cdot, \cdot v), \quad i = \alpha, \ldots, 3, \\ \zeta_{f,\beta}^4 &= \int_{\mathbb{R}^3} \mathrm{d}v\, \frac{1}{2} v^2 v_\beta f(\cdot, \cdot v). \end{aligned} \tag{4.14}$$

All the above identities are in the form of a system of conservation laws expressing the conservation of mass, momentum and energy, but they cannot be considered a closed system of PDEs because, in order to compute the currents, one needs to know the function f. As in the case of particle systems, it is the local equilibrium assumption which transforms the identities (4.12) into hydrodynamical equations, but that requires a rescaling of variables and the considerations we are going to present.

Before doing this, let us point out the consequences of Lemma 4.3. Let us look for the solutions M of the equation

$$Q[M,M]=0. \tag{4.15}$$

If M is a positive solution of (4.15), then, by multiplying it by $\log M$ and integrating with respect to the velocity, one gets that

$$D[M]=0.$$

Therefore, by Lemma 4.3, $\log M$ is a linear combination of collision invariants. The condition that M is integrable and normalized imposes restrictions on the coefficients c_α, so that the solutions M of (4.15) can be parametrized by five real parameters and have the form

$$M(v)=M_{\rho,T,u}(v)=\frac{\rho}{(2\pi T)^{3/2}}\exp\left[-\frac{(v-u)^2}{2T}\right]. \tag{4.16}$$

The above solutions are called *Maxwellians* and the parameters $\rho>0$, $T>0$ and $u\in\mathbb{R}^3$ are the mass density, temperature and mean velocity of the Maxwellian. Indeed, it is immediate to check that

$$\rho_{\mathrm{M}}=\rho, \qquad u_{\mathrm{M}}=u, \qquad e_{\mathrm{M}}=\frac{3}{2}T.$$

The Maxwellians are the analog of the Gibbs states for the ideal gases.

Moreover, since in the above discussion only the v-dependence of M has been discussed, the parameters can be constants as well as functions of q and τ. In the first case they are referred as *global Maxwellians*, while in the second case they are called *local Maxwellians* and they are the analog of the local equilibrium. In conclusion we have proved the following proposition:

PROPOSITION 4.4. *The solutions to* (4.15) *are the local Maxwellians* $M_{\rho,T,u}$, *with* ρ, T *positive functions on* $\Omega\times\mathbb{R}_+$ *and* u *vector valued function on* $\Omega\times\mathbb{R}_+$.

We define *equilibrium solution* for the Boltzmann equation any solution to the Boltzmann equation which is also Maxwellian. If the spatial domain has no particular symmetry, then one can immediately show that all the equilibrium solutions are the global Maxwellians. This is false, for example, if the spatial domain is a ball, because in this case u can be the velocity field of a rigid motion. We restrict ourselves to a space domain which is a three-dimensional torus of size L, $\mathbb{T}^3_L$, thus avoiding such complications as well as the interesting discussion of the interaction of the system with the boundaries, for which we refer to [27].

Another important consequence of Lemma 4.3 is the following: For any positive solution to the Boltzmann equation, denoting by η_f and J_f

$$\eta_f = \int_{\mathbb{R}^3} \mathrm{d}v\, f(\cdot,\cdot,v)\log f(\cdot,\cdot,v),$$
$$J_{f,\alpha} = \int_{\mathbb{R}^3} \mathrm{d}v\, v_\alpha f(\cdot,\cdot,v)\log f(\cdot,\cdot,v), \tag{4.17}$$

the entropy and entropy flux densities, they satisfy the inequality

$$\partial_\tau \eta_f + \nabla_q \cdot J_f \leqslant 0, \tag{4.18}$$

which is an entropy–entropy flux inequality typical of the systems of conservation laws. Inequality (4.18) is obtained by multiplying the Boltzmann equation by $1 + \log f$ and integrating with respect to the velocity and using Lemmas 4.1 and 4.3.

If one further integrates with respect to q, using the fact that the space domain is a torus one obtains

$$\frac{\mathrm{d}}{\mathrm{d}\tau} H\big[f(\tau,\cdot,\cdot)\big] \leqslant 0, \tag{4.19}$$

where

$$H[f] = \int_\Omega \mathrm{d}q\, \eta_f(q,\cdot)$$

is the Boltzmann H function. The inequality (4.19) shows formally the Boltzmann H-theorem and underlines the irreversible character of the Boltzmann equation, to be compared with the reversible character of the Newton equations discussed in Section 3. Thus, in a sense, the Boltzmann equation is much closer to a macroscopic description of the fluid, rather than to a microscopic one.

We conclude this preliminaries by introducing the *linear Boltzmann operator*, which plays a major role in the Hilbert expansion. We fix a Maxwellian M and define the operator

$$Lf = 2Q[M,f]. \tag{4.20}$$

This is an operator acting only on the v-dependence of f. The Maxwellian M may be either global or local, since in the discussion below q and τ play only the role of parameters. When the space–time dependence will become relevant, we shall keep track of it by using the notation $L_{q,\tau}$.

We introduce the Hilbert space $\mathcal{H}$ of the real functions on $\mathbb{R}^3$ with inner product

$$(f,g) = \int_{\mathbb{R}^3} \mathrm{d}v\, M^{-1}(v) f(v) g(v) \tag{4.21}$$

and the corresponding norm $\|f\|_{\mathcal{H}} = (f, f)^{1/2}$. The operator L is well defined on the dense subspace of $\mathcal{D}_L \subset \mathcal{H}$

$$\mathcal{D}_L = \{f \in \mathcal{H} \mid \nu f \in \mathcal{H}\}$$

where

$$\nu(v) = \int_{\mathbb{R}^3} \int_{S_+} \mathrm{d}v_* \, \mathrm{d}\omega \, B(|v - v_*|, \omega) M(v_*). \tag{4.22}$$

The following properties of L are relevant:

(1) $$Lf = 0 \quad \text{if and only if} \quad f = \sum_{\alpha=0}^{4} c_\alpha \psi_\alpha, \qquad \psi_\alpha = M\chi_\alpha, \ \alpha = 0, \ldots, 4;$$
$$(\psi_\alpha, Lf) = 0 \quad \text{for any } f \in \mathcal{D}_L, \ \alpha = 0, \ldots, d; \tag{4.23}$$

(2) $$Lf = -\nu f + Kf, \tag{4.24}$$
where

$$(Kf)(v) = \int_{\mathbb{R}^3} \int_{S_+} \mathrm{d}v_* \, \mathrm{d}\omega \, B(|v - v_*|, \omega) \times \left[M(v') f(v'_*) + M(v'_*) f(v') - M(v) f(v_*)\right] \tag{4.25}$$

is a compact operator on $\mathcal{H}$;

(3) For any $f \in \mathcal{H}$ we denote by $\hat{f}$ its orthogonal projection on

$$\mathrm{Null}\, L = \left\{ f \,\middle|\, f = \sum_{\alpha=0}^{4} c_\alpha \psi_\alpha, c_\alpha \in \mathbb{R}, \alpha = 0, \ldots, 4 \right\}$$

and $\tilde{f} = f - \hat{f}$. The projector on Null L is denoted by P and $P^\perp = 1 - P$.

$$(f, Lg) = (Lg, f) \quad \text{for any } f, g \in \mathcal{D}_L,$$
$$(f, Lf) \leqslant 0 \qquad \text{for any } f \in \mathcal{D}_L; \tag{4.26}$$

Moreover, there exists $\mu > 0$ such that

$$(f, Lf) \leqslant -\mu(f, f) \quad \text{for any } f \in P^\perp \mathcal{D}_L; \tag{4.27}$$

(4) Given $g \in \mathcal{H}$, consider the equation

$$Lf = g \tag{4.28}$$

Because of the orthogonality conditions $(4.23)_2$, g has to satisfy the compatibility conditions

$$(\psi_\alpha, g) = 0, \quad \alpha = 0, \ldots, 4. \tag{4.29}$$

Moreover, by property (2), the Fredholm alternative holds and, if $g \in (\operatorname{Null} L)^\perp$ there are solutions in $\mathcal{H}$ to (4.28) differing from each other by elements of $\operatorname{Null} L$. Therefore, denoting by $L^{-1}g$ the unique solution to (4.28) in $(\operatorname{Null} L)^\perp$, we have proved that

$$g \in (\operatorname{Null} L)^\perp \quad \text{and} \quad Lf = g \quad \text{if and only if} \quad f = L^{-1}g + h,$$

for an arbitrary h belonging to $\operatorname{Null} L$.

Two important special cases of solutions to (4.28) we will use in the sequel are given by special choices of g: Let

$$A_{\alpha,\beta}(v) = v_\alpha v_\beta - \frac{1}{3}v^2\delta_{\alpha,\beta}, \qquad B_\alpha(v) = \frac{1}{2}v^2(v_\alpha - 5T). \tag{4.30}$$

It can be proved [27] that

$$L^{-1}[A_{\alpha,\beta}M] = -\lambda_1(|v|)A_{\alpha,\beta}, \qquad L^{-1}[B_\alpha M] = -\lambda_2(|v|)B_\alpha, \tag{4.31}$$

where the *generalized eigenvalues* λ_γ, $\gamma = 1, 2$, are such that for any $\sigma < 1$ there are constants C_γ such that

$$\lambda_\gamma(|v|) \leqslant C_\gamma\big(M(v)\big)^\sigma. \tag{4.32}$$

A similar statement has been proved in [19] for the soft potentials case.

After this preparation we are now ready to study the hydrodynamical limit for the Boltzmann equation. As in the particle case, we need to rescale the space variables from the kinetic to macroscopic scale. To do this we introduce the a-dimensional parameter ε which represents the ratio between the mean free path and the size of the container. The variable $x = \varepsilon q$ thus varies in the three-dimensional torus of size 1, $\mathbb{T}_1^3 \equiv \mathbb{T}^3$. We choose a smooth function $F^{(0)}$ on $\mathbb{T}^3$ and, for ε very small, consider the initial value

$$f^{(0),\varepsilon}(q, v) = F^{(0)}(\varepsilon q, v)$$

varying slowly on the kinetic scale and look at its evolution according to the Boltzmann equation. Let $f^\varepsilon(q, v, \tau)$ be the solution to the Boltzmann equation with above initial datum. Locally (in the kinetic space scale) the solution is well approximated by the homogeneous Boltzmann equation where the transport term is neglected and only the collisions are taken into account. In order to see significant changes due to the space inhomogeneity, we need to wait for a sufficiently long time $\tau = \varepsilon^{-1}t$. Therefore, we define

$$F^\varepsilon(x, v, t) = f^\varepsilon\big(\varepsilon^{-1}x, v, \varepsilon^{-1}t\big)$$

and write the corresponding rescaled Boltzmann equation

$$(\partial_t + v \cdot \nabla_x) F^\varepsilon = \varepsilon^{-1} Q\big[F^\varepsilon, F^\varepsilon\big], \quad F^\varepsilon(\cdot, \cdot, 0) = F^{(0)}(\cdot, \cdot). \tag{4.33}$$

Hilbert proposed to construct the solution F^ε by looking at the power series

$$F^\varepsilon = \sum_{n=0}^{\infty} \varepsilon^n F_n \tag{4.34}$$

and was able to compute all the F_n. They have to satisfy the recursive relations

$$Q[F_0, F_0] = 0, \qquad 2Q[F_0, F_n] = S_n[F_0, \dots, F_{n-1}], \quad n \geqslant 1, \tag{4.35}$$

with

$$S_n[F_0, \dots, F_{n-1}] \\ = (\partial_t + v \cdot \nabla_x) F_{n-1} - \sum_{(\ell,\ell'):\ \ell,\ell' \geqslant 1,\ \ell+\ell'=n} Q(F_\ell, F_{\ell'}). \tag{4.36}$$

The first of (4.35), by Proposition 4.4 implies

$$F_0 = M_{\rho,T,u}, \tag{4.37}$$

where the hydrodynamical fields ρ, T and u not yet determined and in general depend on x and t. Hence, the second of (4.35) can be written

$$L F_n = S_n[F_0, \dots, F_{n-1}], \quad n \geqslant 1, \tag{4.38}$$

with L denotes $L_{\cdot,\cdot,\cdot}$ the linear Boltzmann operator corresponding to the above still unknown Maxwellian. Equations (4.38) are in the form (4.28). Therefore, in order to solve them one has to fulfill the compatibility conditions

$$P\big(S_n[F_0, \dots, F_{n-1}]\big) = 0, \quad n \geqslant 1,$$

which is equivalent, by the expression (4.36) of S_n and Lemmas 4.1 and 4.2, to:

$$P\big[(\partial_t + v \cdot \nabla_x) F_{n-1}\big] = 0, \quad n \geqslant 1. \tag{4.39}$$

Let us look at the above condition for $n = 1$. Simple Gaussian integrations show that it can be written as

$$\begin{aligned} &\partial_t \rho + \nabla_x \cdot [\rho u] = 0, \\ &\rho(\partial_t + u \cdot \nabla_x) u + \nabla_x P(\rho, T) = 0, \\ &\frac{3}{2} \rho(\partial_t + u \cdot \nabla_x) T + P(\rho, T) \nabla_x \cdot u = 0, \end{aligned} \tag{4.40}$$

with

$$P(\rho, T) = \rho T. \tag{4.41}$$

Equations (4.40) are the Euler equations for an ideal fluid, with the pressure P satisfying the law of perfect gases (4.41). In order to determine fully u, ρ and T we have to specify the initial conditions, which are given in terms of $F^{(0)}$:

$$\begin{aligned}
\rho(\cdot,0) = \rho^{(0)} &= \int_{\mathbb{R}^3} \mathrm{d}v\, F^{(0)}(\cdot, v),\\
u(\cdot,0) = u^{(0)} &= \int_{\mathbb{R}^3} \mathrm{d}v\, vF^{(0)}(\cdot, v),\\
T(\cdot,0) = T^{(0)} &= \int_{\mathbb{R}^3} \mathrm{d}v\, \frac{1}{3}\big(v - u(\cdot,0)\big)^2 F^{(0)}(\cdot, v).
\end{aligned} \tag{4.42}$$

We assume that initial data are smooth enough to provide a unique smooth solution to the Euler equations (4.40), at least for a time interval $0 \leqslant t \leqslant t_1$. With this assumption the lowest order of the expansion F_0 is completely determined as the Maxwellian M with parameters evolving according to the Euler equations.

After this step one can determine the other terms of the series. Since we have fulfilled the condition (4.39) for $n = 1$, we can solve (4.38) with $n = 1$ by the use of the property (4) of the operator L. As already remarked, (4.38) can only fix the part of F_1 orthogonal to $\operatorname{Null} L$. Therefore we have

$$F_1 = L^{-1}\big[(\partial_t + v \cdot \nabla_x)M\big] + \widehat{F}_1, \tag{4.43}$$

with $\widehat{F}_1 \in \operatorname{Null} L$ and hence of the form

$$\widehat{F}_1 = \sum_{\alpha=0}^{4} c_1^\alpha \psi_\alpha, \tag{4.44}$$

and c_1^α five real functions of x and t to be determined. We take advantage of this arbitrariness to satisfy the compatibility condition (4.39) for $n = 2$. In fact, it is a system of five inhomogeneous linear PDEs in the unknown c_1^α's, which can be solved once one fixes the initial conditions $c_1^\alpha(\cdot, 0)$. We will let the initial conditions undetermined for the moment. Then we can solve (4.38) for $n = 2$ and determine F_2 up to its part $\widehat{F}_2$ in $\operatorname{Null} L$. Using the arbitrariness of $\widehat{F}_2$ we fulfill the compatibility condition (4.39) for $n = 3$ and so on, for arbitrary n. In conclusion the Hilbert procedure determines all the terms of the Hilbert expansion, up to the choice of the initial values $c_n^\alpha(\cdot, 0)$ for $n \geqslant 1$. The lowest order is completely determined and is a local equilibrium driven by the Euler equations.

The reasons why the above discussion cannot be considered as a satisfactory proof of the hydrodynamical limit of the Boltzmann equation are the following:

(i) It is not known and not even expected that the Hilbert series has a positive radius of convergence;

(ii) Even if the Hilbert series were convergent, it does not satisfy the initial conditions. Indeed in the procedure presented before, the only track of the initial condition $F^{(0)}$ is in the initial conditions (4.42) for the hydrodynamic fields. The functions F_n, $n \geqslant 1$, obtained by solving (4.38) are determined up to $t = 0$ (only their hydrodynamical part being left undetermined). Therefore the "solution" F^ε constructed via the Hilbert series, at time $t = 0$ satisfies the initial conditions for the hydrodynamic fields and nothing else. Moreover, because we have no control of the sign of the F_n's, it is not a priori obvious that the "solution" one construct in this way is non negative, as it should be because of its probabilistic interpretation.

The first difficulty is by far more serious and it has been solved in the case of one space dimension by Caflisch [18], who has proposed to look at a suitable truncation of the Hilbert series, being able then to evaluate the remainder. The extension to higher space dimension has been given by Lachowitz [63]. The main restriction is that it requires a lot of regularity of the solutions to the Euler equations, which is available only locally in time and with suitable assumptions on the initial data.

The second difficulty has also been fixed in [63] by supplementing the Hilbert expansion with an initial layer expansion. The idea is very simple: For times τ much shorter than the hydrodynamical times $\varepsilon^{-1}t$ the evolution, to the lowest order, is locally given by the homogeneous Boltzmann equation, which, provided the initial datum $F^{(0)}$ is close enough to a Maxwellian, converges exponentially fast to that Maxwellian. The corrections to this can be computed and used to fulfill the initial conditions. In conclusion, one can construct a smooth solution to the initial value problem which, at any positive hydrodynamical time t is close to the truncated Hilbert expansion with remainder, up to terms which are exponentially small in ε. Moreover, since it is initially nonnegative, it remains nonnegative at later times.

Following the approach in [18,63] we show more precisely how to make working rigorously the Hilbert expansion. The solution F^ε to the Boltzmann equation can be expressed as

$$F^\varepsilon = \sum_{n=0}^{N} \varepsilon^n f_n^\varepsilon + \varepsilon^m R_{N,m}^\varepsilon, \tag{4.45}$$

where $R_{N,m}^\varepsilon$ will be determined later on, and

$$f_n^\varepsilon(\cdot,\cdot,t) = F_n(\cdot,\cdot,t) + I_n\big(\cdot,\cdot,\varepsilon^{-1}t\big), \tag{4.46}$$

with F_n the Hilbert expansion terms constructed before and I_n the initial layer expansion terms, depending on the kinetic time τ, defined by the equations:

$$\begin{aligned} \partial_\tau I_0 &= L_0 I_0 + Q[I_0, I_0], \\ \partial_\tau I_n &= L_0 I_n + s_n[I_0, \ldots, I_{n-1}], \qquad n = 1, \ldots, N, \end{aligned} \tag{4.47}$$

which can be obtained with a procedure similar to the one used for the Hilbert expansion. Here L_0 is the linearized Boltzmann operator referred to the Maxwellian at time $t = 0$, but still depending on x and M_0 denotes the Maxwellian $M_{\rho^{(0)},T^{(0)},u^{(0)}}$. The source terms s_n are given by

$$\begin{aligned} s_1[I_0] &= -v\cdot\nabla_x I_0,\\ s_n[I_0,\dots,I_{n-1}] &= -v\cdot\nabla_x I_{n-1}+\varepsilon^{-1}[L_{x,\varepsilon\tau}-L_{x,0}]I_{n-1}\\ &\quad+\sum_{(\ell,\ell'):\,\ell,\ell'\geqslant 1,\ \ell+\ell'=n} Q(I_\ell,I_{\ell'}),\quad n\geqslant 2. \end{aligned} \tag{4.48}$$

Note that the term with ε^{-1} in front in (4.48) is only apparently divergent because $L_{x,\varepsilon\tau}-L_{x,0}$ is of order ε by the regularity of the hydrodynamical solution.

The initial conditions are fixed as follows:

$$I_0(\cdot,\cdot,0)=F^{(0)}-M_0,\qquad I_n(\cdot,\cdot,0)=-F_n(\cdot,\cdot,0),\quad n\geqslant 1, \tag{4.49}$$

thus fulfilling the initial condition in (4.33).

We need to require also that the initial layer terms go to 0 exponentially fast, as τ goes to infinity. From (4.27) one can obtain the time decay provided that the initial data and the source terms are orthogonal to Null L_0. For each fixed x, I_0 is the solution of a spatially homogeneous Boltzmann equation in the near equilibrium regime, so the exponential convergence is automatically satisfied because $F^{(0)}$ and M_0 have the same hydrodynamical moments and $I_0(\cdot,\cdot,0)$ is sufficiently small to control the nonlinearity.

For $n\geqslant 1$ a little more care is required: For each fixed x we have simple linear equations, but it is not generally true that the initial value and the source terms are orthogonal to Null L_0. In fact we have not assumed this for F_n at time $t=0$ and the multiplication by v in the sources produces terms in Null L_0. Hence, it is possible that $I_n(\tau)$ converges to I_n^∞ belonging to Null L_0. Note that for any given positive n, I_n^∞ is finite because, by induction, we assume the exponential decay for I_k, $k<n$. Note, however, that $I_n(\tau)-I_n^\infty$ is still a solution to $(4.47)_2$ (because I_N^∞ belongs to Null L_0 and is constant in time) and goes to 0 exponentially fast. Finally, we use the arbitrariness of the $c_n^\alpha(\cdot,0)$ in the Hilbert expansion to restore the initial conditions which may have been perturbed by the previous subtraction.

Once the f_n have been determined completely from the previous procedure, (4.45) can be interpreted as the definition of $R^\varepsilon_{N,m}$. There is still some arbitrariness in the choice of the integer m which will be fixed below. Since F^ε solves the initial value problem (4.33), the remainder, denoted shortly by R, has to solve the *remainder equation*

$$(\partial_t+v\cdot\nabla_x)R=\varepsilon^{-1}LR+L_1R+\varepsilon^{m-1}\big[Q[R,R]+A^\varepsilon\big], \tag{4.50}$$

with A^ε depending only on the f_n's and given explicitly by

$$\begin{aligned} A^\varepsilon=\varepsilon^{N-2m+1}\Bigg[&-\partial_t F_N-v\cdot\nabla_x f_N\\ &+\sum_{(\ell,\ell'):\,\ell,\ell'\geqslant 1,\ \ell+\ell'>N}\varepsilon^{\ell+\ell'-1-N}Q(f_\ell,f_{\ell'})\Bigg], \end{aligned} \tag{4.51}$$

and L_1 a linear operator also depending only on the f_n and given by

$$L_1 R = 2\sum_{n=1}^{N} \varepsilon^{n-1} Q[f_n, R]. \tag{4.52}$$

The initial condition for R is

$$R(\cdot,\cdot,0) = 0. \tag{4.53}$$

The solution F^ε can be written in the form (4.45) if and only if R is solution to (4.50). We are interested to construct a solution R bounded uniformly in ε, because this would lead to the desired conclusion on the hydrodynamical limit of the Boltzmann equation. If $m > 1$, for ε small the equation for R is weakly nonlinear. Moreover, if $N \geqslant 2m - 1$ the only negative power of ε appears in front of LR. Therefore we choose $N = 2m - 1$. It is also convenient from the technical point of view to take m sufficiently large, rather than $m = 2$.

To take care of the negative power of ε multiplying the linear Boltzmann operator, the only possibility is to use the nonpositivity of L stated in the inequality (4.26). This requires the use of the inner product (4.21). The difficulty is that in this case the Maxwellian depends on x and t, so that, when computing the time derivative of

$$\langle R, R\rangle = \int_\Omega \mathrm{d}x \int_{\mathbb{R}^3} \mathrm{d}v\, R(x,v,t)^2 M_{x,t}^{-1}(v),$$

we obtain a term of the form

$$\int_\Omega \mathrm{d}x \int_{\mathbb{R}^3} \mathrm{d}v\, R(x,v,t)^2 (\partial_t + v\cdot\nabla_x) M_{x,t}^{-1}(v),$$

which cannot be bounded in terms of $\langle R, R\rangle$ because $M^{-1}(\partial_t + v\cdot\nabla_x)M_{x,t}$ grows like a polynomial in v. This technical difficulty was overcome by Caflisch by means of a clever decomposition of the solution into an high and low velocity part which permitted to get bounds on R uniform in ε. We do not enter in such rather technical argument for which we refer to the original papers by Caflisch [18] and Lachowitz [63]. We summarize the results in the following theorem:

THEOREM 4.5. *Assume*:

(1) *For suitable choices of $\gamma > 0$ and of the positive integers ℓ and s, $F^{(0)} > 0$ is in the space $B_{\gamma,\ell,s}$ of the bounded continuous functions with norm*

$$\|f\|_{\gamma,\ell,s} = \sup_{v\in\mathbb{R}^3} \left[\mathrm{e}^{\gamma v^2}(1+v^2)^{\ell/2} \left| f(\cdot,v) \right|_s \right],$$

$|\cdot|_s$ being the Sobolev norm of order s on Ω;

(2) *The initial values for* ρ, T *and* u, *given by* (4.42) *are sufficiently smooth, so that there is a time* t_1 *and a unique solution* (ρ, T, u) *in the time interval* $[0, t_1]$ *to the Euler equations with such initial data, with bounded Sobolev norms up to an order* s *sufficiently large. Furthermore, there are* $\rho_0 > 0$, $T_0 > 0$, ρ_1 *and* T_1 *finite such that*

$$\rho_0 < \rho(x,t) < \rho_1, \qquad T_0 < T(x,t) < T_1 \quad \text{for any } x \in \Omega,\ t \in [0, t_1];$$

(3) *Let* M_t *be the Maxwellian whose parameters are given by the solution to the Euler equation of previous item* (2). *There is* $\delta > 0$ *sufficiently small such that initially*

$$\left\| F^{(0)} - M_0 \right\|_{\gamma,\ell,s} < \delta.$$

Then, there is $\varepsilon_0 > 0$ *such that, for any* $0 < \varepsilon < \varepsilon_0$, *there is a smooth nonnegative solution* F^ε *to* (4.33) *and the solution is unique in the class of the smooth solutions to* (4.33). *Moreover, for any* $t_0 > 0$, *there is* C *such that*

$$\sup_{t_0 \leqslant t \leqslant t_1} \left\| F^\varepsilon(\cdot,\cdot,t) - M_t \right\|_{\gamma,\ell,s} < C\varepsilon.$$

In particular, this proves the validity of the local equilibrium assumption and of the Euler equations for the ideal gas as limiting behavior of a rarefied gas in the hydrodynamical regime.

Remarks and bibliographical notes

The problem of the derivation of the hydrodynamical equations from the kinetic theory was addressed by Maxwell [70] and lately by Hilbert [57] in his sixth question about the derivation of the macroscopic equations from the microscopic dynamics. Hilbert proposed his expansion and used his theory of the integral equations [58] to find the terms of the expansion as explained before. Grad [53,54] studied in great detail the expansion and obtained many bounds on them, including a weaker version of (4.32). The estimates (4.21) are consequences of a result in [31]. The first rigorous result on the hydrodynamical limit is due to Nishida [74] who used an abstract Cauchy–Kovalevskaja theorem to construct the solution to the rescaled Boltzmann equation on a time interval $[0, t_0]$ depending on the initial datum which he assumed analytical. A similar result was obtained with initial data in Sobolev spaces in [4]. Both results are local in time even when the hydrodynamical equations have global solutions, because the time t_0 depends on kinetic properties, not on the hydrodynamical initial data. The validity of the Euler equations on the time interval where they are regular was proved in [21] for the Broadwell model and by Caflisch in [18] in one space dimension, where the high velocity problem represents a major difficulty. The extension to higher dimensions and to initial data close but not necessarily coinciding with M_0 was given by Lachowitz [63]. One should note that the closeness to M_0 was used essentially to ensure the exponential convergence of the solution of the homogeneous Boltzmann equation to the M_0. For initial data not close to a Maxwellian the results available when the paper was written ensured only the convergence without any estimate of the rate. The recent results obtained in [36] provide a rate of convergence which is given by arbitrary high degree polynomials. Therefore it should be possible to carry out the initial layer analysis for smooth arbitrary initial data.

The technique of the Hilbert expansion, possibly modified with initial or boundary or shock layer expansion has been used to discuss hydrodynamical solutions to the Boltzmann equation in different situation: weak shock [20,69] and boundary value problems [5,40–42]. It should be noted however that in such situations the spatial gradients may not be of order ε so that viscosity effects become relevant. The effect of viscosity is discussed in the next section.

5. Navier–Stokes equations from the Boltzmann equation and from particles

5.1. *Navier–Stokes equations from the Boltzmann equation*

The derivation of the Navier–Stokes equations is a much more delicate task than the derivation of the Euler equations, for two main reasons. The first is that the viscosity and the heat conduction are essentially nonequilibrium effects which cannot be accounted of by the local equilibrium and corrections to the local equilibrium are necessary in order to recover them. The second is that a byproduct of the derivation is the relation between the transport coefficients involved in the Navier–Stokes equations and the microscopic interaction. This is quite nontrivial and in fact the derivation of such a relation was one of the first great successes of the kinetic theory. The problem is much harder for particle systems and of course still unsolved rigorously for deterministic systems.

Beyond the above problems, a basic difficulty arises in deriving the Navier–Stokes equations. In fact, in order to derive an equation by a scaling argument, it is necessary that the equation itself be invariant under such a scaling. This is not the case for the Navier–Stokes equations because they involve simultaneously first- and second-order differential operators so that we cannot have any space–time scaling invariance. As we shall see, the space–time scaling invariance confines the derivation to the case of the incompressible equations. We present first the derivation from the Boltzmann equation which is conceptually simpler. Then we discuss the derivation from particles and finally we present a stochastic particle model which has the Navier–Stokes equation as limit under the appropriate scaling. The Hilbert expansion in principle provides not only the local equilibrium, but also the corrections to it at all orders in ε. Let us compute the first correction F_1, by using (4.43). Since the hydrodynamical fields satisfy the Euler equations,

$$PD_t M = 0, \tag{5.1}$$

where we use the short notation

$$D_t F = (\partial_t F + v \cdot \nabla_x F), \tag{5.2}$$

after a straightforward calculation we have

$$D_t M = P^{\perp} D_t M = M \sum_{\alpha,\beta=1}^{3} A_{\alpha,\beta} D_{\alpha,\beta} + MB \cdot \nabla_x T, \tag{5.3}$$

where

$$D_{\alpha,\beta} = \frac{1}{2}(\partial_{x_\alpha} u_\beta + \partial_{x_\beta} u_\alpha) \tag{5.4}$$

and A and B are given by (4.30). Therefore by (4.31), we obtain

$$F_1 = -\lambda_1 \sum_{\alpha,\beta=1}^{3} A_{\alpha,\beta} D_{\alpha,\beta} - \lambda_2 B \cdot \nabla_x T + \widehat{F}_1. \tag{5.5}$$

The above expression is the main tool to compute the viscosity and heat conduction corrections, but the use of the Hilbert expansion hardly could provide the Navier–Stokes equations, because the hydrodynamical fields appearing in it are solutions to the Euler equation, rather than the Navier–Stokes equations. Indeed, the hydrodynamic part $\widehat{F}_1$ of F_1 satisfies a system of linear PDEs which are the compatibility conditions for (4.38) with $n = 2$. Solving them, one can compute the corrections ρ_1, u_1 and T_1 to the solutions to the Euler equations (4.40). As a consequence one can consider the fields

$$\rho_\varepsilon = \rho + \varepsilon\rho_1, \qquad u_\varepsilon = u + \varepsilon u_1, \qquad T_\varepsilon = T + \varepsilon T_1.$$

By computing their time derivative one can check that they do not satisfy a closed system of equations and one has to add some suitable $O(\varepsilon^2)$ terms in order to get such a system. One of the possible procedures to do this provides the Navier–Stokes equations.

In fact the derivation of the Navier–Stokes equations is obtained by a suitable resummation of the Hilbert series which is known as the Chapmann–Enskog expansion. We do not explain its original version in details, but present a simplified version which is sufficient to compute correctly the first order correction and produces the Navier–Stokes equations. For the full version of the Chapmann–Enskog expansion we refer to [27,30] and references quoted therein.

Since we are interested in the corrections to the local equilibrium, we write the solution F^ε to (4.33) (after some initial layer) as

$$F^\varepsilon = M + \varepsilon G^\varepsilon, \tag{5.6}$$

where M is a local Maxwellian with hydrodynamical fields not yet determined and G^ε is a correction to it. There is no loss of generality in assuming

$$P_M G^\varepsilon = 0, \tag{5.7}$$

P_M being the projector on $\mathrm{Null}\, L_M$ and L_M the linear Boltzmann operator with respect to M. In fact, if (5.7) were not satisfied, one could redefine the hydrodynamical fields in M to absorb the hydrodynamical part of G^ε.

The equation in terms of G^ε can be written as

$$D_t M + \varepsilon D_t G^\varepsilon = L_M G^\varepsilon + \varepsilon Q\big(G^\varepsilon, G^\varepsilon\big). \tag{5.8}$$

By projecting it on Null L_M and $(\text{Null}\, L_M)^\perp$, we obtain respectively

$$P_M D_t M + \varepsilon P_M D_t G^\varepsilon = 0, \tag{5.9}$$

$$L_M G^\varepsilon + \varepsilon Q(G^\varepsilon, G^\varepsilon) = P_M^\perp D_t M + \varepsilon P_M^\perp D_t G^\varepsilon. \tag{5.10}$$

But by (5.7), (5.10) is equivalent to

$$G^\varepsilon = L_M^{-1} P_M^\perp D_t M + \varepsilon L_M^{-1}\left(P_M^\perp D_t G^\varepsilon - Q(G^\varepsilon, G^\varepsilon)\right). \tag{5.11}$$

Because of (5.11), in (5.9) we can approximate G^ε to the lowest order by

$$G_1 = L_M^{-1} P_M^\perp D_t M. \tag{5.12}$$

By (5.3) and (5.5), we obtain

$$G_1 = -\lambda_1 \sum_{i,j=1}^{3} A_{i,j} D_{i,j} - \lambda_2 B \cdot \nabla_x T, \tag{5.13}$$

but now the hydrodynamical fields are not solutions to the Euler equations, as in the Hilbert method, but rather to the equation we obtain by replacing G^ε by G_1 in (5.9):

$$P_M D_t M + \varepsilon P_M D_t G_1 = 0. \tag{5.14}$$

It should be noted that G_1 depends on ε as well as M, whose parameters solve equations depending on ε. They follow from some Gaussian integrations, namely it is immediate to obtain the equations

$$\begin{aligned}
&\partial_t \rho + \nabla_x \cdot [\rho u] = 0,\\
&\rho \partial_t u + \rho(u \cdot \nabla_x) u + \nabla_x P = \nabla_x \cdot (\nu_\varepsilon \nabla_x u) + \nabla_x(\sigma_\varepsilon \nabla_x \cdot u),\\
&\frac{3}{2}\rho\left[\partial_t T + \rho(u \cdot \nabla_x) T\right] + P \nabla_x \cdot u = \nabla_x \cdot (\kappa_\varepsilon \nabla_x T) + \nu_\varepsilon (\nabla_x u)^2 + \sigma_\varepsilon (\nabla_x u)^2,
\end{aligned} \tag{5.15}$$

with

$$\mu_\varepsilon = \varepsilon\mu, \qquad \sigma_\varepsilon = \varepsilon\sigma, \qquad \kappa_\varepsilon = \varepsilon\kappa \tag{5.16}$$

and

$$\nu = \int_{\mathbb{R}^3} \mathrm{d}v \left[A_{1,2}(v)\right]^2 \lambda_1(|v|), \qquad \sigma = \int_{\mathbb{R}^3} \mathrm{d}v \left[A_{1,1}(v)\right]^2 \lambda_1(|v|), \tag{5.17}$$

$$\kappa = \int_{\mathbb{R}^3} \mathrm{d}v \left[B_1(v)\right]^2 \lambda_2(|v|). \tag{5.18}$$

Equations (5.15) are the Navier–Stokes–Fourier equations for the ideal gas, with transport coefficients proportional to ε and given by (5.17) and (5.18). Hence one of the aims of the derivation is achieved, namely the relation between the transport coefficients and the microscopic interaction, which appears in the coefficients λ_γ through the scattering cross section.

Notice that, since formally

$$-L^{-1} = \int_0^\infty \mathrm{d}t\, \mathrm{e}^{tL}, \tag{5.19}$$

where e^{tL} is the semigroup with generator L, the formulas for the transport coefficients can be rewritten as

$$\begin{aligned} \nu &= \int_0^\infty \mathrm{d}t \int_{\mathbb{R}^3} \mathrm{d}v\, A_{1,2}\mathrm{e}^{tL}(A_{1,2}M), \\ \sigma &= \int_0^\infty \mathrm{d}t \int_{\mathbb{R}^3} \mathrm{d}v\, A_{1,1}\mathrm{e}^{tL}(A_{1,1}M), \end{aligned} \tag{5.20}$$

$$\kappa = \int_0^\infty \mathrm{d}t \int_{\mathbb{R}^3} \mathrm{d}v\, B_1\mathrm{e}^{tL}(B_1 M). \tag{5.21}$$

The above time integrals are convergent because of (4.31). Formulas of the type (5.20) and (5.21) are known in the physical literature as Green–Kubo formulas [55,62] and the ones described above are probably the first established rigorously.

We notice that the equations we have obtained depend on ε as well as their solutions, so that the above procedure, which is a simplified version of the Chapmann–Enskog expansion, by no means can be interpreted as a power series expansions, but rather as a suitable partial re-summation of the Hilbert expansion, since it contains powers of ε of arbitrary order, not just the first-order terms. Of course, different re-summations are conceivable with different resulting equations, more or less consistent from the physical point of view. The main reason to accept (5.15) as the equations describing the evolution of a viscous, heat conducting ideal gas is in the great success of their use in many applications. On the other hand, it would be desirable to remove such ambiguities from the conceptual point of view, in order to give a solid microscopic foundation to the Navier–Stokes–Fourier equations.

Another remark which is in order about Equations (5.15) is that the transport coefficient appearing there are of order ε, which is a very small parameter. Therefore their effect on the solutions can be appreciated in one of the two following cases:

(1) There are sufficiently large spatial gradients to compensate the small factor ε in front of the second-order operators;
(2) The time interval where we look at the solutions is so large that the cumulative effects of the second-order operators become relevant.

The first case is discussed in the papers quoted at the end of the previous section. The second case is the object of the rest of this section. In both cases the effective transport coefficients are finite.

The above difficulties are related to the lack of scale invariance of the Navier–Stokes equations, a property that is necessary, as remarked before, in order to derive them as a sharp scaling limit, and hence without ambiguities, from an underlying more fundamental model.

To be more explicit, let us write down the equations we would like to derive, namely Equations (5.15) with ν_ε, σ_ε and κ_ε replaced by ν, σ and κ given by (5.17) and (5.18):

$$\begin{aligned}
&\partial_t \rho + \nabla_x \cdot [\rho u] = 0,\\
&\rho \partial_t u + \rho (u \cdot \nabla_x) u + \nabla_x P = \nabla_x \cdot (\nu \nabla_x u) + \nabla_x (\sigma \nabla_x \cdot u),\\
&\frac{3}{2}\rho\big[\partial_t T + \rho (u \cdot \nabla_x) T\big] + P \nabla_x \cdot u = \nabla_x \cdot (\kappa \nabla_x T) + \nu (\nabla_x u)^2 + \sigma (\nabla_x u)^2.
\end{aligned} \tag{5.22}$$

Suppose that we scale the time and space according to the scaling

$$x' = \varepsilon^{-1} x, \qquad t' = \varepsilon^{-1} t, \qquad \textit{hyperbolic scaling.}$$

Since first-order differential operators scale as ε, while second-order ones scale as ε^2, what we obtain are exactly (5.15), namely (5.22) with ε in front of the transport coefficients. This is not surprising, since this is exactly the scaling under which we obtained them.

In order to get finite transport coefficients after the scaling one can try with different powers of ε. Therefore we assume the most general scaling:

$$x' = \varepsilon^{-1} x, \qquad t' = \varepsilon^{-a} t.$$

The equations become:

$$\begin{aligned}
&\partial_{t'} \rho + \varepsilon^{1-a} \nabla_{x'} \cdot [\rho u] = 0,\\
&\rho \partial_{t'} u + \varepsilon^{1-a}\big[\rho (u \cdot \nabla_{x'}) u + \nabla_{x'} P\big]\\
&\quad = \varepsilon^{2-a}\big[\nabla_{x'} \cdot (\nu \nabla_{x'} u) + \nabla_{x'} (\sigma \nabla_{x'} \cdot u)\big],\\
&\frac{3}{2}\rho\big[\partial_{t'} T + \varepsilon^{1-a}\big[\rho (u \cdot \nabla_{x'}) T\big] + P \nabla_{x'} \cdot u\big]\\
&\quad = \varepsilon^{2-a}\big[\nabla_{x'} \cdot (\kappa \nabla_{x'} T) + \nu (\nabla_{x'} u)^2 + \sigma (\nabla_{x'} u)^2\big].
\end{aligned} \tag{5.23}$$

It is clear that there is no choice of a which makes the rescaled equation independent of ε. The case $a = 1$ is the one already discussed. The case $a = 2$, called *parabolic scaling* provides finite transport coefficients, but the Eulerian transport terms are proportional to ε^{-1} and hence very large. This is also true if $1 < a < 2$ where, however, the dissipative terms are still negligible.

On the other hand, if one can make the Eulerian part small with an extra scaling, the choice $a = 2$ has good chances to be the right candidate to get finite transport coefficients.

One can notice that all the Euler terms involve the velocity field u. If u is small of order ε^{a-1}, then this smallness can compensate the diverging factor. Therefore, we assume the extra scaling

$$u = \varepsilon^{a-1} u'. \tag{5.24}$$

This scaling, for $a > 1$, means that we are considering velocities very small compared with the sound speed, that is the only parameter with the dimensions of a velocity appearing in the equations through the pressure. This situation is usually referred to as the *low Mach number* limit, the Mach number being the ratio between the velocity of the flow and the sound speed. By including the scaling (5.24), (5.23) become:

$$\begin{aligned}
&\partial_{t'}\rho + \nabla_{x'}\cdot\left[\rho u'\right] = 0,\\
&\rho\partial_{t'}u' + \rho\left(u'\cdot\nabla_{x'}\right)u' + \varepsilon^{2(1-a)}\nabla_{x'}P\\
&\quad = \varepsilon^{2-a}\left[\nabla_{x'}\cdot\left(\nu\nabla_{x'}u'\right) + \nabla_{x'}\left(\sigma\nabla_{x'}\cdot u'\right)\right],\\
&\frac{3}{2}\rho\left[\partial_{t'}T + \rho\left(u'\cdot\nabla_{x'}\right)T\right] + P\nabla_{x'}\cdot u'\\
&\quad = \varepsilon^{2-a}\left[\nabla_{x'}\cdot\left(\kappa\nabla_{x'}T\right) + \varepsilon^{2(a-1)}\left(\nu\left(\nabla_{x'}u'\right)^2 + \sigma\left(\nabla_{x'}u'\right)^2\right)\right].
\end{aligned} \tag{5.25}$$

The only negative powers of ε are in front of the pressure in the second equation of (5.25). The asymptotics of this system is well known in Hydrodynamics and it has been proved, in particular cases, that it corresponds to the incompressible regime: If $a < 2$ it corresponds to the incompressible Euler regime, while, for $a = 2$, it is the incompressible limit of the Navier–Stokes equations [61].

Let us consider only the case $a = 2$. An inspection of $(5.25)_2$, shows that the pressure has to be constant up to the second order in ε:

$$P = P_0 + \varepsilon P_1 + \varepsilon^2 P_2, \tag{5.26}$$

with ∇P_0 and ∇P_1 vanishing. Since P is a function of ρ and T, assumptions on their initial data are necessary. For the ideal gas, where the pressure P is given by ρT, a particular choice to fulfill the conditions on the pressure is

$$\rho = \bar{\rho} + \varepsilon\rho_1 + \cdots, \qquad T = \overline{T} + \varepsilon\theta + \cdots, \tag{5.27}$$

$$\nabla_x\left(\overline{T}\rho_1 + \bar{\rho}\theta\right) = 0, \tag{5.28}$$

with $\bar{\rho}$ and $\overline{T}$ constants. By suitable choices of the units we set such constants to 1. Equation (5.28) is known as the Boussinesq condition [16]. More general choices are possible and correspondingly a rich phenomenology has been found. Since the only mathematical results, as far as we know, are obtained in the conditions (5.27) and (5.28), we confine ourselves to this case and refer to [82], for the interesting phenomena arising when (5.27) is not assumed.

The above discussion suggests to look at the Boltzmann equation in the parabolic scaling, $a = 2$, with the extra scaling on the velocity field and the assumptions (5.27) and (5.28) at the initial time. Therefore we define

$$F^\varepsilon(x, v, t) = f^\varepsilon\left(\varepsilon^{-1}x, v, \varepsilon^{-2}t\right),$$

which satisfies

$$(\partial_t + \varepsilon^{-1} v \cdot \nabla_x) F^\varepsilon = \varepsilon^{-2} Q[F^\varepsilon, F^\varepsilon], \quad F^\varepsilon(\cdot,\cdot,0) = F_\varepsilon^{(0)}(\cdot,\cdot). \tag{5.29}$$

The assumptions on the initial density and temperature are:

$$\begin{aligned} &\int_{\mathbb{R}^3} \mathrm{d}v\, F_\varepsilon^{(0)}(\cdot, v) = 1 + \varepsilon \rho_1^{(0)} + \mathrm{O}(\varepsilon^2), \\ &\int_{\mathbb{R}^3} \mathrm{d}v\, \frac{v^2}{2} F_\varepsilon^{(0)}(\cdot, v) = 1 + \varepsilon \theta^{(0)} + \mathrm{O}(\varepsilon^2), \end{aligned} \tag{5.30}$$

with

$$\nabla_x \big(\rho_1^{(0)} + \theta^{(0)}\big) = 0. \tag{5.31}$$

The low Mach number assumption is

$$\int_{\mathbb{R}^3} \mathrm{d}v\, v F_\varepsilon^{(0)}(\cdot, v) = \varepsilon u^{(0)} + \mathrm{O}(\varepsilon^2). \tag{5.32}$$

A particular choice of $F_\varepsilon^{(0)}$ satisfying the above conditions is

$$\begin{aligned} &F_\varepsilon^{(0)} = M_0^\varepsilon, \\ &M_0^\varepsilon(\cdot, v) = \frac{1 + \varepsilon \rho_1^{(0)}}{[2\pi(1 + \varepsilon \theta^{(0)})]^{3/2}} \exp\left[-\frac{(v - \varepsilon u^{(0)})^2}{2(1 + \varepsilon \theta^{(0)})} \right]. \end{aligned} \tag{5.33}$$

One can use the same arguments presented for the Euler equations to find a solution to (5.29) in the form of a truncated expansion with remainder:

$$F^\varepsilon(\cdot,\cdot,t) = M_t^\varepsilon(\cdot,\cdot) + \varepsilon^2 f_2(\cdot,\cdot,t) + \varepsilon^3 f_3(\cdot,\cdot,t) + \cdots, \tag{5.34}$$

with

$$M_t^\varepsilon(\cdot, v) = \frac{1 + \varepsilon \rho_1(\cdot, t)}{[2\pi(1 + \varepsilon \theta(\cdot, t))]^{3/2}} \exp\left[-\frac{(v - \varepsilon u(\cdot, t))^2}{2(1 + \varepsilon \theta(\cdot, t))} \right] \tag{5.35}$$

and f_n to be computed by some recursive relations. Note that the term f_1 is missing in (5.34). In fact, even if one includes a nonvanishing f_1, the equation forces it to be in $\operatorname{Null} L$, so it can be absorbed in the Maxwellian M_t^ε by redefining u, θ and ρ_1. The solvability conditions for f_2 require that

$$\begin{aligned} &\nabla_x \cdot u = 0, \\ &\nabla_x (\theta + \rho_1) = 0. \end{aligned} \tag{5.36}$$

The first is the incompressibility condition while the second is the Boussinesq condition. Both have to be satisfied at any time, not just at the initial time. The solvability condition for f_3 requires

$$\begin{aligned} &\partial_t u + u \cdot \nabla_x u + \nabla_x p = \nu \Delta u, \\ &\frac{5}{2}(\partial_t \theta + u \cdot \nabla_x \theta) = \kappa \Delta \theta, \end{aligned} \tag{5.37}$$

where ν and κ are given by (5.17) and (5.18) with the local Maxwellian replaced by the global Maxwellian

$$M_0(v) = \frac{1}{(2\pi)^{3/2}} \exp\left[-\frac{v^2}{2}\right]. \tag{5.38}$$

The first of (5.36) and the first of (5.37) are the *incompressible Navier–Stokes equations* for a fluid. Note that the pressure appearing in it is not the one given by the equation of state of the ideal gases, as in the Euler equation, but rather an unknown multiplier, with the physical meaning of reaction of the fluid due to the constraint of incompressibility. The second of (5.37) is the heat equation with a transport term due to the flow of the gas. Finally, one can satisfy the second of (5.36) for example by choosing $\rho_1 = -\theta$.

Given a smooth solution to above equations in a suitable time interval, one can construct recursively the next terms of the expansion and, by the procedure illustrated in Section 4, it is possible to control the remainder equation (see [32]) thus providing the proof of the following theorem:

THEOREM 5.1. *Assume*:

(1) *The initial datum is given by* (5.33), *with* $\rho_1^{(0)} = -\theta^{(0)}$.

(2) *The initial values for* θ, *and* u, *are sufficiently smooth, so that there is a time* t_1 *and a unique solution* (θ, u, p) *in the time interval* $[0, t_1]$ *to the incompressible Navier–Stokes–Fourier equations with such initial data, with bounded Sobolev norms up to an order* s *sufficiently large.*

Then, there is $\varepsilon_0 > 0$ *such that, for any* $0 < \varepsilon < \varepsilon_0$, *there is a smooth nonnegative solution* F^ε *to* (5.29) *and the solution is unique in the class of the smooth solutions. Moreover, there are* $\gamma > 0$, *and positive integers* ℓ *and* s *such that*

$$\sup_{0 \leqslant t \leqslant t_1} \left\| F^\varepsilon(\cdot, \cdot, t) - M_t^\varepsilon \right\|_{\gamma, \ell, s} < C\varepsilon^2$$

for a suitable C.

Note that, although the result is stated in terms of closeness to the local equilibrium, in order to compute the hydrodynamical parameters determining the local equilibrium, one has to know the transport coefficients which are computed in terms of the first correction to the local equilibrium.

The above result, as the one discussed for the compressible Euler equations, is valid on a time interval where the hydrodynamical equations have a smooth solution, namely locally in time for general initial data and globally in time in two space dimensions or for small three-dimensional initial data. On the other hand, the incompressible Navier–Stokes equations have global weak solutions, although smoothness and uniqueness are not known.

Therefore it is natural to ask whether the validity of the incompressible Navier–Stokes equations can be obtained under weaker assumptions. In particular, one may try to show that the DiPerna–Lions solutions to the Boltzmann equation [37] converge to the Leray–Hopf weak solutions to the incompressible Navier–Stokes equations under the above scaling limit. The problem is quite difficult because we have a very little control of the weak solutions to the Boltzmann equations, just conservation of the L^1 norm and energy and entropy inequalities. In particular, the conservation laws (4.12) are only formal in this framework.

This program was started in [6,7] and progressive improvements have been obtained in a long list of papers by several authors [8–10,49,50,66–68] till the very recent work [51] which provides a complete proof under some restrictions on the scattering cross section.

We sketch here the main ideas of this approach. The Boltzmann equation, after a suitable rescaling, takes the form

$$\tau_\varepsilon \partial_t F^\varepsilon + v \cdot \nabla_x F^\varepsilon = \varepsilon^{-1} Q\big[F^\varepsilon, F^\varepsilon\big], \tag{5.39}$$

with

$$F^\varepsilon = M_0(1 + \delta_\varepsilon g_\varepsilon), \tag{5.40}$$

τ_ε and δ_ε being suitably chosen. To compare (5.39) with (5.29) discussed above, one has to choose $\tau_\varepsilon = \delta_\varepsilon = \varepsilon$. Formally, the solutions to (5.39) converge to different regimes, depending on the choices of δ_ε and τ_ε: If $\tau_\varepsilon = 1$ (Euler time scale) and $\delta_\varepsilon \to 0$, the result is the linear acoustic; if $\tau_\varepsilon = \delta_\varepsilon$ we get as we know, the incompressible regime, Eulerian if $\tau_\varepsilon \varepsilon^{-1} \to \infty$ and Navier–Stokes if $\tau_\varepsilon = \varepsilon$. Finally, the result is the Stokes equations if $\tau_\varepsilon = \varepsilon$ and $\delta_\varepsilon \varepsilon^{-1} \to 0$.

Let us focus on the incompressible Navier–Stokes regime which is the most interesting one. We assume τ_ε and δ_ε coinciding with ε. Hence the deviation from the Maxwellian, g_ε, satisfies the equation

$$\begin{aligned} \varepsilon \partial_t g_\varepsilon + v \cdot \nabla_x g_\varepsilon &= \frac{1}{\varepsilon^2} M_0^{-1} Q\big[F^\varepsilon, F^\varepsilon\big] \\ &= \varepsilon^{-1} \widetilde{L} g_\varepsilon + M_0^{-1} Q[M_0 g_\varepsilon, M_0 g_\varepsilon], \end{aligned} \tag{5.41}$$

where $\widetilde{L} f = 2M_0^{-1} Q[M_0, M_0 f]$. Equation (5.41) implies that, if g_ε converges, its limit $g \in \operatorname{Null} \widetilde{L}$. Thus by (4.23),

$$g = \rho_1 + u \cdot v + \frac{1}{2}\theta\big(v^2 - 3\big). \tag{5.42}$$

Multiplying by M_0 and $M_0 v$ and integrating (5.41) on v, we obtain

$$\nabla \cdot u = 0, \qquad \nabla(\rho_1 + \theta) = 0,$$

which are the incompressibility and Boussinesq conditions.

In order to determine the equations for the hydrodynamic fields, we note that multiplying (5.41) by $M_0 v \tau_\varepsilon^{-1}$ we get

$$\partial_t u_\varepsilon + \varepsilon^{-1} \nabla \cdot \int \mathrm{d}v\, M_0 g_\varepsilon v \otimes v = 0, \tag{5.43}$$

with $u_\varepsilon = \int M_0 g_\varepsilon v\, \mathrm{d}v$. To compute the tensor in the previous equation one notes that its diagonal part produces a pressure term, so one is left with computing the contribution due to $A = v \otimes v - \frac{v^2}{3}\mathbb{1}$. By (4.31) there is φ such that $A = \widetilde{L}\varphi$ and hence

$$\varepsilon^{-1} \int \mathrm{d}v\, M_0 g_\varepsilon A = \varepsilon^{-1} \int \mathrm{d}v\, M_0 g_\varepsilon \widetilde{L}\varphi = \int \mathrm{d}v M_0 \bigl(\varepsilon^{-1} \widetilde{L} g_\varepsilon\bigr)\varphi,$$

because L is self-adjoint. Now $\varepsilon^{-1} L g_\varepsilon$ can be computed from (5.41). The result is

$$\varepsilon^{-1} \nabla \cdot \int \mathrm{d}v\, M_0 g_\varepsilon v \otimes v \to -\nu \Delta u + \nabla \cdot (u \otimes u) + \nabla p.$$

Thus one obtains the incompressible Navier–Stokes. In a similar fashion, we could obtain the incompressible Stokes and Euler equations.

Such a derivation, presented in [7], is of course only formal and obviously provides the same results which follow from the power series expansion. There are several difficulties to make rigorous the present argument, among them the fact that the DiPerna–Lions solutions are not guaranteed to satisfy the conservation laws. Such solutions, however, satisfy the following entropy bound:

$$\varepsilon^{-2} H\bigl(F^\varepsilon \big| M_0\bigr) + \frac{\varepsilon^{-2}}{\varepsilon \tau_\varepsilon} \int_0^t \mathrm{d}s \int \mathrm{d}x\, D\bigl[F^\varepsilon(x, \cdot, s)\bigr] \leqslant \varepsilon^{-2} H\bigl(F^\varepsilon(\cdot, \cdot, 0) \big| M_0\bigr),$$

where $H(\cdot|\cdot)$ is the relative entropy and D is defined in (4.8). Now the initial datum is assumed such that the right-hand side of this inequality is finite, so that weak compactness of the family of solutions is at hand. The convergence, however, is not strong enough to guarantee the limiting procedures. We do not enter into the details necessary to overcome the difficulties one has to face in making rigorous the program. After the first papers by Bardos, Golse and Levermore [6,7], most of the obstructions have been removed in a sequence of papers [8–10,49–51,66–68] and at the moment complete proofs are available for the incompressible Navier–Stokes case, with just a restriction left on the scattering cross section. The theorem can be stated as follows:

THEOREM 5.2. *Assume that the scattering cross section $B(|V|, \omega)$ is such that the following two conditions are verified*:

(i) $0 < c_1 < B(|V|, \omega)/|\cos(\widehat{V\omega})| < c_2$;

(ii) *the functions λ_γ in (4.31) (which depend on B) are such that $\lambda_\gamma M_0^{-1}$ grow at most as polynomials in v.*

Moreover, the spatial domain is $\mathbb{R}^3$ and the initial values of u_ε and θ_ε converge in $L^2(\mathbb{R}^3)$ to u_0 and θ_0. Then there is a subsequence $\varepsilon_n \to 0$ such that $M_0 g_{\varepsilon_n}$ converges weakly to $M_0 g$, g is of the form (5.41) with $\rho_1 + \theta = 0$, $\nabla \cdot u = 0$ and (u, θ) are weak solutions to (5.37) with initial data (u_0, θ_0) and ν and κ given by (5.17) and (5.18) with the local Maxwellian replaced by M_0.

5.2. *Navier–Stokes equations from the Newton equations*

One can try to apply the above ideas to Hamiltonian particle systems. The difficulties here are related to the lack of ergodic properties as for the case of compressible Euler equations (see Section 2). In addition, since in the parabolic scaling we are looking at much larger times, one has to compute the corrections to the local equilibrium which determine the transport coefficients.

As discussed before, one has to restrict the analysis to the low Mach numbers limit. One can look at an initial local Gibbs state of of the type (2.17), with the hydrodynamical fields satisfying the analogues of (5.30) and (5.32)

$$\begin{aligned} &P_0^\varepsilon(x_1, v_1, \ldots, x_N, v_N) \\ &\quad = Z^{-1} \prod_{i=1}^N \exp\Bigg[\big(1 + \varepsilon\mu_0(x_i)\big) - \frac{1}{2(1+\varepsilon\theta_0(x_i))} \\ &\qquad \times \Bigg[\big(v_i - \varepsilon u_0(x_i)\big)^2 + \sum_{j\neq i} \phi\big(\varepsilon^{-1}|x_i - x_j|\big)\Bigg]\Bigg]. \end{aligned} \tag{5.44}$$

Consider the Liouville equation in the parabolic scaling:

$$\partial_t f_t^\varepsilon + \mathcal{L}^\varepsilon f_t^\varepsilon = 0, \tag{5.45}$$

where

$$\mathcal{L}^\varepsilon f = \sum_{i=1}^N \varepsilon^{-1} v_i \cdot \partial_{x_i} f + \varepsilon^{-2} \sum_{i=1}^N \sum_{j\neq i} (\nabla\phi)\big(\varepsilon^{-1}(x_i - x_j)\big) \cdot \partial_{v_i} f \tag{5.46}$$

is the rescaled Liouville operator. In this case the local equilibrium assumption would be the following: Let

$$\begin{aligned} &P_t^\varepsilon(x_1, v_1, \ldots, x_N, v_N) \\ &\quad = Z^{-1} \prod_{i=1}^N \exp\Bigg[\big(1 + \varepsilon\mu(x_i, t)\big) - \frac{1}{2(1+\varepsilon\theta(x_i, t))} \\ &\qquad \times \Bigg[\big(v_i - \varepsilon u(x_i, t)\big)^2 + \sum_{j\neq i} \phi\big(\varepsilon^{-1}|x_i - x_j|\big)\Bigg]\Bigg], \end{aligned} \tag{5.47}$$

then the solution to the Liouville equation, (5.45), can be written as

$$f_t^\varepsilon = P_t^\varepsilon + \varepsilon^2 R_t^\varepsilon. \tag{5.48}$$

While in the Euler case we did not need any information about the deviations from the local equilibrium, R_t^ε, in the Navier–Stokes case we need to solve the analog of (4.43) to find the first correction to the local equilibrium. In this case, however, the situation is much less clear because, while for the linearized Boltzmann operator L we could use the Fredholm alternative to solve the equation and get (5.5), the Liouville operator $\mathcal{L}^\varepsilon$ has no such good spectral properties and the solution of the analog of (4.43) is too difficult to construct. Anyway, although it is not possible to solve rigorously the equation, one can at least compute, at a formal level, the solution in the special case of interest. Using this one can find the limiting equations which are the incompressible Navier–Stokes–Fourier equations

$$\begin{aligned} &\partial_t u + u \cdot \nabla_x u + \nabla_x p = \nu \Delta u, \\ &c_p(\partial_t \theta + u \cdot \nabla_x \theta) = \kappa \Delta \theta, \end{aligned} \tag{5.49}$$

where c_p is the *specific heat at constant pressure* and the transport coefficients are given by the Green–Kubo formulas. In fact the viscosity coefficient is given by

$$\begin{aligned} \nu &= \int_0^\infty \mathrm{d}t \int \mathrm{d}x\, \mathbb{E}\big[\bar{\zeta}_{1,2}(0,0)\bar{\zeta}_{1,2}(x,t)\big] \\ &= \int_0^\infty \mathrm{d}t \int \mathrm{d}x\, \mathbb{E}\big[\bar{\zeta}_{1,2}(0,0)\big(\exp[t\mathcal{L}]\bar{\zeta}_{1,2}\big)(x,0)\big], \end{aligned} \tag{5.50}$$

where the expectation $\mathbb{E}$ is with respect to the global Gibbs state (5.45) with $\varepsilon = 0$, $\mathcal{L}$ is the unscaled Liouville operator and $\bar{\zeta}_{1,2}$ is given by (2.6) after a suitable subtraction. The heat conduction coefficient is given by a similar formula. For the details of this formal derivation we refer to [43] and references quoted therein. Note that it is not obvious (and indeed not known) that the time integral in (5.50) is finite, because the time evolution is the Hamiltonian and the spectrum of $\mathcal{L}$ does not contain any negative real part. Hence the situation in this case is much worse than in the Euler case where, as in [76], one can make a suitable ergodicity assumption and then derive rigorously the limiting equations. In this case it is totally unclear how to overcome the second obstruction, related to the spectral properties of the Liouville operator.

Recently Varadhan [86] has introduced a technique, known as the *nongradient method*, which is designed to solve the above class of problems in stochastic systems. At the moment it is not known if the method works for continuous particle systems with some stochastic noise. On the other hand, it has been successfully applied to several examples of particle systems on the lattice. This technique is the main tool to study stochastic lattice gas models with the aim of deriving the Navier–Stokes equation.

The simplest case to consider is the Simple Exclusion Process, which converges to the Burgers equation in the Euler limit, as discussed in Section 3. We now consider it in the parabolic scaling, where the diffusive effects may become significant. As before, the Eulerian part of the evolution becomes too large in general, so we need to include the analog of the low Mach numbers assumption, namely we assume the initial density ρ_0 of the form

$$\rho_0 = \bar{\rho} + \varepsilon u_0, \tag{5.51}$$

where $\bar{\rho} \in (0, 1)$ is some background constant density, while u_0 represents the deviation from the constant profile, which is assumed of order ε. In the parabolic scaling and with initial conditions fulfilling (5.51) it is possible to prove [45] that, in dimension three or more, at any positive time t the empirical density of the process differs (weakly in probability) from the constant $\bar{\rho}$ by a deviation $\varepsilon u(x, t)$, where u is the solution to the viscous Burgers equation

$$\partial_t u + F \cdot \nabla_x u^2 = \nabla_x \cdot [D \nabla_x u], \qquad u(\cdot, 0) = u_0(\cdot). \tag{5.52}$$

Here the diffusivity matrix D is determined by a variational formula which is equivalent to the Green–Kubo formula for this model. In particular, the convergence of the time integral in the Green–Kubo formula is a consequence of the result. The method employed to prove the result is the entropy method of Yau [89], which compares the nonequilibrium distribution with the local equilibrium. In this case it is necessary to include the correction to the local equilibrium, which is determined by using the Varadhan method [86].

The SEP is an example of a particle system where the program of deriving the Navier–Stokes equation (in the low Mach numbers regime) can be accomplished. Although it captures several features of particle systems, it is far from being a realistic model, namely the dynamics is on a lattice, it is stochastic, the particles do not have velocity and there is only one conserved quantity, the mass. However, we can consider more realistic systems on the lattice including all the conservation laws of a standard fluid.

In order to give a meaning to the notion of velocity on the lattice, we note that particles jumping on the lattice according to SEP with a jump probability whose drift is F, can be interpreted as particles moving with uniform velocity F, and subject to some noise. This provides a possible way to construct a model of particles with velocities. We fix a finite set $\mathcal{V}$ of vectors $v \in \mathbb{R}^3$ and, for each $v \in \mathcal{V}$, we consider a species of particles moving according to a SEP with drift v. The particles of the different species can coexist in the same point of the lattice, while there is exclusion between particles with the same velocity. The evolution of such a system would be just the evolution of each species independently of the others. Then we introduce the interaction by including collisions: Two particles with velocity $v_1, v_2 \in \mathcal{V}$ sitting at the same site x can collide at random exponential times. The result of the collision is the annihilation of the two particles and the creation of two new particles with velocities $v_1', v_2' \in \mathcal{V}$, provided that the process does not violate the exclusion rule and the total momentum is conserved:

$$v_1' + v_2' = v_1 + v_2. \tag{5.53}$$

Of course particular assumptions are necessary on the set of velocities $\mathcal{V}$ in order that all makes sense. Under suitable assumption, it is possible to check that the only conserved quantities are total mass (regardless of the species) and total momentum. The hydrodynamical behavior of such a system can be studied on the Euler time scale, where one can derive a system of conservation laws for the macroscopic mass and momentum densities in the parabolic scaling and the low Mach numbers regime. The above model has been introduced and studied in [46] where it has been proved, with a suitable choice of $\mathcal{V}$ and the parabolic scaling, that in the low Mach numbers regime, the hydrodynamical limit is given by the incompressible Navier–Stokes equations with viscosity coefficient given by the Green–Kubo formula. The result in [46] is restricted to smooth solutions to the Navier–Stokes equations. The extension to the weak solutions, via the study of the "large deviations", has been done in [79]. A model including also the conservation of energy has been introduced in [12] in order to obtain the analog of (5.49) and the results have been extended to this case. For a detailed discussion of the ideas and methods involved in the study of such models we refer to [44].

6. Kinetic and hydrodynamical limits for the Lorentz model

6.1. *Mathematical properties of the linear Boltzmann equation*

In this section we consider the Lorentz model which consists of a single particle moving in a given distribution of scatterers. For this simple model, often used for the study of transport problems and for numerical computations, we derive a linear Boltzmann equation under the low density scaling and, from this, the linear heat equation under the diffusive limit. We also mention the important result by Bunimovich and Sinai, showing that it is possible to derive a diffusion equation directly from the microscopic model under the diffusive scaling.

In other words, all three steps:

(1) From the microscopic model to the Boltzmann description,

(2) From the Boltzmann to the hydrodynamic equations,

(3) From the microscopic model to the hydrodynamical equations,

we have discussed in the previous sections, are fully carried out in this simple linear context. Moreover the first two steps are also technically easy so that we discuss them here in some detail. The third one is however much deeper and technically difficult so that we limit ourselves to establish the result.

This section is devoted to the study of the following linear equation

$$\partial_t f = Lf = (K - I)f, \tag{6.1}$$

where

$$Kf(v') = \int_{S^1} \mathrm{d}v\, k(v|v') f(v) \tag{6.2}$$

and $k(v|v')$ is a transition probability density: $k : S^1 \times S^1 \to \mathbb{R}^+$ and $\int k(v|v')\,\mathrm{d}v' = 1$ for all v belonging to S^1.

The process described by (6.1) is that of a particle with velocity of modulus one, having random transitions (collisions) which preserve the energy. We introduce a physical mechanism for such transitions. Consider a particle, with initial velocity $v \in S^1$, hitting a circular

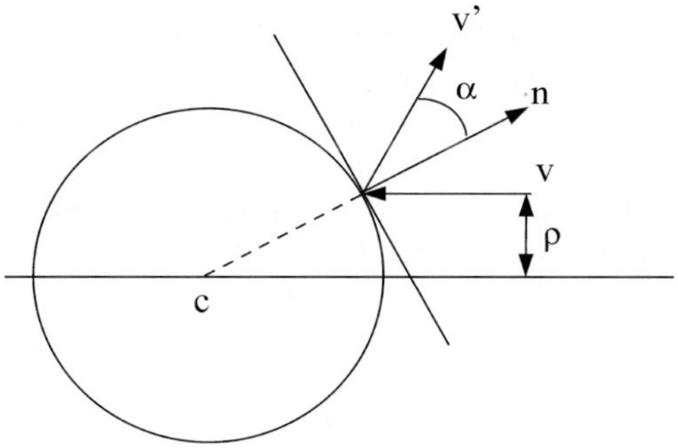

Fig. 6.

obstacle of unit diameter (see Figure 6), whose center c is random. If we denote by ρ the impact parameter and n the outward normal in the collision point, we can compute the outgoing velocity v' (in terms of the conservation of the energy and angular momentum) finding

$$v' = v - 2(v \cdot n)n. \tag{6.3}$$

We want to associate a probability $k(v|v')\,\mathrm{d}v'$ related to the transition $v \to v'$. A reasonable choice is to assume ρ as a random variable uniformly distributed in $[-1/2, 1/2]$, so that the probability to have the transition $v \to v'$ is $\mathrm{d}\rho = \frac{\mathrm{d}\rho}{\mathrm{d}v'}\,\mathrm{d}v'$. However, instead of computing $k(v|v')$ explicitly, we find more convenient to express the operator K, by means of an integration with respect to the angle α. Using

$$\mathrm{d}\rho = \left|\frac{\mathrm{d}\rho}{\mathrm{d}\alpha}\right|\mathrm{d}\alpha = |n \cdot v|\,\mathrm{d}n,$$

we get

$$\begin{aligned} Kf(v') &= \int_{S^1} \mathrm{d}n\,|n \cdot v| f(v)\chi(n \cdot v \leqslant 0) \\ &= \int_{S^1} \mathrm{d}n\,|n \cdot v| f(v)\chi\big(n \cdot v' \geqslant 0\big), \end{aligned} \tag{6.4}$$

where v' is given by (6.3). Note that in (6.3) v' can be considered as incoming or outgoing velocity. What distinguishes if v' incoming or outgoing is the scalar product $n \cdot v$ which is positive or negative if v is outgoing or incoming, respectively.

Other possible transition probabilities can be introduced considering the collision of the particle by an obstacle generating a smooth potential ϕ. In this case an analogous argument leads us to the expression

$$Kf(v) = \int_{S^1} \mathrm{d}n\, B\big(n, |v|\big) f\big(v'\big), \tag{6.5}$$

where $B(n, |v|)$ is proportional to the cross section associated to the potential ϕ (defined by $|\frac{\mathrm{d}\rho}{\mathrm{d}\alpha}|$).

Here we are considering the two-dimensional case for simplicity, however all our considerations can be extended easily to the three-dimensional case.

Up to now we have considered the process in the velocity space. If we are interested as well to the particle position, we introduce the nonhomogeneous equation

$$(\partial_t + v \cdot \nabla) f = Lf = (K - I) f. \tag{6.6}$$

To avoid problems with boundary conditions, we shall assume that the space variable x runs over Λ, where Λ is either $\mathbb{R}^2$ or a two-dimensional torus (i.e. periodic boundary condition).

It is obvious that the Cauchy problem associated to (6.6) has a unique global solution in any reasonable space. Moreover, what is more important, we can give the explicit solution. We shall assume that the initial probability distribution f_0 is a bounded, continuous function expressing the initial distribution of our test particle.

By (6.6), we have

$$f(x,t) = \mathrm{e}^{-t} f_0(x - vt, v) + \int_0^t \mathrm{d}s\, \mathrm{e}^{(t-s)} (Kf)\big(x - v(t-s), v\big). \tag{6.7}$$

To avoid inessential regularity problems, we consider (6.7) as the equation under investigation.

Iterating (6.7) we find the following formal expansion:

$$\begin{aligned} f(x,v,t) &\\ = \mathrm{e}^{-t} & f_0(x - vt, v) \\ + \mathrm{e}^{-t} & \sum_{n>0} \int_0^t \mathrm{d}t_1 \int_0^{t_1} \mathrm{d}t_2 \cdots \int_0^{t_{n-1}} \mathrm{d}t_n \int \mathrm{d}v_1 \cdots \int \mathrm{d}v_n \\ & \times k(v|v_1) k(v_1|v_2) \cdots k(v_{n-1}|v_n) \\ & \times f_0\big(x - v(t - t_1) - v_1(t_1 - t_2) - \cdots - v_n t_n, v_n\big). \end{aligned} \tag{6.8}$$

The physical interpretation of (6.8) is transparent: The value of f in (x, v) at time t is a sum of infinitely many contributions. The term

$$\mathrm{e}^{-t} f_0(x - vt, v)$$

is the value of the initial datum in the backward trajectory, (with the weight e^{-t} which takes into account the mass loss) and is interpreted as the zero collision contribution. The first term of the series in the right-hand side of (6.8),

$$\mathrm{e}^{-t} \int_0^t \mathrm{d}t_1 \int \mathrm{d}v_1 k(v|v_1) f_0\big(x - v(t - t_1) - v_1 t_1, v_1\big),$$

is the contribution to the probability density to be in (x, v) at time t after a collision and is called the one-collision term. The generic n-collision term can be interpreted in an analogous way. We see immediately that the series (6.8) is uniformly convergent in x, v, t in any compact set.

The main feature of (6.6), is that it possesses a large family of Lyapunov functions. Consider the following functional:

$$I_\Phi = \int \mathrm{d}x\,\mathrm{d}v\,\Phi\big(f(x, v, t)\big), \tag{6.9}$$

where Φ is a convex function. Then I_Φ is decreasing along the solutions.

Indeed

$$\begin{aligned}\dot{I}_\Phi &= \int \Phi'(f)(Kf - f)\,\mathrm{d}x\,\mathrm{d}v \\ &= \int \mathrm{d}x \int \mathrm{d}v \int \mathrm{d}v'\, k(v|v')\Phi(f)'(v)\big(f(x, v) - f(x, v')\big) \\ &= -\frac{1}{2}\int \mathrm{d}x \int \mathrm{d}v \int \mathrm{d}v' \\ &\quad \times k(v|v')\big[\big(\Phi(f)'(x, v) - \Phi(f)'(x, v')\big)\big(f(x, v) - f(x, v')\big)\big].\end{aligned} \tag{6.10}$$

The term in square brackets is positive (Φ' is increasing) and the claim is proved.

The asymptotic behavior (in time) of the solutions is expected to be completely determined at least for the homogeneous equation

$$\partial_t f = Kf - f, \tag{6.11}$$

which we study first.

We obviously have a (stationary) solution to $Lf = 0$ which is $f = 1$. We have no other solutions to this equation, which is the eigenvalue problem

$$Kf = f. \tag{6.12}$$

Indeed $Lf = 0$ implies $(f, Lf) = 0$ (here $(\cdot\,,\,\cdot)$ denotes the L^2 norm in S^1) and, using (6.10) with $\Phi(f) = \frac{f^2}{2}$,

$$0 = (f, Lf) = -\frac{1}{2}\int k(v|v')\big|f(v') - f(v)\big|^2\,\mathrm{d}v\,\mathrm{d}v', \tag{6.13}$$

so that if $k : S^1 \times S^1 \to \mathbb{R}^+$ is positive a.e. in $S^1 \times S^1$, then the condition $(f, Lf) = 0$ implies that f is constant.

For the nonhomogeneous situation the result is the same. The stationary condition is

$$v \cdot \nabla f = Lf. \tag{6.14}$$

By taking the inner product of (6.14) by f, since L is symmetric and $v \cdot \nabla$ is antisymmetric we must have simultaneously

$$(f, Lf) = 0 \tag{6.15}$$

and

$$(f, v \cdot \nabla f) = 0. \tag{6.16}$$

By (6.15) we have $f(x, v) = \rho(x)$ for some ρ depending on x only. Therefore, we have also $Lf = 0$ and hence $v \cdot \nabla f = 0$, implying that f is constant. In conclusion, the only solution to (6.14) is $f = \text{const}$. If we want to restrict the solutions to the family of the probability distributions, the constant solution is acceptable only if the domain is bounded.

We now pass to the analysis of the asymptotic behavior of the solutions to the transport equation and we start with the simpler homogeneous case.

A natural question is to see whether the solution $f(v, t)$ of the homogeneous equation converges, as $t \to \infty$, to the stationary solution with an exponential rate. The spectral properties of the operator K give the result. Indeed K is a compact operator (in $L^2(S^1)$), self-adjoint and positive. Since $\|K\| \leqslant 1$, the spectrum is contained in [0, 1]. The value 0 is the only accumulation point and 1 is a simple eigenvalue. As matter of facts, there exists an orthonormal basis $\{e_i\}_{i=0}^{\infty}$ associated to the eigenvalues

$$1 = \alpha_0 > \alpha_1 \geqslant \alpha_2 \geqslant \cdots,$$

for which we can write

$$Kf = \sum_{n=1}^{\infty} \alpha_n f_n e_n, \tag{6.17}$$

where $f_n = (e_n, f)$.

Suppose now that $f \perp \text{const}$. Then a standard computation gives us

$$(f, Lf) \leqslant -(1 - \alpha_1)\|f\|^2. \tag{6.18}$$

Moreover,

$$\frac{1}{2}\frac{\mathrm{d}}{\mathrm{d}t}\|f\|^2 = (f, Lf) \leqslant -(1 - \alpha_1)\|f\|^2 \tag{6.19}$$

and hence

$$\|f(t)\| \leqslant \mathrm{e}^{-(1-\alpha_1)t}\|f\|. \tag{6.20}$$

We now pass to analyze the behavior of the solution to the nonhomogeneous equation. We have to distinguish between the two cases, when $\Lambda = \mathbb{R}^2$ and $\Lambda = T^2$, the two-dimensional torus. In the first case, it is not hard to see that $f \to 0$ as $t \to \infty$ (this is the dispersive behavior where all the mass is dispersed at the infinity). We leave it as an exercise to the reader. In the second case we expect that $f(t)$ converges to a constant as in the homogeneous case. Even in this simple linear case, however, this property is not completely obvious. Even though we have a single stationary solution, if we look at the operator $-v \cdot \nabla + L$ from a spectral point of view, it is not evident that such an operator have the correct spectral properties sufficient for an exponential decay.

On the other hand we can take advantage of the stochastic nature of the Markov semigroup we are interested in. Since the probability transition is positive, by relatively standard probabilistic arguments, the exponential convergence to the unique invariant measure can be established. Since we do not use this property in the sequel, we do not give here the proof of this statement.

6.2. *A mechanical derivation*

In this section we want to analyze how the process we have introduced is indeed obtained from the deterministic time evolution of a mechanical system in the low density limit.

We consider the transport equation we have discussed so far,

$$(\partial_t + v \cdot \nabla) f = (K - I) f, \tag{6.21}$$

with

$$Kf(v) = \int_{S^1} B(n, v) f\big(x, v'\big)\,\mathrm{d}n, \tag{6.22}$$

where

$$v' = v - 2(v \cdot n)n \tag{6.23}$$

and

$$B(n, v) = |n \cdot v| \chi(n \cdot v \geqslant 0). \tag{6.24}$$

We prefer, in the present context, to express the collision operator in terms of the angular integration as in (6.4).

We consider a point particle in $\mathbb{R}^2$ moving in a random distribution of identical disks of radius ε (see Figure 7). The interaction is assumed elastic. This means that the particles move freely up to the first instant of contact with an obstacle. Then it is elastically reflected and so on. Since the modulus of the velocity of the test particle is constant, we assume it to be unity so that the phase space of our system is $\mathbb{R}^2 \times S^1$.

The scatterers are distributed according to a Poisson distribution of parameter $\mu_\varepsilon = \varepsilon^{-1}\mu$, where μ is positive and ε is a small parameter. More explicitly, the probability

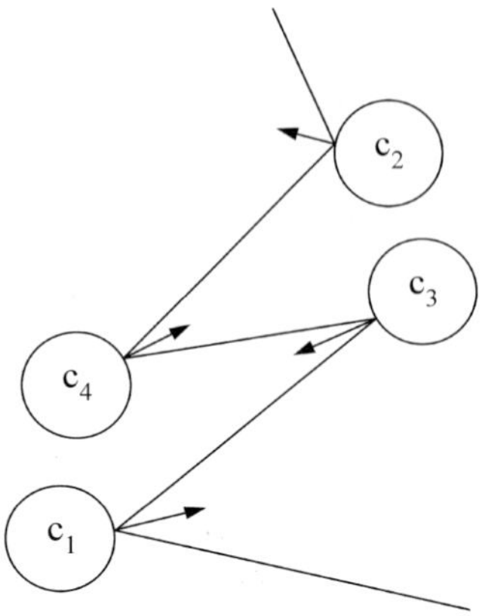

Fig. 7.

distribution of finding exactly N obstacles in a bounded measurable set $\Lambda \subset \mathbb{R}^d$ is given by

$$e^{-\mu_\varepsilon|\Lambda|}\frac{\mu_\varepsilon^N}{N!}\,\mathrm{d}c_1\cdots\mathrm{d}c_N, \tag{6.25}$$

where $(c_1, \ldots, c_N)$ are the positions of the scatterers, denoted also shortly by $\mathbf{c}$ or $\mathbf{c}_N$ and $|\Lambda|$ denotes the Lebesgue measure of Λ.

For a given initial distribution $f_0 = f_0(x, v)$ continuous in both variables we can define

$$f_\varepsilon(t, x, v) = \mathbb{E}^\varepsilon\big[f_0\big(S_{\mathbf{c},\varepsilon}^{-t}(x, v)\big)\big], \tag{6.26}$$

where $S_{\mathbf{c},\varepsilon}^t(x, v)$ is the flow associated to the initial datum (x, v) for a given scatterer configuration $\mathbf{c}$ and $\mathbb{E}^\varepsilon$ denotes the expectation with respect to the Poisson measure of parameter μ_ε in $\mathbb{R}^d$.

We are interested in the asymptotic behavior of f_ε when ε goes to 0. We can prove:

THEOREM 6.1.

$$\lim_{\varepsilon\to 0}\sup_{0\leqslant t\leqslant T}\big|f_\varepsilon(t, x, v) - f(t, x, v)\big| = 0 \quad \textit{a.e. } (x, v) \in \mathbb{R}^2 \times S^1, \tag{6.27}$$

where f *solves* (6.21) *with initial datum* f_0.

REMARK 1. The limit we are considering here is analogous to the Boltzmann–Grad limit we discussed for interacting systems. Indeed $\mu_\varepsilon = \langle N\rangle_{|\Lambda|}$, the average number of scatterers in the region Λ. Hence we are requiring $\varepsilon\langle N\rangle_{|\Lambda|} = \text{const}$ (in three dimensions it would be $\varepsilon^2\langle N\rangle_{|\Lambda|} = \text{const}$). Of course we could also have postulated the equivalent scaling (low density limit) in which the radius of any disk is not scaled but we scale hyperbolically space and time as in the hydrodynamical limit making the density simultaneously vanishing suitably.

REMARK 2. Here we allow the overlapping of scatterers: The Poisson measure is that of a free gas. It would also be possible to consider the Poisson measure restricted to non-overlapping configurations, namely the Gibbs measure for a systems of hard disks in the plane. However the two measures are asymptotically equivalent and the result does hold also in this case.

REMARK 3. The flow $S^{-t}_{\mathbf{c},\varepsilon}(x, v)$ is not defined globally in time if the trajectory hits one of the intersection points of the boundaries of a group of overlapping scatterers. However this happens for a negligible set of initial data. So we do not care of it. We also remark that we are assuming that the initial point is outside the disk configuration.

To prove the above theorem we start by writing explicitly $f_\varepsilon(t, x, v)$. For an arbitrary but fixed time $T > 0$, consider $B(x)$, the ball of center x and radius T. Then for $t \leqslant T$ (using the shorthand notation $d\mathbf{c}_N$ for $dc_1 \cdots dc_N$),

$$f_\varepsilon(t, x, v) = e^{-\mu_\varepsilon |B(x)|} \sum_{N \geqslant 0} \frac{\mu_\varepsilon^N}{N!} \int_{B(x)^N} d\mathbf{c}_N \, f_0\big(S^{-t}_{\mathbf{c}_N}(x, v)\big). \tag{6.28}$$

Here we used that, for $t \leqslant T$, only the scatterers in $B(x)$ may influence the motion of the test particle.

Among the obstacles c belonging to $\mathbf{c} \cap B(x)$, we distinguish between those influencing the motion of the light particle and the others. Indeed we call "external" the obstacles c in $\mathbf{c} \cap B(x)$ such that

$$\inf_{0 \leqslant t \leqslant T} \big| x_{\mathbf{c}}(t) - c \big| > \varepsilon \tag{6.29}$$

and "internal" all the others. Here we used the notation $S^t_{\mathbf{c}_N}(x, v)$ for $(x_{\mathbf{c}}(t), v_{\mathbf{c}}(t))$. Then we decompose a given configuration $\mathbf{c}_N$ of $B(x)^N$ in the following way,

$$\mathbf{c}_N = \mathbf{a}_P \cup \mathbf{b}_Q,$$

where $\mathbf{a}_P$ is the set of all external obstacles and $\mathbf{b}_Q$ is the set of all internal ones.

Realizing then that

$$S^t_{\mathbf{c}_N} = S^t_{\mathbf{b}_Q}, \tag{6.30}$$

we get

$$\begin{aligned} &f_\varepsilon(t, x, v) \\ &\quad = e^{-\mu_\varepsilon |B(x)|} \sum_{Q \geqslant 0} \frac{\mu_\varepsilon^Q}{Q!} \int_{B(x)^Q} d\mathbf{b}_Q \sum_{P \geqslant 0} \frac{\mu_\varepsilon^P}{P!} \int_{B(x)^P} d\mathbf{a}_P \\ &\qquad \times \chi\big(\{\text{the } \mathbf{a}_P \text{ are external and the } \mathbf{b}_Q \text{ are internal}\}\big) \\ &\qquad \times f_0\big(S^{-t}_{\mathbf{a}_P \cup \mathbf{b}_Q}(x, v)\big) \end{aligned}$$

$$= \sum_{Q \geqslant 0} \frac{\mu_\varepsilon^Q}{Q!} \int_{B(x)^Q} \mathrm{d}\mathbf{b}_Q \, \mathrm{e}^{-\mu_\varepsilon |\mathcal{T}(\mathbf{b}_Q)|} \times \chi\big(\{\text{the } \mathbf{b}_Q \text{ are internal}\}\big) f_0\big(S_{\mathbf{b}_Q}^{-t}(x, v)\big). \tag{6.31}$$

The factor $\mathrm{e}^{-\mu_\varepsilon |\mathcal{T}(\mathbf{b}_Q)|}$, where $\mathcal{T}(\mathbf{b}_Q)$ is the tube defined by

$$\mathcal{T}(\mathbf{b}_Q) = \big\{ y \in B(x),\ \exists s \in [0, t],\ \big| y - x_{\mathbf{b}_Q}(s) \big| < \varepsilon \big\}, \tag{6.32}$$

arises from the integration over $\mathrm{d}\mathbf{a}_P$ which has been performed explicitly. Moreover,

$$\chi\big(\{\text{the } \mathbf{b}_Q \text{ are internal}\}\big) = \chi\big(\big\{\mathbf{b}_Q \subset \mathcal{T}(\mathbf{b}_Q)\big\}\big). \tag{6.33}$$

Note also that when $\varepsilon \to 0$,

$$\big|\mathcal{T}(\mathbf{b}_Q)\big| \approx \varepsilon t. \tag{6.34}$$

Using the symmetry of the internal obstacles we can write (6.31) in the more convenient form:

$$f_\varepsilon(t, x, v) = \sum_{Q \geqslant 0} \mu_\varepsilon^Q \int_{B(x)^Q} db_1 \cdots db_Q \, \chi_{\mathrm{ord}}(\mathbf{b}_Q) \times \mathrm{e}^{-\mu_\varepsilon |\mathcal{T}(\mathbf{b}_Q)|} f_0\big(S_{\mathbf{b}_Q}^{-t}(x, v)\big), \tag{6.35}$$

where $\chi_{\mathrm{ord}}(\mathbf{b}_Q)$ is the characteristic function of the fact that all the obstacles are hit and that the collision are ordered: The scatterer b_i is hit before b_j, if $i < j$.

We now look at the series expansion defining $f(x, v, t)$, namely

$$f(x, v, t) = \mathrm{e}^{-t} \sum_{Q \geqslant 0} \int_0^t \mathrm{d}t_1 \int_0^{t_1} \mathrm{d}t_2 \cdots \int_0^{t_{Q-1}} \mathrm{d}t_Q \int_{S_+^1} \mathrm{d}n_1 \cdots \int_{S_+^1} \mathrm{d}n_Q \times \prod_{i=0}^{Q-1} (n \cdot v_i) \, f_0\big(x - v(t - t_1) - v_1(t_1 - t_2) - \cdots - v_n t_n, v_n\big), \tag{6.36}$$

where v_0 coincides with v and the velocities $v_1, v_2, \ldots, v_Q$ are the ingoing velocities of the Q collisions, which are completely determined by the choices of $n_1, \ldots, n_Q$.

We write the above expression in terms of the impact parameters:

$$f(x, v, t) = \mathrm{e}^{-t} \sum_{Q \geqslant 0} \mu_\varepsilon^Q \int_0^t \mathrm{d}t_1 \int_0^{t_1} \mathrm{d}t_2 \cdots \int_0^{t_{Q-1}} \mathrm{d}t_Q \int_{-\varepsilon/2}^{+\varepsilon/2} \mathrm{d}\rho_1 \cdots \int_{-\varepsilon/2}^{+\varepsilon/2} \mathrm{d}\rho_Q \times f_0\big(x - v(t - t_1) - v_1(t_1 - t_2) - \cdots - v_n t_n, v_n\big), \tag{6.37}$$

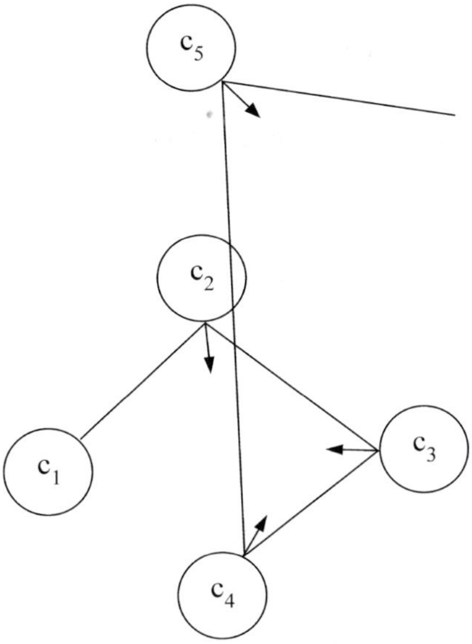

Fig. 8.

where $\rho_1, \ldots, \rho_Q$ are the impact parameters.

Now, given the impact times $t_1, \ldots, t_Q$ and the impact parameters $\rho_1, \ldots, \rho_Q$, we know how the obstacles yielding the collision $v_j \to v_{j+1}$ are located, so that we want to introduce the change of variables

$$\{\rho_i, t_i\}_{i=1}^{Q} \to (\mathbf{b}_Q) \tag{6.38}$$

to make the expression (6.36) as much as possible similar to (6.35).

Since the Jacobian of the change of variables (6.38) is 1, we have

$$f(t,x,v) = \mathrm{e}^{-t} \sum_{Q \geqslant 0} \mu_\varepsilon^Q \int_{B(x)^Q} \mathrm{d}b_1 \cdots \mathrm{d}b_Q \, \chi_{\mathrm{ord}}(\mathbf{b}_Q) \times f_0\big(x - v(t-t_1) - v_1(t_1 - t_2) - \cdots - v_n t_n, v_n\big). \tag{6.39}$$

The two expressions (6.39) and (6.35) look similar, but they are not the same. Indeed the flow $T^t_{\mathbf{b}_Q}(x,v) = (x - v(t-t_1) - v_1(t_1-t_2) - \cdots - v_n t_n, v_n)$ that we have constructed as consequence of the change of variables (6.38) can deliver physical inconsistencies as shown in Figure 8. However, these inconsistencies can happen, for any Q, for a vanishing set of scatterer configurations so that the term by term convergence of the series (6.35) to the series (6.39) is ensured. Moreover, the two series are absolutely convergent, so that the proof of Theorem 2.1 can be easily achieved.

REMARK 4. Equation (6.36) is an evolution equation for the probability density associated to a particle performing random jumps in the velocity variable at random Markov times.

On the contrary, the original system is Hamiltonian, the only stochasticity being that of the positions of the scatterers (and the initial state distributed according f_0). The change of variables (6.38) outlines explicitly that the Poisson distribution of the scatterers induces a distribution of the instants and angles of collisions which, due to possible re-collisions and the fact that some transitions are not geometrically possible, are not independent nor Markov. The long tail memory is however lost in the limit.

NOTES. The randomness of the obstacles is absolutely necessary to obtain the correct kinetic description. Indeed it can be shown (see [15]) that a properly scaled periodic configuration of scatterers cannot give a Poisson distribution of times as requested by a linear Boltzmann equation. The linear Boltzmann equation can however be derived by a periodic lattice gas of scatterers at a low density (see [23]). The derivation we present here is due to Gallavotti (see [47,48]). Subsequent improvements as regards the type of convergence or more general scatterer distributions can be found in [13] and [83], respectively. For an analysis of long range potentials and grazing collisions, see [34] and [39], [35], respectively.

6.3. *The hydrodynamic limit*

In the previous section we have seen how to pass from the particle to the Boltzmann picture for the Lorentz model. In the present section we show how to pass from the linear Boltzmann equation to the heat equation, which is, in the present case, the correct hydrodynamic equation.

As usual we consider a gas obeying the linear Boltzmann equation in a large box Λ_ε of side ε^{-1}, ε being a small parameter to be sent to zero. We use the variables q and τ for the microscopic space and time and f^ε is the solution of the linear Boltzmann equation in Λ_ε:

$$\partial_\tau f^\varepsilon(q, v) + v \cdot \nabla_q f^\varepsilon(qx, v) = Lf^\varepsilon(q, v). \tag{6.40}$$

We want to look at the system not on the typical length scale of the large box (which are ε^{-1}), but on the unit scale. To do this we introduce the new variables

$$x = \varepsilon q, \qquad t = \varepsilon\tau.$$

The time is also scaled because we want to have typical velocities unscaled to see displacements in the unit box. Under this scaling the mean free path is of order ε and (6.40) becomes

$$\partial_t f^\varepsilon(x, v) + v \cdot \nabla_x f^\varepsilon(x, v) = \frac{1}{\varepsilon} Lf^\varepsilon(x, v). \tag{6.41}$$

Note that under this scaling the average number of collisions diverges when $\varepsilon \to 0$. However, although the above hyperbolic scaling is that usually considered in the derivation of the Euler equation for the nonlinear Boltzmann equation as we have seen in Chapter 3, in the present case it gives a trivial result for the following reason. Taking the integral with respect to v in (6.41) we find, using the mass conservation,

$$\partial_t \rho^\varepsilon + \int \mathrm{d}v\, v \cdot \nabla_x f^\varepsilon = 0, \tag{6.42}$$

where $\rho_\varepsilon(x,t) = \int \mathrm{d}v\, f_\varepsilon(x,v,t)$. On the other hand, if the limit $\varepsilon \to 0$ makes sense, we expect that the limit function f^0 must satisfy $Lf^0 = 0$ (it is a local equilibrium), to give a sense to the right-hand side of (6.41). Thus it is of the form $f^0 = \rho^0$, namely it is constant as regards the v dependence. Looking at (6.42) we finally find $\partial_t \rho^0 = 0$. Therefore we do not get any hydrodynamical equation.

The physical reason of this feature is that the momentum is not a locally conserved quantity and, at the equilibrium, the mean value must vanish. In other words we have a trivial Euler description.

To see some nontrivial behavior we have to look at larger times (parabolic or diffusive scaling), that is,

$$x = \varepsilon q, \qquad t = \varepsilon^2 \tau.$$

Under this scaling we find

$$\partial_t f^\varepsilon + \frac{1}{\varepsilon} v \cdot \nabla_x f^\varepsilon = \frac{1}{\varepsilon^2} L f^\varepsilon. \tag{6.43}$$

We now want to study (6.43) with $x \in \Lambda$ which is the unit box. To make a simple choice we assume also periodic boundary conditions.

We approach (6.43) with the Hilbert expansion. Consider the following formal series

$$f^\varepsilon = \sum_{k=0}^{\infty} \varepsilon^k f_k, \tag{6.44}$$

where the coefficients f_k are independent of ε. We hope to determine them recursively, by imposing that f^ε is a solution of (3.4). Of course we do not expect that the series (6.44) is convergent, for the moment we want only to give a sense to each term f_k.

This can be easily done. In facts we have that f_0 must be a function of x and t only to satisfy the equilibrium condition.

At order ε^{-1} we find

$$L f_1 = v \cdot \nabla_x f_0 \tag{6.45}$$

and, at order ε^0

$$L f_2 = \partial_t f_0 + v \cdot \nabla_x f_1. \tag{6.46}$$

More generally,

$$L f_{n+2} = \partial_t f_n + v \cdot \nabla_x f_{n+1}, \quad n \geqslant 1. \tag{6.47}$$

The main problem is to see whether the operator L can be inverted in (6.45)–(6.47). We know that the kernel of L is formed by the constants (we are arguing in the velocity space only, here x is only a parameter). Therefore (6.45)–(6.47) have a solution if the right-hand sides have vanishing integrals with respect to v.

Since f_0 does not depend on v,

$$\int \mathrm{d}v\, v \cdot \nabla_x f_0 = 0, \tag{6.48}$$

so that f_1 can be determined. This allows also a complete determination of f_0. Indeed from (6.46), integrating in $\mathrm{d}v$, we have

$$\partial_t f_0 + \int \mathrm{d}v\, v \cdot \nabla_x f_1 = 0. \tag{6.49}$$

By (6.46), we finally find

$$\partial_t f_0 = \sum_{\alpha,\beta=1}^{2} D_{\alpha,\beta}\, \partial_\alpha \partial_\beta f_0, \tag{6.50}$$

where $\partial_\alpha = \partial_{x_\alpha}$ and

$$D_{\alpha,\beta} = -\int \mathrm{d}v\, v_\alpha L^{-1} v_\beta. \tag{6.51}$$

We are tempted to consider (6.50) as a diffusion equation. We have to show, however, that the matrix D is positive. A better equivalent expression for $D_{\alpha,\beta}$ is

$$D_{\alpha,\beta} = \int \mathrm{d}v \int_0^\infty \mathrm{d}t\, v_\alpha \mathrm{e}^{Lt} v_\beta = \int \mathrm{d}v \int_0^\infty \mathrm{d}t\, v_\alpha \mathbb{E}\big[v_\beta(t,v)\big], \tag{6.52}$$

where $v(t,v)$ is the jump process with generator L described before. We see that $D_{1,1} = D_{2,2} = D > 0$ (due to the isotropy) and $D_{1,2} = D_{2,1} = 0$ (as follows by the change $(v_1 \to -v_1)$. Therefore

$$\partial_t f_0 = D\Delta f_0. \tag{6.53}$$

By the integral expression for L^{-1},

$$L^{-1} g(v) = \int_0^\infty \mathrm{d}t\, \mathrm{e}^{Lt} g(v) = \int_0^\infty \mathrm{d}t\, \mathbb{E}\big[g\big(v(t,v)\big)\big] \tag{6.54}$$

(note that the integral is convergent if g is orthogonal to the constants), we see that L^{-1} preserves the parity, it maps even (odd) functions in even (odd) functions. Therefore f_1 is odd. Moreover, integrating (6.46), the right-hand side is the diffusion equation and hence is vanishing. This implies that f_2 exists and it is even. The procedure can be iterated to an arbitrary n.

Let us now establish the convergence. Instead of considering the Hilbert expansion (3.5), we simply look for a solution to (6.43) of the form

$$f^\varepsilon = f_0 + \varepsilon f_1 + \varepsilon^2 f_2 + \varepsilon R^\varepsilon, \tag{6.55}$$

where f_i are the first three coefficients of the Hilbert expansion and R^ε, the unknown, satisfies the following equation:

$$\partial_t R^\varepsilon + v \cdot \nabla_x R^\varepsilon = \frac{1}{\varepsilon^2} L R^\varepsilon + A_\varepsilon. \tag{6.56}$$

Here

$$A_\varepsilon = \partial_t (f_1 + \varepsilon f_2) + v \cdot \nabla_x f_2. \tag{6.57}$$

We assume now that the initial condition $f^\varepsilon(x, v, 0) = g(x)$ is a given local equilibrium profile. We require only that g be in $L^1(\Lambda)$. By the smoothness of the solution of the heat equation it is easy to verify that A_ε is uniformly bounded (with respect to ε) in $C([0,T]; L^2(\Lambda \times S^1))$. Moreover, by (6.56), denoting by $(\cdot,\cdot)$ and $\|\cdot\|$ scalar product and norm in $L^2(\Lambda \times S^1)$, we find

$$\frac{1}{2}\frac{\mathrm{d}}{\mathrm{d}t}\|R^\varepsilon(t)\|^2 = \frac{1}{\varepsilon^2}\big(R^\varepsilon(t), LR^\varepsilon\big) + \big(R^\varepsilon(t), A_\varepsilon(t)\big). \tag{6.58}$$

By the positivity of L and the Cauchy–Schwarz inequality, we deduce that

$$\|R^\varepsilon(t)\| \leqslant \int_0^t \|A_\varepsilon(s)\|\, \mathrm{d}s, \tag{6.59}$$

which implies that R^ε is uniformly bounded in $C([0,T]; L^2(\Lambda \times S^1))$.

As a consequence we have proved the following theorem.

THEOREM 6.2. *Let $g \in L^1(\Lambda)$ be an initial datum for* (6.43) *whose solution is denoted by $f^\varepsilon(x, v, t)$. Let $f_0(x, t)$ be the solution of the heat equation with diffusion constant D. Then f^ε converges to f_0 in $L^2(x, v)$ uniformly in $t \in [0, T]$.*

A sequence of remarks are in order.

REMARK 1. We have assumed that the initial condition for the evolution given by the linear Boltzmann equation is given by a local equilibrium state (independent of v). Strictly speaking this is not necessary as we have already seen in the more difficult nonlinear case: There is an initial regime (the initial layer) in which a general state $F_0(x, v)$ thermalize very fast in time and locally in space, namely,

$$\lim_{\tau \to +\infty} F(x, v, \tau) = g(x) = \int \mathrm{d}v\, F_0(x, v),$$

where $F(x, v, \tau)$ denotes the solution of the unrescaled equation (6.40) with initial datum $F_0(x, v)$. Then the convergence in Theorem 6.2 holds for any positive macroscopic t (uniformly in all the intervals not containing the origin).

REMARK 2. Note that the diffusion coefficient D depends only on the jump process in the velocity. D depends on the physical parameters by means of the transition probabilities, or, by means of the cross section. In other words, different potentials give rise to different diffusion coefficients because they have different cross sections.

REMARK 3. A more complete description of the above convergence can be done by looking at the processes. It is possible to show the convergence of the jump process (in space and velocity) to the diffusion process underlying the heat equation.

6.4. *The diffusive limit for the Lorentz model*

A much more difficult problem is to obtain the diffusive behavior starting directly from the particle dynamics we have introduced in Section 2. Actually we would expect that, under the diffusive scaling, the distribution density of the test particle converges to that of a diffusion process. This is actually a very difficult and still unsolved problem. However, Bunimovich and Sinai (see [17]) showed that such diffusive limit holds in the case when the scatterers are periodically distributed. One can formulate the result in the following way.

We consider a point particle elastically colliding with a periodic distribution of scatterers in the case of finite horizon, that is the path of the test particle between two successive collision is bounded. For instance we consider the two-dimensional triangular lattice configuration in Figure 9.

We denote by $q(t)$ the trajectory of the test particle.

The dynamics of the particle can be projected in a suitable torus $\mathbb{T}^2$ and the phase space of the system is $\Gamma = Q \times S^1$ where Q, the configuration space is $\mathbb{T}^2$ deprived, in our cases, by two scatterers (see Figure 10).

Therefore the motion is a dynamical system with continuous time in Γ with an invariant probability measure $\mathrm{d}\mu = C\,\mathrm{d}q\,\mathrm{d}\theta$, where $\mathrm{d}q$ and $\mathrm{d}\theta$ are the Lebesgue measure on

Fig. 9.

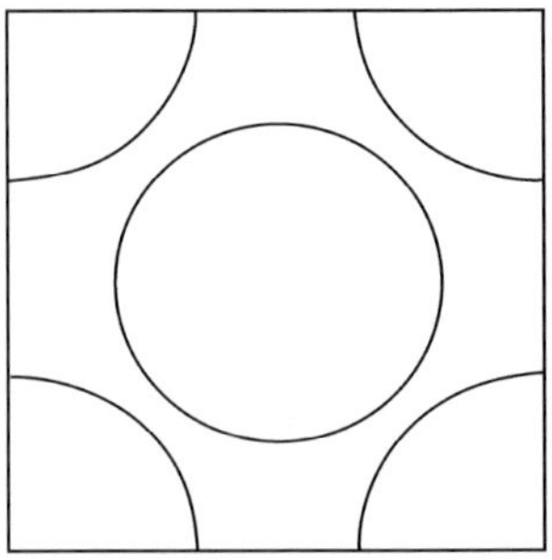

Fig. 10.

Q and S^1 respectively and C is a normalization factor. We denote by $\Phi^t(q,\theta)$ the flow. This is the famous Sinai billiard model which has been proved, in the seventies, to enjoy strong ergodic properties.

Using the dispersive property of the billiard, Bunimovich and Sinai were able to prove the following result. They assume that the initial conditions $q(0)$, $v(0)$ are distributed according to a probability measure μ_0, absolutely continuous with respect to the Lebesgue measure. Then for $t \in [0,1]$ we consider the rescaled trajectory

$$q_\varepsilon(t) = \sqrt{\varepsilon} q\left(\frac{t}{\varepsilon}\right),$$

which is an element of $C([0,1];\mathbb{R}^2)$. The initial measure induces a probability distribution μ_ε on the space of all continuous trajectories $C([0,1];\mathbb{R}^2)$. In other words, q_ε is a stochastic process. Note that the dynamics is completely deterministic the only stochasticity being confined on the uncertainty of the initial condition. Then:

THEOREM 6.3 [17]. *The measure μ_ε converges weakly to a Wiener measure as $\varepsilon \to 0$.*

In other words the distribution of $q_\varepsilon(t)$ obey to the heat equation in the limit $\varepsilon \to 0$.

We do not enter into the details of the proof which is much beyond the purposes of the present work.

References

[1] R.K. Alexander, *The infinite hard sphere system*, Ph.D. Thesis, Dept. of Math., University of California, Berkeley (1975).

[2] L. Arkeryd and A. Nouri, *The stationary Boltzmann equation in the slab with given weighted mass for hard and soft forces*, Ann. Sc. Norm. Sup. Pisa Cl. Sci. (4) **27** (1998), 533–556.

[3] L. Arkeryd and A. Nouri, *On the Milne problem and the hydrodynamic limit for a steady Boltzmann equation model*, J. Statist. Phys. **99** (3/4) (2000), 993–1019.

[4] K. Asano and S. Ukai, *The Euler limit and the initial layer of the nonlinear Boltzmann equation*, Hokkaido Math. J. **12** (1983), 303–324.

[5] C. Bardos, R.E. Caflisch and B. Nicolaenko, *Thermal layer solutions of the Boltzmann equation*, Random Fields: Rigorous Results in Statistical Physics, Koszeg 1984, J. Fritz, A. Jaffe and D. Szasz, eds, Birkhäuser, Boston (1985).

[6] C. Bardos, F. Golse and D. Levermore, *Sur les limites asymptotiques de la théorie cinétique conduisant à la dynamique des fluides incompressibles*, C. R. Acad. Sci. Paris Sér. I Math. **309** (1989), 727–732.

[7] C. Bardos, F. Golse and D. Levermore, *Fluid dynamical limits of kinetic equations, I. Formal derivations*, J. Statist. Phys. **63** (1991), 323–344.

[8] C. Bardos, F. Golse and D. Levermore, *Fluid dynamical limits of kinetic equations, II. Convergence proofs for the Boltzmann equation*, Comm. Pure Appl. Math. **46** (1993), 667–753.

[9] C. Bardos, F. Golse and D. Levermore, *Acoustic and Stokes limits for the Boltzmann equation*, C. R. Acad. Sci. Paris Sér. I Math. **327** (1998), 323–328.

[10] C. Bardos, F. Golse and D. Levermore, *The acoustic limit for the Boltzmann equation*, Arch. Ration. Mech. Anal. **153** (2000), 177–204.

[11] C. Bardos and B. Nicolaenko, *Navier–Stokes equations and dynamical systems*, Handbook of Dynamical Systems, Vol. 2, B. Fieldler, ed., North-Holland, Amsterdam (2002), 503–597.

[12] O. Benois, R. Esposito and R. Marra, *Navier–Stokes limit for a thermal stochastic lattice gas*, J. Statist. Phys. **96** (1999), 653–713.

[13] C. Boldrighini, L.A. Bunimovich and Ya.G. Sinai, *On the Boltzmann equation for the Lorentz gas*, J. Statist. Phys. **32** (1983), 477–501.

[14] C. Boldrighini, R.L. Dobrushin and Y.M. Suhov, *One-dimensional hard rod caricatures of hydrodynamics*, J. Statist. Phys. **31** (1983), 577–616.

[15] J. Bourgain, F. Golse and B. Wennberg, *On the distribution of free path length for the periodic Lorentz gas*, Comm. Math. Phys. **190** (1998), 491–508.

[16] J. Boussinesq, *Theorie analytique de la chaleur*, Gauthier-Villars, Paris (1903).

[17] L.A. Bunimovich and Ya.G. Sinai, *Statistical properties of the Lorentz gas with periodic configuration of scatterers*, Comm. Math. Phys. **78** (1981), 478–497.

[18] R.E. Caflisch, *The fluid dynamical limit of the nonlinear Boltzmann equation*, Comm. Pure Appl. Math. **33** (1980), 651–666.

[19] R.E. Caflisch, *The Boltzmann equation with a soft potential*, Comm. Math. Phys. **74** (1980), 71–95.

[20] R.E. Caflisch and B. Nicolaenko, *Shock waves the Boltzmann equation. Nonlinear partial differential equations* (Durham, N.H., 1982), Contemp. Math. **17** (1983), 35–44.

[21] R.E. Caflisch and G. Papanicolaou, *The fluid dynamical limit of a nonlinear Boltzmann equation*, Comm. Pure Appl. Math. **32** (1979), 589–616.

[22] E. Caglioti, C. Marchioro and M. Pulvirenti, *Non-equilibrium dynamics of three-dimensional infinite particle systems*, Comm. Math. Phys. **215** (2000), 25–43.

[23] E. Caglioti, M. Pulvirenti and V. Ricci, *Derivation of a linear Boltzmann equation for a periodic Lorentz gas*, Markov Process. Related Fields **3** (2000), 265–285.

[24] S. Caprino, M. Pulvirenti and W. Wagner, *A particle systems approximating stationary solutions to the Boltzmann equation*, SIAM J. Math. Anal. **4** (1998), 913–934.

[25] C. Cercignani, *On the Boltzmann equation for rigid spheres*, Transport Theory Statist. Phys. **2** (1972), 211–225.

[26] C. Cercignani, *The Grad limit for a system of soft spheres*, Comm. Pure Appl. Math. **36** (1983), 479–494.

[27] C. Cercignani, *The Boltzmann Equation and Its Applications*, Springer-Verlag, New York (1988).

[28] C. Cercignani, *Ludwig Boltzmann. The Man Who Trusted Atoms*, Oxford University Press, Oxford (1998).

[29] C. Cercignani, R. Illner and M. Pulvirenti, *The Mathematical Theory of Dilute Gases*, Springer-Verlag, New York (1994).

[30] S. Chapman and T.G. Cowling, *The Mathematical Theory of Nonuniform Gases*, Cambridge University Press, Cambridge (1952).

[31] C. Costantini and R. Marra, *Hydrodynamic limits for the Boltzmann process*, J. Statist. Phys. **67** (1992), 229–249.

[32] A. De Masi, R. Esposito and J.L. Lebowitz, *Incompressible Navier–Stokes and Euler limits of the Boltzmann equation*, Comm. Pure Appl. Math. **42** (1989), 1189–1214.

[33] A. De Masi, N. Ianiro, S. Pellegrinotti and E. Presutti, *A survey of the hydrodynamical behavior of many-particle systems*, Nonequilibrium Phenomena, II. From Stochastics to Hydrodynamics, J.L. Lebowitz and E.W. Montroll, eds, North-Holland, Amsterdam (1984), 123–294.

[34] L. Desvillettes and M. Pulvirenti, *The linear Boltzmann equation for long-range forces: A derivation from particle systems*, Models Methods Appl. Sci. **9** (1999), 1123–1145.

[35] L. Desvillettes and V. Ricci, *Rigorous derivation of a linear kinetic equation of Fokker–Plank type in the limit of grazing collisions*, J. Statist. Phys. **104** (2001), 1173–1189.

[36] L. Desvillettes and C. Villani, *Entropic methods for the study of the long time behavior of kinetic equations*, The Sixteenth International Conference on Transport Theory, Part I (Atlanta, GA, 1999), Transport Theory Statist. Phys. **30** (2001), 155–168.

[37] R.J. DiPerna and P.-L. Lions, *On the Cauchy problem for the Boltzmann equations: Global existence and weak stability*, Ann. Math. **130** (1989), 321–366.

[38] R.L. Dobrushin and J. Fritz, *Non-equilibrium dynamics of two-dimensional infinite particle systems with a singular interaction*, Comm. Math. Phys. **57** (1977), 67–81.

[39] D. Dürr, S. Goldstein and J.L. Lebowitz, *Asymptotic motion of a classical particle in a random potential in two dimensions: Landau Model*, Comm. Math. Phys. **113** (1987), 209–230.

[40] R. Esposito, J.L. Lebowitz and R. Marra, *Hydrodynamic limit of the stationary Boltzmann equation in a slab*, Comm. Math. Phys. **160** (1994), 49–80.

[41] R. Esposito, J.L. Lebowitz and R. Marra, *The Navier–Stokes limit of stationary solutions of the nonlinear Boltzmann equation*, J. Statist. Phys. **78** (1995), 389–412.

[42] R. Esposito, J.L. Lebowitz and R. Marra, *Solutions to the Boltzmann equation in the Boussinesq regime*, J. Statist. Phys. **90** (1998), 1129–1178.

[43] R. Esposito and R. Marra, *On the derivation of the incompressible Navier–Stokes equation for Hamiltonian particle systems*, J. Statist. Phys. **74** (1993), 981–1004.

[44] R. Esposito and R. Marra, *Hydrodynamics as scaling limit of kinetic systems and stochastic particle systems*, Trends Statist. Phys. **3** (2000), 1–43.

[45] R. Esposito, R. Marra and H.T. Yau, *Diffusive limit of asymmetric simple exclusion*, Rev. Math. Phys. **6** (1994), 1233–1267.

[46] R. Esposito, R. Marra and H.T. Yau, *Navier–Stokes equations for stochastic particle systems on the lattice*, Comm. Math. Phys. **182** (1996), 395–456.

[47] G. Gallavotti, *Divergences and approach to equilibrium in the Lorentz and Wind-tree models*, Phys. Rev. **185** (1969), 308–322.

[48] G. Gallavotti, *Rigorous theory of the Boltzmann equation in the Lorentz gas*, Nota interna, n. 358, Istituto di Fisica Università di Roma (1972); reprinted in G. Gallavotti, *Meccanica statistica*, Quaderni del CNR, n. 50 (1995), 191–204.

[49] F. Golse and C.D. Levermore, *The Stokes–Fourier limit for the Boltzmann equation*, C. R. Acad. Sci. Paris Sér. I Math. **333** (2001), 145–150.

[50] F. Golse and C.D. Levermore, *The Stokes and acoustic limits for the Boltzmann equation*, Comm. Pure Appl. Math. **55** (2002), 336–393.

[51] F. Golse and L. Saint-Raymond, *The Navier–Stokes limit of the Boltzmann equation: Convergence proof*, Preprint; *The Navier–Stokes limit for the Boltzmann equation*, C. R. Acad. Sci. Paris Sér. I Math. **333** (2001), 897–902.

[52] H. Grad, *On the kinetic theory of rarefied gases*, Comm. Pure Appl. Math. **2** (1949), 331–407.

[53] H. Grad, *Asymptotic theory of the Boltzmann equation, II*, Rarefied Gas Dynamics, Vol. I, J.A. Laurmann, ed., Academic Press, New York (1963), 26–59.

[54] A. Grad, *Asymptotic equivalence of the Navier–Stokes and nonlinear Boltzmann operator*, Proc. Symp. Appl. Math. **17** (1965), 154–183.

[55] M.S. Green, *Markoff random processes and the statistical mechanics of time-dependent phenomena, II. Irreversible processes in fluids*, J. Chem. Phys. **22** (1954), 398–413.

[56] D. Hilbert, *Mathematical Problems*, Göttinges Nachr. (1900), 253–297.

[57] D. Hilbert, *Begründung der kinetischen Gastheorie*, Math. Ann. **72** (1916/17), 331–407.

[58] D. Hilbert, *Grundzügeiner allgemeinen Theorie der linearen Integralgleichungen*, Chelsea, New York, (1953).

[59] R. Illner and M. Pulvirenti, *Global validity of the Boltzmann equation for two and three-dimensional rare gas in vacuum*, Comm. Math. Phys. **121** (1989), 143–146.

[60] M. Kac, *Probability and Related Topics*, Interscience, New York (1959).

[61] S. Klainerman and A. Majda, *Compressible and incompressible fluids*, Comm. Pure Appl. Math. **35** (1982), 629–651.

[62] R. Kubo, *Statistical–mechanical theory of irreversible processes, I. General theory and simple applications in magnetic and conduction problems*, J. Phys. Soc. Jap. **12** (1957), 570–586.

[63] M. Lachowicz, *On the initial layer and the existence theorem for the nonlinear Boltzmann equation*, Math. Methods Appl. Sci. **9** (3) (1987), 27–70.

[64] M. Lachowicz and M. Pulvirenti, *A stochastic system of particles modelling the Euler equation*, Arch. Ration. Mech. Anal. **109** (1990), 81–93.

[65] O.E. Lanford, *Time Evolution of large classical systems*, Lecture Notes in Phys., Vol. 38, J. Moser, ed., Springer-Verlag, Berlin–Heidelberg (1976), 1–111.

[66] P.-L. Lions and N. Masmoudi, *Une approche locale de la limite incompressible*, C. R. Acad. Sci. Paris Sér. I Math. **329** (1999), 387–392.

[67] P.-L. Lions and N. Masmoudi, *From Boltzmann equations to incompressible fluid mechanics equations, I*, Arch. Ration. Mech. Anal. **158** (2001), 173–193.

[68] P.-L. Lions and N. Masmoudi, *From Boltzmann equations to incompressible fluid mechanics equations, II*, Arch. Ration. Mech. Anal. **158** (2001), 195–211.

[69] T.-P. Liu and S.-H. Yu, *Boltzmann equation: Micro–macro decompositions and positivity of shock profiles*, Comm. Math. Phys. **246** (2004), 133–179.

[70] J.C. Maxwell, *On the dynamical theory of gases*, Philos. Trans. Roy. Soc. London **157** (1867), 49–88.

[71] S. Méléard, *Probabilistic models for nonlinear partial differential equations*, Lecture Notes in Math., Vol. 1627, D. Talay and L. Tubaro, eds, Springer-Verlag, New York (1995), 42–95.

[72] C.B. Morrey, *On the derivation of the equations of hydrodynamics from statistical mechanics*, Comm. Pure Appl. Math. **8** (1955), 279–326.

[73] M.G. Mürmann, *The hydrodynamic limit of a deterministic particle system with conservation of mass and momentum*, J. Statist. Phys. **105** (2001), 483–510.

[74] T. Nishida, *Fluid dynamical limit of the nonlinear Boltzmann equation to the level of the compressible Euler equation*, Comm. Math. Phys. **61** (1978), 119–148.

[75] K. Oelschlager, *On the connection between Hamiltonian many-particle systems and the hydrodynamical equations*, Arch. Ration. Mech. Anal. **115** (1991), 297–310.

[76] S. Olla, S.R.S. Varadhan and H.T. Yau, *Hydrodynamical limit for a Hamiltonian system with weak noise*, Comm. Math. Phys. **155** (1993), 523.

[77] M. Pulvirenti, *Kinetic limits for stochastic particle systems*, Lecture Notes in Math., Vol. 1627, D. Talay and L. Tubaro, eds, Springer-Verlag, New York (1995), 96–126.

[78] M. Pulvirenti, W. Wagner and M.B. Zavelani Rossi, *Convergence of particle schemes for the Boltzmann equation*, Eur. J. Mech. B Fluids **13** (1994), 339–351.

[79] J. Quastel and H.T. Yau, *Lattice gases, large deviations and the incompressible Navier–Stokes equations*, Ann. Math. **148** (1998), 51–108.

[80] F. Rezakhanlou, *Hydrodynamic limit for attractive particle systems on* Z^d, Comm. Math. Phys. **140** (1991), 417–448.

[81] D. Ruelle, *Statistical Mechanics*, Benjamin, Reading, MA (1969).

[82] Y. Sone, K. Aoki, S. Tabata, H. Sugimoto and A.V. Bobylev, *Inappropriateness of the heat-conduction equation for the description of a temperature field of a stationary gas in the continuum limit: Examination by asymptotic analysis and numerical computation of the Boltzmann equation*, Phys. Fluids **8** (1996), 628–638.

[83] H. Spohn, *The Lorentz flight process converges to a random flight process*, Comm. Math. Phys. **60** (1978), 277–290.

[84] M. Toda, R. Kubo and N. Saitô, *Statistical Physics I: Equilibrium Statistical Mechanics*, Springer-Verlag, Berlin–Heidelberg (1992).

[85] K. Uchiyama, *On the Boltzmann–Grad limit for the Broadwell model of the Boltzmann equation*, J. Statist. Phys. **52** (1988), 331–3557.

[86] S.R.S. Varadhan, *Nonlinear diffusion limit for a system with nearest neighbor interactions, II*, Asymptotic Problems in Probability Theory: Stochastic Models and Diffusion on Fractals, K.D. Elworthy and N. Ikeda, eds, Pitman Research Notes in Math., Vol. 283, Wiley, New York (1994), 75–128.

[87] C. Villani, *A review of mathematical topics in collisional kinetic theory*, Handbook of Mathematical Fluid Dynamics, Vol. 1, S. Friedlander and D. Serre, eds, North-Holland, Amsterdam (2002), 71–306.

[88] W. Wagner, *A convergence proof for Bird's direct simulation Monte Carlo method for the Boltzmann equation*, J. Statist. Phys. **66** (1992), 1011–1044.

[89] H.T. Yau, *Relative entropy and hydrodynamics of Ginsburg–Landau models*, Math. Phys. Lett. **22** (1991), 63–80.

CHAPTER 2

Two-Dimensional Euler System and the Vortex Patches Problem

Jean-Yves Chemin

Centre de Mathématiques de l'École Polytechnique
91 128 Palaiseau Cedex, France
E-mail: chemin@math.polytechnique.fr

Contents

HANDBOOK OF MATHEMATICAL FLUID DYNAMICS, VOLUME III
Edited by S.J. Friedlander and D. Serre

Introduction

The vortex patches problem comes from two-dimensional fluid mechanics. Let us consider the Euler system related to a two-dimensional incompressible inviscid fluid with constant density, namely

$$\begin{cases} \partial_t v + v \cdot \nabla v = -\nabla p, \\ \operatorname{div} v = 0, \\ v_{|t=0} = v_0. \end{cases} \tag{E}$$

Here v is a time-dependent divergence free vector field on $\mathbb{R}^2$. This vector field $v(t,x)$ is assumed to be the description of the speed of a particle of the fluid located in x at time t. The choice of $\mathbb{R}^2$ instead of the more realistic case of a bounded domain is motivated by simplicity.

In dimension two, the curl of the vector field v, called the vorticity, is the scalar usually denoted by ω with

$$\omega \stackrel{\text{def}}{=} \partial_1 v^2 - \partial_2 v^1.$$

With suitable hypotheses on ω, we have the familiar Biot–Savart law, which allows us to recover the vector field from the vorticity

$$v(x) = \int_{\mathbb{R}^2} \frac{(x-y)^\perp}{|x-y|^2} \omega(y)\, dy \quad \text{with } z^\perp \stackrel{\text{def}}{=} \left(-z^2, z^1\right). \tag{1}$$

The key point of (E) is that the vorticity is constant along the particle paths because of the equation

$$\partial_t \omega + v \cdot \nabla \omega = 0. \tag{V}$$

So, when a is a real number greater than 1, a classical result of harmonic analysis (essentially the Marcinkiewicz theorem, see Theorem 1.2) implies that the gradient of v is in L^a if the vorticity is in L^a. Moreover, (V) implies that the L^a norm of ω is preserved. When $a = \infty$ of course the L^∞ norm of the vorticity ω is preserved, but the Marcinkiewicz theorem is known to be false. Let us give here a very simple example: as proved in Proposition 1.3, when ω is the characteristic function of the square $0 \leqslant x_1, x_2 \leqslant 1$, then ∇v is not bounded.

It is well known that in quasilinear evolution of hyperbolic type, the control of the Lipschitz norm is crucial. Thus it is not possible to get straightforward Lipschitz estimates on the vector field v from the conservation relation (V). In spite of that, Wolibner proved in 1933 (see [87]) that global regular solutions of (E) exist for initial data in C^r with $r \in]1,2[$, which means that a constant C exists such that

$$\left|\nabla v(x) - \nabla v(y)\right| \leqslant C|x-y|^{r-1}.$$

This global solution is of course Lipschitz with respect to the space variable. When an evolution system of partial differential equations has global existence for regular enough data, it is classical to play the following game: take a rough initial data, regularize this rough initial data, then consider the family of global smooth solutions of the system and try to pass to the limit.

It is a natural question to investigate the case when the initial vorticity ω_0 is simply bounded, an interesting case in applications being the case when ω is the characteristic function of a domain. As the example mentioned above shows us, it is hopeless to remain in the framework of Lipschitz solutions.

Let us notice that the case when the initial vorticity ω_0 is simply a measure has been the purpose of many works. Among them, let us mention the work of Delort (see [45]) who proved that it is possible to pass to the limit in the case when the initial vorticity belongs to the Sobolev space H^{-1} and has a nonnegative singular part.

In this text we shall focus on two-dimensional Euler equation in the case when the initial vorticity ω_0 is bounded and compactly supported. We shall also pay a lot of attention to the case when this ω_0 is the characteristic function of a bounded (connected) domain.

In Section 1, we shall study the way we can deduce information about the vector field from informations about the vorticity. As (V) tells us that the vorticity is preserved along the trajectories it is obviously a crucial problem. As we shall see, if the vorticity is in L^p, so is the gradient of the associate divergence free vector field. As shown by an explicit counterexample, this turns out to be false when $p = \infty$. This point is in fact the core of most of this text.

In Section 2, we shall prove that the incompressible Euler system is locally well posed for regular enough initial data. In dimension 2 the conservation (V) implies that it is globally well posed (this is the Wolibner theorem). Moreover, we shall prove that if the initial vorticity is only bounded and compactly supported, then the two-dimensional incompressible Euler system is globally well posed (this is the Yudovich theorem).

Those two sections recall briefly results and sketch of proofs of very basic (and old) results on two-dimensional incompressible Euler equations.

Section 3 does not investigate regularity properties. In this section, we investigate the growth of the diameter of the support of a bounded compactly supported vorticity. This is a quantitative question. When the vorticity is bounded and compactly supported, it is easy to see that the associated divergence free vector field is bounded uniformly in time. Thus it is clear, thanks to (V) that the diameter of the support of $\omega(t)$ is less or equal to $C(t+1)$.

The case when the sign of the vorticity changes is very different from the case when the vorticity is nonnegative. The reason why is the following. Thanks to (V) and (1), we have

$$\frac{\mathrm{d}}{\mathrm{d}t}\int_{\mathbb{R}^2} x\omega(t,x)\,\mathrm{d}x = 0 \quad \text{and} \quad \frac{\mathrm{d}}{\mathrm{d}t}\int_{\mathbb{R}^2} |x|^2\omega(t,x)\,\mathrm{d}x = 0.$$

It is clear that the efficiency of the second conservation is not same when the vorticity ω is nonnegative and when the sign of the vorticity changes. In the first case the rate of growth is as bad as possible (i.e., like Ct). It is proved by an explicit counterexample. In the case when the vorticity is nonnegative, the growth of the diameter is less than $(t\log t)^{1/4}$. The question of the optimality of such a rate remains an open problem.

Section 4 is devoted to a detailed study of the Biot–Savart law. In particular, we prove that, when the vorticity is the characteristic function of a bounded domain with $C^{1+\varepsilon}$ boundary, the associated divergence free vector field is Lipschitz. The main point of this section is to generalize this result. In fact we define the concept of tangential regularity with respect to a family of vector fields. It means that if we differentiate a function with respect to a vector field of this family, the result is a little bit better than expected.

For instance when we differentiate the characteristic function of a bounded domain D_0 with smooth boundary ∂D_0 with respect a vector field Y which is tangent to ∂D_0, we find 0. The following property will be more important for us:

Let us consider a family of regularization of the characteristic function of D_0. When we differentiate with respect to Y this family of functions, we find a bounded family of distributions in the Hölder space $C^{-\alpha}$ for some $\alpha \in]0, 1[$. A precise definition of this space will be given in Definition 4.1. Let us simply say that this space $C^{-\alpha}$ is the space of distributions which are the sum of a smooth bounded function and of the divergence of a vector field with coefficients in the classical Hölder space $C^{1-\alpha}$.

Section 5 studies the problem of the regularity of the boundary of a vortex patch. The original problem consists in the investigation of the precise regularity of the solution in the case when the vorticity of the initial data is the characteristic function of a bounded domain with a smooth enough boundary. As seen in Section 4 (Proposition 4.1), the initial data is Lipschitz in this case. The question is the following: Is this property preserved by the flow of the equation which does exist and is unique thanks to the Yudovich theorem (Theorem 2.2)?

This theorem shows that the global unique solution is log-Lipschitzian and that, thanks to conservation of the vorticity (V), the vorticity of the solution at time t is the characteristic function of a bounded domain the topology of which is preserved along the evolution. Then two very natural questions can be stated: does the boundary of the domain remain regular for all time? If so, what happens for large time?

Let us imagine that the vorticity of the solution at time t is the characteristic function of the interior D_t of a closed simple curve of class $C^{1+\varepsilon}$. Let us denote by $\gamma(t)$ a $C^{1+\varepsilon}$ one-to-one map from $\mathbb{S}^1$ into $\mathbb{R}^2$, the range of which is the boundary of D_t. Thanks to the Biot–Savart law, the solution $v(t,\cdot)$ is given by

$$v(t) = \nabla^{\perp} f(t) \quad \text{with } f(t,x) = \frac{1}{2\pi}\int_{D_t} \log|x-y|\,\mathrm{d}y.$$

Using Green's formula, we get that

$$v(t,x) = \frac{1}{2\pi}\int_0^{2\pi} \log\bigl|x-\gamma(t,\sigma)\bigr|\partial_\sigma\gamma(t,\sigma)\,\mathrm{d}\sigma.$$

It looks reasonable to seek for a γ which satisfies

$$\partial_t\gamma(t,s) = v\bigl(t,\gamma(t,s)\bigr). \tag{2}$$

Thus we want to solve

$$\partial_t \gamma(t,s) = \frac{1}{2\pi}\int_0^{2\pi} \log\left|\gamma(t,s)-\gamma(t,\sigma)\right|\partial_\sigma\gamma(t,\sigma)\,\mathrm{d}\sigma. \tag{3}$$

The problem of the propagation of the regularity of the boundary of the domain of a vortex patch has been stated by Yudovich in his fundamental paper [88].

In [68], Majda conjectured that a breakdown of the regularity of the boundary happens. This conjecture was based on two facts. First numerical simulations done by Zabusky, Hughes and Roberts (see [89]), Dritschel (see [48]), Dritschel and MacIntyre (see [49]) and Buttke (see [19,20]) led to think that this breakdown should occur.

Let us consider the problem through (43). We can study small perturbations of a circular patch which is obviously a stationary solution of the equation. Then considering the asymptotic expansion up to quadratic terms we get a simplified equation. This approach has been developed by Constantin and Titi in [38]. Alinhac proved in [3] that this simplified equation was very close to breakdown in finite time.

In this section, we present the structure of the proof of the global regularity for the boundary of vortex patches. This means that if we consider an initial data which is the characteristic function of a bounded domain with $C^{1+\varepsilon}$ boundary, then the vorticity remains for any time the characteristic function of a bounded domain which remains $C^{1+\varepsilon}$.

Section 6 is a brief introduction to paradifferential calculus which is the foundation of the proof of global regularity of vortex patches. Introduced by Bony in order to study the propagation of singularities in nonlinear hyperbolic equations (see [17]), paradifferential calculus is now a tool often used in the study of equations of fluid mechanics. The long series of works inspired by the paper by Cannone, Meyer and Planchon (see [22]) on incompressible Navier–Stokes system is a good example is this. For an introduction to this subject, the reader can consult the book of Lemarié-Rieusset [65].

The basic idea of paradifferential calculus is the following one. Thanks to the inverse Fourier formula, we have

$$uv(x) = (2\pi)^{-d}\int \mathrm{e}^{\mathrm{i}x\cdot(\xi+\eta)}\hat{u}(\xi)\hat{v}(\eta)\,\mathrm{d}\xi\,\mathrm{d}\eta.$$

This integral can be cut into three parts:

- one where $|\xi|$ is small be respect to $|\eta|$, which means that $|\xi| \leqslant \varepsilon|\eta|$ for some small ε,
- one where $|\eta|$ is small be respect to $|\xi|$,
- and the last one where $|\xi|$ and $|\eta|$ are equivalent (i.e., $\varepsilon|\xi| \leqslant |\eta| \leqslant \varepsilon^{-1}|\xi|$).

Roughly speaking, the basic facts are the following ones:

- the first term, called the paraproduct, is defined for any couple of tempered distributions and its regularity is essentially determined by the regularity of v,
- the last term, called the remainder, is not always defined but the regularity of u and v sum up.

Precise definitions and statements will be given in this section. This is also the opportunity to give complete proofs of statements of the preceding two sections.

In Section 7 we state some generalizations and mention other approaches to the vortex patches problem. The generalizations concern mainly the regularity of the boundary of the domain. As mentioned previously, when the vorticity is the characteristic function of the square $0 \leqslant x_1, x_2 \leqslant 1$, the gradient of the associated divergence free vector field is not bounded. But when the vorticity is the characteristic function of a bounded domain with smooth boundary, then the gradient of the associated divergence free vector field is bounded.

This fact in general. It is possible to prove that, when the vorticity is the characteristic function of a bounded domain, the boundary of which is a $C^{1+\varepsilon}$ curve with a singular set Σ then the gradient of the associated divergence free vector field in bounded outside any neighborhood of Σ. Moreover, the blow up of the gradient is less that the logarithm of the distance to the singular set Σ.

The other part of this section mentions an other approach to this problem. Incompressible Euler equations can be seen as an ordinary differential equation of order two with respect to the flow ψ of v. In the spacial case of dimension two, using Biot–Savart law (1) and the conservation of vorticity (V), we have by definition of the flow

$$\frac{\mathrm{d}\psi}{\mathrm{d}t}(t,x) = \int_{\mathbb{R}^2} \frac{(\psi(t,x)-y)^{\perp}}{|\psi(t,x)-y|^2} \omega_0\big(\psi^{-1}(t,y)\big)\,\mathrm{d}y.$$

This approach, developed in the context of the vortex patches problem in particular by Serfati, gives shorter proofs of the global regularity of the boundary. But this approach seems too rigid to be adapted to the problem we shall investigate in the next section.

In Section 8 we investigate the stability of vortex patches with smooth boundary under viscous perturbations. Let us explain what it is. Let us consider an initial vorticity which is the characteristic function of a bounded domain with smooth enough boundary and the solution v_ν of incompressible Navier–Stokes equations with viscosity v_ν. Then when ν tends to 0, what happens? Let us notice that, in all the previous sections, the fact that we work in the whole space $\mathbb{R}^2$, i.e. in a domain without boundary was a (sometimes huge) technical simplification. In this last section, the fact that we work in the whole space $\mathbb{R}^2$ is fundamental. This means that no boundary layer phenomena appear.

In this section, we shall prove that, when the viscosity ν tends to 0, the vorticity of the vector field v_ν converges to the vorticity of the solution of two-dimensional incompressible Euler equations. Moreover, if we consider the range of the initial domain by the flow of v_ν, it converges to the range of this domain by the flow of the solution of Euler equation. These results are based on a remarkable estimate about the gradient of v_ν which is independent of ν. All this uses in a crucial way the flexibility of the concept of tangential regularity.

1. The Biot–Savart law

In all this section, the divergence free vector field v will be computed from the scalar ω with the formula (1).

The first property is trivial but will be useful in the proof of the Yudovich theorem.

PROPOSITION 1.1. *If* $1 < a < 2 < b < \infty$, *then*

$$\|v\|_{L^\infty} \leqslant C\|\omega\|_{L^a \cap L^b}.$$

The second one, much harder to prove, is a classical corollary of the well-known Marcinkiewicz theorem (see, for instance, [84]).

PROPOSITION 1.2. *A constant C exists which satisfies the following property: For any a belonging to $]1,\infty[$ and for any divergence free vector field v the gradient of which belongs to L^a, we have*

$$\|\nabla v\|_{L^a} \leqslant C\frac{a^2}{a-1}\|\omega\|_{L^a}.$$

As said in the Introduction, the inequality

$$\|\nabla v\|_{L^\infty} \leqslant C\|\omega\|_{L^\infty\cap L^a}$$

turns out to be wrong. The above inequality is violated in particular by the following simple example. Let us consider the function ω defined by

$$\omega(x) = H(x_1)H(1-x_1)H(x_2)H(1-x_2),$$

where H is the characteristic function of $\mathbb{R}^+$. The fact that the vector field v associated with ω is not Lipschitz is proved next.

PROPOSITION 1.3. *A neighborhood of the origin V and a positive real number α exist such that v satisfies*

$$\forall x \in V, \quad |\nabla v(x)| \geqslant \alpha\bigl(1-\log|x|\bigr).$$

PROOF. Let θ be a smooth function which vanishes outside the unit ball and the value of which is 1 near the origin. It is clear that the function

$$x \mapsto \int_{\mathbb{R}^d} \frac{(x-y)^\perp}{|x-y|^2}\bigl(1-\theta(y)\bigr)\omega(y)\,dy$$

is smooth near the origin. So we have to state near the origin the vector field $\tilde v$ defined by

$$\tilde v(x) = \nabla^\perp E_2 \star (\theta\omega) \quad \text{with } E_2 \stackrel{\text{def}}{=} \frac{1}{2\pi}\log r.$$

By definition of $\tilde v$, we have that

$$\partial_1 \tilde v^1 = -\partial_1\partial_2\bigl(E_2 \star (\theta\omega)\bigr) = -E_2 \star \bigl(\partial_1\partial_2(\theta\omega)\bigr).$$

Moreover, the Leibniz formula implies that

$$\begin{aligned}&\partial_1\partial_2(\theta\omega)(x)\\ &\quad = \delta_0 + H(x_1)H(x_2)\,\partial_1\partial_2\theta(x) + H(x_1)\,\partial_1\theta(x)\delta_{(x_2=0)} + H(x_2)\,\partial_2\theta(x)\delta_{(x_1=0)}.\end{aligned}$$

So it turns out that

$$\partial_1 \tilde{v}^1(x) = -\frac{1}{2\pi} \log r - E_2 \star \big(H(x_1)H(x_2)\, \partial_1 \partial_2 \theta\big) \\ + E_2 \star H(x_1)\, \partial_1 \theta(x) \delta_{(x_2=0)} + E_2 \star H(x_2)\, \partial_2 \theta(x) \delta_{(x_1=0)}.$$

The last three functions of the above identity are obviously locally bounded. Thus the proposition is proved. □

In fact, the vector field is close to be Lipschitz. This is made clear by the following two propositions which will be proved in Section 6.

PROPOSITION 1.4. *Let us define*

$$\|v\|_{LL} \stackrel{def}{=} \sup_{0<|x-x'|\leqslant 1} \frac{|v(t,x) - v(t,x')|}{|x-x'|(1-\log|x-x'|)} < \infty.$$

Then we have that

$$\|v\|_{LL} \leqslant C\|\omega\|_{L^a \cap L^\infty}.$$

PROPOSITION 1.5. *For any positive ε and for any $a \in]1, \infty[$, a constant C exists such that*

$$\|\nabla v\|_{L^\infty} \leqslant C\left(\|\omega\|_{L^a} + \|\omega\|_{L^\infty} \log\left(e + \frac{\|\omega\|_\varepsilon}{\|\omega\|_{L^\infty}}\right)\right)$$

and

$$\|\nabla v\|_{L^\infty} \leqslant C\left(\|v\|_{L^2} + \|\omega\|_{L^\infty} \log\left(e + \frac{\|\omega\|_\varepsilon}{\|\omega\|_{L^\infty}}\right)\right),$$

where $\|a\|_\varepsilon$ denotes the Hölder norm of a, i.e., $\|a\|_\varepsilon \stackrel{def}{=} \|a\|_{L^\infty} + \sup_{0<|x-y|\leqslant 1} \frac{|a(x)-a(y)|}{|x-y|^\varepsilon}$.

2. Wolibner and Yudovich theorems

2.1. *A problem at infinity*

We are interested in initial data the vorticity of which is the characteristic function of a bounded domain. Let us consider a nonnegative and compactly supported function ω. It is obvious, using the Biot–Savart law (1) that if $|x|$ is large enough, we have

$$\big|v(x)\big| \geqslant \frac{C}{|x|} \int_{\mathbb{R}^2} \omega(y)\, dy.$$

It is obvious that v does not belong to L^2. As we also want to use somewhere energy estimate, let us introduce the following definition.

DEFINITION 2.1. We call stationary vector field and we denote by σ, any vector field of the form

$$\sigma = \left(-\frac{x^2}{r^2} \int_0^r \rho g(\rho)\, \mathrm{d}\rho, \frac{x^1}{r^2} \int_0^r \rho g(\rho)\, \mathrm{d}\rho \right), \quad \text{where } g \in \mathcal{D}\big(\mathbb{R} \setminus \{0\}\big).$$

It is a very easy exercise left to the reader to prove that such vector fields are stationary solutions of the incompressible Euler system and that the vorticity of σ is $g(|x|)$. Let us notice that if the function g is nonnegative, if $|x|$ is large enough, we have

$$\left|\sigma(x)\right| \geqslant \frac{1}{2|x|} \int g\big(|x|\big)\, \mathrm{d}x.$$

This of course implies that if σ is not identically 0, then σ does not belong to L^2. We shall consider L^2 perturbation of vector fields of type σ. As the following lemma shows us, such a class of vector fields is large enough. In particular it contains vector fields, the vorticity of which is the characteristic function of a bounded domain. It also contains vector fields, the vorticity of which is the length measure of a smooth curve.

LEMMA 2.1. *Let μ be a measure such that $(1 + |x|)\mu$ is a bounded measure. Let us also assume that μ belongs to $H^{-1}(\mathbb{R}^2)$. Then a unique divergence free vector field v exists in $\sigma + L^2(\mathbb{R}^2; \mathbb{R}^2)$ for some stationary vector field σ and such that $\omega(v) = \mu$.*

PROOF. Let us choose some stationary vector field σ such that

$$\int_{\mathbb{R}^2} \omega(\sigma)\, \mathrm{d}x = \int_{\mathbb{R}^2} \mathrm{d}\mu.$$

Let us consider the vector field v defined by

$$\begin{aligned} v = \sigma + \mathcal{F}^{-1}\big(\chi(\xi)\xi^{\perp}|\xi|^{-2}\big(\hat{\mu}(\xi) - \hat{\omega}(\sigma)(\xi)\big) \\ + \big(\mathrm{Id} - \chi(\xi)\big)\xi^{\perp}|\xi|^{-2}\big(\hat{\mu}(\xi) - \hat{\omega}(\sigma)(\xi)\big)\big). \end{aligned}$$

The fact that $\hat{\mu}$ is Lipschitzian and belongs to $H^{-1}(\mathbb{R}^2)$ implies the lemma because the uniqueness of v is an exercise left to the reader. □

Now we can state the following definition.

DEFINITION 2.2. Let m be a real number. We shall denote by E_m the space of divergence free vector fields v such that a stationary vector field σ exists such that

$$\int_{\mathbb{R}^2} \omega(\sigma) = m \quad \text{and} \quad v - \sigma \in L^2.$$

Let us remark that Lemma 2.1 implies in particular that $E_0 = L^2$.

2.2. *Wolibner theorem*

Let us recall the definition of Hölder spaces.

DEFINITION 2.3. Let r be a positive noninteger real number. If r belongs to the interval $]0, 1[$, we denote by $C^r(\mathbb{R}^d)$, or by C^r when no confusion is possible, the space of bounded functions u on $\mathbb{R}^d$ such that a constant C exists such that, for any x and y in $\mathbb{R}^d$,

$$\left|u(x) - u(y)\right| \leqslant C|x - y|^r.$$

If $r > 1$, we denote $C^r(\mathbb{R}^d)$, or by C^r when no confusion is possible, the space of functions u such that, for any multiinteger α of length less than or equal to the integral part of r, denoted by $[r]$, we have

$$\partial^\alpha u \in C^{r-[r]}.$$

It is obvious that

$$\left\|\tilde{u}\right\|_r \stackrel{\text{def}}{=} \sum_{\alpha/|\alpha|\leqslant[r]} \left(\left\|\partial^\alpha u\right\|_{L^\infty} + \sup_{x\neq y} \frac{|\partial^\alpha u(x) - \partial^\alpha u(y)|}{|x - y|^{r-[r]}}\right)$$

gives to C^r a structure of Banach space.

THEOREM 2.1. *Let m be a real number and v_0 a divergence free vector field belonging the space to $E_m \cap C^r$. A unique global solution (v, p) exists in the space*

$$L^\infty_{\text{loc}}(\mathbb{R}; C^r) \cap C(\mathbb{R}; E_m) \times L^\infty_{\text{loc}}(\mathbb{R}; H^1 \cap C^{r+1}).$$

SKETCH OF PROOF. A detailed modern proof of this basic result relies on estimates which will be proved in Section 6. We shall give here only the sketch of the proof. The first thing to do is to determine precisely the pressure from the vector field v. Applying the operator "divergence" to the Euler equation we find that the pressure p must satisfy

$$-\Delta p = \sum_{j,k} \partial_j v^k \, \partial_k v^j.$$

For any v and w in $C^r \cap E_m$, let us state

$$\Pi(v, w) \stackrel{\text{def}}{=} \mathcal{F}^{-1} \sum_{j,k} \chi(\xi) \frac{\xi_j \xi_k}{|\xi|^2} \mathcal{F}\left(v^j w^k\right) + \sum_{j,k} \left(1 - \chi(\xi)\right) \frac{1}{|\xi|^2} \mathcal{F}\left(\partial_j v^k \, \partial_k w^j\right). \tag{4}$$

As the two vector fields v and w are divergence free, we have that $-\Delta p = \Pi(v,v)$. We shall prove in Section 6 that

$$\|\Pi(a,b)\|_{\widetilde{C}^r} \leqslant C_r\big(\|a\|_{\widetilde{C}^r}\|\nabla b\|_{L^\infty} + \|\nabla a\|_{L^\infty}\|b\|_{\widetilde{C}^r}\big) \tag{5}$$

and

$$\|\Pi(a,b)\|_{\widetilde{C}^r} \leqslant C_r\big(\|a\|_{\widetilde{C}^r}\|\nabla b\|_{L^\infty} + \|\nabla a\|_{L^\infty}\|\nabla b\|_{C^{r-1}}\big). \tag{6}$$

Moreover, it is obvious that the only harmonic function in the space $H^1 \cap C^{r+1}$ is the null function. Thus the pressure p is totally determined in $H^1 \cap C^{r+1}$ by the relation $-\Delta p = \Pi(v,v)$.

REMARK. In the above inequality the gradient of the pressure appears as a nonlinear operator of order 0 (i.e., without any loss of derivative). Let us notice that if we think about the pressure p as

$$p = \mathcal{F}^{-1}\sum_{j,k}\frac{\xi_j\xi_k}{|\xi|^2}\mathcal{F}\big(v^j v^k\big),$$

the gradient of the pressure appears as a nonlinear operator with the same loss of derivative as $\operatorname{div} v \otimes v$. This observation in one of the key point of the proof of Delort's theorem about vortex sheets (see [45]).

Now let us go back to the proof of Theorem 2.1. Let us choose, according to Lemma 2.2, a stationary vector field σ such that $E_m = \sigma + L^2$. We look for a solution of the form

$$v = \tilde{v} + \sigma \quad \text{with } \tilde{v} \in \widetilde{C}^r \stackrel{\text{def}}{=} L^2 \cap C^r.$$

The two inequalities, (5) and (6), allow us to prove that the classical iterative scheme

$$\partial_t \tilde{v}_{n+1} + (\sigma + \tilde{v}_n)\cdot\nabla\tilde{v}_{n+1} = \nabla\Pi(\tilde{v}_n,\tilde{v}_n) + 2\nabla\Pi(\tilde{v}_n,\sigma) - \tilde{v}_n\cdot\nabla\sigma$$

converges in $L^\infty([0,T];\widetilde{C}^r)$ for some positive time which is greater than $C_r\|\tilde{v}_0\|_{\widetilde{C}^r}^{-1}$. Of course, the solution is unique. Thus we have a maximal time $T^\star$ such that a unique solution of (E) exists in the space $\sigma + L^\infty_{\text{loc}}([0,T^\star[;\widetilde{C}^r)$. Thanks to the above lower bound of the life span, we have that, if $T^\star$ is finite,

$$\|\tilde{v}(t)\|_{\widetilde{C}^r} \geqslant \frac{C}{T^\star - t}\cdot \tag{7}$$

Moreover, the energy estimate shows us that

$$\|\tilde{v}(t)\|_{L^2} \leqslant \|\tilde{v}_0\|_{L^2}\exp\big(t\|\nabla\sigma\|_{L^\infty}\big). \tag{8}$$

So using (5) and (6), we get that, for any t less than $T^\star$, we have

$$\left\|\tilde{v}(t)\right\|_{\widetilde{C}^r} \leqslant \left\|\tilde{v}_0\right\|_{\widetilde{C}^r} \exp\left(Ct\|\nabla\sigma\|_{C^{r-1}} + C\int_0^t \left\|\nabla v(t')\right\|_{L^\infty} \mathrm{d}t'\right). \tag{9}$$

So using (7), we infer that, if $T^\star$ is finite, then

$$\int_0^{T^\star} \left\|\nabla\tilde{v}(t)\right\|_{L^\infty} \mathrm{d}t = +\infty. \tag{10}$$

At this step, nothing crucial is really specific to dimension two. To get any global information, it seems necessary to use a global conservation at a much higher level of regularity than L^2, namely the conservation of vorticity along the trajectories given by (V) which is specific to dimension two. To do so let us use Proposition 1.5 which shows us that

$$\left\|\nabla v(t)\right\|_{L^\infty} \leqslant \|\nabla\sigma\|_{L^\infty} + \left\|\tilde{v}(t)\right\|_{L^2} + \left\|\omega(t)\right\|_{L^\infty} \log\left(e + \frac{\|\tilde{v}(t)\|_{C^r}}{\|\omega(t)\|_{L^\infty}}\right). \tag{11}$$

As the L^∞ norm of ω is preserved, which is the main fact specific the dimension two, we have, using the energy estimate (8),

$$\begin{aligned}
&\left\|\nabla v(t)\right\|_{L^\infty} \\
&\quad\leqslant \|\nabla\sigma\|_{L^\infty} + \left\|\tilde{v}_0\right\|_{L^2} \exp\left(t\|\nabla\sigma\|_{L^\infty}\right) + \|\omega_0\|_{L^\infty} \log\left(e + \frac{\|\tilde{v}(t)\|_{C^r}}{\|\omega_0\|_{L^\infty}}\right).
\end{aligned}$$

Plugging (9) into the above inequality gives

$$\begin{aligned}
&\left\|\nabla v(t)\right\|_{L^\infty} \\
&\quad\leqslant \|\nabla\sigma\|_{L^\infty} + \left\|\tilde{v}_0\right\|_{L^2} \exp\left(t\|\nabla\sigma\|_{L^\infty}\right) \\
&\qquad + \|\omega_0\|_{L^\infty}\log\left(e + \frac{\|\tilde{v}_0\|_{C^r}}{\|\omega_0\|_{L^\infty}} \exp\left(Ct\|\nabla\sigma\|_{C^{r-1}} + C\int_0^t \left\|\nabla v(t')\right\|_{L^\infty} \mathrm{d}t'\right)\right) \\
&\quad\leqslant \|\nabla\sigma\|_{L^\infty} + \left\|\tilde{v}_0\right\|_{L^2} \exp\left(t\|\nabla\sigma\|_{L^\infty}\right) + \|\omega_0\|_{L^\infty} \log\left(e + \frac{\|\tilde{v}_0\|_{C^r}}{\|\omega_0\|_{L^\infty}}\right) \\
&\qquad + \|\omega_0\|_{L^\infty} \exp\left(Ct\|\nabla\sigma\|_{C^{r-1}} + C\int_0^t \left\|\nabla v(t')\right\|_{L^\infty} \mathrm{d}t'\right).
\end{aligned}$$

Gronwall lemma implies immediately that

$$\left\|\nabla v(t)\right\|_{L^\infty} \leqslant C_0 \exp(C_1 t) \quad \text{and} \quad \left\|v(t)\right\|_{\widetilde{C}^r} \leqslant \|v_0\|_{\widetilde{C}^r} \exp\left(C\mathrm{e}^{Ct}\right). \tag{12}$$

Thus $T^\star = +\infty$ and Theorem 2.1 is proved. □

2.3. *Yudovich theorem*

As said in the Introduction we are going to regularize the initial data, the vorticity of which is simply bounded and then we are going to pass to the limit.

THEOREM 2.2. *Let m be a real number and v_0 a divergence free vector field belonging to E_m. Let us assume that ω_0 belongs to $L^\infty \cap L^a$ with $1 < a < +\infty$. Then a unique solution (v, p) of (E) exists in the space $C(\mathbb{R}; E_m) \times L^\infty_{\rm loc}(\mathbb{R}; L^2)$ such that the vorticity ω of v belongs to $L^\infty(\mathbb{R}^3) \cap L^\infty_{\rm loc}(\mathbb{R}; L^a(\mathbb{R}^2))$.*

Moreover, this vector field v has a flow. This means that a unique map ψ continuous from $\mathbb{R} \times \mathbb{R}^2$ into $\mathbb{R}^2$ exists such that

$$\psi(t, x) = x + \int_0^t v\big(s, \psi(s, x)\big)\, \mathrm{d}s \tag{13}$$

and a constant C exists such that

$$\psi(t) - \mathrm{Id} \in C^{\exp(-Ct\|\omega_0\|_{L^\infty \cap L^a})}.$$

PROOF. Let us prove the uniqueness of the solution (v, p). As the pressure is uniquely determined by p, the key point is the following lemma which estimates the L^2-distance between two solutions using only a control on the $L^\infty \cap L^a$ norm of the vorticity.

LEMMA 2.2. *Let a be a real number greater than 1, a constant C exists which satisfies the following property.*

Let (v_1, p_1) and (v_2, p_2) be two solutions of the system (E) which belong both to the same space $L^\infty_{\rm loc}(\mathbb{R}; E_m) \times L^\infty_{\rm loc}(\mathbb{R}; L^2)$ and such that their vorticity ω_i belongs to $L^\infty \cap L^a$. Let us define

$$\alpha(t) \stackrel{\text{def}}{=} \Big(C \max_i \big\| v_i(0) - \sigma \big\|_{L^2} \mathrm{e}^{t\|\nabla\sigma\|_{L^\infty}} + \max \|\omega_i\|_{L^\infty \cap L^a} + 1\Big)^{2/a}$$

and

$$\beta(t) = e \int_0^t \alpha(s)\, \mathrm{d}s.$$

If the time t and the two initial values $v_1(0)$ and $v_2(0)$ satisfy

$$\big\| v_1(0) - v_2(0) \big\|_{L^2}^2 \leqslant \mathrm{e}^{-a(\exp\beta(t) - 1)},$$

then we have

$$\big\| v_1(t) - v_2(t) \big\|_{L^2}^2 \leqslant \big\| v_1(0) - v_2(0) \big\|_{L^2}^{2\exp(-\beta(t))} \mathrm{e}^{a(1 - \exp(-\beta(t)))}.$$

All the following computations of this proof assume that the functions are regular enough. They can be made rigorous only under the hypothesis of the lemma (see, for instance, [35]). Let us define

$$I(t) \stackrel{\text{def}}{=} \|(v_1 - v_2)(t)\|_{L^2}^2.$$

We can write that

$$\begin{aligned} I'(t) = &-\sum_j \int_{\mathbb{R}^2} v_1^j(t,x)\, \partial_j \big|(v_1 - v_2)(t,x)\big|^2 \, \mathrm{d}x \\ &+ 2\sum_{i,j} \int_{\mathbb{R}^2} (v_1 - v_2)^i(t,x)(v_1 - v_2)^j(t,x)\, \partial_j v_2^i(t,x)\, \mathrm{d}x \\ &+ \sum_i \int_{\mathbb{R}^2} (v_1 - v_2)^i(t,x)\, \partial_i p(t,x)\, \mathrm{d}x. \end{aligned}$$

As the vector fields v_i are divergence free, we get by integrations by parts

$$I'(t) \leqslant J(t) \stackrel{\text{def}}{=} 2\int_{\mathbb{R}^2} \big|(v_1 - v_2)(t,x)\big|^2 \big|\nabla v_2(t,x)\big|\, \mathrm{d}x.$$

Hölder inequality implies that, for any $b \geqslant a$, we have

$$I'(t) \leqslant 2\bigg(\int_{\mathbb{R}^2} \big|v_1(t,x) - v_2(t,x)\big|^{2b/(b-1)} \mathrm{d}x\bigg)^{1-1/b} \bigg(\int_{\mathbb{R}^2} \big|\nabla v_2(t,x)\big|^b \mathrm{d}x\bigg)^{1/b}.$$

So for any b greater than a,

$$I_\varepsilon'(t) \leqslant 2\big\|v_1(t) - v_2(t)\big\|_{L^\infty}^{2/b} I_\varepsilon(t)^{1-1/b} \big\|\nabla v_2(t)\big\|_{L^b} + R_\varepsilon(t).$$

Thanks to the Biot–Savart law, Theorem 1.2 and the conservation of the vorticity along the trajectories, we have, for any $b \geqslant a$,

$$\big\|\nabla v_2(t)\big\|_{L^b} \leqslant Cb\big\|\omega_2(0)\big\|_{L^a \cap L^\infty}.$$

Moreover, we have

$$\|v_i\|_{L^\infty} \leqslant C\Big(\big\|v_i(0) - \sigma\big\|_{L^2} \mathrm{e}^{t\|\nabla\sigma\|_{L^\infty}} + \|\sigma\|_{L^\infty} + \big\|\omega_i(0)\big\|_{L^\infty}\Big).$$

So it turns out that for any $b \geqslant a$, we have

$$I'(t) \leqslant \alpha(t) b I(t)^{1-1/b}. \tag{14}$$

Now let us assume that $\|v_1(0) - v_2(0)\|_{L^2} < 1$. Let η be in $]0, 1 - I(0)[$. Let us state

$$J_\eta(t) = \eta + I(t).$$

From (14) we infer that $J'_\eta(t) \leqslant \alpha(t) b J_\eta(t)^{1-1/b}$. Now let us choose $b = a - \log J_\eta(t)$. We get

$$\begin{aligned} J'_\eta(t) &\leqslant \alpha(t)\big(a - \log J_\eta(t)\big) J_\eta(t) \exp\left(\frac{-\log J_\eta(t)}{a - \log J_\eta(t)}\right) \\ &\leqslant e\alpha(t)\big(a - \log J_\eta(t)\big) J_\eta(t). \end{aligned}$$

Thus by integration, we get that

$$J_\eta(t) \leqslant J_\eta(0) + \int_0^t e\alpha(t')\big(a - \log J_\eta(t')\big) J_\eta(t')\,\mathrm{d}t'. \tag{15}$$

Now the key point is the classical Osgood lemma which can be understood as a generalization of Gronwall lemma.

LEMMA 2.3. *Let ρ be a measurable nonnegative function, γ a nonnegative locally integrable function and μ a continuous and nondecreasing function such that $\mu(r)$ is positive when r is positive. Let us assume that, for a nonnegative real number c, the function ρ satisfies*

$$\rho(t) \leqslant c + \int_{t_0}^t \gamma(s)\mu\big(\rho(s)\big)\,\mathrm{d}s. \tag{16}$$

If c is positive, then we have

$$-\mathcal{M}\big(\rho(t)\big) + \mathcal{M}(c) \leqslant \int_{t_0}^t \gamma(s)\,\mathrm{d}s \quad \textit{with } \mathcal{M}(x) = \int_x^1 \frac{\mathrm{d}r}{\mu(r)}. \tag{17}$$

If $c = 0$ and if μ satisfies

$$\int_0^1 \frac{\mathrm{d}r}{\mu(r)} = +\infty \tag{18}$$

then the function ρ is identically 0.

PROOF. In order to prove this lemma, let us state

$$R_c(t) \stackrel{\text{def}}{=} c + \int_{t_0}^t \gamma(s)\mu\big(\rho(s)\big)\,\mathrm{d}s.$$

The function R_c is a continuous nondecreasing function. So we have

$$\begin{aligned}\frac{\mathrm{d}R_c}{\mathrm{d}t} &= \gamma(t)\mu\big(\rho(t)\big)\\ &\leqslant \gamma(t)\mu\big(R_c(t)\big).\end{aligned} \tag{19}$$

Let us assume that c is positive. The function R_c is also positive. So we infer from (19) that

$$-\frac{\mathrm{d}}{\mathrm{d}t}\mathcal{M}\big(R_c(t)\big) = \frac{\mathrm{d}R_c}{\mathrm{d}t}\frac{1}{\mu(R_c(t))} \leqslant \gamma(t).$$

Thus we get (17) by integration. Let us assume now that $c = 0$ and that ρ is not identically 0 near t_0. As the function μ is nondecreasing, we can consider the function $\tilde{\rho}(t) \stackrel{\text{def}}{=} \sup_{s\in[t_0,t]} \rho(s)$ instead of ρ. A real number t_1 greater than t_0 exists such that $\rho(t_1)$ is positive. As the function ρ satisfies (16) with $c = 0$, it also satisfies this inequality for any positive c' less than $\rho(t_1)$. Then it comes from (17) that

$$\forall c' \in \big]0, \rho(t_1)\big], \qquad \mathcal{M}\big(c'\big) \leqslant \int_{t_0}^{t_1} \gamma\big(t'\big)\,\mathrm{d}\tau + \mathcal{M}\big(\rho(t_1)\big),$$

which implies that $\int_0^1 \frac{\mathrm{d}r}{\mu(r)} < +\infty$. Thus the lemma is proved. □

The existence of the solution is proved in a very classical way by using this lemma for the solutions with regularized initial data.

Our purpose is now to prove the Lagrangian part of Theorem 2.2. Let us define the space of vector fields we are going to work with.

DEFINITION 2.4. The set of log-Lipschitzian vector fields on $\mathbb{R}^d$, denoted by *LL*, is the set of bounded vector fields v such that

$$\|v\|_{LL} \stackrel{\text{def}}{=} \sup_{0<|x-x'|\leqslant 1} \frac{|v(t,x) - v(t,x')|}{|x - x'|(1 - \log|x - x'|)} < \infty.$$

As Proposition 1.4 shows us, a divergence free vector field with bounded vorticity is a log-Lipschitzian vector field. The following theorem implies the Lagrangian part of Theorem 2.2.

THEOREM 2.3. *Let v be a vector field in $L^1_{\mathrm{loc}}(\mathbb{R}; LL)$, a unique continuous map ψ from $\mathbb{R}\times\mathbb{R}^d$ into $\mathbb{R}^d$ exists such that*

$$\psi(t,x) = x + \int_0^t v\big(s, \psi(s,x)\big)\,\mathrm{d}s.$$

Moreover, ψ is such that, for any t,

$$\psi(t) - \mathrm{Id} \in C^{\exp(-\int_0^t \|v(s)\|_{LL}\,\mathrm{d}s)}.$$

More precisely, we have

$$|x-y|\leqslant \mathrm{e}^{1-\exp(\int_0^t \|v(s)\|_{LL}\,\mathrm{d}s)}$$
$$\Longrightarrow\quad \bigl|\psi(t,x)-\psi(t,y)\bigr|\leqslant |x-y|^{\exp(-\int_0^t \|v(s)\|_{LL}\,\mathrm{d}s)}\mathrm{e}^{1-\exp(-\int_0^t \|v(s)\|_{LL}\,\mathrm{d}s)}.$$

Let us digress a little on ordinary differential equations associated to non-Lipschitzian vector fields. This will lead us to the above theorem. Here the function μ is supposed to be continuous and nondecreasing from $\mathbb{R}^+$ into $\mathbb{R}^+$ such that $\mu(0)=0$.

DEFINITION 2.5. Let (X,d) be a metric space and $(E,\|\cdot\|)$ a Banach space. Let us denote by $\mathcal{C}_\mu(X,E)$ the space of bounded functions u from X into E such that

$$\|u\|_\mu \stackrel{\text{def}}{=} \|u\|_{L^\infty}+\sup_{(x,y)\in X\times X, x\neq y}\frac{\|u(x)-u(y)\|}{\mu(d(x,y))}<\infty.$$

The theorem about ordinary differential equations which generalizes the classical Cauchy–Lipschitz theorem is the following.

THEOREM 2.4. *Let E be a Banach space, Ω an open subset of E, I an open interval of $\mathbb{R}$ and (t_0,x_0) an element of $I\times\Omega$. Let us consider a function $F\in L^1_{\mathrm{loc}}(I;\mathcal{C}_\mu(\Omega;E))$. Let us assume that μ satisfies* (18). *Then an open interval J exists such that $t_0\in J\subset I$ and such that the equation*

$$x(t)=x_0+\int_{t_0}^t F\bigl(s,x(s)\bigr)\,\mathrm{d}s \qquad (EDO)$$

has a unique continuous solution defined on J.

PROOF. Let us begin by proving the uniqueness of the trajectories. Let $x_1(t)$ and $x_2(t)$ two solutions of (*EDO*) defined in an open interval $\widetilde{J}$ which contains t_0 with the same initial data x_0. Let us define $\rho(t)=\|x_1(t)-x_2(t)\|$. It is obvious that

$$0\leqslant \rho(t)\leqslant \int_{t_0}^t \gamma(s)\mu\bigl(\rho(s)\bigr)\,\mathrm{d}s \quad \text{with } \gamma\in L^1_{\mathrm{loc}}(I) \text{ and } \gamma\geqslant 0. \qquad (20)$$

Thanks to Lemma 2.3 we immediately get that $\rho\equiv 0$.

Now let us prove the existence by considering the classical Picard scheme

$$x_{k+1}(t)=x_0+\int_{t_0}^t F\bigl(s,x_k(s)\bigr)\,\mathrm{d}s.$$

We skip the fact that for J small enough, the sequence $(x_k)_{k\in\mathbb{N}}$ is well defined and bounded in the space $C_b(J,\Omega)$. Let us state $\rho_{k,n}(t)=\sup_{s\leqslant t}\|x_{k+n}(s)-x_k(s)\|$. We have that

$$0\leqslant \rho_{k+1,n}(t)\leqslant \int_{t_0}^t \gamma(s)\mu\bigl(\rho_{k,n}(s)\bigr)\,\mathrm{d}s.$$

Let us state $\rho_k(t) \stackrel{\text{def}}{=} \sup_n \rho_{k,n}(t)$. From the fact that the function μ is a nondecreasing function we deduce that

$$0 \leqslant \rho_{k+1}(t) \leqslant \int_{t_0}^{t} \gamma(s)\mu\big(\rho_k(s)\big)\,\mathrm{d}s.$$

Fatou's lemma implies now that

$$\tilde{\rho}(t) \stackrel{\text{def}}{=} \limsup_{k\to+\infty} \rho_k(t) \leqslant \int_{t_0}^{t} \gamma(s)\mu\big(\tilde{\rho}(s)\big)\,\mathrm{d}s.$$

Lemma 2.3 implies that $\tilde{\rho}(t) \equiv 0$ near t_0; this concludes the proof of Theorem 2.4. □

To prove the whole Yudovich theorem, we shall apply Lemma 2.3. Let us study the regularity of the flow with respect to the variable x. Let us consider two trajectories $x_1(t)$ and $x_2(t)$ such that

$$\big\|x_1(0) - x_2(0)\big\| < 1.$$

As long as $\|x_1(t) - x_2(t)\| < 1$, we have

$$\begin{aligned}\big\|x_1(t) - x_2(t)\big\| &\leqslant \|x_1 - x_2\| + \int_0^t \big\|v\big(s, x_1(s)\big) - v\big(s, x_2(s)\big)\big\|\,\mathrm{d}s \\ &\leqslant \|x_1 - x_2\| + \int_0^t \big\|v(s)\big\|_{LL}\mu\big(\big\|x_1(s) - x_2(s)\big\|\big)\,\mathrm{d}s.\end{aligned}$$

Thanks to Lemma 2.3, it turns out that

$$-\log\big(1 - \log\big\|x_1(t) - x_2(t)\big\|\big) + \log\big(1 - \log\|x_1 - x_2\|\big) \leqslant \int_0^t \big\|v(s)\big\|_{LL}\,\mathrm{d}s.$$

An obvious computation implies that

$$\big\|x_1(t) - x_2(t)\big\| \leqslant \|x_1 - x_2\|^{\exp(-\int_0^t \|v(s)\|_{LL}\,\mathrm{d}s)} \mathrm{e}^{1-\exp(-\int_0^t \|v(s)\|_{LL}\,\mathrm{d}s)}. \tag{21}$$

Thus the Yudovich theorem is proved. □

2.4. *An example of decay of regularity*

The purpose of this section is to exhibit a solution of the incompressible Euler system which shows that the regularity of the flow given by the Yudovich theorem is essentially optimal. The solution we are going to construct satisfies the following properties:

- the vorticity ω of the solution v is at any time t, bounded and compactly supported,

- at any time t, the flow $\psi(t)$ of v does not belong to the Hölder class $C^{\exp(-t)}$.

Let us define the initial data. The function ω_0 is the function null outside $[-1, 1] \times [-1, 1]$, odd with respect to each variables x_1 and x_2 and the value of which is 2π on $[0, 1] \times [0, 1]$. So the vector field v_0 is defined by

$$v_0(x_1, x_2) = \begin{cases} -\frac{1}{2\pi} \int \frac{x_2 - y_2}{|x-y|^2} \omega_0(y)\,\mathrm{d}y, \\ \frac{1}{2\pi} \int \frac{x_1 - y_1}{|x-y|^2} \omega_0(y)\,\mathrm{d}y. \end{cases}$$

We shall prove the following theorem.

THEOREM 2.5. *Let v be the solution of (E) associated with the initial data defined above. At time t, the flow $\psi(t)$ of v does not belong to C^α for $\alpha > \exp(-t)$.*

The symmetry properties are extremely useful for the computations. We have the following estimate.

PROPOSITION 2.1. *A constant C exists such that if $x_1 \in [0, C]$, then*

$$v_0^1(x_1, 0) \geqslant -2x_1 \log x_1.$$

PROOF. Let us state $\tilde{\omega}_0(x_1) \stackrel{\text{def}}{=} 2H(x_1) - 1$ (H denotes Heavyside function). We have, thanks to symmetry properties,

$$\begin{aligned} v_0^1(x_1, 0) &= \frac{1}{2\pi} \int \frac{y_2}{|x-y|^2} \omega_0(y)\,\mathrm{d}y \\ &= \int_{-1}^{1} \mathrm{d}y_1\, \tilde{\omega}_0(y_1) \int_0^1 \frac{2y_2}{(x_1 - y_1)^2 + y_2^2}\,\mathrm{d}y_2 \\ &= \tilde{v}_0^1(x_1, 0) + \bar{v}_0^1(x_1, 0) \end{aligned}$$

with

$$\tilde{v}_0^1(x_1, 0) = -\int_0^1 \log(x_1 - y_1)^2\,\mathrm{d}y_1 + \int_0^1 \log(x_1 + y_1)^2\,\mathrm{d}y_1$$

and

$$\bar{v}_0^1(x_1, 0) = \int_0^1 \log \frac{1 + (x_1 - y_1)^2}{1 + (x_1 + y_1)^2}\,\mathrm{d}y_1.$$

It is clear that $x_1 \mapsto \bar{v}_0^1(x_1, 0)$ is an odd smooth function. Moreover, straightforward computations give that, for $0 \leqslant x_1 < 1$,

$$\tilde{v}_0^1(x_1, 0) = -4x_1 \log x_1 + 2(1 + x_1)\log(1 + x_1) - 2(1 - x_1)\log(1 - x_1).$$

Thus if $0 \leqslant x_1 < 1$, we have

$$v_0^1(x_1, 0) = -4x_1 \log x_1 + f(x_1),$$

where f denotes a smooth odd function on $]-1, 1[$. Thus the proposition is proved. □

PROOF OF THEOREM 2.5. Let us go back to incompressible Euler system. Thanks to the Yudovich theorem (Theorem 2.2), the flow of v is a continuous function of the variable (t, x). Moreover, at any time t, the vector field v is symmetric with respect to the coordinates axes. Thus both axes are globally invariant by the flow. Thus the origin also. This means that for any t,

$$\psi(t, 0) = 0, \qquad \psi^1(t, 0, x_2) = 0 \quad \text{and} \quad \psi^2(t, x_1, 0) = 0. \tag{22}$$

Let T be any fixed positive time. The vorticity is preserved along the trajectories. Thus a neighborhood of the origin W exists such that, for any time $t \in [0, T]$,

$$\omega(t)_{|W} = \omega_{0|W}.$$

The divergence free vector field $\tilde{v}(t) = v(t) - v_0$ is symmetric with respect to the two coordinates axis and its vorticity is identically 0 on W. Thus a constant A exists such that, for any $t \in [0, T]$,

$$\left|v(t, x) - v_0(x)\right| \leqslant A|x|.$$

From Proposition 2.1 we infer the existence of a constant C' such that for any (t, x_1) belonging to $[0, T] \times [0, C']$, we have

$$v(t, x_1, 0) \geqslant -x_1 \log x_1.$$

A constant α exists such that if $x_1 \in [0, \alpha[$ then for any $t \in [0, T]$, we have $\psi^1(t, x_1, 0) \in [0, C']$. It comes from the above inequality that

$$\psi^1(t, x_1, 0) \geqslant x_1(t) \quad \text{with } \dot{x}_1(t) = -x_1(t) \log x_1(t).$$

Then it turns out that

$$\psi^1(t, x_1, 0) \geqslant x_1^{\exp(-t)}.$$

As $\psi(t, 0) = 0$, Theorem 2.5 is proved. □

2.5. *References and remarks*

The first result of local well-posedness in Hölder spaces has been proved in 1928 by Lichtenstein in the impressive series of works [67]. Thanks to the Lagrangian point of view, the Euler system is seen as an ordinary differential equation in Hölder spaces. This point of view has been rediscovered by Serfati in [75,76]. One of the main ideas of these two papers is that the ordinary differential equation turns out to be analytic. Serfati proved in particular the following theorem.

THEOREM 2.6. *Let v be a solution of (E) given by Theorem* 2.1. *Let us denote by ψ its flow. Then the map*

$$t \mapsto \psi(t,\cdot)$$

is analytic with value in the Banach space C^r.

This theorem has been proved independently by Gamblin in [52] following the Eulerian point of view. Moreover, in this work Gamblin establishes a remarkable microlocal interpretation of the above theorem.

THEOREM 2.7. *Let v be a solution of (E) given by Theorem* 2.1. *The analytic wave front set of v is included in the characteristic variety*

$$\{(t,x,\tau,\xi) \in \mathbb{R}^{1+d} \times \mathbb{R}^{1+d} \setminus \{0\}, \tau + \langle v, \xi\rangle = 0\}.$$

For the definition of the analytic wave front set, we refer to [56], Section 8.4.

Theorem 2.2 has been proved in 1964 by Yudovich in [88]. In this seminal paper, he also stated the problem of vortex patches which will be the purpose of Sections 4–8 of this chapter.

In this framework of solutions with bounded vorticity, Gamblin proved in [52] the following theorem of regularity of the trajectories.

THEOREM 2.8. *Let v be a solution of (E) given by Theorem* 2.2. *Let us denote by ψ its flows. Then for any x in $\mathbb{R}^2$, the map*

$$t \mapsto \psi(t,x)$$

is in the class G^3 which is the Gevrey class of order 3.

This means that this map is C^∞ and that $\|\partial_t^k \psi\|_{L^\infty([0,T]\times\mathbb{R}^2)} \leqslant C_T^{k+1}(k!)^3$. The following microlocal statement has been proved by Gamblin in [52].

THEOREM 2.9. *Let v be a solution of (E) given by Theorem* 2.2. *The Gevrey* 3 *wave front set of v is included in the characteristic variety*

$$\{(t,x,\tau,\xi) \in \mathbb{R}^{1+d} \times \mathbb{R}^{1+d} \setminus \{0\}, \tau + \langle v, \xi\rangle = 0\}.$$

For the definition of the Gevrey 3 wave front set, we refer again to [56], Section 8.4. The above theorems were established with the weaker conclusion C^∞ instead of analytic or G^3 by the author in [31].

Let us point out a recent generalization of Theorem 2.2 done by Vishik in [86]. In this impressive work, Vishik proves the existence and uniqueness of solution of the two-dimensional Euler system (E) for a class of initial data the vorticity of which is not in L^p for p greater than some p_0. We shall say more about this surprising result at the end of Section 6.

The example of Section 2.4 has been exhibited by Bahouri and the author in [5].

3. The growth of the diameter of a vortex patch

The problem addressed in this section is the following: Let us consider a global solution of two-dimensional incompressible Euler equation, is it possible to get bounds on the time evolution of the diameter of the support of the vorticity?

As proved in Proposition 1.1, the velocity field v is globally bounded in $\mathbb{R} \times \mathbb{R}^2$. So through an immediate time integration, we get that the diameter of the support of $\omega(t)$ is bounded by

$$\delta(0) + C\|\omega_0\|_{L^1 \cap L^\infty} t.$$

3.1. *The case when the sign of the vorticity changes*

In spite of the fact that this estimate seems rather rough, it turns out to be optimal when the sign of the vorticity changes. More precisely, we have the following proposition.

PROPOSITION 3.1. *Let ω_0 be a bounded compactly supported function on $\mathbb{R}^2$ such that*

$$\omega_0(x) = -\omega_0(\bar{x}) = -\omega_0(-\bar{x}) = \omega_0(-x), \tag{23}$$

which means that ω_0 is odd with respect to each variables x_1 and x_2. Let us assume that stating $Q \stackrel{def}{=} \{x \in \mathbb{R}^2 / x_1 \geqslant 0 \text{ and } x_2 \geqslant 0\}$, we have $\omega_{0|Q} \geqslant 0$. Then if ω_0 is not identically 0, then a constant C_0 exists such that $\delta(t) \geqslant C_0 t$.

PROOF. In order to prove this, let us first observe that the relation (23) holds for any time t and that the Biot–Savart law can be written in this case

$$\begin{aligned} v^1(x) = \int_Q \bigg(& (x_2 - y_2)\left(\frac{1}{|x-y|^2} - \frac{1}{|x+\bar{y}|^2}\right) \\ & + (x_2 + y_2)\left(\frac{1}{|x-\bar{y}|^2} - \frac{1}{|x+y|^2}\right)\bigg)\omega(y)\,dy, \\ v^2(x) = \int_Q \bigg(& (x_1 - y_1)\left(\frac{1}{|x-y|^2} - \frac{1}{|x-\bar{y}|^2}\right) \\ & + (x_1 + y_1)\left(\frac{1}{|x+y|^2} - \frac{1}{|x+\bar{y}|^2}\right)\bigg)\omega(y)\,dy. \end{aligned} \tag{24}$$

Let us define now the center of mass of the vorticity restricted on Q by

$$\mathbf{P}(t) \stackrel{\text{def}}{=} \frac{1}{m_0} \int_Q \omega(t,x) x \, dx.$$

Using the above version of the Biot–Savart law and the fact that

$$\mathbf{P}'(t) = \frac{1}{m_0} \int_Q \omega(t,x) v(t,x) \, dx,$$

we get

$$\frac{d\mathbf{P}^1}{dt}(t) = \frac{1}{m_0} \int_{Q\times Q} \frac{4x_1 y_1 (x_2 + y_2)}{|x-\bar{y}|^2 |x+y|^2} \omega(x)\omega(y) \, dx \, dy,$$

$$\frac{d\mathbf{P}^2}{dt}(t) = -\frac{1}{m_0} \int_{Q\times Q} \frac{4x_2 y_2 (x_1 + y_1)}{|x+\bar{y}|^2 |x+y|^2} \omega(x)\omega(y) \, dx \, dy.$$

It is obvious that $\mathbf{P}^1$ is increasing and that $\mathbf{P}^2$ is decreasing. The energy conservation will be used to estimate $\mathbf{P}^1$ from below. As seen in Section 2.1, v_0 is in L^2 because $\int_{\mathbb{R}^2} \omega = 0$. Using the Biot–Savart law (1), we have, for any positive t,

$$E_0 \stackrel{\text{def}}{=} \frac{1}{2} \left\| v(t) \right\|_{L^2}^2 = -\frac{1}{2\pi} \int_{\mathbb{R}^2 \times \mathbb{R}^2} \log|x-y| \omega(t,x) \omega(t,y) \, dx \, dy.$$

Thanks to symmetries, we get

$$E_0 = \frac{1}{\pi} \int_{Q\times Q} \log\left(1 + \frac{16 x_1 y_1 x_2 y_2}{|x-y|^2 |x+y|^2}\right) \omega(t,x) \omega(t,y) \, dx \, dy.$$

Using the conservation of the integral of the vorticity

$$m_0 \stackrel{\text{def}}{=} \int_{\mathbb{R}^2} \omega_0(x) \, dx = \int_{\mathbb{R}^2} \omega(t,x) \, dx,$$

Hölder's inequality in the above integral with $q = 6/5$ implies that

$$E_0^6 \leqslant C m_0 \mathbf{P}_1'(t) I^5(t) \tag{25}$$

with

$$I(t) \stackrel{\text{def}}{=} \int_{Q\times Q} \left(\frac{|x-\bar{y}|^2 |x+y|^2}{x_1 y_1 (x_2 + y_2)} \right)^{1/5} \times \left(\log\left(\frac{|x-\bar{y}|^2 |x+\bar{y}|^2}{|x-y|^2 |x+y|^2} \right) \right)^{1/5} \omega(t,x) \omega(t,y) \, dx \, dy. \tag{26}$$

Using the fact that $\log(1+z) \leqslant Cz(1+z)^{-5/6}$, we get, after computations detailed in [59] that

$$I(t) \leqslant C \int_{Q\times Q} \frac{(x_2+y_2)^{3/5}}{|x-y|^{2/5}} \omega(t,x)\omega(t,y)\,dx\,dy.$$

Applying Hölder's inequality we infer that

$$I^5(t) \leqslant I_2^2(t) I_3^3(t)$$

with

$$I_2(t) \stackrel{\text{def}}{=} \int_{Q\times Q} \frac{1}{|x-y|} \omega(t,x)\omega(t,y)\,dx\,dy$$

and

$$I_3(t) \stackrel{\text{def}}{=} \int_{Q\times Q} (x_2+y_2)\omega(t,x)\omega(t,y)\,dx\,dy.$$

Let us observe that $I_3(t) = 2m_0\mathbf{P}_2(t)$. As $\mathbf{P}_2(t)$ is a decreasing function, we get

$$I_3(t) \leqslant 2m_0^2\mathbf{P}_2(0).$$

Moreover, by easy interpolation arguments we have that

$$I_2(t) \leqslant m_0^{3/2} \|\omega_0\|_{L^\infty}^{1/2}.$$

Then we get that $I^5(t) \leqslant Cm_0^9\mathbf{P}_2^3(0)\|\omega\|_{L^\infty}$ and then using (26), we get that

$$\mathbf{P}_1'(t) \geqslant E_0 m_0^{-9}\mathbf{P}_2^{-3}(0).$$

Thus Proposition 3.1 is proved. □

3.2. *The case when the vorticity is nonnegative*

The case when the vorticity is nonnegative is very different. In this case, the growth of the diameter of the vortex patch is much slower. More precisely, we have the following theorem.

THEOREM 3.1. *Let v be a solution of (E) the initial data of which has a bounded, nonnegative and compactly supported vorticity ω_0. Then a constant C_0 exists such that, for any time t, the diameter of the support of $\omega(t)$ is less than*

$$4d_0 + C_0\big(t\log(2+t)\big)^{1/4}.$$

The proof of this theorem is essentially based on conservation laws of the two-dimensional Euler equation. First, the conservation of the mass of the vorticity

$$\forall t \in \mathbb{R}, \quad m_0 \stackrel{\text{def}}{=} \int_{\mathbb{R}^2} \omega_0(x)\,\mathrm{d}x = \int_{\mathbb{R}^2} \omega(t,x)\,\mathrm{d}x. \tag{27}$$

Then the conservation of the supremum of ω, namely

$$\forall t \in \mathbb{R}, \quad M_0 \stackrel{\text{def}}{=} \sup_{x\in\mathbb{R}^2} \omega_0(x) = \sup_{x\in\mathbb{R}^2} \omega(x). \tag{28}$$

Moreover, the center of mass is preserved because thanks to equation (V) and the Biot–Savart law, we have

$$\begin{aligned}\frac{\mathrm{d}}{\mathrm{d}t}\int_{\mathbb{R}^2} x\omega(t,x)\,\mathrm{d}x &= -\int_{\mathbb{R}^2} x(v\cdot\nabla\omega)(x)\,\mathrm{d}x\\ &= \int_{\mathbb{R}^2} \omega v\,\mathrm{d}x\\ &= \frac{1}{2\pi}\int_{\mathbb{R}^2\times\mathbb{R}^2} \frac{(x-y)^{\perp}}{|x-y|^2}\omega(x)\omega(y)\,\mathrm{d}x\,\mathrm{d}y\\ &= 0.\end{aligned}$$

If we assume that the center of mass at time 0 is the origin (i.e., $\int_{\mathbb{R}^2} x\omega_0(x)\,\mathrm{d}x = 0$), we have

$$\forall t \in \mathbb{R}, \quad \int_{\mathbb{R}^2} x\omega(t,x)\,\mathrm{d}x = 0. \tag{29}$$

Then let us observe that

$$\frac{\mathrm{d}}{\mathrm{d}t}\int_{\mathbb{R}^2} |x|^2\omega(t,x)\,\mathrm{d}x = 2\int_{\mathbb{R}^2} \big(v(x)\cdot x\big)\omega(x)\,\mathrm{d}x.$$

By definition of the vorticity and after integration by parts we have

$$\begin{aligned}\int_{\mathbb{R}^2} v(x)\cdot x\omega(x)\,\mathrm{d}x = &-\int_{\mathbb{R}^2} \big(\partial_1 v^1 x_1 v^2 + v^1 v^2 + x_2\,\partial_1 v^2 v^2\big)\,\mathrm{d}x\\ &+\int_{\mathbb{R}^2} \big(\partial_2 v^1 x_1 v^1 + v^1 v^2 + x_2\,\partial_2 v^2 v^1\big)\,\mathrm{d}x\\ = &-\int_{\mathbb{R}^2} v(x)\cdot x\,\omega(x)\,\mathrm{d}x.\end{aligned}$$

We infer that

$$\forall t \in \mathbb{R}, \quad i_0 \stackrel{\text{def}}{=} \int_{\mathbb{R}^2} |x|^2\omega_0(x)\,\mathrm{d}x = \int_{\mathbb{R}^2} |x|^2\omega(t,x)\,\mathrm{d}x. \tag{30}$$

Let us assume proven the following assertion

$$|x| \geqslant 4d_0 + C_0\big(t\log(2+t)\big)^{1/4} \quad \Longrightarrow \quad \left|\frac{x}{|x|}\cdot v(t,x)\right| \leqslant \frac{C_0}{|x|^3}. \tag{31}$$

The above assertion claims that the vector field $(1, v(t,x))$ points inward the domain

$$\mathcal{D} \stackrel{\text{def}}{=} \big\{(t,x) \in \mathbb{R}^+ \times \mathbb{R}^2,\, |x| < 4d_0 + C_0\big(t\log(2+t)\big)^{1/4}\big\}.$$

Thus the proof of Theorem 3.1 reduces to the proof of (31). Using that $x\cdot(x-y)^\perp = -x\cdot y^\perp$, and that the center of mass of the vorticity is the origin (Equation (29)), we get

$$\frac{x}{|x|}\cdot v(t,x) = -\frac{1}{2\pi}\int_{\mathbb{R}^2} \frac{x\cdot y^\perp}{|x|}\left(\frac{1}{|x-y|^2} - \frac{1}{|x|^2}\right)\omega(t,y)\,dy.$$

Cutting the integral into the two parts $|x-y| \leqslant \frac{|x|}{2}$ and $|x-y| > \frac{|x|}{2}$, we get that

$$\frac{x}{|x|}\cdot v(t,x) = J_1(x) + J_2(x)$$

with

$$J_1(t,x) \stackrel{\text{def}}{=} \int_{|x-y|>|x|/2} \frac{x\cdot y^\perp}{|x|}\left(\frac{1}{|x-y|^2} - \frac{1}{|x|^2}\right)\omega(t,y)\,dy$$

and

$$J_2(t,x) \stackrel{\text{def}}{=} \int_{|x-y|\leqslant|x|/2} \frac{x\cdot y^\perp}{|x|}\left(\frac{1}{|x-y|^2} - \frac{1}{|x|^2}\right)\omega(t,y)\,dy.$$

It is easily seen that as $|x-y| > |x|/2$, we have that $|2x-y| \leqslant 3|x-y|$. Thus as ω is nonnegative, we infer that

$$\begin{aligned} J_1(t,x) &\leqslant \frac{C}{|x|^3}\int_{|x-y|>|x|/2} |y|^2\omega(t,y)\,dy \\ &\leqslant \frac{C}{|x|^3} i_0. \end{aligned}$$

The second part J_2 is in fact harder to estimate. As ω is nonnegative , we have

$$\big|J_2(t,x)\big| \leqslant C\int_{|x-y|\leqslant|x|/2} \frac{\omega(y)}{|x-y|}\,dy.$$

Using the fact that

$$\left\{y\in\mathbb{R}^2,\, |x-y| \leqslant \frac{|x|}{2}\right\} \subset \left\{y\in\mathbb{R}^2,\, |y| \geqslant \frac{|x|}{2}\right\},$$

we have that

$$\left|J_2(x)\right| \leqslant C \int_{|y| \geqslant |x|/2} \frac{\omega(y)}{|x-y|} \, \mathrm{d}y.$$

Then using the fact that for any domain S of $\mathbb{R}^2$ and any point x of $\mathbb{R}^2$, we have

$$\int_S \frac{h(y)}{|x-y|} \, \mathrm{d}y \leqslant C \|h\|_{L^1(S)}^{1/2} \|h\|_{L^\infty(S)}^{1/2},$$

we have that

$$\left|J_2(t,x)\right| \leqslant C M_0^{1/2} \left(\int_{|y| \geqslant |x|/2} \omega(y) \, \mathrm{d}y \right)^{1/2}.$$

Assertion (31) and thus Theorem 3.1 will follow from the following proposition.

PROPOSITION 3.2. *For any k, a constant C_0 (depending of course of m_0, M_0, d_0 and i_0) exists such that*

$$\forall (t,x) \in \mathbb{R}^+ \times \left(\mathbb{R}^2 \setminus \mathcal{D}\right), \quad \int_{|y| \geqslant |x|/2} \omega(y) \, \mathrm{d}y \leqslant \frac{C_0}{|x|^k}.$$

In fact $k = 6$ is enough for our purpose. As done in [59] this proposition can be proved using only conservation laws (27)–(30). Here we shall follow the Appendix of [59] where high-order momenta of the vorticity are introduced. Let us define

$$m_n(t) \stackrel{\text{def}}{=} \int_{\mathbb{R}^2} |x|^{4n} \omega(t,x) \, \mathrm{d}x.$$

Although those quantities are not preserved along the flow, something relevant can be said about their large time behavior. More precisely, we have the following lemma.

LEMMA 3.1. *A constant C_0 exists such that*

$$\forall t \in \mathbb{R}^+, \ \forall n \in \mathbb{N}, \quad m_n(t) \leqslant m_0 \left(d_0^4 + C_0 i_0 n t\right)^n,$$

where d_0 is the one which appears in the statement of Theorem 3.1.

Let us prove that the above lemma implies Proposition 3.2 for $k = 6$. Let t be large enough and let n be the integer greater that $3/2$ such that

$$6 \log(2+t) - 1 < n \leqslant 6 \log(2+t). \tag{32}$$

Using again the nonnegativity of the vorticity, we have

$$\int_{|y|\geqslant r} \omega(t,y)\,\mathrm{d}y \leqslant \frac{m_n(t)}{r^{4n}}$$
$$\leqslant \frac{m_0}{r^{4n}}\big(d_0^4 + C_0 i_0 n t\big)^n.$$

Thus we have

$$\int_{|y|\geqslant r} \omega(t,y)\,\mathrm{d}y \leqslant \frac{m_0}{r^6} 2^{3/2-n}\big(d_0^4 + C_0 i_0 t \log(2+t)\big)^{3/2}.$$

By definition of n, we have that $2^{n+1} \geqslant (2+t)^{6\log 2}$. Thus

$$\int_{|y|\geqslant r} \omega(t,y)\,\mathrm{d}y \leqslant \frac{C_0}{r^6}.$$

PROOF OF LEMMA 3.1. Now let us prove Lemma 3.1. Using the conservation of the vorticity along the flow and the Biot–Savart law (1), we have

$$\frac{\mathrm{d}m_n}{\mathrm{d}t} = \frac{2n}{\pi} \int_{\mathbb{R}^2\times\mathbb{R}^2} K(x,y)|x|^{4n-2}\omega(t,x)\omega(t,y)\,\mathrm{d}x\,\mathrm{d}y$$

with

$$K(x,y) \stackrel{\text{def}}{=} x\cdot(x-y)^{\perp}\left(\frac{1}{|x-y|^2} - \frac{1}{|x|^2}\right).$$

Let us decompose the above integral into three parts

$$\frac{\mathrm{d}m_n}{\mathrm{d}t} = \sum_{i=1}^{3} \alpha_i(t)$$

with

$$\alpha_i(t) \stackrel{\text{def}}{=} \frac{2n}{\pi} \int_{A_i} K(x,y)|x|^{4n-2}\omega(t,x)\omega(t,y)\,\mathrm{d}x\,\mathrm{d}y$$

and

$$A_1(t) \stackrel{\text{def}}{=} \left\{(x,y)\in\mathbb{R}^2\times\mathbb{R}^2/|y| \leqslant \left(1-\frac{1}{2n}\right)|x|\right\},$$

$$A_2(t) \stackrel{\text{def}}{=} \left\{(x,y)\in\mathbb{R}^2\times\mathbb{R}^2\Big/\left(1-\frac{1}{2n}\right)|x| < |y| \leqslant \left(1-\frac{1}{2n}\right)^{-1}|x|\right\}$$

and

$$A_3(t) \overset{\text{def}}{=} \left\{(x, y) \in \mathbb{R}^2 \times \mathbb{R}^2 / |x| \leqslant \left(1 - \frac{1}{2n}\right)|y|\right\}.$$

Let us estimate $K(x, y)$ when $(x, y) \in A_1$. As $|x - y| > (2n)^{-1}|x|$ and $|2x - y| \leqslant 3|x|$, we have

$$\left|K(x, y)\right| \leqslant \frac{|y|^2|2x - y|}{|x|^2|x - y|} \leqslant 6n\frac{|y|^2}{|x|^2}.$$

Then we get

$$\begin{aligned}\alpha_1(t) &\leqslant \frac{12n^2}{\pi}\int_{A_i} |y|^2\omega(t, x)\omega(t, y)\,\mathrm{d}x\,\mathrm{d}y\\ &\leqslant \frac{12n^2}{\pi} i_0 m_{n-1}(t).\end{aligned} \tag{33}$$

If (x, y) belongs to A_3, then we have

$$\begin{aligned}K(x, y) &\leqslant \frac{x \cdot (x - y)^\perp}{|x - y|^2} + \frac{x \cdot y^\perp}{|x|^2}\\ &\leqslant \frac{|x|}{|x - y|} + \frac{|y|}{|x|}\\ &\leqslant 2n\frac{|y|^2}{|x|^2}.\end{aligned}$$

From this we immediately infer that

$$\alpha_3(t) \leqslant \frac{4n^2}{\pi} i_0 m_{n-1}(t). \tag{34}$$

The term α_2 is harder to estimate. Let us write

$$\alpha_2(t) = I_1(t) + I_2(t)$$

with

$$I_1(t) \overset{\text{def}}{=} -\frac{2n}{\pi}\int_{A_2} |x|^{4n-2}\frac{x \cdot y^\perp}{|x - y|^2}\omega(t, x)\omega(t, y)\,\mathrm{d}x\,\mathrm{d}y$$

and

$$I_2(t) \overset{\text{def}}{=} \frac{2n}{\pi}\int_{A_2} |x|^{4(n-1)} x \cdot y^\perp\omega(t, x)\omega(t, y)\,\mathrm{d}x\,\mathrm{d}y.$$

As in region A_2 we have $|x| \leqslant 2|y|$, we can write that

$$\left|I_2(t)\right| \leqslant \frac{4n^2}{\pi} i_0 m_{n-1}(t). \tag{35}$$

To estimate I_1, let us observe that A_2 is symmetric with respect to the diagonal and that

$$H(x, y) \stackrel{\text{def}}{=} \frac{x \cdot y^\perp}{|x-y|^2} = H(y, x).$$

Thus we have that

$$I_1(t) = -\frac{n}{\pi} \int_{A_2} H(x, y)\big(|x|^{4n-2} - |y|^{4n-2}\big) \frac{x \cdot y^\perp}{|x-y|^2} \omega(t, x)\omega(t, y)\, \mathrm{d}x\, \mathrm{d}y.$$

But in region A_2, we have

$$\big||x|^{4n-2} - |y|^{4n-2}\big| \leqslant 6n|y||x-y||x|^{4(n-1)}.$$

But as $|H(x, y)| \leqslant \frac{|y|}{|x-y|}$, we have that

$$\left|I_1(t)\right| \leqslant \frac{6n^2}{\pi} i_0 m_{n-1}(t).$$

Now plugging estimates (33)–(35) together gives

$$\frac{\mathrm{d}m_n}{\mathrm{d}t} \leqslant C_0 i_0 m_{n-1}(t).$$

But, using Hölder inequality we infer

$$\frac{\mathrm{d}m_n}{\mathrm{d}t} \leqslant C_0 i_0 m_0^{1/n} m_{n-1}(t)^{1-1/n}.$$

This gives the result by integration. □

3.3. *References and remarks*

This section is based on the work [59] of Iftimie, Sideris and Gamblin. The problem of the rate of growth of the diameter of a vortex patch has been addressed previously by Marchioro in [69]. He proved that the rate of growth in the case of positive vorticity is less than $t^{1/3}$. Let us point out that the problem of the optimality of the rate $(t \log t)^{1/4}$ proved by Iftimie, Sideris and Gamblin remains open.

More generally we can say that our understanding of the large time behavior of solutions of the two-dimensional Euler system is, for the time being, rather poor. For instance, the problem of the effective growth of $\|v(t)\|_{\widetilde{C}^r}$ for solutions of the two-dimensional Euler system given by Theorem 2.1 is totally open. In this framework of large time behavior of the vorticity, let us mention the pioneer work of Iftimie, Lopes Filho and Nussenzveig (see [60]). They prove that in the case of the incompressible Euler equation if the upper half plane, if the rescaled vorticity defined by

$$\tilde{w}(t,x) \stackrel{\text{def}}{=} t^2 \omega(t,tx)$$

convergences weakly to $\tilde{\mu}$, then this measure μ satisfy $\tilde{\mu} = \mu \otimes \delta_0$ where the measure μ is a finite sum of Dirac masses.

In the case of the three dimensional Euler system, let us also mention the results of Serre (see [80]) about the (at least) linear growth of the vorticity in the case when a global smooth solution might exist.

4. More about the Biot–Savart law

4.1. *The regularity of the vector field when the boundary of the vortex patch is smooth*

In this section, we shall prove, using basic techniques of singular integral of curves, that if the vorticity is the characteristic function of a domain with a smooth boundary, then the associate divergence free vector field is Lipschitzian. Let us denote by γ a smooth function from the unit circle $\mathbb{S}^1$ into $\mathbb{R}^2$ which is a parametrization of the boundary. Easy computations imply that

$$v(x) = \frac{1}{2\pi} \int_{-\pi}^{\pi} T(\sigma) \log|x - \gamma(\sigma)| \, d\sigma \quad \text{with } T(\sigma) \stackrel{\text{def}}{=} \left(\frac{d\gamma_1}{d\sigma}, \frac{d\gamma_2}{d\sigma} \right). \tag{36}$$

We shall prove the following proposition.

PROPOSITION 4.1. *The vector field v is Lipschitzian and a constant C_γ which depends only on $c_\gamma \stackrel{def}{=} \inf_{\sigma \neq \sigma'} \frac{|\gamma(\sigma)-\gamma(\sigma')|}{|\sigma-\sigma'|}$ and $\|\gamma\|_{1+\varepsilon}$ exists such that*

$$\|\nabla v\|_{L^\infty} \leqslant C_\gamma.$$

PROOF. Let x be a point of $\mathbb{R}^2 \setminus \gamma(\mathbb{S}^1)$. We can suppose that $d(x,\gamma) = |x - \gamma(0)|$. We have

$$x = \gamma(0) \pm d(x,\gamma) N(0).$$

Taylor inequality implies that

$$|x - \gamma(\sigma)| \geqslant \frac{1}{2}\bigl(d(x,\gamma) + c_\gamma|\sigma|\bigr) - \|\gamma\|_{1+\varepsilon}|\sigma|^{1+\varepsilon}.$$

Then we infer that

$$|\sigma| \leqslant \alpha_\gamma \quad \Longrightarrow \quad \big|x-\gamma(\sigma)\big| \geqslant \frac{d(x,\gamma)+c_\gamma|\sigma|}{4} \quad \text{with}$$
$$\alpha_\gamma = \left(\frac{c_\gamma}{4\|\gamma\|_{1+\varepsilon}}\right)^{1/\varepsilon}. \tag{37}$$

As $|x-\gamma(\sigma)| \geqslant |\gamma(\sigma)-\gamma(0)| - |x-\gamma(0)|$, we have

$$d(x,\gamma) \leqslant \frac{1}{2}c_\gamma\alpha_\gamma \quad \Longrightarrow \quad \left(|\sigma| \geqslant \alpha_\gamma \Longrightarrow \big|x-\gamma(\sigma)\big| \geqslant \frac{1}{2}c_\gamma\alpha_\gamma\right). \tag{38}$$

On $\mathbb{R}^2 \setminus \gamma(\mathbb{S}^1)$ we have

$$\nabla v(x) = \frac{1}{2\pi}\int_{-\pi}^{\pi} T(\sigma)\frac{x-\gamma(\sigma)}{|x-\gamma(\sigma)|^2}\,\mathrm{d}\sigma. \tag{39}$$

Denoting by $\ell(\gamma)$ the length of γ, it turns out that

$$d(x,\gamma) \geqslant \frac{\alpha_\gamma c_\gamma}{2} \quad \Longrightarrow \quad \big|\nabla v(x)\big| \leqslant \frac{2\ell(\gamma)}{\alpha_\gamma c_\gamma}. \tag{40}$$

From now on let us assume that $d(x,\gamma) \leqslant \alpha_\gamma c_\gamma/2$ and let us estimate

$$I_{i,j}(x) = J_{i,j}(x) + \widetilde{J}_{i,j}(x)$$

with

$$J_{i,j}(x) = \int_{\sigma\in]-\alpha_\gamma,\alpha_\gamma[} \frac{x^j-\gamma^j(\sigma)}{|x-\gamma(\sigma)|^2} T^i(\sigma)\,\mathrm{d}\sigma$$

and

$$\widetilde{J}_{i,j}(x) = \int_{\sigma\notin]-\alpha_\gamma,\alpha_\gamma[} \frac{x^j-\gamma^j(\sigma)}{|x-\gamma(\sigma)|^2} T^i(\sigma)\,\mathrm{d}\sigma.$$

From (38) it comes $|\widetilde{J}_{i,j}(x)| \leqslant \frac{2\ell(\gamma)}{\alpha_\gamma c_\gamma}$. Let us estimate $J_{i,j}(x)$. It is easily checked that

$$\left|J_{i,j}(x) - T^i(0)\int_{-\alpha_\gamma}^{\alpha_\gamma} \frac{x^j-\gamma^j(0)-\sigma T^j(0)}{d(x,\gamma)^2+\sigma^2|T(0)|^2}\,\mathrm{d}\sigma\right| \leqslant C_\gamma.$$

But using the symmetry with respect to σ, we have that

$$\int_{-\alpha_\gamma}^{\alpha_\gamma} \frac{x^j-\gamma^j(0)-\sigma T^j(0)}{d(x,\gamma)^2+\sigma^2|T(0)|^2}\,\mathrm{d}\sigma = \int_{-\alpha_\gamma}^{\alpha_\gamma} \frac{x^j-\gamma^j(0)}{d(x,\gamma)^2+\sigma^2|T(0)|^2}\,\mathrm{d}\sigma$$

and then it is obvious that

$$\int_{-\alpha_\gamma}^{\alpha_\gamma} \frac{x^j - \gamma^j(0) - \sigma T^j(0)}{d(x,\gamma)^2 + \sigma^2 |T(0)|^2}\, d\sigma \leqslant 2\int_0^{\alpha_\gamma} \frac{d(x,\gamma)}{d(x,\gamma)^2 + c_\gamma^2 \sigma^2}\, d\sigma \leqslant \frac{2\pi}{c_\gamma}.$$

So Proposition 4.1 is proved. □

If we keep in mind Proposition 1.3, it is clear that the fact that the gradient of a divergence free vector field, the vorticity of which is the characteristic function of a bounded domain is bounded or not depends on the regularity of the boundary of this domain.

4.2. *The statement of the basic estimate for vortex patches*

The aim of this section is to find a flexible enough condition on a bounded function u which implies that the functions $\partial_i \partial_j \Delta^{-1} u$ are also bounded. The above two examples suggest that tangential regularity with respect to some given geometrical structure has something to do with the problem.

This leads to the following definition. In all this section, ε denotes a number in $]0,1[$. Let us first define the Hölder spaces of negative index (between 0 and -1).

DEFINITION 4.1. Let ε be in the interval $]0,1[$. The space $C^{\varepsilon-1}$ is the set of tempered distributions u such that a sequence $(u_j)_{0\leqslant j\leqslant d}$ of C^ε functions exists such that

$$u = u_0 + \sum_{j=1}^{d} \partial_j u_j.$$

It is a Banach space for the norm

$$\|u\widetilde{\|}_{\varepsilon-1} \stackrel{\text{def}}{=} \inf_{\substack{(u_j)_{0\leqslant j\leqslant d} \in (C^\varepsilon)^{1+d} \\ u = u_0 + \sum_{j=1}^d \partial_j u_j}} \sup_{0\leqslant j\leqslant d} \|u_j\widetilde{\|}_\varepsilon.$$

Roughly speaking, the space $C^{\varepsilon-1}$ is the space of derivatives of C^ε functions. Now, we can define precisely the type of regularity of the vorticity which will describe the vortex patches situation.

DEFINITION 4.2. Let $X = (X_\lambda)_{\lambda\in\Lambda}$ be a family of vector fields. This family is C^ε-admissible if and only if its coefficients and its divergence are C^ε and

$$I(X) = \inf_{x\in\mathbb{R}^2} \sup_{\lambda\in\Lambda} |X_\lambda(x)| > 0.$$

Now let us define the concept of tangential regularity with respect to such a family.

DEFINITION 4.3. Let X be a C^ε-admissible family of vector fields. Let us denote by $C^\varepsilon(X)$ the space of bounded functions u such that, for any $\lambda \in \Lambda$,

$$X_\lambda(x, D)u \stackrel{\text{def}}{=} \operatorname{div}(uX_\lambda) - u \operatorname{div} X_\lambda \in C^{\varepsilon-1}.$$

Let us define

$$N_\varepsilon(X) \stackrel{\text{def}}{=} \frac{1}{\varepsilon} \sup_{\lambda\in\Lambda} \frac{\|X_\lambda\|_\varepsilon + \|\operatorname{div} X_\lambda\|_\varepsilon}{I(X)}$$

and

$$\|u\|_X^\varepsilon \stackrel{\text{def}}{=} N_\varepsilon(X)\|u\|_{L^\infty} + \sup_{\lambda\in\Lambda} \frac{\|X_\lambda(x, D)u\|_{\varepsilon-1}}{I(X)}.$$

EXAMPLE. Let us consider a compact curve γ_0 of class $C^{1+\varepsilon}$ and a defining function f_0 of γ_0, i.e., a function of class $C^{1+\varepsilon}$ such that in a neighborhood of γ_0 exists such that $f^{-1}(0) = V_0 \cap \gamma_0$ and the gradient of f_0 does not vanish in V_0. Now let us define the following vector fields

$$X_{0,0} = \nabla^\perp f_0, \qquad X_{0,1} = (1-\alpha)\partial_1 \quad \text{and} \quad X_{0,2} = (1-\alpha)\partial_2,$$

where α is a smooth function supported in a sufficiently small compact neighborhood of the curve γ_0 and has value 1 near this curve γ_0. It is obvious that the above family is C^ε-admissible. Moreover, if ω_0 is the characteristic function of the interior of γ_0, then $X_{0,i}(x, D)\omega_0 = 0$. Finally we have

$$N_\varepsilon(X_0) = \frac{\|\nabla f_0\|_\varepsilon}{\min_{x\in V_0} |\nabla f(x)|}.$$

Now let us state the main theorem of this section which is a generalization of Proposition 4.1.

THEOREM 4.1. *Let ε be in $]0,1[$ and a in $]1,+\infty[$. A constant C exists which satisfies the following properties.*

Let X be a C^ε-admissible family of vector fields and let us consider a function ω belonging to $C^\varepsilon(X) \cap L^a$. If v is the divergence free vector field the gradient of which is L^a and the vorticity of which is ω, then the gradient of v is bounded and

$$\|\nabla v\|_{L^\infty} \leqslant C\|\omega\|_{L^a} + C\|\omega\|_{L^\infty} \log\left(e + \frac{\|\omega\|_X^\varepsilon}{\|\omega\|_{L^\infty}}\right).$$

PROOF. Let us prove this theorem in the very particular case when the family X is reduced to the vector field ∂_1. Let us also assume that the Fourier transform of ω does not contain the origin. It is obvious that $\|X\|_\varepsilon = I(X) = 1$.

Thanks to Proposition 6.6, we have for $j \in \{1, 2\}$,

$$\left\|\partial_1 \partial_j \Delta^{-1} \omega\right\|_{L^\infty} \leqslant C\|\omega\|_{L^\infty} \log\left(e + \frac{\|\partial_1 \partial_j \Delta^{-1} \omega\|_\varepsilon}{\|\omega\|_{L^\infty}}\right).$$

Using Proposition 6.5, we get

$$\left\|\partial_1 \partial_j \Delta^{-1} \omega\right\|_{L^\infty} \leqslant C\|\omega\|_{L^\infty} \log\left(e + \frac{\|\partial_1 \omega\|_{\varepsilon-1}}{\|\omega\|_{L^\infty}}\right). \tag{41}$$

To estimate $\|\partial_2^2 \Delta^{-1} \omega\|_{L^\infty}$, let us observe that $\partial_2^2 = \Delta - \partial_1^2$. Thus

$$\left\|\partial_i \partial_j \Delta^{-1} \omega\right\|_{L^\infty} \leqslant C\|\omega\|_{L^\infty} \log\left(e + \frac{\|\partial_1 \omega\|_{\varepsilon-1}}{\|\omega\|_{L^\infty}}\right).$$

So the theorem is proved in this very particular case. □

This theorem will be proved in the general case in Section 6. The two main difficulties we shall meet to prove it are, first the fact that the vector fields X_λ are not smooth, and second those vector fields X_λ can have a tendency to vanish (controlled by $N_\varepsilon(X)$). Both difficulties will be bypassed thanks to paradifferential calculus (see Section 6.1).

4.3. *References and remarks*

The idea of using regularity properties with respect to a family of vector field goes back to the work of Bony about propagation of singularities in semilinear equations (see [16]). In this work, vector fields were smooth. In the context of quasilinear equations (which is of course the case of Euler equation), it is not possible to assume these vector fields to be smooth. Tangential regularity with respect to vector fields of low regularity has been introduced independently by Alinhac in [1] and by the author in [24] to study the problem of propagation of regularity in quasilinear hyperbolic systems.

5. The global regularity of vortex patches

5.1. *An introduction to the problem*

The original problem is to investigate the precise regularity of the solution in the case when the vorticity of the initial data is the characteristic function of a bounded domain with a smooth enough boundary. As seen in Proposition 4.1, the initial data is Lipschitzian in this case. The question is the following: is this property preserved by the flow of the equation which does exist and is unique thanks to the Yudovich theorem (Theorem 2.2)?

That theorem shows that the global unique solution is log-Lipschitzian and that, thanks to conservation of the vorticity (V) the vorticity of the solution at time t is the characteristic function of a bounded domain the topology of which is preserved along the evolution. Then two very natural questions can be stated: does the boundary of the domain remain regular for all time? If so, what happens for large time?

Let us imagine that the vorticity of the solution at time t is the characteristic function of the interior D_t of a closed simple curve of class $C^{1+\varepsilon}$. Let us denote by $\gamma(t)$ a $C^{1+\varepsilon}$ one-to-one map from $\mathbb{S}^1$ into $\mathbb{R}^2$, the range of which is the boundary of D_t. Thanks to the Biot–Savart law, the solution $v(t,\cdot)$ is given by

$$v(t) = \nabla^\perp f(t) \quad \text{with } f(t,x) = \frac{1}{2\pi}\int_{D_t} \log|x-y|\,\mathrm{d}y.$$

Using Green's formula, we get that

$$v(t,x) = \frac{1}{2\pi}\int_0^{2\pi} \log\bigl|x-\gamma(t,\sigma)\bigr|\partial_\sigma\gamma(t,\sigma)\,\mathrm{d}\sigma.$$

It looks reasonable to seek for a γ which satisfies

$$\partial_t\gamma(t,s) = v\bigl(t,\gamma(t,s)\bigr). \tag{42}$$

Thus we want to solve

$$\partial_t\gamma(t,s) = \frac{1}{2\pi}\int_0^{2\pi} \log\bigl|\gamma(t,s)-\gamma(t,\sigma)\bigr|\partial_\sigma\gamma(t,\sigma)\,\mathrm{d}\sigma. \tag{43}$$

5.2. *The statement of the result*

THEOREM 5.1. *Let ε be in the interval $]0,1[$ and γ_0 a $C^{1+\varepsilon}$ function from $\mathbb{S}^1$ into $\mathbb{R}^2$ which is one-to-one and the differential of which does not vanish. A unique solution $\gamma(t,s)$ of* (43) *exists which belongs to $L^\infty_{\mathrm{loc}}(\mathbb{R};C^{1+\varepsilon}(\mathbb{S}^1;\mathbb{R}^2))$ and which is for any time an embedding of the circle.*

PROOF. In fact, we shall prove a more general result which involves the concept of tangential regularity introduced in the preceding section in Definition 4.3.

THEOREM 5.2. *Let ε be in $]0,1[$, a in $]1,+\infty[$ and $X_0 = (X_{0,\lambda})_{\lambda\in\Lambda}$ a C^ε-admissible family of vector fields. Let us consider a divergence free vector field v_0 on $\mathbb{R}^2$, the gradient of which is in L^a. If ω_0 belongs to $C^\varepsilon(X_0)$, then a unique solution v of (E) exists such that*

$$v \in L^\infty_{\mathrm{loc}}(\mathbb{R};\mathit{Lip}) \quad \text{and} \quad \nabla v \in L^a.$$

Moreover, if ψ denotes the flow of v, then for any λ,

$$X_{0,\lambda}(x,D)\psi(t,x) \in L^\infty_{\mathrm{loc}}\bigl(\mathbb{R};C^\varepsilon\bigr).$$

Then if $X_{t,\lambda} = \psi(t)_\star X_{0,\lambda}$, *then the family* $X_t = (X_{t,\lambda})_{\lambda\in\Lambda}$ *is* C^ε*-admissible and we have*

$$N_\varepsilon(X_t) \in L^\infty_{\mathrm{loc}}(\mathbb{R}) \quad \textit{and} \quad \|\omega(t)\|_{\varepsilon,X_t} \in L^\infty_{\mathrm{loc}}(\mathbb{R}).$$

Before proving Theorem 5.2, let us check that it implies Theorem 5.1. The example shown just after Definition 4.3 allows us to apply Theorem 5.2. Let us check that the conclusions of Theorem 5.2 imply the ones of Theorem 5.1. Let σ_0 be a point of the unit circle $\mathbb{S}^1$ and let us consider $\tilde{\gamma}_0$ the solution of the ordinary differential equation

$$\begin{cases} \partial_\sigma \tilde{\gamma}_0(\sigma) = X_{0,0}(\tilde{\gamma}_0(\sigma)), \\ \tilde{\gamma}_0(\sigma_0) = x_0 \in f_0^{-1}(0) \end{cases}$$

and the function $\tilde{\gamma}(t)$ defined by

$$\tilde{\gamma}(t,\sigma) = \psi(t, \tilde{\gamma}_0(\sigma)). \tag{44}$$

Theorem 5.2 claims that $X_{0,0}(x,D)\psi$ belongs to $L^\infty_{\mathrm{loc}}(\mathbb{R}; C^\varepsilon)$. By differentiation of (44), we get

$$\partial_\sigma \tilde{\gamma}(t,\sigma) = \big(X_{0,0}(x,D)\psi\big)(t, \tilde{\gamma}_0(\sigma)).$$

Thus $\partial_\sigma \tilde{\gamma}$ belongs to $L^\infty_{\mathrm{loc}}(\mathbb{R}; C^\varepsilon)$ and obviously does not vanish. Theorem 5.1 is proved. □

5.3. *The structure of the proof*

Now let us prove Theorem 5.2. We shall only prove a priori estimates on smooth solutions We skip the problems coming from regularization and passing to the limit (see [35] for the details). The key estimate is described by the following proposition.

PROPOSITION 5.1. *Let* ε *be in* $]0,1[$ *and* a *in* $[1,+\infty[$, *a constant* C *exists which satisfies the following property. Let* $X_0 = (X_{0,\lambda})_{\lambda\in\Lambda}$ *be a* C^ε*-admissible family and* v *be a solution of the Euler system which belongs to* $L^\infty_{\mathrm{loc}}(\mathbb{R}; C^\infty_b)$. *Then, for any time* t,

$$\|\nabla v(t)\|_{L^\infty} \leqslant \widetilde{N}(X_0,\varepsilon,\omega_0)\exp\big(Ct\|\omega_0\|_{L^\infty}\big)$$

with

$$\widetilde{N}(X_0,\varepsilon,\omega_0) \overset{def}{=} C\left(\|\omega_0\|_{L^a} + \|\omega_0\|_{L^\infty}\log\frac{\|\omega_0\|_{\varepsilon,X_0}}{\|\omega_0\|_{L^\infty}}\right).$$

The idea of the proof is to propagate the geometrical structure (i.e., the family X_0) by the flow of the solution. More precisely, we define the family $X_t = (X_{t,\lambda})_{\lambda\in\Lambda}$ by

$$X_{t,\lambda}(x) = \psi(t)_\star X_{0,\lambda}(x) := \big(X_{0,\lambda}(x,D)\psi(t)\big)\big(\psi^{-1}(t,x)\big). \tag{45}$$

In an Eulerian way, this is written

$$\partial_t X_{t,\lambda} + v \cdot \nabla X_{t,\lambda} = X_{t,\lambda}(x, D)v \tag{46}$$

which means also that the two vector fields $\partial_t + v \cdot \nabla$ and $X_{t,\lambda}$ commute. The key point of the proof consists in estimating $\|\omega(t)\|_{\varepsilon, X_t}$.

From (46) and the conservation of the vorticity (V) we get, as the vector field v is divergence free,

$$\partial_t \operatorname{div} X_{t,\lambda} + v \cdot \nabla \operatorname{div} X_{t,\lambda} = 0, \tag{47}$$

$$\partial_t X_{t,\lambda}(x, D)\omega + v \cdot \nabla X_{t,\lambda}(x, D)\omega = 0. \tag{48}$$

Let us admit for the time being the following inequality which will be proved in Section 6 as a particular case of Proposition 6.11.

PROPOSITION 5.2. *Two bilinear operators W_1 and W_2 exists such that for any divergence free vector field v and any vector field Y, we have*

$$Y(x, D)v = W_1(Y, v) + W_2(Y, v)$$

and such that, for any positive ε, we have

$$\left\| W_1(Y, v) \right\|_\varepsilon \leqslant C \left\| Y(x, D)\omega \right\|_{\varepsilon - 1}$$

and

$$\left\| W_2(Y, v) \right\|_\varepsilon \leqslant C\big(\|Y\|_\varepsilon + \|\operatorname{div} Y\|_\varepsilon\big)\|\nabla v\|_{L^\infty}.$$

Now let us state the following propagation lemma which will be proved in Section 6 as a particular case of Theorem 6.2.

LEMMA 5.1. *Let r be in $]-1, 1[\backslash\{0\}$. A constant C exists which satisfies the following properties. Let v be a divergence free vector field, the order 1 derivatives of which are bounded on $[0, T] \times \mathbb{R}^d$. Let us state $V(t) \stackrel{\text{def}}{=} \|\nabla v(t)\|_{L^\infty}$. Let (f, g) be in $L^\infty([0, T]; C^r) \times L^1([0, T]; C^r)$ such that*

$$g = g_1 + g_2 \quad \text{with } \left\|g_2(t)_r\right\| \leqslant C V(t) \left\|f(t)_r\right\|.$$

If $\partial_t f + v \cdot \nabla f = g$, then we have

$$\begin{aligned} \left\|f(t)\right\|_r &\leqslant \left\|f(0)\right\|_r \exp\left(C \int_0^t V(t')\, \mathrm{d}t'\right) \\ &\quad + \int_0^t \left\|g_1(t')\right\|_r \exp\left(C \int_{t'}^t V(t'')\, \mathrm{d}t''\right). \end{aligned} \tag{49}$$

Then using identities (46)–(48), we get

$$I(X_t) \geqslant I(X_0)\exp\biggl(-\int_0^t \|\nabla v(t')\|_{L^\infty}\,\mathrm{d}t'\biggr), \tag{50}$$

$$\|X_{t,\lambda}(x,D)\omega(t)\|_{\varepsilon-1} \leqslant C\|X_{0,\lambda}(x,D)\omega_0\|_{\varepsilon-1}\exp\biggl(C\int_0^t \|\nabla v(t')\|_{L^\infty}\,\mathrm{d}t'\biggr), \tag{51}$$

$$\|\operatorname{div} X_{t,\lambda}\|_\varepsilon \leqslant \|\operatorname{div} X_{0,\lambda}\|_\varepsilon \exp\biggl(C\int_0^t \|\nabla v(t')\|_{L^\infty}\,\mathrm{d}t'\biggr), \tag{52}$$

$$\begin{aligned}\|X_{t,\lambda}\|_\varepsilon \leqslant{}& C\biggl(\|X_{0,\lambda}\|_\varepsilon + \|\operatorname{div} X_{0,\lambda}\|_\varepsilon + \frac{\|X_{0,\lambda}(x,D)\omega_0\|_{\varepsilon-1}}{\|\omega_0\|_{L^\infty}}\biggr)\\ &\times \exp\biggl(C\int_0^t \|\nabla v(t')\|_{L^\infty}\,\mathrm{d}t'\biggr).\end{aligned} \tag{53}$$

Those inequalities can be summarized in the following one

$$\|\omega(t)\|^\varepsilon_{X_t} \leqslant \|\omega_0\|^\varepsilon_{X_0}\exp\biggl(C\int_0^t \|\nabla v(t')\|_{L^\infty}\,\mathrm{d}t'\biggr).$$

Then applying the stationary estimate stated in Theorem 4.1, using the conservation of vorticity (V) and the fact that $x \mapsto x\log(e+\alpha x^{-1})$ is an increasing function for nonnegative α, we get

$$\begin{aligned}&\|\nabla v(t)\|_{L^\infty}\\ &\quad\leqslant C\biggl(\|\omega_0\|_{L^a} + \|\omega_0\|_{L^\infty}\log\biggl(e+\frac{\|\omega_0\|^\varepsilon_{X_0}}{\|\omega_0\|_{L^\infty}}\exp\biggl(C\int_0^t \|\nabla v(t')\,\mathrm{d}t'\|_{L^\infty}\biggr)\biggr)\biggr)\\ &\quad\leqslant \widetilde{N}(X_0,\varepsilon,\omega_0) + C\int_0^t \|\nabla v(t')\,\mathrm{d}t'\|_{L^\infty}\,\mathrm{d}t' \quad \text{with, as in Proposition 5.1,}\end{aligned}$$

$$\widetilde{N}(X_0,\varepsilon,\omega_0) = C\biggl(\|\omega_0\|_{L^a} + \|\omega_0\|_{L^\infty}\log\biggl(e+\frac{\|\omega_0\|^\varepsilon_{X_0}}{\|\omega_0\|_{L^\infty}}\biggr)\biggr).$$

Then Gronwall lemma allows to conclude.

6. A short introduction to Littlewood–Paley theory and paradifferential calculus

The key estimates used in the proof of Theorem 5.2 rely on the Littlewood–Paley theory and one of its developments: paradifferential calculus. The purpose of this section in to give an idea of those methods.

6.1. *Localization in frequency space*

The very basic idea of this theory consists in a localization procedure in the frequency space. The interest of this method is that the derivatives (or more generally the Fourier multipliers) act in a very special way on distributions the Fourier transform of which is supported in a ball or a ring. More precisely, we have the following lemma.

LEMMA 6.1. *Let $\mathcal{C}$ be a ring, B a ball centered at 0, k a positive integer and σ any smooth homogeneous function of degree $m \in \mathbb{R}$. A constant C exists such that, for any positive real number λ and any function u in L^a, we have*

$$\operatorname{Supp}\hat{u} \subset \lambda B \quad \Longrightarrow \quad \sup_{\alpha=k} \|\partial^\alpha u\|_{L^b} \leqslant C\lambda^{k+d(1/a-1/b)}\|u\|_{L^a};$$

$$\operatorname{Supp}\hat{u} \subset \lambda \mathcal{C} \quad \Longrightarrow \quad C^{-1}\lambda^k\|u\|_{L^a} \leqslant \sup_{\alpha=k} \|\partial^\alpha u\|_{L^a} \leqslant C\lambda^k\|u\|_{L^a}.$$

Moreover, if σ is a smooth function on $\mathbb{R}^d$ which is homogeneous of degree m outside a fixed ball, then we have

$$\operatorname{Supp}\hat{u} \subset \lambda \mathcal{C} \quad \Longrightarrow \quad \|\sigma(D)u\|_{L^b} \leqslant C\lambda^{m+d(1/a-1/b)}\|u\|_{L^a}.$$

PROOF. Using a dilation of size λ, we can assume in all along the proof that $\lambda = 1$. Let ϕ be a function of $\mathcal{D}(\mathbb{R}^d)$ the value of which is 1 near B. As $\hat{u}(\xi) = \phi(\xi)\hat{u}(\xi)$, we can write, if g denotes the inverse Fourier transform of ϕ,

$$\partial^\alpha u = \partial^\alpha g \star u.$$

Young inequalities imply the results. To prove the second assertion, let us consider a function $\tilde{\phi}$ of $\mathcal{D}(\mathbb{R}^d \setminus \{0\})$ the value of which is identically 1 near the ring $\mathcal{C}$. Using the following algebraic identity

$$\sum_{|\alpha|=k} (\mathrm{i}\xi)^\alpha(-\mathrm{i}\xi)^\alpha = |\xi|^{2k}, \tag{54}$$

and stating $g_\alpha \stackrel{\text{def}}{=} \mathcal{F}^{-1}(\mathrm{i}\xi)^\alpha|\xi|^{-2k}\tilde{\phi}(\xi)$, we can write, as $\hat{u} = \tilde{\phi}\hat{u}$ that

$$\tilde{u} = \sum_{|\alpha|=k} (-\mathrm{i}\xi)^\alpha \hat{g}_\alpha \tilde{u},$$

which implies that $u = \sum_{|\alpha|=k} g_\alpha \star \partial^\alpha u$ and the result using Young inequalities. Let us denote by g_σ the inverse Fourier transform of $\phi\sigma$. As $\hat{u}(\xi) = \phi(\xi)\hat{u}(\xi)$, we have $\sigma(D)u = g_\sigma \star u$ and thus the lemma is proved using again Young inequalities. □

The following lemma describes properties of power of functions the Fourier transform of which is supported in a ring.

LEMMA 6.2. *Let $\mathcal{C}$ be a ring. A constant C exists such that for any positive real number λ, for any $a \in [1, \infty[$ and any function u in L^{2a} such that* $\operatorname{Supp} \hat{u} \subset \lambda\mathcal{C}$, *we have*

$$\|u^a\|_{L^2} \leqslant C\lambda^{-1} \|\nabla(u^a)\|_{L^2}.$$

REMARK. This lemma is in some sense surprising. Let us take for instance $a = 2$. The function u^2 has a Fourier transform the support of which is of course not supported in a ring, but in a ball. Despite of that, we can control its L^2 norm with the L^2 norm of its gradient.

PROOF OF LEMMA 6.2. Let us prove this inequality only in the case when a is an integer. The identity (54) applied with $k = 1$ implies that

$$\begin{aligned}\int_{\mathbb{R}^d} u^{2a}\,\mathrm{d}x &= \int_{\mathbb{R}^d} u u^{2a-1}\,\mathrm{d}x \\ &= \sum_{j=1}^{d} \partial_j u_j u^{2a-1}\,\mathrm{d}x\end{aligned}$$

with

$$u_j \stackrel{\text{def}}{=} \mathcal{F}^{-1}\big(\mathrm{i}\xi_j |\xi|^{-2} \hat{u}(\xi)\big).$$

By integration by parts, we infer that

$$\begin{aligned}\int_{\mathbb{R}^d} u^{2a}\,\mathrm{d}x &= -(2a-1) \sum_{j=1}^{d} \int_{\mathbb{R}^d} u_j\, \partial u_j u^{2a-2}\,\mathrm{d}x \\ &= -\frac{2a-1}{a} \sum_{j=1}^{d} \int_{\mathbb{R}^d} u_j\, \partial_j(u^a) u^{a-1}\,\mathrm{d}x.\end{aligned}$$

By Cauchy–Schwarz inequality, we have that

$$\int_{\mathbb{R}^d} u^{2a}\,\mathrm{d}x \leqslant C \|\nabla(u^a)\|_{L^2} \left(\sum_{j=1}^{d} \int_{\mathbb{R}^d} |u_j|^2 u^{2(a-1)}\,\mathrm{d}x\right)^{1/2}.$$

As in the proof of Lemma 6.1, let us consider a function $\tilde{\phi}$ of $\mathcal{D}(\mathbb{R}^d \setminus \{0\})$ the value of which is identically 1 near the ring $\mathcal{C}$ and let us state $g_j \stackrel{\text{def}}{=} \mathcal{F}^{-1}(\mathrm{i}\xi_j)^\alpha |\xi|^{-2k} \tilde{\phi}(\xi)$, we have that g_j is in $\mathcal{S}$ and that $u_j = g_j \star u$. Thus $\|u_j\|_{L^{2a}} \leqslant C\|u\|_{L^{2a}}$. By Hölder inequality, we have

$$\int_{\mathbb{R}^d} u^{2a}\,\mathrm{d}x \leqslant C \|\nabla(u^a)\|_{L^2} \|u\|_{L^{2a}}^a$$

and thus the result is proved. □

Now, let us define a dyadic partition of unity. We shall use it all along this text.

PROPOSITION 6.1. *Let us define by $\mathcal{C}$ the ring of center* 0, *of small radius* 3/4 *and great radius* 8/3. *Two radial functions χ and φ, the values of which are in the interval* [0, 1] *exist respectively in $\mathcal{D}(B(0,4/3))$ and in $\mathcal{D}(\mathcal{C})$ so that*

$$\forall \xi \in \mathbb{R}^d, \quad \chi(\xi) + \sum_{q\geqslant 0} \varphi(2^{-q}\xi) = 1, \tag{55}$$

$$\forall \xi \in \mathbb{R}^d \setminus \{0\}, \quad \sum_{q\in\mathbb{Z}} \varphi(2^{-q}\xi) = 1, \tag{56}$$

$$|p-q| \geqslant 2 \quad \Longrightarrow \quad \operatorname{Supp}\varphi(2^{-q}\cdot) \cap \operatorname{Supp}\varphi(2^{-p}\cdot) = \emptyset, \tag{57}$$

$$q \geqslant 1 \quad \Longrightarrow \quad \operatorname{Supp}\chi \cap \operatorname{Supp}\varphi(2^{-q}\cdot) = \emptyset. \tag{58}$$

If $\widetilde{\mathcal{C}} = B(0,2/3) + \mathcal{C}$, then $\widetilde{\mathcal{C}}$ is a ring and we have

$$|p-q| \geqslant 5 \quad \Longrightarrow \quad 2^p\widetilde{\mathcal{C}} \cap 2^q\mathcal{C} = \emptyset, \tag{59}$$

$$\forall \xi \in \mathbb{R}^d, \quad \frac{1}{3} \leqslant \chi^2(\xi) + \sum_{q\geqslant 0} \varphi^2(2^{-q}\xi) \leqslant 1, \tag{60}$$

$$\forall \xi \in \mathbb{R}^d \setminus \{0\}, \quad \frac{1}{2} \leqslant \sum_{q\in\mathbb{Z}} \varphi^2(2^{-q}\xi) \leqslant 1. \tag{61}$$

This proposition is proved for instance in [35]. We shall consider all along this text two fixed functions χ and φ satisfying the assertions (55)–(60). Now let us fix the notations that will be used in all the following of this text.

NOTATIONS.

$$h = \mathcal{F}^{-1}\varphi \quad \text{and} \quad \tilde{h} = \mathcal{F}^{-1}\chi,$$

$$\Delta_{-1}u = \chi(D)u = \mathcal{F}^{-1}\big(\chi(\xi)\hat{u}(\xi)\big),$$

$$\text{if } q \geqslant 0, \quad \Delta_q u = \varphi(2^{-q}D)u = 2^{qd}\int_{\mathbb{R}^d} h(2^q y)u(x-y)\,dy,$$

$$\text{if } q \leqslant -2, \quad \Delta_q u = 0,$$

$$S_q u = \sum_{p\leqslant q-1} \Delta_p u = \chi(2^{-q}D)u = \int_{\mathbb{R}^d} \tilde{h}(2^q y)u(x-y)\,dy,$$

$$\text{if } q \in \mathbb{Z}, \quad \dot{\Delta}_q u = \varphi(2^{-q}D)u = 2^{qd}\int_{\mathbb{R}^d} h(2^q y)u(x-y)\,dy,$$

$$\text{if } q \in \mathbb{Z}, \quad \dot{S}_q u = \sum_{p \leqslant q-1} \dot{\Delta}_p u.$$

Now let us have a look at the case when we can write

$$\mathrm{Id} = \sum_q \Delta_q \quad \text{or} \quad \mathrm{Id} = \sum_q \dot{\Delta}_q.$$

This is described by the following proposition, proved in [35].

PROPOSITION 6.2. *Let u be an element of $\mathcal{S}'(\mathbb{R}^d)$. Then, we have, in the sense of the convergence in the space $\mathcal{S}'(\mathbb{R}^d)$,*

$$u = \lim_{q \to \infty} S_q u.$$

6.2. *Hölder spaces and Littlewood–Paley theory*

In this section, we shall describe Hölder spaces in terms of Littlewood–Paley theory. For spaces of positive index, we have the following proposition

PROPOSITION 6.3. *For any r in $\mathbb{R}^+ \setminus \mathbb{N}$, a constant C exists such that for any function u in C^r, we have*

$$\sup_q 2^{qr} \|\Delta_q u\|_{L^\infty} \leqslant C_r \|\tilde{u}\|_r.$$

Let B be a ball of $\mathbb{R}^d$. For any r in $\mathbb{R}^+ \setminus \mathbb{N}$, a constant C exists such that we have the following properties:

Let u be a tempered distribution such that

$$u = \sum_{q \geqslant 0} u_q \quad \textit{with } \operatorname{Supp} \hat{u}_q \subset 2^q B.$$

The distribution u belongs to C^r if and only if the sequence $(2^{qr}\|u_q\|_{L^\infty})_{q\in\mathbb{N}}$ is bounded, and we have

$$C_r^{-1} 2^{qr} \|\Delta_q u\|_{L^\infty} \leqslant \|\tilde{u}\|_r \leqslant C_r \sup_{q \geqslant 0} 2^{qr} \|u_q\|_{L^\infty}.$$

PROOF. For sake of simplicity, we shall prove the above proposition only in the case when $r \in]0, 1[$ (see [35] for complete details). By definition of the operator Δ_q we have

$$\Delta_q u(x) = 2^{qd} \int_{\mathbb{R}^d} h\big(2^q(x-y)\big) u(y)\, \mathrm{d}y.$$

The fact that the function φ is identically 0 near the origin implies we have, in particular, that the integral of h is 0. So it turns out that

$$\Delta_q u(x) = 2^{qd} \int_{\mathbb{R}^d} h\big(2^q(x-y)\big)\big(u(y)-u(x)\big)\,\mathrm{d}y. \tag{62}$$

But as the function u is in C^r we have

$$\big|\Delta_q u(x)\big| \leqslant 2^{qd}\|\tilde{u}\|_r \int |x-y|^r \big|h\big(2^q(x-y)\big)\big|\,\mathrm{d}y.$$

Conversely, let us notice that, as $\|u_q\|_{L^\infty} \leqslant C2^{-qr}$, the series $(u_q)_{q\in\mathbb{N}}$ is convergent in L^∞ and if u denotes its sum, we have

$$\|u\|_{L^\infty} \leqslant C_r \sup_{q\geqslant 0} 2^{qr}\|u_q\|_{L^\infty}. \tag{63}$$

Now let us write

$$\big|u(x)-u(y)\big| \leqslant \sum_{q=0}^{N-1} \big|u_q(x)-u_q(y)\big| + 2\sum_{q\geqslant N} \|u_q\|_{L^\infty},$$

where the integer N will be chosen later on. We estimate the first term by

$$\big|u_q(x)-u_q(y)\big| \leqslant C|x-y|\|\nabla u_q\|_{L^\infty}.$$

Thanks to Lemma 6.1, we have

$$\big|u_q(x)-u_q(y)\big| \leqslant C|x-y|2^{-q(r-1)} \sup_{q\geqslant 0} 2^{qr}\|u_q\|_{L^\infty}. \tag{64}$$

Using (64), it turns out that

$$\big|u(x)-u(y)\big| \leqslant C\Big(\sup_{q\geqslant 0} 2^{qr}\|u_q\|_{L^\infty}\Big)\left(\sum_{q=0}^{N} 2^{-q(r-1)}|x-y| + \sum_{q\geqslant N+1} 2^{-qr}\right).$$

Using (63), we can assume that $|x-y| \leqslant 1$. Choosing $N = [-\log_2|x-y|]+1$, the above inequality concludes the proof of the proposition. □

As seen in Section 4.2, it is of interest to consider Hölder spaces of negative index. The following proposition is the analogue of Proposition 6.3 in the case of index between 0 and -1.

PROPOSITION 6.4. *For any r in $]-1,0[$ a constant C_r exists such that for any function u in C^r, we have*

$$\sup_q 2^{qr}\|\Delta_q u\|_{L^\infty} \leqslant C_r\|\tilde{u}\|_r.$$

Let $\widetilde{C}$ be a ring of $\mathbb{R}^d$ and r a real number in $]-1,0[$. A constant C_r exists which satisfies the following property.

Let u be a tempered distribution such that

$$u = \sum_{q\geqslant 0} u_q \quad \textit{with } \operatorname{Supp}\hat{u}_q \subset 2^q\widetilde{C}.$$

This distribution u belongs to C^r if and only if the sequence $(2^{qr}\|u_q\|_{L^\infty})_{q\in\mathbb{N}}$ is bounded, and we have

$$C_r^{-1}2^{qr}\|\Delta_q u\|_{L^\infty} \leqslant \|\tilde{u}\|_r \leqslant C_r \sup_{q\geqslant 0} 2^{qr}\|u_q\|_{L^\infty}.$$

In order to prove the proposition, let us notice that if $u = u_0 + \sum_{j=1}^d \partial_j u_j$ we have, thanks to Lemma 6.1,

$$\begin{aligned}\|\Delta_q u\|_{L^\infty} &\leqslant \|\Delta_q u_0\|_{L^\infty} + C\sum_{j=1}^d 2^q\|\Delta_q u_j\|_{L^\infty}\\ &\leqslant C2^{-qr}\sup_{0\leqslant j\leqslant d}\|u_j\|_{C^{r+1}}\end{aligned}$$

which proves the first assertion of the lemma. Then, as the support of u_q is included in the ring $2^q C$, we have that

$$\Delta_q u = \sum_{j=1}^d \partial_j u_{q,j}$$

with

$$u_{q,j} \stackrel{\text{def}}{=} \mathcal{F}^{-1}\big(-i\xi_j|\xi|^{-2}\hat{u}_q\big).$$

Using Lemma 6.1, we have that

$$\begin{aligned}\|u_{q,j}\|_{L^\infty} &\leqslant C2^{-q}\|u_q\|_{L^\infty}\\ &\leqslant C2^{-q(r+1)}\sup_q 2^{qr}\|u_q\|_{L^\infty}.\end{aligned}$$

For any j the series $(u_{q,j})_{q\in\mathbb{N}}$ is convergent in L^∞ and its sum belongs to C^{r+1}. Thus $(u_q)_{q\in\mathbb{N}}$ is a convergent series in $\mathcal{S}'(\mathbb{R}^d)$ the sum of which belongs to C^r with $\|\widetilde{u}\|_r \leqslant C \sup_q 2^{qr} \|u_q\|_{L^\infty}$.

Now let us state the following definition.

DEFINITION 6.1. For any real number r the space C^r is the space of the tempered distributions u such that

$$\|u\|_r \stackrel{\text{def}}{=} \sup_q 2^{qr} \|\Delta_q\|_{L^\infty}.$$

The following proposition is an immediate consequence of Lemma 6.1.

PROPOSITION 6.5. *If σ is a smooth function on $\mathbb{R}^d$ which is homogeneous of degree m outside a fixed ball, then for any $r \in \mathbb{R}$, the operator $\sigma(D)$ maps continuously C^r into C^{r-m}.*

If we consider a divergence free vector field v such that the support of $\hat{v}$ does not contain the origin then ∇v belongs to $C^0_\star$ (in the very special case when r is an integer, those spaces are traditionally denoted by $C^r_\star$) and we have

$$\|\nabla v\|_0 \leqslant C\|\omega\|_0 \leqslant C\|\omega\|_{L^\infty}. \tag{65}$$

An interesting example for our purpose has been given by Proposition 1.3. But this non-inclusion of $C^0_\star$ in L^∞ is not too strong as described by the following proposition.

PROPOSITION 6.6. *Let ε be in $]0, 1[$, a constant C exists such that, for any f is C^ε,*

$$\|f\|_{L^\infty} \leqslant C\|f\|_0\left(1 + \log \frac{\|f\|_\varepsilon}{\|f\|_0}\right).$$

To prove this, let us write the function f as the sum of $\Delta_q f$. For any positive integer N, we have that

$$\begin{aligned}
\|f\|_{L^\infty} &\leqslant \sum_{q\leqslant N-1} \|\Delta_q f\|_{L^\infty} + \sum_{q\geqslant N} \|\Delta_q f\|_{L^\infty} \\
&\leqslant (N+1)\|f\|_0 + \frac{2^{-(N-1)\varepsilon}}{2^\varepsilon - 1}\|f\|_\varepsilon.
\end{aligned}$$

We get the result choosing

$$N = 1 + \left[\frac{1}{\varepsilon} \log_2 \frac{\|f\|_\varepsilon}{\|f\|_0}\right].$$

The following corollary is an immediate consequence of (65) and Proposition 6.6.

COROLLARY 6.1. *Let ε be in $]0,1[$ and a in $]1,+\infty[$. A constant C exists which satisfies the following properties. If v is a divergence free vector field which belongs to $C^{1+\varepsilon}$, the vorticity of which belongs to L^a, we have*

$$\|\nabla v\|_{L^\infty} \leqslant C\left(\|\omega\|_{L^a} + \|\omega\|_{L^\infty} \log\left(e + \frac{\|\omega\|_\varepsilon}{\|\omega\|_{L^\infty}}\right)\right).$$

As seen above, a vector field the vorticity of which belongs to $L^a \cap L^\infty$ belongs to $C^1_\star$ (traditionally called the Zygmund class). Then Proposition 1.4 will be implied by the following proposition.

PROPOSITION 6.7. *A constant C exists such that, for any function u in $C^1_\star$ and any x and y in $\mathbb{R}^d$ such that $|x-y| \leqslant 1$, we have*

$$\left|u(x) - u(y)\right| \leqslant C\|u\|_1 |x-y|\left(1 - \log|x-y|\right).$$

PROOF. The proof is very similar to the above ones. Let us write, using Lemma 6.1,

$$\begin{aligned}\left|u(x) - u(y)\right| &\leqslant C|x-y| \sum_{q<N} 2^q \|\Delta_q u\|_{L^\infty} + C \sum_{q\geqslant N} \|\Delta_q u\|_{L^\infty} \\ &\leqslant C\|u\|_1\left((N+1)|x-y| + 2^N\right).\end{aligned}$$

Choosing as above $N = [-\log_2|x-y|] + 1$, we get the result. □

6.3. *Besov spaces*

In Section 8, we shall need Besov spaces. Let us define a restricted class of them which will be enough for our purpose.

DEFINITION 6.2. Let a be $[1,\infty]$ and r a real number. The space B^r_a is the space of tempered distributions u such that

$$\|u\|_{B^r_a} \stackrel{\text{def}}{=} \sup_q 2^{qr} \|\Delta_q u\|_{L^a} < \infty.$$

It is an exercise left to the reader to prove that B^r_a equipped with the norm $\|\cdot\|_{B^r_a}$ is a Banach space. Hölder spaces are a particular case of Besov spaces which are presented here. Obviously we have $B^r_\infty = C^r$. In fact, up to an arbitrary small error about the index of regularity, they look very much like Sobolev spaces but in our context, the use of Besov space is much easier. For our purpose, the most important property is the following Sobolev type embedding.

PROPOSITION 6.8. *For any a and b in $[1, +\infty]$ such that $b \geqslant a$ and any real number r, the space B^r_a is continuously embedded in $B^{r-d(1/a-1/b)}_b$. In particular, B^r_a is continuously embedded in $C^{r-d/a}$.*

This proposition is an immediate consequence of Lemma 6.1 because for any u in B^r_a, we have

$$\|\Delta_q u\|_{L^b} \leqslant C 2^{qd(1/a-1/b)} \|\Delta_q u\|_{L^a}.$$

THEOREM 6.1. *Let σ be a smooth function on $\mathbb{R}^d$ which coincides with an homogeneous function of degree m outside a compact neighborhood of the origin. Then the operator $\sigma(D)$ defined by*

$$\sigma(D)u \stackrel{def}{=} \mathcal{F}^{-1}\big(\sigma(\xi)\hat{u}\big)$$

maps continuously B^r_a into B^{r-m}_a.

6.4. *Paradifferential calculus*

In this section, we shall give a very sketchy version of Bony's paradifferential calculus introduced in [17]. The basic idea is the following. Let us decompose the product uv in the following way, known as Bony' decomposition:

$$uv = T_u v + T_v u + R(u, v) \tag{66}$$

with

$$T_u v \stackrel{\text{def}}{=} \sum_q S_{q-1} u \Delta_q v \tag{67}$$

and

$$R(u, v) \stackrel{\text{def}}{=} \sum_{q;\, j\in\{-1,0,1\}} \Delta_q u \Delta_{q-j} v. \tag{68}$$

Because of the definition of S_q and Δ_q given in Proposition 6.1, the support of the Fourier transform of $S_{q-1} u \Delta_q v$ is included in $2^q \widetilde{\mathcal{C}}$. Thus we have the following proposition.

PROPOSITION 6.9. *For any a in $[1, +\infty]$, for any positive real number ε and for any real number r, the bilinear operator T maps continuously $L^\infty \times B^r_a$ into B^r_a and $C^{-\varepsilon} \times B^r_a$ into $B^{r-\varepsilon}_a$.*

PROOF. The proof consists in observing that thanks to (59) we have

$$\Delta_q T_u v = \sum_{|q'-q|\leqslant 4} \Delta_q(S_{q'-1}u\Delta_{q'}v).$$

Thus we infer that

$$\begin{aligned}\|\Delta_q T_u v\|_{L^a} &\leqslant C \sum_{|q'-q|\leqslant 4} \big\|\Delta_q(S_{q'-1}u\Delta_{q'}v)\big\|_{L^a}\\ &\leqslant C2^{-qr} \sum_{|q'-q|\leqslant 4} \|\Delta_q S_{q'-1}u\|_{L^\infty} 2^{q'r}\|\Delta_{q'}v\|_{L^a}.\end{aligned}$$

As $\|S_{q'-1}u\|_{L^\infty} \leqslant C\|u\|_{L^\infty}$, we get by definition of the B^r_a-norm that

$$\|\Delta_q T_u v\|_{L^a} \leqslant C2^{-qr}\|u\|_{L^\infty}\|v\|_{B^r_a}.$$

The case of the space $C^{-\varepsilon}$ is strictly analogous if we notice that

$$\begin{aligned}\|S_{q'-1}u\|_{L^\infty} &\leqslant \sum_{p\leqslant q'} 2^{p\varepsilon}2^{-p\varepsilon}\|\Delta_p u\|_{L^\infty}\\ &\leqslant \sum_{p\leqslant q'} 2^{p\varepsilon}\|u\|_{C^{-\varepsilon}}\\ &\leqslant \frac{1}{1-2^{-\varepsilon}} 2^{q\varepsilon}\|u\|_{C^{-\varepsilon}}.\end{aligned}$$

Thus the proposition is proved. □

Let us point out the fact that $T_u v$ is defined on any product $B^r_a \times B^{r'}_{a'}$. This is of course not the case for the product itself. Thus it will not be the case of the term $R(u,v)$. The case when the operator R is well defined in Besov spaces we consider here is described by the following proposition.

PROPOSITION 6.10. *Let r and r' be two real numbers the sum of which is positive. Let a, a' and b in $[1,+\infty]$ such that*

$$\frac{1}{a} + \frac{1}{a'} \geqslant \frac{1}{b}.$$

Then R maps continuously $B^r_a \times B^{r'}_{a'}$ into $B_b^{r+r'-d(1/a+1/a'-1/b)}$.

PROOF. As the support of the Fourier transform of $\Delta_{q'}u\Delta_{q'-j}v$ is included in a ball of the type $2^q\widetilde{\mathcal{B}}$ where $\widetilde{\mathcal{B}}$ is a fixed ball, we have

$$\Delta_q R(u,v) = \Delta_q \sum_{q'\geqslant q-N_0;\, j\in\{-1,0,1\}} \Delta_q u\Delta_{q-j}v,$$

where N_0 is some fixed integer. Using Lemma 6.1, we have thanks to Hölder inequality,

$$\begin{aligned}
&\left\| \Delta_q R(u,v) \right\|_{L^b} \\
&\quad \leqslant C 2^{qd(1/a+1/a'-1/b)} \sum_{q' \geqslant q-N_0;\, j \in \{-1,0,1\}} \| \Delta_{q'} u \Delta_{q'-j} v \|_{L^{aa'/(a+a')}} \\
&\quad \leqslant C 2^{qd(1/a+1/a'-1/b)} \\
&\qquad \times \sum_{q' \geqslant q-N_0;\, j \in \{-1,0,1\}} 2^{-q'(r+r')} 2^{q'r} \| \Delta_q u \|_{L^a} 2^{(q'-j)r} \| \Delta_{q'-j} v \|_{L^{a'}}.
\end{aligned}$$

As the sum $r + r'$ is positive, we get by definition of $\| \cdot \|_{B^r_a}$,

$$\begin{aligned}
\left\| \Delta_q R(u,v) \right\|_{L^b} &\leqslant C 2^{qd(1/a+1/a'-1/b)} \biggl(\sum_{q' \geqslant q-N_0} 2^{-q'(r+r')} \biggr) \|u\|_{B^r_a} \|v\|_{B^{r'}_{a'}} \\
&\leqslant C 2^{-q(r+r'-d(1/a+1/a'-1/b))} \|u\|_{B^r_a} \|v\|_{B^{r'}_{a'}}
\end{aligned}$$

which proves the proposition. □

The above two propositions imply easily the following corollary.

COROLLARY 6.2. *For any $(r,a) \in]0,+\infty[\times [1,+\infty]$, if f and g are two functions in $B^r_a \cap L^\infty$ then the product ab also belongs to $B^r_a \cap L^\infty$. Moreover, a constant C exists such that*

$$\|ab\|_{B^r_a} \leqslant C \bigl(\|a\|_{L^\infty} \|b\|_{B^r_a} + \|a\|_{B^r_a} \|b\|_{L^\infty} \bigr).$$

It is obvious that, thanks to Proposition 6.5, the estimates (5) and (6) holds.

The following proposition describes the regularity required on Y in order to be able to deduce tangential regularity of v with respect to Y from tangential regularity of ω with respect to Y.

PROPOSITION 6.11. *Two bilinear operators W_1 and W_2 exist such that for any divergence free vector field v and any vector field Y, we have*

$$Y(x,D)v = W_1(Y,v) + W_2(Y,v)$$

and such that, for any a in $[1,\infty]$ and any positive r, we have

$$\left\| W_1(Y,v) \right\|_{B^r_a} \leqslant C \left\| Y(x,D)\omega \right\|_{B^{r-1}_a}$$

and

$$\left\| W_2(Y,v) \right\|_{B^r_a} \leqslant C \bigl(\|Y\|_{B^r_a} + \|\operatorname{div} Y\|_{B^r_a} \bigr) \|\nabla v\|_{L^\infty}.$$

PROOF. In order to prove this, let us write that (see [35] for details)

$$Y(x, D)v = \sum_{i=1}^{5} V_i$$

with

$$V_1 = \big(\mathrm{Id} - \chi(D)\big)\nabla^\perp \Delta^{-1} Y(x, D)\omega,$$

$$V_2 = \sum_{j=1}^{2} \big[T_{Y^j}, \nabla^\perp \Delta^{-1}\big]\partial_j \omega,$$

$$V_3 = \big(\mathrm{Id} - \chi(D)\big)\nabla^\perp \Delta^{-1} R(\omega, \operatorname{div} Y),$$

$$V_4 = -\big(\mathrm{Id} - \chi(D)\big)\nabla^\perp \Delta^{-1} \sum_{j=1}^{2} \big(T_{\partial_j \omega} Y^j + \partial_j R(\omega, Y_j)\big)$$

and

$$V_5 = \sum_{j=1}^{2} \big(T_{\partial_j v} Y^j + R\big(\partial_j v, Y^j\big)\big).$$

Then let us state

$$W_1(Y, v) \stackrel{\text{def}}{=} V_1 \quad \text{and} \quad W_2(Y, v) = \sum_{j=2}^{5} V_j.$$

Thanks to Theorem 6.1, we have

$$\big\| W_1(Y, v)\big\|_{B^r_a} \leqslant C \big\| Y(x, D)\omega\big\|_{B^{r-1}_a}.$$

The fact that for $j \in \{3, 4, 5\}$ we have

$$\| V_j\|_{B^r_a} \leqslant C\big(\|Y\|_{B^r_a} + \|\operatorname{div} Y\|_{B^r_a}\big)\|\nabla v\|_{L^\infty}$$

comes easily from Propositions 6.9 and 6.10. □

6.5. *A propagation theorem*

THEOREM 6.2. *Let r be in $]-1, 1[\setminus\{0\}$ and $a \in [1, +\infty]$. A constant C exists which satisfies the following properties. Let v be a divergence free vector field, the derivatives* *of order* 1 *of which are bounded on* $[0,T]\times\mathbb{R}^d$. *Let* (f,g) *be in* $L^\infty([0,T];B^r_a)\times L^1([0,T];B^r_a)$ *such that*

$$g=g_1+g_2\quad\text{with } \|g_2(t)\|_{B^r_a}\leqslant CV(t)\|f(t)\|_{B^r_a}.$$

If $\partial_t f+v\cdot\nabla f=g$, *then we have*

$$\|f(t)\|_{B^r_a}\leqslant\|f(0)\|_{B^r_a}\mathrm{e}^{C\int_0^t V(t')\,\mathrm{d}t'}+\int_0^t\|g_1(t')\|_{B^r_a}\mathrm{e}^{C\int_{t'}^t V(t'')\,\mathrm{d}t''}\,\mathrm{d}t'. \tag{69}$$

The key point of the proof of this theorem lies in the following commutation lemma.

LEMMA 6.3. *Let* r *be in* $]-1,1[\setminus\{0\}$ *and* a *in* $[1,+\infty]$. *A constant* C *exists such that the following property holds. If* v *is a divergence free Lipschitz vector field,* f *is a function in* B^r_a, *then we have*

$$\|[v\cdot\nabla,\Delta_q]f\|_{L^a}\leqslant C2^{-qr}\|\nabla v\|_{L^\infty}\|f\|_{B^r_a}.$$

PROOF. Let us admit this lemma for a while. Let us apply the operator Δ_q to the transport equation. This gives

$$\partial_t\Delta_q f+v\cdot\nabla\Delta_q f=\Delta_q g_1+\Delta_q g_2+[v\cdot\nabla,\Delta_q]f.$$

As the vector field v is divergence free, using Lemma 6.3 we infer

$$\begin{aligned}&\|\Delta_q f(t)\|_{L^a}\\&\quad\leqslant\|\Delta_q f(0)\|_{L^a}\\&\qquad+C2^{-qr}\int_0^t\big(\|g_1(t')\|_{B^r_a}+\|g_2(t')\|_{B^r_a}+\|\nabla v(t')\|_{L^\infty}\|f(t')\|_{B^r_a}\big)\,\mathrm{d}t'.\end{aligned}$$

Multiplying this inequality by 2^{qr} and taking the supremum over q we get

$$\begin{aligned}\|f(t)\|_{B^r_a}&\leqslant\|f(0)\|_{B^r_a}\\&\quad+C\int_0^t\big(\|g_1(t')\|_{B^r_a}+\|g_2(t')\|_{B^r_a}+\|\nabla v(t')\|_{L^\infty}\|f(t')\|_{B^r_a}\big)\,\mathrm{d}t'.\end{aligned}$$

Then Gronwall's lemma implies the result of Theorem 6.2 provided we prove Lemma 6.3.

In order to do so, let us use Bony's decomposition (66). This gives

$$v\cdot\nabla f=\sum_{j=1}^d\big(T_{v^j}\partial_j f+T_{\partial_j f}v^j+\partial_j R(v^j,f)\big)-R(\operatorname{div}v,f). \tag{70}$$

As the "commuting" operator is linear, it is enough to study the commutation with each of the above operators. Using definition (67) of the paraproduct and properties (55)–(60), we get

$$[\Delta_q, T_{v^j}\partial_j] = \sum_{j=1}^{d} \sum_{|q-q'|\leqslant 4} [\Delta_q, S_{q'-1}(v^j)]\Delta_{q'}\partial_j.$$

In order to estimate this quantity, let us observe that

$$\begin{aligned}\big([\Delta_q, a]b\big)(x) &= \Delta_q(ab)(x) - a(x)\Delta_q b(x)\\ &= 2^{qd}\int_{\mathbb{R}^d} h\big(2^q(x-y)\big)\big(a(y)-a(x)\big)b(y)\,\mathrm{d}y.\end{aligned}$$

Using Taylor inequality we have

$$\big|a(y)-a(x)\big| \leqslant \|\nabla a\|_{L^\infty}|y-x|.$$

It turns out that

$$\big|\big([\Delta_q, a]b\big)(x)\big| \leqslant 2^{qd}\|\nabla a\|_{L^\infty}\int_{\mathbb{R}^d} \big|h\big(2^q(x-y)\big)\big|\,|y-x|\,\big|b(y)\big|\,\mathrm{d}y.$$

Then Young inequality implies that

$$\big\|[\Delta_q, a]b\big\|_{L^2} \leqslant 2^{-q}\big\|h(\cdot)|\cdot|\big\|_{L^1}\|\nabla a\|_{L^\infty}\|b\|_{L^p}.$$

Applying the above inequality with $b = \Delta_{q'}\partial_j f(t)$ and $a = S_{q'-1}v^j$, we infer that

$$\big\|\big[\Delta_q, S_{q'-1}(v^j)\big]\Delta_{q'}\partial_j f\big\|_{L^\infty} \leqslant C^{\rho+1}2^{-q\rho}\|f\|_{B^\rho_a}\|\nabla v\|_{L^\infty}.$$

Thus we have

$$\big\|[\Delta_q, T_{v^j(t)}\partial_j]f(t)\big\|_{L^\infty} \leqslant C^{\rho+1}2^{-q\rho}c_q\|f\|_{B^\rho_a}\|\nabla v\|_{L^\infty}. \tag{71}$$

Now let us study the commutation between Δ_q and $T_{\partial_j \cdot v^j}(t)$. As there is no low frequencies of b in the paraproduct $T_a b$, using Proposition 6.9 we have

$$\|T_a b\|_\rho \leqslant C^{|\rho|+1}\|a\|_{L^\infty}\|\nabla b\|_{B^{\rho-1}_a},$$

and if ρ is negative

$$\|T_a b\|_{B^{\sigma+\rho}_a} \leqslant \frac{C^{|\rho+\sigma|+1}}{-\rho}\|a\|_{B^\rho_a}\|\nabla b\|_{B^\rho_a}.$$

Thus we infer

$$\left\|\Delta_q T_{\partial_j f(t)} v^j(t)\right\|_{L^\infty} \leqslant C^{|\rho|+1} 2^{-q\rho} \left\|\nabla f(t)\right\|_{L^\infty} \left\|\nabla v(t)\right\|_{B^\rho_a}, \tag{72}$$

and if $\rho < 1$,

$$\left\|\Delta_q T_{\partial_j f} v^j\right\|_{L^\infty} \leqslant \frac{C^{|\rho|+1}}{1-\rho} 2^{-qr} \|\nabla f\|_{B^{\rho-1}_a} \|\nabla v\|_{B^0_\infty}. \tag{73}$$

Using the properties of the dyadic partition of unity, in particular (59), we have

$$T_{\partial_j \Delta_q u} v^j = \sum_{q' \geqslant q} S_{q'-1} \Delta_q \partial_j u \Delta_{q'} v^j.$$

Thus using Lemma 6.1, we get

$$\begin{aligned}
2^{q\rho} \left\|T_{\partial_j \Delta_q f} v^j\right\|_{L^a} &\leqslant \sum_{q' \geqslant q} 2^{q\rho} \|\Delta_q \nabla f\|_{L^a} 2^{-q'} \|\nabla \Delta_{q'} v\|_{L^\infty} \\
&\leqslant \sum_{q' \geqslant q} 2^{q\rho} \|\Delta_q f\|_{L^a} 2^{q-q'} \|\nabla \Delta_{q'} v\|_{L^\infty}.
\end{aligned}$$

Using the now standard argument on convolution of the series, we get

$$\left\|T_{\partial_j \Delta_q f} v^j\right\|_{L^a} \leqslant C \|f\|_{B^\rho_a} \|\nabla v\|_{B^0_\infty}. \tag{74}$$

From estimates (72)–(74) we infer

$$\left\|\left[\Delta_q, T_{\partial_j \cdot} v^j\right] f\right\|_{L^\infty} \leqslant C^{\rho+1} 2^{-q\rho} \|\nabla f\|_{L^\infty} \|\nabla v\|_{B^{\rho-1}_a} \quad \text{if } \rho > 1, \tag{75}$$

$$\left\|\left[\Delta_q, T_{\partial_j \cdot} v^j\right] f\right\|_{L^\infty} \leqslant \frac{C^{\rho+1}}{1-\rho} 2^{-q\rho} \|f\|_{B^\rho_a} \|\nabla v\|_{B^0_\infty} \quad \text{if } \rho < 1. \tag{76}$$

The terms of the type $[\Delta_q, R(\operatorname{div} v, \cdot)]$ are very easy to estimate. For any positive ρ, we have, thanks to Proposition 6.10,

$$\begin{aligned}
\left\|\Delta_q R(\operatorname{div} v, u)\right\|_{L^a} &\leqslant 2^{-q\rho} \left\|R(\operatorname{div} v, f)\right\|_{B^\rho_a} \\
&\leqslant \frac{C^{\rho+1}}{\rho} 2^{-q\rho} \|\operatorname{div} v\|_{B^0_\infty} \|f\|_{B^\rho_a}.
\end{aligned}$$

By definition of R, we have

$$R(\operatorname{div} v, \Delta_q u) = \sum_{|q'-q| \leqslant 1, |q''-q'| \leqslant 1} \Delta_{q''} \operatorname{div} v \Delta_{q'} \Delta_q u.$$

Thus for any positive ρ we infer

$$\big\|\big[\Delta_q, R(\operatorname{div} v, \cdot)\big]u\big\|_{L^\infty} \leqslant \frac{C^{\rho+1}}{\rho} 2^{-q\rho} \|\nabla v\|_{B^0_\infty} \|f\|_{B^\rho_a}. \tag{77}$$

If the vector field v is divergence free, this term of course disappears.

Let us now estimate $[\Delta_q, \partial_j R(v^j, \cdot)]$. For any ρ greater than -1, we have

$$\begin{aligned}\big\|\Delta_q \partial_j R\big(v^j, f\big)\big\|_{L^a} &\leqslant \frac{C^{|\rho|+2}}{1+\rho} 2^{-q\rho} \big\|R\big(v^j, f\big)\big\|_{B^\rho_a}\\ &\leqslant C_\rho 2^{-q\rho} \|v\|_{B^1_\infty} \|f\|_{B^\rho_a}.\end{aligned}$$

Moreover, it is obvious that

$$\partial_j R\big(v^j, \Delta_q u\big) = \sum_{|q'-q|\leqslant 2, |q''-q'|\leqslant 1} (\Delta_{q''} \operatorname{div} v) \Delta_{q'} \Delta_q f.$$

The support of the Fourier transform of $\Delta_{q''} \operatorname{div} v \Delta_{q'} \Delta_q u$ is included in a ball of radius $C2^q$, then we infer from Lemma 6.1

$$\big\|\partial_j R\big(v^j, \Delta_q f\big)\big\|_{L^a} \leqslant C 2^{-q\rho} \|v\|_{B^1_\infty} \|f\|_{B^\rho_a}.$$

It turns out that for any real r greater than -1, we have

$$\big\|\big[\Delta_q, \partial_j R\big(v^j, \cdot\big)\big] f\big\|_{L^a} \leqslant \frac{C^{|\rho|+2}}{1+\rho} 2^{-q\rho} \|v\|_{B^1_\infty} \|f\|_{B^\rho_a}. \tag{78}$$

To get exactly the lemma we need to substitute $\|\nabla v\|_{B^0_\infty}$ to $\|v\|_{B^1_\infty}$ in the above estimate. Let us observe that using Lemma 6.1 we have, for any $q \geqslant 0$,

$$\|\Delta_q v\|_{L^\infty} \leqslant C 2^{-q} \|\Delta_q \nabla v\|_{L^\infty}.$$

This implies that

$$\big\|(\mathrm{Id} - S_{-1}) v\big\|_{B^1_\infty} \leqslant C \big\|(\mathrm{Id} - S_{-1}) \nabla v\big\|_{B^0_\infty} \leqslant C \|\nabla v\|_{B^0_\infty}.$$

Thus we have

$$\big\|\big[\Delta_q, \partial_j R\big((\mathrm{Id} - S_{-1}) v^j, \cdot\big)\big] f\big\|_{L^a} \leqslant \frac{C^{|\rho|+2}}{1+\rho} 2^{-q\rho} \|\nabla v\|_{B^0_\infty} \|f\|_{B^\rho_a}. \tag{79}$$

Let us estimate $\|[\Delta_q, \partial_j R(\chi(D) v^j, \cdot)] f\|_{L^a}$. We have

$$R\big(S_{-1} v^j, f\big) = \sum_{|k|\leqslant 1, q'\leqslant 0} \big(\Delta_{q'} S_{-1} v^j\big) \Delta_{q'-j} f.$$

Thus by definition of Δ_q, we have

$$\big[\Delta_q, \partial_j R\big(\chi(D)v^j, \cdot\big)\big] f = 0 \quad \text{when } q \geqslant 3.$$

When $q \leqslant 2$, let us write that

$$\begin{aligned}
&\big[\Delta_q, \partial_j R\big(\chi(D)v^j, \cdot\big)\big] f(x) \\
&\quad = 2^{qd} \sum_{|k|\leqslant 1, q'\leqslant 0} \partial_j \int_{\mathbb{R}^d} h\big(2^q(x-y)\big) \\
&\qquad\qquad \times \big(\Delta_{q'} S_{-1} v^j(x) - \Delta_{q'} S_{-1} v^j(y)\big) \Delta_{q'-j} f(y)\, \mathrm{d}y.
\end{aligned}$$

Like in the proof of (71), we have as $q \leqslant 2$,

$$\big\| \big[\Delta_q, \partial_j R\big(\chi(D)v^j, \cdot\big)\big] f \big\|_{L^a} \leqslant C \|\nabla v\|_{L^\infty} \|f\|_{B_a^\rho}.$$

This concludes the proof of the lemma. □

Using definition (67) of the paraproduct and properties (55)–(60), we get

$$[\Delta_q, T_{v^j} \partial_j] = \sum_{j=1}^{d} \sum_{|q-q'|\leqslant 4} \big[\Delta_q, S_{q'-1}(v^j)\big] \Delta_{q'} \partial_j.$$

Applying the above lemma with $b = \Delta_{q'} \partial_j f(t)$ and $a = S_{q'-1}(v^j)$, we infer that

$$\big\| \big[\Delta_q, S_{q'-1}(v^j)\big] \Delta_{q'} \partial_j f \big\|_{L^\infty} \leqslant C^{r+1} 2^{-qr} \|f\|_r \|\nabla v\|_{L^\infty}$$

and thus that

$$\big\| [\Delta_q, T_{v^j(t)} \partial_j] f(t) \big\|_{L^\infty} \leqslant C^{|r|+1} 2^{-qr} \big\| f(t) \big\|_r \big\| \nabla v(t) \big\|_{L^\infty}. \tag{80}$$

6.6. *The proof of Theorem* 4.1

As low frequencies are smooth, we can assume that the Fourier transform of ω is identically 0 near the origin.

The first step of the proof consists in estimating $\|Y(x, D)v\|_{L^\infty}$ for any vector field Y the coefficients of which are C^ε. To do so, we use the following lemma which we shall admit (see [35] for the complete details).

LEMMA 6.4. *Let ε be in $]0,1[$. A constant C exists which satisfies the following property. Let us consider w_1 and w_2 two vector fields on $\mathbb{R}^d$ and let us assume that w_1 is C^ε and that w_2 is bounded. If the scalar product $(w_1|w_2)$ is a C^ε function, then*

$$\|(w_1|w_2)\|_{L^\infty} \leqslant C\|w_1\|_{L^\infty}\|w_2\|_0 \log\left(e + \frac{\|w_1\|_\varepsilon\|w_2\|_0 + \|(w_1|w_2)\|_\varepsilon}{\|w_1\|_{L^\infty}\|w_2\|_0}\right).$$

Let us apply this lemma with $w_1 = Y$ and $w_2 = \nabla v^i$. This gives

$$\|Y(x,D)v\|_{L^\infty} \leqslant C\|Y\|_{L^\infty}\|\nabla v\|_0 \log\left(e + \frac{\|Y\|_\varepsilon\|\nabla v\|_0 + \|Y(x,D)v\|_\varepsilon}{\|Y\|_{L^\infty}\|\nabla v\|_0}\right).$$

Using (65) and the fact that the function $x \mapsto x\log(e+\alpha x^{-1})$ is nondecreasing, we get

$$\|Y(x,D)v\|_{L^\infty} \leqslant C\|Y\|_{L^\infty}\|\omega\|_{L^\infty} \log\left(e + \frac{\|Y\|_\varepsilon}{\|Y\|_{L^\infty}} + \frac{\|Y(x,D)v\|_\varepsilon}{\|Y\|_{L^\infty}\|\omega\|_{L^\infty}}\right). \tag{81}$$

The above inequality is the analogue of (41). The analogue of the identity $\partial_2^2 = \Delta - \partial_1^2$ is

$$|Y(x)|\partial_1^2 = \frac{Y^1(x)(Y^1\partial_1 + Y^2\partial_2)\partial_1 - Y^2(x)(Y^1\partial_1 + Y^2\partial_2)\partial_2 + (Y^2(x))^2\Delta}{|Y(x)|},$$

$$|Y(x)|\partial_2^2 = \frac{Y^2(x)(Y^1\partial_1 + Y^2\partial_2)\partial_2 - Y^1(x)(Y^1\partial_1 + Y^2\partial_2)\partial_1 + (Y^1(x))^2\Delta}{|Y(x)|},$$

$$|Y(x)|\partial_1\partial_2 = \frac{Y^1(x)(Y^1\partial_1 + Y^2\partial_2)\partial_2 + Y^1(x)(Y^1\partial_1 + Y^2\partial_2)\partial_2 - Y^1(x)Y^2(x)\Delta}{|Y(x)|}.$$

As $|Y^i(x)| \leqslant |Y(x)|$ we have, for any point x of the plane,

$$\begin{aligned}|Y(x)||\nabla v(x)| &\leqslant \|Y\|_{L^\infty}\|\omega\|_{L^\infty} + C\|Y\|_{L^\infty}\|\omega\|_{L^\infty} \\ &\quad \times \log\left(e + \frac{\|Y\|_\varepsilon}{\|Y\|_{L^\infty}} + \frac{\|Y(x,D)v\|_\varepsilon}{\|Y\|_{L^\infty}\|\omega\|_{L^\infty}}\right).\end{aligned} \tag{82}$$

If we apply directly with inequality to the vector fields X_λ of the family X, the quantity $N(X)$ will appear outside the logarithm which will prevent global regularity as we have seen in Section 5. We shall apply the above inequality (82) to vector fields Y_λ deduced from the vector fields X_λ. Let us introduce

$$U_\lambda = \{x \in \mathbb{R}^2 \mid 2|X_\lambda(x)| > I(X)\} \quad \text{and} \quad \delta = \left(\frac{I(X)}{4\|X_\lambda\|_\varepsilon}\right)^{1/\varepsilon}.$$

Let ρ be a nonnegative function in $\mathcal{D}(B(0,1))$ the integral of which is 1. Let us state

$$\theta_\lambda \stackrel{\text{def}}{=} \delta^{-2}\rho(\delta^{-1}\cdot) \star \mathbf{1}_{U_\lambda} \quad \text{and} \quad Y_\lambda \stackrel{\text{def}}{=} \frac{\theta_\lambda}{|X_\lambda|}X_\lambda.$$

It is easily checked that

$$|X_\lambda(x)| \geqslant \frac{3I(X)}{4} \quad \Longrightarrow \quad \theta_\lambda(x) = 1. \tag{83}$$

Now let us apply inequality (82) to the vector field Y_λ with $\varepsilon/2$. As $\|Y_\lambda\|_{L^\infty} = 1$, we infer that if x belongs to V_λ with

$$V_\lambda \stackrel{\text{def}}{=} \left\{ x \in \mathbb{R}^2 \;\Big|\; |X_\lambda(x)| \geqslant \frac{3I(X)}{4} \right\},$$

we have

$$|\nabla v(x)| \leqslant C\|\omega\|_{L^\infty} \log\left(e + \|Y_\lambda\|_{\varepsilon/2} + \frac{\|Y_\lambda(x, D)v\|_{\varepsilon/2}}{\|\omega\|_{L^\infty}} \right). \tag{84}$$

The fact that $\|Y_\lambda\|_\varepsilon \leqslant CN_\varepsilon(X)$ is easy (see [35] for the details). The estimate about the quantity $\|Y_\lambda(x, D)v\|_{\varepsilon/2}$ is more delicate. It is easily seen that

$$\text{if} \quad \theta_{X,\lambda}(x) = \frac{\theta_\lambda(x)}{|X_\lambda(x)|},$$

$$\text{then} \quad \|\theta_{X,\lambda}\|_{L^\infty} \leqslant \frac{1}{I(X)} \quad \text{and} \quad \|\theta_{X,\lambda}\|_\varepsilon \leqslant C\frac{\|X_\lambda\|_\varepsilon}{I(X)^2} \leqslant \frac{CN_\varepsilon(X)}{I(X)}.$$

Moreover, using Bony's decomposition and applying Propositions 6.9 and 6.10, we have

$$\begin{aligned} \|Y_\lambda(x, D)v\|_{\varepsilon/2} &\leqslant \|\theta_{X,\lambda}\|_{L^\infty} \|X_\lambda(x, D)v\|_{\varepsilon/2} + C\|\theta_{X,\lambda}\|_\varepsilon \|X_\lambda(x, D)v\|_{-\varepsilon/2} \\ &\leqslant C\frac{\|X_\lambda(x, D)v\|_{\varepsilon/2} + N_\varepsilon(X)\|X_\lambda(x, D)v\|_{-\varepsilon/2}}{I(X)}. \end{aligned} \tag{85}$$

But the vector field v is divergence free, so we have

$$X_\lambda(x, D)v = \sum_j \left(T_{X_\lambda^j}\partial_j v + T_{\partial_j v}X^j\right) + \sum_j \partial_j R\left(v^j, X_\lambda^j\right).$$

Applying again Propositions 6.9 and 6.10, we get

$$\begin{aligned} \|X_\lambda(x, D)v\|_{-\varepsilon/2} &\leqslant C\|\nabla v\|_{-\varepsilon/2}\|X_\lambda\|_\varepsilon \\ &\leqslant C\|X_\lambda\|_\varepsilon\|\omega\|_{L^\infty}. \end{aligned}$$

Using (85) and (84), we get

$$\|\nabla v\|_{L^\infty(V_\lambda)} \leqslant \|\omega\|_{L^\infty} + C\|\omega\|_{L^\infty} \log\left(e + N_\varepsilon(X) + \frac{\|X_\lambda(x, D)v\|_{\varepsilon/2}}{I(X)\|\omega\|_{L^\infty}} \right). \tag{86}$$

Thanks to Proposition 6.9, we have

$$\big\|T_{\partial_j v} X_\lambda^j\big\|_{\varepsilon/2} \leqslant C\|\nabla v\|_{-\varepsilon/2}\|X_\lambda\|_\varepsilon \leqslant \frac{C}{\varepsilon}\|\omega\|_{L^\infty}\|X_\lambda\|_\varepsilon.$$

From Proposition 6.10 we infer that, for any λ, we have

$$\big\|X_\lambda(x,D)v\big\|_{\varepsilon/2} \leqslant C\big\|X_\lambda(x,D)\omega\big\|_{\varepsilon-1} + C\big(\|X_\lambda\|_\varepsilon + \|\operatorname{div} X_\lambda\|_\varepsilon\big)\|\omega\|_{L^\infty}. \quad (87)$$

By definition of $\|\omega\|_X^\varepsilon$ and from (86) we infer that

$$\|\nabla v\|_{L^\infty(V_\lambda)} \leqslant C\|\omega\|_{L^\infty}\log\left(e + N_\varepsilon(X) + \frac{\|X_\lambda(x,D)\omega\|_{\varepsilon-1}}{I(X)\|\omega\|_{L^\infty}}\right). \quad (88)$$

The family $(X_\lambda)_{\lambda\in\Lambda}$ is C^ε-admissible, thus the union of the V_λ is the whole plane $\mathbb{R}^2$. So we get

$$\|\nabla v\|_{L^\infty} \leqslant C\|\omega\|_{L^\infty}\log\left(e + \frac{\|\omega\|_X^\varepsilon}{\|\omega\|_{L^\infty}}\right),$$

which proves the theorem.

6.7. *References and remarks*

Littlewood–Paley theory is a classical tool of harmonic analysis. It appears as a central method in the context of nonlinear partial differential equations in the famous paper [17] by Bony. This paper constructs the paradifferential calculus. It has been the source of inspiration of a lot of works in nonlinear partial differential equations for instance in incompressible Navier–Stokes equations by Cannone, Meyer and Planchon (see [22] and the book [21] by Cannone) or in Yang–Mills equations by Klainerman and Tataru (see [64]) or in quasilinear waves equations by Bahouri and the author (see [6–8]) and then by Tataru (see [82,83]), and by Klainerman and Rodnianski (see [63]).

For a detailed and complete exposition of the Littlewood–Paley theory, we refer to the book [72] by Runts and Sickel.

As announced in Section 2.5, we shall give a precise statement (in a particular case) of a theorem of Vishik proved in [86]. Of course, we use the notations of Section 2.

THEOREM 6.3. *Let v_0 be a vector field in E_m the vorticity of which ω_0 satisfies*

$$\forall N \geqslant -1, \quad \sum_{q\leqslant N}\|\Delta_q\omega_0\|_{L^\infty} \leqslant C\log(2+N). \quad (89)$$

Then a unique solution exists in $L^\infty(\mathbb{R}^+; E_m)$ such that, for any positive T, we have

$$\sup_{t\in[0,T],N}\frac{1}{(2+N)\log(2+N)}\sum_{q\leqslant N}\|\Delta_q\omega_0\|_{L^\infty} < +\infty.$$

This result is surprising because in [86] Vishik exhibits a function ω_0 which satisfies the above condition (89) such that ω_0 does not belongs to L^p for any p greater than some p_0. Thus we are in some sense far away from the Yudovich theorem (Theorem 2.2). The proof consists in a very sophisticated generalization of Osgood's lemma.

7. Some generalizations and other approaches for vortex patches problem

In this section, we want to mention different generalizations of Theorem 5.2 and also another approach of this problem.

7.1. *Generalization*

The generalizations mainly go into three directions. First the result about vortex patches, namely Theorem 5.2 has been generalized for three-dimensional incompressible fluid locally in time. The main theorem in this direction has been proved by Gamblin and Saint-Raymond in [53]. In order to state it, let us extend the concept of admissible family in the context of dimension three.

DEFINITION 7.1. Let $X = (X_j)_{1\leqslant j\leqslant N}$ a family of vector fields with coefficients and divergence in C^ε. This family is admissible if and only if

$$I(X) \stackrel{\text{def}}{=} \inf_{j\neq k;x} \left|X_j(x) \wedge X_k(x)\right| > 0.$$

This definition says that the system of vector fields in of rank at least 2 in a uniform way. The theorem proved in [53] can be stated as follows.

THEOREM 7.1. *Let v_0 be a divergence free vector field in $L^2(\mathbb{R}^3)$ such that its vorticity Ω_0 is bounded. Let X_0 be an admissible family in the sense of the above definition such that, for any $j \in \{1, \dots, N\}$,*

$$X_{0,j}(x, D)\Omega_0 \in C^{\varepsilon-1}.$$

Then a positive time T exists such that a unique solution v exists in $L^\infty([0, T; Lip \cap L^2)$. Moreover, if ψ denotes the flow of v, then for any j,

$$X_{0,j}(x, D)\psi \in L^\infty\bigl([0, T]\bigr).$$

If $X_{t,j} = \psi(t)_\star X_{0,j}$, then the family $X_t = (X_{t,\lambda})_{\lambda\in\Lambda}$ is admissible and we have

$$\left\|X_{0,j}(x, D)\Omega(t)\right\|_{\varepsilon-1} \in L^\infty\bigl([0, T]\bigr).$$

The second type of generalizations goes back to two-dimensional fluids and considers vortex patches, the boundary of which has singular points in the sense of the geometry (namely corners or cusps). We saw in Section 1 that when there are corners, there is no hope for the vector field to be Lipschitz. Nevertheless it is possible to control the set of the singularities along the evolution of the corresponding Yudovich solution. More precisely, the following theorem is proved in Chapter 9 of [35].

THEOREM 7.2. *Let D_0 be a bounded domain of the plane the boundary of which Γ_0 is a $C^{1+\varepsilon}$ curve outside a closed set Σ_0. Let us consider the Yudovich solution associated with the vector field v_0 in the space E_m the vorticity of which is $\mathbf{1}_{D_0}$. At time t, the domain $D(t) = \psi(t, D_0)$ has a boundary $\Gamma(t)$ which is a curve $C^{1+\varepsilon}$ outside the closed subset $\Sigma(t) = \psi(t, \Sigma_0)$.*

But a surprising observation as been done by Danchin in [42]. If the vorticity ω_0 is the characteristic function of a domain with cusp singularities then the associated vector field is Lipschitz. This observation has been possible thanks to numerical simulation done by Cohen and Danchin in [36]. A precise statement is a little bit technical. Thus we refer to [42] for further details.

The third type of generalizations considers the Euler system (E) in a bounded domain of $\mathbb{R}^2$ or $\mathbb{R}^3$ with boundary condition $v \cdot n = 0$. When the vortex patch does not meet the boundary, the results are like in $\mathbb{R}^d$. This has been proved by Depauw in [46] and Dutrifoy in [50].

When the vortex patch is tangent to the boundary, Depauw proved in [46] that the boundary of the vortex patch remains smooth for finite time. This situation is technically close to the situation treated by Danchin in [42].

These generalizations are definitely nontrivial and use an other approach we are going to explain now.

7.2. *The Lagrangian approach*

This approach already mentioned in Section 2.5 leads to consider the incompressible Euler system as an ordinary differential equation on the flow of the solution. This is the Lagrangian point of view. This has been followed in this context in particular by Huang in [57,58] and by Serfati in [74–77].

For instance, if ω_0 is the initial vorticity and ψ the flow, incompressible Euler system becomes

$$\frac{\mathrm{d}}{\mathrm{d}t}\psi(t,x) = \int_{\mathbb{R}^2} \frac{(\psi(t,x)-y)^\perp}{|\psi(t,x)-y|^2}\omega_0\big(\psi^{-1}(t,y)\big)\,\mathrm{d}y.$$

This allows Serfati to recover the global regularity for vortex patches in a very simple way in [74] in dimension two. In dimension three, Huang proved in [58] an extension of the Beale–Kato–Majda criteria in the case of vortex patch type solution of incompressible Euler system.

This approach related to Lagrangian point of view is simpler but more rigid. The fact that it follows particle paths will make this method inefficient for the study of viscous perturbations of incompressible Euler system. This is the purpose of the following last section.

8. The case of incompressible fluids with vanishing viscosity

In this section, we shall investigate the following problem. Let us consider v_ν $(\nu > 0)$, a time dependent vector field on $\mathbb{R}^2$ which satisfies the incompressible Navier–Stokes system

$$\begin{cases} \partial_t v_\nu + v_\nu \cdot \nabla v_\nu - \nu \Delta v_\nu = -\nabla p_\nu, \\ \operatorname{div} v_\nu = 0, \\ v_{\nu|t=0} = v^0. \end{cases} \tag{NS_ν}$$

Here, we look for the convergence of v_ν when ν tends to 0 and expect to find a result of strong convergence to the solution v of the incompressible Euler system (E) with the same initial data v^0. Of course, the results presented here are valid only in the case of the whole plane $\mathbb{R}^2$ because the problem of totally different in a domain with boundary because of boundary layers phenomena.

8.1. *The case when the vorticity is bounded*

Let us first recall the following classical Leray's theorem about bidimensional incompressible Navier–Stokes system.

THEOREM 8.1. *Let v_0 be in the space E_m for some real m. A unique solution v_ν of (NS_ν) exists in the space*

$$C(\mathbb{R}_+; E_m) \cap L^2_{\mathrm{loc}}(\mathbb{R}_+; H^1).$$

There is a fundamental theorem, proved by Delort in [45] about weak convergence for such families (v_ν).

THEOREM 8.2. *Assume that v_0 belongs to E_m and that the singular part (with respect to the Lebesgue measure) of ω^0 is a positive measure. Then some sequence $(\nu_n)_{n\in\mathbb{N}}$ converging to 0 and a solution v of the Euler system (E) belonging to the space $L^\infty_{\mathrm{loc}}(\mathbb{R}^+; E_m)$ exist so that*

$$\lim_{n\to\infty} v_{\nu_n} = v \quad \textit{weakly} \star \textit{ in the space } L^\infty_{\mathrm{loc}}(\mathbb{R}^+; E_m).$$

When the initial vorticity is bounded, the convergence is of course much stronger. It is described by the following theorem.

THEOREM 8.3. *Let v^0 in E_m. Let us denote by $(v_\nu)_{\nu\in\mathbb{R}_+^\star}$ the family of solution of (NS_ν) associated to the initial datum v^0 and by v the solution of (E) associated to v^0. Moreover, assume that $\omega^0 \in L^\infty \cap L^2$. Then*

$$\lim_{\nu\to 0} v_\nu = v \quad \textit{in the space } L^\infty_{\mathrm{loc}}(\mathbb{R}_+; E_m).$$

More precisely, we have the following estimate. Let T be a positive number; if

$$\nu \leqslant \frac{\mathrm{e}^{2-2\exp(C\|\omega^0\|_{L^\infty\cap L^2}T)}}{4T},$$

then we have

$$\|v_\nu - v\|^2_{L^\infty([0,T];L^2)} \leqslant (4\nu T)^{\exp(-C\|\omega^0\|_{L^2\cap L^\infty}T)}\|\omega^0\|_{L^2\cap L^\infty}\mathrm{e}^{2-2\exp(-C\|\omega^0\|_{L^2\cap L^\infty}T)}.$$

PROOF. The simplest idea is the good one to prove Theorem 8.3. Let us state $w_\nu = v_\nu - v$. By difference between (NS_ν) and (E), we get

$$\partial_t w_\nu + v_\nu \cdot \nabla w_\nu = -\nabla\tilde{p}_\nu + \nu\Delta v_\nu - w_\nu\cdot\nabla v.$$

So we have to estimate the L^2 norm of w_ν. Classical energy estimate in L^2 says that

$$\frac{\mathrm{d}}{\mathrm{d}t}\|w_\nu(t)\|^2_{L^2} = 2\nu\int \Delta v_\nu(t,x)\cdot w_\nu(t,x)\,\mathrm{d}x - 2\int (w_\nu\cdot\nabla v)(t,x)\cdot w_\nu(t,x)\,\mathrm{d}x. \tag{90}$$

Let us examine now the estimate about the family (v_ν). In the context of the two-dimensional Navier–Stokes system, the conservation of vorticity (V) turns out to be

$$\partial_t\omega_\nu + v_\nu\cdot\nabla\omega_\nu - \nu\Delta\omega_\nu = 0. \tag{V_ν}$$

So, by usual energy estimate in L^2, we have

$$\frac{\mathrm{d}}{\mathrm{d}t}\|\omega_\nu(t)\|^2_{L^2} + 2\nu\|\nabla\omega_\nu\|^2_{L^2} = 0.$$

In particular, we infer that, for any $\nu \geqslant 0$,

$$\|\omega_\nu(t)\|_{L^2} \leqslant \|\omega^0\|_{L^2}. \tag{91}$$

Then, thanks to the maximum principle, we have that

$$\|\omega_\nu(t)\|_{L^\infty} \leqslant \|\omega^0\|_{L^\infty}. \tag{92}$$

To estimate the right term in (90), we write that

$$\left|\int_{\mathbb{R}^2} \Delta v_\nu(t,x) w_\nu(t,x)\,\mathrm{d}x\right| \leqslant \|\nabla v_\nu(t)\|_{L^2}\|\nabla w_\nu(t)\|_{L^2}$$
$$\leqslant \|\omega_\nu(t)\|_{L^2}\big(\|\omega_\nu(t)\|_{L^2} + \|\omega(t)\|_{L^2}\big)$$
$$\leqslant 2\|\omega^0\|_{L^2}^2. \tag{93}$$

In order to prove Theorem 8.3, we have to estimate the term

$$J(t) \stackrel{\text{def}}{=} \int_{\mathbb{R}^2} |\nabla v(t,x)||w_\nu(t,x)|^2\,\mathrm{d}x$$

exactly as in the proof of Yudovich theorem as done in Section 2.3. Thus we get

$$\frac{\mathrm{d}}{\mathrm{d}t}\|w_\nu\|_{L^2}^2 \leqslant Ca\|\omega^0\|_{L^\infty\cap L^2}\|w_\nu(t)\|_{L^2}^2$$
$$+ Ca\|\omega^0\|_{L^\infty\cap L^2}^{1+2/a}\|w_\nu(t)\|_{L^2}^{2(1-1/a)} + 4\nu\|\omega^0\|_{L^2}^2. \tag{94}$$

Let us define the function $\delta_\nu(t)$ by

$$\delta_\nu(t) = \frac{\|w_\nu(t)\|_{L^2}^2}{\|\omega^0\|_{L^\infty\cap L^2}^2} + \delta,$$

where δ belongs to $]0,1[$. The inequality (94) can be written in the following way

$$\frac{\mathrm{d}}{\mathrm{d}t}\delta_\nu(t) \leqslant Ca\|\omega^0\|_{L^\infty\cap L^2}\big(\delta_\nu(t) + \delta_\nu(t)^{1-1/a}\big) + 4\nu.$$

Now, we use the fact that this inequality is valid for any a greater than 2. The following sequence of computations will be valid as long as $\delta_\nu(t) \leqslant 1$. If so, we have that

$$\delta_\nu(t) \leqslant \delta_\nu(t)^{1-1/a}.$$

Thus the above inequality becomes

$$\frac{\mathrm{d}}{\mathrm{d}t}\delta_\nu(t) \leqslant Ca\|\omega^0\|_{L^\infty\cap L^2}\delta_\nu(t)^{1-1/a} + 4\nu.$$

Now choosing $a = 2 - \log\delta_\nu(t)$, we get that

$$\delta_\nu'(t) \leqslant 4\nu + C\big(2 - \log\delta_\nu(t)\big)\|\omega^0\|_{L^\infty\cap L^2}\big(1 + \delta_\nu(t)^{-1/\log\delta_\nu(t)}\big)\delta_\nu(t).$$

It is obvious that $\delta_\nu(t)^{-1/\log\delta_\nu(t)} \leqslant \mathrm{e}^{-1}$. Then it turns out that

$$\delta_\nu'(t) \leqslant 4\nu + C\|\omega^0\|_{L^\infty\cap L^2}\big(2 - \log\delta_\nu(t)\big)\delta_\nu(t). \tag{95}$$

Let us define the function μ by $\mu(r) = r(2 - \log r)$. By integration, we get, as long as $\delta_\nu(t)$ is smaller than 1,

$$\delta_\nu(t) \leqslant 4\nu t + C\|\omega^0\|_{L^\infty \cap L^2} \int_0^t \mu(\delta_\nu(t'))\, d\tau.$$

Osgood's lemma (Lemma 2.3) allows to conclude the proof. □

Let us remark that if the solution of the Euler system (E) belongs to the space $L^1_{\mathrm{loc}}(\mathbb{R}; \mathit{Lip})$ (which is of course the case of the initial data of vortex patch type), we have very easily, for any t, the better estimate

$$\|v_\nu(t) - v(t)\|_{L^2}^2 \leqslant 4\nu\|\omega^0\|_{L^2}^2 \int_0^t \exp\left(2\int_\tau^t \|\nabla v(\tau')\|_{L^\infty}\, d\tau'\right) d\tau. \tag{96}$$

To prove this inequality, we simply observe that

$$J(t) \stackrel{\mathrm{def}}{=} \int_{\mathbb{R}^2} |\nabla v(t,x)| |w_\nu(t,x)|^2\, dx \leqslant 2\int_{\mathbb{R}^2} \|\nabla v(t,\cdot)\|_{L^\infty} \|w_\nu(t,\cdot)\|_{L^2}^2\, dt,$$

and then use Gronwall lemma.

In fact, as we shall see in the next section, much more can be done in this case considering the description of the convergence of the vorticity ω_ν to ω.

8.2. *Statement of the results*

THEOREM 8.4. *Let Ω_0 be a bounded domain of the plane, the boundary of which is a $C^{1+\varepsilon}$ curve. Let v_0 be the divergence free vector field in $E_{|\Omega_0|}$ the vorticity of which is the characteristic function of Ω_0. Then, for any positive ν, the equation (NS_ν) (resp. (E)) has a unique solution v_ν (resp. v) in the space $L^\infty_{\mathrm{loc}}(\mathbb{R}^+; \mathit{Lip}(\mathbb{R}^2) \cap E_{|\Omega_0|})$. A constant C exists such that*

$$\forall t \geqslant 0, \quad \sup_{\nu > 0} \|\nabla v_\nu(t)\|_{L^\infty} \leqslant C\exp(Ct) \quad \text{and} \quad \|\nabla v(t)\|_{L^\infty} \leqslant C\exp(Ct).$$

Let γ_0 in $C^{1+\varepsilon}(\mathbb{S}^1; \mathbb{R}^2)$ be a regular $C^{1+\varepsilon}$-parametrization of the boundary of Ω_0; let us state $\gamma_\nu(t) \stackrel{def}{=} \psi_\nu(t, \gamma_0)$ and $\gamma(t) \stackrel{def}{=} \psi(t, \gamma_0)$ where ψ_ν (resp. ψ) is the flow of v_ν (resp. v). Then we have:

- *for any positive ε' less than ε, the family $(\gamma_\nu)_{\nu > 0}$ is bounded in $L^\infty_{\mathrm{loc}}(\mathbb{R}^+; C)^{1+\varepsilon'}(\mathbb{S}^1; \mathbb{R}^2)$;*
- *for any positive ν and t positive, for any positive ε' less than ε, the function $\gamma_\nu(t)$ is a regular $C^{1+\varepsilon'}$-parametrization of the boundary of $\Omega_{t,\nu} \stackrel{def}{=} \psi_\nu(t, \Omega_0)$;*
- *for any positive T and any positive ε' less than ε, we have*

$$\lim_{\nu \to 0} \sup_{t \in [0,T]} \|\gamma_\nu(t) - \gamma(t)\|_{C^{1+\varepsilon'}} = 0.$$

Moreover, the vorticity ω_ν of v_ν converges to the vorticity ω of v is the following sense: for any positive h, t and ν, we have

$$\left\|\omega_\nu(t)\right\|_{L^2((\Omega_{t,\nu})^c_h)} \leqslant \|\omega_0\|_{L^2} \mathrm{e}^{-\frac{h^2}{4\nu t}\exp(-4(\mathrm{e}^{Ct}-1))} \tag{97}$$

and

$$\begin{aligned}&\left\|\omega_\nu(t) - \mathbf{1}_{\Omega_{t,\nu}}\right\|_{L^2((\Omega^c_{t,\nu})_h)}\\ &\quad\leqslant 2\|\omega_0\|_{L^2}\min\left\{1, C\frac{(\nu t)^{1/2}}{h}\mathrm{e}^{-\frac{h^2}{32\nu t}\exp(-4(\mathrm{e}^{Ct}-1))}\right\}\end{aligned} \tag{98}$$

with the definitions $(F)^c_h \stackrel{def}{=} \{x \in \mathbb{R}^2 / d(x, F) > h\}$ and $(F^c)_h \stackrel{def}{=} \{x \in \mathbb{R}^2, d(x, \partial F) > h)\}$.

As in the case of Euler equation, this theorem will be the consequence of another one, formulated in a much more geometrical way.

In all that follows, we shall consider (a, r) in $]2, +\infty] \times]0, 1[$ such that $r > \frac{2}{a}$. Let us first generalize a little bit Definitions 4.2 and 4.3.

DEFINITION 8.1. Let $X = (X_\lambda)_{\lambda \in \Lambda}$ be a family of vector fields. This family is B^r_a-admissible if and only if its coefficients and its divergence are B^r_a and

$$I(X) = \inf_{x \in \mathbb{R}^2} \sup_{\lambda \in \Lambda} \left|X_\lambda(x)\right| > 0.$$

If $a \leqslant b$, the space B^r_a is continuously embedded in $B^{r-2(1/a-1/b)}_b$, a B^r_a-admissible family is a $B^{r-2(1/a-1/b)}_b$-admissible family for any b greater than a; in particular, it is a $C^{r-2/a}$-admissible family. As in Section 4.2 we can define the concept of tangential regularity with respect to such a family in the framework of Besov spaces.

DEFINITION 8.2. Let X be a B^r_a-admissible family of vector fields. Let us denote by $B^r_a(X)$ the space of bounded functions u such that, for any $\lambda \in \Lambda$,

$$X_\lambda(x, D)u \stackrel{\text{def}}{=} \operatorname{div}(uX_\lambda) - u \operatorname{div} X_\lambda \in B^{r-1}_a.$$

Let us define the following quantities

$$N_{r,a}(X) \stackrel{\text{def}}{=} \sup_{\lambda \in \Lambda} \frac{\|X_\lambda\|_{B^r_a} + \|\operatorname{div} X_\lambda\|_{B^r_a}}{I(X)}$$

and

$$\|u\|^{r,a}_X \stackrel{\text{def}}{=} N_{r,a}(X)\|u\|_{L^\infty} + \sup_{\lambda \in \Lambda} \frac{\|X_\lambda(x, D)u\|_{B^{r-1}_a}}{I(X)}.$$

Let us notice that $N_{r,a}(X) \leqslant C N_{r-2/a}(X)$ and $\|u\|_X^{r-2/a} \leqslant C\|u\|_X^{r,a}$. Exactly as in Section 5.2, we can prove that Theorem 8.4 is included in the following one.

THEOREM 8.5. *Let v_0 be a divergence free vector field in E_m the vorticity of which ω_0 is in $B^r_a(X_0)$ for a given B^r_a-admissible family X_0 (with $a \in]2, +\infty[$). Let v_ν (resp. v) be the solution of (NS_ν) (resp. (E)) with initial data v_0 and ψ_ν (resp. ψ) the flow of v_ν (resp. v). Let us state $X_{t,\nu,\lambda} \stackrel{def}{=} \psi_\nu(t)^\star X_{0,\lambda}$ and $X_{t,\lambda} \stackrel{def}{=} \psi(t)^\star X_{0,\lambda}$. Then the families $X_\nu(t)$ and $X(t)$ are B^r_a-admissible, the vorticity $\omega_\nu(t)$ (resp. $\omega(t)$) belongs to $B^r_a(X_\nu(t)$ (resp. $B^r_a(X(t))$. Moreover, a constant C (which depends only on the initial data v_0 and the family X_0) exists such that*

$$\begin{aligned}
&\|X_{0,\lambda}(x, D)\psi_\nu(t)\|_{B^r_a} + \|X_{0,\lambda}(x, D)\psi(t)\|_{B^r_a} \leqslant C e^{\exp(Ct)},\\
&N^r_a(X_\nu(t)) + N^r_a(X(t)) \leqslant C e^{\exp(Ct)},\\
&\|\omega_\nu(t)\|^{r,a}_{X_\nu(t)} + \|\omega(t)\|^{r,a}_{X(t)} \leqslant C e^{\exp(Ct)}
\end{aligned}$$

and thus

$$\|\nabla v_\nu(t)\|_{L^\infty} + \|\nabla v(t)\|_{L^\infty} \leqslant C e^{Ct}.$$

We also have the following convergences:

$$\forall \rho < 1,\ \forall T > 0, \quad \lim_{\nu\to 0} \sup_{t\in[0,T],\lambda\in\Lambda} \|v_\nu(t) - v(t)\|_{C^\rho} = 0,$$

$$\begin{aligned}
\forall r' < r,\ \forall T > 0, \quad \lim_{\nu\to 0} \sup_{t\in[0,T],\lambda\in\Lambda} &\|X_{0,\lambda}(x, D)\psi_\nu(t)\\
&- X_{0,\lambda}(x, D)\psi(t)\|_{B^{r'}_a} = 0,
\end{aligned}$$

$$\begin{aligned}
\forall r' < r,\ \forall T > 0, \quad \lim_{\nu\to 0} \sup_{t\in[0,T],\lambda\in\Lambda} &\big(\|X_{t,\nu,\lambda} - X_{t,\lambda}\|_{B^{r'}_a}\\
&+ \|\operatorname{div} X_{t,\nu,\lambda} - \operatorname{div} X_{t,\lambda}\|_{B^{r'}_a}\big) = 0
\end{aligned}$$

and

$$\begin{aligned}
\forall r' < r,\ \forall T > 0, \quad \lim_{\nu\to 0} \sup_{t\in[0,T],\lambda\in\Lambda} &\|X_{t,\nu,\lambda}(x, D)\omega_\nu(t)\\
&- X_{t,\lambda}(x, D)\omega(t)\|_{B^{r'}_a} = 0.
\end{aligned}$$

Inequalities (97) and (98) come from estimates about transport–diffusion equation for vanishing viscosity because of the equation

$$\partial_t\omega_\nu + v_\nu \cdot \nabla\omega_\nu - \nu\Delta\omega_\nu = 0 \tag{V_ν}$$

satisfied by the vorticity of v_ν. The study of equations of this type is also crucial for the proof of Theorem 8.5. This is the purpose of the next section.

8.3. *Transport–diffusion equations*

The aim of this section is the proof of estimates about the solutions of transport–diffusion equations, namely about solution of equation of the type

$$\begin{cases} \partial_t u + v \cdot \nabla u - \nu \Delta u = 0, \\ u_{|t=0} = u_0, \end{cases} \qquad (T_\nu)$$

when v is a divergence free vector field which belongs to $L^1_{\mathrm{loc}}(\mathbb{R}^+; Lip)$. More precisely, we have the following theorem.

THEOREM 8.6. *A constant C exists which satisfies the following properties. Let v be a divergence free vector field which belongs to $L^1_{\mathrm{loc}}(\mathbb{R}^+; Lip)$ and ν a positive real number. Let us consider a function u which is solution of the equation (T_ν), i.e.,*

$$\begin{cases} \partial_t u + v \cdot \nabla u - \nu \Delta u = 0, \\ u_{|t=0} = u_0, \end{cases} \qquad (T_\nu)$$

where u_0 is a compactly supported function in $L^2(\mathbb{R}^2)$. Let us denote by ψ the flow of v and let us state

$$F_t \stackrel{def}{=} \psi\big(t, \mathrm{Supp}(u_0)\big),$$

$$(F_t)^c_h \stackrel{def}{=} \big\{x \in \mathbb{R}^2 \mid d(x, F_t) > h\big\},$$

$$\big(F^c_t\big)_h \stackrel{def}{=} \big\{x \in F_t \mid d(x, \partial F_t) > h\big\}$$

and

$$V(t) \stackrel{def}{=} \int_0^t \big\|\nabla v(t')\big\|_{L^\infty} \, \mathrm{d}t'.$$

Then we have, for any $(t, h) \in \mathbb{R}^+ \times \mathbb{R}^+$,

$$\big\|u(t)\big\|_{L^2((F_t)^c_h)} \leqslant \|u_0\|_{L^2} \mathrm{e}^{-\frac{h^2}{4\nu t}\exp(-4V(t))}. \qquad (99)$$

Moreover, if ω_0 is the characteristic function of a bounded domain F_0, then

$$\big\|u(t) - \mathbf{1}_{F_t}\big\|_{L^2((F^c_t)_h)}$$
$$\leqslant \|u_0\|_{L^2} \min\left\{1, C\left(\frac{\nu t}{h^2}\right)^{1/2} \mathrm{e}^{2V(t) - \frac{h^2}{4\nu t}\exp(-4V(t))}\right\}. \qquad (100)$$

PROOF. The proof of this theorem relies on energy estimates. Using regularization arguments we may assume that the vector field v and the function u are smooth. Let us consider a smooth function Φ_0, let us denote by ψ the flow of v and let us state

$$\Phi(t,x) \stackrel{\text{def}}{=} \Phi_0\big(\psi^{-1}(t,x)\big).$$

It is obvious that

$$\partial_t(\Phi u) + v\cdot\nabla(\Phi u) - \nu\Delta(\Phi u) = -\nu u\Delta\Phi - 2\nu\nabla\Phi\cdot\nabla u.$$

Taking the L^2 scalar product with Φu and doing integrations by parts gives

$$\frac{1}{2}\frac{\mathrm{d}}{\mathrm{d}t}\|\Phi u\|_{L^2}^2 + \nu\big\|\nabla(\Phi u)\big\|_{L^2}^2 = \nu\|u\nabla\Phi\|_{L^2}^2.$$

Let us choose $\Phi(t,x) = \exp(\phi(t,x))$ with $\phi(t,x) = \phi_0(\psi^{-1}(t,x))$. From the above relation, we get that

$$\frac{\mathrm{d}}{\mathrm{d}t}\|\Phi u\|_{L^2}^2 \leqslant 2\nu\|\nabla\phi\|_{L^\infty}^2\|\Phi u\|_{L^2}^2.$$

Using Gronwall lemma we infer that

$$\big\|(\Phi u)(t)\big\|_{L^2} \leqslant \big\|(\Phi u)(0)\big\|_{L^2}\exp\left(\|\nabla\phi_0\|_{L^\infty}^2\int_0^t \big\|\nabla\psi^{-1}(t',\cdot)\big\|_{L^\infty}\,\mathrm{d}t'\right).$$

Let us choose

$$\phi_0(x) \stackrel{\text{def}}{=} \alpha\min\big\{R, d\big(x, \mathrm{Supp}(u_0)\big)\big\}.$$

Let us remark that with this choice, we have $(\Phi u)(0,x) = u_0(x)$. Using that $\|\nabla\phi(t,\cdot)\|_{L^\infty} = \alpha\exp V(t)$, we get by Gronwall lemma that

$$\big\|\Phi u(t)\big\|_{L^2} \leqslant \|u_0\|_{L^2}\mathrm{e}^{\nu\alpha^2 t\exp(V(t))}.$$

By definition of Φ, it turns out that, if $0 < \eta \leqslant R$, we have

$$\mathrm{e}^{\alpha\eta}\big\|u(t)\big\|_{L^2(\psi_t(F_0)_h^c} \leqslant \|u_0\|_{L^2}\mathrm{e}^{\nu\alpha^2 t\exp(V(t))}.$$

But obviously,

$$(F_t)_h^c \subset \phi_t(F_0)_{\delta(t,h)}^c \quad \text{with } \delta(t,h) \stackrel{\text{def}}{=} \frac{h}{\|\nabla\psi_t\|_{L^\infty}}. \tag{101}$$

Thus if $\delta(t,h) \leqslant R$, we have

$$\|u(t)\|_{L^2((F_t)_h^c)} \leqslant \|u_0\|_{L^2} e^{\nu\alpha^2 t \exp(V(t)) - \alpha h \exp(-V(t))}.$$

As the above inequality is independent of R, it is true for any (t,h); then the best choice for α gives inequality (99).

The proof of (100) follows essentially the same lines. Let us state $w(t,x) = u(t,x) - \mathbf{1}_{F_t}(x)$ and $\Phi(t,x) = \Phi_0(\psi^{-1}(t,x))$ with Φ_0 in $\mathcal{D}(F_0)$. Then we have

$$(\partial_t + v\cdot\nabla - \nu\Delta)(\Phi w) = -\nu w\Delta\Phi - 2\nu\nabla\Phi\cdot\nabla w.$$

As above, we get by energy estimate that

$$\frac{1}{2}\frac{\mathrm{d}}{\mathrm{d}t}\|\Phi w\|_{L^2}^2 + \nu\|\nabla(\Phi w)\|_{L^2}^2 = \nu\|w\nabla\Phi\|_{L^2}^2.$$

A constant C exists such that for any positive h_0, a function χ exists in $\mathcal{D}(F_0)$ such that χ is identically 1 on $(F_0^c)_{h_0}$ and $\|\nabla\chi\|_{L^\infty} \leqslant Ch_0^{-1}$. Then choosing $\Phi_0 = \chi e^{\phi_0}$, where ϕ_0 is equal to (a regularization of) the function $x \mapsto d(x, F_0)$, we get that

$$\frac{\mathrm{d}}{\mathrm{d}t}\|\Phi w\|_{L^2}^2 \leqslant \nu e^{2V(t)}\left(4\alpha^2\|\Phi w\|_{L^2}^2 + \frac{C}{h_0^2}\|u_0\|_{L^2}^2\right).$$

Using (101) with $F_t = F_t^c$, we get that for any t and h such that $he^{-V(t)} \geqslant h_0$, we have

$$\|w(t)\|_{L^2((F_t^c)_h)}^2 \leqslant C\|u_0\|_{L^2}^2 \frac{e^{2\alpha h_0 - h\exp(-V(t))}}{\alpha^2 h_0}\big(e^{4\alpha^2 t\exp(2V(t))} - 1\big).$$

Now using the fact that $e^{-x}(e^{x/2} - 1) \leqslant e^{-x/2}$ and choosing

$$h_0 = \frac{he^{-V(t)}}{2} \quad \text{and} \quad \alpha = \frac{he^{-3V(t)}}{8\nu t}$$

gives the result. □

The propagation of regularity with a control on ν is described by the following theorem.

THEOREM 8.7. *Let r be in the interval $]-1,1[$ and a be a real number greater than or equal to 2. A constant C exists which satisfies the following properties. Let ν be any positive real number, v a divergence free (time dependent) smooth vector field the derivatives of which are bounded, u a function belonging to $L^\infty_{\mathrm{loc}}(\mathbb{R}^+; B^r)$, f (resp. g) a function belonging to $L^\infty_{\mathrm{loc}}(\mathbb{R}^+; B^r_a)$ (resp. to $L^\infty_{\mathrm{loc}}(\mathbb{R}^+; B^{r-2}_a)$). Let us assume that*

$$\partial_t u + v\cdot\nabla u - \nu\Delta u = f + \nu g. \qquad (T'_\nu)$$

Then we have

$$\|u(t)\|_{B^r_a} \leqslant C\mathrm{e}^{CV(t)}\Big(\|u(0)\|_{B^r_a} + \|g\|_{L^\infty([0,T];B^{r-2}_a)} + \int_0^t \mathrm{e}^{-V(t')}\big(\|f(t')\|_{B^r_a} + \nu\|g(t')\|_{B^{r-2}_a}\big)\,\mathrm{d}t'\Big)$$

with $V(t) \stackrel{\text{def}}{=} \int_0^t \|\nabla v(t',\cdot)\|_{L^\infty}\,\mathrm{d}t'$.

If we have, in addition, that for any t*, the support of the Fourier transform of* $g(t,\cdot)$ *is included in* $\mathbb{R}^d \setminus B(0,\lambda_0)$ *for some fixed* λ_0*, we have*

$$\|u(t)\|_{B^r_a} \leqslant C\mathrm{e}^{CV(t)}\Big(\|u(0)\|_{B^r_a} + \|g\|_{L^\infty([0,T];B^{r-2}_a)} + \int_0^t \mathrm{e}^{-V(t')}\|f(t')\|_{B^r_a}\,\mathrm{d}t'\Big).$$

PROOF. The proof of this theorem relies of course on localization in frequency space. For the sake of simplicity let us assume that a is an even integer and let us state $a = 2p$. Let us apply the operator Δ_q to the equation (V'_ν). Stating $u_q \stackrel{\text{def}}{=} \Delta_q u$, we get that

$$\begin{aligned}
&(\partial_t + v\cdot\nabla - \nu\Delta)u_q = f_q + \nu g_q \quad \text{with}\\
&f_q \stackrel{\text{def}}{=} \Delta_q f + [v\cdot\nabla, \Delta_q]u \quad \text{and}\\
&g_q \stackrel{\text{def}}{=} \Delta_q g.
\end{aligned} \tag{102}$$

Using the chain rule, we can write that

$$(\partial_t + v\cdot\nabla - \nu\Delta)u_q^p = pu_q^{p-1}(f_q + \nu g_q) + \nu p(p-1)|\nabla u_q|^2 u_q^{p-2}.$$

By energy estimate we infer that

$$\frac{1}{2}\frac{\mathrm{d}}{\mathrm{d}t}\|u_q(t)\|^{2p}_{L^{2p}} + \frac{\nu}{p}\|\nabla(u_q(t))^p\|^2_{L^2} \leqslant p\big(\|f_q\|_{L^{2p}} + \nu\|g_q(t)\|_{L^{2p}}\big)\|u_q(t)\|^{2p-1}_{L^{2p}}.$$

Applying Lemma 6.2, we get that

$$\frac{1}{2}\frac{\mathrm{d}}{\mathrm{d}t}\|u_q(t)\|^{2p}_{L^{2p}} + c_p\nu 2^{2q}\|u_q(t)\|^{2p}_{L^{2a}} \leqslant p\big(\|f_q\|_{L^{2p}} + \nu\|g_q(t)\|_{L^{2p}}\big)\|u_q(t)\|^{2p-1}_{L^{2p}}.$$

Using Gronwall-type inequality, we get that

$$\|u_q(t)\|_{L^a} \leqslant \|\Delta_q u_0\|_{L^a} + \int_0^t e^{c_p \nu 2^{2q}(t-t')}\left(\|f_q(t')\|_{L^a} + \nu\|g_q(t')\|_{L^a}\right) dt'.$$

Lemma 6.3 implies that

$$\|f_q(t)\|_{L^a} \leqslant 2^{-qr}\left(\|f(t)\|_{B^r_a} + C\|\nabla v(t)\|_{L^\infty}\|u(t)\|_{B^r_a}\right). \tag{103}$$

Thus we can write that

$$\begin{aligned}\|u_q(t)\|_{L^a} \leqslant{}& \|\Delta_q u_0\|_{L^a} + C\nu 2^{-2q}\|g_q(t')\|_{L^\infty([0,t];L^a)} \\ &+ 2^{-qr}\int_0^t \left(\|f(t')\|_{B^r_a} + \|\nabla v(t',\cdot)\|_{L^\infty}\|u(t')\|_{B^r_a}\right) dt'.\end{aligned}$$

Multiplying by 2^{qr} and taking the supremum in q gives the results by Gronwall lemma (skipping the case of low frequencies). □

8.4. *The structure of the proof*

The basic idea is the same as in the case when $\nu = 0$. For the convenience of the reader, we shall recall definition (45) and properties (46), (47), (50) and (52) in the framework of Besov spaces

$$X_{t,\lambda}(x) \stackrel{\text{def}}{=} \psi_\star(t)X_{0,\lambda}(x) = \left(X_{0,\lambda}(x,D)\psi(t)\right)\left(\psi^{-1}(t,x)\right), \tag{104}$$

$$\partial_t X_{t,\lambda} + v\cdot\nabla X_{t,\lambda} = X_{t,\lambda}(x,D)v, \tag{105}$$

$$I(X_t) \geqslant I(X_0)\exp\left(-\int_0^t \|\nabla v(t')\|_{L^\infty} dt'\right), \tag{106}$$

$$\|\operatorname{div} X_{t,\lambda}\|_{B^r_a} \leqslant \|\operatorname{div} X_{0,\lambda}\|_{B^r_a}\exp\left(C\int_0^t \|\nabla v(t')\|_{L^\infty} dt'\right). \tag{107}$$

Of course, we want to use Theorem 4.1 which claims thanks to Proposition 6.8 that

$$\|\nabla v\|_{L^\infty} \leqslant C\|\omega\|_{L^a} + C\|\omega\|_{L^\infty}\log\left(e + \frac{\|\omega\|^{r,a}_X}{\|\omega\|_{L^\infty}}\right).$$

The main difference (and difficulty) is that (48) is not valid anymore. The major problem is the following: If we apply the vector field $X_{t,\nu,\lambda}$ to the equation (V_ν), we get

$$\partial_t X_{t,\lambda}(x,D)\omega + v\cdot\nabla X_{t,\lambda}(x,D)\omega - \nu X_{t,\lambda}(x,D)\Delta\omega = 0.$$

But the coefficients of the vector field $X_{t,\nu,\lambda}$ are only in B^r_a. Thus it is impossible to estimate (or even to define) their product with second-order derivatives of ω which is simply bounded. The way to bypass this difficulty is the use of paradifferential calculus. Propositions 6.9 and 6.10 imply that

$$\left\|X_{t,\nu,\lambda}(x,D)\omega - T_{X_{t,\nu,\lambda}}\cdot\nabla\omega\right\|_{B^{r-1}_a} \leqslant C\|\omega\|_{L^\infty}\big(\|X_{t,\nu,\lambda}\|_{B^r_a} + \|\operatorname{div} X_{t,\nu,\lambda}\|_{B^r_a}\big)$$

with

$$T_{X_{t,\nu,\lambda}}\cdot\nabla\omega \overset{\text{def}}{=} \sum_{j=1}^{2} T_{X^j_{t,\nu,\lambda}}\partial_j\omega.$$

Thus it is enough to estimate $\|\sum_{j=1}^2 T_{X^j_{t,\nu,\lambda}}\partial_j\omega\|_{B^{r-1}_a}$. The following difficult lemma is proved in [40].

LEMMA 8.1. *We have the following identity*

$$\partial_t(T_{X_{t,\nu,\lambda}}\cdot\nabla\omega) + v\cdot\nabla(T_{X_{t,\nu,\lambda}}\cdot\nabla\omega) - \nu X_{t,\lambda}(x,D)\Delta(T_{X_{t,\nu,\lambda}}\cdot\nabla\omega) = f + \nu g,$$

where f and g satisfy the following inequalities

$$\|f\|_{B^{r-1}_a} \leqslant C\big(\|T_{X_{t,\nu,\lambda}}\cdot\nabla\omega\|_{B^{r-1}_a} + \|\nabla v\|_{L^\infty} N_{r,a}(X_{t,\nu}) I(X_{t,\nu})\big)$$

and

$$\|g\|_{B^{r-3}_a} \leqslant C N_{r,a}(X_{t,\nu})\|\omega\|_{L^\infty}.$$

Then the proof goes on as in the case of inviscid fluid. The complete details are done in [40].

8.5. *Remarks and references*

Theorem 8.3 has been proved by the author in [34]. Inequality (96) has been proved by Constantin and Wu in [39] in the very particular case when the initial datum is of vortex patches type, i.e., initial datum the vorticity of which is the characteristic function of a bounded domain with smooth boundary. This case is a very particular case of general vortex patches.

Let us notice that when the initial data is smooth enough (i.e., belongs to some Sobolev space H^s with s strictly greater than 2), then the rate of convergence of the L^2 norm is ν instead of $\nu^{1/2}$ in (96) (see [70]).

Theorems 8.5–8.7 have been proved by Danchin in [40]. Then it has been generalized in the case of three-dimensional axisymmetric flows by Ben Ameur and Danchin in [14] and (locally in time) by Danchin in [41].

References

[1] S. Alinhac, *Interaction d'ondes simples pour des équations complètement non linéaires*, Ann. Sci. École Norm. Sup. **21** (1988), 91–133.

[2] S. Alinhac, *Un problème de concentration évanescente pour les flots non stationnaires et incompressibles en dimension deux*, Comm. Math. Phys. **127** (1990), 585–596.

[3] S. Alinhac, *Remarques sur l'instabilité du problème des poches de tourbillon*, J. Funct. Anal. **98** (1991), 361–379.

[4] S. Alinhac and P. Gérard, *Opérateurs pseudo-différentiels et théorème de Nash–Moser*, *Savoirs actuels*, Interéditions, Paris (1991).

[5] H. Bahouri and J.-Y. Chemin, *Équations de transport relatives à des champs de vecteurs non-lipschitziens et mécanique des fluides*, Arch. Ration. Mech. Anal. **127** (1994), 159–182.

[6] H. Bahouri and J.-Y. Chemin, *Équations d'ondes quasilinéaires et inégalités de Strichartz et équations d'ondes quasilinéaires*, Amer. J. Math. **121** (1999), 1337–1377.

[7] H. Bahouri and J.-Y. Chemin, *Équations d'ondes quasilinéaires et effet dispersif*, Int. Math. Res. Not. **21** (1999), 1141–1178.

[8] H. Bahouri and J.-Y. Chemin, *Microlocal analysis, bilinear estimates and cubic quasilinear wave equation*, Astéristique **284** (2003), 93–141.

[9] H. Bahouri and B. Dehman, *Remarques sur l'apparition de singularités dans les écoulements Eulériens incompressibles à données initiales Hölderiennes*, J. Math. Pures Appl. **73** (1994), 335–346.

[10] S. Baouendi and C. Goulaouic, *Problèmes de Cauchy pseudo-différentiels analytiques*, Séminaire Goulaouic–Schwartz, École polytechnique (1975–1976).

[11] C. Bardos and S. Benachour, *Domaine d'analyticité des solutions de l'équation d'Euler dans un ouvert de* $\mathbb{R}^n$, Ann. Sc. Norm. Sup. Pisa **4** (1977), 647–687.

[12] J. Beale, T. Kato and A. Majda, *Remarks on the breakdown of smoothness for the 3-D Euler equations*, Comm. Math. Phys. **94** (1984), 61–66.

[13] S. Benachour, *Analyticité des solutions des équations d'Euler*, Arch. Ration. Mech. Anal. **71** (1979), 271–299.

[14] J. Ben Ameur and R. Danchin, *Limite non visqueuse pour des fluides incompressibles axisymétriques*, Nonlinear Partial Differential Equations and Their Applications. Collège de France Seminar, Vol. XIV (Paris, 1997/1998), Stud. Math. Appl., Vol. 31, North-Holland, Amsterdam (2002), 29–55.

[15] A. Bertozzi and P. Constantin, *Global regularity for vortex patches*, Comm. Math. Phys. **152** (1993), 19–28.

[16] J.-M. Bony, *Propagation des singularités pour des équations aux dérivées partielles non linéaires*, Sémin. Équ. Dériv. Partielles (1979–1980).

[17] J.-M. Bony, *Calcul symbolique et propagation des singularités pour les équations aux dérivées partielles non linéaires*, Ann. Sci. École Norm. Sup. **14** (1981), 209–246.

[18] J.-M. Bony, *Analyse microlocale et singularités non linéaires*, Nonlinear Hyperbolic Problems, Proceedings Bordeaux, Lecture Notes in Math., Vol. 1042, Springer-Verlag, New York (1988), 114–127.

[19] T. Buttke, *The observation of singularities in the boundary of patches of vorticity*, Phys. Fluids A **1** (7) (1989), 1283–1285.

[20] T. Buttke, *A fast adaptive vortex method for patches of constant vorticity in two dimensions*, J. Comput. Phys. **89** (1990), 161–186.

[21] M. Cannone, *Ondelettes, paraproduits et Navier–Stokes*, Diderot, Paris (1995).

[22] M. Cannone, Y. Meyer and F. Planchon, *Solutions autosimilaires des équations de Navier–Stokes*, Séminaire Équations aux Dérivées Partielles de l'École polytechnique (1993–1994).

[23] E. Chacón Vera, *Analyticity for vortex patches*, Adv. Math. Sci. Appl. **9** (1999), 571–580.

[24] J.-Y. Chemin, *Calcul paradifférentiel précisé et application à des équations aux dérivées partielles non linéaires*, Duke Math. J. **56** (1988), 431–469.

[25] J.-Y. Chemin, *Régularité de la solution d'un problème de Cauchy non linéaire à données singulières en un point*, Ann. Inst. Fourier **39** (1989), 101–122.

[26] J.-Y. Chemin, *Évolution d'une singularité ponctuelle dans des équations strictement hyperboliques non linéaires*, Amer. J. Math. **112** (1990), 805–860.

[27] J.-Y. Chemin, *Évolution d'une singularité ponctuelle dans un écoulement compressible*, Comm. Partial Differential Equations **15** (1990), 1237–1263.

[28] J.-Y. Chemin, *Remarques sur l'apparition de singularités fortes dans les écoulements compressibles*, Comm. Math. Phys. **133** (1990), 323–329.

[29] J.-Y. Chemin, *Sur le mouvement des particules d'un fluide parfait, incompressible, bidimensionnel*, Invent. Math. **103** (1991), 599–629.

[30] J.-Y. Chemin, *Persistance des structures géométriques liées aux poches de tourbillon*, Sémin. Équ. Dériv. Partielles (1990–1991).

[31] J.-Y. Chemin, *Régularité des trajectoires des particules d'un fluide incompressible remplissant l'espace*, J. Math. Pures Appl. **71** (1992), 407–417.

[32] J.-Y. Chemin, *Persistance de structures géométriques dans les fluides incompressibles bidimensionnels*, Ann. Sci. Ecole Norm. Sup. **26** (1993), 1–26.

[33] J.-Y. Chemin and N. Lerner, *Flot de champs de vecteurs non-lipschitziens et équations de Navier–Stokes*, J. Differential Equations **121** (1995), 314–328.

[34] J.-Y. Chemin, *A remark on the inviscid limit for two-dimensional incompressible fluid*, Comm. Partial Differential Equations **21** (1996), 1771–1779.

[35] J.-Y. Chemin, *Fluides parfaits incompressibles*, Astérisque **230** (1995); English transl.: *Incompressible Perfect Fluids*, Oxford Univ. Press (1998).

[36] A. Cohen and R. Danchin, *Multiscale approximation of vortex patches*, SIAM J. Appl. Math. **60** (2000), 477–502.

[37] R. Coifman and Y. Meyer, *Au delà des opérateurs pseudo-différentiels*, Astérisque **57** (1978).

[38] P. Constantin and E. Titi, *On the evolution of nearly circular vortex patches*, Comm. Math. Phys. **119** (1988), 177–198.

[39] P. Constantin and J. Wu, *Inviscid limit of vortex patches*, Nonlinearity **8** (1995), 735–742.

[40] R. Danchin, *Poches de tourbillon visqueuses*, J. Math. Pures Appl. **76** (1997), 609–647.

[41] R. Danchin, *Persistance de structures géométriques et limite non visqueuse pour les fluides incompressibles en dimension quelconque*, Bull. Soc. Math. France **127** (1999), 179–227.

[42] R. Danchin, *Évolution d'une singularité de type cusp dans une poche de tourbillon*, Rev. Mat. Iberoamericana **16** (2000), 281–329.

[43] P. Delacour, *Poches de tourbillon régulières en dimension* 3, Ph.D. Thesis, University of Nantes (2001).

[44] J.-M. Delort, *Estimations fines pour des opérateurs pseudo-differentiels analytiques sur un ouvert à bord de* $\mathbb{R}^n$. *Application aux équations d'Euler*, Comm. Partial Differential Equations **10** (1985), 1465–1525.

[45] J.-M. Delort, *Existence de nappes de tourbillon en dimension deux*, J. Amer. Math. Soc. **4**. (1991), 553–586.

[46] N. Depauw, *Poche de tourbillon pour Euler* 2*D dans un ouvert à bord*, J. Math. Pures Appl. **78** (1999), 313–351.

[47] R. DiPerna and P.-L. Lions, *Ordinary differential equations, transport theory and Sobolev spaces*, Invent. Math. **98** (3) (1989), 511–549.

[48] D. Dritschel, *Contour dynamics and contour surgery: Numerical algorithms for extended, high-resolution modeling of vortex dynamics in two-dimensional inviscid incompressible flow*, Comput. Phys. Reports **10** (1989), 77–146.

[49] D. Dritschel and M. MacIntyre, *Does contour dynamics go singular*? Phys. Fluids A **2** (1990), 748–753.

[50] A. Dutrifoy, *On* 3*-D vortex patches in bounded domains*, submitted.

[51] T.M. Fleet, *Differential Analysis*, Cambridge Univ. Press (1980).

[52] P. Gamblin, *Système d'Euler incompressible et régularité microlocale analytique*, Ann. Inst. Fourier **44** (1994), 1449–1475.

[53] P. Gamblin and X. Saint-Raymond, *On three-dimensional vortex patches*, Bull. Soc. Math. France **123** (1995), 375–424.

[54] P. Gérard, *Résultats récents sur les fluides parfaits incompressibles bidimensionnels* (d'après J.-Y. Chemin et J.-M. Delort), Séminaire Bourbaki **757** (1992).

[55] N. Hanges and F. Trêves, *On the analyticity of solutions of first order non linear partial differential equations*, Trans. Amer. Math. Soc. **331** (1992), 627–638.

[56] L. Hörmander, *The Analysis of Linear Partial Differential Equations*, Springer-Verlag, New York (1983).

[57] C. Huang, *Remarks on regularity of non-constant vortex patches*, Comm. Appl. Anal. **3** (1999), 449–459.

[58] C. Huang, *Singular integral system approach to regularity of* 3*D vortex patches*, Indiana Univ. Math. J. **50** (2001), 509–552.

[59] D. Iftimie, P. Gamblin and T. Sideris, *On the evolution of compactly supported planar vorticity*, Comm. Partial Differential Equations **24** (1999), 1709–1730.

[60] D. Iftimie, M.C. Lopes Filho and H.J. Nussenzveig Lopes, *Comportement en temps grand pour les écoulements parfaits incompressibles dans un demi-plan*, Sém. Équ. Dériv. Partielles (2001–2002).

[61] T. Kato and G. Ponce, *Well-posedness of the Euler and Navier–Stokes equations in the Lebesgue spaces* $L_s^p(\mathbb{R}^2)$, Rev. Mat. Iberoamericana **2** (1986), 73–88.

[62] S. Klainerman and I. Rodnianski, *Rough solutions of the Einstein vacuum equations*, C. R. Math. Acad. Sci. Paris **334** (2) (2002), 125–130.

[63] S. Klainerman and I. Rodnianski, *Improved regularity results for the quasilinear wave equations*, Duke Math. J. **117** (2003), 1–124.

[64] S. Klainerman and D. Tataru, *On the optimal local regularity for the Yang–Mills equations in* $\mathbb{R}^{4+1}$, J. Amer. Math. Soc. **12** (1999), 93–116.

[65] P.-G. Lemarié-Rieusset, *Recent Developments in the Navier–Stokes Problem*, Chapman and Hall/CRC Res. Notes Math., Vol. 431, Chapman and Hall/CRC, Boca Raton, FL (2002).

[66] J. Leray, *Étude de diverses équations intégrales non linéaires et quelques problèmes que pose l'hydrodynamique*, J. Math. Pures Appl. **9** (12) (1933), 1–82.

[67] L. Lichtenstein, *Über einige Existenzprobleme der Hydrodynamik homogenerunzusammendrückbarer, reibungsloser Flüßigkeiten und die Helmoltzochen Wirbelsatze*, Math. Z. **23** (1925), 89–154; **26** (1927), 196–323; **28** (1928), 387–415; **32** (1930), 608–725.

[68] A. Majda, *Vorticity and the mathematical theory of an incompressible fluid flow*, Comm. Pure Appl. Math. **38** (1986), 187–220.

[69] C. Marchioro, *Bounds on the growth of the support of a vortex patch*, Comm. Math. Phys. **164** (1994), 507–524.

[70] F. McGrath, *Nonstationary plane flow of viscous and ideal fluids in* $\mathbb{R}^3$, Arch. Ration. Mech. Anal. **4** (1968), 329–438.

[71] Y. Meyer, *Ondelettes et Opérateurs*, Tome 2, Herman, Paris (1991).

[72] T. Runts and W. Sickel, *Sobolev Spaces of Fractionnal Order, Nemytskij Operators and Non Linear Partial Differential Equations*, de Gruyter Series in Non Linear Analysis and Applications, Vol. 3, de Gruyter, Berlin (1996).

[73] X. Saint-Raymond, *Remarks on axisymmetric solutions of the incompressible Euler system*, Comm. Partial Differential Equations **19** (1994), 321–334.

[74] P. Serfati, *Une preuve directe d'existence globale des vortex patches 2D*, C. R. Acad. Sci. Paris Sér. I **318** (1994), 515–518.

[75] P. Serfati, *Régularité stratifiée et équation d'Euler à temps grand*, C. R. Acad. Sci. Paris Sér. I **318** (1994), 925–928.

[76] P. Serfati, *Équation d'Euler et holomorphies à faible régularité spatiale*, C. R. Acad. Sci. Paris Sér. I **320** (1995), 175–180.

[77] P. Serfati, *Structures holomorphes à faible régularité spatiale en mécanique des fluides*, J. Math. Pures Appl. **74** (1995), 95–104.

[78] D. Serre, *Écoulement parfait incompressible à rotationnel croissant avec le temps*, C. R. Acad. Sci. Paris Sér. I **312** (1991), 315–318.

[79] D. Serre, *Sur le principe variationnel des équations de la mécanique des fluides parfaits*, RAIRO Modél. Math. Anal. Numér. **27** (1993), 739–758.

[80] D. Serre, *La croissance de la vorticité dans les écoulements parfaits incompressibles*, C. R. Acad. Sci. Paris Sér. I **328** (1999), 549–552.

[81] E.M. Stein, *Singular Integrals and Differentiability Properties of Functions*, Princeton Univ. Press (1970).

[82] D. Tataru, *Strichartz estimates for operators with nonsmooth coefficients and the nonlinear wave equation*, Amer. J. Math. **122** (2000), 349–376.

[83] D. Tataru, *Strichartz estimates for second order hyperbolic operators with nonsmooth coefficients, II*, Amer. J. Math. **123** (2001), 385–423.

[84] A. Torchinsky, *Real Variable Methods in Harmonic Analysis*, Pure Appl. Math., Vol. 123, Academic Press, Florida, CA (1986).

[85] H. Triebel, *Interpolation Theory, Function Spaces, Differential Operators*, North-Holland, Amsterdam (1978).

[86] M. Vishik, *Incompressible flows of an ideal fluid with vorticity in borderline spaces of Besov type*, Ann. Sci. École Norm. Sup. **32** (1999), 769–812.
[87] W. Wolibner, *Un théorème d'existence du mouvement plan d'un fluide parfait, homogène, incompressible, pendant un temps infiniment long*, Math. Z. **37** (1933), 698–726.
[88] V. Yudovich, *Non stationary flows of an ideal incompressible fluid*, Zh. Vychisl. Math. **3** (1963), 1032–1066.
[89] N. Zabusky, M. Hughes and K. Roberts, *Contour dynamics for the Euler equations in two dimensions*, J. Comput. Phys. **30** (1979), 96–106.
[90] P. Zhang and Q.-J. Qiu, *Propagation of higher-order regularities of the boundaries of 3-D vortex patches*, Chinese Ann. Math. Ser. A **18** (1997), 381–390.

CHAPTER 3

Harmonic Analysis Tools for Solving the Incompressible Navier–Stokes Equations

Marco Cannone

Laboratoire d'Analyse et de Mathématiques Appliquées, Université de Marne-la-Vallée, Cité Descartes–5, bd Descartes, Champs-sur-Marne, 77454 Marne-la-Vallée Cedex 2, France

E-mail: cannone@math.univ-mlv.fr

Contents

HANDBOOK OF MATHEMATICAL FLUID DYNAMICS, VOLUME III
Edited by S.J. Friedlander and D. Serre

Introduction

Formulated and intensively studied at the beginning of the nineteenth century, the classical partial differential equations of mathematical physics represent the foundation of our knowledge of waves, heat conduction, hydrodynamics and other physical problems. Their study prompted further work by mathematical researchers and, in turn, benefited from the application of new methods in pure mathematics. It is a vast subject, intimately connected to various sciences such as Physics, Mechanics, Chemistry, Engineering Sciences, with a considerable number of applications to industrial problems.

Although the theory of partial differential equations has undergone a great development in the twentieth century, some fundamental questions remain unresolved. They are essentially concerned with the global existence, regularity and uniqueness of solutions, as well as their asymptotic behavior.

The immediate object of this chapter is to review some improvements achieved in the study of a celebrated nonlinear partial differential system, the incompressible Navier–Stokes equations. The nature of a turbulent motion of a fluid, an ocean for instance, or the creation of a vortex inside it, are two typical problems related to the Navier–Stokes equations, and they are still far from being understood.

From a mathematical viewpoint, one of the most intriguing unresolved questions concerning the Navier–Stokes equations and closely related to turbulence phenomena is the regularity and uniqueness of the solutions to the initial value problem. More precisely, given a smooth datum at time zero, will the solution of the Navier–Stokes equations continue to be smooth and unique for all time? This question was posed in 1934 by Leray [148,149] and is still without answer, neither in the positive nor in the negative. Smale includes the uniqueness and regularity question for the Navier–Stokes equations as one of the 18 open problems of the twentieth century [212].

There is no uniqueness proof except for over small time intervals and it has been questioned whether the Navier–Stokes equations really describe general flows. But there is no proof for nonuniqueness either.

Maybe a mathematical ingenuity is the reason for the missing (expected) uniqueness result. Or maybe the methods used so far are not pertinent and the refractory Navier–Stokes equations should be approached with a different strategy.

Uniqueness of the solutions of the equations of motion is the cornerstone of classical determinism [74]. If more than one solution were associated to the same initial data, the committed determinist will say that the space of the solutions is too large, beyond the real physical possibility, and that uniqueness can be restored if the unphysical solutions are excluded. On the other hand, anarchists will be happy to conclude that the laws of motion are not verified and that chaos reigns. More precisely, a nonuniqueness result would represent such an insulting paradox to classical determinism, that the introduction of a more sophisticated model for the study of the motion of a viscous fluid would certainly be justified [39,42,84,119].

Thirty years ago, Shinbrot wrote [211]:

> Without the d'Alembert (and other paradoxes), who would have thought it necessary to study more intricate models than the ideal fluid? However, it is usually through paradoxes that mathematical

work has the greatest influence on physics. In terms of existence and uniqueness theory, this means that the most important thing to discover is what is not true. When one proves the Navier–Stokes equations have solutions, the physicist yawns. If one can prove these solutions are not unique (say), he opens his eyes instead of his mouth. Thus, when we prove existence theorems, we are only telling the world where paradoxes are not and perhaps sweeping away some of the mist that surrounds the area where they are.

If the problem of uniqueness relates to the predictive power aspect of the theory, the existence issue touches the question of the self-consistency of the physical model involved in the Navier–Stokes equations; if no solution exists, then the theory is empty.

In the nineteenth century, the existence problems arising from mathematical physics were studied with the aim of finding exact solutions to the corresponding equations. This is only possible in particular cases. For instance, very few exact solutions of the Navier–Stokes equations were found and, except for some exact stationary solutions, almost all of them do not involve the specifically nonlinear aspects of the problem, since in general the corresponding nonlinear terms in the Navier–Stokes equations vanish.

In the twentieth century, the strategy changed. Instead of explicit formulas in particular cases, the problems were studied in all their generality. This led to the concept of weak solutions. The price to pay is that only the existence of the solutions can be ensured. In fact, the construction of weak solutions as the limit of a subsequence of approximations leaves open the possibility that there is more than one distinct limit, even for the same sequence of approximations.

The uniqueness question is among the most important unsolved problems in fluid mechanics: "Instant fame awaits the person who answers it. (Especially if the answer is negative!)" [211]. Moreover, as for the solutions of the Navier–Stokes equations, such a uniqueness result is not available for the solutions of the Euler equations of ideal fluids, or the Boltzmann equation of rarefied gases, or the Enskog equation of dense gases either.

A question intimately related to the uniqueness problem is the regularity of the solution. Do the solutions to the Navier–Stokes equations blow-up in finite time? The solution is initially regular and unique, but at the instant T when it ceases to be unique (if such an instant exists), the regularity could also be lost.

One may imagine that blow-up of initially regular solutions never happens, or that it becomes more likely as the initial norm increases, or that there is blow-up, but only on a very thin set of probability zero. Nobody knows the answer and the Clay Mathematical Institute is offering a prize for it [80]. As Fefferman [80] remarks, finite blow-up in the Euler equation of an "ideal" fluid is an open and challenging mathematical problem as it is for the Navier–Stokes equations. Constantin [67] suggests that it is finite time blow-up in the Euler equations that is the physically more important problem, since blow-up requires large gradients in the limit of zero viscosity. The best result in this direction concerning the possible loss of smoothness for the Navier–Stokes equations was obtained by Caffarelli, Kohn and Nirenberg [31,151], who proved that the one-dimensional Hausdorff measure of the singular set is zero.

After providing such a pessimistic scenery, revealing our lack of comprehension in the study of the Navier–Stokes equations, let us briefly recall here some more encouraging, even if partial, research directions. Roughly speaking, we can summarize the discussion by saying that if "some quantity" turns out to "be small", then the Navier–Stokes equations are well posed in the sense of Hadamard (existence, uniqueness and stability of the corresponding solutions).

For instance, a unique global solution exists provided the data – the initial value and the exterior force – are small, and the solution is smooth depending on smoothness of the data. Another quantity that can be small is the dimension. If we are in dimension $n = 2$, the situation is easier than in dimension $n = 3$ and completely understood [152,218]. Finally, if the domain $\Omega \subset \mathbb{R}^3$ is small, in the sense that Ω is thin in one direction, say $\Omega = \omega \times (0, \varepsilon)$, then the question is also settled [235].

Other good news is contained in the following pages. They reflect the progress achieved in the last seven years by approaching the Navier–Stokes equations with mathematical tools directly taken from the harmonic analysis world. We mean the use of the Fourier transform and its natural heirs, better suited for the study of nonlinear problems: the Littlewood–Paley decomposition, the paraproducts, the Besov spaces and the wavelets.

Motivated by a somewhat esoteric paper of Federbush entitled "Navier and Stokes meet the wavelets" [78], in 1995 we launched an ambitious program [34]: solve the *nonlinear Navier–Stokes* equations by means of *wavelet transform* and *Besov spaces*. Of course, at the origin of our hopes was the remark that it is possible to solve the *linear heat* equation by *Fourier transform* in *Sobolev spaces*, a very tempting comparison indeed.

Following these ideas and this program, some important results were obtained. They concern the existence of a global solution for highly oscillating data (Section 4), the uniqueness of this solution (Section 5) and its asymptotic behavior, via the existence of self-similar solutions (Section 6).

In the following pages, after recalling these results, we will realize, a posteriori, that the harmonic analysis tools were not necessary at all for their discovery. In fact, each proof of the previous theorem (existence, uniqueness, self-similar solutions) originally found by means of 'Fourier analysis methods', more precisely, by using 'Besov spaces', was followed, shortly after its publication, by a 'real variable methods' proof.

Temam [217] was able to construct a global solution with highly oscillating data by using a classical Sobolev space. This solution was shown to be unique by Meyer [166], with a proof that makes use of a Lorentz space, instead of a Besov one. Finally, Le Jan and Sznitman [138] discovered an elementary space for the existence of self-similar solutions.

The historical details that led to each theorem and each proof are contained in the paper entitled "Viscous flows in Besov spaces" [37], that should be considered as a companion to this article.

1. Preliminaries

1.1. *The Navier–Stokes equations*

We study the Cauchy problem for the Navier–Stokes equations governing the time evolution of the velocity $v(t, x) = (v_1(t, x), v_2(t, x), v_3(t, x))$ and the pressure $p(t, x)$ of an incompressible viscous fluid (whose viscosity coefficient is given by the positive constant ν) filling all of $\mathbb{R}^3$ and in the presence of an external force $\phi(t,x)$:

$$\begin{aligned} &\frac{\partial v}{\partial t} - \nu\Delta v = -(v\cdot\nabla)v - \nabla p + \phi, \\ &\nabla\cdot v = 0, \\ &v(0) = v_0, \quad x\in\mathbb{R}^3, t\geqslant 0. \end{aligned} \tag{1}$$

Here, the external force $\phi(t,x)$ will be considered as arising from a potential $V(t,x)$ in such a way that

$$\phi = \nabla\cdot V \tag{2}$$

which means, that

$$\phi_j = \sum_{k=1}^{3} \partial_k V_{kj}, \quad j = 1,2,3. \tag{3}$$

As we will describe in Section 6.4, more general types of forces can be considered, this is done for instance in the recent paper [42,40] (for other examples see also [130,133]).

We will also assume that the viscosity ν is equal to one. This can be done, without loss of generality, because of the invariant structure of the Navier–Stokes equations and we will return to this issue in Section 3.2.

Finally, thanks to the divergence-free property $\nabla\cdot v = 0$, expressing the incompressibility of the fluid, we can write $(v\cdot\nabla)v = \nabla\cdot(v\otimes v)$. This remark is important because the product of two tempered distributions is not always defined, whereas it is always possible to take the derivative (in the distribution sense) of an L^1_{loc} function. Thus, it will be enough to require $v\in L^2_{\mathrm{loc}}$ in order to make the quadratic term $\nabla\cdot(v\otimes v)$ well defined.

Here and in the following, we say that a vector $a = (a_1,a_2,a_3)$ belongs to a function space X if $a_j\in X$ holds for every $j=1,2,3$, and we put $\|a\| = \max_{1\leqslant j\leqslant 3}\|a_j\|$. To be more precise, we should write $X(\mathbb{R}^3)$ instead of X (for instance $v=(v_1,v_2,v_3)\in L^2_{\mathrm{loc}}$ means $v_j\in L^2_{\mathrm{loc}}(\mathbb{R}^3)$ for every $j=1,2,3$). In order to avoid any confusion, if the space is not $\mathbb{R}^3$ (for example, if the dimension is 2 or if the space is a bounded domain Ω_{b} as considered at the end of Section 5.1) we will write it explicitly (say $X(\mathbb{R}^2)$ or $X(\Omega_{\mathrm{b}})$). The reason why we are mainly interested in the whole space $\mathbb{R}^3$ (or more generally $\mathbb{R}^n$, $n\geqslant 2$) is that we will make constant use of Fourier transform tools, that are easier to handle in the case of the whole space (or a bounded space with periodic conditions, as in [222]) than that of a domain with boundaries. A detailed analysis of the problems that can occur if the Navier–Stokes (or more general) equations are supplemented by the homogeneous Dirichlet (no-slip) boundary conditions is contained in [83].

Our attention will be focused on the existence of solutions $v(t,x)$ to (1) in the space $\mathcal{C}([0,T);X)$ that are strongly continuous functions of $t\in[0,T)$ with values in the Banach space X of vector distributions. Depending on whether T will be *finite* ($T<\infty$) or *infinite* ($T=\infty$) we will obtain respectively *local* or *global* (in time) solutions.

Before introducing the appropriate functional setting, let us transform the system (1) into the operator equation [30,87,117]:

$$\begin{aligned} &\frac{\mathrm{d}v}{\mathrm{d}t} - Av = -\mathbb{P}\nabla\cdot(v\otimes v) + \mathbb{P}\phi, \\ &v(0) = v_0, \quad x\in\mathbb{R}^3, t\geqslant 0, \end{aligned} \tag{4}$$

where A is formally defined as the operator $A = -\mathbb{P}\Delta$ and $\mathbb{P}$ is the Leray–Hopf orthogonal projection operator onto the divergence-free vector field defined as follows.

We let

$$D_j = -\mathrm{i}\frac{\partial}{\partial x_j}, \quad j = 1, 2, 3; \mathrm{i}^2 = -1, \tag{5}$$

and we denote the Riesz transforms by

$$R_j = D_j(-\Delta)^{-1/2}, \quad j = 1, 2, 3. \tag{6}$$

For an arbitrary vector field $v(x) = (v_1(x), v_2(x), v_3(x))$ on $\mathbb{R}^3$, we set

$$z(x) = \sum_{k=1}^{3}(R_k v_k)(x) \tag{7}$$

and define the Leray–Hopf operator $\mathbb{P}$ by

$$(\mathbb{P}v)_j(x) = v_j(x) - (R_j z)(x) = \sum_{k=1}^{3}(\delta_{jk} - R_j R_k)v_k, \quad j = 1, 2, 3. \tag{8}$$

Another equivalent way to define $\mathbb{P}$ is to make use of the properties of the Fourier transform and write

$$\widehat{(\mathbb{P}v)}_j(\xi) = \sum_{k=1}^{3}\left(\delta_{jk} - \frac{\xi_j\xi_k}{|\xi|^2}\right)\hat{v}_k(\xi), \quad j = 1, 2, 3. \tag{9}$$

As such, $\mathbb{P}$ is a pseudo-differential operator of degree zero and is an orthogonal projection onto the kernel of the divergence operator. In other words the pressure p in (1) ensures that the incompressibility condition $\nabla\cdot v = 0$ is satisfied.

Finally, making use of this projection operator $\mathbb{P}$ and the semigroup

$$S(t) = \exp(-tA), \tag{10}$$

it is a straightforward procedure to reduce the operator equation (4) into the following integral equation

$$v(t) = S(t)v_0 - \int_0^t S(t-s)\mathbb{P}\nabla \cdot (v \otimes v)(s)\,\mathrm{d}s + \int_0^t S(t-s)\mathbb{P}\nabla \cdot V(s)\,\mathrm{d}s. \tag{11}$$

On purpose, we are being a little cavalier here: we shall not justify the formal transition (1) → (4) → (11). We shall rather start from (11) and prove the existence and uniqueness of a solution $v(t, x)$ for it. Then, we shall prove that this solution is regular enough to form, with an appropriate pressure $p(t, x)$, a classical solution of the system (1).

Since our attention will essentially be devoted to the study of the integral equation (11) and since we will only consider the case of the all space $\mathbb{R}^3$, so that the semigroup $S(t)$ reduces to the well-known heat semigroup $\exp(t\Delta)$, we will separate the different contributions in (11) in the following way: the linear term containing the initial data

$$S(t)v_0 =: \exp(t\Delta)v_0, \tag{12}$$

the bilinear operator expressing the nonlinearity of the equation

$$B(v, u)(t) =: -\int_0^t \exp\bigl((t-s)\Delta\bigr)\mathbb{P}\nabla \cdot (v \otimes u)(s)\,\mathrm{d}s \tag{13}$$

and finally the linear operator L involving the external force

$$L(V)(t) =: \int_0^t \exp\bigl((t-s)\Delta\bigr)\mathbb{P}\nabla \cdot V(s)\,\mathrm{d}s. \tag{14}$$

The precise meaning of the integral defined by (13) in different function spaces is one of the main problems arising from this approach and will be discussed carefully in the following section.

Let us note here that there is a kind of competition in this integral term between the regularizing effect represented by the heat semigroup $S(t-s)$ and the loss of regularity that comes from the differential operator ∇ and from the pointwise multiplication $v \otimes u$. This loss of regularity is illustrated by the following simple example: if two (scalar) functions f and g are in H^1, their product only belongs to $H^{1/2}$ and their derivative $\partial(fg)$ is even less regular as it belongs to $H^{-1/2}$.

1.2. *Classical, mild and weak solutions*

As yet the existence of a global solution in time has not been proved nor disproved for a three-dimensional flow and sufficiently general initial conditions; but as we will see in the following pages, a global, regular solution does exist whenever the initial data are highly oscillating or sufficiently small in certain function spaces.

To begin with, it is necessary to clarify the meaning of "solution of the Navier–Stokes equations", because, since the appearance of the pioneer papers of Leray, the word "solution" has been used in a more or less generalized sense. Roughly speaking, two main types of solutions can be distinguished: "strong solutions" (for which existence and uniqueness are known) and "weak solutions" (for which only the existence is known).

In the following pages, we will take the term "solution" in the generic sense of classical ordinary differential equations in t with values in the space of tempered distributions $\mathcal{S}'$, in order to be able to use the Fourier transforms tools. This interpretation is suggested by the notion of solution in the sense of distribution used in evolution equations. Moreover, we will ask that the function space X, to which the initial data v_0 belong, is such that $X \hookrightarrow L^2_{\mathrm{loc}}$, in order to be able to give a (distributional) meaning to the nonlinear term $(v \cdot \nabla)v = \nabla \cdot (v \otimes v)$. More generally, we will ask $v \in L^2_{\mathrm{loc}}([0, T); \mathbb{R}^3)$.

In the recent papers of Amann [1] and of Lemarié [142,145], we can count many different definitions of solutions (see also [71]) distinguished only by the class of functions to which they are supposed to belong: *classical, strong, mild, weak, very weak, uniform weak* and *local Leray* solutions of the Navier–Stokes equations!

We will not present all the possible definitions here but concentrate our attention on three cases, respectively classical (Hadamard), weak (Leray) and mild (Yosida) solutions.

DEFINITION 1 (Classical). A classical solution $(v(t,x), p(t,x))$ of the Navier–Stokes equations is a pair of functions $v : t \to v(t)$ and $p : t \to p(t)$ satisfying the system (1), for which all the terms appearing in the equations are continuous functions of their arguments. More precisely, a classical solution is a solution to the system (1) that verifies:

$$v(t,x) \in \mathcal{C}\big([0,T); E\big) \cap \mathcal{C}^1\big([0,T); F\big), \tag{15}$$

$$E \hookrightarrow F \quad \text{(continuous embedding)}, \tag{16}$$

$$v \in E \quad \Longrightarrow \quad \Delta v \in F \quad \text{(continuous operator)}, \tag{17}$$

$$v \in E \quad \Longrightarrow \quad \nabla \cdot (v \otimes v) \in F \quad \text{(continuous operator)}, \tag{18}$$

where E and F are two Banach spaces of distributions.

For example, if E is the Sobolev space H^s and $s > 3/2$ (thus giving H^s the structure of an algebra when endowed with the usual product of functions), we can chose $F = H^{s-2}$, because $\Delta v \in H^{s-2}$ and $\nabla \cdot (v \otimes v) \in H^{s-1} \hookrightarrow H^{s-2}$.

As we recalled in the Introduction, it is very difficult to ensure the existence of classical solutions, unless we look for exact solutions (that do not involve the specific aspects of the problem, since in general the corresponding nonlinear terms in the equations vanish), or we impose very restrictive conditions on the initial data (see Section 3). This is not the case when we take the word solution in the weak sense given by Leray.

DEFINITION 2 (Weak). A *weak solution* $v(t,x)$ of the Navier–Stokes equations in the sense of Leray and Hopf is supposed to have the following properties:

$$v(t,x) \in L^\infty\big([0,T); \mathbb{P}L^2\big) \cap L^2\big([0,T); \mathbb{P}H^1\big) \tag{19}$$

and

$$\int_0^T \left(-\langle v, \partial_t \varphi\rangle + \langle \nabla v, \nabla \varphi\rangle + \langle (v \cdot \nabla)v, \varphi\rangle\right) \mathrm{d}s = \langle v_0, \varphi(0)\rangle + \int_0^T \langle \phi, \varphi\rangle \,\mathrm{d}s \tag{20}$$

for any $\varphi \in \mathcal{D}([0, T); \mathbb{P}\mathcal{D})$. The symbol $\langle \cdot, \cdot \rangle$ denotes the L^2-inner product, whereas $\mathbb{P}X$ denotes the subspace of X (here $X = L^2$, H^1 or $\mathcal{D}$) of solenoidal functions,[1] characterized by the divergence-free condition $\nabla \cdot v = 0$. Finally, such a weak solution is supposed to verify the following energy inequality

$$\frac{1}{2}\|v(t)\|_2^2 + \int_0^t \|\nabla v(s)\|_2^2 \,\mathrm{d}s \leqslant \frac{1}{2}\|v(0)\|_2^2 + \int_0^t \langle \phi, v\rangle \,\mathrm{d}s, \qquad t > 0. \tag{21}$$

Sometimes this inequality is satisfied not only on the interval $(0, t)$ but on all intervals $(t_0, t_1) \subset (0, T)$ except possibly for a set of measure zero. Such a solution is called *turbulent* in Leray's papers.

Finally, after the papers of Kato and his collaborators, we got used to calling *mild solutions* a third category of solutions, whose existence is obtained by a fixed point algorithm applied to the integral equation (11). In other words, the Navier–Stokes equations are studied by means of semigroup techniques as in the pioneering papers of Yosida [238]. More precisely, mild solutions are defined in the following way.

DEFINITION 3 (Mild). A *mild solution* $v(t, x)$ of the Navier–Stokes equations satisfies the integral equation (11) and is such that

$$v(t, x) \in \mathcal{C}\big([0, T); \mathbb{P}X\big), \tag{22}$$

where X is a Banach space of distributions on which the heat semigroup $\{\exp(t\Delta); t \geqslant 0\}$ is strongly continuous and the integrals in (11) are well defined in the sense of Bochner.

Historically, the introduction of the term "mild" in connection with the integral formulation for the study of an arbitrary evolution equation goes back to Browder [30]. We do not expect to use the energy inequality, but we hope to ensure in this way the uniqueness of the solution, in other words that the solution is strong. This is in contrast with Leray's construction of *weak solutions*, relying on compactness arguments and *a priori* energy estimates. Moreover, the fixed point algorithm is stable and constructive. Thus the problem of defining mild solutions is closely akin to the question of knowing whether the Cauchy problem for Navier–Stokes equations is well posed in the sense of Hadamard. This question will be discussed in Section 7 in connection with the theory of stability and Lyapunov functions.

Let us recall that for a function $u(t, \cdot)$ that takes values in a Banach space E, the integral $\int_0^T u(t, \cdot)\,\mathrm{d}t$ exists either because $\int_0^T \|u(t, \cdot)\|_E \,\mathrm{d}t < \infty$ (in this case we say that the

[1] In the literature this space is usually denoted by X_σ.

integral is defined in the sense of Bochner) or because $\int_0^T |\langle u(t,\cdot), y\rangle|\, dt$ converges for any vector y of the dual (or pre-dual) E' of E (the integral is said to be weakly convergent). The weak convergence is ensured by the oscillatory behavior of $u(t,\cdot)$ in the Banach space E.

Now, the oscillatory property of the bilinear term arising from the Navier–Stokes equations is systematically taken into account in all papers that are based on the energy inequality, in particular $\langle B(v,v), v\rangle = 0$ as long as $\nabla \cdot v = 0$. In the following pages, we will *never* take advantage of this remarkable property, for we will only consider functional spaces where it is *not* possible to write $\langle B(v,v), v\rangle$. In fact, $B(v,v)$ will *never* belong to a space that is a dual of the one to which v belongs. This is the reason why our works ([46,47] excepted) are *not* based on the innermost structure of the Navier–Stokes equations and can be easily extended to other nonlinear partial differential equations [14–17,52–54, 89,90,113,156,183,184,190–192,196–202,222,223].

More explicitly, in the literature concerning the existence and uniqueness of mild solutions for the Navier–Stokes equations as inaugurated by Fujita and Kato's celebrated papers [87,117], the oscillatory behavior of $B(v,u)$ is lost from the very beginning because, by definition, mild solutions require *strong* estimates in the *strong* topology, so that $B(v,u)$ can be replaced by $|B(v,u)|$ without affecting the corresponding existence and uniqueness results.

On the other hand, as far as the *weak* solutions are concerned, introduced in the pioneering papers by Leray [148–150], the oscillatory behavior of $B(v,u)$ is frequently analyzed by means of the well-known identity

$$\big\langle \nabla \cdot (u \otimes v), v \big\rangle = 0, \tag{23}$$

where $\nabla \cdot u = 0$. In that case the problem is different, for the above identity does not allow a great flexibility in the choice of the functional setting, that is forced to be defined in terms of an energy norm (e.g., $L^2, H^1, \ldots$).

1.3. *Navier meets Fourier*

The title of this section is borrowed from a paper by Federbush "Navier and Stokes meet the wavelets" [78,79] that will be dealt with in Section 2.4.

The Navier–Stokes equations did not yet exist when Fourier gave the explicit solution of the heat equation

$$\begin{aligned} &\frac{\partial u}{\partial t} - \Delta u = f, \\ &u(0) = u_0. \end{aligned} \tag{24}$$

This equation, governing the evolution of temperature $u(x,t)$, in the presence of an exterior source of heat $f(x,t)$, at a point x and time t of a body assumed here to fill the whole space $\mathbb{R}^3$, becomes, when we consider its partial Fourier transform with respect to x, an ordinary differential equation in t, whose solution is given by

$$u(t,x) = S(t)u_0 + \int_0^t S(t-s)f(s)\,ds, \tag{25}$$

$S(t)$ being the convolution operator defined as in (12) by the heat semigroup

$$S(t) = \exp(t\Delta) = \left(\frac{1}{4\pi t}\right)^{3/2} \exp\left(-\frac{|x|^2}{4t}\right). \tag{26}$$

The Navier–Stokes equations, that describe the motion of a viscous fluid, were introduced by Navier in 1822 [178], the same year that, by a curious coincidence, Fourier published the celebrated treatise "Théorie analytique de la chaleur" [86], in which he developed in a systematic way the ideas contained in a paper of 1807.

But this is not only a mere coincidence. In fact Navier, engineer of the *Ecole Nationale des Ponts et Chaussées*, was also a very close friend of many mathematicians, in particular Fourier. Fourier had a strong influence on Navier's life and career, both as a friend and as a teacher. In turn, Navier was a noticeable proponent of the important mathematical techniques developed by Fourier.[2]

In this section we want to show how to take advantage of the Fourier transform in order to study the Navier–Stokes equations.

We have already remarked that, following Fourier's method to solve the Navier–Stokes equations for a viscous incompressible fluid, we obtain the integral equation (11), very similar to (25), that led to the concept of a mild equation and a mild solution.

If we want to make use of the Fourier transform again, the second idea that comes to mind is to rewrite (11) componentwise ($j = 1, 2, 3$) in Fourier variables

$$\begin{aligned}\hat{v}_j(\xi) &= \exp(-t|\xi|^2)\hat{v}_{0j} \\ &\quad - \int_0^t \exp(-(t-s)|\xi|^2) \sum_{l,k=1}^{3} \left(\delta_{jk} - \frac{\xi_j\xi_k}{|\xi|^2}\right)(\mathrm{i}\xi_l)\hat{v}_l(\xi) * \hat{v}_k(\xi) \\ &\quad + \int_0^t \exp(-(t-s)|\xi|^2) \sum_{l,k=1}^{3} \left(\delta_{jk} - \frac{\xi_j\xi_k}{|\xi|^2}\right)(\mathrm{i}\xi_l)\widehat{V_{lk}}(\xi)\,ds.\end{aligned}$$

We use the notations introduced by Miyakawa in [173] and denote by $F(t,x)$ the tensor kernel associated with the operator $\exp(t\Delta)\mathbb{P}\nabla\cdot$, say

$$\widehat{F_{l,k,j}}(t,\xi) = \exp(-t|\xi|^2)\left(\delta_{jk} - \frac{\xi_j\xi_k}{|\xi|^2}\right)\mathrm{i}\xi_l. \tag{27}$$

[2]This was not the case for most other engineers of his period. Navier's interests in more mathematical aspects of physics, mechanics and engineers sciences were so deep that, when the suspension bridge across the Seine he had designed collapsed, sarcastic articles appeared in the press against Navier, who was referred to as "that eminent man of science whose calculations fail in Paris" (see [38]).

It is easy to see that the kernel $F(t,x) = \{F_{l,k,j}(t,x)\}$ defined in this way verifies

$$|F(t,x)| \lesssim |x|^{-\alpha} t^{-\beta/2}, \quad \alpha \geqslant 0, \beta \geqslant 0, \alpha + \beta = 4, \tag{28}$$

and

$$\|F(t,x)\|_p \lesssim t^{-(4-3/p)/2}, \quad 1 \leqslant p \leqslant \infty. \tag{29}$$

In the following pages we will not take advantage of these general estimates. In fact, we will never use the full structure of the operator $\exp(t\Delta)\mathbb{P}\nabla\cdot$ and our analysis will apply to a more general class of evolution equations.

Let us be more explicit. Our existence and uniqueness theorems for the mild Navier–Stokes equations will be obtained by using the Banach fixed point theorem. The continuity of the bilinear term B as well as the continuity of the linear term L defined in (13) and (14) will be the main ingredients of the proofs. The functional spaces where the initial data will be considered are such that the Riesz transforms operate continuously. The conclusion is easy: we will get rid of the Riesz transforms from the very beginning and limit ourselves to the study of a simplified version of the operator $\exp(t\Delta)\mathbb{P}\nabla\cdot$ giving rise to simplified versions of the operators B and L.

We denote with the letters B_s and L_s these operators defined by

$$B_s(f,g)(t) =: -\int_0^t \big[S(t-s)\dot{\Lambda}\big](fg)(s)\,\mathrm{d}s \tag{30}$$

and

$$L_s(h)(t) =: \int_0^t \big[S(t-s)\dot{\Lambda}\big]h(s)\,\mathrm{d}s, \tag{31}$$

where $f = f(t,x)$, $g = g(t,x)$ and $h(t,x)$ are generic *scalar* fields and

$$\dot{\Lambda} =: \sqrt{-\Delta} \tag{32}$$

denotes the well-known Calderón's homogeneous pseudo-differential operator whose symbol in Fourier transform is $|\xi|$.

In order to obtain such simplified *scalar* versions of the operators B and L, we have not taken into account all Riesz transforms contained in the full *vectorial* operators. For example, as far as the continuity of the bilinear operator is concerned in a certain function space, we can pass from the full vectorial operator B,

$$j \in \{1,2,3\},$$

$$B(u,v)_j = -\mathrm{i}\sum_{m=1}^{3} R_m B_s(u_m, v_j) + \mathrm{i}\sum_{k=1}^{3}\sum_{l=1}^{3} R_j R_k R_l B_s(u_l, v_k), \tag{33}$$

to its scalar simplified version B_s just by using the continuity of the Riesz transforms in this space.

With this simplification in mind, and by recalling the elementary properties of the Fourier transform, we finally get an even simpler expression for the bilinear term (that by abuse of notation will be always denoted by the letter B):

$$B(f,g) = -\int_0^t (t-s)^{-2}\Theta\left(\frac{\cdot}{\sqrt{t-s}}\right) * (fg)(s)\,\mathrm{d}s, \tag{34}$$

where $f = f(t,x)$ and $g = g(t,x)$ are two scalar fields and $\Theta = \Theta(x)$ is a function of x whose Fourier transform is given by

$$\widehat{\Theta}(\xi) = |\xi|\mathrm{e}^{-|\xi|^2}. \tag{35}$$

As such, Θ is analytic, behaves like $\mathrm{O}(|x|^{-4})$ at infinity (this can also be deduced by (28) for $\alpha = 4$ and $\beta = 0$) and its integral is zero.

In the same way, the linear operator L involving the external force will be treated in the simplified scalar form

$$L(h) = \int_0^t (t-s)^{-2}\Theta\left(\frac{\cdot}{\sqrt{t-s}}\right) * h(s)\,\mathrm{d}s. \tag{36}$$

In particular, we notice that

$$B(f,g) = -L(fg) \tag{37}$$

which allows to treat both the bilinear and the linear terms in exactly the same way. This is why, for the sake of simplicity, in the following pages we will only consider the case when there is no external force and refer the reader to [39,42,40,47] for the general case.

2. Functional setting of the equations

2.1. *The Littlewood–Paley decomposition*

Let us start with the Littlewood–Paley decomposition in $\mathbb{R}^3$. To this end, we take an arbitrary function φ in the Schwartz class $\mathcal{S}$ and whose Fourier transform $\hat{\varphi}$ is such that

$$0 \leqslant \hat{\varphi}(\xi) \leqslant 1, \quad \hat{\varphi}(\xi) = 1 \quad \text{if } |\xi| \leqslant \frac{3}{4}, \qquad \hat{\varphi}(\xi) = 0 \quad \text{if } |\xi| \geqslant \frac{3}{2}, \tag{38}$$

and let

$$\psi(x) = 8\varphi(2x) - \varphi(x), \tag{39}$$

$$\varphi_j = 2^{3j}\varphi(2^j x), \qquad j \in \mathbb{Z}, \tag{40}$$

$$\psi_j(x) = 2^{3j}\psi(2^j x), \quad j \in \mathbb{Z}. \tag{41}$$

We denote by S_j and Δ_j, respectively, the convolution operators with φ_j and ψ_j. Finally, the set $\{S_j, \Delta_j\}_{j\in\mathbb{Z}}$ is the Littlewood–Paley decomposition, so that

$$I = S_0 + \sum_{j\geqslant 0} \Delta_j. \tag{42}$$

To be more precise, we should say "a decomposition", because there are different possible (equivalent) choices for the function φ. On the other hand, for an arbitrary tempered distribution f, the last identity gives

$$f = \lim_{j\to\infty} S_0 f + \sum_{j\geqslant 0} \Delta_j f. \tag{43}$$

The interest in decomposing a tempered distribution into a sum of dyadic blocks $\Delta_j f$, whose support in Fourier space is localized in a corona, comes from the nice behavior of these blocks with respect to differential operations. This fact is illustrated by the following celebrated Bernstein's lemma in $\mathbb{R}^3$, whose proof can be found in [162].

LEMMA 1. *Let* $1 \leqslant p \leqslant q \leqslant \infty$ *and* $k \in \mathbb{N}$, *then one has*

$$\sup_{|\alpha|=k} \|\partial^\alpha f\|_p \simeq R^k \|f\|_p \tag{44}$$

and

$$\|f\|_q \lesssim R^{3(1/p-1/q)} \|f\|_p \tag{45}$$

whenever f *is a tempered distribution in* $\mathcal{S}'$ *whose Fourier transform* $\hat{f}(\xi)$ *is supported in the corona* $|\xi| \simeq R$.

In the case of a function whose support is a ball (as, for instance, for $S_j f$) the lemma reads as follows:

LEMMA 2. *Let* $1 \leqslant p \leqslant q \leqslant \infty$ *and* $k \in \mathbb{N}$, *then one has*

$$\sup_{|\alpha|=k} \|\partial^\alpha f\|_p \lesssim R^k \|f\|_p \tag{46}$$

and

$$\|f\|_q \lesssim R^{3(1/p-1/q)} \|f\|_p \tag{47}$$

whenever f *is a tempered distribution in* $\mathcal{S}'$ *whose Fourier transform* $\hat{f}(\xi)$ *is supported in the ball* $|\xi| \lesssim R$.

Let us go back to the decomposition of the unity (42) and (43). It was introduced in the early 1930s by Littlewood and Paley to estimate the L^p norm of trigonometric Fourier series when $1 < p < \infty$. If we omit the trivial case $p = 2$, it is not possible to ensure the belonging of a generic Fourier series to the Lebesgue space L^p by simply using its Fourier coefficients, but this becomes true if we consider instead its dyadic blocks. In the case of a function f (not necessarily periodic), this property is given by the following equivalence

$$\text{if } 1 < p < \infty \quad \text{then } \|f\|_p \simeq \|S_0 f\|_p + \left\| \left(\sum_{j=0}^{\infty} |\Delta_j f(\cdot)|^2 \right)^{1/2} \right\|_p. \tag{48}$$

It is even easier to prove that the classical Sobolev spaces $H^s = H^s_2$, $s \in \mathbb{R}$, can be characterized by the following equivalent norms

$$\|f\|_{H^s} \simeq \|S_0 f\|_2 + \left(\sum_{j=0}^{\infty} 2^{2js} \|\Delta_j f\|_2^2 \right)^{1/2}. \tag{49}$$

As far as the more general norms $\|f\|_{H^s_p} = \|(I - \Delta)^{s/2} f\|_p$, $s \in \mathbb{R}$, $1 < p < \infty$, corresponding to the Sobolev–Bessel spaces H^s_p (that is, when s is an integer, reduce to the well-known Sobolev spaces $W^{s,p}$ whose norm are given by $\|f\|_{W^s_p} = \sum_{|\alpha| \leqslant s} \|\partial^\alpha f\|_p$) we will see in the next section how (49) has to be modified.

Another easier case we wish to present here is provided by the Hölder–Zygmund spaces C^s, $s \in \mathbb{R}$, that can be characterized by the following norms

$$\|f\|_{C^s} \simeq \|S_0 f\|_\infty + \sup_{j \geqslant 0} 2^{js} \|\Delta_j f\|_\infty. \tag{50}$$

We will not prove this property here and we refer the reader to [82]. Let us just remind the reader of the usual definition of these spaces, in order to better appreciate the simplicity of (50). If $0 < s < 1$ we denote the Hölder space by

$$\|f\|_{C^s} = \|f\|_\infty + \sup_{x \neq y} \frac{|f(x) - f(y)|}{|x - y|^s}. \tag{51}$$

As it is well known, this definition has to be modified in the case $s = 1$ in the following way

$$\|f\|_{C^1} = \|f\|_\infty + \sup_{x \neq y} \frac{|f(x+y) + f(x-y) - 2f(x)|}{|x - y|} \tag{52}$$

and defines the Zygmund class C^1. It is now easy to define, for any $s > 0$, the quantities

$$\|f\|_{C^s} = \|f\|_\infty + \sum_{i=1}^{n} \|\partial_i f\|_{C^{s-1}}. \tag{53}$$

In the case $s < 0$ we define the Hölder–Zygmund spaces by the following rule:

$$\mathcal{C}^{s-1} = \left\{ f = \sum_{i=1}^{n} \partial_i g_i,\, g_i \in \mathcal{C}^s \right\},$$
$$\|f\|_{\mathcal{C}^{s-1}} = \inf \sup_{i=1,2,\ldots,n} \|g_i\|_{\mathcal{C}^s}, \tag{54}$$

the infimum being taken over the set of g_i such that $f = \sum_{i=1}^{n} \partial_i g_i$.

Before defining the Besov spaces that will play a key role in our study of the Navier–Stokes equations, let us recall the homogeneous decomposition of the unity, analogous to (42), but containing also all the low frequencies ($j < 0$), say

$$I = \sum_{j \in \mathbb{Z}} \Delta_j. \tag{55}$$

If we apply this identity to an arbitrary tempered distribution f, we may be tempted to write

$$f = \sum_{j \in \mathbb{Z}} \Delta_j f, \tag{56}$$

but, at variance with (43), this identity has no meaning in $\mathcal{S}'$ for several reasons. First of all, the sum in (56) does not necessarily converge in $\mathcal{S}'$ as we can see if we consider a test function $g \in \mathcal{S}$ whose Fourier transform is equal to 1 near the origin, because in this case the quantity $\langle \Delta_j f, g \rangle$ is, for all $j \ll 0$, a positive constant not depending on j. And, even when the sum is convergent, the convergence has to be understood modulo polynomials, because, for these particular functions P, we have $\Delta_j P = 0$ for all $j \in \mathbb{Z}$.

A way to restore the convergence is to "sufficiently" derive the formal series $\sum_{j \in \mathbb{Z}}$ as it stated in the following lemma (see [21,22,183] for a simple proof).

LEMMA 3. *For any tempered distribution f, there exists an integer d such that, for any α, $|\alpha| \geqslant d$, the series $\sum_{j<0} \partial^\alpha (\Delta_j f)$ converges in $\mathcal{S}'$.*

The following corollary, whose proof follows from the previous lemma, gives the correct meaning to the convergence (56), that is modulo polynomials.

COROLLARY 1. *For any integer N, there exists a polynomial P_N of degree $< d$ such that the quantity $\sum_{j=-N}^{\infty} \Delta_j f - P_N$ converges in $\mathcal{S}'$ when $N \to \infty$.*

In such a way, the series $\Delta_j f$ is always well defined; furthermore, it is not difficult to prove that the difference $f - \sum_{j \in \mathbb{Z}} \Delta_j f$ has its spectrum reduced to zero; in other words, it is a polynomial. In this way, the convergence in (56), that fails to be valid in $\mathcal{S}'$, is ensured in the quotient space $\mathcal{S}'/\mathcal{P}$.

2.2. *The Besov spaces*

The Littlewood–Paley decomposition is very useful because we can define (independently of the choice of the initial function φ) the following (inhomogeneous) Besov spaces [82,185].

DEFINITION 4. Let $0 < p, q \leqslant \infty$ and $s \in \mathbb{R}$. Then a tempered distribution f belongs to the (inhomogeneous) Besov space $B_q^{s,p}$ if and only if

$$\|S_0 f\|_q + \Big(\sum_{j>0} (2^{sj} \|\Delta_j f\|_q)^p\Big)^{1/p} < \infty. \tag{57}$$

For the sake of completeness, we also define the (inhomogeneous) Triebel–Lizorkin spaces, even if we will not make a great use of them in the study of the Navier–Stokes equations.

DEFINITION 5. Let $0 < p \leqslant \infty$, $0 < q < \infty$ and $s \in \mathbb{R}$. Then a tempered distribution f belongs to the (inhomogeneous) Triebel–Lizorkin space $F_q^{s,p}$ if and only if

$$\|S_0 f\|_q + \Big\|\Big(\sum_{j>0} (2^{sj} |\Delta_j f|)^p\Big)^{1/p}\Big\|_q < \infty. \tag{58}$$

It is easy to see that the above quantities define a norm if $p, q \geqslant 1$ and a quasi-norm in general, with the usual convention that $p = \infty$ in both cases corresponds to the usual L^∞ norm. On the other hand, we have not included the case $q = \infty$ in the second definition because the L^∞ norm has to be replaced here by a more complicated Carleson measure (see [82]).

As we have already remarked before for some particular values of s, p, q, see (48)–(50), the Besov and Triebel–Lizorkin spaces generalize the usual Lebesgue ones, for instance,

$$L^q = F_q^{0,2}, \quad 1 < q < \infty, \tag{59}$$

and more generally the Sobolev–Bessel spaces,

$$H_q^s = F_q^{s,2}, \quad s \in \mathbb{R}, 1 < q < \infty, \tag{60}$$

and the Hölder ones,

$$C^s = B_\infty^{s,\infty}. \tag{61}$$

Another interesting case is given by the space $F_q^{0,2}$ with $0 < q \leqslant 1$ that corresponds to a local version of the Hardy space, whereas $F_\infty^{0,2}$ gives the local version *bmo* of the John and Nirenberg space *BMO* of Bounded Mean Oscillation functions[3] whose norm is defined by

$$\|f\|_{BMO} = \sup_B \left(\frac{1}{\mu(B)} \int_B |f - f_B|^2 \,\mathrm{d}x \right)^{1/2}, \tag{62}$$

where B stands for the set of Euclidean balls, $\mu(B)$ the volume of B and f_B denotes the average of the function f over B, say $f_B = \frac{1}{|B|} \int_B f(x)\,\mathrm{d}x$. It is clear that this quantity is in general a seminorm, unless we argue modulo constant functions (whose *BMO*-norm is zero). Moreover, it is evident that $L^\infty \hookrightarrow BMO$ but these spaces are different, because the functions $f(x) = \ln|p(x)|$, for all polynomials $p(x)$, belong to *BMO* but not to L^∞.

A space that will be useful in the following pages is provided by the set of functions which are derivatives of functions in *BMO*. More precisely, we are talking about the space introduced by Koch and Tataru in [123], that is denoted by BMO^{-1} (or by ∇BMO) and is defined as the space of tempered distributions f such that there exists a vector function $g = (g_1, g_2, g_3)$ belonging to *BMO* such that

$$f = \nabla \cdot g. \tag{63}$$

The norm in BMO^{-1} is defined by

$$\|f\|_{BMO^{-1}} = \inf_{g \in BMO} \sum_{j=1}^{3} \|g_j\|_{BMO}. \tag{64}$$

At this point, in order to provide the reader with the dyadic decomposition of the classical Hardy $\mathcal{H}^q$, *BMO* and BMO^{-1} spaces, we have to recall that their norms, at variance with the local ones, are "homogeneous".

Let us be more explicit and consider some familiar examples. The Lebesgue space L^p is "homogeneous", because its norm satisfies, with respect to the dilatation group, the following invariance $\|f(\lambda\cdot)\|_p = \lambda^{-3/p}\|f\|_p$ for all $\lambda > 0$. On the other hand, the Sobolev space H^1 normed with $\|f\|_{H^1} = \|f\|_2 + \|\nabla f\|_2$ does not verify a property of this type because the two terms composing the norm have different homogeneity (resp. $\lambda^{-3/2}$ and $\lambda^{1-3/2}$). A possible way to restore the scaling invariance would be to forget the L^2 part and define the "homogeneous" Sobolev space $\dot{H}^1$ simply by $\|f\|_{\dot{H}^1} = \|\nabla f\|_2$. Of course the attentive reader, armed with the discussion that follows (56), will protest that this quantity is not a norm, unless we work in $\mathcal{S}'$ modulo polynomials (in the case of $\dot{H}^1$, modulo constants would be sufficient). A very simple condition that prevents constant functions to belong to $\dot{H}^1$ is given by [166]:

$$\int_{|x| \leqslant R} |f(x)|\,\mathrm{d}x = o(R^3), \qquad R \to +\infty. \tag{65}$$

[3]For a different interpretation of the acronym... see [185], page 175!

A stronger, but more natural condition is provided by the celebrated Sobolev embedding in $\mathbb{R}^3$

$$\|f\|_6 \lesssim \|\nabla f\|_2, \tag{66}$$

thus suggesting the following definition: A function f belongs to $\dot{H}^1$ if and only if ∇f belongs to L^2 and f belongs to L^6, the norm of f in $\dot{H}^1$ being $\|\nabla f\|_2$. Indeed, this definition is equivalent to defining $\dot{H}^1$ as the closure of the test functions space C_0^∞ for the norm $\|f\|_{\dot{H}^1} = \|\nabla f\|_2$. In the same way, we define the space $\dot{H}_p^s$ when $s < 3/p$ as the closure of the space

$$\mathcal{S}_0 = \{f \in \mathcal{S}, 0 \notin \operatorname{Supp} \hat{f}\} \tag{67}$$

for the norm

$$\|f\|_{\dot{H}_p^s} = \|\dot{\Lambda}^s f\|_p, \tag{68}$$

where, as usual, $\dot{\Lambda} = \sqrt{-\Delta}$ denotes the homogeneous Calderón pseudo-differential operator (see Section 1.3). Finally, when $3/p + d \leqslant s < 3/p + d + 1$ and d is an integer, $\dot{H}_p^s$ is a space of distributions modulo polynomials of degree $\leqslant d$.

We are now ready to define the homogeneous version of the Besov and Triebel–Lizorkin spaces [21,22,82,185].

If $m \in \mathbb{Z}$, we denote by $\mathcal{P}_m$ the set of polynomials of degree $\leqslant m$ with the convention that $\mathcal{P}_m = \emptyset$ if $m < 0$. If $q = 1$ and $s - 3/p \in \mathbb{Z}$, we put $m = s - 3/p - 1$; if not, we put $m = [s - 3/p]$, the brackets denoting the integer part function.

DEFINITION 6. Let $0 < p, q \leqslant \infty$ and $s \in \mathbb{R}$. Then a tempered distribution f belongs to the (homogeneous) Besov space $\dot{B}_q^{s,p}$ if and only if

$$\Big(\sum_{j\in\mathbb{Z}} (2^{sj}\|\Delta_j f\|_q)^p\Big)^{1/p} < \infty \quad \text{and} \quad f = \sum_{-\infty}^{\infty} \Delta_j f \quad \text{in } \mathcal{S}'/\mathcal{P}_m. \tag{69}$$

DEFINITION 7. Let $0 < p \leqslant \infty$, $0 < q < \infty$ and $s \in \mathbb{R}$. Then a tempered distribution f belongs to the (homogeneous) Triebel–Lizorkin space $\dot{F}_q^{s,p}$ if and only if

$$\Big\|\Big(\sum_{j\in\mathbb{Z}} (2^{sj}|\Delta_j f|)^p\Big)^{1/p}\Big\|_q < \infty \quad \text{and} \quad f = \sum_{-\infty}^{\infty} \Delta_j f \quad \text{in } \mathcal{S}'/\mathcal{P}_m, \tag{70}$$

with an analogous modification as in the inhomogeneous case when $q = \infty$.

As expected, we have the following identifications:

$$L^q = \dot{F}_q^{0,2}, \quad 1 < q < \infty, \tag{71}$$

and, more generally,

$$\dot{H}^s_q = \dot{F}^{s,2}_q, \quad s \in \mathbb{R}, 1 < q < \infty, \tag{72}$$

$$\dot{C}^s = \dot{B}^{s,\infty}_\infty, \quad s \in \mathbb{R}, \tag{73}$$

$$\dot{F}^{0,2}_q = \mathcal{H}^q, \quad 0 < q \leqslant 1, \tag{74}$$

$$\dot{F}^{0,2}_\infty = BMO, \tag{75}$$

and

$$\dot{F}^{-1,2}_\infty = BMO^{-1}. \tag{76}$$

Moreover, we have the following continuous embedding (see [34]):

$$\begin{aligned} &3 \leqslant q_1 \leqslant q_2 < \infty, \\ &L^3 \hookrightarrow \dot{B}^{-1+3/q_1,\infty}_{q_1} \hookrightarrow \dot{B}^{-1+3/q_2,\infty}_{q_2} \hookrightarrow \dot{F}^{-1,2}_\infty \hookrightarrow \dot{B}^{-1,\infty}_\infty. \end{aligned} \tag{77}$$

We will come back on the "maximal" space $\dot{B}^{-1,\infty}_\infty$ in Proposition 7.

The next four propositions are of paramount importance because they give definitions for the Besov and Triebel–Lizorkin norms in terms of the heat semigroup $S(t)$ (that appears in (12)) and in terms of the function Θ (that appears in (34) and (36)). The first two equivalences given hereafter, are very natural. The idea is that the convolution operators Δ_j can be interpreted as a discrete subset ($j \in \mathbb{Z}$) of the continuous set ($t > 0$) of convolution operators Θ_t where

$$\Theta_t = \frac{1}{t^3}\Theta\left(\frac{\cdot}{t}\right) \tag{78}$$

and, as in (35), Θ is defined by its Fourier transform $\widehat{\Theta}(\xi) = |\xi|\mathrm{e}^{-|\xi|^2}$. If the function Θ were smooth and compactly supported on the Fourier side, this would indeed be the usual characterization for Besov and Triebel–Lizorkin spaces without any restriction on the third (regularity) index s that appears in Definitions 1 and 2. This would also be the case if the function Θ had all its moments equal to zero [185]. In the case we are dealing with, we only know that Θ has its first moment (the integral) equal to zero. This is why we have to require $s < 1$ (see [185]). The reader can consult [185] for the detailed proofs and [82,225,226] for a more general characterization.

PROPOSITION 1. *Let $1 \leqslant p, q \leqslant \infty$ and $s < 1$, then the quantities*

$$\left(\sum_{j\in\mathbb{Z}} (2^{sj}\|\Delta_j f\|_q)^p\right)^{1/p} \tag{79}$$

and

$$\left(\int_0^\infty (t^{-s}\|\Theta_t f\|_q)^p \frac{\mathrm{d}t}{t}\right)^{1/p} \tag{80}$$

are equivalent and will be referred to in the sequel by $\|f\|_{\dot{B}_q^{s,p}}$.

PROPOSITION 2. *Let* $1 \leqslant p \leqslant \infty$, $1 \leqslant q < \infty$ *and* $s < 1$, *then the quantities*

$$\left\| \left(\sum_{j\in\mathbb{Z}} (2^{sj} |\Delta_j f|)^p \right)^{1/p} \right\|_q \tag{81}$$

and

$$\left\| \left(\int_0^\infty (t^{-s} |\Theta_t f|)^p \frac{\mathrm{d}t}{t} \right)^{1/p} \right\|_q \tag{82}$$

are equivalent and will be referred to in the sequel by $\|f\|_{\dot{F}_q^{s,p}}$.

The next two equivalences are even more useful because they allow us to pass from Δ_j to S_j (and from the discrete set S_j to the continuous $S(t)$ one). Here a restriction in the range of exponents also appears and we will be forced to assume that $s < 0$. More precisely, the reason why the equivalences under consideration are not true if $s \geqslant 0$ is essentially the following: even if we can easily estimate any quantity involving Δ_j from above with one only involving S_j, because of the identity

$$\Delta_j = S_{j+1} - S_j, \tag{83}$$

passing from Δ_j to S_j, via the relation

$$S_{j+1} = \sum_{k\leqslant j} \Delta_k, \tag{84}$$

it is not possible when $s \geqslant 0$ (see [185]). In the context of the Navier–Stokes equations, an explicit counter-example for $s = 0$ was given in [34] for the Besov spaces. A second one for the Triebel–Lizorkin spaces (always with $s = 0$) will be given in the following pages.

But let us state the equivalences we are talking about (for a proof see [225], p. 192).

PROPOSITION 3. *Let* $1 \leqslant p, q \leqslant \infty$ *and* $s < 0$, *then the quantities*

$$\left(\sum_{j\in\mathbb{Z}} (2^{sj} \|\Delta_j f\|_q)^p \right)^{1/p}, \tag{85}$$

$$\left(\sum_{j\in\mathbb{Z}} (2^{sj} \|S_j f\|_q)^p \right)^{1/p}, \tag{86}$$

$$\left(\int_0^\infty (t^{-s/2} \|S(t) f\|_q)^p \frac{\mathrm{d}t}{t} \right)^{1/p} \tag{87}$$

and

$$\left(\int_0^\infty \left(t^{-s}\|\Theta_t f\|_q\right)^p \frac{\mathrm{d}t}{t}\right)^{1/p} \tag{88}$$

are equivalent and will be referred to in the sequel by $\|f\|_{\dot{B}_q^{s,p}}$.

PROPOSITION 4. *Let* $1 \leqslant p \leqslant \infty$, $1 \leqslant q < \infty$ *and* $s < 0$, *then the quantities*

$$\left\|\left(\sum_{j\in\mathbb{Z}} \left(2^{sj}|\Delta_j f|\right)^p\right)^{1/p}\right\|_q, \tag{89}$$

$$\left\|\left(\sum_{j\in\mathbb{Z}} \left(2^{sj}|S_j f|\right)^p\right)^{1/p}\right\|_q, \tag{90}$$

$$\left\|\left(\int_0^\infty \left(t^{-s/2}|S(t) f|\right)^p \frac{\mathrm{d}t}{t}\right)^{1/p}\right\|_q \tag{91}$$

and

$$\left\|\left(\int_0^\infty \left(t^{-s}|\Theta_t f|\right)^p \frac{\mathrm{d}t}{t}\right)^{1/p}\right\|_q \tag{92}$$

are equivalent and will be referred to in the sequel by $\|f\|_{\dot{F}_q^{s,p}}$.

The next propositions will be also useful in the following pages. Of course the embeddings are also valid for inhomogeneous spaces.

PROPOSITION 5.

$$\textit{If } s_1 > s_2 \textit{ and } s_1 - \frac{3}{q_1} = s_2 - \frac{3}{q_2}, \textit{ then } \dot{B}_{q_1}^{s_1,p_1} \hookrightarrow \dot{B}_{q_2}^{s_2,p_2} \textit{ and } \dot{F}_{q_1}^{s_1,p_1} \hookrightarrow \dot{F}_{q_2}^{s_2,p_2}. \tag{93}$$

$$\textit{If } p_1 < p_2, \textit{ then } \dot{B}_q^{s_1,p_1} \hookrightarrow \dot{B}_q^{s_2,p_2} \textit{ and } \dot{F}_q^{s_1,p_1} \hookrightarrow \dot{F}_q^{s_2,p_2}. \tag{94}$$

$$\textit{For any } p, q \textit{ and } s,\ \dot{B}_p^{s,\min(p,q)} \hookrightarrow \dot{F}_q^{s,p} \hookrightarrow \dot{F}_q^{s,\max(p,q)}. \tag{95}$$

2.3. *The paraproduct rule*

In order to study how the product acts on Besov spaces, we need to recall Bony's paraproduct algorithm [20], one of the most celebrated tools of paradifferential calculus. The Greek prefix "para" is added here in front of *product* and *differential* to underline that the new operations "go beyond" the usual ones. In particular, the new calculus enables us to define a new product between distributions which turns out to be continuous in many functional spaces where the usual product does not even make sense.

More precisely, let us consider two tempered distributions f and g and write, in terms of a Littlewood–Paley decomposition,

$$f = \sum_j \Delta_j f, \tag{96}$$

$$g = \sum_j \Delta_j g \tag{97}$$

so that, formally,

$$fg = \sum_n [S_{n+1} f S_{n+1} g - S_n f S_n g] + S_0 f S_0 g. \tag{98}$$

Now, after some simplifications, we get

$$\begin{aligned} fg &= \sum_n [\Delta_n f S_n g + \Delta_n g S_n f + \Delta_n f \Delta_n g] \\ &= \sum_n \Delta_n f S_{n-2} g + \sum_n \Delta_n g S_{n-2} f + \sum_{|n-n'|\leqslant 2} \Delta_{n'} f \Delta_n g. \end{aligned} \tag{99}$$

In other words, the product of two tempered distributions is decomposed into two paraproducts, respectively,

$$\pi(f, g) = \sum_n \Delta_n f S_{n-2} g \tag{100}$$

and

$$\pi(g, f) = \sum_n \Delta_n g S_{n-2} f, \tag{101}$$

plus a remainder. Finally, if we want to analyze the product fg by means of the frequency filter Δ_j we deduce from (101), modulo some nondiagonal terms that we are neglecting for simplicity,

$$\Delta_j(fg) = \Delta_j f S_{j-2} g + \Delta_j g S_{j-2} f + \Delta_j \Bigl(\sum_{k \geqslant j} \Delta_k f \Delta_k g \Bigr). \tag{102}$$

Usually, the first two contributions are easier to treat than the third remainder term.

2.4. *The wavelet decomposition*

The Littlewood–Paley decomposition allows us to describe an arbitrary tempered distribution into the sum of regular functions that are well localized in the frequency variable.

The wavelet decomposition allows us to obtain an even better localization for these functions, say in both space and frequency. Of course, the ideal case of functions that are compactly supported in space as well as in frequency is excluded by Heisenberg's principle. Wavelets were discovered at the beginning of the 1980s and the best reference is Meyer's work [162,163].

The idea of using a wavelet decomposition to study turbulence questions was advocated from the very beginning, at about the same time when wavelets tools were available. In fact, due to the strong impact that wavelets had in several important scientific and technological discoveries, many people started dreaming that wavelets could provide the "golden rule" to attack the Navier–Stokes equations, from both mathematical and numerical points of view (see for instance the paper of Farge [77] and the references therein).

We do not discuss here the relevance of wavelets in numerical simulations of the Navier–Stokes equations and refer the reader to Meyer's conclusion in [166]. From the point of view of nonlinear partial differential equations, the situation is a little disappointing. The first attempt to approach the Navier–Stokes equations, by expanding the unknown velocity field $v(t, x)$ into a wavelet basis in space variable, came from Federbush, who wrote an intriguing paper in 1993 [78]. The techniques and insights employed arose from the theory of phase cell analysis used in constructive quantum field theory, and were the starting point and the first source of inspiration of our work [34].

The disappointing note is that, as we will see in the following sections, Federbush's program can be realized as well by using the less sophisticated Littlewood–Paley decomposition. On the other hand, the good news is that the systematic use of harmonic analysis tools (Littlewood–Paley and wavelets decomposition and their natural companions, Besov spaces and Bony's paraproducts techniques) paved the way for important discoveries for Navier–Stokes: the existence of a global solution for highly oscillating data, the uniqueness of this solution and its asymptotic behavior, via the existence of self-similar solutions.

As we have already announced in the Introduction, our story is full of surprises and bad news follows here at once. In fact, each proof of the previous results originally discovered by means of 'Fourier analysis methods', more precisely, by using 'Besov spaces', was followed shortly after its publication by a 'real variable methods' proof.

We will come back to these questions – existence, uniqueness, self-similar solutions – and treat them in detail in three separate sections (resp. Sections 4–6). Before doing this and in order to clarify the previous discussion, let us briefly recall here, for the convenience of the reader, some definitions taken from the wavelet world. Roughly speaking, a wavelet decomposition is a decomposition of the type

$$f = \sum_{\lambda} \langle f, \psi_\lambda \rangle \psi_\lambda, \tag{103}$$

where ψ_λ is "essentially" localized in frequency in a dyadic annulus 2^j and "essentially" localized in space in a dyadic cube 2^{-j}. More precisely, following Meyer [162], we have the following definition:

DEFINITION 8. A wavelet decomposition of regularity $m > 0$ is a set of $2^3 - 1 = 7$ functions ψ_ε, $\varepsilon \in \{0, 1\}^3 \setminus \{0, 0, 0\}$ verifying the following properties:

1. *Regularity*: ψ_ε belongs to $\mathcal{C}^m$.
2. *Localization*:

$$\forall\alpha,\ |\alpha|\leqslant m,\ \forall N\in\mathbb{N}, \exists C:\quad \left|\partial^\alpha\psi_\varepsilon\right|(x)\leqslant C\big(1+|x|\big)^{-N}. \tag{104}$$

3. *Oscillation*:

$$\forall\alpha,\ |\alpha|\leqslant m:\qquad \int x^\alpha\psi_\varepsilon(x)\,\mathrm{d}x=0. \tag{105}$$

4. *Orthogonality*: The set

$$\big\{2^{3j/2}\psi_\varepsilon\big(2^j x-k\big)\big/ j\in\mathbb{Z}^3, \varepsilon\in\{0,1\}^3\backslash\{0,0,0\}\big\} \tag{106}$$

is an orthogonal basis of L^2.

If we denote $\psi_{j,k}(x)=2^{3j/2}\psi(2^j x-k)$ (where, for the sake of simplicity, the parameter ε is neglected), then we obtain the following "homogeneous" decomposition

$$f=\sum_{j\in\mathbb{Z}}\sum_{k\in\mathbb{Z}^3}\langle f,\psi_{j,k}\rangle\psi_{j,k}=\sum_{j,k}c_{j,k}\psi_{j,k} \tag{107}$$

that, as in the case of the homogeneous Littlewood–Paley decomposition, has to be understood in $\mathcal{S}'$ modulo polynomials.

Formally, a Littlewood–Paley decomposition Δ_j gives a wavelet decomposition $\psi_{j,k}$ by letting

$$c_{j,k}\simeq\Delta_j f\big(2^{-j}k\big) \tag{108}$$

and, vice versa, from a wavelet decomposition we can recover a Littlewood–Paley one just by taking

$$\Delta_j f\simeq\sum_{k\in\mathbb{Z}^3}c_{j,k}\psi_{j,k}. \tag{109}$$

Finally, the wavelets coefficients $c_{j,k}$ of a function f allow us to obtain an equivalent definition of the Besov and Triebel–Lizorkin spaces. For example, we have the following proposition [162]:

PROPOSITION 6. *If ψ is a wavelet of regularity $m>0$, then, for any $|s|<m$ and any $1\leqslant p,q\leqslant\infty$, we have the equivalence of norms*

$$\|f\|_{\dot{B}_q^{s,p}}\simeq\bigg(\sum_{j\in\mathbb{Z}}2^{jp(s+3(1/2-1/q))}\bigg(\sum_{k\in\mathbb{Z}^3}|c_{j,k}|^q\bigg)^{p/q}\bigg)^{1/p}. \tag{110}$$

REMARK. In the study of the Navier–Stokes equations and other *incompressible* fluid equations, one would expect that the wavelets functions ψ_ε in Definition 8 have an additional property:

5. *Divergence-free*: Divergence-free basis of wavelets were first discovered by Battle and Federbush [6,7] and their construction was improved later by Lemarié [139,140,145]. A simple presentation of these basis is contained in the paper by Meyer [166].

2.5. *Other useful function spaces*

Before we enter the heart of the paper, devoted to existence and uniqueness theorems for the Navier–Stokes equations, we wish to end this section by presenting other functional spaces, that will be useful in the following pages.

2.5.1. *Morrey–Campanato spaces.* For $1 \leqslant q \leqslant p \leqslant \infty$, the inhomogeneous Morrey–Campanato space M_q^p is defined as the space of functions f which are locally in L^q and such that

$$\sup_{x\in\mathbb{R}^3, 0<r\leqslant 1} R^{3/p}\left(R^{-3}\int_{|x-y|\leqslant r} |f(x)|^q \,\mathrm{d}y\right)^{1/q} < \infty, \tag{111}$$

where the left-hand side of this inequality is the norm of f in M_q^p. The homogeneous Morrey–Campanato space $\dot{M}_q^p$ is defined in the same way, by taking the supremum over all $r \in (0,\infty)$ instead of $r \in (0,1]$.

2.5.2. *Lorentz spaces.* Let $1 \leqslant p, q \leqslant \infty$, then a function f belongs to the Lorentz space $L^{(p,q)}$ if and only if 'the quantity'

$$\|f\|_{L^{(p,q)}} = \left(\frac{q}{p}\int_0^\infty \left[t^{1/p} f^*(t)\right]^q \frac{\mathrm{d}t}{t}\right)^{1/q} < \infty, \tag{112}$$

of course, if $q = \infty$ this means

$$\|f\|_{L^{(p,\infty)}} = \sup_{t>0} t^{1/p} f^*(t) < \infty, \tag{113}$$

where f^* is the decreasing rearrangement of f:

$$f^*(t) = \inf\{s \geqslant 0;\ \left|\{|f| > s\}\right| \leqslant t\}, \quad t \geqslant 0. \tag{114}$$

We know [215] that for $p > 1$, a norm on $L^{(p,q)}$ equivalent to 'the quantity' $\|f\|_{L^{(p,q)}}$ exists such that $L^{(p,q)}$ becomes a Banach space. If $p = q$, the space $L^{(p,p)}$ is nothing else than the Lebesgue space L^p. Moreover, generalization versions of Hölder and Young inequalities hold for the Morrey–Campanato spaces [111]. Finally, for these spaces, the theory of real interpolation gives the equivalence (see [10])

$$\left(L^{p_0}, L^{p_1}\right)_{(\theta,q)} = L^{(p,q)}, \tag{115}$$

where $1 < p_0 < p < p_1 < \infty$ and $0 < \theta < 1$ satisfy $1/p = (1-\theta)/p_0 + \theta/p_1$ and $1 \leqslant q \leqslant \infty$.

2.5.3. *Le Jan–Sznitman spaces.* Recently, Le Jan and Sznitman [137,138] considered the space of tempered distributions f whose Fourier transform verifies

$$\sup_{\mathbb{R}^3} |\xi|^2 \left|\hat{f}(\xi)\right| < \infty. \tag{116}$$

Now, if in the previous expression we consider $\int_{\xi\in\mathbb{R}^3}$ instead of $\sup_{\xi\in\mathbb{R}^3}$, we obtain the (semi)-norm of a homogeneous Sobolev space. This is not the case: the functions whose Fourier transform is bounded define the pseudo-measure space $\mathcal{PM}$ of Kahane. In other words, a function f belongs to the space introduced by Le Jan and Sznitman if and only if $\Delta f \in \mathcal{PM}$, Δ being the Laplacian (in three dimensions). A simple calculation (see [48]) shows that condition (116) is written, in the dyadic decomposition Δ_j of Littlewood and Paley in the form $4^j \|\Delta_j f\|_{\mathcal{PM}} = 4^j \|\widehat{\Delta_j f}\|_\infty \in \ell^\infty(\mathbb{Z})$ and defines in this way "the homogeneous Besov space" $\dot{B}^{2,\infty}_{\mathcal{PM}}$.

Let us note that this quantity is not a norm, unless we work in $\mathcal{S}'$ modulo polynomials, as we did in Section 2.2 in the case of homogeneous Besov spaces (for example, if f is a constant or, more generally a polynomial of degree 1, it is easy to see that $|\xi|^2|\hat{f}(\xi)| = 0$). Another possibility to avoid this technical point is to ask that $\hat{f} \in L^1_{\mathrm{loc}}$. In other words, the Banach functional space relevant to our study is defined by

$$\mathcal{PM}^2 = \left\{ v \in \mathcal{S}' : \hat{v} \in L^1_{\mathrm{loc}}, \|v\|_{\mathcal{PM}^2} \equiv \sup_{\xi\in\mathbb{R}^3} |\xi|^2 \left|\hat{v}(\xi)\right| < \infty \right\}. \tag{117}$$

A generalization of this functional space was recently introduced in the paper by Bhattacharya, Chen, Dobson, Guenther, Orum, Ossiander, Thomann and Waymire (see [8]).

3. Existence theorems

3.1. *The fixed point theorem*

We will recall here two classical results concerning the existence of fixed point solution to abstract functional equations. These theorems are known under the name of Picard in France, Caccioppoli in Italy, and Banach in Poland and ... in the rest of the world!

LEMMA 4. *Let X be an abstract Banach space with norm $\|\cdot\|$ and $B : X \times X \to X$ a bilinear operator such that, for any $x_1, x_2 \in X$,*

$$\big\|B(x_1, x_2)\big\| \leqslant \eta\|x_1\|\|x_2\|, \tag{118}$$

then, for any $y \in X$ such that

$$4\eta\|y\| < 1, \tag{119}$$

the equation

$$x = y + B(x, x) \tag{120}$$

has a solution x in X. In particular, the solution is such that

$$\|x\| \leqslant 2\|y\| \tag{121}$$

and it is the only one such that

$$\|x\| < \frac{1}{2\eta}. \tag{122}$$

The following lemma is a generalization of the previous one ($\lambda = 0$) and will be useful when treating the mild Navier–Stokes equations in the presence of a nontrivial external force (11).

LEMMA 5. *Let X be an abstract Banach space with norm $\|\cdot\|$, $L : X \to X$ a linear operator such that, for any $x \in X$,*

$$\big\|L(x)\big\| \leqslant \lambda\|x\| \tag{123}$$

and $B : X \times X \to X$ a bilinear operator such that, for any $x_1, x_2 \in X$,

$$\big\|B(x_1, x_2)\big\| \leqslant \eta\|x_1\|\|x_2\|, \tag{124}$$

then, for any λ, $0 < \lambda < 1$, and for any $y \in X$ such that

$$4\eta\|y\| < (1 - \lambda)^2, \tag{125}$$

the equation

$$x = y + B(x, x) + L(x) \tag{126}$$

has a solution x in X. In particular, the solution is such that

$$\|x\| \leqslant \frac{2\|y\|}{1 - \lambda} \tag{127}$$

and it is the only one such that

$$\|x\| < \frac{1-\lambda}{2\eta}. \tag{128}$$

For an elementary proof of the above mentioned lemmata the reader is referred to [34] and to [3] where a different proof is given that also applies to the (optimal) case where the equality sign holds in (119), (122), (125) and (128).

3.2. *Scaling invariance*

The Navier–Stokes equations are invariant under a particular change of time and space scaling. More exactly, assume that, in $\mathbb{R}^3 \times (0, \infty)$, $v(t, x)$ and $p(t, x)$ solve the system

$$\begin{aligned} &\frac{\partial v}{\partial t} - \nu \Delta v = -(v \cdot \nabla) v - \nabla p, \\ &\nabla \cdot v = 0, \end{aligned} \tag{129}$$

then the same is true for the rescaled functions

$$v_\lambda(t, x) = \lambda v\big(\lambda^2 t, \lambda x\big), \qquad p_\lambda(t, x) = \lambda^2 p\big(\lambda^2 t, \lambda x\big). \tag{130}$$

On the other hand, the functions $v(\lambda t, \lambda x)$ and $p(\lambda t, \lambda x)$ solve a different Navier–Stokes system, where ν is replaced by $\lambda \nu$, thus allowing us to assume that viscosity is equal to unity, as we did in Section 1.1 (because, if not, it is possible to find a $\lambda > 0$ such that $\lambda \nu = 1$). The above scaling invariance leads to the following definition.

DEFINITION 9. *Critical space.* A translation invariant Banach space of tempered distributions X is called a critical space for the Navier–Stokes equations if its norm is invariant under the action of the scaling $f(x) \to \lambda f(\lambda x)$ for any $\lambda > 0$. In other words, we require the embedding

$$X \hookrightarrow \mathcal{S}' \tag{131}$$

and that, for any $f \in X$,

$$\big\| f(\cdot) \big\| = \big\| \lambda f(\lambda \cdot - x_0) \big\| \quad \forall \lambda > 0, \forall x_0 \in \mathbb{R}^3. \tag{132}$$

Critical spaces are all embedded in a same function space, as stated in the following proposition.

PROPOSITION 7 (A remarkable embedding). *If X is a critical space, then X is continuously embedded in the Besov space $\dot{B}_\infty^{-1,\infty}$.*

The proof of this result is so simple that we would like to present it here. We argue as in the proof of the "minimality of $\dot{B}_1^{0,1}$" given by Frazier, Jawerth and Weiss in [82] (see also [3,161,166]).

To begin, we note that if X satisfies (131), then there exists a constant C such that

$$\left|\left\langle \exp\left(-|x|^2/4\right), f\right\rangle\right| \leqslant C\|f\|_X \quad \forall f \in X. \tag{133}$$

Now, using the translation invariance of X we obtain

$$\left\|\exp(\Delta) f\right\|_{L^\infty} \leqslant C\|f\|_X \tag{134}$$

and, by the invariance under the scaling $f(x) \to \lambda f(\lambda x)$, we get

$$t^{1/2}\left\|\exp(t\Delta) f\right\|_{L^\infty} \leqslant C\|f\|_X. \tag{135}$$

It is now easy to conclude if we recall Proposition 3, say

$$\sup_{t>0} t^{1/2}\left\|\exp(t\Delta) f\right\|_{L^\infty} \simeq \|f\|_{\dot{B}_\infty^{-1,\infty}}. \tag{136}$$

As we will see in the following pages, it is a remarkable feature that the Navier–Stokes equations are well posed in the sense of Hadamard (existence, uniqueness and stability) when the initial data is divergence-free and belongs to certain *critical* function spaces. Actually, it is unclear whether this property is true for either a generic critical space or for the bigger critical space $\dot{B}_\infty^{-1,\infty}$ (see the conjecture formulated in [166], Chapter 8, and [160]), but it happens to be the case for most of the critical functional spaces we have described so far.

For example, in the Lebesgue family $L^p = L^p(\mathbb{R}^3)$ the critical invariant space corresponds to the value $p = 3$ (more generally in $\mathbb{R}^n$, $p = n$) and we will see how to construct mild solutions to the Navier–Stokes equations with data in L^3. The same argument applies to the critical Sobolev space $\dot{H}^{1/2}$, to the Morrey–Campanato $\dot{M}_p^3$ $(1 \leqslant p \leqslant 3)$, the Lorentz $L^{(3,q)}$ $(1 \leqslant q \leqslant \infty)$, the pseudo-measure space of Le Jan and Sznitman $\mathcal{PM}^2$, the Besov $\dot{B}_p^{3/p-1,q}$ $(1 \leqslant q \leqslant \infty, 1 \leqslant p < \infty)$ as well as the Triebel–Lizorkin spaces $\dot{F}_p^{3/p-1,q}$ $(1 \leqslant q < \infty, 1 \leqslant p < \infty)$. The reader is referred to [3] for a precise and exhaustive analysis of the Navier–Stokes equations in critical spaces. Here we will only treat the case of the Lebesgue space L^3 in detail.

Another (equivalent) way of defining critical spaces for the Navier–Stokes equations is to note that in this case the nonlinear term $\nabla \cdot (v \otimes v)$ has the same strength as the Laplace operator; that is $\nabla \cdot (v \otimes v)$ is not subordinate to $-\Delta v$. For instance, if $v \in L^p$ $(p \geqslant 2)$, then $\nabla \cdot (v \otimes v) \in W^{p/2,-1}$ whereas $-\Delta v \in W^{p,-2}$ and, by Sobolev embedding, $W^{p/2,-1} \hookrightarrow W^{p,-2}$ as long as $p \geqslant 3$.

Before recalling the main steps of the proof for the existence of mild solution with initial data in L^3, let us begin with an easier case, the so-called 'supercritical' space L^p, $p > 3$. We will not give a precise definition of 'critical', 'supercritical', or 'subcritical' spaces. The meaning of their names should be clear enough to any reader (for more details and examples see [34,43]). Let us just notice that what we call here 'supercritical' spaces are called 'subcritical' spaces (and vice versa) in the paper by Klainerman [122].

3.3. *Supercritical case*

The main theorem of the existence of mild solutions in L^p, $3 < p < \infty$, was known since the papers of Fabes, Jones and Rivière [76] (1972) and Giga [100] (1986). Concerning the space L^∞, let us note that the existence was obtained only recently in [34,43] by using the simplified structure of the bilinear term we introduced in (34). In fact, as pointed out in a different proof by Giga and his students in Sapporo [104], the difficulty comes from the fact that the Leray–Hopf projection $\mathbb{P}$ is not bounded in L^∞, nor in L^1. The proof we are going to present applies to $3 < p \leqslant \infty$ and is contained in [34,43]. The idea is of course to use the fixed point theorem by means of the following two lemmata, whose proofs are obtained by a simple application of the Young inequality.

LEMMA 6. *Let X be a Banach space, whose norm is translation invariant. For any $T > 0$ and any $v_0 \in X$, we have*

$$\sup_{0<t<T} \left\| S(t)v_0 \right\|_X = \|v_0\|_X. \tag{137}$$

Of course this lemma applies for example when X is a Lebesgue space, in our case $X = L^p$ with $3 < p \leqslant \infty$.

LEMMA 7. *Let $3 < p \leqslant \infty$ be fixed. For any $T > 0$ and any functions $f(t), g(t) \in C([0,T); L^p)$, then the bilinear term $B(f,g)(t)$ also belongs to $\in C([0,T); L^p)$ and we have*

$$\sup_{0<t<T} \left\| B(f,g)(t) \right\|_p \lesssim \frac{T^{1/2(1-3/p)}}{1-3/p} \sup_{0<t<T} \left\| f(t) \right\|_p \sup_{0<t<T} \left\| g(t) \right\|_p. \tag{138}$$

Combining these lemmata with the fixed point algorithm Lemma 4 we obtain the following existence result (see Section 5.2 for its uniqueness counterpart).

THEOREM 1. *Let $3 < p \leqslant \infty$ be fixed. For any $v_0 \in L^p$, $\nabla \cdot v_0 = 0$, there exists a $T = T(\|v_0\|_p)$ such that the Navier–Stokes equations has a solution in $C([0,T); L^p)$.*

To be more precise, according to the notations introduced in Definition 2, we should write $v \in C([0,T); \mathbb{P}L^p)$, because the solution constructed so far is of course a solenoidal (i.e., divergence-free) vector field. To simplify the discussion, we prefer not to use such notation in the following.

We should also remark that the strong continuity at $T = 0$ is not ensured in the case L^∞, because this space is nonseparable. In other words, if it is true that

$$\lim_{t\to 0} \left\| v(t) - v_0 \right\|_p = 0, \quad 3 < p < \infty, \tag{139}$$

this is not the case if $p = \infty$, for the heat semigroup $S(t)$ is not strongly continuous as $t \to 0$.

There are two ways to restore continuity in the case of a nonseparable Banach space X. The first is to restrict the attention to X_*, the closure of C_0^∞ in X. Then, $S(t)$ is strongly continuous and the existence theorem applies as stated. On the other hand, if X is nonseparable, but instead X is the dual of a separable space Y (here $X = L^\infty$ and $Y = L^1$), it is natural to replace $\mathcal{C}([0,T]; X)$ with the space we will denote $\mathcal{C}_*([0,T]; X)$ consisting of *bounded* functions $v(t)$ with values in X which have the property that v is continuous in t with values in X, when X is endowed with the $\sigma(X,Y)$ topology (see [34,43,104,166,216]).

Finally, we will see in the next section that the solution constructed so far is always regular, unique and stable. This means that the Cauchy problem is *locally in time* well posed if the data belong to the supercritical space L^p, $3 < p \leqslant \infty$. It is an open question to know whether the solution is actually global in time. The noninvariance of the L^p norm, $p \neq 3$ ensures that such a global result would not depend on the size of the initial data, say the quantity $\|v_0\|_p$ (or, more generally, if $\nu \neq 1$, the quantity $\|v_0\|_p/\nu$).

3.4. *Critical case*

By means of the critical Lebesgue space L^3 we will see how to construct the existence not only of local solutions for arbitrary initial data, but also of global ones, for small or highly oscillating data (this property will be described in detail in Section 4).

Let us begin with an unpleasant remark. If we try to apply the fixed point theorem to the integral Navier–Stokes equation

$$v(t) = S(t)v_0 - \int_0^t S(t-s)\mathbb{P}\nabla \cdot (v \otimes v)(s)\,\mathrm{d}s \tag{140}$$

in the (natural) function space

$$\mathcal{N} = \mathcal{C}\big([0,T]; L^3\big), \tag{141}$$

we are faced with a difficulty that did not appear in the supercritical case: the bilinear term $B(v,u) = -\int_0^t S(t-s)\mathbb{P}\nabla \cdot (v \otimes u)(s)\,\mathrm{d}s$ is not continuous from $\mathcal{N} \times \mathcal{N} \to \mathcal{N}$.

Of course, the fact that the estimate (138) diverges when $p = 3$ is not enough to show the noncontinuity: first, we would expect a reverse inequality, second, this reverse inequality should apply to the full vectorial bilinear term (in fact, in a way reminiscent of the so-called "div–curl" lemma [66], one can imagine that the full bilinear operator is continuous even if its simplified scalar version is not).

In his unpublished doctoral thesis [183], Oru proved the noncontinuity of the full vectorial term not only in the Lebesgue space L^3, but also in any Lorentz space $L^{(3,q)}$, for any $q \in [1,\infty)$:

PROPOSITION 8. *The (vectorial) bilinear operator B is not continuous from $\mathcal{C}([0,T]; L^{(3,q)}) \times \mathcal{C}([0,T]; L^{(3,q)}) \to \mathcal{C}([0,T]; L^{(3,q)})$, whatever $0 < T \leqslant \infty$ and $q \in [1,\infty)$ are.*

At about the same time Meyer [166] showed that the critical space $L^{(3,\infty)}$ is very different since:

PROPOSITION 9. *The bilinear operator B is continuous from $\mathcal{C}([0,T); L^{(3,\infty)}) \times \mathcal{C}([0,T); L^{(3,\infty)}) \to \mathcal{C}([0,T); L^{(3,\infty)})$ for any $0 < T \leqslant \infty$.*

Oru's theorem is based on the following remark (see also [145]):

LEMMA 8. *If X is a critical space in the sense of Definition 9 and if the bilinear operator B is continuous in the space $\mathcal{C}([0,T); X)$ for a certain T, then X contains a function of the form*

$$\frac{\omega(x)}{|x|} + \phi(x), \tag{142}$$

where ω does not vanish identically, is homogeneous of degree 0, is C^∞ outside the origin and ϕ is a C^∞ function.

In fact, it is possible to prove that functions of the type (142) do not belong to $L^{(3,q)}$, if $q \neq \infty$ but can be in $L^{(3,\infty)}$, thus not contradicting Proposition 9.

Let us note, in passing, that it is very surprising that for a generic critical space we cannot be sure whether the bilinear term is continuous or not. Another example where it is quite easy to prove the continuity of the bilinear term (and thus the existence of a solution) is provided by the critical space $\mathcal{PM}^2$ introduced by Le Jan and Sznitman [138]. We will describe some important consequences of the continuity of the bilinear term in the spaces $L^{(3,\infty)}$ and $\mathcal{PM}^2$ in Sections 6.2 and 6.4.

Let us go back to L^3. If we want to find a mild solution with initial data in this space, there are (at least) three ways to circumvent the obstacle arising from Proposition 8 and are all based on the following remark: the fixed point algorithm in $\mathcal{N}$ is only a *sufficient* condition to ensure the existence of a solution in $\mathcal{N}$ and a different strategy can be considered.

To be more explicit, another sufficient condition leading to the existence of a solution in $\mathcal{N}$ is to find a function space $\mathcal{F}$ (whose elements are functions $v(t,x)$ with $0 < t < T$ and $x \in \mathbb{R}^3$) such that:

(1) the bilinear term $B(u,v)(t)$ is continuous from $\mathcal{F} \times \mathcal{F} \to \mathcal{F}$,

(2) if $v_0 \in L^3$, then $S(t)v_0 \in \mathcal{F}$, and

(3) the bilinear term $B(u,v)(t)$ is continuous from $\mathcal{F} \times \mathcal{F} \to \mathcal{N}$.

In fact, the first two conditions ensure the existence of a (mild) solution $v(t,x) \in \mathcal{F}$, via the fixed point algorithm and, thanks to the third condition, this solution belongs to $\mathcal{N}$ as well (if $\mathcal{F} \hookrightarrow \mathcal{N}$, the third condition being of course redundant).

The three ways known in the literature to obtain a solution $v(t,x) \in \mathcal{N}$ with data in L^3 correspond to three different choices of spaces $\mathcal{F}$ [47]. For the convenience of the reader we will briefly recall in the following sections these spaces leading to the same existence theorem in $\mathcal{N}$ that reads as follows.

THEOREM 2. *For any $v_0 \in L^3$, $\nabla \cdot v_0 = 0$, there exists a $T = T(v_0)$ such that the Navier–Stokes equations have a local solution in $\mathcal{C}([0,T); L^3)$. Moreover, there exists $\delta > 0$ such that if $\|v_0\|_3 < \delta$, then the solution is global, i.e., we can take $T = \infty$.*

As it will be clear in the following pages, here at variance with Theorem 1 we cannot say that $T = T(\|v_0\|_3)$. Again, as far as the uniqueness of the solution, the situation is more delicate and will be revealed in Section 5.3.

3.4.1. *Weissler's space.* In 1981, Weissler [234] gave the first existence result of mild solutions in the half space $L^3(\mathbb{R}^3_+)$, then Giga and Miyakawa [106] generalized the proof to $L^3(\Omega_b)$, Ω_b an open bounded domain in $\mathbb{R}^3$. Finally, in 1984, Kato [114] obtained, by means of a purely analytical proof (involving only Hölder and Young inequalities and without using any estimate of fractional powers of the Stokes operator), an existence theorem in the whole space $L^3(\mathbb{R}^3)$.

In [34,35,47] we showed how to simplify Kato's proof. The idea is to take advantage of the structure of the bilinear operator in its scalar form, as in (34) and (36). In particular, the divergence $\nabla\cdot$ and heat $S(t)$ operators can be treated as a single convolution operator [34]. This is why no explicit conditions on the gradient of the unknown function v and no restriction on q (namely $3 < q < 6$) will be required here, as they were indeed in Kato's original paper [114]. In a different context [34,43] and by using the same simplified scalar structure, it was possible to show the existence of a solution with data in the Lebesgue space L^∞ (Section 3.3), even if the Leray–Hopf operator $\mathbb{P}$ is not bounded in L^∞.

In order to proceed, we have to recall the definition of the auxiliary space $\mathcal{K}_q$ ($3 \leqslant q \leqslant \infty$) introduced by Weissler and systematically used by Kato. More exactly, this space $\mathcal{K}_q$ is made up by the functions $v(t,x)$ such that

$$t^{\alpha/2} v(t,x) \in \mathcal{C}\big([0,T); L^q\big) \tag{143}$$

and

$$\lim_{t\to 0} t^{\alpha/2} \big\|v(t)\big\|_q = 0, \tag{144}$$

with q being fixed in $3 < q \leqslant \infty$ and $\alpha = \alpha(q) = 1 - 3/q$. In the case $q = 3$, it is also convenient to define the space $\mathcal{K}_3$ as the natural space $\mathcal{N}$ with the additional condition that its elements $v(t,x)$ satisfy

$$\lim_{t\to 0} \big\|v(t)\big\|_3 = 0. \tag{145}$$

The theorem in question, that implies Theorem 2, is the following [34]:

THEOREM 3. *Let $3 < q < \infty$, and $\alpha = 1 - 3/q$ be fixed. There exists a constant $\delta_q > 0$ such that, for any initial data $v_0 \in L^3$, $\nabla \cdot v_0 = 0$ in the sense of distributions such that*

$$\sup_{0<t<T} t^{\alpha/2} \big\|S(t)v_0\big\|_q < \delta_q, \tag{146}$$

then there exists a mild solution $v(t,x)$ to the Navier–Stokes equations belonging to $\mathcal{N}$, which tends strongly to v_0 as time goes to zero. Moreover, this solution belongs to all spaces $\mathcal{K}_q$ for all $3 < q < \infty$. In particular, (146) *holds for arbitrary $v_0 \in L^3$ provided we consider $T(v_0)$ small enough, and as well if $T = \infty$, provided the norm of v_0 in the Besov space $\dot{B}_q^{-\alpha,\infty}$ is smaller than δ_q.*

The existence part of the proof of this theorem is a consequence of the following lemmata that we recall here.

LEMMA 9. *If $v_0 \in L^3$, then $S(t)v_0 \in \mathcal{K}_q$ for any $3 < q \leqslant \infty$. In particular this implies (when $T = \infty$) the continuous embedding*

$$L^3 \hookrightarrow \dot{B}_q^{-\alpha,\infty}, \quad 3 < q \leqslant \infty. \tag{147}$$

In particular, this lemma implies that the conclusion of Theorem 3 holds not only in the general case of arbitrary $v_0 \in L^3$ when $T = \infty$, provided the norm of v_0 in the Besov space $\dot{B}_q^{-\alpha,\infty}$ is smaller than δ_q, but also in the more restrictive case of $v_0 \in L^3$ and small enough in L^3, as we recalled in the statement of Theorem 2 and originally proved in the papers of Weissler, Giga and Miyakawa, and Kato. In other words, a function in L^3 can be arbitrarily large in the L^3 norm but small in $\dot{B}_q^{-\alpha,\infty}$. This remark will play a key role in Section 4. Another important consequence of this lemma is that L^3 and $B_q^{-\alpha,\infty}$ are different spaces, for $|x|^{-1} \in \dot{B}_q^{-\alpha,\infty}$ and $|x|^{-1} \notin L^3$ and this will allow the construction of self-similar solutions in Section 6.

The second lemma we need in order to prove Theorem 3 is the following:

LEMMA 10. *The bilinear operator $B(f,g)(t)$ is bicontinuous from $\mathcal{K}_q \times \mathcal{K}_q \to \mathcal{K}_q$ for any $3 < q < \infty$.*

Once these two lemmata are applied for a *certain* q, $3 < q < \infty$, one can easily deduce, provided (146) is satisfied and via the fixed point algorithm, the existence of *a solution* $v(t,x) \in \mathcal{N}$ that tends strongly to v_0 at zero and belongs to $\mathcal{K}_q$ for *all* $3 < q < \infty$.

The latter properties are a consequence of the following generalization of Lemma 10, applied to the bilinear B term.

LEMMA 11. *The bilinear operator $B(f,g)(t)$ is bicontinuous from $\mathcal{K}_q \times \mathcal{K}_q \to \mathcal{K}_p$ for any $3 \leqslant p < \frac{3q}{6-q}$ if $3 < q < 6$; any $3 \leqslant p < \infty$ if $q = 6$; and $q/2 \leqslant p \leqslant \infty$ if $6 < q < \infty$.*

The proof of the uniqueness of the solution in $\mathcal{N}$ requires a more careful study of the bilinear term as it will be explained in Section 5.3.

Before moving on to a different strategy to prove Theorem 2, let us mention here that the limit value $q = \infty$ cannot be considered in the statement of Lemma 10 because, if we use the standard approach to prove the continuity in L^∞, we are led to a divergent integral (see [166], Chapter 19). Thus, a priori, it is not possible to deduce the existence of a mild solution in $\mathcal{N}$ when the condition expressed by (146) is satisfied for $q = \infty$, say

$$\sup_{0<t<T} t^{1/2} \| S(t)v_0 \|_\infty < \delta, \tag{148}$$

(which means, when $T = \infty$, that the norm of v_0 in the Besov space $\dot{B}_\infty^{-1,\infty}$ is small enough). If, instead, we just require the strongest condition

$$\sup_{0<t<T} t^{1/2} \| S(t)v_0 \|_\infty + \sup_{0<t<T} \| S(t)v_0 \|_3 < \delta \tag{149}$$

(which means that, when $T = \infty$, the norm of v_0 in L^3 is small enough), then the existence of a mild solution $v(t, x)$ belonging to $\mathcal{N}$ can be ensured. Moreover, this solution belongs to $\mathcal{K}_\infty$.

Once again, it is obvious that this result implies Theorem 2, at least when $T = \infty$. At difference with the proof of Theorem 3, here we cannot apply the fixed point theorem directly in $\mathcal{K}_\infty$, but in the space $\mathcal{K}$ whose elements are functions $v(t, x)$ belonging to the intersection $\mathcal{K}_\infty \cap \mathcal{N}$ and whose norm is given by $\sup_{0<t<T} t^{1/2} \|v(t)\|_\infty + \sup_{0<t<T} \|v(t)\|_3$. In fact, the following lemma:

LEMMA 12. *The bilinear operator $B(f, g)(t)$ is bicontinuous from $\mathcal{K} \times \mathcal{K} \to \mathcal{K}$,*

whose proof is contained, for example, in [166], holds true and allows us to conclude.

3.4.2. *Calderón's space.* Another way to prove the existence of a solution with data in L^3 was discovered by Calderón [32] in 1990 and was independently proposed five years later in [34] (see [37] for more details).

Here the auxiliary function space will be denoted by the letter $\mathcal{M}$. Its elements $v(t, x)$ are such that

$$\| |v| \|_{\mathcal{M}} = \Big\| \sup_{0<t<T} |v(t, x)| \Big\|_3 \tag{150}$$

is finite.

It is easy to see that $\mathcal{M}$ is continuously embedded in $\mathcal{N}$, because of the following elementary inequality

$$\sup_{0<t<T} \|v(t, x)\|_3 \leqslant \Big\| \sup_{0<t<T} |v(t, x)| \Big\|_3. \tag{151}$$

The method we will pursue here is to solve the mild Navier–Stokes equations in $\mathcal{M}$. This will be possible because, at variance with $\mathcal{N}$, the bilinear operator is bicontinuous in $\mathcal{M}$. More precisely, the following two lemmata hold true [32–34].

LEMMA 13. *$S(t)v_0 \in \mathcal{M}$ if and only if $v_0 \in L^3$.*

This lemma, whose proof follows from Hardy–Littlewood maximal function, shows that the equivalence stated in Proposition 4 is not true if for example $s = 0$, $p = \infty$ and $q = 3$. In fact, the equivalence under consideration can be seen as a consequence of the well-known result that the Hardy space $\mathcal{H}^3$ is equivalent to L^3, which in turn is equivalent to the Triebel–Lizorkin space $\dot{F}_3^{0,2}$. For a more detailed explanation on this subject we refer the reader to [225,226].

The following lemma concerns the bilinear term [32–34].

LEMMA 14. *The bilinear operator $B(f, g)(t)$ is bicontinuous from $\mathcal{M} \times \mathcal{M} \to \mathcal{M}$.*

Before proceeding, we want to make an additional comment here. The fact that the bilinear operator $B(f, g)$ is bicontinuous both in $\mathcal{M}$ (that *is included* in $\mathcal{N}$) and, as it was announced by Meyer [166], bicontinuous in the Lorentz space $\mathcal{C}([0, T); L^{(3,\infty)})$ (that *includes* $\mathcal{N}$), is very peculiar, since Oru showed in [183] that $B(f, g)$ is not bicontinuous in the natural space $\mathcal{N}$.

This remark being made, let us see how, by a simple variant of the proof above, one can generalize Lemma 14. In order to do that, let us introduce the space $\mathcal{H}_p^s$ whose elements $v(t, x)$ are such that

$$\|v\|_{\mathcal{H}_p^s} = \Big\| \sup_{0<t<T} |\dot{\Lambda}^s v(t, x)| \Big\|_p < \infty. \tag{152}$$

Here $\dot{\Lambda}^s$ is as usual the pseudo-differential operator whose symbol in Fourier transform is given by $|\xi|^s$ and $\dot{\Lambda} = \sqrt{-\Delta}$ is the Calderón operator.

In other words, $\mathcal{H}_p^s$ is the subspace of the natural space $\mathcal{C}([0, T); \dot{H}_p^s)$ obtained by interchanging the time and space norms. Here, $\dot{H}_p^s = \dot{F}_p^{s,2}$ corresponds to the so-called Sobolev–Bessel or homogeneous Lebesgue space. In particular, for $p < 3$, we have the following continuous embedding,

$$\dot{H}_p^{3/p-1} \hookrightarrow L^3 = \dot{H}_3^0 \tag{153}$$

which, in turn, gives ($p < 3$)

$$\mathcal{H}_p^{3/p-1} \hookrightarrow \mathcal{M} \hookrightarrow \mathcal{N}. \tag{154}$$

We are now ready to generalize Lemma 14 ($p = 3$) in the following:

LEMMA 15. *Let $3/2 < p < 3$ be fixed. The bilinear operator $B(f, g)(t)$ is bicontinuous from $\mathcal{M} \times \mathcal{M} \to \mathcal{H}_p^{3/p-1}$.*

This lemma should be interpreted as a supplementary regularity property of the bilinear term as it was extensively analyzed in [34,43,48,186]. By means of a more accurate study of the cancellation properties of the bilinear term, the limit case $p = 3/2$ (with the natural norm in time and space variables) can be included as well (see [48]).

This remark being made, let us observe that, just by using Lemmas 13 and 14, we are in a position, via the fixed point algorithm, to prove the existence of a global solution in $\mathcal{M}$ with initial data v_0 sufficiently small in L^3, say

$$\|v_0\|_3 < \delta. \tag{155}$$

However, because the bicontinuity constant arising in Lemma 14 does not depend on T and the condition (corresponding to (144) in the definition of $\mathcal{K}_q$)

$$\lim_{T\to 0}\Big\|\sup_{0<t<T}\big|S(t)v_0\big|\Big\|_3 = 0 \tag{156}$$

is *not verified* if $v_0 \in L^3$, $v_0 \not\equiv 0$, there is no evidence to guarantee that such a global solution is strongly continuous at the origin (and thus unique as we will see in the following pages), and, which is intimately related, that such a solution exists locally in time for an arbitrary initial data v_0 in L^3.

We use here the same trick introduced in [34]. More precisely, instead of looking for a mild solution $v(t,x) \in \mathcal{M}$, via the point fixed Lemma 4, we will look for a solution

$$w(t,x) = v(t,x) - S(t)v_0 \in \mathcal{M} \tag{157}$$

via the point fixed Lemma 5. More precisely, we will solve the equation

$$w(t,x) = \widetilde{B}\big(S(t)v_0, S(t)v_0\big) + 2\widetilde{B}\big(w, S(t)v_0\big) + \widetilde{B}(w,w), \tag{158}$$

where the symmetric bilinear operator $\widetilde{B}$ is defined, in terms of B, by

$$\widetilde{B}(v,u)(t) = \frac{B(v,u)(t) + B(u,v)(t)}{2}. \tag{159}$$

We can now take advantage of the particular structure of the heat semigroup appearing in (158). More exactly, we can generalize the previous lemmata and obtain the following ones:

LEMMA 16. *Let $\alpha = 1 - 3/q$ and $3 < q < \infty$ be fixed. Then*

$$\Big\|\sup_{0<t<T} t^{\alpha/2}\big|S(t)v_0\big|\Big\|_q \leqslant C_q\|v_0\|_3, \tag{160}$$

and in particular, if $v_0 \in L^3$, the left-hand side of (160) *tends to zero as T tends to zero.*

Now $\alpha > 0$, so (160) is a direct consequence of Proposition 4 and the following Sobolev embedding (see [225,226])

$$L^3 = \dot{F}_3^{0,2} \hookrightarrow \dot{F}_q^{-\alpha,2} \hookrightarrow \dot{F}_q^{-\alpha,\infty}. \tag{161}$$

LEMMA 17. *Let $\alpha = 1 - 3/q$, $3 < q < 6$, and $f(t,x) = S(t)f_0$, with $f_0 = f_0(x)$, then the following estimate holds for the bilinear operator*

$$\| B(S(t)f_0, S(t)f_0) \|_{\mathcal{M}} \leqslant C_q \Big\| \sup_{0<t<T} t^{\alpha/2} |S(t)f_0| \Big\|_q^2. \tag{162}$$

LEMMA 18. *Let $\alpha = 1 - 3/q$, $3 < q < \infty$, and $f(t,x) = S(t)f_0$, with $f_0 = f_0(x)$, and $g = g(t,x)$ then the following estimate holds for the bilinear operator*

$$\| B(S(t)f_0, g) \|_{\mathcal{M}} \leqslant C'_q \|g\|_{\mathcal{M}} \Big\| \sup_{0<t<T} t^{\alpha/2} |S(t)f_0| \Big\|_q. \tag{163}$$

We can now state the following existence and uniqueness theorem of [34,47] as:

THEOREM 4. *Let $3 < q < 6$ and $\alpha = 1 - 3/q$. There exists a constant $\delta_q > 0$ such that, for any initial data $v_0 \in L^3$, $\nabla \cdot v_0 = 0$ in the sense of distributions such that*

$$\Big\| \sup_{0<t<T} t^{\alpha/2} |S(t)v_0| \Big\|_q < \delta_q, \tag{164}$$

then there exists a mild solution $v(t,x)$ belonging to $\mathcal{N}$, which tends strongly to v_0 as time goes to zero. Moreover, this solution belongs to the space $\mathcal{M}$ and the function $w(t)$ defined in (157) belongs to $\mathcal{H}_p^{3/p-1}$ $(3/2 < p < 3)$. In particular, (164) holds for arbitrary $v_0 \in L^3$ provided we consider $T(v_0)$ small enough, and as well if $T = \infty$, provided the norm of v_0 in the Triebel–Lizorkin space $\dot{F}_q^{-\alpha,\infty}$ is smaller than δ_q.

The existence part of the proof is now a consequence of Lemma 5, while its uniqueness will be treated in Section 5.3.

In order to appreciate the result we have just stated, let us now concentrate on comparing the hypotheses that arise in the statements of Theorems 3 and 4.

It is not difficult to see that, for any $T > 0$ and $3 \leqslant q \leqslant \infty$, $\alpha = 1 - 3/q$,

$$\sup_{0<t<T} t^{\alpha/2} \| S(t)v_0 \|_q \leqslant \Big\| \sup_{0<t<T} t^{\alpha/2} |S(t)v_0| \Big\|_q \tag{165}$$

which corresponds, for $T = \infty$, to the well-known embedding

$$\dot{F}_q^{-\alpha,\infty} \hookrightarrow \dot{B}_q^{-\alpha,\infty}. \tag{166}$$

This circumstance indicates that, as far as the initial data v_0 is concerned, condition (146) is stronger than (164). However, with regard to the Navier–Stokes equations in the presence of a nontrivial external force (e.g., the gravity) as described in (11) with $\phi \not\equiv 0$, Calderón's method allows us to obtain some better estimates, in particular, as explained in [47], to improve the results contained in [56].

Before ending this section, we would like to remark that the idea of interchanging time and space in the mixed norms can also be adapted in the case of different spaces for the

Navier–Stokes equations. Explicit calculations were performed in [34] in the case of the above defined Sobolev-type space $\mathcal{H}_2^s$ ($s \geqslant 1/2$). In fact, Lemma 15 would be enough to derive such a result when $s = 1/2$. However, other less trivial examples can be obtained.

3.4.3. *Giga's space.* As we recalled in the previous section, the method for finding a strongly continuous solution with values in L^3 makes use of an *ad hoc* auxiliary subspace of functions that are continuous in the t-variable and take values in a Lebesgue space in the x-variable. Moreover, Giga proved in [99] that not only does the solution under consideration belong to $L_t^\infty(L_x^3)$ and $\mathcal{K}_q$ but also, for all q in the interval $3 < q \leqslant 9$, it belongs to the space $\mathcal{G}_q = L_t^{2/\alpha}(L_x^q)$, whose elements $f(t,x)$ are such that

$$\|f\|_{\mathcal{G}_q} =: \left(\int_0^T \|f(t,x)\|_q^{2/\alpha}\,\mathrm{d}t\right)^{\alpha/2} < \infty, \tag{167}$$

T being, as usual, either finite or infinite, and $\alpha = \alpha(q) = 1 - 3/q$.

At this point, one can naturally ask whether these spaces $\mathcal{G}_q$ can be used, independently, as auxiliary *ad hoc* subspaces to prove the existence of a solution with data in L^3. This question arises also in view of the fact that $L_t^p(L_x^q)$ estimates (and, more generally, the so-called Strichartz estimates) are frequently used for the study of other well-known nonlinear partial differential equations, like the Schrödinger one or the wave equation. Even if this does not lead here to a breakthrough as in the case of the Schrödinger equation, making *direct* use of $L_t^p(L_x^q)$ estimates for Navier–Stokes is indeed possible. This was proved by Kato and Ponce in [118], where, in fact, the authors consider the case of a much larger functional class, including the $\mathcal{G}_q$ one.

In what follows, we will focus our attention only on the latter case and prove an existence theorem of local (resp. global) strong solutions in $\mathcal{C}([0,T); L^3)$ with initial data (resp. small enough) in a certain Besov space.

The "Besov language" will provide a very convenient and powerful tool, needed to overcome difficulties which were absent in the previous section.

As in the previous cases, we will start with an estimate of the linear term $S(t)v_0$ in the auxiliary space $\mathcal{G}_q$. We have following lemma.

LEMMA 19. *Let $3 < q \leqslant 9$ and $\alpha = 1 - 3/q$ be fixed. Then*

$$\left(\int_0^T \|S(t)v_0\|_q^{2/\alpha}\,\mathrm{d}t\right)^{\alpha/2} \leqslant C_q\|v_0\|_3, \tag{168}$$

where the integral in the left-hand side of (168) *tends to zero as T tends to zero provided $v_0 \in L^3$.*

Keeping Proposition 4 in mind, this lemma can be proved if we recall the well-known Sobolev embedding [225,226]

$$L^3 \hookrightarrow \dot{B}_q^{-\alpha,2/\alpha}, \tag{169}$$

which holds true as long as $3 < q \leqslant 9$. Here the restriction $q \leqslant 9$ appears as a limit exponent in the Sobolev embedding for Besov spaces. A direct proof of (168) is contained in the papers by Giga [99], Kato [114] and Kato and Ponce [118] and makes use of the Marcinkiewicz interpolation theorem. In short, our lemma says that if $v_0 \in L^3$, then $S(t)v_0$ is in $\mathcal{G}_q$, and therefore we are allowed to work within that functional framework.

The fact that the left-hand side of (168) tends to zero as T tends to zero can be easily checked by using the Banach–Steinhaus theorem. What we would like to stress here, is that this property is of paramount importance, because it will ensure (as in Theorems 3 and 4) the strong continuity at the origin of the solution given by the fixed point scheme. Once we get a solution in $\mathcal{C}([0,T); L^3)$ that tends in the strong L^3 topology to v_0 as time tends to zero, this solution will automatically be *unique*, as we will see in Section 5.3.

Let us now concentrate on the bilinear operator [186].

LEMMA 20. *The bilinear operator $B(f,g)(t)$ is bicontinuous from $\mathcal{G}_q \times \mathcal{G}_q \to \mathcal{G}_p$ for any $3 < p < \frac{3q}{6-q}$ if $3 < q < 6$; any $3 < p < \infty$ if $q = 6$; and $q/2 \leqslant p \leqslant \infty$ if $6 < q < \infty$.*

In the case $q = p$ this result was originally proved by Fabes, Jones and Rivière [76] and represents the equivalent of Lemma 10 in the space $\mathcal{K}_q$.

This lemma can be proved by duality (in the t-variable) in a way reminiscent of Giga's method introduced in [99] and based on the Hardy–Littlewood–Sobolev inequality (see [47]). Here the restrictions on the exponents p and q come from the Young and Hardy–Littlewood–Sobolev inequalities. In particular, the value $\beta = 0$ corresponding to $p = 3$ is excluded. This is why Lemma 20 *cannot be used* directly to get (as in Lemma 11) an $L^\infty_t(L^3_x)$ estimate. That appears to be the main difference with the methods involving the Besov $\dot{B}_q^{-(1-3/q),\infty}$ and Triebel–Lizorkin spaces $\dot{F}_q^{-(1-3/q),\infty}$ that were considered in the previous cases. As a matter of fact, the estimates obtained in those spaces, having their third index equal to ∞, are essentially based on the scaling invariance of the Navier–Stokes equations, which is a very crude property of the nonlinear term. Here, on the contrary, we need to investigate further and to explicitly take into account the oscillatory property of the bilinear term, say

$$\int_{\mathbb{R}^3} \Theta(x)\,\mathrm{d}x = 0 \tag{170}$$

or, equivalently, the fact that the Fourier transform of Θ is zero at the origin. Of course, we are still far away from exploiting the full structure of the bilinear term.

This remark being made, let us now see how to use (170) in the proof of the following lemma.

LEMMA 21. *The bilinear operator $B(f,g)(t)$ is bicontinuous from $\mathcal{G}_6 \times \mathcal{G}_6 \to \mathcal{N}$. In fact, $B(f,g)$ takes its values in $\mathcal{C}(0,T; \dot{B}_3^{0,2})$, which is a proper subset of $\mathcal{N}$.*

We would like to mention here that a variant of this result was applied in [91,92] in the proof of the uniqueness theorem for strong L^3 solutions (see also [48] for more comments) as we will see in Section 5.3.

Let us now outline the proof of Lemma 21, by using once again a duality argument: first we show that $B(f,g)$ is bicontinuous from $L^4_t(L^6_x) \times L^4_t(L^6_x)$ into $L^\infty_t(\dot{B}^{0,2}_3)$ and then we conclude by a usual argument in order to restore the strong continuity in time [47].

To prove the proposition by duality (in the x-variable), let us consider an arbitrary test function $h(x) \in \mathcal{C}^\infty_0$ and let us evaluate

$$I_t = \int_{\mathbb{R}^3} \int_0^t (t-s)^{-2} \Theta\left(\frac{\cdot}{\sqrt{t-s}}\right) * (fg)(s) h(x)\, \mathrm{d}s\, \mathrm{d}x. \tag{171}$$

It is useful here to see the t variable as a fixed parameter. After interchanging the integral over $\mathbb{R}^3$ with the convolution with $h(x)$, and after applying the Hölder inequality (in x) and the Cauchy–Schwarz inequality (in t), we get

$$|I_t| \lesssim \left(\int_0^t \|fg\|_3^2\, \mathrm{d}s\right)^{1/2} \left(\int_0^t \|\Theta_u * h\|_{3/2}^2 \frac{\mathrm{d}u}{u}\right)^{1/2}, \tag{172}$$

where

$$\Theta_u = \frac{1}{u^3} \Theta\left(\frac{\cdot}{u}\right). \tag{173}$$

In order to conclude, we only remark that the oscillatory property of Θ, say (170), allows us to consider the quantity

$$\left(\int_0^\infty \|\Theta_u * h\|_{3/2}^2 \frac{\mathrm{d}u}{u}\right)^{1/2} \tag{174}$$

as an (equivalent) norm on the homogeneous Besov space $\dot{B}^{0,2}_{3/2}$. As we observed in Section 2.2, if the function Θ were smooth and compactly supported on the Fourier side, this would indeed be the usual characterization. Removing the band-limited condition is trivial, and it turns out that smoothness is not a critical assumption, thus allowing a greater flexibility in the definition of the Besov space. What is certainly *not possible* is to get such an equivalence if, as is the case for $S(t)$, the function Θ does not have a zero integral. More explicitly, a property analogous to the one stated in Proposition 3 would not apply here and, in general, does not apply for a Besov space of the type $\dot{B}^{s,p}_q$, with $s \geqslant 0$. A counterexample for $s = 0$, $p = \infty$ and $q = 3$ can be found, for instance, in [34] (Lemma 4.2.10). The reader should refer to [185] for a very enlightening discussion of the definition of Besov spaces, and to [82,225,226] for precise results.

Let us go back to the Besov space $\dot{B}^{0,2}_{3/2}$. A standard argument shows that the dual space of $\dot{B}^{0,2}_{3/2}(\mathbb{R}^3)$ is exactly $\dot{B}^{0,2}_3$. All this finally implies that the bilinear operator $B(f,g)$ is bicontinuous from $L^4_t(L^6_x) \times L^4_t(L^6_x)$ into $L^\infty_t(\dot{B}^{0,2}_3)$, which completes the proof of Lemma 21. Moreover, as

$$\dot{B}^{0,2}_3 \hookrightarrow L^3, \tag{175}$$

we have obtained our $L_t^\infty(L_x^3)$ estimation, and even improved it. As in [34,43,48,186], this provides another example in which the regularity of the bilinear term is better than the linear one.

We are now in position to prove the following theorem [186]:

THEOREM 5. *Let* $3 < q < 9$ *and* $\alpha = 1 - 3/q$ *be fixed. There exists a constant* δ_q *such that for any initial data* $v_0 \in L^3$, $\nabla \cdot v_0 = 0$ *in the sense of distributions such that*

$$\left(\int_0^T \|S(t)v_0\|_q^{2/\alpha} \, dt \right)^{\alpha/2} < \delta_q \tag{176}$$

and then there exists a mild solution $v(t,x)$ *belonging to* $\mathcal{N}$, *which tends strongly to* v_0 *as time goes to zero. Moreover, this solution belongs to all the spaces* $\mathcal{G}_q$ $(3 < q < 9)$ *and is such that the fluctuation* $w(t,x)$ *defined in* (157) *satisfies*

$$w \in \mathcal{C}\big([0,T); \dot{B}_3^{0,2}\big) \tag{177}$$

and

$$w \in L^2\big((0,T); L^\infty\big). \tag{178}$$

Finally, (176) *holds for arbitrary* $v_0 \in L^3$ *provided we consider* $T(v_0)$ *small enough, and as well if* $T = \infty$, *provided the norm of* v_0 *in the Besov space* $\dot{B}_q^{-\alpha,2/\alpha}$ *is smaller than* δ_q.

Keeping in mind the previous propositions and remarks, the proof of that theorem is easily carried out as follows (see [186] for more details).

First, we apply the fixed point algorithm in the space $\mathcal{G}_q = L^{2/\alpha}([0,T); L^q)$ (q and α being assigned in the statement) to get, by means of Lemma 20, a mild solution $v(t,x) \in \mathcal{G}_q$. Then, again using Lemma 20, we find that $v(t,x) \in \mathcal{G}_q$ for all $3 < q < 9$. In particular $v(t,x) \in \mathcal{G}_6 = L^4([0,T); L^6)$, which gives $v(t,x) \in \mathcal{N}$ and (177) (once Lemma 21 is taken into account).

As we presented in [48], this regularity result can even be improved to get $w(t) \in \mathcal{C}([0,T); \dot{F}_{3/2}^{1,2})$, which means that the gradient of $w(t)$ belongs uniformly in time to $L^{3/2}$ and we observe that $\dot{F}_{3/2}^{1,2} \hookrightarrow \dot{B}_3^{0,2}$. The latter regularity result can be seen in connection with an estimate derived by Kato [114] that assures that the gradient of $v(t)$, solution of Theorem 3 in $\mathcal{N}$, is such that $t^{1-3/(2q)} \nabla v(t) \in \mathcal{C}([0,T); L^q)$ for any $q \geqslant 3$. We proved in [48] that the function $w(t)$ satisfies the last estimate for the optimal exponent $q = 3/2$.

Finally, as the bilinear term is bicontinuous from $\mathcal{G}_q \times \mathcal{G}_q$ into $L_t^2(L_x^\infty)$, and arguing by duality, ($\mu(s)$ being a test function), we can obtain the estimate (178), say

$$\left| \int_0^T \|B(f,g)(s)\|_\infty \mu(s) \, ds \right| \lesssim \int_0^T \int_0^t \frac{\|fg\|_{q'/2}(s)\mu(t)}{(t-s)^{1/2+3/q'}} \, ds \, dt \lesssim \|\mu\|_2. \tag{179}$$

4. Highly oscillating data

At difference with Leray's well-known weak approach, the method described in the previous pages – the so-called "Tosio Kato's method" (see the book [88] for many examples of applications of this method to nonlinear PDEs) – also implies the uniqueness of the corresponding solution, as it will be explained in Section 5. However, the existence of the solution holds under a restrictive condition on the initial data, that is required to be small, which is not the case for Leray's weak solutions. In Section 7 we will make the link between this property, the smallness of the Reynolds number associated with the flow, the stability of the corresponding global solution and the existence of Lyapunov functions for the Navier–Stokes equations.

The aim of this section is to give an interpretation of the smallness of the initial data in terms of an oscillation property. The harmonic analysis tools we developed so far will play a crucial role here.

Let us recall that, as stated in Theorem 3, a global solution in $\mathcal{C}([0,\infty); L^3)$ exists, provided that the initial data v_0 is divergence-free and belongs to L^3, and that its norm is small enough in L^3, or more generally, small in the Besov space $\dot{B}_q^{-\alpha,\infty}$ (for a certain $3 < q < \infty$ and $\alpha = 1 - 3/q$ fixed). In other words, a function v_0 in L^3 whose norm is arbitrarily large in L^3 but small enough in $\dot{B}_q^{-\alpha,\infty}$ (or in a Triebel–Lizorkin space $\dot{F}_q^{-\alpha,\infty}$ as in Theorem 4, or in the Besov space $\dot{B}_q^{-\alpha,2/\alpha}$ as in Theorem 5) also ensures the existence of a global mild solution in $\mathcal{C}([0,\infty); L^3)$.

The advantage of using a Besov norm instead of a Lebesgue one is that the condition of being small enough in a Besov space is satisfied by highly oscillating data (Section 4.1). A second remarkable property is that these spaces contain homogeneous functions of degree -1, leading to global self-similar solutions (Section 6). Moreover, Besov spaces led to the (first) proof of the uniqueness for solutions in $\mathcal{C}([0,\infty); L^3)$ (Section 5.3).

The a posteriori disappointing observation is that… Besov spaces were not necessary at all in any of these discoveries!

4.1. *A remarkable property of Besov spaces*

In order to appreciate the formulation of Kato's theorem in terms of the Besov space $\dot{B}_q^{-\alpha,\infty}$ given in Theorem 3, we shall devote ourselves here to illustrating that the condition $\|v_0\|_{\dot{B}_q^{-\alpha,\infty}} < \delta$ is satisfied in the particular case of a sufficiently oscillating function v_0.

A typical situation will be given by the following example. Let v_0 be an arbitrary (not identically vanishing) function belonging to L^3. If we multiply v_0 by an exponential, say the function $w_k = \exp[ix \cdot k]$, we obtain, for any $k \in \mathbb{R}^3$, a function $w_k v_0$ such that (Lemma 22)

$$\lim_{|k|\to\infty} \|w_k v_0\|_{\dot{B}_q^{-\alpha,\infty}} = 0, \tag{180}$$

in spite of the fact that

$$\lim_{|k|\to\infty} \|w_k v_0\|_3 = \|v_0\|_3. \tag{181}$$

In other words, the smallness condition $\|w_k v_0\|_{\dot{B}_q^{-\alpha,\infty}} < \delta$, is verified as long as we choose a sufficiently high frequency k. At this point, it is tempting to consider $w_k v_0$ as the new initial data of the problem and to affirm that Kato's solution exists globally in time, provided we consider sufficiently oscillating data. One can argue that $w_k v_0$ is no longer a divergence-free function. Nevertheless, the function $w_k v_0$ is divergence-free *asymptotically* for $|k| \to \infty$, which is exactly the situation we are dealing with. More precisely, it turns out that (Lemma 23)

$$\lim_{|k|\to\infty} \left\|\nabla \cdot (w_k v_0) - w_k \nabla \cdot v_0\right\|_3 = 0. \tag{182}$$

LEMMA 22. *Let v be an arbitrary function in L^3 and let $w_k(x)$, $k \in \mathbb{N}$, be a sequence of functions such that $\|w_k\|_\infty \leqslant C$ and $w_k \rightharpoonup 0$ (as $k \to \infty$) in the distributional sense. Then, the products $w_k v$ tend to 0 in the strong topology of $\dot{B}_q^{-\alpha,\infty}$* ($\alpha = 1 - 3/q > 0$).

The proof of this lemma is quite easy and we wish to present the main components here (for more details see [34,35]).

We will make use of a density argument. To this end, let us introduce the following decomposition of the function v:

$$v = h + g, \tag{183}$$

where $h \in L^3$ and

$$\|h\|_3 \leqslant \varepsilon \tag{184}$$

and $g \in C_0^\infty$. The next step is to recall the continuous embedding (Lemma 9) $L^3 \hookrightarrow \dot{B}_q^{-\alpha,\infty}$ to infer the following inequality ($k \geqslant 0$)

$$\|w_k h\|_{\dot{B}_q^{-\alpha,\infty}} \lesssim \|w_k h\|_3 \lesssim \varepsilon. \tag{185}$$

On the other hand, Young's inequality gives ($j \in \mathbb{Z}$)

$$\left\|S_j(w_k g)\right\|_q \leqslant \left\|2^{3j}\varphi(2^j\cdot)\right\|_r \|w_k g\|_p, \tag{186}$$

where

$$\frac{1}{q} = \frac{1}{r} + \frac{1}{p} - 1. \tag{187}$$

This implies

$$2^{-\alpha j}\left\|S_j(w_k g)\right\|_q \lesssim 2^{-j(1-3/q)} 2^{-j(1-3/r)} \|g\|_p = 2^{-j(1-3/p)} \|g\|_p \tag{188}$$

so that, for any $k \geqslant 0$, any $j \geqslant j_1 > 0$ and any $j \leqslant j_0 < 0$, we have

$$2^{-\alpha j}\left\|S_j(w_k g)\right\|_q \lesssim \varepsilon \tag{189}$$

(in fact, if $j \geqslant j_1$ we let $p = q > 3$ and if $j \leqslant j_0$ we let $1 \leqslant p < 3$).

We are now left with the terms $S_j(w_k g)$ for $j_0 < j < j_1$. Making use of the hypothesis $m_k \rightharpoonup 0$ together with the Lebesgue dominated convergence theorem, we finally find, for any $k \geqslant k_0$ and $j_0 < j < j_1$,

$$2^{-\alpha j} \left\| S_j(w_k g) \right\|_q \lesssim \varepsilon \tag{190}$$

which concludes the proof of the lemma.

LEMMA 23. *Let $m(\xi) \in C^\infty(\mathbb{R}^3 \setminus \{0\})$ be a homogeneous function of degree 0 and let M be the convolution operator associated with the multiplier $m(\xi)$. If we consider $|\xi_0| = 1$, $v \in L^p$ and $1 < p < \infty$, then*

$$\lim_{\lambda \to \infty} \sup_{|\xi_0|=1} \left\| M\big(\exp(\mathrm{i}\lambda\xi_0 \cdot x)v(x)\big) - \exp(\mathrm{i}\lambda\xi_0 \cdot x)m(\xi_0)v(x) \right\|_p = 0. \tag{191}$$

In the case we are interested in, this lemma will be used for $p = 3$ and with M replaced by the Leray–Hopf projection operator $\mathbb{P}$ onto the divergence-free vector fields and $m(\xi)$ replaced by a 3×3 matrix whose entries are homogeneous symbols of degree 0.

In order to prove the lemma in its general form, we remark that the symbol of the operator $\exp(-\mathrm{i}\lambda\xi_0 \cdot x)M(\exp(\mathrm{i}\lambda\xi_0 \cdot x)v) - m(\lambda\xi_0)v(x)$ is given by $m(\xi + \lambda\xi_0) - m(\lambda\xi_0)$, this by virtue of the homogeneity of m.

Equation (191) will now be proved by means of a density argument. In fact, it is sufficient to limit ourselves to functions $v \in \mathcal{V} \subset L^p$, where $\mathcal{V}$ is the dense subspace of L^p defined by $v \in \mathcal{S}$ and the Fourier transform $\hat{v}$ of v has compact support. Now, we put

$$v_\lambda = \exp(-\mathrm{i}\lambda\xi_0 \cdot x)M\big(\exp(\mathrm{i}\lambda\xi_0 \cdot x)v\big) - m(\lambda\xi_0)v, \tag{192}$$

then the Fourier transform of v_λ is given by

$$\hat{v}_\lambda(\xi) = \big[m(\xi + \lambda\xi_0) - m(\lambda\xi_0)\big]\hat{v}(\xi). \tag{193}$$

Finally, $\hat{v}$ has compact support, say in $|\xi| \leqslant R$, and then

$$m(\xi + \lambda\xi_0) - m(\lambda\xi_0) = r_\lambda(\xi), \tag{194}$$

where, on $|\xi| \leqslant R$, $r_\lambda(\xi) \to 0$ together with all its derivatives in the L^∞ norm. We thus have $v_\lambda \to 0$ in $\mathcal{S}$ when $\lambda \to \infty$. A fortiori, $\|v_\lambda\|_p \to 0$ when $\lambda \to \infty$, and the lemma is proved.

4.2. *Oscillations without Besov norms*

Some years after the publication of [34,35] Temam [217] informed us that the property we described in the previous pages, that highly oscillating data lead to global solutions to Navier–Stokes, was implicitly contained in the pioneering papers of Kato and Fujita [87,117] of 1962.

These papers deal with mild solutions to Navier–Stokes that are continuous in time and take values in the Sobolev space $\dot{H}^s$, say $v \in \mathcal{C}([0, T); \dot{H}^s)$. It is easy to see, in the three-dimensional case, that the critical Sobolev space corresponds to the value $s = 1/2$. More precisely, the Sobolev spaces $\dot{H}^s$, $s > 1/2$ are super-critical. In other words, as far as the scaling is concerned, they have the same invariance as the Lebesgue spaces L^p if $p > 3$. This means that, using the simplified version of the bilinear operator, one can easily prove the existence of a local mild solution for arbitrary initial data [34], that is, the theorem.

THEOREM 6. *Let $1/2 < s < \infty$ be fixed. For any $v_0 \in \dot{H}^s$, $\nabla \cdot v_0 = 0$, there exists a $T = T(\|v_0\|_s)$ such that the Navier–Stokes equations have a mild solution in $\mathcal{C}([0, T); \dot{H}^s)$.*

On the other hand, in the critical case $s = 1/2$, one can ensure the existence of a local solution, that turns out to be global when the initial data are small enough:

THEOREM 7. *There exists a constant $\delta > 0$ such that for any initial data $v_0 \in \dot{H}^{1/2}$, $\nabla \cdot v_0 = 0$ in the sense of distributions, such that*

$$\|v_0\|_{\dot{H}^{1/2}} < \delta, \tag{195}$$

then there exists a mild solution $v(t, x)$ to the Navier–Stokes equations belonging to $\mathcal{C}([0, \infty); \dot{H}^{1/2})$.

In the particular case $s = 1$, we also have at our disposal a persistence result, namely:

THEOREM 8. *There exists a constant $\delta > 0$ such that if the initial data $v_0 \in \dot{H}^{1/2} \cap \dot{H}^1$, $\nabla \cdot v_0 = 0$ in the sense of distributions and satisfies*

$$\|v_0\|_{\dot{H}^{1/2}} < \delta, \tag{196}$$

then the mild solution $v(t, x)$ to the Navier–Stokes equations, whose existence is ensured by Theorem 7, also belongs to $\mathcal{C}([0, \infty); \dot{H}^1)$.

To prove such a result, it is enough to show that the $\dot{H}^1$ norm of the solution is a Lyapunov function, which means that it is decreasing in time. The study of the Lyapunov functions for the Navier–Stokes equations will be examined in detail in Section 7.1.

Actually, to obtain a global mild solution in the space $\mathcal{C}([0, \infty); \dot{H}^1)$ it would be enough to get a uniform estimate of the kind

$$\|v(t)\|_{\dot{H}^1} \leqslant \|v_0\|_{\dot{H}^1} \quad \forall t > 0, \tag{197}$$

because a classical "bootstrap" argument will allow to pass from a local solution to a global one.

This property turns out to be satisfied when the initial data $v_0 \in \dot{H}^1$ has a sufficiently small norm in the space $\dot{H}^{1/2}$. More precisely, as we will describe in detail in Section 7.1, the following inequality is proven in the celebrated papers by Kato and Fujita [87,117]:

$$\frac{\mathrm{d}}{\mathrm{d}t}\|v(t)\|_{\dot{H}^1}^2 \leqslant -2\|v(t)\|_{\dot{H}^2}^2\left(\nu - C\|v(t)\|_{\dot{H}^{1/2}}\right). \tag{198}$$

This immediately implies the aforementioned property of decrease in time of the homogeneous norm $\|v\|_{\dot{H}^1}$, as long as $\|v_0\|_{\dot{H}^{1/2}}$ is small enough. On the other hand, it is easy to show that the L^2 norm of the solution v also decreases in time, say

$$\frac{\mathrm{d}}{\mathrm{d}t}\|v(t)\|_2^2 = -2\nu\|\nabla v(t)\|_2^2 < 0, \tag{199}$$

which allows us to deduce the decreasing of the nonhomogeneous norm $\|v\|_{H^1}$ as well.

Now, Temam's remark is very simply and reads as follows. Suppose $v_0 \in \mathcal{S}'$ is such that $\hat{v}_0(\xi) = 0$ if $|\xi| \leqslant R$, then

$$\|v_0\|_{\dot{H}^{1/2}} \leqslant R^{-1/2}\|v_0\|_{\dot{H}^1} \tag{200}$$

and thus one can get the existence of a global mild solution in $\mathcal{C}([0,\infty); \dot{H}^1)$ provided the initial data is concentrated at high frequencies ($R \gg 1$), say highly oscillating!

4.3. *The result of Koch and Tataru*

In his doctoral thesis [186,187], Planchon gave the precise interpretation of the persistence result stated in Theorem 8, replacing the smallness of the $\dot{H}^{1/2}$ norm of the initial data, with the smallness (or oscillation) in a Besov space. Everything takes place as in [34] for the critical space L^3: there exists an absolute constant $\beta > 0$ such that if $\|v_0\|_{\dot{B}_4^{-1/4,\infty}} < \beta$ and $v_0 \in \dot{H}^1$, then there exists a global solution in $\mathcal{C}([0,\infty); \dot{H}^1)$. What make things work here is that, even if $\dot{H}^1$ is not a critical space, it is embedded in $\dot{H}^{1/2}$ (which is not the case for any Lebesgue space L^p, $p \geqslant 3$, when working in unbounded domains as $\mathbb{R}^3$). The importance of such a result is that it allows us to obtain *global* and *regular* solutions in the energy space $\dot{H}^1$, under the hypothesis of oscillation of the initial data. In other words, at variance with the L^3 setting, we can establish a link between Leray's weak solutions and Kato's mild ones.

This approach was generalized first by Koch and Tataru [123] and then by Furioli, Lemarié, Zahrouni and Zhioua [89,93,145,240]. Both of these results seem optimal.

Roughly speaking the theorem by Koch and Tataru says that if the norm of the initial data is small enough in the critical space BMO^{-1}, then there is a global mild solution for the Navier–Stokes equations. Again, the norm of the product of a fixed function in L^3 times an oscillating function, say $w_k = \exp[ix \cdot k]$, tends to zero as $|k|$ tends to infinity. It is not clear whether this theorem is optimal, because, if it is true that it generalizes the results of the previous section (in fact BMO^{-1} contains L^3 as well as $\dot{B}_q^{-\alpha,\infty}$, for any $3 < q < \infty$ and $\alpha = 1 - 3/q$), we should recall that BMO^{-1} is contained in the biggest critical space $\dot{B}_{\infty}^{-1,\infty}$ (as stated in (77) and Proposition 7) and nobody knows whether the Navier–Stokes system is well posed in this space (see [166]). Incidentally, we wish to remind the reader that Montgomery-Smith proved a blow-up result in the space $\dot{B}_{\infty}^{-1,\infty}$ for a modified (with respect to the nonlinear term) Navier–Stokes equations [176]. Moreover, his result also shows there is initial data that exists in every Triebel–Lizorkin or Besov space (and hence in every Lebesgue and Sobolev space), such that after a finite time, the solution of the Navier–Stokes-like equation is in no Triebel–Lizorkin or Besov space (and hence in no Lebesgue or Sobolev space).

On the other hand, the persistence result by Furioli, Lemarié, Zahrouni and Zhioua says that if the initial data is not only small in BMO^{-1}, but also belongs to the Banach space X, where X can be either the Lebesgue space L^p, $1 \leqslant p \leqslant \infty$, or the inhomogeneous Besov space $B_q^{s,p}$ with $1 \leqslant p \leqslant \infty$, $1 \leqslant q \leqslant \infty$ and $s > -1$, or the homogeneous Besov space $\dot{B}_q^{s,p}$ with $1 \leqslant p \leqslant \infty$, $1 \leqslant q \leqslant \infty$ and $s > -1$, then the corresponding solution also belongs to $L^{\infty}((0,\infty); X)$.

In order to simplify the discussion, we will limit ourselves to present only global solutions. However, solutions which are local in time as we previously constructed in the critical space L^3 are also available. More exactly, we are talking about the following results.

THEOREM 9. *There exists a constant $\delta > 0$ such that, for any initial data $v_0 \in BMO^{-1}$ that verifies*

$$\|v_0\|_{BMO^{-1}} < \delta, \tag{201}$$

then there exists a global mild solution $v(t,x)$ to the Navier–Stokes equations such that

$$\sqrt{t}\,v(t,x) \in L^{\infty}\big((0,\infty), \mathbb{R}^3\big) \tag{202}$$

and

$$\sup_{t>0, x_0\in\mathbb{R}^3} \frac{1}{t^{3/2}} \int_{0<\tau<t} \int_{|x-x_0|<\sqrt{t}} \left|v(\tau,x)\right|^2 \mathrm{d}\tau\, \mathrm{d}x < \infty. \tag{203}$$

The proof of this theorem is contained in the paper of Koch and Tataru [123]. The condition expressed by (203), comes from the fact that a Carleson measure characterization of BMO^{-1} (see [214]) says that a function v_0 belongs to BMO^{-1} if and only if

$$\sup_{t>0, x_0\in\mathbb{R}^3} \frac{1}{t^{3/2}} \int_{0<\tau<t} \int_{|x-x_0|<\sqrt{t}} \left|S(\tau)v_0\right|^2 \mathrm{d}\tau\, \mathrm{d}x < \infty, \tag{204}$$

$S(\tau) = \exp(\tau\Delta)$ denoting, as usual, the heat semigroup. On the other hand, this condition seems the weaker possible one, say BMO^{-1} seems the largest space where local or global solutions exist. In fact, as we recalled in Section 1.2, in order to give a sense to the Navier–Stokes equations we want to have at least

$$v(t,x) \in L^2_{\mathrm{loc}}\big([0,\infty); \mathbb{R}^3\big). \tag{205}$$

Now the Navier–Stokes equations are invariant with respect to scaling, hence we want a scale and translation invariant version of L^2-boundedness, say

$$\sup_{t>0,x_0\in\mathbb{R}^3} \frac{1}{|B_t(x)|} \iint_{B_t(x)\times[0,t^2]} \left|v(\tau,x)\right|^2 \mathrm{d}\tau\,\mathrm{d}x < \infty \tag{206}$$

(where $|B_t(x)|$ denotes the Lebesgue measure of the ball $B_t(x)$ centered at x and radius t), which is precisely the condition expressed by (203).

Finally, let us quote the persistence result announced in [93].

THEOREM 10. *Let v_0 verify the condition of Theorem 9 and $v(t,x)$ the corresponding global solution, then if X is one of the following Banach spaces*:

$$\textit{Lebesgue } L^p, \quad 1 \leqslant p \leqslant \infty, \tag{207}$$

or

$$\textit{inhomogeneous Besov } B_q^{s,p}, \quad 1 \leqslant p \leqslant \infty, 1 \leqslant q \leqslant \infty, s > -1, \tag{208}$$

or

$$\textit{homogeneous Besov } \dot{B}_q^{s,p}, \quad 1 \leqslant p \leqslant \infty, 1 \leqslant q \leqslant \infty, s > -1, \tag{209}$$

then the corresponding solution also belongs to $L^\infty((0,\infty); X)$.

From the sketch of the proof contained in [93] it is clear that this result applies more generally to any Banach space X such that the following condition is satisfied

$$\|fg\|_X \lesssim \big(\|f\|_X\|g\|_\infty + \|g\|_X\|f\|_\infty\big), \tag{210}$$

as is the case for the spaces quoted above as well as for the Sobolev space H^s, $s \geqslant 1/2$.

5. Uniqueness theorems

In 1994 Jean Leray summarized the state of the art for the Navier–Stokes equations in the following way [150]:

> A fluid flow initially regular remains so over a certain interval of time; then it goes on indefinitely; but does it remain regular and well-determined? We ignore the answer to this double question. It was addressed sixty years ago in an extremely particular case [149]. At that time H. Lebesgue, questioned, declared: "Don't spend too much time for such a refractory question. Do something different!"

This is not the case for Kato's mild solutions for which a general uniqueness theorem, that is the subject of this section, is available. In order to appreciate the simplicity of its proof, let us start by recalling why the uniqueness of weak solutions remains a challenging question.

5.1. *Weak solutions*

Before dealing with the uniqueness of weak solutions for Navier–Stokes, let us examine a more general case. We consider the difference $v_1 - v_2$ of two weak solutions v_1 and v_2 that, for the moment, may take different initial values (i.e., $v_1(0) - v_2(0)$ is not necessarily zero), but with the same boundary conditions, say $v_1(t,x) - v_2(t,x) = 0$ if $x \in \partial\Omega$ for all $t > 0$ (this is always the case if we suppose the no-slip boundary conditions, $v_1 = v_2 = 0$ on $(0,T) \times \partial\Omega$)). Of course, if Ω is unbounded, this condition concerns the behavior of the solutions at infinity.

We obtain

$$\frac{\partial}{\partial t}(v_1 - v_2) + v_1 \cdot \nabla(v_1 - v_2) + (v_1 - v_2) \cdot \nabla v_2 = \Delta(v_1 - v_2) - \nabla(p_1 - p_2) \tag{211}$$

and if we take the inner product $\langle\cdot,\cdot\rangle$ of $L^2(\Omega)$ with $(v_1 - v_2)$ we finally get

$$\frac{1}{2}\frac{\mathrm{d}}{\mathrm{d}t}\|v_1 - v_2\|_2^2 + \big\|\nabla(v_1 - v_2)\big\|_2^2 = -\big\langle(v_1 - v_2) \cdot \nabla v_2, v_1 - v_2\big\rangle. \tag{212}$$

In fact, since $(v_1 - v_2)(t,x) = 0$ if $x \in \partial\Omega$ for all $t > 0$, Green's formula gives

$$\begin{aligned}&\big\langle v_1 \cdot \nabla(v_1 - v_2), v_1 - v_2\big\rangle \\ &\quad = -\big\langle\nabla \cdot v_1, |v_1 - v_2|^2\big\rangle - \big\langle v_1 \cdot \nabla(v_1 - v_2), v_1 - v_2\big\rangle = 0\end{aligned} \tag{213}$$

and

$$\big\langle\nabla(p_1 - p_2), v_1 - v_2\big\rangle = -\big\langle p_1 - p_2, \nabla \cdot (v_1 - v_2)\big\rangle = 0. \tag{214}$$

Thus, we obtain

$$\frac{1}{2}\frac{\mathrm{d}}{\mathrm{d}t}\|v_1 - v_2\|_2^2 + \big\|\nabla(v_1 - v_2)\big\|_2^2 \leqslant \|\nabla v_2\|_\infty \|v_1 - v_2\|_2^2 \tag{215}$$

which finally gives, via Gronwall's lemma, the estimate

$$\begin{aligned}&\big\|(v_1 - v_2)(s)\big\|_2^2 + 2\int_0^s \big\|\nabla(v_1 - v_2)\big\|_2^2\,\mathrm{d}t \\ &\quad \leqslant \big\|(v_1 - v_2)(0)\big\|_2^2 \exp\biggl(\int_0^s 2\|\nabla v_2\|_\infty\,\mathrm{d}t\biggr)\end{aligned} \tag{216}$$

and implies uniqueness of weak solutions as long as the (formal) manipulations we have performed are justified and the quantity $\int_0^s \|\nabla v_2\|_\infty\,\mathrm{d}t$ remains bounded. In particular, this argument shows the uniqueness of classical smooth solutions. More precisely, if *one* smooth weak solution, say v_2, exists and is such that $\int_0^s \|\nabla v_2\|_\infty\,\mathrm{d}t$ remains bounded, then *all* weak solutions have to coincide with it.

But there is another way to estimate the term $-\langle (v_1 - v_2) \cdot \nabla v_2, v_1 - v_2 \rangle$, say

$$\left|\langle (v_1 - v_2) \cdot \nabla v_2, v_1 - v_2 \rangle\right| \leqslant \|\nabla v_2\|_2 \|v_1 - v_2\|_4^2 \tag{217}$$

which suggests the use of the Sobolev inequality

$$\|v_1 - v_2\|_4 \leqslant c \|v_1 - v_2\|_2^{1-n/4} \left\|\nabla(v_1 - v_2)\right\|_2^{n/4} \tag{218}$$

where $n = 2$ or $n = 3$ denotes, as usual, the space dimension. Now, if we consider the two cases separately, we obtain after some straightforward calculations (see [110,235])

$$\left\|(v_1 - v_2)(s)\right\|_2^2 \leqslant \left\|(v_1 - v_2)(0)\right\|_2^2 \exp\left(c \int_0^s \|\nabla v_2\|_2^2 \,\mathrm{d}t\right) \tag{219}$$

if $n = 2$, and

$$\left\|(v_1 - v_2)(s)\right\|_2^2 \leqslant \left\|(v_1 - v_2)(0)\right\|_2^2 \exp\left(c \int_0^s \|\nabla v_2\|_2^4 \,\mathrm{d}t\right) \tag{220}$$

if $n = 3$.

If we make use of the energy inequality (21), which is the only information on weak solutions we can (and should) use here, it is easy to conclude and get a uniqueness result only in the case $n = 2$. In fact, nothing can be said if $n = 3$ because, at variance with the case $n = 2$, the energy inequality does not allow us here to treat the term $\int_0^s \|\nabla v_2\|_2^4 \,\mathrm{d}t$. If we could, we would of course not only obtain uniqueness, but also continuous dependence on initial data and the full regularity of the solution.

A third way to obtain uniqueness was suggested by Serrin [208,209] and improved later on by many authors. The idea is that if some additional integrability property is satisfied by *at least one* weak solution, more exactly, if $v_2 \in L^s((0, T); L^r)$ and if $2/s + n/r = 1$ with $n < r \leqslant \infty$, then *all* weak solutions have to coincide with it (recently, Kozono and Taniuchi in [126] considered the marginal case $s = 2$, $r = \infty$ in a larger class, say $v_2 \in L^2((0, T); BMO)$, see also [127,194]). In general, if v_2 is a weak solution, it is possible to prove that there exist s_0 and r_0 such that $2/s_0 + n/r_0 = n/2$ so that $v_2 \in L^{s_0}((0, T); L^{r_0})$. In particular, from this remark and Serrin's criterion we can recover, in the two-dimensional case, the uniqueness result shown above. But, again, in three dimensions this is not enough to conclude.

Finally, concerning the critical exponents $n = r$ and $s = \infty$, Serrin's result was adapted by von Wahl [232] (resp. by Kozono and Sohr [125]) to obtain the following result. Suppose that *one* weak solution, say v_2, satisfies $v_2 \in \mathcal{C}([0, T]; L^n)$ (resp. $v_2 \in L^\infty((0, T); L^n)$), then *all* weak solutions have to coincide with it (for a different proof see the papers of Lions and Masmoudi [152–154]). More recently, the smoothness of such a weak solution was proved by Escauriaza, Seregin and Sverák [75]. On the other hand, Montgomery-Smith announced in [177] a logarithmic improvement over the usual Serrin condition.

These types of results are known under the equivalence "weak = strong". In other words it is possible to show that if there exists a more regular weak solution, then the usual one (whose existence was proved by Leray) and such a regular solution necessarily coincide. The moral of the story is that if we *postulate* more regularity on weak solutions, then the uniqueness follows. In particular this argument shows that the uniqueness, the continuous dependence on the initial data and the regularity problems for the Navier–Stokes equations are closely related. In other words, any global weak solution coincides with a more regular one as long as such a solution exists.

It is also clear from this remark and from the analysis performed in Section 3, that if a weak solution v exists and if the initial data $v_0 \in L^3$, then the solution is a strong one on some interval $[0, T)$ with $T > 0$ (hence $v(t)$ is smooth for $0 < t < T$). Moreover, we may take $T = \infty$ if $\|v_0\|_3$ is small enough. In fact, as we recalled in Section 3, there exists a strong solution $u \in \mathcal{C}([0, T); L^3)$ with $T > 0$, with $u_0 = v_0$ and satisfying Serrin's criterion. This is a simple consequence of (167) and follows directly from the result by von Wahl [232] and by Kozono and Sohr [125] (see [114]).

On the other hand, we cannot apply the uniqueness result of von Wahl to prove the uniqueness of mild solutions in $\mathcal{C}([0, T); L^p)$ (neither for the critical case $p = 3$ nor for the supercritical one $p > 3$) because the initial data only belong to L^p and, in general, not to L^2. There are of course two exceptions: the case of a bounded domain and the case of the space dimension two. As a matter of fact, if Ω_b is a bounded domain in $\mathbb{R}^3$, by means of the embedding $L^p(\Omega_b) \hookrightarrow L^2(\Omega_b)$, if $p > 2$ (rather $p \geqslant 3$ so that the existence of a solution is guaranteed, as we have seen in Section 3) and von Wahl's uniqueness theorem, it is possible to prove that Leray's weak solutions coincide with Kato's mild ones, so that their uniqueness follows in a straightforward manner [92]. In the same way, if we consider $\mathbb{R}^2$ instead of $\mathbb{R}^3$, it is obvious that the uniqueness criterion of von Wahl gives uniqueness of mild solutions with data in the critical space $L^2(\mathbb{R}^2)$ (the supercritical case $L^q(\mathbb{R}^2)$, $q \geqslant 2$, always being easier to treat as we are going to see in the following section). In other words, once again, in two dimensions there is no mystery concerning uniqueness: Leray's theory, based on the energy space $L^2(\mathbb{R}^n)$, is in a perfect agreement with Kato's one, based on the invariant space $L^n(\mathbb{R}^n)$, because the two spaces involved coincide if $n = 2$.

5.2. *Supercritical mild solutions*

From the previous discussion it is clear that we will limit ourselves to the case of the whole three-dimensional space $\mathbb{R}^3$. Of course, *mutatis mutandis*, the results of this and the following sections apply as usual to $\mathbb{R}^n$, $n \geqslant 2$, as well. A very simple case is provided by the uniqueness of mild solutions in supercritical spaces. For example, in the case of the Lebesgue spaces L^p, $p > 3$, the following result holds true:

THEOREM 11. *Let $3 < p \leqslant \infty$ be fixed. For any $v_0 \in L^p$, $\nabla \cdot v_0 = 0$, and any $T > 0$, there exists at most a mild solution in $\mathcal{C}([0, T); L^p)$ to the Navier–Stokes equations. In other words, the solution $v(t, x)$ given by Theorem* 1 *is unique in the space $\mathcal{C}([0, T); L^p)$.*

The proof of this property is so simple that we wish to sketch it here. Let us suppose that $v_1(t, x) \in \mathcal{C}([0, T); L^p)$ and $v_2(t, x) \in \mathcal{C}([0, T); L^p)$ solve the mild integral equation

$$v_i(t) = S(t)v_0 + B(v_i, v_i)(t), \quad i = 1, 2, \tag{221}$$

with the same initial data v_0. Then, by taking the difference between these equations

$$v_1 - v_2 = B(v_1, v_1 - v_2) + B(v_1 - v_2, v_2) \tag{222}$$

and using (138), we get

$$\sup_{0<t<T} \|(v_1 - v_2)(t)\|_p \lesssim \eta(T, p)\Big(\sup_{0<t<T} \|v_1(t)\|_p + \sup_{0<t<T} \|v_2(t)\|_p\Big) \sup_{0<t<T} \|(v_1 - v_2)(t)\|_p, \tag{223}$$

where

$$\eta(T, p) = \frac{T^{1/2(1-3/p)}}{1 - 3/p}. \tag{224}$$

We can always take $T = T'$ small enough in order to obtain

$$\eta(T', p)\Big(\sup_{0<t<T'} \|v_1(t)\|_p + \sup_{0<t<T'} \|v_2(t)\|_p\Big) < 1 \tag{225}$$

which obviously implies $v_1 = v_2$ in $\mathcal{C}([0, T'); L^p)$. Now, it is also easy to see that this argument can be iterated to get uniqueness up to time $2T'$ (and so on to $3T'$, etc.). In other words, as explained in the papers by Kato and Fujita [117] (p. 254) and [87] (p. 290), the iteration scheme is well posed and leads to uniqueness up to time T.

5.3. *Critical mild solutions*

In this section we are interested in the proof of the uniqueness of the solution given by Theorem 2. The historical details describing the achievement of this result are contained in [37] and for a systematic approach of the existence and uniqueness problem for mild solutions, the reader is also referred to the papers of Amann [1].

Let us note from the very beginning that, by a simple application of Lemma 4 and Theorem 3, it is always possible to ensure the uniqueness of a mild local solution $v(t, x)$ in a critical space (e.g., $\mathcal{C}([0, T); L^3)$) associated with an initial datum (resp. $v_0 \in L^3$, $\nabla \cdot v_0 = 0$), if we just require that it belongs to one of the auxiliary spaces described before (introduced by Weissler, Calderón and Giga) and if the norm of the solution $v(t, x)$ in such a space is smaller than a given constant (for example, smaller than $2\|v_0\|_3$, as follows directly from (119), (121) and (122)). Even if this remark is trivial and despite the fact that the condition under which the uniqueness is satisfied is very restrictive, we will use this elementary uniqueness result in Section 6 devoted to the proof of existence of self-similar solutions for the Navier–Stokes equations.

Since the introduction at the beginning of the 1960s of the mild formulation of the Navier–Stokes equations by Kato and Fujita [87,117], other results were discovered, ensuring the uniqueness of the corresponding solution under several regularity hypotheses near $t = 0$. In the simplest case, when the solutions belong to $\mathcal{C}([0,T); L^3)$, these additional conditions are written [114] $\lim_{t\to 0} t^{\frac{\alpha}{2}} \|v(t)\|_q = 0$, $\alpha = 1 - \frac{3}{q}$, $3 < q < \infty$, or [100], for the same values of α and q, $v \in L^{\frac{2}{\alpha}}((0,T), L^q)$. In fact, as we described in detail in Section 3.4, the use of one of these two auxiliary norms (corresponding to two auxiliary subspaces of $\mathcal{C}([0,T); L^3)$) makes it possible to apply the fixed point algorithm to obtain the existence of mild solutions in $\mathcal{C}([0,T); L^3)$. This is the reason why, in one of these subspaces, the uniqueness of the solution can be guaranteed as well. For example, the following result was known since the fundamental papers by Kato and Fujita.

THEOREM 12. *Let $3 < q \leqslant \infty$ be fixed. For any $v_0 \in L^3$, $\nabla \cdot v_0 = 0$, and any $T > 0$, there exists at most a mild solution to the Navier–Stokes equations such that $v(t,x) \in \mathcal{C}([0,T); L^3)$, $t^{1/2(1-3/q)} v(t,x) \in \mathcal{C}([0,T); L^q)$ and the following condition is satisfied*

$$\lim_{t\to 0} t^{1/2(1-3/q)} \|v(t)\|_q = 0. \tag{226}$$

In other words, using the notation of Section 3.4, Theorem 12 guarantees uniqueness (only) in the subspace $\mathcal{N} \cap \mathcal{K}_q$, $3 < q \leqslant \infty$. If $q = \infty$, the uniqueness is treated in detail in [166]. If $3 < q < \infty$, the proof follows directly from Lemma 10. In fact, if v_i, $i = 1, 2$ are two solutions that verify $t^{1/2(1-3/q)} v_i(t,x) \in \mathcal{C}([0,T); L^q)$ and $\lim_{t\to 0} t^{1/2(1-3/q)} \|v_i(t)\|_q = 0$ we have by Lemma 10 (here $3 < q < \infty$)

$$\begin{aligned} &\sup_{0<t<T} t^{1/2(1-3/q)} \|(v_1 - v_2)(t)\|_q \\ &\quad \lesssim \sup_{0<t<T} t^{1/2(1-3/q)} \|(v_1 - v_2)(t)\|_q \\ &\qquad \times \Bigl(\sup_{0<t<T} t^{1/2(1-3/q)} \|v_1(t)\|_q + \sup_{0<t<T} t^{1/2(1-3/q)} \|v_2(t)\|_q \Bigr) \end{aligned} \tag{227}$$

and it is possible to chose $T = T'$ small enough so that

$$\Bigl(\sup_{0<t<T'} t^{1/2(1-3/q)} \|v_1(t)\|_q + \sup_{0<t<T'} t^{1/2(1-3/q)} \|v_2(t)\|_q \Bigr) < 1, \tag{228}$$

thus implying uniqueness (again, let us state that this argument can be iterated in time as in the proof of Theorem 11).

Of course, the previous result is not satisfactory and one would expect that the following result holds true.

THEOREM 13. *For any $v_0 \in L^3$, $\nabla \cdot v_0 = 0$ and any $T > 0$, there exists at most a mild solution to the Navier–Stokes equations such that $v(t,x) \in \mathcal{C}([0,T); L^3)$.*

The first proof of Theorem 13, say of the uniqueness in $\mathcal{C}([0,T); L^3)$ without any additional hypothesis (that was followed by at least five different other proofs [143,147]), was obtained in 1997 and was based on two well-known ideas. The first one is that it is more simple to study the bilinear operator $B(v,u)(t)$ in a Besov frame [34]; the second is that it is helpful to distinguish in the solution v the contribution from the tendency $\exp(t\Delta)v_0$ and from the fluctuation $B(v,v)(t)$, the latter function always being more regular than the former [34]. More precisely, Furioli, Lemarié and Terraneo in [91,92] were able to prove the uniqueness theorem in its optimal version, say Theorem 13, by using the bicontinuity of the scalar operator $B(f,g)(t)$ (and thus the vectorial as well) respectively from $L^\infty((0,T);L^3)\times L^\infty((0,T);L^3)\to L^\infty((0,T);\dot{B}_2^{1/2,\infty})$ and from $L^\infty((0,T);\dot{B}_2^{1/2,\infty})\times L^\infty((0,T);L^3)\to L^\infty((0,T);\dot{B}_2^{1/2,\infty})$.

What is remarkable is that, contrary to what one would expect, the spaces L^3 and $\dot{B}_2^{1/2,\infty}$ are not comparable. The fact that the Besov space of the positive regularity index played only a minor role in the paper [92] led naturally to the question whether one could do without it. Some months after the announcement of the uniqueness theorem of Lemarié and his students, Meyer showed how to improve this result. The distinction between the fluctuation and the tendency was not used, the time–frequency approach was unnecessary and the Besov spaces did not play any role. Meyer's proof shortened the problem to the bicontinuity of the bilinear term $B(f,g)(t)$ in the Lorentz space $L^{(3,\infty)}$ and more precisely, as stated in Proposition 9, in $\mathcal{C}([0,T);L^{(3,\infty)})$ [166]. This result by itself is even more surprising because, as we recalled in Section 3, Oru proved otherwise that, in spite of all the cancellations that it contains, the full vectorial bilinear term $B(v,u)(t)$ is not continuous in $\mathcal{C}([0,T);L^3)$ [183].

Let us now see how Proposition 9 simply implies Theorem 13. Let v_1 and v_2 two mild solutions in $\mathcal{C}([0,T);L^3)$ with same initial data $v_0\in L^3$ and consider their difference

$$\begin{aligned} v_1-v_2 &= B(v_1,v_1-v_2)+B(v_1-v_2,v_2)\\ &= B\big(v_1-S(t)v_0,v_1-v_2\big)+B\big(S(t)v_0,v_1-v_2\big)\\ &\quad + B\big(v_1-v_2,v_2-S(t)v_0\big)+B\big(v_1-v_2,S(t)v_0\big). \end{aligned}\tag{229}$$

Now, by means of Proposition 9 (via the embedding $L^3\hookrightarrow L^{(3,\infty)}$) and of a slight modification of Lemma 18, we get the following estimate

$$\begin{aligned} &\sup_{0<t<T}\big\|(v_1-v_2)(t)\big\|_{L^{3,\infty}}\\ &\quad\lesssim \sup_{0<t<T}\big\|(v_1-v_2)(t)\big\|_{L^{3,\infty}}\Big(\sup_{0<t<T}t^{1/2(1-3/q)}\big\|S(t)v_0\big\|_q\\ &\qquad + \sup_{0<t<T}\big\|v_1-S(t)v_0\big\|_{L^3}+\sup_{0<t<T}\big\|v_2-S(t)v_0\big\|_{L^3}\Big), \end{aligned}\tag{230}$$

where q can be chosen in the interval $3<q\leqslant\infty$ (for instance $q=\infty$ in the proof contained in [166]). Finally, it is possible to chose $T=T'$ small enough so that

$$\begin{aligned} &\Big(\sup_{0<t<T'}t^{1/2(1-3/q)}\big\|S(t)v_0\big\|_q\\ &\quad + \sup_{0<t<T'}\big\|v_1-S(t)v_0\big\|_{L^3}+\sup_{0<t<T'}\big\|v_2-S(t)v_0\big\|_{L^3}\Big)<1, \end{aligned}\tag{231}$$

this property being a direct consequence of Lemma 9 and of the strong continuity in time of the L^3 norm of the solutions v_1 and v_2. From this estimate we deduce that locally in time $v_1 - v_2$ is equal to zero in the sense of distribution, thus $v_1 - v_2$ is equal to zero in L^3 in the interval $0 \leqslant t \leqslant T'$ and the argument can of course be iterated in the time variable.

The proof of the uniqueness of the solution in the more general cases given by Theorems 3–5 (say, when the initial data belongs to a Besov space) is contained in [92].

To conclude, we wish to present a different proof of the uniqueness result from the one contained in [166], based on Proposition 9. In fact, following [36,48], we will give here a more precise result.

PROPOSITION 10. *Let $3/2 < q < \infty$ and $0 < T \leqslant \infty$ be fixed. The bilinear operator $B(f, g)(t)$ is bicontinuous from $L^\infty((0,T); L^{(3,\infty)}) \times L^\infty((0,T); L^{(3,\infty)}) \to L^\infty((0,T); \dot{B}_q^{3/q-1,\infty})$.*

We will prove this proposition by duality, as we did in the proof of Lemmas 20 and 21. Let us consider a test function $\chi(x) \in C_0^\infty$ and evaluate the duality product in $\mathbb{R}^3$ with the bilinear term. We get

$$\left|\langle B(f,g)(t), \chi\rangle\right| \leqslant \int_0^t \left|\left\langle s^{-2}\Theta\left(\frac{\cdot}{\sqrt{s}}\right) * \chi, (fg)(t-s)\right\rangle\right| \mathrm{d}s. \tag{232}$$

If we had at our disposal a generalization of the classical Young's inequality

$$\|a * b\|_\infty \leqslant \|a\|_{3/2}\|b\|_3, \tag{233}$$

we could hope to modify the following argument that gives the continuity of $B(f, g)$ from $L^\infty((0,T); L^3) \times L^\infty((0,T); L^3) \to L^\infty((0,T); \dot{B}_{3/2}^{1,\infty})$, that is,

$$\begin{aligned}
&\left|\langle B(f,g)(t), \chi\rangle\right| \\
&\quad \leqslant \Big(\sup_{0<t<T} \|fg(t)\|_{3/2}\Big) \int_0^t \left\| s^{-2}\Theta\left(\frac{\cdot}{\sqrt{s}}\right) * \chi\right\|_3 \mathrm{d}s \\
&\quad \leqslant 2\Big(\sup_{0<t<T} \|f(t)\|_3\Big)\Big(\sup_{0<t<T} \|g(t)\|_3\Big) \int_0^\infty u \left\|\frac{1}{u^3}\Theta\left(\frac{\cdot}{u}\right) * \chi\right\|_3 \frac{\mathrm{d}u}{u} \\
&\quad \lesssim \Big(\sup_{0<t<T} \|f(t)\|_3\Big)\Big(\sup_{0<t<T} \|g(t)\|_3\Big)\|\chi\|_{\dot{B}_3^{-1,1}},
\end{aligned} \tag{234}$$

the last estimate being a consequence of the equivalence of Besov norms given in Proposition 3.

Now, the generalized Young's inequality applied to the Lorentz spaces [111]

$$\|a * b\|_r \leqslant C_{p,q}\|f\|_p\|g\|_{(q,\infty)} \tag{235}$$

holds only if $1 < p, q, r < \infty$ and $p^{-1} + q^{-1} = 1 + r^{-1}$. Thus, there is no hope of modifying (233).

To circumvent such a difficulty, we will decompose the kernel Θ in two parties Θ_1 and Θ_2 defined by their Fourier transforms as

$$\widehat{\Theta}_1(\xi) =: |\xi| e^{-|\xi|^2/2} \tag{236}$$

and

$$\widehat{\Theta}_2(\xi) =: e^{-|\xi|^2/2}, \tag{237}$$

in such a way that

$$|\xi| \exp[-s|\xi|^2] = \frac{1}{\sqrt{s}} \widehat{\Theta}(\sqrt{s}\xi) = \frac{1}{\sqrt{s}} \widehat{\Theta}_1(\sqrt{s}\xi) \widehat{\Theta}_2(\sqrt{s}\xi). \tag{238}$$

With this decomposition, we can write, by taking the inverse Fourier transform (p and q being conjugate exponents)

$$\begin{aligned} &\left|\langle B(f,g)(t), \chi\rangle\right| \\ &\quad \leqslant \int_0^t \left|\left\langle s^{-2}\Theta_1\left(\frac{\cdot}{\sqrt{s}}\right) * \chi, \left(\frac{1}{\sqrt{s}}\right)^3 \Theta_2\left(\frac{\cdot}{\sqrt{s}}\right) * fg(t-s)\right\rangle\right| \mathrm{d}s \\ &\quad \leqslant \int_0^t \left\| \left(\frac{1}{\sqrt{s}}\right)^3 \Theta_2\left(\frac{\cdot}{\sqrt{s}}\right) * fg(t-s)\right\|_q \left\| s^{-2}\Theta_1\left(\frac{\cdot}{\sqrt{s}}\right) * \chi\right\|_p \mathrm{d}s \end{aligned} \tag{239}$$

and Young's generalized inequality ($3/2 < q < \infty$, $q^{-1} + 1 = \alpha^{-1} + 2/3$)

$$\begin{aligned} \left\| \left(\frac{1}{\sqrt{s}}\right)^3 \Theta_2\left(\frac{\cdot}{\sqrt{s}}\right) * fg(t-s)\right\|_q &\lesssim \left\| \left(\frac{1}{\sqrt{s}}\right)^3 \Theta_2\left(\frac{\cdot}{\sqrt{s}}\right)\right\|_\alpha \| fg(t-s)\|_{(3/2,\infty)} \\ &\lesssim s^{-3/2(2/3-1/q)} \| fg(t-s)\|_{(3/2,\infty)} \end{aligned} \tag{240}$$

allows to conclude

$$\begin{aligned} &\left|\langle B(f,g)(t), \chi\rangle\right| \\ &\quad \leqslant \Big(\sup_{0<t<T} \|fg(t)\|_{(3/2,\infty)}\Big) \int_0^t \frac{\|s^{-2}\Theta_1(\frac{\cdot}{\sqrt{s}}) * \chi\|_p}{s^{\frac{3}{2}(2/3-1/q)}} \mathrm{d}s \\ &\quad \leqslant 2\Big(\sup_{0<t<T} \|f(t)\|_{(3,\infty)}\Big)\Big(\sup_{0<t<T} \|g(t)\|_{(3,\infty)}\Big) \int_0^\infty \frac{\|\frac{1}{u^3}\Theta_1(\frac{\cdot}{u}) * \chi\|_p}{u^{1-3/q}} \frac{\mathrm{d}u}{u} \\ &\quad \lesssim \Big(\sup_{0<t<T} \|f(t)\|_{(3,\infty)}\Big)\Big(\sup_{0<t<T} \|g(t)\|_{(3,\infty)}\Big) \|\chi\|_{\dot{B}_p^{1-3/q,1}}. \end{aligned} \tag{241}$$

In order to make use of Proposition 10 in the proof of Theorem 13 we need a classical result (see [10]).

LEMMA 24. *The following embedding are continuous*: $\dot{B}_q^{3/q-1,\infty} \hookrightarrow L^{(3,\infty)}$ *for any* $0 < q < 3$ *and* $L^{(3,\infty)} \hookrightarrow \dot{B}_q^{3/q-1,\infty}$ *for any* $3 < q < \infty$.

Without losing generality, let us prove this lemma only when $q = 2$. In order to do this, we make use of the characterization of Besov and Lorentz spaces given by the interpolation theory as stated in (115) (see [10])

$$\left(L^2, L^4\right)_{(2/3,\infty)} = L^{(3,\infty)} \tag{242}$$

and

$$\left(\dot{B}_2^{0,1}, \dot{B}_2^{3/4,1}\right)_{(2/3,\infty)} = \dot{B}_2^{1/2,\infty}. \tag{243}$$

Now, as

$$\dot{B}_2^{0,1} \hookrightarrow L^2 \tag{244}$$

and

$$\dot{B}_2^{3/4,1} \hookrightarrow \dot{B}_4^{0,1} \hookrightarrow L^4, \tag{245}$$

we get the required result

$$\dot{B}_2^{1/2,\infty} \hookrightarrow L^{(3,\infty)}. \tag{246}$$

Proposition 9 is proved and Theorem 13 follows (see [166]).

6. Self-similar solutions

The viscous flows for which the profiles of the velocity field at different times are invariant under a scaling of variables are called self-similar. More precisely, we are talking about solutions to the Navier–Stokes equations

$$\begin{aligned} &\frac{\partial v}{\partial t} - \nu\Delta v = -(v\cdot\nabla)v - \nabla p,\\ &\nabla\cdot v = 0,\\ &v(0) = v_0 \end{aligned} \tag{247}$$

such that

$$v(t,x) = \lambda(t)V\bigl(\lambda(t)x\bigr), \qquad p(t,x) = \lambda^2(t)P\bigl(\lambda(t)x\bigr), \tag{248}$$

$\lambda(t)$ being a function of time, $P(x)$ a function of x and $V(x)$ a divergence-free vector field.

Two possibilities arise in what follows.

DEFINITION 10 (Backward). A backward self-similar solution is a solution of the form (248), where $\lambda(t) = 1/\sqrt{2a(T-t)}$, $a > 0$, $T > 0$ and $t < T$. As such, $V(x)$ and $P(x)$ solve the system

$$\begin{aligned} &-\nu\Delta V + aV + a(x\cdot\nabla)V + (V\cdot\nabla)V + \nabla P = 0, \\ &\nabla\cdot V = 0. \end{aligned} \tag{249}$$

DEFINITION 11 (Forward). A forward self-similar solution is a solution of the form (248), where $\lambda(t) = 1/\sqrt{2a(T+t)}$, $a > 0$, $T > 0$ and $t > -T$. As such, $V(x)$ and $P(x)$ solve the system

$$\begin{aligned} &-\nu\Delta V - aV - a(x\cdot\nabla)V + (V\cdot\nabla)V + \nabla P = 0, \\ &\nabla\cdot V = 0. \end{aligned} \tag{250}$$

6.1. *Backward: Singular*

The motivation for studying backward self-similar solutions is that, if they exist, they would possess a singularity when $t = T$; indeed $\lim_{t\nearrow T}\|\nabla v(t)\|_2 = \infty$. In 1933, Leray remarked that if a weak solution v becomes "turbulent" at a time T, then the quantity $u(t) = \sup_{x\in\mathbb{R}^3}\sqrt{v\cdot v}$ has to blow-up like $\frac{1}{\sqrt{2a(T-t)}}$ when t tends to T. Furthermore, he suggested, without proving their existence, to look for backward self-similar solutions. His conclusion was the following [148]:

> [...] unfortunately I was not able to give an example of such a singularity [...]. If I had succeeded in constructing a solution to the Navier equations that becomes irregular, I would have the right to claim that turbulent solutions not simply reducing to regular ones do exist. But if this position were wrong, the notion of turbulent solution, that for the study of viscous fluids will not play a key role any more, would not lose interest: there have to exist some problems of Mathematical Physics such that the physical causes of regularity are not sufficient to justify the hypothesis introduced when the equations are derived; to these problems we can apply similar considerations of the ones advocated so far.

The first proof of the nonexistence of backward self-similar solutions sufficiently decreasing at infinity seems to have been given by a physicist at the beginning of the 1970s in a somewhat esoteric paper, written by Rosen [203]. Another argument for the nonexistence of nontrivial solutions to the system (249) was given by Foias and Temam in [81].

But the mathematical proof for the nonexistence of backward self-similar solutions as imagined by Leray was available in functional spaces only later, in 1996, thanks to the works of the Czech school of J. Nečas.

In a paper published in the French Academy "Comptes Rendus" [179] – the last one to be presented by Leray (1906–1998) – Nečas, Růžička and Šverák announced that any weak solution V to the Navier–Stokes equations (249) belonging to the space $L^3 \cap W^{1,2}_{\mathrm{loc}}$ reduces to the zero solution. The proof of this remarkable statement [180] is based on asymptotic estimates at infinity (in the Caffarelli–Kohn–Nirenberg sense) for the functions

V and P as well as for their derivatives, and on the maximum principle for the function $\Pi(x) = \frac{1}{2}|V(x)|^2 + P(x) + ax \cdot V(x)$ on a bounded domain of $\mathbb{R}^3$. A different approach to obtain the same result, without using the Caffarelli–Kohn–Nirenberg theory, but under the more restrictive condition $V \in W^{1,2}$ was proposed afterwards by Málek, Nečas, Pokorný and Schonbek [155] (see also [170] for a generalization of the method to the proof of nonexistence of pseudo self-similar solutions).

Now, if we impose that the norms of v that appear naturally in the energy equality derived from (247) are finite, we get the estimates $\int_{\mathbb{R}^3} |V|^2 < \infty$ and $\int_{\mathbb{R}^3} |\nabla V|^2 < \infty$, i.e., $V \in W^{1,2}$ which implies $V \in L^3$, by Sobolev embedding. But if, on the contrary, we only impose that the local version of the energy equality is finite, in other words $V \in W^{1,2}_{\mathrm{loc}}$, we get some conditions that do not imply $V \in L^3$. This case, left open in [155,180], was solved by Tsai and gave origin to the following theorem [227,146]:

THEOREM 14. *Any weak backward self-similar solution V to the Navier–Stokes equations* (249) *belonging either to the space L^q, $3 < q < \infty$ or to $W^{1,2}_{\mathrm{loc}}$ reduces to the zero solution.*

6.2. *Forward: Regular or singular*

As we will see in this section, the situation is more favorable in the case of mild forward self-similar solutions. In fact, since the pioneering paper of Giga and Miyakawa [107], we know of the existence of many mild forward self-similar solutions of the type (248) with $\lambda(t) = 1/\sqrt{t}$. These solutions cannot be of finite energy. In fact, if we consider the inner product between V and the equation (250) and integrate by parts in the whole space, we get, if V is sufficiently decreasing at infinity

$$\int_{\mathbb{R}^3} |\nabla V|^2 + a \int_{\mathbb{R}^3} |V|^2 = 0. \tag{251}$$

Finally, this equality results in the conclusion that $V = 0$, in particular when $V \in W^{1,2}$. (It is important to stress here that such a conclusion is not true for backward self-similar solutions because of the difference of signs in (249) and (250).)

This is why Giga and Miyakawa suggested, as an alternative to Sobolev spaces, to consider the Morrey–Campanato ones. They succeeded in proving the existence and the uniqueness of mild forward self-similar solutions to the Navier–Stokes equations written in terms of the vorticity as unknown, without applying their method to the Navier–Stokes equations in terms of the velocity. Four years later, Federbush [78,79] considered the super-critical Morrey–Campanato spaces $\dot{M}_2^q$, $3 < q < \infty$, for these equations. The critical space $\dot{M}_2^3$ was treated shortly after by Taylor [216] who, surprisingly, did not take advantage of this space which contains homogeneous functions of degree -1, to get the existence of self-similar solutions as shown in [34].

As pointed out in the previous section, a remarkable property of the Besov spaces is that they contain homogeneous functions of degree -1 among their elements, such as, e.g., $|x|^{-1}$. This is a crucial point if we look for solutions to the Navier–Stokes equations which satisfy the scaling property

$$v(t,x) = v_\lambda(t,x) = \lambda v(\lambda^2 t, \lambda x) \quad \forall \lambda > 0 \tag{252}$$

or, equivalently, taking $\lambda^2 t = 1$, such that

$$v(t,x) = \frac{1}{\sqrt{t}} V\left(\frac{x}{\sqrt{t}}\right). \tag{253}$$

In fact, whenever they exist, these particular ($a = 1/2$ and $T = 0$) forward self-similar solutions $v(t,x)$ are such that their initial value $v(0,x)$ is a homogeneous function of degree -1.

We will show here how to obtain, by using a generalization of Kato's celebrated Theorem 3, the existence of mild forward self-similar solutions $v(t,x)$ with initial data v_0 homogeneous of degree -1, divergence-free and sufficiently small in a Besov space. In [34,35,44,45], we showed how to construct mild forward self-similar solutions for the Navier–Stokes equations (247), by using Besov spaces. In particular, the existence of regular forward self-similar solutions of the form $\frac{1}{\sqrt{t}} V(\frac{x}{\sqrt{t}})$ with $V \in L^q$ and $3 < q < \infty$ is contained as a corollary in [34]. The main idea of the aforementioned papers is to study the Navier–Stokes equations by the fixed point algorithm in a critical space containing homogeneous functions of degree -1. Furthermore, as noted by Planchon [186], the equivalence between the integral mild equation and the elliptic problem (250) is completely justified.

The result we are talking about is the following theorem.

THEOREM 15. *Let $3 < q < \infty$, and $\alpha = 1 - 3/q$ be fixed. There exists a constant $\delta_q > 0$ such that for any initial data $v_0 \in \dot{B}_q^{-\alpha,\infty}$, homogeneous of degree -1, $\nabla \cdot v_0 = 0$ in the sense of distributions and such that*

$$\|v_0\|_{\dot{B}_q^{-\alpha,\infty}} < \delta_q, \tag{254}$$

then there exists a global mild forward self-similar solution $v(t,x)$ to the Navier–Stokes equations such that

$$v(t,x) = \frac{1}{\sqrt{t}} V\left(\frac{x}{\sqrt{t}}\right) \tag{255}$$

where $V(x)$ is a divergence-free function belonging to $\dot{B}_q^{-\alpha,\infty} \cap L_q$.

The proof of these results follows by a simple modification of Theorem 3, once we recall that it is always possible to ensure the uniqueness of a mild solution $v(t,x)$ in a critical space, if the norm of the solution $v(t,x)$ in such a space is smaller than a given constant (see Section 5.3). In fact, suppose that $v(t,x)$ solves Navier–Stokes with a datum $v_0 \in \dot{B}_q^{-\alpha,\infty}$

such that $v_0 = \lambda v_0(\lambda x)\ \forall \lambda > 0$, then the corresponding solution $v(t,x)$, whose uniqueness is ensured if $\sup_{0<t<\infty} t^{\alpha/2} \|v(t,x)\|_q \leqslant C$, has to coincide with $\lambda v(\lambda^2 t, \lambda x)\ \forall \lambda > 0$ for the latter inequality is invariant under the same self-similar scaling.

Since 1995, Barraza has suggested replacing the Besov spaces with the Lorentz ones $L^{(3,\infty)}$ (see also Kozono and Yamazaki's results [131,133,236]), always with the aim of proving the existence of forward self-similar solutions [4], but he did not achieve the bicontinuity of the bilinear operator in this space. This result was proven later by Meyer (see Proposition 9), and was applied, not only to obtain the uniqueness of Kato's mild solutions (Theorem 13), but also to prove the existence of forward self-similar solutions. More precisely:

THEOREM 16. *There exists a constant $\delta > 0$ such that for any initial data $v_0 \in L^{(3,\infty)}$, homogeneous of degree -1, $\nabla \cdot v_0 = 0$ in the sense of distributions and such that*

$$\|v_0\|_{L^{(3,\infty)}} < \delta, \tag{256}$$

then there exists a global mild forward self-similar solution $v(t,x)$ to the Navier–Stokes equations such that

$$v(t,x) = \frac{1}{\sqrt{t}} V\left(\frac{x}{\sqrt{t}}\right), \tag{257}$$

where $V(x)$ is a divergence-free function belonging to $L^{(3,\infty)}$.

Once again, the proof of this theorem is trivial if we recall the bicontinuity of the bilinear term $B(f,g)(t)$ in $\mathcal{C}([0,T); L^{(3,\infty)})$ [166] (see Proposition 9). This result shows that there is no need for Fourier transform or Besov spaces to prove the existence of self-similar solutions for Navier–Stokes.

As we have already pointed out, Le Jan and Sznitman [137,138] gave an even simpler ad hoc setting to prove such a result. The space they introduced is defined, however, by means of a Fourier transform condition. More exactly, following the notations of Section 2.5.3,

$$\psi \in \mathcal{PM}^2 \quad \text{if and only if} \quad \hat{\psi} \in L^1_{\text{loc}} \text{ and } \|\psi\|_{\mathcal{PM}^2} = \sup_{\xi} |\xi|^2 \left|\hat{\psi}(\xi)\right| < \infty. \tag{258}$$

Now, according to the simplified version of Le Jan and Sznitman's result contained in [48], we have:

THEOREM 17. *The bilinear operator $B(f,g)$ is bicontinuous from $L^\infty_t(\mathcal{PM}^2) \times L^\infty_t(\mathcal{PM}^2)$ into $L^\infty_t(\mathcal{PM}^2)$. Therefore there exists a unique global mild solution to the Navier–Stokes equations in $L^\infty_t(\mathcal{PM}^2)$ provided the initial data is divergence-free and sufficiently small in $\mathcal{PM}^2$.*

Note that the authors made use of some probabilistic tools in [137,138] requiring rather subtle techniques to obtain the continuity of the bilinear operator. More precisely, the main idea contained in these papers is to study the non linear integral equation verified by the Fourier transform of the Laplacian of the velocity vector field associated with the "deterministic equations" of Navier–Stokes. This integral representation involves a Markovian kernel K_ξ, associated to the branching process, called stochastic cascades, in which each particle located at $\xi \neq 0$, after an exponential holding time of parameter $|\xi|^2$, with equal probability either dies out or gives birth to two descendants, distributed according to K_ξ. By taking the inverse Fourier transform one can thus obtain a solution to the Navier–Stokes equations... arising from a sequence of cascades!

However, as pointed out in [48], in the particular case of the pseudo-measures, Theorem 17 is a straightforward consequence of the fixed point algorithm and it is enough to show why the bilinear operator is bicontinuous. We work in Fourier space, with $\hat{f}$ and $\hat{g}$ instead of f and g. A standard argument (rotational invariance and homogeneity) shows that [214,215]

$$\frac{1}{|\xi|^2} * \frac{1}{|\xi|^2} \simeq \frac{C}{|\xi|}. \tag{259}$$

Thus

$$\widehat{B(f,g)}(t,\xi) = \int_0^t |\xi| \mathrm{e}^{-(t-s)|\xi|^2} \hat{f}(s) * \hat{g}(s)\,\mathrm{d}s, \tag{260}$$

and, upon using (259),

$$\sup_{t,\xi}\big(|\xi|^2 \big|\widehat{B}(t)\big|\big) \lesssim \sup_{t,\xi}\big(|\xi|^2 \big|\hat{f}(t)\big|\big) \sup_{t,\xi}\big(|\xi|^2 \big|\hat{g}(t)\big|\big) \sup_{t,\xi} \int_0^t |\xi|^2 \mathrm{e}^{-(t-s)|\xi|^2}\,\mathrm{d}s. \tag{261}$$

This last integral is in turn less than unity, which concludes the proof once the fixed point algorithm is recalled.

Finally, the norm of the space $\mathcal{PM}^2$ being critical in the sense of Definition 9, the following result can be easily deduced from the previous estimate.

THEOREM 18. *There exists a constant $\delta > 0$ such that for any initial data $v_0 \in \mathcal{PM}^2$, homogeneous of degree -1, $\nabla \cdot v_0 = 0$ in the sense of distributions and such that*

$$\|v_0\|_{\mathcal{PM}^2} < \delta, \tag{262}$$

then there exists a global mild forward self-similar solution $v(t,x)$ to the Navier–Stokes equations such that

$$v(t,x) = \frac{1}{\sqrt{t}} V\left(\frac{x}{\sqrt{t}}\right), \tag{263}$$

where $V(x)$ is a divergence-free function belonging to $\mathcal{PM}^2$.

REMARK. As far as backward self-similar solutions are concerned, we can exclude the existence of singularities for the Navier–Stokes equations simply by using Nečas, Růžička and Šverák and Tsai's results. However, singular forward self-similar solutions may exist. More precisely, there is a substantial difference between the self-similar solutions constructed in Theorem 15 and those constructed in Theorems 16 and 18. Both have a singularity at time $t = 0$ (of the type $\sim 1/|x|$), but the solution constructed in Theorem 15 becomes *instantaneously* smooth for $t > 0$, whereas this property cannot be ensured a priori for the other two families of self-similar solutions. The reason is the following. Even if they are both issued from the fixed point algorithm, the solutions in Theorem 15, and in Theorems 16 and 18 are constructed in a very different way. In the first case, in order to overcome the difficulty (and sometimes the impossibility) of proving the continuity of the bilinear estimate in the so-called critical spaces, we had to make use of Kato's celebrated idea of considering *two norms* at the same time, the so-called natural norm and the auxiliary regularizing norm. As such, Kato's approach imposes a priori a regularization effect on the solutions we look for. In other words, they are considered as fluctuations around the solution of the heat equation with the same initial data. In the case of the self-similar solution arising from Theorem 15, this regularity condition is imposed by the Lebesgue norm. More explicitly, not only does the divergence-free function $V(x)$ belong to the Besov space $\dot{B}_q^{-\alpha,\infty}$, but also to L_q, which is not a priori the case for the solutions in Theorems 16 and 18.

For people who believe in blow-up and singularities, this a priori condition coming from the *two norms* approach is indeed very strong. In other words, at variance with Leray's approach, Kato's algorithm does not seem to provide a framework for studying a priori singular solutions. However, as we have seen in the previous pages, two exceptions exist, i.e., two critical spaces where Kato's method applies with just *one norm*: the Lorentz space $L^{(3,\infty)}$ (considered independently by Kozono and Yamazaki [131,133,236], Barraza [4,5], Meyer [166]) and the pseudo-measure space of Le Jan and Sznitman [137,138]. The approach with only *one norm* gives the existence of a solution in a larger space which, in our case, contains genuinely singular solutions that are not smoothened by the action of the nonlinear semigroup associated.

The importance of this remark will be clear in Section 6.4, where we will construct explicit forward self-similar solutions, singular for any time $t \geqslant 0$, and we will suggest how to obtain loss of smoothness for solutions with large data.

If the debate concerning singularities is still open, as far as Besov spaces and harmonic analysis tools are concerned, it is clear that they have nothing to do with the existence (Theorem 16) nor the nonexistence (Theorem 14) of self-similar solutions.

6.3. *Asymptotic behavior*

Finding self-similar solutions is important because of their possible connection with attractor sets. In other words, they are related to the asymptotic behavior of global solutions of the Navier–Stokes equations. A heuristic argument is the following: let $v(t, x)$ be a global solution to the Navier–Stokes system, then, for any $\lambda > 0$, the function

$v_\lambda(t,x) =: \lambda v(\lambda^2 t, \lambda x)$ is also a solution to the same system. Now, if in a "certain sense" the limit $\lim_{\lambda\to\infty} v_\lambda(t,x) =: u(t,x)$ exists, then it is easy to see that $u(t,x)$ is a self-similar solution and that $\lim_{t\to\infty} \sqrt{t}v(t,\sqrt{t}x) = u(1,x)$. In [186,188,189], Planchon gave the precise mathematical frame to explain the previous heuristic argument (see also Meyer [164], Barraza [5] and, for more general nonlinear equations, Karch [113]).

As we suggested among the open problems in [34], the existence of self-similar solutions also evokes the study of exact solutions for Navier–Stokes. In the following section, we will describe the result of Tian and Xin, who gave an explicit one-parameter family of self-similar solutions, singular in a single point [221], and we will show how to interpret their result as a loss of smoothness for large data.

We would like to mention here the papers of Okamoto [181,182] that contain a systematic study of exact solutions of the systems (249) and (250). These results merit attention, especially since the resolution of these elliptic equations seems very difficult. One could image to apply these results to the study of mild solutions in the subcritical case, for which neither the existence nor the uniqueness is known (see also [37]) unless some restriction are required (see [32,33,145]).

More precisely, let us suppose that we can prove the existence of a nontrivial self-similar solution $v(t,x) = \frac{1}{\sqrt{t}}V(\frac{x}{\sqrt{t}})$ – in other words a solution V of (250) – with $V \in L^p$ and $1 \leqslant p < 3$. Then the Cauchy problem associated to the zero initial data would allow two different solutions, viz. v and 0, both belonging to $\mathcal{C}([0,T); L^p)$. In fact, $\lim_{t\to 0} \|\frac{1}{\sqrt{t}}V(\frac{x}{\sqrt{t}})\|_p = 0$, provided $1 \leqslant p < 3$. And the Cauchy problem would be ill-posed in $\mathcal{C}([0,T); L^p)$, $1 \leqslant p < 3$ in the same way that it is ill posed for a semilinear partial differential equation studied in 1985 by Haraux and Weissler [108].

This point of view should confirm the conjecture formulated by Kato [116], according to which the Cauchy problem is ill posed in the sense of Hadamard when $1 \leqslant p < 3$. In the case $p = 2$, for example, we will not obtain a unique, global, regular and stable solution and the scenario imagined by Leray would be possible. We will come back to this question in Section 7.2.

Finally, let us quote the book of Giga and Giga [102] "*Nonlinear Partial Differential Equations – Asymptotic Behavior of Solutions and Self-Similar Solutions*", whose English translation should be available soon, that contains one of the most comprehensive and self-contained state of the art of the results available in this direction for the Navier–Stokes and other partial differential equations (e.g., the porous medium, the nonlinear Schrödinger and the KdV equations).

6.4. *Loss of smoothness for large data?*

As we recalled in the Introduction, a question intimately related to the uniqueness problem is the regularity of the solutions to the Navier–Stokes equations. Several possibilities can be conjectured. One may imagine that blow-up of initially regular solutions never happens, or that it becomes more likely as the initial norm increases, or that there is blow-up, but only on a very thin set of probability zero. Or it is "possible that singular solutions exist but are unstable and therefore difficult to construct analytically and impossible to detect numerically [...], which would contradict the almost universal assumption that these equations are globally regular"[122].

As we have seen in Section 3, when using a fixed point approach, existence and uniqueness of global solutions are guaranteed only under restrictive assumptions on the initial data, that is required to be small in some sense, i.e., in some functional space. In Section 4 we pointed out that fast oscillations are sufficient to make the fixed point scheme work, even if the norm in the corresponding function space of the initial data is arbitrarily large (in fact, a different auxiliary norm turns out to be small). Here we would like to suggest how some particular data, arbitrarily large (not oscillating) could give rise to singular solutions. It is extremely unpleasant that we have no criteria to decide whether for arbitrarily large data the corresponding solution is regular or singular.

As observed by Heywood in [110], in principle "it is easy to construct a singular solution of the NS equations that is driven by a singular force. One simply constructs a solenoidal vector field u that begins smoothly and evolves to develop a singularity, and then defines the force to be the residual".

Recently, Tian and Xin [221] found explicit formulas for a one-parameter family of stationary "solutions" of the three-dimensional Navier–Stokes system (1) "with $\phi \equiv 0$" which are regular except at a given point. These explicit "solutions" agree with those previously obtained by Landau for special values of the parameter (see [135,136]). Due to the translation invariance of the Navier–Stokes system, one can assume that the singular point corresponds to the origin. More exactly, the main theorem from [221] reads as follows.

All solutions to the Navier–Stokes system (with $\phi \equiv 0$*)* $u(x) = (u_1(x), u_2(x), u_3(x))$ *and* $p = p(x)$ *which are steady, symmetric about* x_1*-axis, homogeneous of degree* -1*, regular except* $(0,0,0)$ *are given by the following explicit formula*:

$$
\begin{aligned}
u_1(x) &= 2\frac{c|x|^2 - 2x_1|x| + cx_1^2}{|x|(c|x| - x_1)^2}, \\
u_2(x) &= 2\frac{x_2(cx_1 - |x|)}{|x|(c|x| - x_1)^2}, \\
u_3(x) &= 2\frac{x_3(cx_1 - |x|)}{|x|(c|x| - x_1)^2}, \\
p(x) &= 4\frac{cx_1 - |x|}{|x|(c|x| - x_1)^2},
\end{aligned}
\tag{264}
$$

where $|x| = \sqrt{x_1^2 + x_2^2 + x_3^2}$ *and* c *is an arbitrary constant such that* $|c| > 1$.

It is clear that these stationary "solutions" are self-similar, because they do not depend on time and they are homogeneous of degree -1 in the space variable. Moreover, there is no hope of describing the "solutions" given by (264) in Leray's theory, because they are not globally of finite energy; in other words, they do not belong to L^2. However, they do belong to L^2_{loc} and this is at least enough to allow us to give a (distributional) meaning to the nonlinear term $(v \cdot \nabla)v = \nabla \cdot (v \otimes v)$. Finally, as pointed out at the end of Section 6.2, the "solutions" discovered by Tian and Xin cannot be analyzed by Kato's two norms method either, because they are global but not smooth. More precisely, they are singular at the origin with a singularity of the kind $\sim 1/|x|$ for all time.

There are at least two ad hoc frameworks for studying such singularity within the fixed point scheme and without using the two norms approach. We are thinking of the Lorentz space $L^{(3,\infty)}$ ([42]) and the pseudo-measure space $\mathcal{PM}^2$ ([40]), because they both contain singularities of the type $\sim 1/|x|$. However, the latter space has the advantage that not only the definition of its norm is very elementary and simplifies the calculations, it will also allow us to treat singular (Delta type) external force, that precisely arise from Tian and Xin's "solutions".

More exactly, by straightforward calculations performed in [40], one can check that, indeed, $(u_1(x), u_2(x), u_3(x))$ and $p(x)$ given by (264) satisfy the Navier–Stokes equations with $\phi \equiv 0$ in the *pointwise sense* for every $x \in \mathbb{R}^3 \setminus \{(0,0,0)\}$. On the other hand, if one treats $(u(x), p(x))$ as a *distributional or generalized* solution to the Navier–Stokes equations in the whole $\mathbb{R}^3$, they correspond to the very singular external force $\phi = (b\delta_0, 0, 0)$, where δ_0 stands for the Dirac delta and the parameter b depends on c and $\lim_{|c|\to\infty} b(c) = 0$. As such, if c is small enough, the existence of these solutions can be ensured as well via the fixed point algorithm as in [42,40].

The stationary solutions defined in (264) are singular with singularity of the kind $\mathcal{O}(1/|x|)$ as $|x| \to 0$. This is a critical singularity, because as it was shown by Choe and Kim [65], every pointwise stationary solution to the Navier–Stokes system with $F \equiv 0$ in $B_R \setminus \{0\} = \{x \in \mathbb{R}^3 : 0 < |x| < R\}$ satisfying $u(x) = o(1/|x|)$ as $|x| \to 0$ is also a solution in the sense of distributions in the whole B_R. Moreover, it is shown in [65] that under the additional assumption $u \in L^q(B_R)$ for some $q > 3$, then the stationary solution $u(x)$ is smooth in the whole ball B_R. In other words, if $u(x) = o(1/|x|)$ as $|x| \to 0$ and $u \in L^q(B_R)$ for some $q > 3$, then the singularity at the origin is removable.

We are now ready to state our remark about a possible loss of regularity of solutions with large data (see [40]).

REMARK. Let us consider the Navier–Stokes equations with external force $\phi \equiv 0$. Then, if one defines the functions $u_\varepsilon(x, 0) = \varepsilon u(x)$, where $u(x)$ is the (divergence-free, homogeneous of degree -1) function given by (264) as the initial data, then for small ε the system has a global regular (self-similar) solution which is even more regular than a priori expected and for $\varepsilon = 1$ the system has a singular "solution" for any time. The fact that, for small ε and external force $\phi \equiv 0$ for every $x \in \mathbb{R}^3$, the solution is smooth follows from a parabolic regularization effect analyzed in [40]. On the other hand, if $\varepsilon > 1$ nothing can be said in general and the corresponding solution can be regular or singular.

However, after a more careful analysis, one realizes that this possible loss of smoothness result does not apply in the "distributional" sense, but, as we explained before, only "pointwise" for every $x \in \mathbb{R}^3 \setminus \{(0,0,0)\}$. On the other hand, as explained in a forthcoming paper [18], this loss of smoothness for large data holds in the distributional sense for a model equation of gravitating particles (for which, moreover, blow-up is known).

7. Stability

As we have seen in the previous sections, when using a fixed point approach, existence and uniqueness of global solutions are guaranteed only under restrictive assumptions on the initial data, that is required to be small in some sense, i.e., in some functional space. In Section 4 we pointed out that fast oscillations are sufficient to make the fixed point scheme work, even if the norm in the corresponding function space of the initial data is arbitrarily large (in fact, a different auxiliary norm turns out to be small). On the other hand, in Section 6 we suggested how arbitrarily large data (not oscillating) could give rise to irregular solutions: in general, we do not know whether for arbitrarily large data the corresponding solution is regular or singular.

For the Navier–Stokes equations one might consider the entire question irrelevant, for the solution is unique and regular for small initial data and no viscous flow can be considered incompressible if the initial data are too large. The problem here is different: the set ($\delta > 0$) of initial data for which one can ensure the existence and the uniqueness ($\|v_0\| < \delta$) is not known precisely and could be too small, and the result meaningless from a physical point of view. In other words, the initial data as well as the unique corresponding solution would be "physically" zero! The "physical" role played by the smallness assumption on the initial data will be dealt with in this section. More precisely, we will make the link between this property, the stability of the corresponding global solution and the existence of Lyapunov functions.

First of all, let us note that the smallness condition is not absolute, but relative to the viscosity ν and, if we do not rescale the variables as we did in Section 3.2, this condition is written $\|v_0\|/\nu < \delta$. Now, if we interpret $\|v_0\|$ as the characteristic velocity of the problem and we suppose (in the whole space $\mathbb{R}^3$ or $\mathbb{T}^3$) the characteristic length is normalized to unity, then the quotient $R =: \|v_0\|/\nu$ can be interpreted as a Reynolds number associated with the problem. More precisely, the complexity of the Navier–Stokes equations is essentially due to the competition between the nonlinear convection term $\rho(v \cdot \nabla)v$, and the linear term of viscous diffusion, $\mu\Delta v$. The order of magnitude of the quotient between these terms (dimension equation)

$$\frac{|\rho(v \cdot \nabla)v|}{|\mu\Delta v|} \equiv \frac{\rho}{\mu}\frac{V^2/L}{V/L^2} = \frac{LV}{\nu} =: R \tag{265}$$

defines a dimensionless quantity R, called Reynolds number, that allows a comparison of the inertial forces and the viscosity ones.

Thus, the condition giving the existence and uniqueness of Kato's (global and regular) solution is nothing but by the smallness of a dimensionless Reynolds number associated with the problem. At this point it would be tempting to prove that for Reynolds numbers that are too large, the solution does not exist, or is not regular, or not unique or simply not stable. This point of view would be confirmed by the image of developed turbulence formulated in 1944 by Landau [136]:

> Yet not every solution of the equations of motion, even if exact, can actually occur in Nature. The flows that occur in Nature must not only obey the equations of fluid dynamics, but also be stable. For the flow to be stable it is necessary that small perturbations, if they arise, should decrease with

time. If, on the contrary, the small perturbations which inevitably occur in the flow tend to increase with time, then the flow is absolutely unstable. Such a flow unstable with respect to infinitely small perturbations cannot exist.

The criteria to find the critical Reynolds numbers above which solutions of the Navier–Stokes could not necessarily be stable under small perturbations are a matter for the theory of hydrodynamics stability and we refer the reader to [36,210] for a more comprehensive discussion and accurate bibliography on the subject. In the following pages we would like to concentrate only on the results that are closely related to the approach for the Navier–Stokes equations introduced in [34].

Let us start with the L^3-valued mild solutions. First of all, we should note that the application that associates with the initial value $v_0 \in L^3$ the corresponding solution $v(t,x) \in \mathcal{C}([0,T); L^3)$ constructed, as in Kato's theory, by the fixed point theory, is analytical in a neighborhood of zero, as a functional acting on L^3 with values in $\mathcal{C}([0,T); L^3)$, as recalled for instance in [3]. Accordingly, the stability of mild solutions follows immediately because, by virtue of the uniqueness theorem (Section 5), any mild solution arises from the fixed point algorithm. As we will see in Section 7.2, this does not hold the case for the subcritical case $2 \leqslant p < 3$ [41,165].

Generalizing previous stability results in L^p (see [195,228]), Kawanago proceeds in the opposite direction [120,121]. First, he obtains a stability estimate, then makes use of it to establish a uniqueness theorem for mild solution. His result concerns global solutions $v \in \mathcal{C}([0,\infty); L^3)$ and reads as follows. For any $v_0 \in L^3$, there exist two constants $\delta(v_0) > 0$ and $C > 0$ such that, if $\|v_0 - \tilde{v}_0\|_3 < \delta$, then $\tilde{v} \in \mathcal{C}([0,\infty); L^3)$ and

$$\left\|v(t) - \tilde{v}(t)\right\|_3 \leqslant \left\|v(0) - \tilde{v}(0)\right\|_3 \exp\left\{C \int_0^t \left\|v(s)\right\|_5^5 \mathrm{d}s\right\} \tag{266}$$

for any $t > 0$. Finally, Barraza obtains some stability and uniqueness results for solutions in $L^{(3,\infty)}$ [5]. But, as we have already remarked, the theorem by Meyer in the same Lorentz space [166] allows a considerable simplification of these results.

As pointed out by Yudovich in [239], the choice of the norm for proving the stability of an infinite-dimensional system (e.g., a viscous fluid) is crucial because the Banach norms are not necessarily equivalent therein. To be more explicit, let us recall the simple example of the linear Cauchy problem [85,239]

$$\begin{aligned} &\frac{\partial v}{\partial t} = x\frac{\partial v}{\partial x}, \\ &v(0,x) = \varphi(x), \end{aligned} \tag{267}$$

whose unique (for an arbitrary smooth initial function φ) explicit solution $v(t,x) = \varphi(x\exp(t))$ is exponentially asymptotically stable in $L^p(\mathbb{R})$ for $1 \leqslant p < \infty$, stable but not asymptotically stable in $L^\infty(\mathbb{R})$ or $W^{1,1}(\mathbb{R})$ and exponentially unstable in any $W^{k,p}(\mathbb{R})$ for $k > 1$, $p \geqslant 1$ or $k = 1$, $p > 1$.

7.1. *Lyapunov functions*

A sufficient condition for a solution to be stable for a given norm is that $\|v(t,x)-\tilde{v}(t,x)\|$, the norm of the difference between the solution v and a perturbation $\tilde{v}$, is a decreasing-in-time function. This leads to the following definition.

DEFINITION 12. Let v be a solution of the Navier–Stokes equations, then any decreasing-in-time function $\mathcal{L}(v)(t)$ is called a Lyapunov function associated to v.

The most well-known example is certainly provided by energy

$$E(v)(t)=\frac{1}{2}\|v(t)\|_2^2, \tag{268}$$

because, a calculation similar to the one performed in (251), gives

$$\frac{\mathrm{d}}{\mathrm{d}t}E(t)=-\nu\|\nabla v(t)\|_2^2<0. \tag{269}$$

This result can easily be generalized in the homogeneous Sobolev spaces $\dot{H}^s$, for $0\leqslant s\leqslant 1$. For example, in the case $s=\frac{1}{2}$, by means of Hölder and Sobolev inequalities in $\mathbb{R}^3$, we get ([117], p. 258)

$$\|\mathbb{P}(v\cdot\nabla)v\|_2\leqslant C\|v\|_6\|\nabla v\|_3\leqslant C\|v\|_{\dot{H}^1}\|v\|_{\dot{H}^{3/2}}. \tag{270}$$

From this estimate we easily deduce the decreasing property for the function $v=v(t)$ that reads as follows

$$\frac{\mathrm{d}}{\mathrm{d}t}\|v(t)\|_{\dot{H}^{1/2}}^2\leqslant -2\|v(t)\|_{\dot{H}^{3/2}}^2\left(\nu-C\|v(t)\|_{\dot{H}^{1/2}}\right) \tag{271}$$

and thus, if the Reynolds number $\|v_0\|_{\dot{H}^{1/2}}/\nu$ is sufficiently small, we get a Lyapunov function associated with the norm $\dot{H}^{1/2}$. As already stated in Section 4.2, a similar argument allows us to obtain for the $\dot{H}^1$ norm:

$$\frac{\mathrm{d}}{\mathrm{d}t}\|v(t)\|_{\dot{H}^1}^2\leqslant -2\|v(t)\|_{\dot{H}^2}^2\left(\nu-C\|v(t)\|_{\dot{H}^{1/2}}\right). \tag{272}$$

This estimate shows that the smallness of the number $\|v_0\|_{\dot{H}^{1/2}}/\nu$ also implies the decrease in time of $\|v\|_{\dot{H}^1}$. Now, the Sobolev spaces $\dot{H}^s$, $s>1/2$ are super-critical. In other words, as far as the scaling is concerned, they have the same invariance as the Lebesgue spaces L^p if $p>3$. This means that one can prove the existence of a local mild solution for arbitrary initial data (Theorem 1). In the case of $\dot{H}^1$, this solution turns out to be global, provided the quantity $\|v_0\|_{\dot{H}^{1/2}}/\nu$ is sufficiently small, thanks to the uniform estimate

$$\|v(t)\|_{\dot{H}^1}\leqslant\|v_0\|_{\dot{H}^1}\quad\forall t>0, \tag{273}$$

that is derived directly from (272).

In other words, this property establishes a direct link between the Lyapunov functions, the existence of global regular solutions in an energy space and the oscillatory behavior of the corresponding initial data.

In a paper that seems to have been completely ignored [115], Kato, after treating the classical cases $\dot{H}^s$, $0 \leqslant s \leqslant 1$, derives new Lyapunov functions for the Navier–Stokes equations not necessarily arising from an energy norm. More precisely: there exists $\delta > 0$ such that if the Reynolds number $R_3(v_0) = \|v_0\|_3/\nu < \delta$, then the quantity $R_3(v)(t) = \|v(t)\|_3/\nu$ is a Lyapunov function associated with v. The importance of this result comes from its connection with the stability theory. In fact, as explained by Joseph [112]:

> It is sometimes possible to find positive definite functionals of the disturbance of a basic flow, other than energy, which decrease on the solutions when the viscosity is larger than a critical value. Such functionals, which may be called generalized energy functionals of the Lyapunov type, are of interest because they can lead to a larger interval of viscosities on which global stability of the basic flow can be guaranteed.

As we proved in [49,50], Kato's result also applies to other functional norms, in particular the Besov ones. See also [2,95,96,145] for related results in this direction. Not only do these properties show the stability for Navier–Stokes in very general functional frames (and imply in particular that set of global regular solutions is open), but as we have noted above, they could shed some light on the research of global Navier–Stokes solutions in supercritical spaces.

7.2. *Dependence on the initial data*

Before leaving this section, we would like to recall a result obtained by Meyer and announced at the Conference in honor of Jacques-Louis Lions held in Paris in 1998 [165]. The full proof will appear in detail in [167]. The theorem in question expresses the dependence on the initial data of the solutions to Navier–Stokes in the subcritical case and could shed some light on the conjecture formulated by Kato in [116], that we recalled in Section 6.3. The result is the following:

THEOREM 19. *There is no application of class $\mathcal{C}^2$ that associates a (mild or weak) solution $v(t,x) \in \mathcal{C}([0,T); L^p)$, $2 \leqslant p < 3$ to the corresponding initial condition $v_0 \in L^p$.*

Note that $p = 2$ corresponds to the most interesting case of weak solutions by Leray. In particular:

1. There is no application of class $\mathcal{C}^2$ that associates Leray's weak solution $v(t,x) \in L^\infty((0,T); L^2)$, to the initial condition $v_0 \in L^2$.
2. If a mild solution exists in the subcritical case ($2 \leqslant p < 3$), it does not arise from a fixed point algorithm. On the other hand, as we have seen in Section 7, the application that associates Kato's mild solution $v(t,x) \in \mathcal{C}([0,T); L^3)$ to the initial data $v_0 \in L^3$ is analytical in a neighborhood of zero as a functional acting on L^3 and taking values in $\mathcal{C}([0,T); L^3)$. In the subcritical case, the regularity of the flow-map changes drastically.

The proof of Theorem 19 is based on a contradiction argument. Briefly stated, it is assumed that for the initial data λv_0, the solution $v_\lambda(t,x)$, whose existence is supposed in Theorem 19, could be written in the form $\lambda v^{(1)}(t,x) + \lambda^2 v^{(2)}(t,x) + o(\lambda^2)$, where little o corresponds to the norm $L^\infty([0,T); L^p)$ and $\lambda \to 0$. Then, the idea is to evaluate (by calculations analogous to that performed in Section 3.4.2) the norm of the bilinear operator that defines $v^{(2)}(t,x)$ in terms of v_0 in order to prove that $v^{(2)}(t,x)$ cannot belong to $\mathcal{C}([0,T); L^p)$. As usual, the main point will be to evaluate not the "exact" value of the symbol of the operator, but its "homogeneity scaling".

This kind of ill-posedness results for solutions arising from the Banach fixed point theorem in the case of the Navier–Stokes equations can be easily generalized to the nonlinear heat equation, the viscous Hamilton–Jacobi equation and the convection–diffusion equation, as it is illustrated in the paper [41].

Conclusion

Should we conclude from the three examples given in this paper (oscillations, uniqueness and self-similarity) that real variable methods are *always* better suited for the study of Navier–Stokes, and that wavelets, paraproducts, Littlewood–Paley decomposition, Besov spaces and harmonic analysis tools in general have nothing to do with these equations?

In order to analyze this question, we list here a series of bad and good news, that will be summarized by a prophetic wish.

For the Navier–Stokes equations, there are other examples in which Fourier methods do not gain against real variable methods. For example, by using Fourier transform in [109], Heywood was hoping to get a better global estimate for $\|\nabla v(t)\|_2$, in order to improve the key inequality analyzed in Section 5.1, Equation (220). However, as he remarks in [110]:

> We give Fourier transform estimates for solutions of the Navier–Stokes equations, without using Sobolev's inequalities, getting again global existence in two dimensions but only local existence in three dimensions. [...] Unfortunately, because of a dimensional dependence in the evaluation of a singular integral, the final result is only a local existence theorem in the three-dimensional. [...] This adds another failure to an already long list of failures to prove global existence in the three-dimensional case, which may reinforce the feeling that singularities really exist.

In practical applications, one never looks for a solution in $\mathbb{R}^3$, yet solid bodies (e.g., the surface of a container), limit the region of space where the flow takes place. However, in the physically more interesting case when boundaries are present, it is very difficult to generalize the methods based on Fourier transform techniques (see [51,68,69,153,154,175,240]), unless some periodicity conditions are considered, like, e.g., the torus $\mathbb{T}^3$ (see [222]).

The situation seems more favorable to Fourier methods in the case of decay as $t \to \infty$ of solutions of the Navier–Stokes equations (see [23–27,95,96,237]). So far, no better techniques than the Fourier splitting introduced by Schonbek and the Hardy spaces considered by Miyakawa [171–173] are known to study the decay at infinity of solutions to the Navier–Stokes equations.

Finally, in the case of the Euler equations, there is a rich literature that makes use of paradifferential tools (see [55,61,229–231]). However, in the case of vortex patches, whose regularity was proved in 1993 by Chemin using Bony's paraproduct rule (see [57,61]), a much simpler proof that does not make use of the paradifferential machinery was discovered by Bertozzi and Constantin [11,12] and by Serfati [207].

The discussion seems endless, the examples innumerable and it is difficult to conclude. As announced, we will to do it with a messianic hope of Federbush [79]:

> One should be able to do more than we have accomplished so far using wavelets: make a dent in the question of the existence of global strong solutions, find a theoretical formalism for turbulence [...]. Someone (perhaps smarter than me, perhaps working harder than me, perhaps luckier than me, perhaps younger than me) should get much further on turbulence and the Navier–Stokes equations with the ideas in wavelet analysis.

Acknowledgments

This work was written during a three months (September–December 2001) visit in Tokyo (National Aerospace Laboratory) and Kyoto (Division of Aeronautics and Astronautics, Graduate School of Engineering, Kyoto University, Kyoto) supported by a STA–JSPS fellowship. The hospitality received at both Institutions and, more general, throughout Japan is gratefully acknowledged. The author wishes also to express his gratitude to Grzegorz Karch, Nataša Pavlović and Fabrice Planchon for their careful rereading of the manuscript.

References

[1] H. Amann, *On the strong solvability of the Navier–Stokes equations*, J. Math. Fluid Mech. **2** (2000), 16–98.

[2] P. Auscher, S. Dubois and P. Tchamitchian, *On the stability of global solutions to Navier–Stokes equations in the space*, J. Math. Pures Appl. **83** (6) (2004), 673–697.

[3] P. Auscher and P. Tchamitchian, *Espaces critiques pour le système des équations de Navier–Stokes incompressibles*, preprint, Université de Picardie Jules Verne (1999); available online at http://www.mathinfo.u-picardie.fr/auscher/.

[4] O.A. Barraza, *Self-similar solutions in weak L^p spaces of the Navier–Stokes equations*, Rev. Mat. Iberoamericana **12** (1996), 411–439.

[5] O.A. Barraza, *Regularity and stability for the solutions of the Navier–Stokes equations in Lorentz spaces*, Nonlinear Anal. **35** (6) (1999), 747–764.

[6] G. Battle and P. Federbush, *Divergence-free vector wavelets*, Michigan Math. J. **40** (1995), 181–195.

[7] G. Battle, P. Federbush and P. Uhlig, *Wavelets for quantum gravity and divergence-free wavelets*, Appl. Comput. Harmon. Anal. **1** (1994), 295–297.

[8] R.N. Bhattacharya, L. Chen, S. Dobson, R.B. Guenther, C. Orum, M. Ossiander, E. Thomann and E.C. Waymire, *Majorizing kernels & stochastic cascades with applications to incompressible Navier–Stokes equations*, Trans. Amer. Math. Soc. **355** (12) (2003), 5003–5040 (electronic).

[9] M. Ben-Artzi, *Global solutions of two-dimensional Navier–Stokes and Euler equation*, Arch. Ration. Mech. Anal. **128** (1994), 329–358.

[10] J. Bergh and J. Löfstrom, *Interpolation Spaces, an Introduction*, Springer-Verlag, Berlin (1976).

[11] A.L. Bertozzi and P. Constantin, *Global regularity of vortex patches*, Comm. Math. Phys. **152** (1993), 19–28.

[12] A.L. Bertozzi and A.J. Majda, *Vorticity and Incompressible Flow*, Cambridge Texts Appl. Math., Cambridge Univ. Press (2002).

[13] O.V. Besov, *Investigation of a Family of Functional Spaces Connected with Embedding and Extension Theorems*, Trudy Mat. Inst. Steklov **60** (1961) (in Russian).

[14] H.A. Biagioni, L. Cadeddu and T. Gramchev, *Semilinear parabolic equations with singular initial data in anisotropic weighted spaces*, Differential Integral Equations **12** (1999), 613–636.

[15] H.A. Biagioni and T. Gramchev, *Evolution PDE with elliptic dissipative terms: Critical index for singular initial data, self-similar solutions and analytic regularity*, C. R. Acad. Sci. Paris Sér. I Math. **327** (1) (1998), 41–46.

[16] P. Biler, *The Cauchy problem and self-similar solutions for a nonlinear parabolic equation*, Stud. Math. **114** (1995), 181–205.

[17] P. Biler, *Local and global solutions of a nonlinear nonlocal parabolic problem* (Warsaw, 1994), Gakuto Internat. Ser. Math. Sci. Appl., Vol. 7, Gakkōtosho, Tokyo (1996), 49–66.

[18] P. Biler, M. Cannone, I. Guerra and G. Karch, *Global regular and singular solutions for a model of gravitating particles*, Math. Ann. (2004), to appear.

[19] P. Biler, M. Cannone and G. Karch, *Asymptotic Stability of the Navier–Stokes Flow Past an Obstacle*, Banach Center Publications, Institute of Mathematics, Polish Academy of Sciences, Warszawa (2004), to appear.

[20] J.-M. Bony, *Calcul symbolique et propagation des singularités pour les équations aux dérivées partielles non linéaires*, Ann. Sci. Ecole Norm. Sup. **14** (4) (1981), 209–246.

[21] G. Bourdaud, *Réalisation des espaces de Besov homogènes*, Ark. Mat. **26** (1) (1988), 41–54.

[22] G. Bourdaud, *Ce qu'il faut savoir sur les espaces de Besov*, Prépublication de l'Université de Paris 7 (Janvier 1993).

[23] L. Brandolese, *Loss of the localization of asymmetric solutions of the Navier–Stokes equations in $\mathbb{R}^n$*, C. R. Acad. Sci. Paris Sér. I Math. **332** (2) (2001), 125–130.

[24] L. Brandolese, *Localisation, Oscillations et Comportament Asymptotique pour les Equations de Navier–Stokes*, Doctoral Thesis, Ecole Normale Supérieure de Cachan, France (2001).

[25] L. Brandolese, *Asymptotic behavior of the energy and pointwise estimates for solutions to the Navier–Stokes equations*, Rev. Mat. Iberoamericana **20** (1) (2004), 223–256.

[26] L. Brandolese, *Atomic decomposition for the vorticity of a viscous fluid*, Math. Nachr., to appear.

[27] L. Brandolese and Y. Meyer, *On the instantaneous spreading for the Navier–Stokes system in the whole space*, ESAIM Control Optim. Calc. Var. **8** (2002), 273–285 (electronic).

[28] H. Brezis, *Remarks on the preceding paper by M. Ben-Artzi*, Arch. Ration. Mech. Anal. **128** (4) (1994), 359–360.

[29] H. Brezis and T. Cazenave, *A nonlinear heat equation with singular initial data*, J. Anal. Math. **68** (1996), 277–304.

[30] F.E. Browder, *Non-linear equations of evolution*, Ann. Math. **80** (1964), 485–523.

[31] L. Caffarelli, R. Kohn and L. Nirenberg, *Partial regularity of suitable weak solutions of the Navier–Stokes equations*, Comm. Pure Appl. Math. **35** (1982), 771–837.

[32] C.P. Calderón, *Existence of weak solutions for the Navier–Stokes equations with initial data in L^p*, Trans. Amer. Math. Soc. **318** (1) (1990), 179–200; Addendum, ibid., 201–207.

[33] C.P. Calderón, *Initial values of solutions of the Navier–Stokes equations*, Proc. Amer. Math. Soc. **117** (3) (1993), 761–766.

[34] M. Cannone, *Ondelettes, paraproduits et Navier–Stokes*, Diderot, Paris (1995).

[35] M. Cannone, *A generalization of a theorem by Kato on Navier–Stokes equations*, Rev. Mat. Iberoamericana **13** (3) (1997), 515–541.

[36] M. Cannone, *Nombres de Reynolds, stabilité et Navier–Stokes*, Evolution Equations: Existence and Singularities, Vol. 52, Banach Center Publications, Institute of Mathematics, Polish Academy of Sciences, Warszawa (2000), 29–59.

[37] M. Cannone, *Viscous flows in Besov spaces*, Proc. Sixth Paseki School: Mathematical Theory in Fluid Mechanics, J. Málek, J. Nečas and M. Rokyta, eds, Springer-Verlag, Berlin (2000), 1–34.

[38] M. Cannone and S. Friedlander, *Navier: Blow-up and collapse*, Amer. Math. Soc. Notices **50** (1) (2003), 7–13.

[39] M. Cannone and G. Karch, *Incompressible Navier–Stokes equations in abstract Banach spaces*, Tosio Kato's Method and Principle for Evolution Equations in Mathematical Physics, H. Fujita, S.T. Kuroda and H. Okamoto, eds, Yurinsha, Tokyo (2002), 27–41; appeared also in Sūrikaisekikenkyūsho Kōkyūroku **1234** (2001).

[40] M. Cannone and G. Karch, *Smooth or singular solutions to the Navier–Stokes system?*, J. Differential Equations **197** (2) (2004), 247–274.

[41] M. Cannone and G. Karch, *On the validity of Picard algorithm for nonlinear parabolic equations*, Preprint (2003).

[42] M. Cannone and G. Karch, *About the generalized Navier–Stokes equations*, J. Math. Fluid Mech. (2004), to appear.
[43] M. Cannone and Y. Meyer, *Littlewood–Paley decomposition and the Navier–Stokes equations*, Meth. Appl. Anal. **2** (1995), 307–319.
[44] M. Cannone, Y. Meyer and F. Planchon, *Solutions auto-similaires des équations de Navier–Stokes in* $\mathbb{R}^3$, Exposé n. VIII, Séminaire X-EDP, Ecole Polytechnique (1994).
[45] M. Cannone and F. Planchon, *Self-similar solutions of the Navier–Stokes equations in* $\mathbb{R}^3$, Comm. Partial Differential Equations **21** (1996), 179–193.
[46] M. Cannone and F. Planchon, *Maximal function inequalities and Navier–Stokes equations*, *Investigations on the structure of solutions to partial differential equations* (in Japanese) (Kyōto, 1997), Sūrikaisekikenkyūsho Kōkyūroku **1036** (1998), 139–159.
[47] M. Cannone and F. Planchon, *On the non stationary Navier–Stokes equations with an external force*, Adv. Differential Equations **4** (5) (1999), 697–730.
[48] M. Cannone and F. Planchon, *On the regularity of the bilinear term for solutions of the incompressible Navier–Stokes equations in* $\mathbb{R}^3$, Rev. Mat. Iberoamericana **1** (2000), 1–16.
[49] M. Cannone and F. Planchon, *Fonctions de Lyapunov pour les équations de Navier–Stokes*, Exposé n. VIII, Séminaire X-EDP, Ecole Polytechnique (2000), 1–7.
[50] M. Cannone and F. Planchon, *More Lyapunov functions for the Navier–Stokes equations*, Navier–Stokes equations: Theory and Numerical Methods, R. Salvi, ed., Lectures Notes in Pure and Appl. Math., Vol. 223, Dekker, New York (2001), 19–26.
[51] M. Cannone, F. Planchon and M.E. Schonbek, *Navier–Stokes equations in the half space*, Comm. Partial Differential Equations **25** (5/6) (2000), 903–924.
[52] T. Cazenave and F.B. Weissler, *Asymptotically self-similar global solutions of the nonlinear Schrödinger and heat equations*, Math. Z. **228** (1) (1998), 83–120.
[53] T. Cazenave and F.B. Weissler, *More self-similar solutions of the nonlinear Schrödinger equation*, NoDEA Nonlinear Differential Equations Appl. **5** (3) (1998), 355–365.
[54] T. Cazenave and F.B. Weissler, *Scattering theory and self-similar solutions for the nonlinear Schrödinger equation*, SIAM J. Math. Anal. **31** (2000), 625–650 (electronic).
[55] D. Chae, *On the well-posedness of the Euler equations in the Triebel–Lizorkin spaces*, Comm. Pure Appl. Math. **55** (5) (2002), 654–678.
[56] J.-Y. Chemin, *Remarques sur l'existence globale pour le système de Navier–Stokes incompressible*, SIAM J. Math. Anal. **23** (1992), 20–28.
[57] J.-Y. Chemin, *Persistence de structures géometriques dans les fluides incompressibles bidimensionnels*, Ann. Sci. École Norm. Sup. **26** (1993), 1–16.
[58] J.-Y. Chemin, *About Navier–Stokes system*, Publ. Lab. Anal. Num. (Univ. Paris 6) **R 96023** (1996), 1–43.
[59] J.-Y. Chemin, *Sur l'unicité dans le système de Navier–Stokes tridimensionnel*, Exposé n. XXIV, Séminaire X-EDP, Ecole Polytechnique (1996–1997).
[60] J.-Y. Chemin, *Théorèmes d'unicité pour le système de Navier–Stokes tridimensionnel*, J. Anal. Math. **77** (1997), 27–50.
[61] J.-Y. Chemin, *Fluides Parfaits Incompressibles*, Astérisque **230** (1995); *Perfect Incompressible Fluids*, transl. by I. Gallagher and D. Iftimie, Oxford Lecture Series in Mathematics and Its Applications, Vol. 14, Clarendon Press–Oxford University Press, New York (1998).
[62] Z.-M. Chen, L_n *solutions to the stationary and nonstationary Navier–Stokes equations in* $\mathbb{R}^n$, Pacific J. Math. **158** (1993), 293–303.
[63] Z.-M. Chen and Z. Xin, *Homogeneity criterion on the Navier–Stokes equations in the whole spaces*, J. Math. Fluid Mech. **3** (2) (2001), 152–182.
[64] Z.-M. Chen, Y. Zhao and Y. Yang, *Hölder–Zygmund techniques to the Navier–Stokes equations in the whole spaces*, J. Partial Differential Equations **13** (2000), 89–96.
[65] H.J. Choe and H. Kim, *Isolated singularity for the stationary Navier–Stokes system*, J. Math. Fluid Mech. **2** (2) (2000), 151–184.
[66] R. Coifman, P.-L. Lions, Y. Meyer and S. Semmes, *Compensated compactness and Hardy spaces*, J. Math. Pures Appl. **72** (1993), 247–286.
[67] P. Constantin, *A few results and open problems regarding incompressible fluids*, Notices Amer. Math. Soc. **42** (6) (1995), 658–663.

[68] N. Depauw, *Solutions peu régulières des equations d'Euler et Navier–Stokes incompressibles sur un domaine à bord*, Doctoral Thesis, Université de Paris Nord, France (1998).

[69] N. Depauw, *Solutions des équations de Navier–Stokes incompressibles dans un domaine extérieur*, Rev. Mat. Iberoamericana **17** (1) (2000), 21–68.

[70] R.J. DiPerna, P.-L. Lions and Y. Meyer, *L^p-regularity of velocity averages*, Ann. Inst. H. Poincaré Anal. Non Linéaire **8** (1991), 271–287.

[71] S. Dubois, *What is a solution to the Navier–Stokes equations?*, C. R. Acad. Sci. Paris **335** (2002), 27–32.

[72] S. Dubois, *Equations de Navier–Stokes dans l'Espace: Espaces Critiques et Solutions d'Énergie Finie*, Doctoral Thesis, Université de Picardie Jules Verne (2002).

[73] S. Dubois, *Mild solutions to the Navier–Stokes equations and energy equalities*, J. Differential Equations **189** (1) (2003), 99–147.

[74] J. Earman, *A Primer on Determinism*, Reidel, Dordrecht (1986).

[75] L. Escauriaza, G. Seregin and V. Sverák, *On $L^{3,\infty}$-solutions to the Navier–Stokes equations and backward uniqueness*, Uspekhi Mat. Nauk **58** (2) (2003), 3–44.

[76] E.B. Fabes, B.F. Jones and N.M. Rivière, *The initial value problem for the Navier–Stokes equations with data in L^p*, Arch. Ration. Mech. Anal. **45** (1972), 222–240.

[77] M. Farge, *Wavelets transform and their applications to turbulence*, Ann. Rev. Fluid Mech. **24** (1992), 395–457.

[78] P. Federbush, *Navier and Stokes meet the wavelet*, Comm. Math. Phys. **155** (1993), 219–248.

[79] P. Federbush, *Navier and Stokes meet the wavelet, II*, Mathematical Quantum Theory. I. Field Theory and Many-Body Theory, Vancouver, BC, 1993, CRM Proc. Lecture Notes **7** (1994), 163–169.

[80] C.L. Fefferman, *Existence and uniqueness of the Navier–Stokes equation* (2002); available online at http://www.claymath.org/millennium/Navier–Stokes_Equations.

[81] C. Foias and R. Temam, *Self-similar universal homogeneous statistical solutions of the Navier–Stokes equations*, Comm. Math. Phys. **90** (1983), 187–206.

[82] M. Frazier, B. Jawerth and G. Weiss, *Littlewood–Paley Theory and the Study of Function Spaces*, CBMS Reg. Conf. Ser. Math., Vol. 79, Amer. Math. Soc., Providence, RI (1991).

[83] J. Frehse and J. Málek, *Problems due to the no-slip boundary in incompressible fluid dynamics*, Geometric Analysis and Nonlinear Partial Differential Equations, S. Hildebrandt and H. Karcher, eds, Springer-Verlag (2003), 559–571.

[84] S. Friedlander and N. Pavlović, *Remarks concerning a modified Navier–Stokes equation*, Discrete Contin. Dyn. Syst. **10** (1/2) (2004), 269–288.

[85] S. Friedlander and V. Yudovich, *Instabilities in fluid motion*, Notices Amer. Math. Soc. **46** (11) (1999), 1358–1367.

[86] J. Fourier, *Théorie analytique de la chaleur*, 1822, published in Œuvres de Fourier **1** Gauthiers-Villar, Paris (1888).

[87] H. Fujita and T. Kato, *On the Navier–Stokes initial value problem, I*, Arch. Ration. Mech. Anal. **16** (1964), 269–315.

[88] H. Fujita, S.T. Kuroda and H. Okamoto, eds, *Tosio Kato's Method and Principle for Evolution Equations in Mathematical Physics*, Yurinsha, Tokyo (2002); appeared also in Sūrikaisekikenkyūsho Kōkyūroku **1234** (2001).

[89] G. Furioli, *Applications de l'analyse harmonique réelle à l'etude des equations de Navier–Stokes et de Schrödinger non linéaire*, Doctoral Thesis, Université de Paris Sud, France (1999).

[90] G. Furioli, *On the existence of selfsimilar solutions of the nonlinear Schrödinger equation with power nonlinearity between* 1 *and* 2, Differential Integral Equations **14** (10) (2001), 1259–1266.

[91] G. Furioli, P.-G. Lemarié-Rieusset and E. Terraneo, *Sur l'unicité dans $L^3(\mathbb{R}^3)$ des solutions mild des équations de Navier–Stokes*, C. R. Acad. Sci. Paris Sér. I Math. **325** (12) (1997), 1253–1256.

[92] G. Furioli, P.-G. Lemarié-Rieusset and E. Terraneo, *Unicité dans $L^3(\mathbb{R}^3)$ et d'autres espaces fonctionnels limites pour Navier–Stokes*, Rev. Mat. Iberoamericana **16** (3) (2000), 605–667.

[93] G. Furioli, P.-G. Lemarié-Rieusset, E. Zahrouni and A. Zhioua, *Un théorème de persistence de la régularité en norme d'espaces de Besov pour les solutions de Koch and Tataru des équations de Navier–Stokes dans $\mathbb{R}^3$*, C. R. Acad. Sci. Paris Sér. I Math. **330** (2000), 339–342.

[94] G. Furioli and E. Terraneo, *Molecules of the Hardy space and the Navier–Stokes equations*, Funkcial. Ekvac. **45** (1) (2002), 141–160.

[95] I. Gallagher, D. Iftimie and F. Planchon, *Non-explosion en temps grand et stabilité de solutions globales des équations de Navier–Stokes*, C. R. Acad. Sci. Paris Sér. I Math. **334** (4) (2002), 289–292.

[96] I. Gallagher, D. Iftimie and F. Planchon, *Asymptotics and stability for global solutions to the Navier–Stokes equations*, Ann. Inst. Fourier **53** (5) (2003), 1387–1424.

[97] I. Gallagher and F. Planchon, *On global infinite energy solutions to the Navier–Stokes equations in two dimensions*, Arch. Ration. Mech. Anal. **161** (2002), 307–337.

[98] P. Gérard, Y. Meyer and F. Oru, *Inégalités de Sobolev précisées, Exposé n. VIII*, Séminaire X-EDP, Ecole Polytechnique (1996–1997).

[99] Y. Giga, *Regularity criteria for weak solutions of the Navier–Stokes system*, Proc. Symp. Pure Math. **45** (1) (1986), 449–453.

[100] Y. Giga, *Solutions for semilinear parabolic equations in L^p and regularity of weak solutions of the Navier–Stokes system*, J. Differential Equations **62** (1986), 186–212.

[101] Y. Giga, *Review of the paper by H. Brezis*, "Remarks on the preceding paper by M. Ben-Artzi", Math. Rev. **96h:35149** (1996); see, for the revised form, at the MathSciNet http address.

[102] Y. Giga and M.-H. Giga, *Nonlinear Partial Differential Equations – Asymptotic Behaviour of Solutions and Self-Similar Solutions* (in Japanese), Kyōritsu Shuppan, Tokyo (1999), English transl. to appear.

[103] Y. Giga, K. Inui, J. Kato and S. Matsui, *Remarks on the uniqueness of bounded solutions of the Navier–Stokes equations*, Proceedings of the Third World Congress of Nonlinear Analysts, Part 6 (Catania, 2000), Nonlinear Anal. **47** (6) (2001), 4151–4156.

[104] Y. Giga, K. Inui and S. Matsui, *On the Cauchy Problem for the Navier–Stokes Equations with Nondecaying Initial Data*, Adv. Fluid Dynamics, Quad. Mat., Vol. 4, P. Maremonti, ed., Aracne, Roma (1999), 28–68.

[105] Y. Giga, S. Matsui and O. Sawada, *Global existence of two-dimensional Navier–Stokes flow with nondecaying initial velocity*, J. Math. Fluid Mech. **3** (2001), 302–315.

[106] Y. Giga and T. Miyakawa, *Solutions in L^r of the Navier–Stokes initial value problem*, Arch. Ration. Mech. Anal. **89** (1985), 267–281.

[107] Y. Giga and T. Miyakawa, *Navier–Stokes flows in $\mathbb{R}^3$ with measures as initial vorticity and the Morrey spaces*, Comm. Partial Differential Equations **14** (1989), 577–618.

[108] A. Haraux and F.B. Weissler, *Non-uniqueness for a semi-linear initial value problem*, Indiana Univ. Math. J. **31** (1985), 167–189.

[109] J.G. Heywood, *Fourier transform estimates for the Navier–Stokes equations*, Proc. Internat. Conf. Advances in Geometric Analysis and Continuum Mechanics, Standford University, International Press, Cambridge (1993), 109–115.

[110] J.G. Heywood, *Remarks on the possible global regularity of solutions of the three-dimensional Navier–Stokes equations*, Progress in Theoretical and Computational Fluid Mechanics, Paseky 1993, Pitman Res. Notes Math. Ser., Vol. 308, Longman, Harlow (1994), 1–32.

[111] R. Hunt, *On $L(p,q)$ spaces*, L'Enseignement Mathématique **12** (4) (1966), 249–276.

[112] D.D. Joseph, *Stability of Fluid Motions* (2 Vols), Springer Tracts in Natural Philosophy, Springer-Verlag, Berlin (1976).

[113] G. Karch, *Scaling in nonlinear parabolic equations*, J. Math. Anal. Appl. **234** (1999), 534–558.

[114] T. Kato, *Strong L^p solutions of the Navier–Stokes equations in $\mathbb{R}^m$ with applications to weak solutions*, Math. Z. **187** (1984), 471–480.

[115] T. Kato, *Liapunov functions and monotonicity in the Navier–Stokes equation*, Lecture Notes in Math., Vol. 1450, Springer-Verlag, Berlin (1990), 53–63.

[116] T. Kato, *Well-posedness nitsuite* (in Japanese), Sūgaku **48** (3) (1996), 298–300.

[117] T. Kato and H. Fujita, *On the non-stationary Navier–Stokes system*, Rend. Sem. Mat. Univ. Padova **32** (1962), 243–260.

[118] T. Kato and G. Ponce, *The Navier–Stokes equations with weak initial data*, Int. Math. Res. Notes **10** (1994), 435–444.

[119] N.H. Katz and N. Pavlović, *A cheap Caffarelli–Kohn–Nirenberg inequality for Navier–Stokes equations with hyper-dissipation*, Geom. Funct. Anal. **12** (2) (2002), 355–379.

[120] T. Kawanago, *Stability of global strong solutions of the Navier–Stokes equations, Nonlinear evolution equations and their applications* (in Japanese), Sūrikaisekikenkyūsho Kōkyūroku **913** (1995), 141–147.

[121] T. Kawanago, *Stability estimate of strong solutions for the Navier–Stokes system and its applications*, Electron. J. Differential Equations (electronic) **15** (1998), 1–23; available online at http://ejde.math.swt.edu/Volumes/1998/15-Kawanago/abstr.html.

[122] S. Klainerman, *PDE as a unified subject*, GAFA 2000 (Tel Aviv, 1999), Geom. Funct. Anal., Special Vol., Part I (2000), 279–315.

[123] H. Koch and D. Tataru, *Well-posedness for the Navier–Stokes equations*. Adv. Math. **157** (1) (2001), 22–35.

[124] H. Kozono, *Well-posedness of the Navier–Stokes equations*, EVEQ 2000 lectures notes (manuscript), Parts I and II, July 3–7 Prague (2000).

[125] H. Kozono and H. Sohr, *Remark on uniqueness of weak solutions to the Navier–Stokes equations*, Analysis **16** (3) (1996), 255–271.

[126] H. Kozono and Y. Taniuchi, *Bilinear estimates in BMO and the Navier–Stokes equations*, Math. Z. **235** (1) (2000), 173–194.

[127] H. Kozono and Y. Taniuchi, *Limiting case of the Sobolev inequality in BMO, with application to the Euler equation*, Comm. Math. Phys. **214** (1) (2000), 191–200.

[128] H. Kozono and M. Yamazaki, *Semilinear heat equations and the Navier–Stokes equation with distributions as initial data*, C. R. Acad. Sci. Paris Sér. I Math. **317** (1993), 1127–1132.

[129] H. Kozono and M. Yamazaki, *Semilinear heat equations and the Navier–Stokes equation with distributions in new function spaces as initial data*, Comm. Partial Differential Equations **19** (1994), 959–1014.

[130] H. Kozono and M. Yamazaki, *The stability of small stationary solutions in Morrey spaces of the Navier–Stokes equation*, Indiana Univ. Math. J. **44** (1995), 1307–1336.

[131] H. Kozono and M. Yamazaki, *The exterior problem for the non-stationary Navier–Stokes equation with data in the space* $L^{n,\infty}$, C. R. Acad. Sci. Paris Sér. I Math. **320** (6) (1995), 685–690.

[132] H. Kozono and M. Yamazaki, *The Navier–Stokes equation with distributions as initial data and application to self-similar solutions*, New Trends in Microlocal Analysis (Tokyo, 1995) (1997), 125–141.

[133] H. Kozono and M. Yamazaki, *Exterior problem for the stationary Navier–Stokes equations in the Lorentz space*, Math. Ann. **310** (2) (1998), 279–305.

[134] H. Kozono and M. Yamazaki, *On a larger class of stable solutions to the Navier–Stokes equations in exterior domains*, Math. Z. **228** (1998), 751–785.

[135] L.D. Landau, *A new exact solution of Navier–Stokes equations*, C. R. (Doklady) Acad. Sci. USSR (N.S.) **43** (1944), 286–288.

[136] L.D. Landau and E.M. Lifchitz, *Fluid Dynamics* (in Russian), Nauka, Moscow, 1986; English transl. by J.B. Sykes and W.H. Reid: Course of Theoretical Physics, Vol. 6, Pergamon Press, London–Paris–Frankfurt; Addison-Wesley, Reading, MA (1959).

[137] Y. Le Jan and A.-S. Sznitman, *Cascades aléatoires et équations de Navier–Stokes*, C. R. Acad. Sci. Paris **324** (7) (1997), 823–826.

[138] Y. Le Jan and A.-S. Sznitman, *Stochastic cascades and 3-dimensional Navier–Stokes equations*, Probab. Theory Related Fields **109** (3) (1997), 343–366.

[139] P.-G. Lemarié, *Analyses multi-résolutions non othogonales, commutation entre projecteurs et dérivations et ondelettes vecteurs à divergence nulle*, Rev. Mat. Iberoamericana **8** (1992), 221–237.

[140] P.-G. Lemarié, *Un théorème d'inxistence pour les ondelettes vecteurs à divergence nulle*, C. R. Acad. Sci. Paris **319** (8) (1994), 811–813.

[141] P.-G. Lemarié-Rieusset, *Some remarks on the Navier–Stokes equations in* $\mathbb{R}^3$, J. Math. Phys. **39** (8) (1998), 4108–4118.

[142] P.-G. Lemarié-Rieusset, *Solutions faibles d'énergie infinie pour les équations de Navier–Stokes dans* $\mathbb{R}^3$, C. R. Acad. Sci. Paris Sér. I **328** (8) (1999), 1133–1138.

[143] P.-G. Lemarié-Rieusset, *Cinq petits théorèmes d'unicité* L^3 *des solutions des équations de Navier–Stokes dans* $\mathbb{R}^3$, Colloque equations de Navier–Stokes et analyse microlocale, CIRM, Luminy, Marseille lectures notes (manuscript) (9–13 Novembre 1999), 1–9.

[144] P.-G. Lemarié-Rieusset, *Une remarque sur l'analyticité des solutions milds des équations de Navier–Stokes dans* $\mathbb{R}^3$, C. R. Acad. Sci. Paris Sér. I **330** (2000), 183–186.

[145] P.-G. Lemarié-Rieusset, *Recent Developments in the Navier–Stokes Problem*, Chapman & Hall/CRS Res. Notes Math., Chapman & Hall/CRC, Boca Raton, FL (2002).

[146] P.-G. Lemarié-Rieusset, *Espaces de Lorentz et Navier–Stokes: Le problème des solutions auto-similaires de Leray*, Preprint, Université d'Evry Val d'Essonne (2002); available online at http://www.maths.univ-evry.fr/pages_perso/lemarie/mypub.html.

[147] P.-G. Lemarié-Rieusset and R. May, *Uniqueness for the Navier–Stokes equations and multipliers between Sobolev spaces*, Preprint of the Université d'Evry Val d'Essonne (2002).

[148] J. Leray, *Etudes de diverses équations intégrales non linéaires et de quelques problèmes que pose l'hydrodynamique*, J. Math. Pures Appl. **12** (1933), 1–82.

[149] J. Leray, *Sur le mouvement d'un liquide visqueux emplissant l'espace*, Acta Math. **63** (1934), 193–248.

[150] J. Leray, *Aspects de la mécanique théorique des fluides*, La Vie des Sciences, C. R. Sér. Générale II **11** (4) (1994), 287–290.

[151] F.-H. Lin, *A new proof of the Caffarelli–Kohn–Nirenberg theorem*, Comm. Pure Appl. Math. **51** (1998), 240–257.

[152] P.-L. Lions, *Mathematical Topics in Fluid Mechanics*, Clarendon Press, Oxford, Vol. 1 (1996); Vol. 2 (1998).

[153] P.-L. Lions and N. Masmoudi, *Unicité des solutions faibles de Navier–Stokes dans $L^N(\Omega)$*, C. R. Acad. Sci. Paris Sér. I Math. **327** (1998), 491–496.

[154] P.-L. Lions and N. Masmoudi, *Uniqueness of mild solutions of the Navier–Stokes system in L^N*, Comm. Partial Differential Equations **26** (11–12) (2001), 2211–2226.

[155] J. Málek, J. Nečas, M. Pokorný and M.E. Schonbek, *On possible singular solutions to the Navier–Stokes equations*, Math. Nachr. **199** (1999), 97–114.

[156] A.L. Mazzucato, *Analysis of the Navier–Stokes and other non-linear evolution equations with initial data in Besov-type spaces*, Ph.D. Thesis, University of North Carolina at Chapel Hill (2000).

[157] A.L. Mazzucato, *Besov–Morrey spaces: Function space theory and applications to non-linear PDE*, Trans. Amer. Math. Soc. **355** (4) (2003), 1297–1364 (electronic).

[158] A.L. Mazzucato, *Decomposition of Besov–Morrey spaces*, Harmonic analysis at Mount Holyoke (South Hadley, MA, 2001), Contemp. Math., Vol. 320, Amer. Math. Soc., Providence, RI (2003), 279–294.

[159] R. May, *Existence, Unicité et Régularité des Solutions Faibles des Équations de Navier–Stokes*, Doctoral Thesis, Université d'Evry Val d'Essonne, France (1999).

[160] R. May, *Rôle de l'espace de Besov $B_\infty^{-1,\infty}$ dans le contrôle de l'explosion éventuelle en temps fini des solutions régulières des équations de Navier–Stokes*, C. R. Acad. Sci. Paris Sér. I Math. **336** (9) (2003), 731–734.

[161] Y. Meyer, *La Minimalité de l'Espace de Besov $\dot{B}_1^{0,1}$ et la Continuité des Opérateurs Définis par des Intégrales Singulières*, Univ. Autonoma de Madrid, Monografias de Matematicas, Vol. 4 (1986).

[162] Y. Meyer, *Wavelets and Operators*, Cambridge Studies in Advanced Mathematics, Vol. 37, Cambridge Univ. Press, Cambridge (1993).

[163] Y. Meyer, *Wavelets and fast numerical algorithms*, Handbook of Numerical Analysis, Vol. 5, P.-G. Ciarlet and J.-L.Lions, eds, Elsevier, Amsterdam (1997), 639–713.

[164] Y. Meyer, *Large time behavior and self-similar solutions of some semi-linear diffusion equations*, Harmonic Analysis and Partial Differential Equations (Chicago, IL, 1996), Chicago Lectures in Math., Univ. Chicago Press, Chicago, IL (1999), 241–261.

[165] Y. Meyer, *New estimates for Navier–Stokes equations (two manuscripts)*, Colloque en l'Honneur de J.-L. Lions à l'Occasion de Son 70e Anniversaire, Paris 26–27 mai 1998, Auditorium du CNRS, 3 rue Michel-Ange, Paris XVIe (1998).

[166] Y. Meyer, *Wavelets, paraproducts and Navier–Stokes equations*, Current Developments in Mathematics 1996, International Press, Cambridge, MA (1999), 105–212.

[167] Y. Meyer, *The role of oscillations in some nonlinear problems*. European Congress of Mathematics, Vol. I (Barcelona, 2000), Progr. Math., Vol. 201, Birkhäuser, Basel (2001), 75–99.

[168] Y. Meyer, *Oscillating Patterns in Image Processing and Nonlinear Evolution Equations*, The Fifteenth Dean Jacqueline B. Lewis Memorial Lectures, University Lecture Series, Vol. 22, Amer. Math. Soc., Providence, RI (2001).

[169] Y. Meyer, *Oscillating patterns in some nonlinear evolution equations*, Mathematical Foundation of Turbulent Viscous Flows, Lectures Notes in Math., M. Cannone and T. Miyakawa, eds, Springer-Verlag, New York (2004), to appear.

[170] J.R. Miller, M. O'Leary and M.E. Schonbek, *Nonexistence of singular pseudo-self-similar solutions of the Navier–Stokes system*, Math. Ann. **319** (2001), 809–815.
[171] T. Miyakawa, *Hardy spaces of solenoidal vector fields, with applications to the Navier–Stokes equations*, Kyushu J. Math. **50** (1) (1996), 1–64.
[172] T. Miyakawa, *Application of Hardy space techniques to the time-decay problem for incompressible Navier–Stokes flows in* $\mathbb{R}^n$, Funkcial. Ekvac. **41** (3) (1998), 383–434.
[173] T. Miyakawa, *On space-time decay properties of nonstationary incompressible Navier–Stokes flows in* $\mathbb{R}^n$, Funkcial. Ekvac. **43** (3) (2000), 541–557.
[174] S. Monniaux, *Uniqueness of mild solutions of the Navier–Stokes equation and maximal* L^p*-regularity*, C. R. Acad. Sci. Paris Sér. I Math. **328** (8) (1999), 663–668.
[175] S. Monniaux, *On Uniqueness for the Navier–Stokes system in* $3D$*-bounded Lipschitz domains*, J. Funct. Anal. **195** (1) (2002), 1–11.
[176] S. Montgomery-Smith, *Finite time blow-up for a Navier–Stokes like equation*, Proc. Amer. Math. Soc. **129** (10) (2001), 3025–3029 (electronic).
[177] S. Montgomery-Smith, *A condition implying regularity of the three dimensional Navier–Stokes equation*, Preprint (2003).
[178] C.L.M.H. Navier, *Mémoire sur les lois du mouvement des fluides*, Mém. Acad. Sci. Inst. France **6** (1822), 389–440.
[179] J. Nečas, M. Růžička and V. Šverák, *Sur une remarque de J. Leray concernant la construction de solutions singulières des équations de Navier–Stokes*, C. R. Acad. Sci. Paris Sér. I Math. **323** (3) (1996), 245–249.
[180] J. Nečas, M. Růžička and V. Šverák, *On Leray self-similar solutions of the Navier–Stokes equations*, Acta Math. **176** (2) (1996), 283–294.
[181] H. Okamoto, *Exact solutions of the Navier–Stokes equations via Leray's scheme*, Proc. Miniconf. Partial Differential Equations and Applications, Seoul Nat. Univ., Seoul (1997); Lecture Notes Ser., Vol. 38, Seoul Nat. Univ. (1997), 1–30.
[182] H. Okamoto, *Exact solutions of the Navier–Stokes equations via Leray's scheme*, Japan J. Indust. Appl. Math. **14** (2) (1997), 169–197.
[183] F. Oru, *Rôle des oscillations dans quelques problèmes d'analyse non linéaire*, Doctoral Thesis, Ecole Normale Supérieure de Cachan, France (1998).
[184] H. Pecher, *Self-similar and asymptotically self-similar solutions of nonlinear wave equations*, Math. Ann. **316** (2) (2000), 259–281.
[185] J. Peetre, *New Thoughts on Besov Spaces*, Duke Univ. Press, Durham, NC (1976).
[186] F. Planchon, *Solutions globales et comportement asymptotique pour les equations de Navier–Stokes*, Doctoral Thesis, Ecole Polytechnique, France (1996).
[187] F. Planchon, *Global strong solutions in Sobolev or Lebesgue spaces for the incompressible Navier–Stokes equations in* $\mathbb{R}^3$, Ann. Inst. H. Poincaré Anal. Non Linéaire **13** (1996), 319–336.
[188] F. Planchon, *Convergence de solutions des équations de Navier–Stokes vers des solutions auto-similaires*, Exposé n. III, Séminaire X-EDP, Ecole Polytechnique (1996).
[189] F. Planchon, *Asymptotic behavior of global solutions to the Navier–Stokes equations in* $\mathbb{R}^3$, Rev. Mat. Iberoamericana **14** (1) (1998), 71–93.
[190] F. Planchon, *Solutions auto-similaires et espaces de données initiales pour l'équation de Schrödinger*, C. R. Acad. Sci. Paris **328** (1999), 1157–1162.
[191] F. Planchon, *Self-similar solutions and Besov spaces for semi-linear Schrödinger and wave equations*, Journées "Equations aux Dérivées Partielles", Saint-Jean-de-Monts, 1999, Exp. No. IX, 11 pp., Univ. Nantes, Nantes (1999).
[192] F. Planchon, *On the Cauchy problem in Besov space for a non-linear Schrödinger equation*, Commun. Contemp. Math. **2** (2) (2000), 243–254.
[193] F. Planchon, *Sur un inégalité de type Poincaré*, C. R. Acad. Sci. Paris Sér. I Math. **330** (1) (2000), 21–23.
[194] F. Planchon, *An extension of the Beale–Kato–Majda criterion for the Euler equations*, Comm. Math. Phys. **232** (2) (2003), 319–326.
[195] G. Ponce, R. Racke, T.C. Sideris and E.S. Titi, *Global Stability of Large Solutions to the* $3D$ *Navier–Stokes equations*, Comm. Math. Phys. **159** (1994), 329–341.
[196] F. Ribaud, *Analyse de Littlewood–Paley pour la résolution d'equations paraboliques semi-linéaires*, Doctoral Thesis, Université de Paris Sud, France (1996).

[197] F. Ribaud, *Problème de Cauchy pour les équations paraboliques semi-linéaires avec données dans* $H_p^s(\mathbb{R}^n)$, C. R. Acad. Sci. Paris **322** (1) (1996), 25–30.

[198] F. Ribaud, *Semilinear parabolic equations with distributions as initial data*, Discrete Contin. Dyn. Syst. **3** (3) (1997), 305–316.

[199] F. Ribaud, *Cauchy problem for semilinear parabolic equations with initial data in* $H_p^s(\mathbb{R}^n)$ *spaces*, Rev. Mat. Iberoamericana **14** (1) (1998), 1–46.

[200] F. Ribaud and A. Youssfi, *Regular and self-similar solutions of nonlinear Schrödinger equations*, J. Math. Pures Appl. **77** (10) (1998), 1065–1079.

[201] F. Ribaud and A. Youssfi, *Solutions globales et solutions auto-similaires de l'équation des ondes non linéaire*, C. R. Acad. Sci. Paris **329** (1) (1999), 33–36.

[202] F. Ribaud and A. Youssfi, *Global solutions and self-similar solutions of semilinear wave equation*, Math. Z. **239** (2) (2002), 231–262.

[203] G. Rosen, *Navier–Stokes initial value problem for boundary-free incompressible fluid flow*, J. Phys. Fluid **13** (1970), 2891–2903.

[204] O. Sawada, *On time-local solvability of the Navier–Stokes equations in Besov spaces*, Adv. Differential Equations **8** (4) (2003), 385–412.

[205] V. Scheffer, *A solution to the Navier–Stokes inequality with an internal singularity*, Comm. Math. Phys. **101** (1) (1985), 47–85.

[206] M.E. Schonbek, *L^2 decay for weak solutions to the Navier–Stokes equations*, Arch. Ration. Mech. Anal. **88** (1985), 209–222.

[207] P. Serfati, *Une preuve directe d'existence globale des vortex patches $2D$*, C. R. Acad. Sci. Paris Sér. I Math. **318** (6) (1994), 515–518.

[208] J. Serrin, *On the interior regularity of weak solutions of the Navier–Stokes equations*, Arch. Ration. Mech. Anal. **9** (1962), 187–195.

[209] J. Serrin, *The initial value problem for the Navier–Stokes equations*, 1963 Nonlinear Problems, Proc. Sympos., Madison, WI, Univ. Wisconsin Press, Madison, WI (1963), 69–98.

[210] Y. Shibata, *Global solutions of nonlinear evolution equations and their stability* (in Japanese), Sūgaku **51** (1999), 1–17.

[211] M. Shinbrot, *Lectures on Fluid Mechanics*, Gordon and Breach, New York–London–Paris (1973).

[212] S. Smale, *Mathematical problems for the next century*, Math. Intelligencer **20** (2) (1998), 7–15.

[213] H. Sohr, *The Navier–Stokes Equations, An Elementary Functional Analytic Approach*, Birkhäuser, Basel–Boston–Berlin (2001).

[214] E.M. Stein, *Harmonic Analysis*, Princeton Math. Ser., Vol. 43, Princeton Univ. Press (1993).

[215] E.M. Stein and G. Weiss, *Introduction to Fourier Analysis on Euclidean Spaces*, Princeton Univ. Press (1971).

[216] M. Taylor, *Analysis on Morrey spaces and applications to the Navier–Stokes and other evolution equations*, Comm. Partial Differential Equations **17** (1992), 1407–1456.

[217] R. Temam, *Some developments on Navier–Stokes equations in the second half of the 20th century*, Development of Mathematics 1950–2000, J.-P. Pier, ed., Birkhäuser, Basel (2000), 1049–1106.

[218] R. Temam, *Navier–Stokes Equations, Theory and Numerical Analysis*, Reprint of the 1984 edition, AMS Chelsea Publishing, Providence, RI (2001).

[219] E. Terraneo, *Applications de certains espaces fonctionnels de l'analyse harmonique réelle aux equations de Navier–Stokes et de la chaleur non linéaire*, Doctoral Thesis, Université d'Evry Val d'Essonne, France (1999).

[220] E. Terraneo, *Sur la non-unicité des solutions faibles de l'équation de la chaleur non linéaire avec non-linéarité u^3*, C. R. Acad. Sci. Paris Sér. I Math. **328** (9) (1999), 759–762.

[221] G. Tian and Z. Xin, *One-point singular solutions to the Navier–Stokes equations*, Topol. Methods Nonlinear Anal. **11** (1) (1998), 135–145.

[222] S. Tourville, *An analysis of numerical methods for solving the two-dimensional Navier–Stokes equations*, Ph.D. Thesis, Washington University of St. Louis, MO (1998).

[223] S. Tourville, *Existence and uniqueness of solutions for a modified Navier–Stokes equation in* $\mathbb{R}^2$, Comm. Partial Differential Equations **23** (1–2) (1998), 97–121.

[224] S. Tourville, *Existence and uniqueness of solutions for the Navier–Stokes equations with hyperdissipation*, J. Math. Anal. Appl. **281** (1) (2003), 62–75.

[225] H. Triebel, *Theory of Function Spaces*, Monograph in Mathematics, Vol. 78, Birkhäuser, Basel (1983).

[226] H. Triebel, *Theory of Function Spaces, II*, Monograph in Mathematics, Vol. 84, Birkhäuser, Basel (1992).

[227] T.P. Tsai, *On Leray's self-similar solutions of the Navier–Stokes equations satisfying local energy estimates*, Arch. Ration. Mech. Anal. **143** (1) (1998), 29–51; Erratum, ibid. **147** (1999), 363.

[228] H.B. Veiga and P. Secchi, *L^p-Stability for the strong solutions of the Navier–Stokes equations in the whole space*, Arch. Ration. Mech. Anal. **98** (1987), 65–70.

[229] M. Vishik, *Hydrodynamics in Besov Spaces*, Arch. Ration. Mech. Anal. **145** (1998), 197–214.

[230] M. Vishik, *Incompressible flows of an ideal fluid with vorticity in borderline spaces of Besov type*, Ann. Sci. Ecole Norm. Sup. **32** (6) (1999), 769–812.

[231] M. Vishik, *Incompressible flows of an ideal fluid with unbounded vorticity*, Comm. Math. Phys. **213** (3) (2000), 697–731.

[232] W. von Wahl, *The Equations of Navier–Stokes and Abstract Parabolic Equations*, Aspect der Mathematik, Vieweg und Sohn, Wiesbaden (1985).

[233] K.A. Voss, *Self-similar solutions of the Navier–Stokes equations*, Ph.D. Thesis, Yale University (1996).

[234] F.B. Weissler, *The Navier–Stokes initial value problem in L^p*, Arch. Ration. Mech. Anal. **74** (1981), 219–230.

[235] M. Wiegner, *The Navier–Stokes equations – A neverending challenge?* Jahresber. Deutsch. Math.-Verein. **101** (1999), 1–25.

[236] M. Yamazaki, *The Navier–Stokes equations in the weak-L^n spaces with time-dependent external force*, Math. Ann. **317** (2000), 635–675.

[237] M. Yamazaki, *Shuju no kansūkūkan ni okeru Navier–Stokes houteishiki* (*Navier–Stokes equations in various function spaces*) (in Japanese), Sūgaku **51** (3) (1999), 291–308; English transl.: Amer. Math. Soc. Transl. Ser. 2, Vol. 204, Selected Papers on Classical Analysis, K. Nomizu, ed., Amer. Math. Soc., Providence, RI (2001), 111–132.

[238] K. Yosida, *Functional Analysis*, Reprint of the sixth edition (1980), Classics in Mathematics, Springer-Verlag, Berlin (1995).

[239] V.I. Yudovich, *The Linearization Method in Hydrodynamical Stability Theory*, Transl. Math. Monogr., Vol. 74, Amer. Math. Soc., Providence, RI (1984).

[240] A. Zhioua, *Solutions mild pour le problème de Navier–Stokes dans le demi-espace pour une donnée singulière*, Doctoral Thesis, Université d'Evry Val d'Essonne, France (2003).

CHAPTER 4

Boundary Layers

E. Grenier

U.M.P.A., U.M.R. 5669, E.N.S. Lyon, 46 allée d'Italie, 69364 Lyon, France

Contents

HANDBOOK OF MATHEMATICAL FLUID DYNAMICS, VOLUME III
Edited by S.J. Friedlander and D. Serre

1. Stability of inviscid shear flows

1.1. *Introduction*

Let us first recall the classical Euler equations for incompressible ideal fluids in a domain Ω

$$\partial_t u + (u \cdot \nabla)u + \nabla p = 0, \tag{1}$$

$$\nabla \cdot u = 0, \tag{2}$$

$$u \cdot n = 0 \quad \text{on } \partial\Omega, \tag{3}$$

where u is the velocity, p is the pressure and n is the outer normal to $\partial\Omega$. If $u(t, x)$ is some divergence free vector field, then linearized Euler equations around $u(t, x)$, on (v, q) are

$$\partial_t v + (u \cdot \nabla)v + (v \cdot \nabla)u + \nabla q = 0, \tag{4}$$

$$\nabla \cdot v = 0, \tag{5}$$

$$v \cdot n = 0 \quad \text{on } \partial\Omega. \tag{6}$$

Let u be a stationary solution of (1)–(3). The main problem is to know whether this solution is stable or unstable, and in which sense. Namely there are several notions of stability: spectral, linear, energy, Lyapunov, nonlinear that we will detail in this chapter.

There is little known of the stability of general two-dimensional or three-dimensional flows and most of the theory is devoted to the study of shear layers, which are particular stationary solutions of Euler equations which depends only on one variable (normal to the boundaries). The study of such flows goes back to the pioneering works of Lord Rayleigh at the end of the 19th century. The aim of this section is, first, to make a review of the classical results of Rayleigh, Fjortoft, Drazin, to describe the spectrum of linearized Euler equations near a shear flow and to give examples of such spectra. Then we will describe more recent results and methods related to nonlinear stability and to nonlinear instability.

In all this chapter, Ω, domain of evolution of the fluid, will be either $\mathbb{R}^2$ (whole space), $\mathbb{T}^2$, where $\mathbb{T} = \mathbb{R}/(2\pi\mathbb{Z})$ (periodic setting), $\mathbb{R} \times \mathbb{T}$ (periodic in one direction), $\mathbb{R} \times [0, 1]$ (strip), $\mathbb{T} \times [0, 1]$ (periodic strip), or the analogs in three-dimensional space ($\mathbb{R}^3$, $\mathbb{T}^3$, $\mathbb{R}^2 \times \mathbb{T}$, $\mathbb{R}^2 \times [0, 1]$, $\mathbb{T}^2 \times [0, 1]$). The generic variable will be denoted by (x, z) or (x, y, z).

1.2. *Rayleigh equation*

Let us consider a three-dimensional shear flow $V(z)e_1$, where e_1 is the unit vector in the x direction, and where $z_1 < z < z_2$, with $z_1 \in \mathbb{R}$ or $z_1 = -\infty$ and $z_2 \in \mathbb{R}$ or $z_2 = +\infty$. The linearized Euler equations (4)–(6) on a perturbation (u_1, u_2, u_3) of $V(z)e_1$ are

$$\partial_t u + V(z)\,\partial_x u + u_3 V'(z)e_1 + \nabla p = 0, \tag{7}$$

$$\nabla \cdot u = 0, \tag{8}$$

$$u_3 = 0 \quad \text{on } z = z_1, z_2. \tag{9}$$

To study the linear stability of the shear flow $V(z)$, a first approach is to take the Fourier–Laplace transform of (7)–(9), namely the Fourier transform in the tangential variables, the Laplace transform in time and to look for solutions of (7)–(9) of the form

$$u(t,x,y,z) = \tilde{u}(z)\exp\bigl[\mathrm{i}(\alpha x + \beta y - \alpha c t)\bigr],$$
$$p(t,x,y,z) = \tilde{p}(z)\exp\bigl[\mathrm{i}(\alpha x + \beta y - \alpha c t)\bigr],$$

where $\alpha, \beta \in \mathbb{R}$ and $c \in \mathbb{C}$. Note the $-\mathrm{i}\alpha c t$ factor of the Laplace transform, which is traditional in Fluid Mechanics. If $\mathcal{I}m\, c > 0$, then u has an exponential growth and V is spectrally unstable. Equations (7)–(9) reduce to

$$\mathrm{i}\alpha(V - c)\tilde{u}_1 + V'\tilde{u}_3 = -\mathrm{i}\alpha\tilde{p}, \tag{10}$$
$$\mathrm{i}\alpha(V - c)\tilde{u}_2 = -\mathrm{i}\beta\tilde{p}, \tag{11}$$
$$\mathrm{i}\alpha(V - c)\tilde{u}_3 = -\mathrm{i}\,\partial_z\tilde{p}, \tag{12}$$
$$\mathrm{i}\bigl(\alpha\tilde{u}_1 + \beta\tilde{u}_2\bigr) + \partial_z\tilde{u}_3 = 0, \tag{13}$$
$$u_3 = 0 \quad \text{at } z = z_1, z_2. \tag{14}$$

This system can be reduced to a two-dimensional system by using Squire's transform and by introducing $\hat{\alpha} = (\alpha^2 + \beta^2)^{1/2}$, $\hat{\alpha}\hat{u}_1 = \alpha\tilde{u}_1 + \beta\tilde{u}_2$, $\hat{p} = \hat{a}\tilde{p}/\tilde{\alpha}$, $\hat{u}_2 = 0$ and $\hat{u}_3 = u_3$ system (10)–(14) then gives

$$\mathrm{i}\hat{\alpha}(V - c)\hat{u}_1 + V'\hat{u}_3 = -\mathrm{i}\hat{\alpha}\hat{p}, \tag{15}$$
$$\mathrm{i}\hat{\alpha}(V - c)\hat{u}_3 = -\mathrm{i}\,\partial_z\tilde{p}, \tag{16}$$
$$\mathrm{i}\hat{\alpha}\hat{u}_1 + \partial_z\tilde{u}_3 = 0, \tag{17}$$

with $\hat{u}_3 = 0$ on z_1 and z_2. Now $\hat{u}$ is a two-dimensional perturbation which, using (15)–(17), is a solution of two-dimensional linearized Euler equations. This leads to the following theorem.

THEOREM 1.1 (Squire's theorem, 1933). *To each unstable three-dimensional disturbance there corresponds a more unstable two-dimensional one.*

Therefore as far as spectral instability of shear flows is concerned it is sufficient to work in two space dimensions, since by Squire's theorem a shear flow which is unstable with respect to three-dimensional perturbations is also unstable with respect to two-dimensional perturbations, and obviously if it is unstable with respect to two-dimensional perturbations it is also unstable with respect to three-dimensional perturbations (independent on the last variable).

In two dimensions it is then convenient to introduce the stream function $\psi(t,x,z)$ of the perturbation and to take its Fourier–Laplace transform. This lead to perturbations of the form

$$u(t,x,z) = \begin{pmatrix} \psi'(z) \\ -\mathrm{i}\alpha\psi(z) \end{pmatrix} \exp \mathrm{i}\alpha(x - ct). \tag{18}$$

Putting (18) in linearized Euler equations gives Rayleigh equation

$$(V - c)(\psi'' - \alpha^2\psi) - V''\psi = 0, \tag{19}$$

$$\psi = 0 \quad \text{at } z = z_1, z_2. \tag{20}$$

Note that the vorticity of the disturbance is given by $(\psi'' - \alpha^2\psi)\exp i\alpha(x - ct)$, and for the spectral analysis it is more convenient to define Rayleigh's operator Ray_α as

$$\mathrm{Ray}_\alpha\, \omega = V\omega - V''\psi, \tag{21}$$

where

$$\psi'' - \alpha^2\psi = \omega, \tag{22}$$

with boundary conditions (20). Note that (21) and (22) is equivalent to (7)–(9) up to the Fourier transform in x and to the introduction of the stream function ψ. Spectral stability of the shear flow is therefore directly linked to the spectrum of Ray_α: a shear flow is spectrally unstable if there exists an eigenvalue c of Ray_α such that $\mathcal{I}m\, c \geqslant 0$.

Note that Ray_α has real coefficients, hence if c is an eigenvalue of Ray_α, then $\bar{c}$ is also an eigenvalue, and $\mathrm{Sp}(\mathrm{Ray}_\alpha)$ is invariant under complex conjugation.

1.3. *Instability criteria*

Let us first study the eigenvalues of Ray_α which are not real. By conjugation we can restrict ourselves to the case $\mathcal{I}m\, c > 0$. In this section we recall two necessary conditions for the existence of eigenvalues c and eigenvectors ψ of Ray_α with $\mathcal{I}m\, c > 0$ (that is, conditions for spectral instability, which implies linear instability).

THEOREM 1.2 (Rayleigh, 1880 [56]). *A necessary condition for instability is that the basic velocity profile should have an inflexion point.*

PROOF. Multiply (19) by $\bar{\psi}$ to get

$$\int |\partial_y\psi|^2 + \alpha^2|\psi|^2 + \frac{V''}{V - c}|\psi|^2 = 0 \tag{23}$$

and take the imaginary part of the equation

$$\mathcal{I}m\, c \cdot \int \frac{V''}{|V - c|^2}|\psi|^2 = 0$$

hence if $\mathcal{I}m\, c \neq 0$, V'' must change sign. □

This result has been refined by Fjortoft.

THEOREM 1.3 (Fjortoft, 1950 [21]). *A necessary condition for instability is that* $V''(V - V(y_s)) < 0$ *somewhere in the field of flow, where* y_s *is a point at which* $V''(y_s) = 0$.

PROOF. Take the real part of (23) to get

$$\int \frac{V''(V - \mathcal{R}e\, c)}{|V - c|^2} |\psi|^2 = -\int |\partial_y \psi|^2 + \alpha^2 |\psi|^2.$$

If we add

$$(\mathcal{R}e\, c - V(y_s)) \int \frac{V''}{|V - c|^2} |\psi|^2 = 0$$

we obtain

$$\int \frac{V''(V - V(y_s))}{|V - c|^2} |\psi|^2 = -\int |\partial_y \psi|^2 + \alpha^2 |\psi|^2$$

and the result follows. □

The localization of the eigenvalues is then precised by Howard's semicircle theorem.

THEOREM 1.4 (Howard, 1961 [17]). *Let* u_{m} *and* u_{M} *be the infimum and supremum of* V *and let* c *be an eigenvalue of Rayleigh equation. Then*

$$\left| \mathcal{R}e\, c - \frac{1}{2}(u_{\mathrm{m}} + u_{\mathrm{M}}) \right|^2 + |\mathcal{I}m\, c|^2 \leqslant \left| \frac{1}{2}(u_{\mathrm{m}} - u_{\mathrm{M}}) \right|^2, \tag{24}$$

namely c *belongs to the disk of diameter* $[u_{\mathrm{m}}, u_{\mathrm{M}}]$.

PROOF. Let us recall the proof as in [17]. Let ψ be an eigenvector and c an eigenvalue. The adjoint equation of Rayleigh on Θ is

$$(\partial_{yy}^2 - \alpha^2)(V - c)\Theta - V''\Theta = 0$$

which can be written

$$\partial_y \big[(V - c)^2 \partial_y \Theta \big] - \alpha^2 (V - c)^2 \Theta = 0. \tag{25}$$

Note that $\Theta = \psi/(V - c)$. Let now

$$Q = |\partial_y \Theta|^2 + \alpha^2 |\Theta|^2 > 0.$$

Multiplying (25) by $\overline{\Theta}$ and integrating by parts we get

$$\int (V - c)^2 Q = 0$$

and taking the real and imaginary parts

$$\int \big((V - \mathcal{R}e\, c)^2 - |\mathcal{I}m\, c|^2\big) Q = 0$$

and

$$2\mathcal{I}m\, c \int (V - \mathcal{R}e\, c) Q = 0.$$

In particular as $\mathcal{I}m\, c \neq 0$, $\mathcal{R}e\, c$ must lie in the range of V. Moreover,

$$0 \geqslant \int (u - u_{\mathrm{m}})(u - u_{\mathrm{M}}) Q = \int \big(|c|^2 - (u_{\mathrm{m}} + u_{\mathrm{M}})\, \mathcal{R}e\, c + u_{\mathrm{m}} u_{\mathrm{M}}\big) Q$$

hence

$$|c|^2 - (u_{\mathrm{m}} + u_{\mathrm{M}})\, \mathcal{R}e\, c + u_{\mathrm{m}} u_{\mathrm{M}} \leqslant 0$$

which proves the theorem. □

1.4. *Description of the spectrum of shear flows*

Let us now turn to the complete description of $\mathrm{Sp}(\mathrm{Ray}_\alpha)$ and prove the theorem.

THEOREM 1.5.

$$\mathrm{Sp}(\mathrm{Ray}_\alpha) = \mathrm{Sp}_{\mathrm{ess}}(\mathrm{Ray}_\alpha) \cup \bigcup_{j=1}^{N_\alpha} c_j(\alpha),$$

where

$$\mathrm{Sp}_{\mathrm{ess}}(\mathrm{Ray}_\alpha) = \Big[\inf_{z_1 < z < z_2} V(z),\ \sup_{z_1 < z < z_2} V(z)\Big]$$

is the essential spectrum, and $c_j(\alpha)$ being N_α eigenvalues with finite multiplicity, with $N_\alpha \leqslant +\infty$.

PROOF. Let $z_1 = 0$ and $z_2 = +\infty$ to fix the ideas, the other cases being similar. The resolvent equation is

$$(\mathrm{Ray}_\alpha - c Id)\omega = f, \tag{26}$$

where $f(z)$ is a given function. This equation is simply a second-order ordinary differential equation since it can be written under the form

$$\psi'' = \left(\frac{V''}{V-c} + \alpha^2\right)\psi + \frac{f}{V-c}. \tag{27}$$

Note first that it degenerates when $V - c$ vanishes for some z. This happens when c is real and lies in the range of V, i.e. when there exists $z_1 < z_c < z_2$ such that $V(z_c) = c$. We then say that z_c is a *critical layer*. It turns out and we will prove it below that it is the only possible degeneracy, and the essential spectrum is given by

$$\mathrm{Sp}_{\mathrm{ess}}(\mathrm{Ray}_\alpha) = \Big[\inf_{z_1<z<z_2} V(z), \sup_{z_1<z<z_2} V(z)\Big].$$

We will now prove that outside $\mathrm{Sp}_{\mathrm{ess}}$ there are only isolated eigenvalues of finite multiplicities. Standard ordinary differential equation results give the existence and uniqueness of a solution $\Theta(z,\alpha,c)$ of (27) with $f = 0$ such that $\Theta \to 0$ as $z \to +\infty$ and more precisely such that $\Theta \sim \mathrm{e}^{-\alpha z}$ as $z \to +\infty$. Moreover, Θ is (locally) an holomorphic function in c for $c \notin \mathrm{Sp}_{\mathrm{ess}}(\alpha)$. Moreover, there exists a nonzero solution of (27) with $f = 0$ and boundary conditions (20) if and only if $\Theta(0,\alpha,c) = 0$. For a fixed α this is an holomorphic equation in $c \notin \mathrm{Sp}_{\mathrm{ess}}(\alpha)$, which therefore has only isolated solutions, in finite or at most countable number. Let $c_j(\alpha)$ be these solutions. Note that by construction $\Theta(y,\alpha,c_j(\alpha))$ is then an eigenvector of Ray_α. Moreover, as the equation is real, the $c_j(\alpha)$ are pairwise complex conjugated. Let us now show that the spectrum $\mathrm{Sp}(\alpha)$ of Ray_α is

$$\mathrm{Sp}(\alpha) = \mathrm{Sp}_{\mathrm{ess}}(\alpha) \cup \big\{c_j(\alpha)\big\}. \tag{28}$$

Namely for $f \neq 0$, standard ordinary differential equations arguments give the existence of a solution ψ_1 of (27) with $\psi_1 \to 0$ as $z \to +\infty$, and if $c \notin \mathrm{Sp}(\alpha)$, $\psi_1 - \psi_1(0)\Theta(0,\alpha,c)^{-1}\Theta(y,\alpha,c)$ is the unique solution of (27), which proves (28). □

PROPOSITION 1.6. *There exists a constant C_V depending only on V such that, for every $\alpha \neq 0$ and for every eigenvalue $c_j(\alpha)$, we have*

$$\alpha^2\big|\mathcal{I}m\, c_j(\alpha)\big| \leqslant C_V.$$

PROOF. Just use

$$\left|\frac{V''}{V-c}\right| \leqslant \sup_z \big|V''(z)\big| |\mathcal{I}m\, c|^{-1}$$

together with (23). □

Moreover, the $c_j(\alpha)$ are continuous in α.

In fact using the classical theory of complex ordinary differential equations (Fuchs), it is possible to define ψ_0 even when c lies in the range of V, and it seems possible to obtain that ψ_0 is analytic for $\mathcal{R}e\, c \geqslant 0$ and for $\mathcal{R}e\, c \leqslant 0$. This would be a way to prove that there exists only a finite number of eigenvalues, as claimed in physics books [17].

Note also that this gives continuity of eigenvalues c (with $\mathcal{R}e\, c \neq 0$) as a function of k.

1.5. *Essential spectrum of general flows*

Few results are available on the spectrum for general two- or three-dimensional flows which are not shear flows. The main result has been obtained by Vishik in [70] using pseudo-differential technics.

Let us first recall the definition of the essential spectrum, following Browder [7]. Let B be a Banach space and let $T \in \mathcal{L}(B)$. A point $z \in \mathrm{Sp}(T)$ is called a point of the discrete spectrum if it is an isolated point in $\mathrm{Sp}(T)$, if it has finite multiplicity and if the range of $z - T$ is closed. Other points are points of the essential spectrum $\mathrm{Sp}_{\mathrm{ess}}(T)$. Following Nussbaum [53] we define the essential spectral radius

$$r_{\mathrm{ess}}(T) = \sup\{|z| \,|\, z \in \mathrm{Sp}_{\mathrm{ess}}(T)\}.$$

THEOREM 1.7 (Essential spectrum radius; Vishik [70]). *Let $\Omega = \mathbb{T}^n$ and u_0 be a C^∞ steady solution to the Euler equation $(u_0 \cdot \nabla)u_0 + \nabla p_0 = 0$, $\operatorname{div} u_0 = 0$, $u_0 \in (C^\infty(\Omega))^n$, $p_0 \in C^\infty(\Omega)$. Let*

$$Lv = (u_0 \cdot \nabla)v + (v \cdot \nabla)u_0 + \nabla q$$

be the linearized Euler operator. Then, for any $t > 0$,

$$r_{\mathrm{ess}}(e^{tL}) = e^{\omega t},$$

where

$$\omega = \lim_{t \to \infty} \frac{1}{t} \log \sup_{\substack{x_0, \xi_0, b_0 \\ (b_0, \xi_0) = 0 \\ |\xi_0| = 1, |b_0| = 1}} \left| b(x_0, \xi_0, b_0; t) \right|$$

and where (x, ξ, b) satisfies the following system of ODEs (which we call the bi-characteristic amplitude equations)

$$\dot{x} = u_0(x), \qquad \dot{\xi} = -\left(\frac{\partial u_0}{\partial x}\right)^T \xi,$$

$$\dot{b} = -\left(\frac{\partial u_0}{\partial x}\right) b + 2|\xi|^{-2}\left(\xi, \frac{\partial u_0}{\partial x} b\right)\xi,$$

$$x(0) = x_0, \qquad \xi(0) = \xi_0, \qquad b(0) = b_0.$$

Note that if there exists $z \in \mathrm{Sp}(L)$ such that $|z| > \omega$, then z is a point of discrete spectrum and the flow is *linearly* and *nonlinearly* unstable in the sense of Theorem 1.10 (which can be easily extended to this case using Theorems 1.13 or 1.14).

The Lyapunov exponent ω can be effectively computed in a number of examples [27]. In [27] it is proved that exponential stretching (positivity of the Lyapunov exponent along at least one Lagrangian trajectory) implies that $\omega > 0$. On the contrary, the theorem.

THEOREM 1.8 (Friedlander, Strauss and Vishik [25]). *Let $\Omega = \mathbb{T}^2$, u_0 as in Theorem 1.7. Let $u_0(x) \neq 0$ for all $x \in T^2$. Then $\omega = 0$ and hence $r_{\mathrm{ess}}(\mathrm{e}^{tL}) = 1$ for $t > 0$.*

1.6. *Nonlinear instability*

1.6.1. *Results.* A first class of nonlinear instability results has been obtained by Arnold in [1] in the periodic setting.

THEOREM 1.9 (Instability in Lagrangian variables; Arnold [1]). *In the Lagrangian variables, the only stable flow is the pure rotation.*

However this results deals with Lagrangian variables and not with the physical Eulerian variables. The first instability result in Eulerian variables has been obtained by Friedlander, Strauss and Vishik who proved in [25, 70] nonlinear instability in H^s with $s > d/2 + 1$, where d is the spatial dimension. However their approach is limited to such spaces and cannot go to L^∞ and L^2 instabilities, which are more interesting norms.

THEOREM 1.10 (Linear instability implies nonlinear instability [35]). *If a shear flow $V(z)$ is spectrally unstable (in the sense that there exists a linear exponentially growing mode) then it is also nonlinearly unstable in the following sense: for every s arbitrary large and for every $\delta > 0$ arbitrary small, there exists a solution $u(t,x,z)$ of Euler equations with*

$$\left\| u(0,\cdot,\cdot) - V \right\|_{H^s} \leqslant \delta$$

and

$$\left\| u(T^\delta,\cdot,\cdot) - V \right\|_{L^\infty} \geqslant \sigma > 0,$$
$$\left\| u(T^\delta,\cdot,\cdot) - V \right\|_{L^2} \geqslant \sigma > 0,$$

where σ is independent on δ and where $T^\delta \leqslant C \log \delta^{-1}$.

Using Theorems 1.7 and 1.8 we also have the following theorem.

THEOREM 1.11. *Let u_{stat} be a smooth stationary solution of Euler equations on $\mathbb{T}^2$. Let us assume that u_{stat} does not vanish on $\mathbb{T}^2$ and that it is spectrally unstable. Then it is also nonlinearly unstable in the sense of Theorem 1.10.*

Theorems 1.10 and 1.11 are consequences of the general method developed in the next section.

1.6.2. *A general method: periodic case.* Let us first describe the construction of instabilities in the periodic case (periodic tangential variables). Let u_{stat} be a time independent solution of Euler equations. Let

$$Q(v_1, v_2) = P\big((v_1 \cdot \nabla)v_2\big)$$

and

$$Au = P\big((u_{\text{stat}} \cdot \nabla)u + (u \cdot \nabla)u_{\text{stat}}\big),$$

where P is the Leray projector on divergence free vector fields. Euler equations can be written as

$$\partial_t u + Q(u, u) = 0 \tag{29}$$

and linearized Euler equations near u_{stat}

$$\partial_t u + Au = 0. \tag{30}$$

Let $v = u - u_{\text{stat}}$ which satisfies

$$\partial_t v + Av + Q(v, v) = 0. \tag{31}$$

An important point is to evaluate the spectral radius of A in order to get sharp bounds on solutions of (30) and (31). A crude and straightforward estimate on (30) gives that the solution u satisfies

$$\big\|u(t)\big\|_{L^2} \leqslant \big\|u(0)\big\|_{L^2} \exp(\lambda_0 t), \tag{32}$$

where $\lambda_0 = 2\|\nabla u_{\text{stat}}\|_{L^\infty}$. Thanks to the divergence free condition, a similar bound holds for solutions v of (31)

$$\big\|v(t)\big\|_{L^2} \leqslant \big\|v(0)\big\|_{L^2} \exp(\lambda_0 t). \tag{33}$$

If there exists a solution of (30) of the form $u_1 \exp(\lambda_0 t)$ with u_1 smooth, then to obtain a nonlinear instability result is straightforward: just write that $u_{\text{stat}} + \delta u_1 \exp(\lambda_0 t)$ is an approximate solution of (29) and that the error is of order $\delta^2 \exp(2\lambda_0 t)$ and end by a Gronwall lemma. Nonlinear instability is also easy if one can find a solution of (30) of the form $u_1 \exp(\lambda_1 t)$ with $\lambda_1 > \lambda_0/2$ (the same proof; see condition (ii) of [70]).

In the general case, however, we have to construct a more precise approximate solution. The framework is the following.

Assumptions. Let V_j for $j \in \mathbb{N}$ be a family of subspaces of complex smooth functions ($V_j \subset H^s(\Omega)$ for every $s \geqslant 0$), stable by complex conjugation, and $\|\cdot\|_l$ ($l \geqslant 1$) be a family of norms (controlling the L^∞, C^1 and L^2 norms) such that:

(A1) *Nonlinear energy estimate.* Let u' be some divergence free, time dependent, smooth function. There exists a smooth function $\lambda_0 > 0$, independent on u' such that solutions $v(t)$ of

$$\partial_t v + Q(u', v) + Q(v, u') + Q(v, v) = R \tag{34}$$

satisfy

$$\partial_t \|v(t)\|_{L^2}^2 \leqslant \lambda_0(\|\nabla u'\|_{L^\infty}) \|v(t)\|_{L^2}^2 + 2\|v(t)\|_{L^2} \|R(t)\|_{L^2}. \tag{35}$$

(A2) *Spectral instability.* There exists $\lambda_1 \in \mathbb{C}$ with $\mathcal{R}e\,\lambda_1 > 0$ and a nonzero smooth function $v_1 \in V_1$ such that $v_1 \exp(\lambda_1 t)$ is a solution of (30) and such that

$$u_1 = v_1 \exp(\lambda_1 t) + \bar{v}_1 \exp(\bar{\lambda}_1 t)$$

satisfies

$$\|u_1\|_{L^2(|y| \leqslant A)} \geqslant C_A \exp(\mathcal{R}e\,\lambda_1 t)$$

for some $A > 0$ and some $C_A > 0$. Let N such that

$$2N\,\mathcal{R}e\,\lambda_1 > \lambda_0(\|\nabla u_{\text{stat}}\|_{L^\infty}). \tag{36}$$

(A3) *Algebra property.* If $u \in V_j$ and $u' \in V_{j'}$, then $Q(u, u')$ can be written under the form

$$\sum_{0 \leqslant j'' \leqslant j+j'} u_{j''},$$

where $u_{j''} \in V_{j''}$ and

$$\|u_{j''}\|_{l-\sigma_0} \leqslant C_{j,j',j'',l} \|u\|_l \|u'\|_l,$$

where σ_0 is independent on j, j' and j''.

(A4) *Estimate on the spectral radius of A on V_j.* If $j \leqslant N$ and if $w \in V_j$ satisfies

$$\|w(t)\|_l \leqslant C_w \exp(\lambda t)$$

for some $l \geqslant \sigma_0 + 1$, with

$$\lambda \geqslant 2\,\mathcal{R}e\,\lambda_1,$$

then the solution $u(t)$ of

$$\partial_t u + Au = w$$

with initial data $u(0) \in V_j$ lies in V_j and satisfies

$$\|u(t)\|_{l-\sigma_0} \leqslant C(C_w, \|u(0)\|_l, j, l, \lambda) \exp(\lambda t). \tag{37}$$

Abstract result

DEFINITION 1.12 (Nonlinear instability). We say that u_{stat} is a nonlinearly unstable solution of (29) if there exists $\delta_0 > 0$ and constants $C_s > 0$ such that, for every s arbitrarily large and every $\delta > 0$ arbitrary small, there exists a solution u of (29) with

$$\|u(0, x, y) - u_{\text{stat}}(y)\|_{H^s} \leqslant \delta$$

and

$$\|u(T^\delta, x, y) - u_{\text{stat}}(y)\|_{L^\infty} \geqslant \delta_0,$$
$$\|u(T^\delta, x, y) - u_{\text{stat}}(y)\|_{L^2} \geqslant \delta_0,$$

where

$$T^\delta \leqslant C_s \log(1 + \delta^{-1}) + C_s.$$

THEOREM 1.13 (Spectral instability implies nonlinear instability). *On* $\Omega = \mathbb{T}^2$ *or* $\mathbb{T} \times [-1, 1]$, *or* $\mathbb{T} \times \mathbb{R}_+$ *or* $\mathbb{T} \times \mathbb{R}$, *under assumptions* (A1), (A2), (A3) *and* (A4), u_{stat} *is a nonlinearly unstable solution of* (29).

We refer to the proof for a more precise description of the instability.

PROOF OF THEOREM 1.13. We will construct an approximate solution u^{app} of the form

$$u^{\text{app}} = u_{\text{stat}} + \sum_{j=1}^{N} \delta^j u_j, \tag{38}$$

where u_j solves

$$\partial_t u_j + A(u_j) + \sum_{k=1}^{j-1} \big(Q(u_k, u_{j-k}) + Q(u_{j-k}, u_k)\big) = 0 \tag{39}$$

with $u_j(0) = 0$ for $j \geqslant 2$, and can be decomposed in

$$u_j = \sum_{0 \leqslant k \leqslant j} u_{j,k}, \tag{40}$$

where $u_{j,k} \in V_k$ and

$$\|u_{j,k}\|_{l-2j\sigma_0} \leqslant C_{j,k,l}\big(\|u_1\|_l\big)\exp(j\,\mathcal{R}e\,\lambda_1 t). \tag{41}$$

Let us show by recurrence that we can construct such an approximate solution. First, (A2) gives

$$u_1 = v_1 \exp(\lambda_1 t) + \bar{v}_1 \exp(\bar{\lambda}_1 t). \tag{42}$$

Let us assume now that u_j are built for $j < j_0$ and let us construct u_{j_0}. Then using (A3),

$$R = \sum_{k=1}^{j_0-1} \big(Q(u_k, u_{j_0-k}) + Q(u_{j_0-k}, u_k)\big) \tag{43}$$

is a sum of functions $R_k \in V_k$ with $k \leqslant j_0$ and with

$$\|R_k\|_{l-2j_0\sigma_0+\sigma_0} \leqslant C_{j_0,k,l}\big(\|u_1\|_l\big)\exp(j_0\,\mathcal{R}e\,\lambda_1 t).$$

Let now $u_{j_0,k}$ be the solution of

$$\partial_t u_{j_0,k} + A u_{j_0,k} + R_k = 0$$

with initial data 0. Notice that as $j_0 \geqslant 2$ and $j_0 \leqslant N$, we can use assumption (A4) to get that $u_{j_0,k} \in V_k$ for all t, and satisfies (41), which ends the construction by recurrence of u^{app}. Notice that

$$\partial_t u^{\mathrm{app}} + Q\big(u^{\mathrm{app}}, u^{\mathrm{app}}\big) = R_{\mathrm{app}}, \tag{44}$$

where

$$\|R_{\mathrm{app}}\|_{l-2N\sigma_0-\sigma_0} \leqslant C\delta^{N+1}\exp\big((N+1)\,\mathcal{R}e\,\lambda_1 t\big). \tag{45}$$

Let

$$T_0 = \frac{\log\delta^{-1}}{\mathcal{R}e\,\lambda_1}.$$

Let us now compare u^{app} with u_{stat}. We have, using (A2),

$$\big\|u^{\mathrm{app}} - u_{\mathrm{stat}}\big\|_{L^2(|y|\leqslant A)} \geqslant C_1\delta\exp(\mathcal{R}e\,\lambda_1 t) - \sum_{j=2}^{N} C_j\delta^j \exp(j\,\mathcal{R}e\,\lambda_1 t) \tag{46}$$

for some constants $C_i > 0$, and thus for $t \leqslant T_0 - \sigma_1$ with σ_1 large enough, but independent on δ,

$$\big\|u^{\mathrm{app}} - u_{\mathrm{stat}}\big\|_{L^2(|y|\leqslant A)} \geqslant \frac{C_1\delta}{2}\exp(\mathcal{R}e\,\lambda_1 t). \tag{47}$$

Let now u be the solution of

$$\partial_t u + Q(u,u) = 0 \tag{48}$$

with initial data $u_{\text{stat}} + \delta u_1(0)$. Let $v = u - u^{\text{app}}$. It satisfies

$$\partial_t v + Q\big(u^{\text{app}}, v\big) + Q\big(v, u^{\text{app}}\big) + Q(v,v) = -R_{\text{app}} \tag{49}$$

hence using (A1),

$$\partial_t \|v\|_{L^2}^2 \leqslant \big(\lambda_0\big(\big\|\nabla u^{\text{app}}\big\|_{L^\infty}\big) + \beta\big)\big\|v(t)\big\|_{L^2}^2 + C_\beta\big\|R_{\text{app}}(t)\big\|_{L^2}^2, \tag{50}$$

where $\beta > 0$ is small, such that $2N\,\mathcal{R}e\,\lambda_1 > \lambda_0(\|\nabla u_{\text{stat}}\|_{L^\infty} + \beta) + \beta$. We have

$$\big\|\nabla u^{\text{app}}\big\|_{L^\infty} \leqslant \|\nabla u_{\text{stat}}\|_{L^\infty} + \sum_{j=1}^{N} C_j \delta^j \exp(\mathcal{R}e\,\lambda_1 jt)$$

and thus for $t \leqslant T_0 - \sigma_2$ with $\sigma_2 > \sigma_1$ large enough but independent on δ,

$$\big\|\nabla u^{\text{app}}\big\|_{L^\infty} \leqslant \|\nabla u_{\text{stat}}\|_{L^\infty} + \beta.$$

For $t \leqslant T_0 - \sigma_2$ we therefore have

$$\begin{aligned}
&\partial_t \big\|v(t)\big\|_{L^2}^2 \\
&\quad \leqslant \big(\lambda_0\big(\|\nabla u_{\text{stat}}\|_{L^\infty} + \beta\big) + \beta\big)\big\|v(t)\big\|_{L^2}^2 + C\delta^{2(N+1)} \exp\big(2(N+1)\,\mathcal{R}e\,\lambda_1 t\big).
\end{aligned}$$

As $v(0) = 0$ we have

$$\big\|v(t)\big\| \leqslant C'\delta^{(N+1)} \exp\big((N+1)\,\mathcal{R}e\,\lambda_1 t\big).$$

Now for $t = T_0 - \sigma_3$ with $\sigma_3 \geqslant \sigma_2$ large enough,

$$\|u - u_{\text{stat}}\|_{L^2(|y| \leqslant A)} \geqslant \big\|u^{\text{app}} - u_{\text{stat}}\big\|_{L^2(|y| \leqslant A)} - \big\|v(t)\big\|_{L^2} \geqslant \delta_0$$

with δ_0 independent on δ, which ends the proof. □

1.6.3. *A general method: localized instabilities.*

Assumptions. Instead of looking at periodic instabilities, we study groups of such instabilities, with nearby wave numbers, in order to get space integrability. The framework is essentially the same as in Section 1.6.2.

Let V_k for $k \in \mathbb{R}$ be the vector space of functions of the form $\exp(ikx)u(y)$, where u is in H^s for every s. Let $\|\cdot\|_l$ be a family of norms (controlling the L^∞, C^1 and L^2 norms, and more generally the H^s norm for $s \leqslant l$) and let $\bar{\delta} > 0$. We will consider the following assumptions:

(A1′) *Nonlinear energy estimate*. Let u' be some divergence free smooth function. There exists a smooth function $\lambda_0 > 0$ such that solutions $v(t)$ of

$$\partial_t v + Q(u', v) + Q(v, u') + Q(v, v) = R \tag{51}$$

satisfy

$$\partial_t \|v(t)\|_{L^2}^2 \leqslant \lambda_0(\|\nabla u'\|_{L^\infty}) \|v(t)\|_{L^2}^2 + 2\|v(t)\|_{L^2} \|R(t)\|_{L^2}. \tag{52}$$

(A2′) *Spectral instability*. There exists k_- and k_+ with $0 \leqslant k_- < k_+$, a complex-valued function $\lambda_1(k)$ smooth in $]-k_+, -k_-[\cup]k_-, k_+[$, whose real part is strictly positive for $k_- < |k| < k_+$, which is identically vanishing outside, with $\lambda_1(-k) = \bar{\lambda}_1(k)$, and functions $v_1(k)$, of strictly positive L^2 norms for $k_- < |k| < k_+$, with $v_1(-k) = \bar{v}_1(k)$ such that

$$v_1(k)\exp(\lambda_1(k)t)\exp(ikx)$$

is a solution of (30) for $k_- \leqslant |k| \leqslant k_+$. Moreover, $\mathcal{R}e\,\lambda_1(k)$ has a maximum for positive k, at $k = k_0 > 0$, which is nondegenerate ($\partial_{kk}^2 \mathcal{R}e\,\lambda_1(k_0) < 0$). We also impose that $(k, y) \to v_1(k, y)$ is smooth, and assume that

$$\bar{\delta} < \max_k \mathcal{R}e\,\lambda_1(k). \tag{53}$$

Let N such that

$$2N\,\mathcal{R}e\,\lambda_1(k_0) > \lambda_0(\|u_{\text{stat}}\|_{L^\infty}).$$

(A3′) *Algebra property*. If $u \in V_k$ and $u' \in V_{k'}$ then $Q(u, u')$ can be written under the form $u_{++} + u_{+-} + u_{-+} + u_{--}$, where $u_{\pm\pm} \in V_{\pm k \pm k'}$ and

$$\|u_{\pm\pm}\|_{l-\sigma_0} \leqslant C_{k,k',l} \|u\|_l \|u'\|_l,$$

where σ_0 is independent on k, k' and l, $C_{k,k',l}$ being locally bounded in k and k'.

(A4′) *Estimate on the spectral radius of A on V_k*. If $|k| \leqslant k_+(N+1)$ and if $w \in V_k$ satisfies

$$\|w(t)\|_l \leqslant C_w \exp(\lambda' t)$$

for some $l \geqslant \sigma_0 + 1$ and some $\lambda' > 0$, then the solution $u(t)$ of

$$\partial_t u + Au = w$$

with initial data $u(0) \in V_k$ lies in V_k and satisfies

$$\|u(t)\|_{l-\sigma_0} \leqslant C(C_w, \|u(0)\|_l, k, l)\exp(\lambda' t) \tag{54}$$

provided $\lambda' > \mathcal{R}e\,\lambda_1(k) + \bar{\delta}$, the constants $C(C_w, \|u(0)\|_l, k, l)$ being locally bounded in k.

Abstract result

THEOREM 1.14 (Spectral instability implies nonlinear instability). *On* $\Omega = \mathbb{R} \times \mathbb{T}$ *or* $\mathbb{R} \times [-1, 1]$ *or* $\mathbb{R} \times \mathbb{R}_+$ *or* $\mathbb{R} \times \mathbb{R}$, *under assumptions* (A1′)–(A4′), u_{stat} *is a nonlinearly unstable solution of* (29) *in the following sense*: *there exists* $\delta_0 > 0$ *and constants* $C_s > 0$ *such that for every* s *arbitrarily large and every* $\delta > 0$ *arbitrary small there exists a solution* u *of* (29) *with*

$$\|u(0, x, y) - u_{\text{stat}}(y)\|_{H^s} \leqslant \delta$$

and

$$\|u(T^\delta, x, y) - u_{\text{stat}}(y)\|_{L^\infty} \geqslant \delta_0,$$

$$\lim_{\delta\to 0} \|u(T^\delta, x, y) - u_{\text{stat}}(y)\|_{L^2} = +\infty,$$

where

$$T^\delta \leqslant C_s \log(1 + \delta^{-1}) + C_s.$$

We emphasize that $u - u_{\text{stat}}$ lies in H^s for every s and in particular in L^2. This theorem is therefore a localized version of Theorem 1.13. The proof of this theorem in fact gives much more details on the instability than the theorem itself.

PROOF OF THEOREM 1.4. Let $k_0 > 0$ be the wave number, where $\mathcal{R}e\,\lambda_1$ reaches its maximum. Let $\eta > 0$ be small enough and such that $\eta \leqslant \inf(|k_0 - k_-|, |k_0 - k_+|)$. Let $I = [-k_0 - \eta, -k_0 + \eta] \cup [k_0 - \eta, k_0 + \eta]$, let ϕ be a smooth function with support in I, with $\phi(k_0) = 1$ and $\phi(-k) = \bar{\phi}(k)$. We take

$$u_1(t, x, y) = \int_I \phi(k) v_1(k, y) \exp(\lambda_1(k)t) \exp(ikx)\, \mathrm{d}k \tag{55}$$

which is a real valued solution of (30) by assumption (A2′). We have

$$\|u_1\|_{L^2}^2 = C \int_I |\phi(k)|^2 \|v_1(k)\|_{L^2}^2 \exp(2\,\mathcal{R}e\,\lambda_1(k)t)\, \mathrm{d}k,$$

hence, as for η small enough, λ_1 has a unique nondegenerate maximum in $[k_0 - \eta, k_0 + \eta]$, at $k = k_0$,

$$\|u_1(t)\|_{L^2} \sim \frac{C_1}{t^{1/4}} \exp(\mathcal{R}e\,\lambda_1(k_0)t) \tag{56}$$

as $t \to +\infty$. Notice that

$$\int \exp(-u^2 t - \mathrm{i}\alpha u t) \exp(\mathrm{i}xu)\,\mathrm{d}u = \frac{C}{\sqrt{t}} \exp\left(-\frac{1}{4t}(x - \alpha t)^2\right). \tag{57}$$

Therefore, using a Taylor expansion of λ_1 at $k = k_0$,

$$\left\| u_1(t) \right\|_{L^\infty} \sim \frac{C_1'}{\sqrt{t}} \exp\bigl(\mathcal{R}e\,\lambda_1(k_0)t\bigr). \tag{58}$$

Notice the different powers of t which arise since the support of v_1 grows like $\sqrt{t}$.

We will construct an approximate solution u^{app} of the form

$$u^{\mathrm{app}} = u_{\mathrm{stat}} + \sum_{j=1}^{N} \delta^j u_j, \tag{59}$$

where the u_j can be written under the form

$$u_j = \int_I \cdots \int_I u_{k_1,\dots,k_j} \exp\bigl(\mathrm{i}x(k_1 + \cdots + k_j)\bigr)\,\mathrm{d}k_1 \cdots \mathrm{d}k_j, \tag{60}$$

where $u_{k_1,\dots,k_j}$ solves

$$\partial_t u_{k_1,\dots,k_j} + A(u_{k_1,\dots,k_j}) + R_{k_1,\dots,k_j} = 0 \tag{61}$$

with

$$R_{k_1,\dots,k_j} = Q(u_{k_1}, u_{k_2,\dots,k_j}) + \cdots + Q(u_{k_1,\dots,k_{j-1}}, u_{k_j}). \tag{62}$$

We will show

$$\|u_{k_1,\dots,k_j}\|_{l-2j\sigma_0} \leqslant C_{j,k_1,\dots,k_j}\bigl(\|u_1\|_l\bigr) \exp\bigl(\mathcal{R}e\,\bigl(\lambda_1(k_1) + \cdots + \lambda_1(k_j)\bigr)t\bigr). \tag{63}$$

Notice that under these assumptions we have

$$\begin{aligned}\|u_j\|_{L^\infty} &\le C \int_I \cdots \int_I \exp\bigl(\mathcal{R}e\,\lambda_1(k_1)t\bigr) \cdots \exp\bigl(\mathcal{R}e\,\lambda_1(k_j)t\bigr)\,\mathrm{d}k_1 \cdots \mathrm{d}k_j \\ &\le C\left(\int_I \exp\bigl(\mathcal{R}e\,\lambda_1(k)t\bigr)\,\mathrm{d}k\right)^j,\end{aligned}$$

therefore

$$\|u_j\|_{L^\infty} \leqslant \frac{C_j}{t^{j/2}} \exp\bigl(\mathcal{R}e\,\lambda_1(k_0)jt\bigr).$$

Moreover,

$$u_j = \int_{jk\in jI} \int_{k_1+\cdots+k_j=jk} u_{k_1,\ldots,k_l} \exp(\mathrm{i}jkx)$$

with

$$\begin{aligned}
&\left\| \int_{k_1+\cdots+k_j=jk} u_{k_1,\ldots,k_l} \right\|_{L^2} \\
&\quad \leqslant C \int_{k_1+\cdots+k_j=jk} \exp\left(\mathcal{R}e \sum_{i=1}^{j} \lambda_1(k_i)t \right) \\
&\quad \leqslant C \int_{k_1+\cdots+k_j=jk} \exp\left(\mathcal{R}e\, (j\lambda_1(k_0)) - \beta j(k_0-k)^2 - \beta \sum_j (k-k_i)^2 \right) t \\
&\quad \leqslant \frac{C}{t^{(j-1)/2}} \exp\bigl(j\,\mathcal{R}e\, \lambda_1(k_0) - \beta j(k_0-k)^2 \bigr)
\end{aligned}$$

for some $\beta > 0$, which leads to

$$\|u_j\|_{L^2} \leqslant \frac{C_j}{t^{j/2}} t^{1/4} \exp\bigl(\mathcal{R}e\, \lambda_1(k_0) jt\bigr).$$

The fact that we can build such an approximate solution can be proved exactly as for Theorem 1.13, provided η is small enough. Notice that

$$\partial_t u^{\mathrm{app}} + Q\bigl(u^{\mathrm{app}}, u^{\mathrm{app}}\bigr) = R_{\mathrm{app}}, \tag{64}$$

where

$$\|R_{\mathrm{app}}\|_{L^2} \leqslant \frac{C\delta^{N+1}}{t^{(N+1)/2}} t^{1/4} \exp\bigl((N+1)\,\mathcal{R}e\, \lambda_1(k_0)t\bigr). \tag{65}$$

Let T_0 such that

$$\frac{\delta}{\sqrt{t}} \exp\bigl(\mathcal{R}e\, \lambda_1(k_0) T_0\bigr) = 1.$$

Let us now compare u^{app} with u_{stat}. Let, for $A > 0$,

$$\Omega_A = \bigl\{ \bigl| x + \mathcal{I}m\, \partial_k \lambda_1(k_0) \bigr| \leqslant A\sqrt{t}, |y| \leqslant A \bigr\}.$$

Using (57), there exists $A > 0$ such that, for every $t \geqslant 0$,

$$\bigl\| u_1(t) \bigr\|_{L^2(\Omega_A)} \geqslant \frac{C_1}{t^{1/4}} \exp\bigl(\mathcal{R}e\, \lambda_1(k_0)t\bigr) \tag{66}$$

for some $C_1 > 0$. We have

$$\|u^{\text{app}} - u_{\text{stat}}\|_{L^2(\Omega_A)} \geqslant \frac{C_1 \delta t^{1/4}}{2t^{1/2}} \exp(\mathcal{R}e\, \lambda_1(k_0)t) - \sum_{j=2}^{N} \frac{C_j \delta^j t^{1/4}}{t^{j/2}} \exp(j\, \mathcal{R}e\, \lambda_1(k_0)t) \tag{67}$$

and thus for $t \leqslant T_0 - \sigma_1$ with σ_1 large enough, but independent on δ,

$$\|u^{\text{app}} - u_{\text{stat}}\|_{L^2(\Omega_A)} \geqslant \frac{C_1 \delta}{4} \frac{t^{1/4}}{t^{1/2}} \exp(\mathcal{R}e\, \lambda_1(k_0)t). \tag{68}$$

Let now u be the solution of

$$\partial_t u + Q(u, u) = 0 \tag{69}$$

with initial data $u_{\text{stat}} + \delta u_1$. Let $v = u - u^{\text{app}}$. It satisfies

$$\partial_t v + Q(u^{\text{app}}, v) + Q(v, u^{\text{app}}) + Q(v, v) = -R_{\text{app}}, \tag{70}$$

hence using (A1′),

$$\partial_t \|v\|_{L^2}^2 \leqslant (\lambda_0(\|\nabla u^{\text{app}}\|_{L^\infty}) + \beta)\|v(t)\|_{L^2}^2 + C_\beta \|R_{\text{app}}(t)\|_{L^2}^2, \tag{71}$$

where $\beta > 0$ is small, such that $2N\, \mathcal{R}e\, \lambda_1(k_0) > \lambda_0(\|\nabla u_{\text{stat}}\|_{L^\infty} + \beta) + \beta$. We have

$$\|\nabla u^{\text{app}}\|_{L^\infty} \leqslant \|\nabla u_{\text{stat}}\|_{L^\infty} + \sum_{j=1}^{N} \frac{C''_j \delta^j}{t^{j/2}} \exp(j\, \mathcal{R}e\, \lambda_1 t)$$

and thus for $t \leqslant T_0 - \sigma_2$ with $\sigma_2 > \sigma_1$ large enough but independent on δ,

$$\|\nabla u^{\text{app}}\|_{L^\infty} \leqslant \|\nabla u_{\text{stat}}\|_{L^\infty} + \beta.$$

For $t \leqslant T_0 - \sigma_2$ we therefore have

$$\partial_t \|v(t)\|_{L^2}^2 \leq (\lambda_0(\|\nabla u_{\text{stat}}\|_{L^\infty} + \beta) + \beta)\|v(t)\|_{L^2}^2 + \frac{C\delta^{2(N+1)} t^{1/2}}{t^{(N+1)}} \exp(2(N+1)\, \mathcal{R}e\, \lambda_1(k_0)t).$$

As $v(0) = 0$, we have

$$\|v(t)\|_{L^2} \leqslant \frac{C' \delta^{N+1}}{t^{(N+1)/2}} t^{1/4} \exp((N+1)\, \mathcal{R}e\, \lambda_1(k_0)t),$$

since if $\lambda_2 < \lambda_3$, a function ϕ satisfying $\phi(0) = 0$ and

$$\partial_t \phi \leqslant \lambda_2 \phi + \frac{\exp(\lambda_3 t)}{1 + t^N}$$

verifies

$$\phi \leqslant C \exp(\lambda_3 t) \int_0^t \frac{\exp((\lambda_2 - \lambda_3)(t - \tau))}{1 + \tau^N} \, d\tau \leqslant C \frac{\exp(\lambda_3 t)}{1 + t^N}.$$

Now for $t = T_0 - \sigma_3$, with $\sigma_3 \geqslant \sigma_2$ large enough,

$$\|u - u_{\mathrm{stat}}\|_{L^2(\Omega_A)} \geqslant \left\|u^{\mathrm{app}} - u_{\mathrm{stat}}\right\|_{L^2(\Omega_A)} - \left\|v(t)\right\|_{L^2} \geqslant \sigma_0 T_0^{1/4}$$

with σ_0 independent on δ, which ends the proof, since the Lebesgue measure of Ω_A is of order $C T_0^{1/2}$. □

REMARKS. The proof of Theorem 1.14 in fact says much more than the theorem itself. In particular it gives a precise description of the solution till the instability time T^δ. For t near T^δ, the solution mainly equals

$$u \sim u_{\mathrm{stat}} + v_1(k_0) \exp(ikx_0) \exp\bigl(\lambda_1(k_0)t\bigr)$$
$$+ v_1(-k_0) \exp(-ikx_0) \exp\bigl(\lambda_1(-k_0)t\bigr)$$

for $|x + \mathcal{I}m\, \partial_k \lambda_1(k_0)t| \ll \sqrt{t}$. Therefore near T^δ, only the most unstable modes $\pm k_0$ emerge, after a travel at speed $-\mathcal{I}m\, \partial_k \lambda_1(k_0)$. This description is valid for every ϕ, provided $\phi(k_0) \neq 0$. Physically if one perturbs u_{stat} by a very small noise, waves of wave numbers k_0 grow more rapidly than the other waves, and travel to create the instability. In general, $\mathcal{I}m\, \partial_k \lambda_1(k_0) \neq 0$ which leads to a so called convective instability.

1.7. *Stable profiles*

1.7.1. *Lyapunov functional approach.* This section is devoted to the nonlinear stability of shear layer profiles and is strongly limited to two-dimensional flows. The methods cannot be extended in general to three-dimensional flows. This approach has been developed formally by Fjortoft and rigorously by Arnold [1]. The aim is to prove the theorem.

THEOREM 1.15 (Particular case of Arnold, 1969 [1]). *Let $u_s = (V(y), 0)$ be a shear layer profile in the strip $\mathbb{R} \times [0, 1]$. Then if u_s has no inflexion point (that is if V is strictly convex or strictly concave), it is (nonlinearly) stable in $H^1(\mathbb{R} \times [0, 1])$.*

Note that with Rayleigh's criterium we already know that the shear layer is spectrally stable. The method is in fact more general and can be used for general stationary two-dimensional flows.

PROOF OF THEOREM 1.15. Let us first observe that if u is a solution of Euler equations, then for any function $F:\mathbb{R}\to\mathbb{R}$,

$$H_F(u)=\int\frac{|u^2|}{2}+\int F(\operatorname{curl} u)$$

is a constant independent on t. Let u_s be a stationary solution of Euler equations. To study its stability we can follow the classical method of Lyapunov, namely look for a function F such that H_F has a critical point at u_s and is locally convex (or concave) near u_s. We first compute $\nabla H_F(u_s)$ which equals

$$\begin{aligned}\nabla H_F(u_s)\cdot v&=\int u_s\cdot v+\int F'(\operatorname{curl} u_s)\operatorname{curl} v\\&=\int v\cdot\big(u_s+F''(\operatorname{curl} u_s)\nabla^\perp\operatorname{curl} u_s\big).\end{aligned}$$

Therefore H_F has a critical point at u_s provided

$$u_s+F''(\operatorname{curl} u_s)\nabla^\perp\operatorname{curl} u_s=0. \tag{72}$$

Note that, as u_s is a stationary solution, $(u_s\cdot\nabla)\operatorname{curl} u_s=0$, therefore u_s and $\nabla^\perp\operatorname{curl} u_s$ are co-linear, hence we can define F'' by

$$F''(\operatorname{curl} u_s)=-\frac{u_s}{\nabla^\perp\operatorname{curl} u_s}. \tag{73}$$

Now the Hessian of H_F equals

$$\operatorname{Hess} H_F(u_s)\cdot v\cdot w=\int v\cdot w+\int F''(\operatorname{curl} u_s)\operatorname{curl} v\operatorname{curl} w.$$

It is positive definite if $F''\geqslant 0$ and is negative definite if Ω is bounded and if $F''<-C_0$ where C_0 is linked to the constant appearing in Poincaré's inequality. We refer to the papers of Arnold [1], and to [51] for details, and will only state the results for shear flows in a strip $0\leqslant z\leqslant 1$ ($\Omega=\mathbb{R}\times[0,1]$). In this case, using invariance by translation with respect to the x direction we get that $\int u_1\,dx$ is also independent on t and we can look for more general conserved quantities of the form

$$H_F(u)=\int\frac{|u|^2}{2}+\int F(\operatorname{curl} u)+\bar u\int u_1,$$

where $\bar u\in\mathbb{R}$ is fixed. Stationary condition (72) then becomes

$$u_s+(\bar u,0)+F''(\operatorname{curl} u_s)\nabla^\perp\operatorname{curl} u_s=0$$

and thus, for a shear layer,

$$F''\big(V'(z)\big) = -\frac{V + \bar{u}}{\partial_{zz} V}.$$

If the velocity profile has no inflexion point then $\partial_{zz} V$ does not change sign for $0 \leqslant z \leqslant 1$ and we can choose $\bar{u}$ such that $F'' > 1$. This leads to a convex function F, such that H_F has a critical point at u_s and is convex. Therefore a convex or concave velocity profile is nonlinearly stable under H^1 perturbations. □

REMARKS. (1) This method can be extended to other systems, like barotropic compressible ideal fluids and plasma situations [38]. It is also important to note that this approach can not be extended to three-dimensional flows (except when there is some symmetry [51]) and can not be extended to viscous flows in bounded domains since there is up to now no way to control the production of vorticity on the boundary.

(2) The stability holds with respect to H^1 perturbations, which is the context of classical physical books [17]. However, it can be argued that stability with respect to L^2 perturbations is more relevant. Shnirelman managed to prove the theorem:

THEOREM 1.16 (Shnirelman, 1999 [65]; instability of shear layers). *Let V_1 and V_2 be two shear layer profiles such that*

$$\int V_1(z)\,\mathrm{d}z = \int V_2(z)\,\mathrm{d}z,$$

$$\int V_1^2(z)\,\mathrm{d}z = \int V_2^2(z)\,\mathrm{d}z.$$

Then, for every $\varepsilon > 0$, there exists a time T_ε, a force f_ε satisfying

$$\int_0^{T_\varepsilon} \|f_\varepsilon\|_{L^2}^2\,\mathrm{d}t < \varepsilon$$

and a solution u_ε of Euler equation with force term f_ε such that $u_\varepsilon(0,\cdot) = (V_1(\cdot), 0)$ and $u_\varepsilon(T_\varepsilon,\cdot) = (V_2(\cdot), 0)$.

This bright result says that any shear flow is nonlinearly unstable with respect to L^2 perturbations (except of course the null flow). The physical motivation is that when we put a small obstacle in the flow or when we add a very small viscosity near the boundaries, the induced perturbations are small in energy but have a very large vorticity, and therefore are small in L^2 but not in H^1.

(3) It is important to note that if H_F is preserved by the nonlinear dynamics of Euler equations, and if H_F has a critical point at u_s, stationary solution, then Hess $H_F \cdot v \cdot v$ is constant if v is a solution of linearized Euler equations near u_s.

Let us give a formal proof (a computational proof is of course possible): let v^0 be a divergence free vector field. Let u^ε be the solution of Euler equations with initial data $u_s + \varepsilon v^0$. Then $H_F(u^\varepsilon)$ is constant in time, and so is $H_F(u_s)$, therefore

$$H_F(u^\varepsilon) - H_F(u_s)$$
$$= \nabla H_F(u_s)(u^\varepsilon - u_s) + \frac{1}{2}\nabla^2 H_F(u_s)(u^\varepsilon - u_s)\cdot(u^\varepsilon - u_s) + O(\varepsilon^3)$$

is constant in time. Now as $\varepsilon \to 0$, $\varepsilon^{-1}(u^\varepsilon - u_s)$ converges to the solution v of linearized Euler equations with initial data v^0, and the result follows.

1.7.2. *Boundary layer type solutions.* The aim of this section is to investigate the existence of solutions of Euler equations of the form $u^\eta(t,x,y,y/\eta)$ as $\eta \to 0$, in $\Omega = \mathbb{R} \times [0,+\infty[$, where all the partial derivatives of u^η are bounded uniformly in η as $\eta \to 0$, that is to investigate the existence of solutions which have a boundary layer type behavior near $y = 0$. We recall that viscosity turns out to have a destabilizing role near boundaries, therefore solutions of this form for Euler equations are expected to be more stable than for Navier–Stokes equations which is our goal. It is easy to see that solutions of the form $u^\eta(t,x,y,y/\eta)$ can persist over times of order η simply looking at Euler equations as a "hyperbolic" system. However to get this behavior up to times of order $\mathcal{O}(1)$ is more delicate, and sometimes false. Let us first state an instability result.

THEOREM 1.17 (Cases of complete instability [16]). *Let $u^\eta(t,x,y,y/\eta)$ be a solution of Euler equation on $[0,T]$ such that $u^\eta \in H^s([0,T]\times\mathbb{R}\times\mathbb{R}_+\times\mathbb{R}_+)$ for every s arbitrary large. Let us assume that there exists x_0 such that the (x independent shear layer) $(u_1^\eta(0,x_0,0,y),0)$ is spectrally unstable, with an unstable eigenvalue whose real part is bounded away from 0. Then, for every s and N arbitrary large, there exists a constant C_0, families of solutions v^η of Euler equations and of times T^η such that*

$$\|v^\eta(0,\cdot) - u^\eta(0,\cdot)\|_{H^s} \leqslant \eta^N,$$
$$\|v^\eta(T^\eta,\cdot) - u^\eta(T^\eta,\cdot)\|_{L^\infty} \geqslant C_0$$

with $T^\eta \to 0$ as $\eta \to 0$.

Therefore even if such solutions exist they are meaningless since they are completely unstable and sensitive to any arbitrary small perturbation. Note that using the machinery of [58], if $u^\eta(0,\cdot)$ is analytic in x, y and y/η, it is possible to construct analytic solutions of the form $u^\eta(t,x,y,y/\eta)$ for $0 \leqslant t \leqslant T$ with $T > 0$: again the analyticity rules out all the instabilities and hides the physically relevant destabilizing mechanisms. On the contrary if for every $x_0 \in \mathbb{R}$ the boundary layer profile $u_1^\eta(0,x_0,0,y)$ is concave or convex, then the situation is stable and we can prove an existence theorem of the form:

THEOREM 1.18 (Existence of solutions for convex boundary layers [35]). *Let $u^0(x,y,Y)$ be a given smooth function defined on $\mathbb{R}\times\mathbb{R}_+\times\mathbb{R}_+$. Let us assume that, for every $x \in \mathbb{R}$, $Y \to u_1^0(x,0,Y)$ is strictly concave or convex, and some additional technical assumptions. Then there exists $T > 0$ and a family of solutions of Euler equations of the form*

$u^\eta(t,x,y,y/\eta)$ with initial data u^0, u^η being moreover a bounded family of H^s functions for every s arbitrary large.

We will not detail the "additional technical assumptions" here and refer to [35] for more details. The proof has two parts: first we construct an approximate solution then we use careful energy estimates. We look for approximate solutions of the form

$$u^{\mathrm{app}} \sim \sum_{j=0}^{+\infty} \eta^j u^{\mathrm{int},j}(t,x,y) + \sum_{j=0}^{+\infty} \eta^j u^{\mathrm{b},j}(t,x,y/\eta),$$

where $u^{\mathrm{int},j}$ and $u^{\mathrm{b},j}$ are smooth functions, u_j^{b} being moreover rapidly decreasing in its last variable $Y = y/\eta$. We find that $u^{\mathrm{int},0}$ is solution of Euler equations, with related pressure $p^{\mathrm{int},0}$, and that $u^{\mathrm{b},0}$ is solution of the following "inviscid Prandtl equations":

$$\partial_t u_1^{\mathrm{b},0} + u_1^{\mathrm{b},0}\,\partial_x u_1^{\mathrm{b},0} + u_2^{\mathrm{b},0}\partial_Y u_1^{\mathrm{b},0} + \partial_x p^{\mathrm{int},0}(t,x,0) = 0, \tag{74}$$

$$\partial_x u_1^{\mathrm{b},0} + \partial_Y u_2^{\mathrm{b},0} = 0, \tag{75}$$

$$u_2^{\mathrm{b},0} = 0 \quad \text{as } Y = 0. \tag{76}$$

It turns out that (74)–(76) is easily solved in small time, simply by turning to Lagrangian coordinates. Again it is striking to see that it is the viscosity which is annoying in Prandtl equations. We then fulfill an energy estimate which follows the ideas of the next section (see [35] for more details).

1.7.3. *Flows in a thin channel.* Let us describe a nearby problem, proposed in [49] and first addressed by Brenier [6]. The purpose is to describe the asymptotic behavior of solutions u^ε of Euler equations in a thin domain $(x,y) \in \Omega^\varepsilon = \mathbb{T}\times[0,\varepsilon]$ as the small parameter ε goes to 0. Let $\Omega = \mathbb{T}\times[0,1]$ and let $Y = y/\varepsilon$ be the rescaled second variable, and let U^ε be the rescaled velocity, defined by

$$U_1^\varepsilon(t,x,Y) = u_1^\varepsilon(t,x,\varepsilon Y), \qquad U_2^\varepsilon(t,x,Y) = \varepsilon^{-1}u_2^\varepsilon(t,x,\varepsilon Y).$$

Formally, U_1^ε and U_2^ε converge to U_1 and U_2 where (U_1,U_2) satisfies an inviscid Prandtl's type equation

$$\partial_t U_1 + U_1\,\partial_x U_1 + U_2\,\partial_Y U_2 + \partial_x p = 0, \tag{77}$$

$$\partial_x U_1 + \partial_Y U_2 = 0, \tag{78}$$

$$U_2(t,x,0) = U_2(t,x,1) = 0 \quad \forall t, \forall x, \tag{79}$$

where $p(t,x)$, function independent of Y, is the remaining pressure and is a Lagrange multiplier for the constraint (79) on the normal velocities. Again note the formal analogy of (77)–(79) with the genuine Prandtl's equations. This system can be called "homogeneous hydrostatic equations" [6] since it is exactly the equations obtained by making the hydrostatic assumption $\partial_Y p = -\rho g$ in the case of an homogeneous fluid (ρ and g constant, the $-\rho g$ can be absorbed in a redefinition of the pressure).

Existence for the limit system is not obvious, since instabilities may appear in the limit $\varepsilon \to 0$. More precisely, as in the previous section, we have the theorem.

THEOREM 1.19 (An instability result). *Let $U^0(Y)$ be a given smooth function, such that the corresponding shear layer $(U^0(Y), 0)$ is spectrally unstable. Then, for every s and N arbitrary large, there exists a constant C_0 and a family of solutions u^ε of Euler equations in Ω^ε with corresponding rescaled velocities U^ε and a family of times $T^\varepsilon > 0$ such that*

$$\|U^\varepsilon(0,\cdot,\cdot) - (U^0, 0)\|_{H^s} \leqslant \varepsilon^N,$$

$$\|U^\varepsilon(T^\varepsilon,\cdot,\cdot) - (U^0, 0)\|_{L^2} \geqslant C_0,$$

$$\|U^\varepsilon(T^\varepsilon,\cdot,\cdot) - (U^0, 0)\|_{L^\infty} \geqslant C_0,$$

where $\lim_{\varepsilon\to 0} T^\varepsilon = 0$.

Such a situation occurs for $U^0(Y) = \sin(2\pi Y)$ for instance. To rule out these instabilities we have to assume a convexity assumption, as in the previous section.

THEOREM 1.20 (Existence for convex profiles; Brenier, 1999 [6]). *Let $(U_1^0, U_2^0) \in H^s(\mathbb{T} \times [0,1])$ with s large enough. Let us assume that*

$$\inf_{(x,Y)\in\mathbb{T}\times[0,1]} \left|\partial_{YY} U_0^1(x,Y)\right| > 0$$

and that $\partial_Y U_1^0(x,0)$ and $\partial_Y U_1^0(x,1)$ are independent on x. Then there exists $T > 0$ and a unique solution $U \in C([0,T], H^s(\mathbb{T}\times[0,1]))$ to (77)–(79) with initial data (U_1^0, U_2^0).

Let $\omega(t,x,Y) = \partial_2 U_1(t,x,Y)$. A first observation is that this reduced vorticity is transported with the flow. The key idea of the proof is then to make a "semi-Lagrangian" change of variables, and more precisely to observe that as ω is strictly monotonic as Y goes from 0 to 1 and has constant values on $Y = 0$ and $Y = 1$, it can be used as a coordinate. With the help of the change of variables $(x, Y) \to (x, \omega)$, Brenier manages to transform (77)–(79) into an hyperbolic symmetric system (with an infinite number of functions) and then proves existence for this hyperbolic system (see [6] for more details).

Once we have existence for the limit system we can prove convergence of solutions as $\varepsilon \to 0$.

THEOREM 1.21 (Convergence in the convex case [33]). *Under the assumptions of Theorem 1.20, the rescaled solutions U^ε obtained from solutions u^ε of Euler equations converge to the limit solution U in $L^\infty([0,T], H^{s-2}(\mathbb{T}\times[0,1]))$ as $\varepsilon \to 0$.*

Note that u^ε exists globally in time (global solutions of two-dimensional Euler equations) but a priori converges to U only in small time (in general U does not exists in large times [6]).

SKETCH OF PROOF OF THEOREM 1.21. The proof of this theorem relies first on the construction of an approximate solution to Euler equations, starting from a solution of (77)–(79) and using linearized versions of this system and we will not detail this point. The difficult point is to make an energy estimate, even on linearized systems. Let us focus on the derivation of an useful energy estimate on linearized Euler equations in Ω^ε around an approximate solution u^{app} (depending on ε)

$$\partial_t v + (u^{\mathrm{app}} \cdot \nabla)v + (v \cdot \nabla)u^{\mathrm{app}} + \nabla q = 0, \tag{80}$$

$$\nabla \cdot v = 0 \tag{81}$$

with boundary condition $v \cdot n = 0$ on $\partial\Omega^\varepsilon$. If we try to make an L^2 estimate we get

$$\partial_t \int \frac{|v|^2}{2} \leqslant \|\nabla u^{\mathrm{app}}\|_{L^\infty} \int |v|^2,$$

which is useless since $\|\nabla u^{\mathrm{app}}\|_{L^\infty}$ behaves like $\mathrm{O}(\varepsilon^{-1})$. The problem is thus to find out a norm $N(v)$ such that

$$\partial_t N(v) \leqslant CN(v),$$

where C is a constant independent on ε. Let us now describe how to find such a norm. The method is quite general and can be applied to various other systems (barotropic case, plasmas [34]).

Let us fix some $x_0 \in \mathbb{T}$ and some time t_0. For simplicity we will take $x_0 = 0$ and $t_0 = 0$. To investigate the stability of the flow near (t_0, x_0) the best thing to do is to make a "blow up" in all the variables, that is, to define

$$T = \frac{t}{\varepsilon}, \qquad X = \frac{x}{\varepsilon}, \qquad Y = \frac{y}{\varepsilon}$$

in an isotropic way. We recall that an hyperbolic type systems responds in times of order ε to small spatial scales of order ε, and there is no a priori reason why the x direction should be unaffected by instabilities (on the contrary, when instabilities take place, they do evolve on time scales of order ε and create "rolls" of size ε in the x direction). After this rescaling Euler equations in Ω^ε become Euler equations in $\mathbb{R} \times [0, 1]$, but now the approximate solution u^{app} varies over x scales of order ε^{-1} and time scales of order ε^{-1}. Let us imagine a small perturbation of u^{app} at $T = 0$ and near $X = 0$. We have to prove that this perturbation remains small over times of order ε^{-1}. If we make a classical L^2 estimate we get that this perturbation grows at most exponentially in time, like $\exp(CT)$. It is clearly not sufficient since over times of order ε^{-1} this leads to an amplification coefficient of order $\exp(C\varepsilon^{-1})$. If we want to prove by an energy method that this perturbation remains bounded, we have to find a norm $\widetilde{N}(v)$ such that $\partial_T \widetilde{N}(v) \leqslant 0$ since any term of the form $C\widetilde{N}(v)$ as it is the case for the L^2 estimate would lead to exponential amplification. But now if we make formally $\varepsilon = 0$ we have to find a norm $\widetilde{N}(v)$ which is decreasing in time for the linearized Euler equations near a constant profile $(u_1^{\mathrm{app}}(0,0,Y),0)$. But we do know such norms (following the remarks of Section 1.7.1). Namely

$$\int \frac{|v|^2}{2} + \int \frac{u_1^{\mathrm{app}}(0,0,Y)+\bar{u}}{\partial_{YY} u_1^{\mathrm{app}}(0,0,Y)} |\partial_Y v_1 - \partial_X v_2|^2$$

(where $\bar{u}$ is some constant) are constant if v is a solution of linearized Euler equations near the constant profile $(u_1^{\mathrm{app}}(0,0,Y),0)$. It remains now to reintroduce the time and space dependence, which can be seen as slow variables (the variable Y being the fast variable).

For this we introduce

$$N(v) = \int \frac{|v|^2}{2} + \int \frac{u_1^{\mathrm{app}}(t,x,Y)+\bar{u}}{\partial_{YY} u_1^{\mathrm{app}}(t,x,Y)} |\partial_Y v_1 - \varepsilon\, \partial_x v_2|^2,$$

where $\bar{u}$ is a constant which ensure that $N(v)$ is positive definite. Now as u^{app} depends on t and x, we cannot expect $N(v)$ to be decreasing, however straightforward computations show that $\partial_t N(v) \leqslant CN(v)$, where C is a constant independent on ε. Moreover, N controls the L^2 norms (and even a part of the H^1 norm). Therefore we get a good norm on the linearized system. The end of the proof is now classical. First we derive higher-order estimates and then turn to non linear estimates (see [33] for more details). □

1.8. *Miscellaneous*

1.8.1. *Piecewise linear velocity profiles.* Rayleigh's equation can be explicitly solved for piecewise linear profiles V, since in the intervals where V is linear, it degenerates into $\psi'' - \alpha^2\psi = 0$ which can be easily solved, and on the boundaries it degenerates in jump relations. Let us give some explicit examples (some are detailed in [17]).

In the half space $\mathbb{R} \times \mathbb{R}_+$, profiles of the form

$$u_{\mathrm{stat}}(y) = \begin{cases} \alpha y & \text{for } y \leqslant 1, \\ \beta y + (\alpha - \beta) & \text{for } 1 \leqslant y \leqslant 2, \\ \alpha + \beta & \text{for } y \geqslant 2 \end{cases} \tag{82}$$

with $\beta > \alpha$ are linearly unstable.

In the whole space $\mathbb{R} \times \mathbb{R}$ shears like

$$u_{\mathrm{stat}}(y) = \begin{cases} -1 & \text{for } y \leqslant -1, \\ y & \text{for } -1 \leqslant y \leqslant 1, \\ 1 & \text{for } y \geqslant 1 \end{cases} \tag{83}$$

are linearly unstable. The eigenvalue relation (Rayleigh, 1894) is

$$c^2 = \frac{(1-2\alpha)^2 - e^{-4\alpha}}{4\alpha^2}.$$

Let $\alpha_s \sim 0.64$ be the root of $1 - 2\alpha + e^{-2\alpha} = 0$. Then c is purely imaginary for $0 < \alpha < \alpha_s$ with a maximum for $\alpha \sim 0.40$ and is real for $\alpha > \alpha_s$.

In the periodic strip $\mathbb{R} \times (2\mathbb{T})$ the shear

$$u_{\text{stat}}(y) = \begin{cases} 2y+2 & \text{for } -1 \leqslant y \leqslant -1/2, \\ -2y & \text{for } -1/2 \leqslant y \leqslant 1/2, \\ 2y+2 & \text{for } 1/2 \leqslant y \leqslant 1 \end{cases} \tag{84}$$

extended by periodicity is linearly unstable (see Figure 1). In a strip $\mathbb{R} \times [-1, 1]$ shears like (84) are linearly unstable (Figure 2).

Let us give another completely explicit example. We take $z_1 = -1$, $z_2 = +1$ and $V(z) = 1$ for $b \leqslant z \leqslant 1$, $V(z) = z/b$ for $-b \leqslant z \leqslant b$ and $V(z) = -1$ for $-1 \leqslant z \leqslant -b$. The eigenvalue relation (Rayleigh, 1894) is

$$c^2 = 1 - \frac{\alpha b(1+X^2)Y^2 + 2\alpha bXY - XY^2}{\alpha^2 b^2((1+X^2)Y + X(1+Y^2))},$$

where $X = \tanh \alpha b$ and $Y = \tanh \alpha(1-b)$. The flow is unstable provided $b < 1/2$.

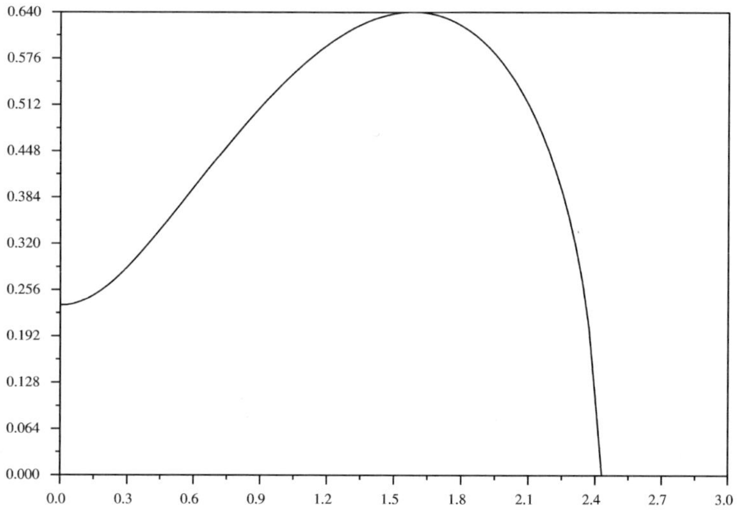

Fig. 1. $k|\mathcal{I}m\, c|$ for the case $\mathbb{R} \times 2\mathbb{T}$.

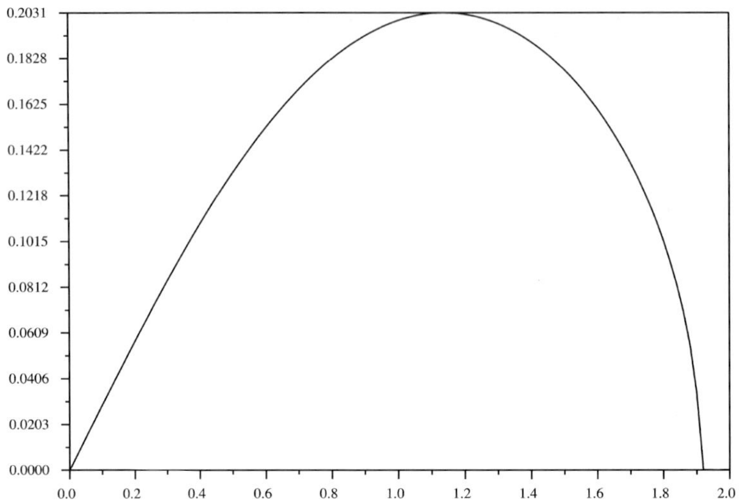

Fig. 2. $|\mathcal{I}m\, c|$ for the case $\mathbb{R} \times [-1, 1]$.

2. Stability of viscous shear flows

2.1. *Introduction*

Let us first recall the classical Navier–Stokes equations for incompressible viscous fluids in a domain Ω, u being the velocity, p the pressure, and $\nu > 0$ the viscosity

$$\partial_t u + (u \cdot \nabla)u - \nu \Delta u + \nabla p = 0, \tag{85}$$

$$\nabla \cdot u = 0, \tag{86}$$

$$u = 0 \quad \text{on } \partial\Omega. \tag{87}$$

Rayleigh's theorem shows that the stability of a shear layer flow in an inviscid fluid is linked to the presence of an inflexion point in its profile. In particular, shears without inflexion points are stable. When we go from inviscid to viscous fluids, we add viscosity, which is a dissipative mechanisms. Therefore we could think that the situation would be easier for viscous fluids than for inviscid ones, and that the viscosity would stabilize profiles with an inflexion point and keep the other stable anyway. This is far from being the case. In fact the viscosity *has a destabilizing role*, and almost all the shears are *unstable* solutions of Navier–Stokes equations. For profiles unstable for Euler equations this simply comes from a perturbative analysis of the unstable eigenvalues, however for profiles stable for Euler equations, the destabilizing mechanism is very subtle and requires a very careful, technical and intricated analysis, which will be only sketched in this review. We define the Reynolds number as the inverse of the viscosity

$$R = \nu^{-1}.$$

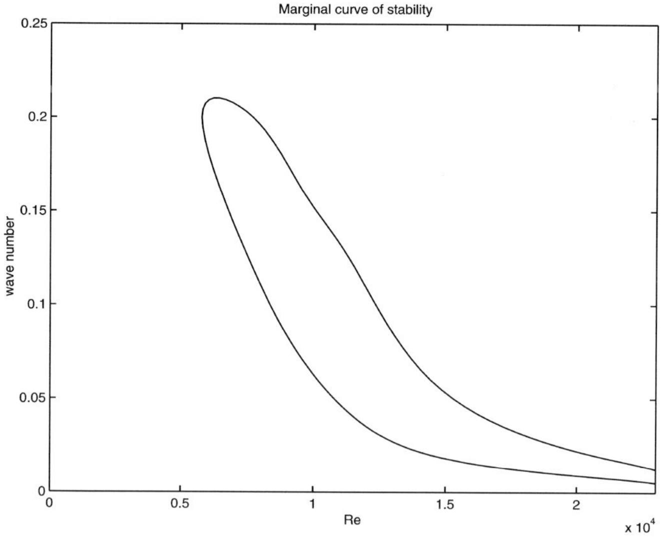

Fig. 3. Stability zone for Poiseuille flow.

Let us begin with the description of the stability of the classical Poiseuille flow $U(z) = 1 - z^2$ between two plates $-1 \leqslant z \leqslant 1$. Poiseuille flow is stable for Reynolds numbers smaller than $R_c = 5772$ and unstable for larger Reynolds numbers. More precisely as shown in Figure 3 it is unstable in the crescent like zone of the α, R plane. This mean that for a given α, all perturbations with this wave number are damped for small Reynolds numbers, are amplified for intermediate Reynolds numbers and then again damped for higher Reynolds numbers, which is a rather strange behavior! At a given Reynolds number larger than 5772, there is only a band of unstable wave numbers, smaller α and larger α being stable.

The situation is a little different for shear layers which are unstable for Euler equations. Let us for instance consider $U(z) = \sin z \exp(-z)$. It is stable for Reynolds numbers smaller than $R_c \sim 100$ and unstable for larger Reynolds numbers. The domain of stability in α, R plane is given by Figure 4. The main difference with Poiseuille flow is that the upper branch of the domain of instability does not go to 0 as the Reynolds number increases. Moreover, the Reynolds number is much smaller.

2.2. *Orr–Sommerfeld equation*

Let $U(z)$ be a given smooth shear layer profile. Squire's theorem (Theorem 1.1) can be extended to Navier–Stokes equations, hence it is sufficient to study stability with respect to two-dimensional perturbations, namely it is sufficient to work in two-dimensional space $(x, z) \in \Omega$, where $\Omega = \mathbb{T} \times [z_1, z_2]$ or $\mathbb{R} \times [z_1, z_2]$ with $z_1 \in \mathbb{R} \cup \{-\infty\}$ and $z_2 \in \mathbb{R} \cup \{+\infty\}$. Linearized Navier–Stokes equations around $(U(z), 0)$ are, on $v = (v_1, v_3)$

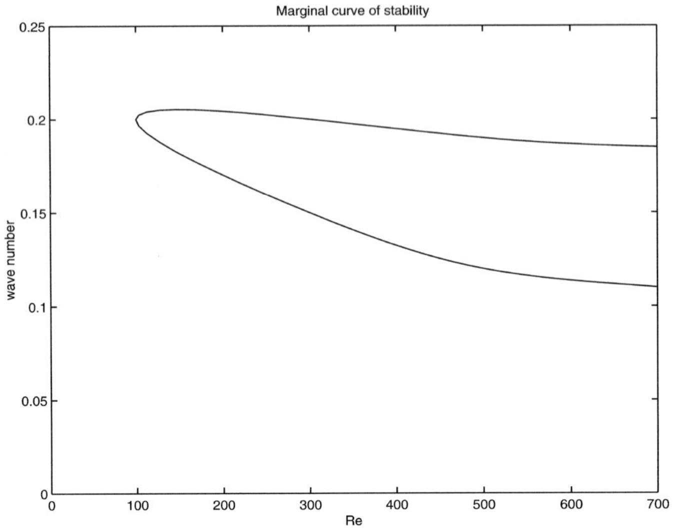

Fig. 4. Stability zone for $\sin z \exp(-z)$ flow.

with $R = \nu^{-1}$,

$$\partial_t v + U(z)\,\partial_z v + v_3\,\partial_z U e_1 = -\nabla p + R^{-1}\Delta v, \tag{88}$$

$$\nabla \cdot v = 0, \tag{89}$$

where $e_1 = (1, 0, 0)$. As for Euler equations we take the Fourier transform of the stream function and look for solutions of the form

$$v = \exp\bigl(\mathrm{i}(\alpha x - \alpha c t)\bigr)\begin{pmatrix} \phi'(z) \\ -\mathrm{i}\alpha\phi(z) \end{pmatrix}.$$

This leads to the famous Orr–Sommerfeld equation

$$(\mathrm{i}\alpha R)^{-1}\bigl(\partial^2 - \alpha^2 r\bigr)^2\phi = (U - c)l\bigl(\partial^2 - \alpha^2 r\bigr)\phi - U''\phi \tag{90}$$

with boundary conditions

$$\phi(0) = \phi'(0) = 0 \quad \text{on } z = z_1, z_2, \tag{91}$$

if we work in $z_1 \leqslant z \leqslant z_2$, or

$$\phi(0) = \phi'(0) = 0 \quad \text{on } z = z_1, \qquad \phi \to 0 \quad \text{as } z \to +\infty, \tag{92}$$

if we work in an half space $z \geqslant z_1$.

Let us define the Orr–Sommerfeld operator

$$\mathrm{Orr}_{\alpha,R}\,\omega = -(\mathrm{i}\alpha R)^{-1}\big(\partial^2 - \alpha^2\big)\omega + U\omega - U''\phi, \tag{93}$$

where ϕ is defined by

$$\big(\partial^2 - \alpha^2\big)\phi = \omega,$$

with

$$\phi(z_1) = \phi(z_2) = 0$$

if we work between two plates and

$$\phi(z_1) = 0, \qquad \phi \to 0 \quad \text{as } z \to +\infty,$$

in the half space. It is a perturbation of $(\partial^2 - \alpha^2)^2$ which is a sectorial operator, hence it is itself a sectorial operator [37].

REMARK. Between two plates, the spectrum of $\mathrm{Orr}_{\alpha,R}$ is purely discrete and composed of eigenvectors. This remark is however useless since we are interested in the limit $R \to +\infty$ where some parts of the spectrum tends to be continuous. In Section 2.3 we will study for which values of α and R there exists an eigenvalue of $\mathrm{Orr}_{\alpha,R}$ with positive imaginary part, which corresponds to unstable modes.

In the half space on the contrary, $\mathrm{Orr}_{\alpha,R}$ has a continuous part, linked with the behavior at infinity, and isolated eigenvalues. The continuous spectrum has a negative imaginary part, therefore, generically, instability of the flow is created by an isolated eigenvalue with strictly positive imaginary part.

2.3. *Curve of marginal stability*

The main physical problem is to compute the curve of marginal stability, that is the real values of c such that (90) and (91) has a nonzero solution ϕ, and in particular the critical Reynolds number. The curve of marginal stability is the limit between the stable area of parameters and the unstable one. It has two typical forms (see Figures 3 and 4), depending on the stability of the profile with respect to Euler equations. The main physical problem is to get an estimation of the critical Reynolds number, smallest Reynolds of instability. The problem has been solved by Tollmien in the 1930s and is a masterpiece of complex analysis.

Let us first describe in an informal way the main difficulties of the analysis. Orr–Sommerfeld operator is a viscous perturbation of the Reynolds operator. Because of the viscous part, a boundary layer appears near the boundary. Its analysis is however straightforward. Moreover, and it is the main difficulty, the Rayleigh part degenerates when $U - c$ vanishes, hence for z such that $U(z) = c$. Such a z is called a “critical layer”. The analysis of critical layers involves complex ordinary differential equations and Airy functions. They are the very cause of the instability. The complexity of the analysis makes the solution quite unexpected (destabilization through viscosity). Let us now give some details.

2.3.1. *Inviscid approximation.* The inviscid limit $R \to +\infty$ of Orr–Sommerfeld operator $\mathrm{Orr}_{\alpha,R}$ is of course the Rayleigh operator Ray_α (21) and (22). All the difficulty lies in the fact that if c is real and lies in the range of U then there exists $z_c \in [z_1, z_2]$ called critical layer, where

$$U(z_c) = c.$$

At z_c the ordinary differential equation (21) and (22) is singular. The classical theory of Fuchs of ordinary differential equations in the complex plane says that there exists two independent solutions called Tollmien's solutions ϕ_1^0 and ϕ_2^0 of (21) and (22) of the form

$$\phi_1^0(z) = (z - z_c) P_1(z) \tag{94}$$

and

$$\phi_2^0(z) = P_2(z) + \frac{U_c''}{U_c} \phi_1^0(z) \log(z - z_c), \tag{95}$$

where P_1 and P_2 are analytic, and $P_1(z_c) = P_2(z_c) = 1$ and $P_2'(z_c) = 0$. All the other solutions of (21) and (22) are linear combinations of ϕ_1^0 and ϕ_2^0.

Note that ϕ_1^0 is smooth and it is easy to construct a sequence of solutions of (90) which converges to ϕ_1^0 as $R \to +\infty$, however ϕ_2^0 is singular at $z = z_c$ and it is more subtle to construct solutions of (90) which converge to it. Note that, as Orr–Sommerfeld is of order 4 we must construct four independent solutions of it. Two of them will converge to solutions of Rayleigh equations, and two will be purely viscous.

2.3.2. *Critical layer.* The vector space of solutions of (90) is of dimension 4 and we want to describe a basis of it, at least asymptotically as $\alpha R \to +\infty$. The main problem lies in $z = z_c$, where the equation degenerates. In fact an interior viscous layer appears near z_c, which is of size $(\alpha R)^{-1/3}$ as will be proven below. To focus on it we introduce the rescaled variable

$$\xi = \frac{z - z_c}{\varepsilon}, \tag{96}$$

where

$$\varepsilon = \left(\mathrm{i}\alpha R U_c'\right)^{-1/3}.$$

Let

$$\phi(z) = \chi(\xi).$$

Equation (90) takes the form

$$\chi'''' - 2\alpha^2\varepsilon^2\chi'' + \alpha^4\varepsilon^4\chi = \mathrm{i}\alpha R\varepsilon^2\big((U - c)\big(\chi'' - \alpha^2\varepsilon^2\chi\big) - \varepsilon^2 U''\chi\big). \tag{97}$$

Let us look for an asymptotic expansion of χ of the form

$$\chi(\xi) = \chi^0(\xi) + \varepsilon\chi^1(\xi) + \cdots. \tag{98}$$

We find on χ^0,

$$\left(\partial_\xi^2 - \xi\right)\partial_\xi^2\chi^0 = 0 \tag{99}$$

and on χ^1

$$\left(\partial_\xi^2 - \xi\right)\partial_\xi^2\chi^1 = \frac{U_c''}{2U_c}\partial_\xi^2\chi^0 - \frac{U_c''}{U_c'}\chi^0. \tag{100}$$

Let us first recall that the solutions of

$$\left(\partial_\xi^2 - \xi\right)\phi = 0$$

are the Airy functions $A_1(\xi)$ and $A_2(\xi)$ which will therefore appear in our analysis. Equation (99) has a vector space of dimension 4 of solutions, which is spanned by

$$\chi_1^0(\xi) = \varepsilon\xi, \qquad \chi_2^0(\xi) = 1,$$
$$\chi_3^0(\xi) = A_1(\xi, 2), \qquad \chi_4^0(\xi) = A_2(\xi, 2),$$

where $A_1(\xi, 2)$ and $A_2(\xi, 2)$ are double primitives of Airy functions. Now for each solution $\chi_1^0, \ldots, \chi_4^0$ of (99) we will construct a corresponding solution of (90) which behaves like it near z_c.

2.3.3. *Construction of* ϕ_1. We take

$$\chi_1^1 = \frac{1}{2}\varepsilon\frac{U_c''}{U_c'}\xi^2$$

in order to match with the second term in the power expansion (98) of ϕ_1^0 at z_c. It is then easy to construct a solution ϕ_1 of (90) which tends to ϕ_1^0 as $R \to +\infty$.

2.3.4. *Construction of* ϕ_2. To χ_2^0 will correspond an approximation of ϕ_2^0. Note that in this case χ_2^1 is a solution of

$$\left(\partial_\xi^2 - \xi\right)\partial_\xi^2\chi_2^1 = -\frac{U_c''}{U_c}. \tag{101}$$

It can be expressed in terms of special functions B_3

$$\chi_2^1(\xi) \sim -\frac{U_c''}{U_c'}\left(B_3(\xi, 2, 1) + A\xi + B\right),$$

where A and B are arbitrary constants which must be chosen so that $\xi_2^0 + \varepsilon\xi_2^1$ is asymptotic to ϕ_2^0 for $|z - z_c| \ll 1$, in order to match ϕ_2^0 with χ_2^1. But for $|z - z_c| \ll 1$, using (95),

$$\phi_2^0(z) = 1 + \frac{U_c''}{U_c'}(z - z_c)\ln(z - z_c) + \mathrm{O}\big((z - z_c)^2 \ln(z - z_c)\big)$$

and, using asymptotic expansions of B_3,

$$\chi_2^1(\xi) \sim -\frac{U_c''}{U_c'}\big(-\xi \ln \xi + \big(\psi(2) + 2\pi \mathrm{i}\big)\xi + A\xi + B + \mathrm{O}\big(\xi^{-3}\big)\big),$$

therefore we must take

$$A = -\big(\ln \varepsilon + \psi(2) + 2\pi \mathrm{i}\big), \qquad B = 0.$$

It is possible to go on with the construction of a solution of (90) which matches χ_2^1 by looking to a complete asymptotic expansion in ε. This gives a solution ϕ_2 of (90) whose leading term is

$$\phi_2(z) \equiv P_2(z) + \varepsilon\chi_2^1(\xi) + \frac{U_c''}{U_c'}\big(P_1(z) - 1\big)(z - z_c)\ln(z - z_c) \tag{102}$$

and such that

$$\begin{aligned}\phi_2'(z) \equiv {}& P_2(z) + \varepsilon\big(\chi_2^1\big)'(\xi) \\ & + \frac{U_c''}{U_c'}\big(\big(P_1(z) - 1\big)\big(\ln(z - z_c) + 1\big) + P_1'(z)(z - z_c)\ln(z - z_c)\big)\end{aligned} \tag{103}$$

up to higher-order terms in ε.

2.3.5. *Construction of ϕ_3 and ϕ_4.* To find ϕ_3 and ϕ_4 it is more convenient to use the so called WKBJ method and to look for ϕ of the form

$$\phi = \exp\bigg(\int_{z_c}^{z} g(z)\,\mathrm{d}z\bigg). \tag{104}$$

Putting (104) in (90) gives

$$\begin{aligned}& g^4 + 6g^2g' + 4gg'' + 3\big(g'\big)^2 + g''' - 2\alpha^2\big(g^2 + g'\big) + \alpha^4 \\ & \quad = \mathrm{i}\alpha R\big((U - c)\big(g^2 + g' - \alpha^2\big) - U''\big),\end{aligned} \tag{105}$$

and we are led to look for an expansion of the form

$$g(z) = (\mathrm{i}\alpha R)^{1/2} g_0(z) + g_1(z) + (\mathrm{i}\alpha R)^{-1/2} g_2(z) + \cdots$$

for z away from z_c. On g_0 we get

$$g_0^2\big(g_0^2 - (U - c)\big) = 0 \tag{106}$$

which has three solutions. The first one is $g_0 = 0$ and in fact leads to ϕ_1^0 and ϕ_2^0 again. We therefore have to investigate the two other solutions

$$g_0 = \mp(U - c)^{1/2}.$$

The equation on g_1 is

$$2g_0\big(2g_0^2 - (U - c)\big)g_1 = -g_0'\big(6g_0^2 - (U - c)\big), \tag{107}$$

and is solved by

$$g_1(z) = -\frac{5}{4}\frac{U'}{U - c}. \tag{108}$$

It is possible to go on with the construction and to compute the following terms g_i $(i \geqslant 2)$. This gives, away from $z = z_c$ a solution of (90) of the form (104), with

$$\phi_{3,4}(z) \sim C_{3,4}(U - c)^{-5/4}\exp\big(\mp(\alpha R)^{1/2}Q(z)\big), \tag{109}$$

where

$$Q(z) = \mathrm{e}^{\mathrm{i}\pi/4}\int_{z_c}^{z}(U - c)^{1/2}\,\mathrm{d}z.$$

Now we have extend these solutions near $z = z_c$ and in fact to match these solutions with χ_3^0 and χ_4^0. For this we use the asymptotic expansions of the Airy functions to get

$$\phi_3(z) \equiv \left(\frac{U - c}{U_c'(z - z_c)}\right)^{-5/4}\exp\left(\frac{2}{3}\xi^{3/2} - (\alpha R)^{1/2}Q(z)\right)A_1(\xi, 2) \tag{110}$$

and similarly

$$\phi_3'(z) \equiv \varepsilon^{-1}\left(\frac{U - c}{U_c'(z - z_c)}\right)^{-3/4}\exp\left(\frac{2}{3}\xi^{3/2} - (\alpha R)^{1/2}Q(z)\right)A_1(\xi, 1). \tag{111}$$

We will not need the expansion of ϕ_4, which is similar.

2.3.6. *Approximate eigenvalue relation in a channel.* The eigenvalue relation is obtained by writing that there exists some non zero combination of $\phi_1, \phi_2, \phi_3, \phi_4$ satisfying the two boundary condition $\phi = \phi' = 0$ on the two boundaries (which makes four boundary conditions). The complete eigenvalue relation is however very heavy to manipulate. Therefore it is useful to look for an approximate eigenvalue relation. We will not justify the approximations in this section. Let us look for a solution

$$\phi = A\phi_1 + \phi_2 + C\phi_3 + D\phi_4 \tag{112}$$

of (90). In a first approximation we neglect ϕ_4 and set $D = 0$ in (112). We also approximate ϕ_1 and ϕ_2 by ϕ_1^0 and ϕ_2^0 and define

$$\Phi = A\phi_1^0 + \phi_2^0.$$

Moreover, at $z = z_2$ we neglect ϕ_3, retaining ϕ_1^0 and ϕ_2^0. To ensure the boundary conditions in z_2 we have to choose A such that $\Phi'(z_2) = 0$ which leads to

$$A = -\frac{(\phi_2^0)'(z_2)}{(\phi_1^0)'(z_2)}.$$

We then write $\Phi(z_1) = \Phi'(z_1) = 0$, and eliminate C to get

$$\frac{\Phi'(z_1)}{\Phi(z_1)} = \frac{\phi_3'(z_1)}{\phi_3(z_1)},$$

or if we introduce

$$W(\alpha, c) = -\frac{U_1'}{c}\frac{\Phi(z_1)}{\Phi'(z_1)}, \tag{113}$$

we get the approximate eigenvalue relation

$$W(\alpha, c) = -\frac{U_1'}{c}\frac{\phi_3(z_1)}{\phi_3'(z_1)}. \tag{114}$$

Let now compute W which depends only on the inviscid solutions ϕ_1^0 and ϕ_2^0. This can be done with the help of the modified Heisenberg solutions and we get (see [17] for the details of the computations)

$$A(\alpha, c) = \frac{(U_1')^2}{I_2}\alpha^{-2}\bigl(1 + \mathrm{O}(\alpha^2, c)\bigr),$$

where

$$I_2 = \int_{z_1}^{z_2} U^2\,\mathrm{d}z.$$

Moreover, $U(z_c) = c$ leads to

$$z_1 - z_c = -\frac{c}{U_1'}\left(1 - \frac{U_1''}{2(U_1')^2}c + \mathrm{O}(c^2)\right).$$

Therefore

$$W(\alpha, c) = 1 - K_1 + o(1) - i\pi \frac{U_1'' I_2}{(U_1')^3} \alpha^2 K_1 \big(1 + o(1)\big), \tag{115}$$

where

$$K_1 = \frac{I_2}{U_1'} \frac{\alpha^2}{c}.$$

2.3.7. *Approximate eigenvalue relation near boundaries.* As previously, we neglect ϕ_4 and replace ϕ_1 and ϕ_2 by their inviscid limits ϕ_1^0 and ϕ_2^0. As $z \to +\infty$, (90) goes to

$$(\partial^2 - \alpha^2)(\partial^2 - \beta^2)\phi = 0,$$

where $\beta^2 = i\alpha R(1 - c) + \alpha^2$ with $\mathcal{R}e\,\beta > 0$. Since β is large like $(\alpha R)^{1/2}$, the viscous solution decays much faster than the inviscid one, and ϕ_3 can be neglected before ϕ_1 and ϕ_2 for large z. Let again $\Phi = A\phi_1 + \phi_2$ be the inviscid solution. We have to choose A such that

$$\Phi \sim B e^{-\alpha z}$$

for some constant B.

To get the eigenvalue relation we then write $\Phi(z_1) = \Phi'(z_1) = 0$, and eliminate C to get

$$\frac{\Phi'(z_1)}{\Phi(z_1)} = \frac{\phi_3'(z_1)}{\phi_3(z_1)},$$

or if we introduce

$$W(\alpha, c) = -\frac{U_1'}{c} \frac{\Phi(z_1)}{\Phi'(z_1)} \tag{116}$$

we get the eigenvalue relation

$$W(\alpha, c) = -\frac{U_1'}{c} \frac{\phi_3(z_1)}{\phi_3'(z_1)}. \tag{117}$$

Let now compute W. To get A and B we have to investigate the asymptotic behavior as $z \to +\infty$ of the inviscid solutions ϕ_1 and ϕ_2. This can be done with the help of the modified Heisenberg solutions and we get

$$A(\alpha, c) = \big(U_1'\big)^2 \alpha^{-1} \big\{1 + O(\alpha, c)\big\}$$

and

$$B(\alpha, c) = U_1' \alpha^{-1} \big\{1 + O(\alpha, c)\big\}.$$

The same methods lead to

$$W(\alpha, c) = 1 - K_2 + \mathrm{o}(1) - \mathrm{i}\pi \frac{U_1''}{(U_1')^3} \alpha K_2 \bigl(1 + \mathrm{o}(1)\bigr), \tag{118}$$

where

$$K_2 = \frac{1}{U_1'} \frac{\alpha}{c}.$$

2.3.8. *Tietjens function.* Using (110) and (111) we get

$$\frac{\phi_3(z_1)}{\phi_3'(z_1)} = (z_1 - z_c) F(Y), \tag{119}$$

where

$$Y = (z_c - z_1)\bigl(\alpha R U_c'\bigr)^{1/3}, \qquad \xi_1 = Y \mathrm{e}^{-5\pi \mathrm{i}/6},$$

and where

$$F(Y) = \frac{A_1(\xi_1, 2)}{\xi_1 A_1(\xi_1, 1)} \tag{120}$$

is the so-called Tietjens function. Introducing $\lambda_1(c)$ defined by

$$1 + \lambda_1(c) = \frac{U_1'}{c}(z_c - z_1),$$

where

$$\lambda_1(c) = -\frac{U_1''}{2(U_1')^2} c + \mathrm{O}\bigl(c^2\bigr),$$

the approximate relation becomes

$$W(\alpha, c) = \bigl(1 + \lambda_1(c)\bigr) F(Y), \tag{121}$$

where the function W is explicitly given by (118) or (115).

The Tietjens function can be easily computed using

$$H(Y) = Y F(Y)$$

which satisfies the second-order nonlinear equation

$$H H'' - 2\bigl(H'\bigr)^2 + 3H' - 1 + \mathrm{i}\bigl(H^3 - Y H^2\bigr) = 0$$

with initial conditions

$$H(0) = \frac{3^{2/3}}{\Gamma(1/3)} \mathrm{e}^{-\mathrm{i}\pi/6} \quad \text{and} \quad H'(0) = 1 - \frac{1}{2} 3^{3/2} \pi^{-1}.$$

Moreover, as $|Y| \to +\infty$, with arguments between $-\pi/6$ and $11\pi/6$, we have

$$F(Y) \sim \mathrm{e}^{\pi\mathrm{i}/4} Y^{-3/2} + \frac{5}{4} \mathrm{e}^{\pi\mathrm{i}/2} Y^{-3} + \frac{151}{32} \mathrm{e}^{3\pi\mathrm{i}/4} Y^{-9/2} + \cdots. \tag{122}$$

The imaginary part of F vanishes at $Y_0 \sim 2.297$ and at this point $F(Y_0) \sim 0.5645$ and $F'(Y_0) \sim -0.1197 + 0.2307\mathrm{i}$.

2.4. *Stability characteristic of classical flows*

2.4.1. *Asymptotics of the lower and upper branches.* The upper branch is obtained by taking $Y \to +\infty$ and approximating $F(Y)$ by $\mathrm{e}^{\pi\mathrm{i}/4} Y^{-3/2}$ in the approximate eigenvalue relation. We get for symmetric flows in a channel

$$c \sim \frac{I_2}{U_1'} \alpha^2, \qquad R \sim \frac{1}{2} \pi^{-2} \frac{(U_1')^{11}}{I_2^5 (U_1'')^2} \alpha^{-11}. \tag{123}$$

For boundary layers this gives

$$c \sim \frac{1}{U_1'} \alpha, \qquad R \sim \frac{1}{2} \pi^{-2} \frac{(U_1')^{11}}{(U_1'')^2} \alpha^{-6} \tag{124}$$

if $U_1'' > 0$ (to be modified in $U_1'' = 0$).

The lower branch is obtained as $Y \to Y_0$. For symmetric flows in a channel we get

$$c \sim 2.296 \frac{I_2}{U_1'} \alpha^2, \qquad R \sim 1.002 \frac{(U_1')^5}{I_2^3} \alpha^{-7}. \tag{125}$$

For boundary layer flows we have

$$c \sim 2.2961 \frac{1}{U_1'} \alpha, \qquad R \sim 1.002 \left(U_1'\right)^5 \alpha^{-4}. \tag{126}$$

2.4.2. *Classical flows.*

- Couette: z for $-1 \leqslant z \leqslant 1$, always stable (it is the only one).
- Poiseuille: $1 - z^2$ for $-1 \leqslant z \leqslant 1$, $R_\mathrm{c} = 5772$.
- Blasius: $f'(z)$ for $z \geqslant 0$, where f solves

$$f''' + f f'' = 0$$

with $f(0) = f'(0) = 0$ and $f'(z) \to 1$ as $z \to +\infty$. $R_c = 520$. Lower branch $R \sim 0.062\alpha^{-4}$ and upper branch $R \sim 2.10^{-5}\alpha^{-10}$ (as $f'(0) = 0$ the case is degenerated).
- Asymptotic suction boundary layer profile: $U = 1 - e^{-z}$ with a suction $-1/R$. Critical Reynolds $R_c \sim 47047$.

2.4.3. *Most unstable mode.* We want now to have a bound on the growth rate of the most unstable mode of Orr–Sommerfeld equation as the Reynolds number increases. Let $(u_s(Y), 0)$ be a given shear layer profile. Let $\sigma(R)$ be the supremum of the imaginary parts of the eigenvalues αc of Orr–Sommerfeld operator, the supremum being taken also with respect to α, the Reynolds number being on the contrary fixed, and let σ_∞ be the supremum of the imaginary parts of the eigenvalues αc of the corresponding Rayleigh operator.

THEOREM 2.1. *If $\sigma_\infty > 0$, then*

$$\sigma(R) \to \sigma_\infty \quad \text{as } R \to +\infty, \tag{127}$$

and the wave number α, where $\sigma(R)$ is reached, converges to α_0 as $R \to +\infty$. On the other hand, if $\sigma_\infty = 0$ and if $V'(0) \neq 0$, then

$$\sigma(R) \sim C_0 R^{-1/2} \quad \text{as } R \to +\infty, \tag{128}$$

for some constant C_0 depending only on $V'(0)$ and $V''(0)$.

PROOF. If this layer is spectrally unstable for Euler equations, let v_{Euler} be the most unstable mode, with corresponding eigenvalue c_{Euler} and wave number α_{Euler}. Near c_{Euler}, α_{Euler} and v_{Euler}, Orr–Sommerfeld operator is a regular perturbation of Rayleigh operator since c is not real (there is no problem any more in the critical layer). Hence it is possible to construct eigenvectors and eigenvalues of Orr–Sommerfeld operator which converge to v_{Euler} and c_{Euler} as $R \to +\infty$. In particular, we can prove that the growth rate of the most unstable mode of Orr–Sommerfeld converges to the growth rate $\alpha_{\text{Euler}} c_{\text{Euler}}$ of the most unstable mode of Rayleigh operator.

If on the contrary the shear layer $(u_s(Y), 0)$ is spectrally stable for Euler equations the situation is more difficult. We know from the previous paragraph that the shear layer is unstable for Navier–Stokes equations and it is necessary to study (121) in the limit $R \to +\infty$. The approximate relation is

$$W(k, c) = F(Y), \tag{129}$$

where

$$Y = \frac{c}{V'(0)^{2/3}} (kR)^{1/3} \big(1 + o(c)\big),$$

$$W(k, c) = 1 - \frac{1}{V'(0)} \frac{k}{c} + o(1) - i\pi \frac{V''(0)}{(V'(0))^4} \frac{k^2}{c} \big(1 + o(1)\big).$$

As $F \to 0$ as $|Y| \to +\infty$ we split the problem in two. First we consider (129) for $|Y| \leqslant C_0$ for some large constant C_0. As F is bounded, k/c is bounded and therefore $k \leqslant C_2|c|$ for some constant C_2. But

$$|Y| = C\frac{|c|}{k}k^{4/3}R^{1/3} \leqslant C_0.$$

This leads to

$$k \leqslant C_0^{3/4}\left(\frac{k}{|c|}\right)^{3/4} R^{-1/4}.$$

But we know that on the domain on instability, $k \geqslant C_1 R^{-1/4}$ therefore there exists a constant C_2 such that $|c| \leqslant C_2 k$. Hence k and c are both of order $R^{-1/4}$, and $k\,\mathcal{I}m\,c$ is of order $R^{-1/2}$.

Let us now consider large Y where (122) can be used. In particular F go to 0, hence $c \sim k/V'(0)$. We therefore make the change of variables

$$c = \frac{k}{V'(0)}(1+\sigma),$$

where $\sigma \in \mathbb{C}$. Relation (129) then becomes

$$\sigma - \sigma^2 + O(\sigma^3) - i\pi\frac{V''(0)}{V'(0)^3}k\bigl(1 - \sigma + O(\sigma^2)\bigr) = F(Y).$$

Moreover, $\mathcal{I}m\,c/\mathcal{R}e\,c \to 0$, therefore the argument of Y goes to 0 and the argument of $F(Y)$ goes to $5\pi/4$. Therefore either $\sigma \gg k$ and

$$\sigma \sim -e^{i\pi/4}V'(0)^{5/2}k^{-2}R^{-1/2},$$

or σ and k are of the same orders. In the first case, as k goes from $R^{-1/4}$ to $R^{-1/6}$, σ ranges from 1 to $R^{-1/6}$ (in this case the assumption $\sigma \gg k$ is no longer valid) and $\mathcal{I}m\,c$ ranges from $R^{-1/4}$ to $R^{-1/3}$. Moreover,

$$k\,\mathcal{I}m\,c \sim \frac{V'(0)^{3/2}}{\sqrt{2}}R^{-1/2}.$$

In the second case, $\mathcal{I}m\,c$ is of order k^2 and ranges from $R^{-1/2}$ to $R^{-1/3}$, and $k\,\mathcal{I}m\,c$ ranges from $R^{-3/4}$ to $R^{-1/2}$, which ends the proof. □

2.5. *Stability and instability results*

The main point is that flows are stable at low Reynolds number, and then unstable above some critical Reynolds number R_c which could be computed in the previous section, through spectral analysis. We can detail a little more the situation: for Reynolds numbers smaller than some critical $R_e < R_c$ number, the flow is stable and this can be proved using energy methods.

More precisely, for $R < R_e$, the energy of any perturbation decreases. For $R_e < R < R_c$, the flow is stable, but there exists perturbations such that their energy first increases then decays to go to 0 as time goes to $+\infty$. In particular these flows are stable but their stability can not be proved using energy estimates. Their linear stability can be proved using spectral analysis, but the problem of their nonlinear stability is mainly open.

For $R > R_c$ the flow is spectrally unstable, and it can be proved that it is also nonlinearly unstable. Let us detail these three regions.

2.5.1. $R < R_e$. Let us compute $\partial_t \int |v|^2$ in the linear case

$$\partial_t \int \frac{|v|^2}{2} + \int v \cdot (v \cdot \nabla u) + \nu \int |\nabla v|^2 = 0$$

therefore provided

$$I(v) = \int v \cdot (v \cdot \nabla u) + \nu \int |\nabla v|^2 > 0$$

for every function v, the flow u is linearly stable. But

$$\int v \cdot (v \cdot \nabla u) = \int v_1 v_3 \, \partial_z u_1$$

$$= \int \left(\int_0^z \partial_z v_1(x, z') \, dz' \right) \left(\int_0^z \partial_z v_3(x, z') \, dz' \right) \partial_z u_1(z) \, dz \, dx$$

which in absolute value is bounded by

$$\left(\int_0^{+\infty} |\partial_z u_1| \, dz \right) \|\partial_z v_1\|_{L^2} \|\partial_z v_3\|_{L^2}.$$

This latest expression is bounded by $\nu \|\nabla v\|_{L^2}^2$ provided that ν is large enough. Note that this computation works also in the nonlinear case (the term $(v \cdot \nabla)v$ disappears in energy estimates). This leads to the proposition:

PROPOSITION 2.2. *Let $u(x, z) = (u_1(z), 0)$ be a shear flow such that*

$$\int_0^{+\infty} |\partial_z u_1| \, dz < +\infty.$$

Then for ν large enough, the energy of any perturbation (in the linear case as well as in the nonlinear case) decreases.

We define R_e as the largest Reynolds number for which this is true. Note that this Reynolds number cannot a priori be computed through the former proof. A first proof of this proposition was given by Serrin [62].

2.5.2. $R_e < R < R_c$. Let us consider the linear case. By definition of R_e we know that there exists a perturbation which at first increases and then decreases. This comes from the fact that Orr–Sommerfeld operator is not self-adjoint.

To explain this fact, let us consider a diagonalizable matrix A, with eigenvalues $\lambda_1, \ldots, \lambda_n$ and related eigenvectors $e_1, \ldots, e_n$. If A is self-adjoint (or more generally normal), then the eigenvectors are orthonormal and we have for instance if the eigenvalues are real

$$\left\| e^{tA} v \right\|^2 = \sum_{i=1}^{n} e^{t\lambda_i} \langle e_i, v \rangle^2,$$

where $\langle e_i, v \rangle$ is the scalar product between e_i and v. If all the λ_i are negative then $\|e^{tA} v\|^2$ is a sum of decreasing terms, and is therefore decreasing with time.

This is no longer true is the eigenvectors are not orthonormal. Assume $n = 2$ to simplify. Let us imagine that the two unit eigenvectors e_1 and e_2 are very close ($\|e_1 - e_2\| = \beta \ll 1$), and that $-\lambda_1 \gg -\lambda_2 > 0$. Then if $v = \alpha_1 e_1 + \alpha_2 e_2$,

$$\left\| e^{tA} v \right\|^2 = \left\| \alpha_1 e^{\lambda_1 t} e_1 + \alpha_2 e^{\lambda_2 t} e_2 \right\|^2.$$

Let $v = \beta^{-1}(e_1 - e_2)$. Then $\|v\| = 1$. As t increases, starting from 0, the first term decreases much faster than the second one. For t of order $-10\lambda_1^{-1}$, it almost vanishes, and $e^{tA} v$ almost equals to $\beta^{-1} e^{\lambda_2 t} e_2$ which has a very large norm (almost β^{-1}). Therefore in a first step, $\|e^{tA} v\|$ increases very rapidly, in spite of the signs of λ_1 and λ_2. In a second step, the norm decreases like $e^{\lambda_2 t}$ and therefore much more slowly.

In particular 0 is stable, but this cannot be proved estimating $\partial_t \|e^{tA} v\|_{L^2}^2$, whose sign changes as time goes on!

This situation does occur in real fluids and is physically very important. For Reynolds numbers between R_e and R_c, numerical computations and physical deductions show that generic linear perturbations grow by a factor of order $O(R)$ before they decrease. If R_c is small (say of order 100), this amplification coefficient is large, but not very relevant. On the contrary if R_c is large (it can be as large as 10^6), then the amplification coefficient is really huge. Of course the flow is always linearly stable, but an initial perturbation of order 10^{-6} leads to perturbations of order $O(1)$ and therefore to nonlinear effects. The initial data for which the linear regime is applicable are therefore very small, typically the energy of the initial perturbation must be of order $O(10^{-7})$. Larger perturbations lead to the development of nonlinear phenomena. Of course $O(10^{-7})$ is purely noise in real experiments! The noise is sufficient to create O(1) flows and to destabilize the flow. In this case the linear approach has a very restrictive range of application. We refer to [60] for more details on this problem, called "nonnormal modes".

2.5.3. $R > R_c$. Shear layers are in general instable for large Reynolds number, except linear flows (Couette profiles). It is possible to prove (see next section) that this linear instability creates a nonlinear instability in Sobolev spaces (H^s with s large enough). The question of the L^∞ and of the L^2 instability is widely open.

3. Inviscid limit of Navier–Stokes equations

This section is devoted to the study of the inviscid limit of Navier–Stokes equations near a boundary. More precisely let $\Omega = \mathbb{R}^2 \times [0, +\infty[$ or $\Omega = \mathbb{R} \times [0, +\infty[$. Do sequences u^ν of solutions of Navier–Stokes equations converge to solutions of Euler equations as $\nu \to 0$? (We have of course to choose carefully the initial data of Navier–Stokes and Euler equations.) Does the convergence holds in H^s?, in L^∞?, in L^2?

We will see that this problem is deeply linked to stability properties of shear layers, and give a negative answer for H^s convergence, conjecture negative results for L^∞ ... and leave the L^2 case open! We will work in two dimensions to fix the ideas (the situation in three dimensions being even worse).

3.1. *Formal derivation of Prandtl's equations*

The first approach has been done by Prandtl a century ago. If we look for asymptotic expansions of the solutions u^ν of Navier–Stokes equations in powers of ν, we are lead to look for u^ν of the form

$$u^\nu(t,x,y) \sim \sum_{j=0}^{+\infty} \nu^j u^{\mathrm{int},j}(t,x,y) + \sum_{j=0}^{+\infty} \nu^j u^{\mathrm{b},j}\big(t,x,\nu^{-1/2}y\big), \tag{130}$$

where $u^{\mathrm{int},j}$ and $u^{\mathrm{b},j}$ are smooth functions, $u^{\mathrm{b},j}$ being moreover rapidly decreasing in its last variable $Y = y/\sqrt{\nu}$. As will be clear below, $\sqrt{\nu}$ is the only natural scale for the boundary layer. Note that the right-hand side of (130) may diverge and is only an asymptotic expansion of u^ν as $\nu \to 0$.

When we put (130) in Navier–Stokes equations we find Euler equations for $u^{\mathrm{int},0}$ with boundary condition $u^{\mathrm{int},0} \cdot n = 0$. Using incompressibility condition in the boundary layer we get at order $\nu^{-1/2}$

$$\partial_Y u_2^{\mathrm{b},0} = 0,$$

hence as $u_2^{\mathrm{b},0}$ goes to 0 as $Y \to +\infty$, $u_2^{\mathrm{b},0} \equiv 0$. Navier–Stokes equations on u_1^ν then give at order ν^0

$$\begin{aligned}
&\partial_t\big(u_1^{\mathrm{int},0}(t,x,0)+u_1^{\mathrm{b},0}(t,x,Y)\big)\\
&\quad+\big(u_1^{\mathrm{int},0}(t,x,0)+u_1^{\mathrm{b},0}(t,x,Y)\big)\,\partial_x\big(u_1^{\mathrm{int},0}(t,x,0)+u_1^{\mathrm{b},0}(t,x,Y)\big)\\
&\quad+\big(Y\,\partial_y u_2^{\mathrm{int},0}(t,x,0)+u_2^{\mathrm{b},1}(t,x,Y)\big)\,\partial_Y\big(u_1^{\mathrm{int},0}(t,x,0)+u_1^{\mathrm{b},0}(t,x,Y)\big)\\
&\quad-\partial_{YY}u_1^{\mathrm{b},0}+\partial_x p^{\mathrm{int},0}(t,x,0)+\partial_x p^{\mathrm{b},0}(t,x,Y)=0
\end{aligned}$$

and Navier–Stokes equations on u_2^ν at order $\nu^{-1/4}$ give $\partial_Y p^{\mathrm{b},0}=0$, that is $p^{\mathrm{b},0}\equiv 0$ (as usual, the pressure does not vary in the boundary layer). Using incompressibility condition we get

$$\partial_x u_1^{\mathrm{b},0}+\partial_Y u_2^{\mathrm{b},1}=0. \tag{131}$$

The equations on $u_1^{\mathrm{b},0}$ can be clarified if we introduce

$$U_1(t,x,Y)=u_1^{\mathrm{int},0}(t,x,0)+u_1^{\mathrm{b},0}(t,x,Y)$$

and

$$U_2(t,x,Y)=Y\,\partial_y u_2^{\mathrm{int},0}(t,x,0)+u_2^{\mathrm{b},1}(t,x,Y).$$

Using (131) we get the classical Prandtl's equations

$$\partial_t U_1+U_1\,\partial_x U_1+U_2\,\partial_Y U_1-\partial_{YY}U_1=f(t,x), \tag{132}$$

$$\partial_x U_1+\partial_Y U_2=0, \tag{133}$$

$$U_1=U_2=0 \quad \text{for } Y=0, \tag{134}$$

where $f(t,x)=-\partial_x p^{\mathrm{int},0}(t,x,0)$ is already known (pressure gradient of the Euler flow $u^{\mathrm{int},0}$).

3.2. *Existence results for Prandtl's equations*

Existence theory for Prandtl's equations is difficult and mainly open, because of the lack of knowledge on U_2. More precisely, U_2 is recovered from U_1 by solving (133)

$$U_2(t,x,Y)=-\int_0^Y \partial_x U_1(t,x,Z)\,\mathrm{d}Z, \tag{135}$$

but in doing this we loose one x derivative (which is of course not recovered by the Y integration). Therefore we loose here one derivative and we cannot close Sobolev-type estimates.

There are mainly three types of results on Prandtl equations: existence in the analytical case [2,58], existence under monotonicity assumption [54], blow up results [20].

3.2.1. *Analytic case.* Let us first describe the existence result in the analytic case. Let $\Sigma_{\sigma,\rho}$ be the complex domain

$$\Sigma_{\sigma,\rho} = \{(x, y) \in \mathbb{C}^2 \mid |\mathcal{I}m\, x| \leqslant \sigma, |\mathcal{I}m\, y| \leqslant \rho\theta(\mathcal{R}e\, y)\},$$

where $\theta(y) = y$ for $y \leqslant 1$ and $\theta(y) = 1$ for $y \geqslant 1$.

THEOREM 3.1 (Existence in the analytic case [2,58]). *Let $\sigma > 0$ and $\rho > 0$ and U_1^0, U_2^0 be two holomorphic functions in $\Sigma_{\sigma,\rho}$. Then there exists $T > 0$ and a solution (U_1, U_2) of Prandtl equations on $0 \leqslant t \leqslant T$ with initial data (U_1^0, U_2^0).*

SKETCH OF THE PROOF. The proof follows the philosophy of the standard Cauchy–Kowalewsky theorem (see [8] for a nice presentation). The idea is to recover the loss of one derivative of (135) by using the analyticity. For holomorphic functions we can bound derivatives in terms of the function by Cauchy's formula. Therefore we can bound U_2 in terms of U_1. The price to pay is that the domain of analyticity shrinks as time goes on. More precisely, we look for $U_1(t,\cdot)$ and $U_2(t,\cdot)$ in $\Sigma_{\sigma-\alpha t,\rho-\beta t}$, where α and β are large enough. This is sufficient to get existence in small time. We refer to [58] for the details. □

3.2.2. *Monotonic case.* The second main existence theorem deals with monotonic function and has been proved by Oleinik in [54]. She obtains, locally in time and space, existence of smooth solution for monotonic profiles.

3.2.3. *A blow up result.* In [20], Engquist and Engquist prove blow up in finite time of smooth solutions of Prandtl equations for particular initial data. More precisely, they consider initial data of the form

$$u_1(0, x, y) = -x b_0(x, y).$$

They prove that under certain conditions on b_0 the solution of Prandtl equation, if it exists, blows up in finite time.

3.3. *Underlying instabilities*

In fact boundary layer of Prandtl's type are highly unstable, and the asymptotic expansion (130) does not describe at all what happens near the boundary. To see this let us take a shear layer profile $(u_s(Y), 0)$ which is linearly unstable for Euler equations and such that $u_s(0) = 0$ (take for instance $u_s(Y) = \sin(Y)\exp(-Y)$). From the analysis of the stability of shear layers of Navier–Stokes equations we already know that there exists a solution of linearized Navier–Stokes equations of the form $v \exp(ct)$, where $\mathcal{R}e\, c \sim C_0/\sqrt{\nu}$ as $\nu \to 0$, namely that the shear layer is unstable on time scales of order $\sqrt{\nu}$. This instability leads to the following result:

THEOREM 3.2 (Instability in strong norm [32]). *Let $(u_s(Y), 0)$ be a smooth shear layer profile, spectrally unstable for Euler equations. Let $u_s(t, Y)$ be the solution of the heat equation*

$$\partial_t u_s - \partial_{YY} u_s = 0, \tag{136}$$

$$u_s(t, 0) = 0 \quad \forall t \geqslant 0, \tag{137}$$

with initial data $u_s(Y)$. Then, for every s and N arbitrarily large, there exist $\sigma > 0$, $C > 0$, a family of times T^ν and a family of solutions u^ν of Navier–Stokes with viscosity ν, such that

$$\left\| u^\nu(0, \cdot) - \left(u_s(Y), 0\right) \right\|_{H^s} \leqslant \varepsilon^N, \tag{138}$$

$$\left\| u^\nu(T^\nu, \cdot) - \left(u_s(T^\nu, Y), 0\right) \right\|_{H^2} \to +\infty, \tag{139}$$

$$\left\| u^\nu(T^\nu, \cdot) - \left(u_s(T^\nu, Y), 0\right) \right\|_{L^\infty} \geqslant C\nu^{1/4}, \tag{140}$$

$$\left\| u^\nu(T^\nu, \cdot) - \left(u_s(T^\nu, Y), 0\right) \right\|_{L^2} \geqslant C\nu^{3/8} \tag{141}$$

as $\nu \to 0$, with

$$T^\nu \leqslant C\nu^{1/4} \log \nu^{-1}.$$

Let us comment the result before going on with the proof. Note that Prandtl equation on a shear layer reduces to the heat equation (136) and (137). The theorem in particular says that $(t, x, Y) \to u^\nu(t, x, \sqrt{\nu} Y)$ will not converge in $L^\infty(H^2)$ norm to a solution of Prandtl. More generally it says that we cannot hope (130) to be true in H^s with $s \geqslant 2$. We cannot even have expansions of the form

$$u^\nu = u^{\mathrm{E}} + u^{\mathrm{P}} + O_{L^\infty}\left(\sqrt{\nu}\right),$$

where u^{E} is a solution of Euler equation and u^{P} of Prandtl equation (assuming that such a solution exists). More generally, we cannot describe what happens in the boundary layer in H^2 or in L^∞ with precision $\nu^{1/4}$.

Up to now we are not able to replace $\nu^{1/4}$ and $\nu^{3/8}$ by ν^0 in (140) and (141) since as $t \to T^\nu$ a second boundary layer of size $\nu^{3/4}$ is created near the boundary. Large gradients appear in this sublayer and we can no longer make an energy estimate.

SKETCH OF PROOF OF THEOREM 3.2. The first step is to make a change of variables: $x \to x/\sqrt{\nu}$, $y \to y/\sqrt{\nu}$ and $t \to \sqrt{\nu}$. The idea is then to start from a shear profile which has an inflexion point and is therefore unstable for Euler equations and to start with a perturbation which is the most unstable linear mode for Euler equations for this shear profile. We build an approximate solution for Navier–Stokes equations, taking in particular into account a boundary layer (in the new variables, which means a sublayer in the original x and y variables) of size $\nu^{1/4}$ which appears near $y = 0$ ($\nu^{3/4}$ in the original variables). We end the proof with an L^2 estimate. This estimate stops when the gradients of the approximate solution become large and go to infinity. The supremum of the gradient is attained in the sublayer, and is of order one as the approximate solution is of order $\nu^{1/4}$ (since the boundary layer is of size $\nu^{1/4}$). When the proof stops, the approximate solution is only of order $\nu^{1/4}$ in supremum norm. Going back to the original variables however, it goes to infinity in H^2 norm. □

3.4. *Inviscid limit in the analytic case*

Note that in Theorem 3.2 the boundary layer is destabilized by perturbations of wave number of order $\mathrm{O}(\nu^{-1/2})$. Such perturbations grow like $\exp(CT\nu^{-1/2})$. If we allow initial perturbations with Sobolev (H^s) regularity, the energy with wave numbers of order $\mathrm{O}(\nu^{-1/2})$ is initially of order $\mathrm{O}(\nu^{-s/2})$, and therefore later on, of size

$$\mathrm{O}\big(\nu^{-s/2}\mathrm{e}^{CT\nu^{-1/2}}\big),$$

which is of order $\mathrm{O}(1)$ for T of order $C'\nu^{1/2}\log\nu^{-1}$.

However if we take initial perturbations with *analytic* regularity, the energy of wave numbers of order $\mathrm{O}(\nu^{-1/2})$ is of order $\exp(-C_0\nu^{-1/2})$ for some constant C_0. Later on this energy is of order

$$\exp\big(-C_0\nu^{-1/2} + CT\nu^{-1/2}\big)$$

which is bounded for T bounded, and small for small T. It is therefore possible to get a convergence result for small time and for analytic initial data. This has been proved by Asano [2], Caflish and Sammartino [58] for analytic initial data, and improved by Cannone [10] for initial data with tangential analytic regularity (and Sobolev regularity in the normal variables). Roughly speaking (with no precise description of the underlying spaces) the following result holds:

THEOREM 3.3 (Convergence for analytic initial data [2,58]). *Let u_0^{E} be an analytic initial data. Then there exists $T > 0$, a solution u^{E} of Euler equation with initial data u_0^{E}, a solution u^{P} of Prandtl equations, and a family of solution u^ν of Navier–Stokes equations, such that*

$$u^\nu = u^{\mathrm{E}} + u^{\mathrm{P}} + \mathrm{o}(1)$$

as ν goes to 0.

Of course the physical instabilities are completely ruled out in this case. Moreover, it is not possible to get global convergence results ($T = +\infty$) even if the limit solution u^{E} is globally analytic.

3.5. *Structure of the viscous layer*

The main point is that a viscous sublayer of size $\mathrm{O}(\nu^{1/4})$ appears near the boundary. This layer creates large gradients which are the main difficulty to get instability results in L^∞. Namely it is possible to construct a Reynolds number with this length scale: the Reynolds number of the sublayer is of order $\mathrm{O}(\nu^{3/4})$ and therefore goes to $+\infty$ as ν goes to 0. This means that the sublayer itself will in turn become instable and get destabilized by some linearly growing mode. As for the original layer of size $\mathrm{O}(\nu^{1/2})$ it will create a sub-sublayer of size $\mathrm{O}(\nu^{7/8})$ and so on. As the Reynolds number increases, more and more sublayers appear. The process stops when the last created sublayer is of size $\mathrm{O}(\nu)$ where the Reynolds number is always $\mathrm{O}(1)$ and remains bounded.

Of course this construction is a purely mathematical one, since these successive layers appear at very large Reynolds numbers, beyond any possible physical experience. However this cascade turns out to be deeply linked with the following beautiful result of Kato [40].

THEOREM 3.4 (Conditional convergence [40]). *Let u^ν be a sequence of solutions of Navier–Stokes equations on some time interval* $[0, T]$. *Let u^{E} be a smooth solution of Euler equations on* $[0, T]$ *such that* $u^\nu(0) \to u^{\mathrm{E}}(0)$ *in* $L^2(\Omega)$. *Then*

$$u^\nu \to u^{\mathrm{E}} \quad \text{in } L^2\big([0, T] \times \Omega\big) \tag{142}$$

if and only if, for every $C > 0$,

$$\nu \int_0^T \int_{d(x,\Omega) \leqslant C\nu} \|\nabla u^\nu\|^2 \to 0 \tag{143}$$

as $\nu \to 0$.

Therefore convergence from Navier–Stokes to Euler is equivalent to the vanishing of the dissipation by viscosity in a strip of size $C\nu$ near the boundary. The same size of layer $\mathrm{O}(\nu)$ appears. Note that Prandtl size $\mathrm{O}(\sqrt{\nu})$ does not enter here.

SKETCH OF PROOF OF THEOREM 3.4. First (142) implies (143) simply using the energy estimates of Navier–Stokes equations, and the conservation of energy for Euler. The converse is obtained by constructing a "fake boundary layer". Let us work in dimension two to fix the ideas. Let Ψ^{E} be the stream function of u^{E}. Let ϕ be a smooth function such that $\phi(0) = 0$ and $\phi(x) = 1$ for $x > 1$. Then we consider

$$\Psi^\nu(x) = \Psi^{\mathrm{E}}(x)\phi\left(\frac{d(x, \partial\Omega)}{\varepsilon}\right)$$

and $v^\nu = \nabla^\perp \Psi^\nu$. Then v^ν vanishes on $\partial\Omega$. An energy estimates on $v^\nu - u^\nu$ ends the proof of Theorem 3.4 under the assumption (143). Note that the boundary layer introduced has nothing to do with Prandtl equation, and is completely arbitrary. □

Physically, it is usual to consider that Prandtl boundary layer is unstable, leads to turbulent motions near the wall. Very close to the wall (distances of order $O(\nu)$), a so-called viscous sublayer appears, where the viscosity dominates the flow. At larger distances, the turbulent behavior is described by log laws.

3.6. *Miscellaneous*

3.6.1. *Other boundary conditions.* Dirichlet conditions are the harder one for Navier–Stokes equations. Other conditions (Navier, inflow) lead to boundary layers of size $O(\nu)$ which are stable under smallness conditions. We refer to [69] and [13] for more details.

3.6.2. *Some physics.* The aim of this subsection is to describe in an informal way the flow created by the motion of a cylinder with constant speed in a fluid at rest.

For very small Reynolds numbers, the flow is completely laminar around the cylinder, with no recirculation and no instability.

For larger Reynolds numbers, a recirculation zone appears behind the obstacle. More precisely, two eddies with opposite signs appear just behind the cylinder and remain close to it.

For larger Reynolds numbers, these eddies are unstable and successively leave the cylinder to propagate in the wake, to form the so called "Von Karman" vortex street.

The situation is then the following: near the front of the obstacle, the boundary layer is stable and laminar. It is well approximated by Prandtl equations (the viscosity is very small, though not zero). The boundary layer splits from the obstacle at a distance well predicted by Prandtl equations. After the splitting the situation is chaotic and the wake may be turbulent. This makes the interest of the Prandtl layers. They do not describe in a good manner the inviscid limit of Navier–Stokes equations, but describe in a pretty good manner a certain domain of parameters, where the viscosity is small, but "not too small". They are not limit of Navier–Stokes equations, but are close to them for a given range of viscosity.

For larger Reynolds number, the boundary layer is laminar near the front of the obstacle, then it becomes turbulent (this is precisely this turbulent behavior which prevents the convergence of Navier–Stokes equations to Prandtl equations), and then it splits from the boundary. The wake is completely turbulent. The main feature of this domain of parameters is the "drag crisis". There exists a critical Reynolds number where the drag force needed to maintain the motion of the cylinder is discontinuous (with respect to the Reynolds number) and is reduced by almost a factor 2. This is linked to the turbulence of the boundary layer where the drag is much less than in a laminar boundary layer.

Note in particular that the convergence of the drag force is a physically (and industrially!) much more relevant question than the convergence from Navier–Stokes to Prandtl or to Euler equations.

4. Ekman's type boundary layers

4.1. *Introduction*

Prandtl's boundary layer is not the only physically interesting boundary layer, and the aim of this section is to describe other types of viscous boundary layers, which appear in particular in Meteorology and in MHD, the basic example being the Ekmann layer. Namely in meteorology or oceanography, with respect to the time scales and lengthscales studied, we cannot neglect the rotation of the Earth, that is the Coriolis force. All the main models are based on the action of the Coriolis force and its balance with the pressure (so called "geostrophic equilibrium"), and some fundamentals models (like quasigeostrophic equations in a bounded domain, or wind driven oceanic circulation) include the effect of the boundary layers near the top and the bottom of the domain, created by the combined action of the rotation, the pressure and the viscous terms. These boundary layers are mainly Ekmann layers that we will now describe precisely. It turns out that Ekmann layers are stable for Reynolds numbers smaller than a critical Reynolds number (of order 54) and unstable for larger Reynolds numbers.

The study of these layers goes back to Ekmann himself [18] and we refer to the monographs of Greenspan [30], Pedlosky [55] and Gill [29] for a physical introduction.

Let us begin by the following model problem: let $\Omega = \mathbb{T}^2 \times [0, 1]$ (periodic setting in x, y, bounded in z) and let us consider the Navier–Stokes Coriolis equations

$$\partial_t u + \nabla(u \otimes u) - \nu \Delta u + \frac{e \times u}{\varepsilon} + \frac{\nabla p}{\varepsilon} = 0, \tag{144}$$

$$\nabla \cdot u = 0, \tag{145}$$

$$u = 0 \quad \text{on } z = 0, 1, \tag{146}$$

where $e = (0, 0, 1)$ is a fixed vector (in the z direction), $\nu > 0$ is the viscosity and $\varepsilon > 0$ is the Rossby number. Physically, ε is a small parameter, no so small in meteorology (of order 10^{-1} to 10^{-2}), much smaller in geomagnetism (10^{-5} and even smaller). Boundary conditions (146) can be replaced by "wind shear" type boundary conditions:

$$u_3 = 0, \qquad \partial_z(u_1, u_2) = \tau(x, y), \tag{147}$$

where τ is a given vector field which models the action of the wind over the free surface of the ocean [55].

Existence results for Navier–Stokes Coriolis system are straightforward since the energy estimates are exactly the same as for genuine Navier–Stokes equations. Leray's proof can then be adapted and we get weak solutions in $L^\infty(L^2) \cap L^2(H^1)$ for any initial data in L^2.

In order to get interesting results we have to link ε and ν, and from now on we assume that $\nu = \beta\varepsilon$ where β is some fixed constant.

4.2. *Formal analysis*

Let us now turn to the formal analysis of the limit $\varepsilon \to 0$. The ε^{-1} terms give the classical geostrophic balance

$$e \times u = -\nabla p.$$

Or, componentwise:

$$\partial_z p = 0, \qquad (u_1, u_2) = -\nabla^{\perp} p,$$

therefore the pressure does not depend on the third component z. This implies that u_1, u_2, the tangential velocity, only depends on t, x, y, and is divergence free. Moreover, u_3 is also constant in z by the divergence free condition. As we shall see later, in fact $u_3 \equiv 0$. In conclusion, formally, solutions u^{ε} of Navier–Stokes Coriolis system converge to a two-dimensional divergence free vector field (this is the famous Taylor–Proudman theorem).

This is not compatible with the boundary conditions (146), therefore boundary layers appear near $z = 0$ and $z = 1$ to ensure $u^{\varepsilon} = 0$ on $\partial\Omega$. In these layers we have equilibrium between Coriolis and viscous forces (the pressure force and the inertia do not appear here, as it will be clear below). If the layer is of size λ, the viscous forces are of order ν/λ^2 and the Coriolis force of order ε^{-1}, therefore to get a balance, we have to take $\lambda = \sqrt{\varepsilon\nu}$. Hence, the typical size of Ekmann layer is $\sqrt{\varepsilon\nu}$. Let

$$Z = \frac{z}{\sqrt{\varepsilon\nu}}.$$

There is also a boundary layer near $z = 1$ and we define

$$\widetilde{Z} = \frac{1 - z}{\sqrt{\varepsilon\nu}}.$$

We have to look for approximate solutions of the form

$$u^{\mathrm{app}}(t, x, y, z) \sim \sum_{j=0}^{+\infty} \sqrt{\varepsilon\nu}^{j} u^{\mathrm{int}, j}(t, x, y, z) + \sum_{j=0}^{+\infty} \sqrt{\varepsilon\nu}^{j} u^{\mathrm{b}, j}(t, x, y, Z)$$
$$+ \sum_{j=0}^{+\infty} \sqrt{\varepsilon\nu}^{j} u^{\mathrm{b}, j, \mathrm{top}}(t, x, y, \widehat{Z}), \tag{148}$$

where $u^{\mathrm{int}, j}$, $u^{\mathrm{b}, j}$ and $u^{\mathrm{b}, j, \mathrm{top}}$ are smooth functions, $u^{\mathrm{b}, j}$ and $u^{\mathrm{b}, j, \mathrm{top}}$ being, moreover, rapidly decreasing in the Z variables, with $u^{\mathrm{int}, j} + u^{\mathrm{b}, j} = 0$ for $z = 0$ and $u^{\mathrm{int}, j} + u^{\mathrm{b}, j, \mathrm{top}} = 0$ for $z = 1$.

To get the equations on $u^{\mathrm{int}, j}$, $u^{\mathrm{b}, j}$ and $u^{\mathrm{b}, j, \mathrm{top}}$ we put (148) in Navier–Stokes Coriolis equations and look at the terms in $\sqrt{\varepsilon\nu}^{j}$ for $j \geqslant -1$.

At order -1 in the boundary layer, (145) gives

$$\partial_Z u_3^{b,0} = 0,$$

which implies $u_3^{b,0} = 0$. As $u_3^{int,0}$ is independent on z this also gives $u_3^{int,0} = 0$. The third equation of (144) in the boundary layer gives $\partial_Z p^{b,0} = 0$, and therefore $p^{b,0} = 0$. As usual the pressure does not vary much in the boundary layer. The other equations of (144) give

$$-u_2^{b,0} = \partial_{ZZ} u_1^{b,0}, \qquad u_1^{b,0} = \partial_{ZZ} u_2^{b,0}$$

with boundary conditions

$$u_1^{b,0}(t,x,y,0) = -u_1^{int,0}(t,x,y,0), \qquad u_2^{b,0}(t,x,y,0) = -u_2^{int,0}(t,x,y,0),$$

and

$$u_1^{b,0}(t,x,y,Z) \to 0, \qquad u_2^{b,0}(t,x,y,Z) \to 0 \quad \text{as } Z \to +\infty.$$

Note that this is an ordinary differential equation, and that t, x, y are only parameters. The solution is given by

$$\begin{aligned} & u_1^{b,0}(t,x,y,Z) \\ & \quad = -e^{-Z/\sqrt{2}}\left(u_1^{int,0}(t,x,y)\cos\frac{Z}{\sqrt{2}} + u_2^{int,0}(t,x,y)\sin\frac{Z}{\sqrt{2}}\right), \end{aligned} \tag{149}$$

$$\begin{aligned} & u_2^{b,0}(t,x,y,Z) \\ & \quad = -e^{-Z/\sqrt{2}}\left(-u_1^{int,0}(t,x,y)\sin\frac{Z}{\sqrt{2}} + u_2^{int,0}(t,x,y)\cos\frac{Z}{\sqrt{2}}\right), \end{aligned} \tag{150}$$

and is called the "Ekmann spiral". Using (145) in the boundary layer at order 0 we have

$$\partial_x u_1^{b,0} + \partial_y u_2^{b,0} + \partial_Z u_3^{b,1} = 0,$$

and using (149) and (150) we get

$$u_3^{b,1}(t,x,y,Z) = e^{-Z/\sqrt{2}}\left(\partial_y u_2^{b,0} - \partial_x u_1^{b,0}\right)\sin\left(Z + \frac{\pi}{4}\right). \tag{151}$$

Now, equations (144) in the interior at order -1 again gives the geostrophic balance, and at order 0, with $u_h^{int,0} = (u_1^{int,0}, u_2^{int,0})$,

$$\partial_t u_h^{int,0} + (u_h^{int,0}\cdot\nabla)u_h^{int,0} + e \times u_h^{int,1} - \nu\Delta u_h^{int,0} + \nabla p^{int,0} = 0,$$

or taking the bi-dimensional curl $\omega^{\text{int},0} = \text{curl}_{2D}\, u_h^{\text{int},0}$,

$$\partial_t \omega^{\text{int},0} + \left(u_h^{\text{int},0} \cdot \nabla\right)\omega^{\text{int},0} - \nu \Delta \omega^{\text{int},0} = \partial_z u_3^{\text{int},1},$$

and integrating in z, keeping in mind that $\omega^{\text{int},0}$ is independent on z,

$$\partial_t \omega^{\text{int},0} + \left(u_h^{\text{int},0} \cdot \nabla\right)\omega^{\text{int},0} - \nu \Delta \omega^{\text{int},0} = u_3^{\text{int},1}(t,x,y,1) - u_3^{\text{int},1}(t,x,y,0).$$

But $u_3^{\text{int},1}(t,x,y,0) = -u_3^{\text{b},1}(t,x,y,0)$ which is already known, therefore

$$\partial_t \omega^{\text{int},0} + \left(u_h^{\text{int},0} \cdot \nabla\right)\omega^{\text{int},0} - \nu \Delta \omega^{\text{int},0} + \sqrt{\frac{\beta}{2}}\omega^{\text{int},0} = 0, \tag{152}$$

or equivalently,

$$\partial_t u^{\text{int},0} + \left(u^{\text{int},0} \cdot \nabla\right)u^{\text{int},0} - \nu \Delta u^{\text{int},0} + \sqrt{\frac{\beta}{2}}u^{\text{int},0} = 0. \tag{153}$$

Hence $u^{\text{int},0}$ satisfies a two-dimensional damped Euler equation ($u_3^{\text{int},0} = 0$). The term $\sqrt{\beta/2}u^{\text{int},0}$ is called the Ekmann pumping term and comes from the energy dissipation by the viscosity in the Ekmann boundary layer. In particular we have existence of a global smooth solution of (153) is the initial data is smooth (with the same regularity). It is possible to go on with the formal analysis and to construct the higher-order profiles for $j \geqslant 1$.

REMARKS.

- We have only done the formal analysis for initial data that in the interior does not depend on z (so-called "well prepared case"). The situation is slightly more complicated in the general case, where waves of high speed (called inertial waves) propagate through Ω (we refer to [31,12] for more details in this "ill prepared case").
- If $\varepsilon \ll \nu$ then "$\beta = +\infty$" and (153) reduces to $u^{\text{int},0} = 0$: the flow is immediately damped to rest by the boundary layers. On the contrary if $\varepsilon \gg \nu$, "$\beta = 0$" and the pumping term disappears. Therefore the only interesting regime is ν of order ε.
- It is usual in meteorology to replace the classical viscosity $-\nu\Delta$ by a "turbulent viscosity" of the form $-\nu_{\text{h}}\Delta_{\text{h}} - \nu_{\text{v}}\partial_{zz}$. The motivation is that the flow is in fact turbulent, which leads to enhanced energy dissipation, and that it is strongly anisotropic, the vertical direction being different from the horizontal directions (vertical motions are very difficult by Taylor–Proudman theorem). Therefore meteorologists often split the viscous term in an horizontal one and a vertical one, with $\nu_{\text{v}} \ll \nu_{\text{h}}$. Typically, ν_{v} is of order ε and ν_{h} is much larger (it can be of order $O(1)$). There is of course no rigorous justification of the "turbulent viscosity".

4.3. *Linear stability*

Let us introduce the Reynolds number associated with Ekmann layers. A Reynolds number is obtained by multiplying a velocity by typical length and then dividing by a viscosity. Here we take $\sqrt{\varepsilon\nu}$ as typical size and $\|u^{\mathrm{int},0}\|_{L^\infty}$ as typical velocity, which gives

$$R = \frac{\|u^{\mathrm{int},0}\|_{L^\infty}\sqrt{\varepsilon\nu}}{\nu} = \|u^{\mathrm{int},0}\|_{L^\infty}\sqrt{\frac{\varepsilon}{\nu}} = \|u^{\mathrm{int},0}\|_{L^\infty}\sqrt{\beta}. \tag{154}$$

Let us consider a pure Ekmann layer of the form

$$u^{\mathrm{E}}(t,x,y,Z) = \mathrm{e}^{-Z/\sqrt{2}}U_\infty\left(\cos\frac{Z}{\sqrt{2}}, -\sin\frac{Z}{\sqrt{2}}, 0\right), \tag{155}$$

where U_∞ is the velocity away from the layer. The corresponding Reynolds number is

$$R = U_\infty\sqrt{\frac{\varepsilon}{\nu}}.$$

Let us now consider Navier–Stokes Coriolis equations linearized around u^{E}

$$\partial_t u + (u_{\mathrm{E}}\cdot\nabla)u + (u\cdot\nabla)u_{\mathrm{E}} - \nu\Delta u + \frac{e\times u}{\varepsilon} + \frac{\nabla q}{\varepsilon} = 0, \tag{156}$$

$$\nabla\cdot u = 0, \tag{157}$$

Then there exists a critical Reynolds number R_{c} such that 0 is a stable solution of (156) and (157) if and only if $R < R_{\mathrm{c}}$. For $R > R_{\mathrm{c}}$ there exists an unstable, exponentially increasing solution of (156) and (157). Up to now there is no complete proof of this fact, and we can only prove the existence of two Reynolds numbers R_1 and R_2 such that (156) and (157) is spectrally stable for $R < R_1$ (using energy estimates) and spectrally unstable for $R > R_2$ (using a perturbative analysis starting from Rayleigh operator). However it is very easy to compute numerically the spectrum of (156) and (157), and we will detail this point now.

As for Euler or Navier–Stokes equations, we take the Fourier transform in the (x, y) plane, introduce a wave number $k \in \mathbb{R}^2$ and look for a solution v of (156) and (157) of the form $\exp(ik\cdot(x,y) - ikct)v_0(z)$, where v_0 is a vector valued function. Note the k factor in $ikct$ which is traditional in Fluid Mechanics. To simplify the equations we first make a change of variables in the (x, y) plane and take $(-k^\perp/\|k\|, k/\|k\|)$ as a new frame, and $(\tilde{x}, \tilde{y})$ as new coordinates. In these new coordinates, k is co-linear with the second vector of the basis. This change is equivalent to a rotation in u^{E} which becomes (dropping the tildes for convenience)

$$u^{\mathrm{E}}(t,x,y,z) = -U_\infty\begin{pmatrix}\cos\gamma - \mathrm{e}^{-Z/\sqrt{2}}\cos(\frac{Z}{\sqrt{2}}+\gamma)\\ -\sin\gamma + \mathrm{e}^{-Z/\sqrt{2}}\sin(\frac{Z}{\sqrt{2}}+\gamma)\\ 0\end{pmatrix}, \tag{158}$$

where γ is the angle of rotation of the frame. Now the v is of the form $\exp(iky - ikct)v_1(z)$ where $k \in \mathbb{R}$. However, as v is independent on x, we can introduce a stream function Ψ and look for v of the form

$$v(t,x,y,Z) = \exp(iky - ikct)\begin{pmatrix} U(Z) \\ \Psi'(Z) \\ -\mathrm{i}k\Psi(Z) \end{pmatrix}. \tag{159}$$

System (156) and (157) then reduces to the following four-by-four system on the two functions (U, Ψ)

$$\partial_Z^2 U - k^2 U + 2\,\partial_Z \Psi = \mathrm{i}kR\big((v_l - c)U - \Psi\,\partial_Z u_l\big), \tag{160}$$

$$\big(\partial_Z^2 - k^2\big)^2 \Psi - \mathrm{i}kR\big((v_l - c)\big(\partial_Z^2 - k^2\big)\Psi - \Psi\,\partial_Z^2 v_l\big) - 2\,\partial_Z U = 0, \tag{161}$$

where

$$u_l = \cos\tilde{\gamma} - \exp\left(-Z\sqrt{\frac{\beta}{2}}\right)\cos\left(\tilde{\gamma} + Z\sqrt{\frac{\beta}{2}}\right)$$

and

$$v_l = -\left(\sin\tilde{\gamma} - \exp\left(-Z\sqrt{\frac{\beta}{2}}\right)\sin\left(\tilde{\gamma} + Z\sqrt{\frac{\beta}{2}}\right)\right),$$

$\tilde{\gamma}$ being an angle between the direction of the flow outside the boundary layer and the direction of k, with boundary conditions

$$U(0) = 0, \qquad \Psi(0) = \partial_Z \Psi(0) = 0$$

on $Z = 0$ and

$$\partial_Z U = \partial_Z^2 \Psi = 0$$

at infinity.

As for Euler and Navier–Stokes equations we can introduce the Ekmann operator defined by

$$\mathrm{Eck}_{k,R}\begin{pmatrix} U \\ \omega \end{pmatrix} = \begin{pmatrix} v_l U - \Psi\,\partial_Z u_l - (\mathrm{i}kR)^{-1}(\partial_Z^2 U - k^2 U + 2\,\partial_Z \Psi) \\ v_l \omega - \psi\,\partial_Z^2 v_l - (\mathrm{i}kR)^{-1}((\partial_Z^2 - k^2)\omega - 2\,\partial_Z U) \end{pmatrix},$$

where as usual $(\partial_Z^2 - k^2)\Psi = \omega$. Note that as $R \to +\infty$,

$$\mathrm{Eck}_{k,R}\begin{pmatrix} U \\ \omega \end{pmatrix} \to \begin{pmatrix} v_l U - \Psi\,\partial_Z u_l \\ v_l \omega - \psi\,\partial_Z^2 v_l \end{pmatrix},$$

the ω part being exactly Rayleigh operator. As v_l as many inflection points, it is not surprising that the corresponding Rayleigh operator is unstable. Making a perturbation analysis as $R \to +\infty$ we see that $\mathrm{Eck}_{k,R}$ has unstable eigenvalues for suitable k and for R large enough.

Numerically it is simple to solve the eigenvalue problem (160) and (161), simply by using finite elements to discretize the derivatives, and brute force to compute the spectrum of the matrix thus obtained. It has been done by Lilly [45] who finds that the profile u_{E} is stable for $R < R_{\mathrm{c}} \sim 54.2$ and unstable for $R > R_{\mathrm{c}}$.

4.4. *Justification for small Reynolds numbers*

This section is devoted to the justification of the formal analysis of section in the case of small Reynolds numbers, using energy methods. More precisely, we will prove the theorem.

THEOREM 4.1 ([36]). *Let $u(0) \in H^4(\mathbb{T}^2)$ and let $u(t)$ be the corresponding global strong solution of the two-dimensional damped Euler equations* (153). *Let $u^\varepsilon(0)$ be a sequence of initial data bounded in L^2 and let $u^\varepsilon(t)$ be the corresponding sequence of weak solutions of* (144)–(146). *If*

$$\left\| u^\varepsilon(0) - u(0) \right\|_{L^2} \to 0,$$

then

$$\sup_{t \in [0,T]} \left\| u^\varepsilon(t) - u(t) \right\|_{L^2} \to 0$$

for every T such that

$$\sup_{0 \leqslant t \leqslant T} R_{\mathrm{BL}}(t) \leqslant R_1, \tag{162}$$

where

$$R_{\mathrm{BL}}(t) = \left\| u(t) \right\|_{L^\infty} \sqrt{\frac{\nu}{\varepsilon}}.$$

PROOF. We will in fact prove a slightly different result, which is typical of the justification of asymptotic equations involving boundary layers: we shall assume that $u(0) \in H^s(\mathbb{T})$ for every s and prove (148). Namely, starting from $u(0)$ it is possible, following Section 4.2 to construct an approximate solution of the form (148), such that if we truncate the right-hand side of (148) to $j \leqslant N$ for some large integer N, the sum v^{app} thus obtained satisfies (144) up to small error terms. However we have to face a small technical problem, since v^{app} does not vanish on $\partial\Omega$ because of exponential tails of the boundary layer terms, moreover, it is not divergence free, but its divergence is of order ε^N. To correct these two points we add a vector field $\tilde{v}$ which is of order ε^N in L^2 and goes to 0 in every H^s, that lifts the boundary conditions and has the right divergence. Set now $u^{\mathrm{app},\varepsilon} = v^{\mathrm{app}} + \tilde{v}$ which satisfies (145) and (146) and

$$\partial_t u^{\mathrm{app},\varepsilon} + (u^{\mathrm{app},\varepsilon} \cdot \nabla) u^{\mathrm{app},\varepsilon} + \frac{e \times u^{\mathrm{app},\varepsilon}}{\varepsilon} - \nu \Delta u^{\mathrm{app},\varepsilon} + \nabla p^{\mathrm{app},\varepsilon} = R^{\mathrm{app},\varepsilon},$$

where

$$\sup_{[0,T]} \|R^{\mathrm{app},\varepsilon}\|_{L^2} \leqslant C\varepsilon^{N-1}.$$

Set now $v^\varepsilon = u^\varepsilon - u^{\mathrm{app},\varepsilon}$ which satisfies

$$\begin{aligned} &\partial_t v^\varepsilon + (u^{\mathrm{app},\varepsilon} \cdot \nabla) v^\varepsilon + (v^\varepsilon \cdot \nabla) u^{\mathrm{app},\varepsilon} \\ &\quad + (v^\varepsilon \cdot \nabla) v^\varepsilon + \frac{e \times v^\varepsilon}{\varepsilon} - \nu \Delta v^\varepsilon + \nabla p^\varepsilon = -R^{\mathrm{app},\varepsilon}, \end{aligned} \tag{163}$$

$$\nabla \cdot v^\varepsilon = 0, \tag{164}$$

$$v^\varepsilon = 0 \quad \text{for } z = 0, 1. \tag{165}$$

A standard L^2 estimates gives

$$\partial_t \int \frac{|v^\varepsilon|^2}{2} + \nu \int |\nabla v^\varepsilon|^2 + \int v^\varepsilon (v^\varepsilon \cdot \nabla) u^{\mathrm{app},\varepsilon} = -\int v^\varepsilon R^{\mathrm{app},\varepsilon} \tag{166}$$

using divergence free condition on v^ε and $u^{\mathrm{app},\varepsilon}$. We have

$$\int v^\varepsilon (v^\varepsilon \cdot \nabla) u^{\mathrm{app},\varepsilon} = \int v^\varepsilon (v_h^\varepsilon \cdot \nabla_h) u^{\mathrm{app},\varepsilon} + \int v^\varepsilon (v_3^\varepsilon \cdot \partial_z) u^{\mathrm{app},\varepsilon}.$$

The first integral of the right-hand side is bounded by

$$\|\nabla_h u^{\mathrm{app},\varepsilon}\|_{L^\infty} \|v^\varepsilon\|^2$$

and $\|\nabla_h u^{\mathrm{app},\varepsilon}\|_{L^\infty}$ is bounded uniformly in ε. The second integral equals

$$\int v_h^\varepsilon (v_3^\varepsilon \cdot \partial_z) u_h^{\mathrm{app},\varepsilon} + \int v_3^\varepsilon (v_3^\varepsilon \cdot \partial_z) u_3^{\mathrm{app},\varepsilon}.$$

The second integral is bounded by

$$\|\partial_z u_3^{\mathrm{app},\varepsilon}\|_{L^\infty} \|v^\varepsilon\|^2$$

and $\|\partial_z u_3^{\mathrm{app},\varepsilon}\|_{L^\infty}$ is bounded uniformly in ε. In the last integral we split $u_h^{\mathrm{app},\varepsilon}$ into $\tilde{u} = u_h^{\mathrm{b},0} + u^{b,0,\mathrm{top}}$ and $u^{\mathrm{app},\varepsilon} - \tilde{u}$. The part involving $u^{\mathrm{app},\varepsilon} - \tilde{u}$ is bounded by $C\|v^\varepsilon\|_{L^2}^2$ and the other part

$$\int v_3^\varepsilon v_h^\varepsilon \cdot \partial_z \tilde{u} = \int \left(\int_0^z \partial_z v_3^\varepsilon(t,x,y,Z)\,\mathrm{d}Z\right)\left(\int_0^z \partial_z v_h^\varepsilon(t,x,y,Z)\,\mathrm{d}Z\right)\partial_z \tilde{u}$$

which is bounded by

$$\begin{aligned}&\int \left(\int_0^1 \left|\partial_z v_3^\varepsilon\right|^2 \mathrm{d}Z\right)^{1/2} \left(\int_0^1 \left|\partial_z v_h^\varepsilon\right|^2 \mathrm{d}Z\right)^{1/2} z\left|\partial_z \tilde{u}\right| \mathrm{d}x\,\mathrm{d}y\,\mathrm{d}z\\ &\quad\leqslant \left\|\partial_z v_3^\varepsilon\right\|_{L^2} \left\|\partial_z v_h^\varepsilon\right\|_{L^2} \sup_{(x,y)\in\mathbb{T}^2} \int z\left|\partial_z \tilde{u}\right| \mathrm{d}z.\end{aligned}$$

Using the explicit expression of $\tilde{u}$ given by (149) and (150), this is bounded by

$$C_0\sqrt{\varepsilon\nu}\left\|\partial_z v_3^\varepsilon\right\|_{L^2}\left\|\partial_z v_h^\varepsilon\right\|_{L^2}\left\|u^{\mathrm{int},0}\right\|_{L^\infty}.$$

Therefore (167) can be rewritten

$$\begin{aligned}&\partial_t \frac{\|v^\varepsilon\|_{L^2}^2}{2} + \nu \int \left|\nabla v^\varepsilon\right|^2\\ &\quad\leqslant C\varepsilon^{N-1}\left\|v^\varepsilon\right\|_{L^2} + C\left\|v^\varepsilon\right\|_{L^2}^2 + C_0\sqrt{\varepsilon\nu}\left\|\partial_z v_3^\varepsilon\right\|_{L^2}\left\|\partial_z v_h^\varepsilon\right\|_{L^2}\left\|u^{\mathrm{int},0}\right\|_{L^\infty}. \qquad (167)\end{aligned}$$

Now if

$$C_0\sqrt{\varepsilon\nu}\left\|u^{\mathrm{int},0}\right\|_{L^\infty} \leqslant \nu$$

that is under assumption (162) we have

$$\partial_t \|v\|_{L^2}^2 \leqslant C\|v\|_{L^2}^2 + C\varepsilon^{2N-2}$$

which ends the proof. □

REMARKS. If we replace $-\nu\Delta$ by the turbulent viscosity $-\nu_h\Delta_h - \nu_v\partial_{zz}$ with ν_{v} or order ε and $\nu_h \gg \nu_v$ then Theorem 4.1 holds true without any size restriction (without (162)). Namely using divergence free condition,

$$C_0\sqrt{\varepsilon\nu_v}\left\|\partial_z v_3^\varepsilon\right\|_{L^2}\left\|\partial_z v_h^\varepsilon\right\|_{L^2}\left\|u^{\mathrm{int},0}\right\|_{L^\infty} = C_0\sqrt{\varepsilon\nu_{\mathrm{v}}}\left\|\partial_h v_h^\varepsilon\right\|_{L^2}\left\|\partial_z v_h^\varepsilon\right\|_{L^2}\left\|u^{\mathrm{int},0}\right\|_{L^\infty}$$

which can be absorbed in

$$\nu_h\left\|\nabla_h v^\varepsilon\right\|_{L^2}^2 + \nu_v\left\|\partial_z v^\varepsilon\right\|_{L^2}^2$$

without any size restriction since $\nu_h \gg \nu_v$. This is not surprising since the turbulent viscosity tries to model the turbulent phenomena, and in particular to handle in a rough way all the instability mechanisms. It is therefore natural that it also handles and models boundary layer instabilities.

4.5. *Instabilities at high Reynolds numbers*

For Reynolds numbers larger than the critical Reynolds number, the flow is linearly unstable. It is then possible to prove that the flow is also nonlinearly unstable. One way (among many others) to express this fact is to prove the nonlinear instability of approximate solutions of high order.

THEOREM 4.2 ([16]). *Let N be arbitrarily large. Let $u^{\mathrm{app},\varepsilon}$ be a sequence of approximate solutions of Navier–Stokes Coriolis equations of the form*

$$u^{\mathrm{app},\varepsilon} = \sum_{j=0}^{N} \varepsilon^j u^{\mathrm{int},j}(t,x,y,z) + \sum_{j=0}^{N} \varepsilon^j u^{\mathrm{b},j}(t,x,y,Z) + \sum_{j=0}^{N} \varepsilon^j u^{\mathrm{b},j,\mathrm{top}}\bigl(t,x,y,\widehat{Z}\bigr) \tag{168}$$

satisfying Navier–Stokes Coriolis equations up to $\mathrm{O}(\varepsilon^N)_{L^2}$ terms on some time interval $[0,T]$. Let

$$R(t) = \bigl\|u^{\mathrm{int},0}\bigr\|_{L^\infty}.$$

Assume that

$$R(0) > R_{\mathrm{c}}. \tag{169}$$

Then $u^{\mathrm{app},\varepsilon}$ is nonlinearly unstable in the following sense: for every s there exists a sequence u^ε of solution of Navier–Stokes Coriolis equations on $[0,T]$ such that

$$\bigl\|u^{\varepsilon,\mathrm{app}}(0) - u^\varepsilon(0)\bigr\|_{H^s} \leqslant \varepsilon^N, \tag{170}$$

and

$$\bigl\|u^{\varepsilon,\mathrm{app}}(T^\varepsilon) - u^\varepsilon(T^\varepsilon)\bigr\|_{L^\infty} > C_0 \tag{171}$$

for some constant C_0 where $T^\varepsilon \to 0$ as $\varepsilon \to 0$.

SKETCH OF THE PROOF. The main idea is to construct a localized instability at the most unstable point of the boundary layer. Careful localization arguments together with energy estimates then end the proof. □

REMARKS.

- Note that between Theorems 4.1 and 4.2 there is a wide range of Reynolds numbers which is not treated (the range 4–54.2). In this range, Ekman layers are linearly stable and it is very likely that they are also nonlinearly stable. However this cannot be proved by energy estimates, since the L^2 norm of an arbitrary perturbation may grow first, before it decays and goes to 0.
- Despite Theorem 4.2 it is possible that the limit of Navier–Stokes Coriolis equations is Euler equation with a damping term (in L^2 sense). The boundary layer is not seen in energy norm, and it may be unstable without entering the core of the flow....
- For supercritical Reynolds numbers, very close to the critical Reynolds number, rolls appear in the boundary layer [39]. The analysis of their dynamics is widely open.

References

[1] V.I. Arnold and B.A. Khesin, *Topological Methods in Hydrodynamics*, Appl. Math. Sci., Vol. 125, Springer-Verlag, New York (1998).

[2] K. Asano, *Zero-viscosity limit of the incompressible Navier–Stokes equation, II, Mathematical analysis of fluid and plasma dynamics, I* (Kyoto, 1986), Sûrikaisekikenkyûsho Kôkyûroku **656** (1988), 105–128.

[3] A. Babin, A. Mahalov and B. Nicolaenko, *Global splitting, integrability and regularity of 3D Euler and Navier–Stokes equations for uniformly rotating fluids*, Preprint (1996).

[4] G.K. Batchelor, *An Introduction to Fluid Dynamics*, Cambridge Univ. Press (1967).

[5] T. Beale and A. Bourgeois, *Validity of the quasigeostrophic model for large scale flow in the atmosphere and ocean*, SIAM J. Math. Anal. **25** (1994), 1023–1068.

[6] Y. Brenier, *Homogeneous hydrostatic flows with convex velocity profiles*, Nonlinearity **12** (3) (1999), 495–512.

[7] F. Browder, *On the spectral theory of elliptic differential operators, I*, Math. Ann. **142** (1961), 22–130.

[8] R. Caflisch, *A simplified version of the abstract Cauchy–Kowalewski theorem with weak singularities*, Bull. Amer. Math. Soc. **23** (2) (1990), 495–500.

[9] M. Cannone, *Ondelettes, Paraproduits et Navier–Stokes*, Diderot, Paris (1995).

[10] M. Cannone, M.C. Lombardo and M. Sammartino, *Existence and uniqueness for the Prandtl equations*, C. R. Acad. Sci. Paris Sér. I Math. **332** (3) (2001), 277–282.

[11] J.-Y. Chemin, *A propos d'un problème de pénalisation de type antisymétrique*, C. R. Acad. Sci. Paris Sér. I Math. **321** (1996), 861–864.

[12] J.-Y. Chemin, B. Desjardins, I. Gallagher and E. Grenier, *Fluids with anisotropic viscosity*, Special issue for R. Temam's 60th birthday, Math. Model. Numer. Anal. **34** (2) (2000), 315–335.

[13] T. Clopeau, A. Mikelić and R. Robert, *On the vanishing viscosity limit for the 2D incompressible Navier–Stokes equations with the friction type boundary conditions*, Nonlinearity **11** (6) (1998), 1625–1636.

[14] T. Colin, *Remarks on an homogeneous model of ocean circulation*, Asympt. Anal. **12** (1996), 153–168.

[15] T. Colin and P. Fabrie, *Rotating fluid at high Rossby number driven by a surface stress: Existence and convergence*, Preprint (1996).

[16] B. Desjardins and E. Grenier, *Linear instability implies nonlinear instability for various types of boundary layers*, Ann. Inst. H. Poincaré, to appear.

[17] P.G. Drazin and W.H. Reid, *Hydrodynamic Stability*, Cambridge Monographs Mech. Appl. Math., Cambridge Univ. Press (1981).

[18] V.W. Ekman, *On the influence of the Earth's rotation on ocean currents*, Ark. Mat. Astron. Fys., Stockholm **2** (11) (1905).

[19] P. Embid and A. Majda, *Averaging over fast gravity waves for geophysical flows with arbitrary potential vorticity*, Comm. Partial Differential Equations **21** (1996), 619–658.

[20] W. Engquist and B. Engquist, *Blowup of solutions of the unsteady Prandtl's equation*, Comm. Pure Appl. Math. **50** (12) (1997), 1287–1293.

[21] R. Fjortoft, *Application of integral theorems in deriving criteria of stability for laminar flows and for the baroclinic circular vortex*, Geofys. Publ., Oslo **17** (1950), 1–52.
[22] S. Friedlander, *Lectures on stability and instability of an ideal fluid*, Hyperbolic Equations and Frequency Interactions, IAS/Park City Math. Ser. 5, Park City, UT (1995), 227–304.
[23] S. Friedlander, *On nonlinear instability and stability for stratified shear flow*, J. Math. Fluid Mech. **3** (1) (2001), 82–97.
[24] S. Friedlander and A. Shnirelman, *Instability of steady flows of an ideal incompressible fluid*, Mathematical Fluid Mechanics, Adv. Math. Fluid Mech., Birkhäuser, Basel (2001), 143–172.
[25] S. Friedlander, W. Strauss and M. Vishik, *Nonlinear instability in an ideal fluid*, Ann. Inst. H. Poincaré Anal. Non Linéaire **14** (2) (1997), 187–209.
[26] S. Friedlander, W. Strauss and M. Vishik, *Robustness of instability for the two-dimensional Euler equations*, SIAM J. Math. Anal. **30** (6) (1999), 1343–1354.
[27] S. Friedlander and M. Vishik, *Instability criteria for steady flows of a perfect fluid*, Chaos **2** (1992), 455–460.
[28] S. Friedlander and V. Yudovich, *Instabilities in fluid motion*, Notices Amer. Math. Soc. **46** (10) (1999), 1358–1367.
[29] A.E. Gill, *Atmosphere Ocean Dynamics*, Academic Press, New York (1982).
[30] H.P. Greenspan, *The theory of rotating fluids*, Reprint of the 1968 original Cambridge Monographs Mech. Appl. Math., Cambridge Univ. Press, Cambridge, NY (1980).
[31] E. Grenier, *Oscillatory perturbations of the Navier–Stokes equations*, J. Math. Pures Appl. **76** (9) (1997), 477–498.
[32] E. Grenier, *Non dérivation des équations de Prandtl*, Séminaire sur les Équations aux Dérivées Partielles, 1997–1998, Exp. No. XVIII, École Polytech., Palaiseau (1998).
[33] E. Grenier, *On the derivation of homogeneous hydrostatic equations*, Math. Model. Numer. Anal. **33** (5) (1999), 965–970.
[34] E. Grenier, *Limite quasineutre en dimension* 1, Journées "Équations aux Dérivées Partielles" (Saint-Jean-de-Monts, 1999), Exp. No. II, Univ. Nantes, Nantes (1999).
[35] E. Grenier, *On the nonlinear instability of Euler and Prandtl equations*, Comm. Pure Appl. Math. **53** (9) (2000), 1067–1091.
[36] E. Grenier and N. Masmoudi, *Ekman layers of rotating fluids, the case of well prepared initial data*, Comm. Partial Differential Equations **22** (5/6) (1997), 953–975.
[37] D. Henry, *Geometric Theory of Semilinear Parabolic Equations*, Lecture Notes in Math., Vol. 840, Springer-Verlag, Berlin–New York (1981).
[38] D.D. Holm, J.E. Marsden and T.S. Raţiu, *The Hamiltonian structure of continuum mechanics in material, inverse material, spatial and convective representations*, Hamiltonian Structure and Lyapunov Stability for Ideal Continuum Dynamics, Sém. Math. Sup., 100, Presses Univ. Montréal, Montreal, QP (1986), 11–124.
[39] G. Iooss, *Existence et stabilité de la solution périodique secondaire intervenant dans les problèmes d'évolution du type Navier–Stokes*, Arch. Ration. Mech. Anal. **47** (1972), 301–329.
[40] T. Kato, *Remarks on zero viscosity limit for nonstationary Navier–Stokes flows with boundary*, Seminar on Nonlinear Partial Differential Equations (Berkeley, CA, 1983), Math. Sci. Res. Inst. Publ., Vol. 2, Springer-Verlag, New York (1984), 85–98.
[41] H. Lamb, *Hydrodynamics*, Cambridge Univ. Press (1932).
[42] J. Leray, *Essai sur les mouvements plans d'un liquide visqueux que limitent des parois*, J. Math. Pures Appl. **13** (1934), 331–418.
[43] D. Levermore and M. Oliver, *Analyticity of solutions for a generalized Euler equation*, J. Differential Equations **133** (1997), 321–339.
[44] D. Levermore, M. Oliver and E. Titi, *Global well-posedness for models of shallow water in a basin with a varying bottom.*, Indiana Univ. Math. J. **45** (1996), 479–510.
[45] H. Lilly, *On the instability of the Ekman boundary layer*, J. Atmos. Sci. **23** (1966), 481–494.
[46] C.C. Lin, *The Theory of Hydrodynamic Stability*, Cambridge Univ. Press (1955).
[47] J.-L. Lions, *Quelques méthodes de résolution des problèmes aux limites non linéaires*, Dunod–Gauther-Villars, Paris (1969).
[48] J.-L. Lions, R. Temam and S. Wang, *Modèles et analyse mathématiques du système Océan/Atmosphère*, C. R. Acad. Sci. Paris Sér. I Math. **316** (1993), 113–119, C. R. Acad. Sci. Paris Sér. I Math. **316** (1993), 211–215, C. R. Acad. Sci. Paris Sér. I Math. **318** (1994), 1165–1171.

[49] P.-L. Lions, *Mathematical Topics in Fluid mechanics*, Vols 1 and 2, Oxford Lecture Ser. Math. Appl., Vol. 3, Oxford Science Publications, Clarendon Press, Oxford Univ. Press, NY (1996).

[50] A. Majda, *Compressible Fluid Flow and Systems of Conservation Laws in Several Space Variables*, Appl. Math. Sci., Vol. 53, Springer-Verlag, New York (1984).

[51] C. Marchioro and M. Pulvirenti, *Mathematical Theory of Incompressible Nonviscous Fluids*, Appl. Math. Sci., Vol. 96, Springer-Verlag, New York (1994).

[52] W.H. Munk and G.F. Carrier, *The wind-driven circulation in ocean basins of various shapes*, Tellus **2** (1950), 158–167.

[53] R. Nussbaum, *The radius of the essential spectrum*, Duke Math. J. **37** (1970), 473–478.

[54] O. Oleinik, *On the system of Prandtl equations in boundary-layer theory*, Dokl. Akad. Nauk SSSR **150** (1963), 28–31.

[55] J. Pedlosky, *Geophysical Fluid Dynamics*, Springer-Verlag, Berlin (1979).

[56] Lord Rayleigh, *On the stability, of instability, of certain fluid motions*, Proc. London Math. Soc. **11** (1880), 57–70.

[57] Lord Rayleigh, *Scientific Papers*, Vols I–VI, Cambridge University Press.

[58] M. Sammartino and R. Caflish, *Zero viscosity limit for analytic solutions, of the Navier–Stokes equation on a half-space, I, II*, Comm. Math. Phys. **192** (2) (1998), 433–461 and 463–491.

[59] H. Schlichting, *Boundary Layer Theory*, McGraw Hill, New York (1968).

[60] P.J. Schmid and D.S. Henningson, *Stability and Transition in Shear Flows*, Appl. Math. Sci., Vol. 142, Springer-Verlag, New York (2000).

[61] S. Schochet, *Fast singular limits of hyperbolic PDEs*, J. Differential Equations **114** (1994), 476–512.

[62] J. Serrin, *On the stability of viscous fluid motions*, Arch. Ration. Mech. Anal. **3** (1959), 1–13.

[63] A. Shnirelman, *On the nonuniqueness of weak solution of the Euler equation*, Comm. Pure Appl. Math. **50** (12) (1997), 1261–1286.

[64] A. Shnirelman, *Weak solution of incompressible Euler equations with decreasing energy*, C. R. Acad. Sci. Paris Sér. I Math. **326** (3) (1998), 329–334.

[65] A. Shnirelman, *On the L^2-instability and L^2-controllability of steady flows of an ideal incompressible fluid*, Journées "Équations aux Dérivées Partielles", Saint-Jean-de-Monts (1999), Exp. No. XIII, 8 pp., Univ. Nantes, Nantes (1999).

[66] H.B. Squire, *On the stability of three-dimensional disturbances of viscous flow between parallel walls*, Proc. Roy. Soc. A **142** (1933), 129–155.

[67] G.I. Taylor, *Experiments on the motion of solid bodies in rotating fluids*, Proc. Roy. Soc. A **104** (1923), 213–218.

[68] R. Temam, *Navier–Stokes Equations. Theory and Numerical Analyses*, Stud. Math. Appl., Vol. 2, North-Holland, Amsterdam (1984).

[69] R. Temam and X. Wang, *Remarks on the Prandtl equation for a permeable wall*, Special issue on the occasion of the 125th anniversary of the birth of Ludwig Prandtl, Z. Angew. Math. Mech. **80** (11/12) (2000), 835–843.

[70] M. Vishik, *Spectrum of small oscillations of an ideal fluid and Lyapunov exponents*, J. Math. Pures Appl. **75** (6) (1996), 531–557.

[71] R.K. Zeytounian, *Modélisation asymptotique en mécanique des fluides newtoniens*, Mathématiques et Applications, Vol. 15, Springer-Verlag, Berlin–Heidelberg (1994).

CHAPTER 5

Stability of Large-Amplitude Shock Waves of Compressible Navier–Stokes Equations

Kevin Zumbrun

Mathematics Department, Indiana University, Bloomington, IN 47405-4301, USA
E-mail: kzumbrun@indiana.edu

with Appendix by

Helge Kristian Jenssen

Department of Mathematics, North Carolina State University, Raleigh, NC 27695, USA
E-mail: hkjensse@unity.ncsu.edu

Gregory Lyng

Department of Mathematics, University of Michigan, Ann Arbor, MI 48109-1109, USA
E-mail: glyng@umich.edu

Contents

HANDBOOK OF MATHEMATICAL FLUID DYNAMICS, VOLUME III
Edited by S.J. Friedlander and D. Serre

Abstract

We summarize recent progress on one-dimensional and multidimensional stability of viscous shock wave solutions of compressible Navier–Stokes equations and related symmetrizable hyperbolic–parabolic systems, with an emphasis on the large-amplitude regime where transition from stability to instability may be expected to occur. The main result is the establishment of rigorous necessary and sufficient conditions for linearized and nonlinear planar viscous stability, agreeing in one dimension and separated in multidimensions by a co-dimension one set, that both extend and sharpen the formal conditions of structural and dynamical stability found in classical physical literature. The sufficient condition in multidimensions is new, and represents the main mathematical contribution of this article. The sufficient condition for stability is always satisfied for sufficiently small-amplitude shocks, while the necessary condition is known to fail under certain circumstances for sufficiently large-amplitude shocks; both are readily evaluable numerically. The precise conditions under and the nature in which transition from stability to instability occurs are outstanding open questions in the theory.

[1]Contributed by H.K. Jenssen and G. Lyng.

Preface

Five years ago, we and co-authors introduced in the pair of papers [GZ] and [ZH] a new, "dynamical systems" approach to stability of viscous shock waves based on Evans function and inverse Laplace transform techniques, suitable for the treatment of large-amplitude and or strongly nonlinear waves such as arise in the regime where physical transition from stability to instability may be expected to occur. Previous results for systems (see, e.g., [MN.1,KMN,Go.1,Go.2,L.1–L.3,SzX,LZ.1,LZ.2] and references therein) had concerned only sufficient conditions for stability, and had been confined almost exclusively to the small-amplitude case.[2] The philosophy of [GZ,ZH] was, rather, to determine useful *necessary and sufficient conditions* for stability in terms of the Evans function, which could then be investigated either numerically [Br.1,Br.2,BrZ,KL] or analytically [MN.1,KMN,Go.1,Go.2,Z.5,HuZ,Hu.1–Hu.3,Li,FreS.2,PZ] in cases of interest.[3]

This approach has proven to be extremely fruitful. In particular, the method has been generalized in [ZS,Z.3,HoZ.1,HoZ.2] and [God.1,Z.3,Z.4,MaZ.1–MaZ.5], respectively, to multidimensions and to real viscosity and relaxation systems, two improvements of great importance in physical applications. Extensions in other directions include generalizations to sonic shocks [H.4,H.5,HZ.3], boundary layers [GG,R], discrete and semi-discrete models [Se.1,God.2,B-G.3,BHR], combustion [Ly,LyZ.1,LyZ.2,JLy.1,JLy.2], and periodic traveling-wave solutions in phase-transitional models [OZ.1,OZ.2]. Most recently, the qualitative connection to inviscid theory pointed out in [ZS] has been made explicit in [MéZ.1,GMWZ.1–GMWZ.4] by the introduction of Kreiss symmetrizer and pseudodifferential techniques, leading to results on existence and stability of curved shock and boundary layers in the inviscid limit.

In this chapter, we attempt to summarize the developments of the past five years as they pertain to the specific context of compressible Navier–Stokes equations and related hyperbolic–parabolic systems. As the theory continues in a state of rapid development, this survey is to some extent a snapshot of a field in motion. For this reason, we have chosen to emphasize those aspects that appear to be complete, in particular, to expose the basic planar theory and technical tools on which current developments are based. It is our hope and expectation that developments over the next five years will substantially change the complexion of the field, from the technical study of stability conditions as carried out here to their exploitation in understanding phenomena of importance in physical applications.

As the list of references makes clear, the development of this program has been the work of several different groups and individuals, each contributing their unique point of view. We gratefully acknowledge the contribution and companionship of our colleagues and co-authors in this venture. Special thanks go to former graduate students Peter Howard, Len Brin, Myungyun Oh, Jeffrey Humpherys and Gregory Lyng (IU Bloomington, advisor K. Zumbrun); Pauline Godillon, Pierre Huot and Frederic Rousset (ENS Lyons, respective advisors D. Serre, S. Benzoni-Gavage and D. Serre); and Ramon Plaza (NYU, advisor J. Goodman); and former postdoctoral researchers Corrado Mascia and Kristian Jenssen

[2]The single exception is the partial, zero-mass stability analysis of [MN.1].

[3]A similar philosophy may be found in the partial, zero-mass stability analysis of [KK.1], which was carried out simultaneously to and independently of [GZ,ZH]; that work, however, concerned only sufficient stability conditions.

(IU Bloomington), whose interest and innovation have changed the Evans function approach to conservation laws from a handful of papers into an emerging subfield.

Special thanks also to Rob Gardner, Todd Kapitula and Chris Jones, whose initial contributions were instrumental to the development of the Evans function approach; to Denis Serre, Sylvie Benzoni-Gavage and Heinrich Freistühler, whose vision and innovation immediately widened the scope of investigations, and whose ideas have contributed still more than is indicated by the record of their joint and individually published articles; to Olivier Guès, Mark Williams and Guy Métivier, whose ideas have not only driven the newest developments in the field, but also profoundly affected our view of the old; and to Arthur Azevedo, Constantin Dafermos, Jonathan Goodman, David Hoff, Tai-Ping Liu, Dan Marchesin, Brad Plohr, Keith Promislow, Jeffrey Rauch, Joel Smoller, Anders Szepessy, Björn Sandstede, Blake Temple and Yanni Zeng for their generously offered ideas and continued support and encouragement outside the bounds of collaboration. Thanks to Denis Serre and the anonymous referee for their careful reading of the manuscript and many helpful corrections. Finally, thanks to my colleagues in Bloomington for their friendship, aid, and stimulating mathematical company, and to my family for their patience, love, and support.

1. Introduction

Fluid- and gas-dynamical flow, though familiar in everyday experience, are sufficiently complex as to defy simple description. Reacting flow, plastic–elastic flow in solids, and magnetohydrodynamic flow are still more complicated, and moreover are removed from direct observation in our daily lives. Historically, therefore, much of the progress in understanding these phenomena has come from the study of simple flows occurring in special situations, and their stability and bifurcation. For example, one may neglect compressibility of the conducting medium for low Mach number flow, or viscosity for high Reynolds number, or both, in each case obtaining a reduction in the complexity of the governing equations. Alternatively, one may consider solutions with special symmetry, reducing the spatial dimension of solutions: for example, laminar (shear) flows exhibiting local symmetry parallel to the direction of flow, or compressive (shock) flows exhibiting local symmetry normal to the flow.

Once existence of such solutions has been ascertained, there remains the interesting question of *stability* of the flow/*validity* of simplifying assumptions; more precisely, one would like to know the parameter range under which such solutions may actually be found in nature and to understand the corresponding bifurcation at the point where stability is lost. These questions comprise the classical subject of *hydrodynamic stability* (see, e.g., [Ke.1,Ke.2,Ra.1–Ra.6,T,Ch,DR,Lin]). With the advent of powerful and affordable processors, this subject has lost its central place to computational fluid dynamics, which in effect brings the most remote parameter regimes into the realm of our direct experience. However, we would argue that this development makes all the more interesting the rigorous understanding of stability/validity. For example, numerical approximation adds another layer of effects between model and experiment that must be carefully separated from actual behavior, and this is an important practical role for rigorous analysis. Likewise, numerical experiments suggest new phenomena for which analysis can yield qualitative understanding.

Stability and the related issue of validity of formal or numerical approximations are difficult questions, and unfortunately analysis has often lagged behind the needs of practical application. However, this area is currently in a period of rapid development, to the extent that practical applications now seem not only realistic but imminent. In this chapter, we give an account of recent developments in the study of stability and behavior of compressive "shock-type" fronts, with an eye toward both mathematical completeness and eventual physical applications. The latter criterion means that we must consider stability in the regime where transition from stability to instability is likely to occur, and this means in most cases that we must consider simplified flow within the context of the full model, without simplifying idealizations, i.e., as occurring in "real media" and usually in the large-amplitude regime. This requirement leads to substantial technical difficulties, and much of the mathematical interest of the analysis.

Consider a general system of *viscous conservation laws*

$$U_t + \sum_j F^j(U)_{x_j} = \nu \sum_{j,k} \big(B^{jk}(U)U_{x_k}\big)_{x_j}, \quad x \in \mathbb{R}^d, U, F^j \in \mathbb{R}^n, B^{jk} \in \mathbb{R}^{n\times n}, \tag{1.1}$$

modeling flow in a compressible medium, where ν is a constant measuring transport effects (e.g., viscosity or heat conduction).

An important class of solutions are *planar viscous shock waves*

$$U(x,t) = \overline{U}\left(\frac{x_1 - st}{\nu}\right), \qquad \lim_{x_1\to\pm\infty} \overline{U}(x_1) = U_\pm, \tag{1.2}$$

satisfying the traveling-wave ordinary differential equation (ODE)

$$B^{11}(\overline{U})\overline{U}' = F^1(\overline{U}) - F^1(U_-) - s(\overline{U} - U_-), \tag{1.3}$$

which reduce in the vanishing-viscosity limit $\nu \to 0$ to "ideal shocks", or planar discontinuities between constant states $U_\pm$ of the corresponding "inviscid" system

$$U_t + \sum_j F^j(U)_{x_j} = 0. \tag{1.4}$$

The shape function $\overline{U}$ is often called a "viscous shock profile". Planar shock waves are archetypal of more general, curved shock solutions, which have locally planar structure (see [M.1–M.4], [GW,GMWZ.2,GMWZ.3] for discussions of existence of curved shocks in the respective contexts (1.1) and (1.4)). The equations of reacting flow, containing additional reaction terms $Q(u)$ on the right-hand sides of (1.1) and (1.4), likewise feature planar solutions (1.2), known variously as strong or weak detonations, or strong or weak deflagrations, depending on their specific physical roles (see, e.g., [FD,CF,GS,Ly] for details).

The theories of stability of shock and detonation waves are historically intertwined. For physical reasons, the more complicated detonation case seems to have driven the study of stability, at least initially. For example, experiments early on indicated several possible types of instabilities that could occur, in particular: *galloping* longitudinal instabilities in which the steadily traveling planar front is replaced by a time-periodic planar solution moving with time-periodic speed; *cellular* instabilities, in which it is replaced by a nonplanar traveling wave with approximately periodic transverse spatial structure and different speed; and *spinning* instabilities, in which it is replaced by a nonplanar, steadily traveling and rotating wave (see, e.g., [FD,Ly,LyZ.1,LyZ.2] for further discussion). The latter two types are the ones most frequently observed; note that these are transverse instabilities, and thus multidimensional in character. The questions of main physical interest seem to be: (1) the origins/mechanism for instability; (2) location of the transition from stability to instability; (3) understanding/prediction of the associated bifurcation to more complicated solutions; and (4) the effect of front geometry (curvature) on stability.

Shock waves, on the other hand, were apparently thought for a long time to be universally stable. Indeed, the possibility of instability was first predicted analytically, through the normal modes analyses of [Kr,D,Er.1,Er.2], in a regime that was at the time not accessible to experiment; see [BE] for an account of these and other interesting details of the early theory. On the other hand, in the inviscid case $\nu = 0$, the stability condition can be explicitly calculated for shock waves, whereas for detonation waves it cannot. For this and other reasons, the shock case is much better understood; in particular, there exists a rather complete analytical multidimensional stability theory for inviscid shock waves, see [M.1–M.4,Mé.1–Mé.5], whereas the corresponding (ZND) theory for detonation waves is mainly numerical, with sensitive numerical issues limiting conclusions (see [LS] for a fairly recent survey).

As we shall discuss further, the inviscid shock theory gives partial answers to question (1), but does not rule out possible additional mechanisms connected with viscosity or other transport effects. Likewise, it gives only a partial answer to question (2) reducing the location of the transition point to an open region in parameter space, and so question (3) cannot even be addressed. Question (4) inherently belongs to the viscous theory, as there is no effect of shock curvature without finite shock width. In this article, we present a mathematical framework for the study of viscous shock stability, based on the Evans function of [E.1–E.4,AGJ,PW], that has emerged through the series of papers [GZ,ZH,BSZ.1,ZS,Z.1–Z.3,MaZ.1–MaZ.5] and references therein, and which features necessary and sufficient conditions for planar viscous stability analogous to but somewhat sharper than the weak and strong (or "uniform") Lopatinski conditions of the inviscid theory [Kr,M.1–M.4,Mé.5]. These are sufficient in principle to resolve questions (1) and (2), and give also a useful starting point for the study of question (3). Question (4) remains for the moment open, and is an important direction for further investigation.

Aside from their intrinsic interest, these results are significant for their bearing on the originally motivating problem of detonation and on related problems associated with nonclassical shocks in phase-transitional or magnetohydrodynamical flows, for which significant stability questions remain even in "typical" regimes occurring frequently in applications. Though we shall not discuss it here, the technical tools we develop in this chapter can be applied also in these more exotic cases; see [ZH,Z.5,Z.3,HZ.2,BMSZ,

FreZ.2,LyZ.2] for discussions in the strictly parabolic case, and [Ly,LyZ.1] in the case of "real", or physical viscosity.

1.1. *Equations and assumptions*

We begin by identifying an abstract class of equations, generalizing the Kawashima class of [Kaw], that isolates those qualities relevant to the investigation of shock stability in compressible flow. The Navier–Stokes equations of compressible gas- or magnetohydrodynamics (MHD) in dimension d, may be expressed in terms of conserved quantities U in the standard form (1.1), with

$$U = \begin{pmatrix} u^I \\ u^{II} \end{pmatrix}, \qquad B^{jk} = \begin{pmatrix} 0 & 0 \\ b_1^{jk} & b_2^{jk} \end{pmatrix}, \tag{1.5}$$

$u^I \in \mathbb{R}^{n-r}$, $u^{II} \in \mathbb{R}^r$, and

$$\operatorname{Re} \sigma \sum \xi_j \xi_k b_2^{jk} \geqslant \theta |\xi|^2, \tag{1.6}$$

$\theta > 0$, for all $\xi \in \mathbb{R}^d$; for details, see Appendix A.1. Since in this work we are interested in time-asymptotic stability for a fixed, finite viscosity rather than the vanishing-viscosity limit, we set $\nu = 1$ from here forward, and suppress the parameter ν.

Near states of stable thermodynamic equilibrium, where there exists an associated convex entropy η, they can be written, alternatively, in terms of the "entropy variable" $W := \mathrm{d}\eta(U)$, in *symmetric hyperbolic–parabolic form*

$$U(W)_t + \sum_j \tilde{F}^j(W)_{x_j} = \sum_{j,k} \big(\tilde{B}^{jk} W_{x_k}\big)_{x_j}, \qquad W = \begin{pmatrix} w^I \\ w^{II} \end{pmatrix}, \tag{1.7}$$

where

$$\mathrm{d}\tilde{F}^j = \mathrm{d}F^j(\partial U/\partial W), \qquad \tilde{B}^{jk} = B^{jk}(\partial U/\partial W) = \begin{pmatrix} 0 & 0 \\ 0 & \tilde{b}^{jk} \end{pmatrix} \tag{1.8}$$

are symmetric, and $(\partial U/\partial W)$ and $\sum \xi_j \xi_k \tilde{b}^{jk}$ are (uniformly) symmetric positive definite,

$$\sum \xi_j \xi_k \tilde{b}^{jk} \geqslant \theta |\xi|^2, \quad \theta > 0, \tag{1.9}$$

for all $\xi \in \mathbb{R}^d$; see [Kaw] and references therein, or Appendix A.1. This fundamental observation generalizes the corresponding observation of Godunov [G] (see also [Fri,M.1,Bo]) in the inviscid setting $B^{jk} \equiv 0$.

In either setting, symmetric form (1.7)–(1.9) has the important property that its structure guarantees local existence under small perturbations of equilibrium, as may be established by standard energy estimates [Fri,Kaw]. The viscous equations, moreover, satisfy the further condition of *genuine coupling*

$$\text{No eigenvector of } \sum_j \xi_j \,\mathrm{d}F^j \text{ lies in the kernel of } \sum \xi_j \xi_k B^{jk}, \tag{1.10}$$

for all nonzero $\xi \in \mathbb{R}^d$ (equivalently, $\sum_j \xi_j \,\mathrm{d}\widetilde{F}^j$, $\sum \xi_j \xi_k \widetilde{B}^{jk}$), which implies that the system is "dissipative" in a certain sense; in particular, it implies, together with form (1.7), *global existence and decay under small perturbations of equilibrium*, via a series of clever energy estimates revealing "hyperbolic compensation" for absent parabolic terms [Kaw]. We refer to equations of form (1.7) satisfying (1.8)–(1.10) as "Kawashima class".

Here, we generalize these ideas to the situation of a viscous shock solution (1.2) of (1.1) consisting of a smooth shock profile connecting two *different* thermodynamically stable equilibria, and small perturbations thereof. Specifically, we assume that, by some invertible change of coordinates $U \to W(U)$, possibly but not necessarily connected with a global convex entropy, followed if necessary by multiplication on the left by a nonsingular matrix function $S(W)$, (1.1) may be written in the *quasilinear, partially symmetric hyperbolic–parabolic form*

$$\widetilde{A}^0 W_t + \sum_j \widetilde{A}^j W_{x_j} = \sum_{j,k} \big(\widetilde{B}^{jk} W_{x_k}\big)_{x_j} + G, \qquad W = \begin{pmatrix} w^I \\ w^{II} \end{pmatrix}, \tag{1.11}$$

$w^I \in \mathbb{R}^{n-r}$, $w^{II} \in \mathbb{R}^r$, $x \in \mathbb{R}^d$, $t \in \mathbb{R}$, where, defining $W_\pm := W(U_\pm)$:

(A1) $\widetilde{A}^j(W_\pm)$, $\widetilde{A}^j_* := \widetilde{A}^j_{11}$, $\widetilde{A}^0$ are symmetric, $\widetilde{A}^0 > 0$.

(A2) No eigenvector of $\sum \xi_j \,\mathrm{d}F^j(U_\pm)$ lies in the kernel of $\sum \xi_j \xi_k B^{jk}(U_\pm)$, for all nonzero $\xi \in \mathbb{R}^d$. (Equivalently, no eigenvector of $\sum \xi_j \widetilde{A}^j (\widetilde{A}^0)^{-1}(W_\pm)$ lies in the kernel of $\sum \xi_j \xi_k \widetilde{B}^{jk}(W_\pm)$.)

(A3) $\widetilde{B}^{jk} = \begin{pmatrix} 0 & 0 \\ 0 & \tilde{b}^{jk} \end{pmatrix}$, $\widetilde{G} = \begin{pmatrix} 0 \\ \tilde{g} \end{pmatrix}$, with $\operatorname{Re} \sum \xi_j \xi_k \tilde{b}^{jk}(W) \geqslant \theta |\xi|^2$ for some $\theta > 0$, for all W and all $\xi \in \mathbb{R}^d$, and $\tilde{g}(W_x, W_x) = \mathcal{O}(|W_x|^2)$.

Here, the coefficients of (1.11) may be expressed in terms of the original equation (1.1), the coordinate change $U \to W(U)$, and the approximate symmetrizer $S(W)$, as

$$\begin{aligned} \widetilde{A}^0 &:= S(W)(\partial U/\partial W), \\ \widetilde{A}^j &:= S(W)\,\mathrm{d}F^j(\partial U/\partial W), \\ \widetilde{B}^{jk} &:= S(W) B^{jk}(\partial U/\partial W), \\ G &= -\sum_{jk} (\mathrm{d}S\, W_{x_j}) B^{jk} (\partial U/\partial W) W_{x_k}. \end{aligned} \tag{1.12}$$

Note that, in accordance with the general philosophy of [ZH,Z.3] in the strictly parabolic case, the conditions of dissipative symmetric hyperbolic–parabolicity connected with stability of equilibrium states are required only at the endstates of the profile, the conditions along the profile – symmetrizable hyperbolicity, $\mathrm{d}\widetilde{F}^j_{11}$, in the first equation and parabolicity, $\sum \xi_j \xi_k \tilde{b}^{jk} > 0$, in the second equation – being concerned with local well-posedness rather than time-asymptotic stability of any intermediate state. This allows for interesting applications to phase-transitional or van der Waals gas dynamics, for which hyperbolicity of the associated ideal system (1.4) may be lost along the profile; see Appendix A.1.

REMARK 1.1. The differential symbol $A^\xi_* = \sum_j \xi_j A^j_*$ defined by (A1) governs convection in "hyperbolic" modes, as described in [MaZ.3]; for general equations of form (1.1), this must be replaced by the pseudodifferential symbol $A^\xi_* := \mathrm{d}F^\xi_{11} - \mathrm{d}F^\xi_{12}(b^{\xi\xi}_2)^{-1}b^{\xi\xi}_1$, where $\mathrm{d}F^\xi := \sum_j \mathrm{d}F^j \xi_j$ and $b^{\xi\xi} := \sum_j b^{jk}\xi_j\xi_k$. That is, the form (4.2) imposes interesting structure also at the linearized symbolic level.

REMARK 1.2. Assumptions (A1)–(A3) were introduced in [MaZ.4] under the restriction $G \equiv 0$, or equivalently $S \equiv I$. This is sufficient to treat isentropic van der Waals gas dynamics, as discussed in the example at the end of the introduction of [MaZ.4]; however, it is not clear whether there exists such a coordinate change for the full equations of gas dynamics with van der Waals equation of state. The improved version $G \not\equiv 0$, introduced in [GMWZ.4], is much easier to check, and in particular holds for the full van der Waals gas equations under the single (clearly necessary) condition $T_e > 0$, where $T = T(\rho, e)$ is temperature, and ρ and e are density and internal energy; see Appendix A.1. Still more general versions of (1.11) (in particular, including zero-order terms with vanishing 1–1 block) might be accommodated in our analysis; however, so far there seems to be no need in applications. Of course, equations in the Kawashima class, in particular systems possessing a global entropy, are included under (A1)–(A3) as a special case.

Along with the above structural assumptions, we make the technical hypotheses:

(H0) F^j, B^{jk}, W, $S \in C^s$, with $s \geqslant 2$ in our analysis of necessary conditions for linearized stability, $s \geqslant 5$ in our analysis of sufficient conditions for linearized stability,[4] and $s \geqslant s(d) := [d/2] + 5$ in our analysis of nonlinear stability.

(H1) The eigenvalues of A^1_* are (i) distinct from the shock speed s; (ii) of common sign relative to s; and (iii) of constant multiplicity with respect to U.[5]

(H2) The eigenvalues of $\mathrm{d}F^1(U_\pm)$ are distinct from s.

(H3) Local to $\overline{U}(\cdot)$, solutions of (1.2)–(1.3) form a smooth manifold $\{\overline{U}^\delta(\cdot)\}$, $\delta \in \mathcal{U} \subset \mathbb{R}^\ell$.

Conditions (H0)–(H2) may be checked algebraically, and are generically satisfied for profiles of the equations of gas dynamics and MHD; see Appendix A.1.[6] We remark that (H1)(i) arises naturally as the condition that the traveling-wave ODE be of nondegenerate type, while (H2) is the condition for normal hyperbolicity of $U_\pm$ as rest points of that ODE; see Section 1.2. The "weak transversality" condition (H3) is more difficult to verify;

[4]As mentioned in [MaZ.3], we suspect that C^3 should suffice, with further work.

[5]In our analysis of necessary conditions for stability, we require only (H1)(i). Condition (H1)(iii) can be dropped, at the expense of some detail in the linear estimates [Z.4].

[6]In particular, $A^j_* = u^j I_{n-r}$, where u^j is the jth component of fluid velocity, and so (H1) is trivially satisfied. That is, hyperbolic modes experience only passive, scalar convection for these equations.

however, it is automatically satisfied for extreme Lax shocks such as arise in gas dynamics; see Remark 2, Section 1.2. More generally, it is implied by but does not imply transversality of the traveling-wave connection $\overline{U}$ as an orbit of (1.3).

In our analysis of sufficient conditions for stability, we make two further hypotheses at the level of the inviscid equations (1.4), analogous to but somewhat stronger than the block structure condition of the inviscid stability theory [Kr,M.1–M.4,Mé.5]. The first is, simply:

(H4) The eigenvalues of $\sum \xi_j \, \mathrm{d}F^j(U_\pm)$ are of constant multiplicity with respect to $\xi \in \mathbb{R}^d \setminus \{0\}$.

This condition was shown in [Mé.4] to imply block structure, and is satisfied for all physical examples for which the block structure condition has currently been verified. In particular, it holds always for gas dynamics, but fails for MHD; see, respectively, Appendix C and [Je,JeT].

The second concerns the structure of the glancing set of the symbol

$$P_\pm(\xi,\tau) := \mathrm{i}\tau + \sum \mathrm{i}\xi_j \, \mathrm{d}F^j(U_\pm) \tag{1.13}$$

of the inviscid equations (1.4) at $x_1 = \pm\infty$. By symmetrizability assumption (A1), together with hypothesis (H4), the characteristic equation $\det P_\pm(\xi,\tau) = 0$ has n roots

$$\tau = -a_r(\xi), \quad r = 1,\dots,n, \tag{1.14}$$

locally analytic and homogeneous degree one in ξ, where a_r are the eigenvalues of $\sum \xi_j \, \mathrm{d}F^j(U_\pm)$.

DEFINITION 1.3. Setting $\xi = (\xi_1,\dots,\xi_d) =: (\xi_1,\tilde{\xi})$, we define the *glancing set* of $P_\pm$ as the set of frequencies $(\tilde{\xi},\tau)$ for which $\tau = a_r(\xi_1,\tilde{\xi})$ and $(\partial a_r/\partial \xi_1) = 0$ for some real ξ_1 and $1 \leqslant r \leqslant n$.

REMARK 1.4. The word "glancing" is used in Definition 1.3 because null bi-characteristics of $P_\pm$ through points (ξ,τ) with $(\partial a_r/\partial \xi_1) = 0$ are parallel to $x_1 = 0$.

The significance of the glancing set for our analysis is that it corresponds to the set of frequencies $\tilde{\xi},\tau$ for which the (rotated) inverse functions $\mathrm{i}\xi_1 = \mu_r(\tilde{\xi},\tau)$ associated with $\tau = a_r(\xi_1,\tilde{\xi})$, corresponding to rates of spatial decay in x_1 for the inviscid resolvent equation, are *pure imaginary* and have *branch singularities* of degree s_r equal to the multiplicity of root ξ_1. As pointed out by Kreiss [Kr], these are the frequencies for which resolvent estimates become delicate, and structural assumptions become important (e.g., strict hyperbolicity as in [Kr], or, more generally, block structure as in [M.1–M.4,Mé.5]).

Our final technical hypothesis is, then:

(H5) The glancing set of $P_\pm$ is the union of k (possibly intersecting) smooth curves $\tau = \eta_q(\tilde{\xi})$, $0 \leqslant k \leqslant n$, defined as the locii on which $\tau = -a_q(\xi_1,\tilde{\xi})$ and $(\partial a_q/\partial \xi_1) = 0$ for some real ξ_1, on which root ξ_1 has *constant multiplicity* s_q, defined as the order of the first nonvanishing partial derivative $(\partial^s a_q/\partial \xi_1^s)$ with respect to ξ_1, i.e., the associated inverse function $\xi_1^q(\tilde{\xi},\tau)$ has constant degree of singularity s_q.

Condition (H5) is automatic in dimensions $d = 1, 2$ and in any dimension for rotationally invariant problems. In one dimension, the glancing set is empty. In the two-dimensional case, the homogeneity of a_r and its derivatives implies that the ray through $(\tilde{\xi}, \tau)$ is the graph of $\tau(\tilde{\xi})$ and that (H5) holds there. By the implicit function theorem, (H5) holds also in the case that all branch singularities are of square root type, degree $s_q = 2$, with η_q defined implicitly by the requirement $(\partial a_q/\partial \xi_1) = 0$ (indeed, η_q is in this case analytic). *In particular, it holds in the case that all eigenvalues $a_r(\xi)$ are either linear or else strictly convex/concave in ξ_1 for $\tilde{\xi} \neq 0$.*[7] Thus, (H5) is satisfied in all dimensions for the equations of gas dynamics, for which the characteristic eigenvalues a_r are are linear combinations of (ξ, η) and $|\xi, \eta|$ (see Appendix C), hence clearly linear or else strictly concave/convex for $\tilde{\xi} \neq 0$. Similarly, it may be calculated (see [MéZ.3]) that all characteristic eigenvalues of the equations of MHD are linear or else strictly concave/convex for $\tilde{\xi} \neq 0$, with the exceptions of the two "slow" magnetoacoustic characteristics, which together may produce a singularity of degree at most two. Thus, again (H5) is always satisfied.

REMARK 1.5. Assuming symmetrizability, (A1), condition (H5) like (H4) implies block structure; see the direct matrix perturbation calculations of [Z.3], Section 5, or Section 5 of this chapter. As was the block structure condition in the inviscid case, conditions (H4) and (H5) are used in the viscous case to make convenient certain estimates on the resolvent, and should be viewed as "first-order nondegeneracy conditions". There is ample motivation in the example of MHD to carry out a refined analysis at the next level of degeneracy, with these conditions relaxed: in particular, the constant multiplicity requirement (H4), which presently restricts the stability analysis for MHD to the one-dimensional case. See [MéZ.3] for some work in this direction.

1.2. *Structure and classification of profiles*

We next state some general results from [MaZ.3,Z.3] regarding structure and classification of profiles, analogous to those proved in [MP] in the strictly parabolic case. Proofs may be found in Appendix A.2; see also [MaZ.3,Z.3,Ly,LyZ.1].

Observe first, given $\det b_2^{11} \neq 0$, (A2), that (H1)(i) is equivalent, by determinant identity $\det\begin{pmatrix} dF_{11}^1 - s & dF_{12}^1 \\ b_1^{11} & b_2^{11}\end{pmatrix} = \det(dF_{11}^1 - s - b_1^{11}(b_2^{11})^{-1} dF_{12}^1) \det b_2^{11}$, to the condition

$$\det\begin{pmatrix} dF_{11}^1 - s & dF_{12}^1 \\ b_1^{11} & b_2^{11}\end{pmatrix} \neq 0. \tag{1.15}$$

Writing the traveling-wave ODE (1.3) as

$$f^I(U) - su^I \equiv f^I(U_-) - su_-^I, \tag{1.16}$$

$$b_1^{11}(u^I)' + b_2^{11}(u^{II})' = f^{II}(U) - f^{II}(U_-) - s(u^{II} - u_-^{II}), \tag{1.17}$$

[7] More generally, just those involved in glancing. Likewise, constant multiplicity (H4) is only needed in our analysis for eigenvalues involved in glancing.

where $F^1 =: \binom{f^I}{f^{II}}$, we find that (1.15) is in turn the condition that (1.17) describe a nondegenerate ODE on the r-dimensional manifold described by (1.16): well defined by the implicit function theorem plus full rank of $(\mathrm{d}F^1_{11} - s, \mathrm{d}F^1_{12})$.

LEMMA 1.6 [Z.3,MaZ.3]. *Given* (H1)–(H3), *the endstates* $U_\pm$ *are hyperbolic rest points of the ODE determined by* (1.17) *on the* r*-dimensional manifold* (1.16), *i.e., the coefficients of the linearized equations about* $U_\pm$, *written in local coordinates, have no center subspace. In particular, under regularity* (H0), *standing-wave solutions* (1.2) *satisfy*

$$\left|(\mathrm{d}/\mathrm{d}x_1)^k(\bar{U} - U_\pm)\right| \leqslant C\mathrm{e}^{-\theta|x_1|}, \quad k = 0, \ldots, 4, \tag{1.18}$$

as $x_1 \to \pm\infty$.

Lemma 1.6 verifies the hypotheses of the gap lemma, [GZ,ZH],[8] on which we shall rely to obtain the basic ODE estimates underlying our analysis in the low-frequency/large-time regime.

Let i_+ denote the dimension of the stable subspace of $\mathrm{d}F^1(U_+)$, i_- denote the dimension of the unstable subspace of $\mathrm{d}F^1(U_-)$, and $i := i_+ + i_-$. Indices $i_\pm$ count the number of incoming characteristics from the right/left of the shock, while i counts the total number of incoming characteristics toward the shock. Then, the *hyperbolic classification* of $\bar{U}(\cdot)$, i.e., the classification of the associated hyperbolic shock (U_-, U_+), is:

$$\begin{cases} \text{Lax type} & \text{if } i = n+1, \\ \text{Undercompressive (u.c.)} & \text{if } i \leqslant n, \\ \text{Overcompressive (o.c.)} & \text{if } i \geqslant n+2. \end{cases}$$

In case all characteristics are incoming on one side, i.e., $i_+ = n$ or $i_- = n$, a shock is called *extreme*.

As in the strictly parabolic case, there is a close connection between the hyperbolic type of a shock and the nature of the corresponding connecting profile. Considering the standing-wave ODE as an r-dimensional ODE on manifold (1.16), let us denote by $1 \leqslant d_\pm \leqslant r$ the dimensions of stable manifold at U_+ and unstable manifold at U_-, and $d := d_+ + d_-$. Then, we have:

LEMMA 1.7 [MaZ.3,LyZ.1]. *Given* (H1)–(H3), *there hold relations*

$$\begin{aligned} n - i_+ &= r - d_+ + \dim\mathcal{U}(A_{*+}), \\ n - i_- &= r - d_- + \dim\mathcal{S}(A_{*-}), \end{aligned} \tag{1.19}$$

where $A_* := \mathrm{d}F^1_{11} - \mathrm{d}F^1_{12}(b^{11}_2)^{-1}b^{11}_1$, *and* $\mathcal{U}(M)$ *and* $\mathcal{S}(M)$ *denote unstable and stable subspaces of a matrix* M. *In particular, existence of a connecting profile implies* $n - i = r - d$.

[8] See also the version established in [KS], independently of and simultaneously to that of [GZ].

The final assertion implies that the "viscous" and "hyperbolic" types of shock connections agree, i.e., Lax, undercompressive and overcompressive designations imply corresponding information about connections. The more detailed information (1.19) implies that "extreme" shocks have "extreme" connections:

COROLLARY 1.8 [MaZ.3,LyZ.1]. *For (right) extreme shocks, $i_+ = n$, there holds also $d_+ = r$, i.e., the connection is also extreme, and also* $\dim\mathcal{U}(A_*) \equiv 0$.

PROOF. Immediate from (1.19), using $d_+ \leqslant r$ and $\dim\mathcal{U}(A_{*+}) \geqslant 0$. □

A complete description of the connection, of course, requires the further index ℓ defined in (H4) as the dimension ℓ of the connecting manifold between $(u_\pm, v_\pm)$ in the traveling-wave ODE. Generically, one expects that ℓ should be equal to the surplus $d - r = i - n$. In case the connection is "dimensionally transverse" in this sense, i.e.:

$$\ell = \begin{cases} 1, & \text{undercompressive or Lax case,} \\ i - n, & \text{overcompressive case,} \end{cases} \tag{1.20}$$

we call the shock "pure" type and classify it according to its hyperbolic type; otherwise, we call it "mixed" under/overcompressive type. Throughout this chapter, we shall assume that (1.20) holds, so that all viscous profiles are of *pure, hyperbolic type*. Indeed, we shall restrict attention mainly to the classical case of *pure, Lax-type profiles* such as occur in the standard gas-dynamical case, confining our discussion of other cases to brief remarks. For more detailed discussion of the nonclassical, overcompressive and undercompressive cases, we refer the reader to [Z.3] and especially [ZS].

REMARKS 1.9. 1. Transverse Lax and overcompressive connections of (1.3) persist under change of parameters (U_-, s), while transverse undercompressive connections are of co-dimension $q := n + 1 - i$ (the "degree of undercompressivity") in parameter space (U_-, s).

2. For extreme indices $d_+ = r$ or $d_- = r$, profiles if they exist are always transverse, and likewise for indices that are "minimal" in the sense that $d_+ = \ell$ or $d_- = \ell$. In particular, *extreme Lax or overcompressive connections (if they exist) are always transverse*, hence satisfy (H3). Undercompressive shocks are never extreme.

3. The relation $0 \leqslant \dim\mathcal{U}(A_+) - \dim\mathcal{U}(A_{*+}) = (r - d_+) \leqslant r$, may bc viewed as a sort of subcharacteristic relation, by analogy with the relaxation case. Indeed, as described in [LyZ.1], it may be used to obtain the full subcharacteristic condition

$$a_j < a_j^* < a_{j+r},$$

as a consequence of linearized stability of constant states, where a_j denote the eigenvalues of $A := \mathrm{d}F^1$ and a_j^* those of A_*; this is closely related to arguments given by Yong (see, e.g., [Yo]) in the relaxation case. The subcharacteristic relation goes the "wrong way" in the relaxation case, and so one cannot conclude a result analogous to Corollary 1.8; for further discussion, see [MaZ.1,MaZ.5].

4. In the special case $r = 1$, there holds $d_+ = d_- = 1$ whenever there is a connection, and so profiles are always of Lax type $n - i = r - d = -1$. This recovers the observation of Pego [P.2] in the case of one-dimensional isentropic gas dynamics that smooth (undercompressive type) phase-transitional shock profiles cannot occur under the effects of viscosity alone, even for a van der Waals-type equation of state. (They can occur, however, when dispersive, capillary pressure effects are taken into account; see, e.g., [Sl.1–Sl.5,B-G.1,B-G.2].)

1.3. *Classical (inviscid) stability analysis*

The classical approach to the study of stability of shock waves, as found in the mathematical physics literature, consists mainly in the study of the related problems of *structural stability*, or existence and stability of profiles (1.2) as solutions of ODE (1.3), and *dynamical stability*, or hyperbolic stability of the corresponding ideal shock

$$U(x,t) = \overline{\overline{U}}(x_1 - st), \qquad \overline{\overline{U}}(x_1) := \begin{cases} U_- & \text{for } x_1 < 0, \\ U_+ & \text{for } x_1 \geqslant 0, \end{cases} \tag{1.21}$$

as a solution of the inviscid equations (1.4); see, for example, the excellent survey articles [BE] and [MeP]. These may be derived formally by matched asymptotic expansion, either in the vanishing-viscosity limit $\nu \to 0$, or, as discussed in Section 1.3, [Z.3], in the low frequency, or large space–time limit

$$\bar{x} := \varepsilon x, \qquad \bar{t} := \varepsilon t, \qquad \varepsilon \to 0, \tag{1.22}$$

with ν held fixed and $\bar{t}$ varying on a bounded interval $[0, T]$. We summarize the relevant results below.

1.3.1. *Structural stability.* The viscous profile problem, or "inner" problem from the matched asymptotic expansion point of view, yields first of all the *Rankine–Hugoniot conditions*

$$s[U] = \left[F^1(U)\right], \tag{RH}$$

where $[H] := H(U_+) - H(U_-)$ denotes jump across the shock in quantity H, as the requirement that $U_\pm$ must both be rest points of the traveling-wave ODE (1.3). These arise more directly in the inviscid theory, as the conditions that mass be conserved across an ideal shock discontinuity (1.21) (see, e.g., [La,Sm]).

In matched asymptotic expansion, solvability of the inner problem serves as the boundary (or free boundary, in this case) condition for the "outer" problem, which in this case (see Section 1.3, [Z.3]) is just the inviscid equations (1.4). By Remark 1.9.1, the Rankine–Hugoniot conditions (RH) are sufficient for existence in the vicinity of a transverse Lax-type or overcompressive-type shock profile, while in the vicinity of a transverse undercompressive shock profile, existence is equivalent to (RH) plus an additional q boundary conditions, sometimes known as "kinetic conditions", where $q = n + 1 - i$ (in the notation of the previous section) is the degree of undercompressivity. *Thus, local to an existing transverse profile, the outer, inviscid problem may in the Lax or overcompressive case be discussed without reference to the inner problem, appealing only to conservation of mass,* (RH). As discussed cogently in [vN] (round table discussion), this remarkable fact is in sharp contrast to the situation in the case of a solid boundary layer, and is at the heart of hyperbolic shock theory.

Of course, there remains the original question of existence in the first place of *any* transverse smooth shock profile, which is a priori far from clear given the incomplete parabolicity of Equations (1.1). Here, the genuine coupling condition (A2) plays a key role. For example, in the "completely decoupled" case that $dF^j_{12} \equiv 0$, the first equation of (1.1) is first-order hyperbolic, so supports only discontinuous traveling-wave solutions (1.21) [Sm]. On the other hand, (A2) precludes such discontinuous solutions, at least for small shock amplitude $[U] := U_+ - U_-$. For, a distributional solution (1.21) of (1.1), $\nu = 1$, may be seen to satisfy not only (RH), but also, using the block structure assumption (1.8) in the symmetric, W coordinates, the condition $[w^{II}] = 0$, or $[W] \in \ker \widetilde{B}^{11}$.[9] For $[U]$ small, however, (RH) implies that $[U]$ is approximately parallel to an eigenvector of dF^1, hence by (A2) cannot lie in $\ker B^{11}(U_-)$, and therefore (using smallness again) $[W]$ cannot lie in $\ker \widetilde{B}^{11}$, a contradiction. Viewed another way, this is an example of the *time-asymptotic smoothing* principle introduced in [HoZi.1,HoZi.2] in the specific context of compressible Navier–Stokes equations, which states that (A2) implies smoothness of time-asymptotic states as in the strictly parabolic case, despite incomplete smoothing for finite times: in particular, of stationary or traveling-wave solutions.

Indeed, Pego [P.1] has shown by a center-manifold argument generalizing that of [MP] in the strictly parabolic case that the small-amplitude existence theory for hyperbolic–parabolic systems satisfying assumption (A2) is identical to that of the strictly parabolic case, as we now describe.

First, recall the Lax structure theorem:

PROPOSITION 1.10 [La,Sm]. *Let $F^1 \in C^2$ be hyperbolic ($\sigma(dF^1)$ real and semisimple) for U in a neighborhood of some base state U_0, and let $a_p(U_0)$ be a simple eigenvalue, where $a_1 \leqslant \cdots \leqslant a_k$ denote the eigenvalues of $dF^1(U)$ and $r_j = r_J(U)$ the associated eigenvectors. Then, there exists a smooth function $H_p : (U_-, \theta) \to U_+, \theta \in \mathbb{R}^1$, with $H_p(U_-, 0) \equiv U_-$ and $(\partial/\partial\theta) H_p(U_-, 0) \equiv r_p(U_-)$, such that, for $U_\pm$ lying sufficiently close to U_0 and $[U] := U_+ - U_-$ lying sufficiently close to $|U_+ - U_-| r_p(U_0)$, the triple (U_-, U_+, s) satisfies the Rankine–Hugoniot conditions* (RH) *if and only if $U_+ = H_p(U_-, \theta)$ for θ sufficiently small.*

PROOF. See [Sm], pp. 328–329, proofs of Theorem 17.11 and Corollary 17.12. □

Note that, for $[U]$ sufficiently small and U sufficiently near U_0, (RH) implies that $[U]$ lies approximately parallel to some $r_j(U_0)$; thus, the assumption that $[U]$ lie nearly parallel

[9] Without symmetric, block form, we cannot make conclusions regarding the product $B^{11} U_{x_1}$ of a discontinuous function and a possibly point measure. However, (1.8) and (1.9) together allow us to eliminate the possibility of a jump in w^{II}.

to $r_p(U_0)$ is no real restriction. For fixed U_-, the image of H_p is known as the (pth) Hugoniot curve through U_-. For gas dynamics, under mild assumptions on the equation of state[10] the Hugoniot curves extend *globally*, describing the full solution set of (RH).

DEFINITION 1.11. For U lying on the pth Hugoniot curve through U_-, denote by $\sigma(U_-, U)$ the associated shock speed defined by (RH). Then, the shock triple $(U_-, U_+ = H_p(U_-, \theta_+), s = \sigma(U_-, U_+))$ is said to satisfy the *strict Liu–Oleinik admissibility condition* if and only if

$$\sigma\big(U_-, U(\theta_*)\big) > \sigma(U_-, U_+) = s \quad \text{for all } \theta_*(\theta_* - \theta_+) < 0, \tag{LO}$$

i.e., on the segment of the Hugoniot curve between U_- and U_+, $\sigma(U_-, U)$ takes on its minimum value at U_+.

A general small-amplitude existence theory may now be concisely stated as follows, with the strictly parabolic theory subsumed as a special case.

PROPOSITION 1.12 [P.1]. *Let $F^1 \in C^2$ hyperbolic and $B^{11} \in C^2$ of form* (1.5)–(1.6) *satisfy the genuine coupling condition* (1.10) *for U in a neighborhood of some base state U_0, and let $a_p(U_0)$ be a simple eigenvalue, where $a_1 \leqslant \cdots \leqslant a_k$ denote the eigenvalues of $\mathrm{d}F^1(U)$. Then, for a sufficiently small neighborhood $\mathcal{U}$ of U_0 and for (U_-, U_+, s) lying sufficiently close to $(U_0, U_0, a_p(U_0))$ and satisfying the noncharacteristic condition $a_p^*(U_\pm) \neq s$, there exists a traveling-wave connection* (1.2) *lying in $\mathcal{U}$ if and only if the triple (U_-, U_+, s) satisfies both the Rankine–Hugoniot conditions* (RH) *and the strict Liu–Oleinik admissibility condition* (LO). *Moreover, such a local connecting orbit, if it exists, is transverse, hence unique up to translation.*

PROOF. See [P.1], or the related [Fre.4]. □

In the case of "standard" gas dynamics, for which the equation of state admits a global convex entropy, there is a correspondingly simple *global existence theory*. In the contrary case of "nonstandard" or "real" (e.g., van der Waals) gas dynamics, the global existence problem so far as we know is open.

PROPOSITION 1.13 [Gi,MeP]. *For gas dynamics with a global convex entropy, a triple (U_-, U_+, s) admits a traveling-wave connections (viscous profile)* (1.2) *if and only if it satisfies both* (RH) *and* (LO). *Moreover, connections if they exist are transverse.*

PROOF. See [Gi] in the "genuinely nonlinear" case $\nabla a_p r_p \neq 0$, [MeP], Appendix C in the general case. Transversality is automatic, by Remark 1.9.1. □

[10] For example, existence of a global convex entropy, plus the "weak condition" of [MeP], satisfied for all known examples possessing a convex entropy (in particular, ideal gas dynamics and the larger classes of equations considered by Bethe, Weyl, Gilbarg, Wendroff and Liu [Be,We,Gi,Wen.1,Wen.2,L.4]); see [MeP].

REMARK 1.14. 1. In the case that eigenvalues a_j retain their order relative to a_p, e.g., a_p remains simple, along the Hugoniot curve, the Liu–Oleinik condition (LO) may be seen to be a strengthened version of the classical *Lax characteristic conditions* [La]

$$a_{p-1}(U_-) < 0 < a_p(U_-), \qquad a_p(U_+) < 0 < a_{p+1}(U_+), \tag{1.23}$$

where $1 \leqslant p \leqslant n$ is the principal characteristic speed associated with the shock, or, equivalently (by preservation of order),

$$a_p(U_-) > s > a_p(U_+), \tag{1.24}$$

by $a_p(U_-) = \sigma(U_-, U_-) > s$ and symmetry with respect to $\pm$ of (LO). *In particular, the profiles described in Propositions* 1.12 *and* 1.13 *are always of classical, Lax type*. Small-amplitude profiles bifurcating from a multiple eigenvalue a_p, or large-amplitude profiles of general hyperbolic–parabolic systems may be of arbitrary type; see [FreS.1] and [CS], respectively.

In the genuinely nonlinear case $\nabla a_p r_p \neq 0$, (LO) is equivalent along the Hugoniot curve $U_+ = H_p(U_-, \theta)$ through U_- to condition (1.24), which in turn is equivalent to $\operatorname{sgn}\theta = -\operatorname{sgn}\nabla a_p r_p(U_-) \neq 0$ [La,Sm]. Here, H_p denotes an arbitrary nondegenerate extension of the local parametrization defined in Proposition 1.10.

2. Condition (LO) yields existence and uniqueness of admissible scale-invariant solutions of (1.4) for *Riemann initial data*

$$U_0 = \begin{cases} U_- & \text{for } x < 0, \\ U_+ & \text{for } x \geqslant 0, \end{cases} \tag{1.25}$$

for small amplitudes $|U_+ - U_-|$ and general equations [L.4], and, under mild additional assumptions (the "medium condition" of [MeP]), for large amplitudes in the case of gas dynamics with a global convex entropy [MeP]. Such *Riemann solutions* represent possible time-asymptotic states (after renormalization (1.22)) for the viscous equations (1.1) with "asymptotically planar" initial data in the sense that $\lim_{x_1 \to \pm\infty} U_0(x) = U_\pm$; see, e.g., [AMPZ.1] for further discussion.

3. For large amplitudes, the genuine coupling condition (A2) does not in general appear to preclude discontinuous traveling-wave solutions of (1.1) analogous to "subshocks" in the relaxation case [L.5,Wh], but rather must be replaced by the nonlinear version $[W] \notin \ker \widetilde{B}^{11}(U_-) = \ker \widetilde{B}^{11}(U_+)$ for (U_-, U_+, s) satisfying (RH). For compressible Navier–Stokes equations, this condition is satisfied globally for classical equations of state. Without loss of generality taking shock speed $s = 0$, we find that failure corresponds to a jump in density with velocity, pressure, and temperature held fixed; see the discussion of gas-dynamical structure (Corollary A.5, Appendix A.1). Thus, it can occur only for a nonmonotone, van der Waals-type pressure law, and corresponds to an undercompressive, phase-transitional type shock. Recall that such shocks do not possess smooth profiles under the influence of viscosity alone, Remark 1.9.4.

1.3.2. *Dynamical stability.* Hyperbolic stability concerns the *bounded-time* stability of (1.21) as a solution of inviscid equations (1.4) augmented with the free-boundary condition of solvability of the viscous profile problem, i.e., well-posedness of the "outer problem" arising through formal matched asymptotic expansion. That is, it is Hadamard-type stability, or well-posedness, that is the relevant notion in this case, in keeping with the homogeneous nature of (1.4). In view of Remark 1.14.1, we focus our attention on the classical, Lax case, for which the free-boundary conditions are simply the Rankine–Hugoniot conditions (RH) arising through conservation of mass.

Postulating a perturbed front location $X(\tilde{x}, t)$, $x =: (x_1, \tilde{x})$, following the standard approach to stability of fluid interfaces (see, e.g., [Ri,Er.1,FD,M.1–M.4]), and centering the discontinuity by the change of independent variables

$$z := x_1 - X(\tilde{x}, t), \tag{1.26}$$

we may convert the problem to a more conventional fixed-boundary problem at $z \equiv 0$, with front location X carried as an additional dependent variable. This device eliminates the difficulty of measures arising in the linearized equations (and eventual nonlinear iteration) through differentiation about a moving discontinuity.

Necessary inviscid stability condition. Linearizing the resulting equations about (1.21), assuming without loss of generality $s = 0$, we obtain constant-coefficient problems

$$U_t + A^1_\pm U_z + \sum_2^d A^j_\pm U_{x_j} = 0, \tag{1.27}$$

$A^j_\pm := \mathrm{d}F^j(U_\pm)$, on $z \gtrless 0$, linked by n transmission conditions given by the linearized Rankine–Hugoniot conditions, which involve also t- and $\tilde{x}$-derivatives of the (scalar) linearized front location.

Linearized inviscid stability analysis by Laplace–Fourier transform of (1.27) yields, in the case of a Lax p-shock (defined by (1.23), or $p := i_+$ in the notation of Section 1.2), the *weak Kreiss–Sakamoto–Lopatinski stability condition*

$$\Delta(\tilde{\xi}, \lambda) \neq 0, \quad \tilde{\xi} \in \mathbb{R}^{d-1}, \mathrm{Re}\,\lambda > 0, \tag{1.28}$$

where $\tilde{\xi} := (\xi_2, \ldots, \xi_d)$ is the Fourier wave number in transverse spatial directions $\tilde{x} := (x_2, \ldots, x_d)$, λ is the Laplace frequency in temporal direction t, and

$$\Delta := \det\bigl(r_1^-, \ldots, r_{p-1}^-, \mathrm{i}\bigl[F^{\tilde{\xi}}\bigr] + \lambda[U], r_{p+1}^+, \ldots, r_n^+\bigr), \tag{1.29}$$

with $F^{\tilde{\xi}}(U) := \sum_{j\neq 1} \xi_j F^j(U)$, $[g] := g(U_+) - g(U_-)$, and $\{r_{p+1}^+, \ldots, r_n^+\}(\tilde{\xi}, \lambda)$ and $\{r_1^-, \ldots, r_{p-1}^-\}(\tilde{\xi}, \lambda)$ denoting (analytically chosen) bases for the unstable (resp. stable) subspaces of matrices

$$A_\pm(\tilde{\xi}, \lambda) := \bigl(\lambda I + \mathrm{i}\,\mathrm{d}F^{\tilde{\xi}}(U_\pm)\bigr)\bigl(\mathrm{d}F^1(U_\pm)\bigr)^{-1}; \tag{1.30}$$

for details, see, e.g., [Se.2–Se.4,ZS,Mé.5]. Condition (1.28) is obtained from the requirement that the inhomogeneous equation

$$\left(r_1^-, \dots, r_{n-i_-}^-, \mathrm{i}\left[F^{\tilde{\xi}}\right] + \lambda[U], r_{i_++1}^+, \dots, r_n^+\right)\alpha = \hat{f}, \tag{1.31}$$

arising through Laplace–Fourier transform of the linearized problem have a unique solution for every datum $\hat{f}$, i.e., the columns of the matrix on the left-hand side form a basis of $\mathbb{R}^n$.

This result was first obtained by Erpenbeck in the case of gas dynamics[11] [Er.1,Er.2], and in the general case by Majda [M.1–M.4]. Condition (1.28) is an example of a general type of stability condition introduced by Kreiss and Sakamoto in their pioneering work on hyperbolic initial-boundary value problems [Kr,Sa.1,Sa.2]; as pointed out by Majda, it plays a role analogous to that of the Lopatinski condition in elliptic theory. Zeroes of Δ correspond to normal modes $e^{\lambda t} e^{i\tilde{\xi}\cdot\tilde{x}} w(x_1)$ of the linearized equations, hence (1.28) is *necessary* for linearized inviscid stability. Evidently, Δ is positive homogeneous (i.e., homogeneous with respect to positive reals), degree one, as are the inviscid equations; thus, instabilities if they occur are of Hadamard type, corresponding to ill-posedness of the linearized problem.

Sufficient inviscid stability condition. A fundamental contribution of Majda [M.1–M.4], building on the earlier work of Kreiss [Kr], was to point out the importance of the *uniform* (or "strong") *Kreiss–Sakamoto–Lopatinski stability condition*

$$\Delta(\tilde{\xi}, \lambda) \neq 0, \quad \tilde{\xi} \in \mathbb{R}^{d-1}, \operatorname{Re}\lambda \geqslant 0, (\tilde{\xi}, \lambda) \neq (0, 0), \tag{1.32}$$

extending (1.28) to imaginary λ, as a *sufficient* condition for nonlinear stability. Assuming (1.32), a certain "block structure" condition on matrix $A(\tilde{\xi}, \tau)$ defined in (1.30) (recall: implied by (H4)), and appropriate compatibility conditions on the perturbation at the shock (ensuring local structure of a single discontinuity, i.e., precluding the formation of shocks in other characteristic fields), Majda established local existence of perturbed ideal shock solutions

$$\tilde{U}(x, t) =: \overline{\overline{U}}\left(x_1 - X(\tilde{x}, t)\right) + U\left(x_1 - X(\tilde{x}, t), \tilde{x}, t\right), \tag{1.33}$$

$\overline{\overline{U}}$ as in (1.21), for initial perturbation sufficiently small in H^s, s sufficiently large, with optimal bounds on $\|U\|_{H^s(x,t)}$ and $\|X\|_{H^s(\tilde{x},t)}$. Moreover, using pseudodifferential techniques, he established a corresponding result for *curved shock fronts* under the local version of the planar uniform stability condition, establishing both existence and stability of curved shock solutions.

These result have since been significantly sharpened and extended by Métivier and co-workers; see [Mé.5], and references therein. In particular, Métivier [Mé.1] has obtained *uniform local existence and stability for vanishing shock amplitudes* under assumptions (A1) and (H2) plus the additional nondegeneracy conditions that principal characteristic $a_p(U, \xi)$ be: (i) simple and genuinely nonlinear in the normal direction to the shock,

[11] See also related, earlier work in [Be,D,Ko.1,Ko.2,Free.1,Free.2,Ro,Ri].

without loss of generality $\xi = (1, 0, \ldots, 0)$, and (ii) strictly convex in $\tilde{\xi}$ in a vicinity of $\xi = (1, 0, \ldots, 0)$. In the existence result of Majda, the time of existence vanished with the shock strength, due to the singular (i.e., characteristic) nature of the small-amplitude limit.

REMARK 1.15. Failure of (1.32) on the boundary $\lambda = i\tau$, τ real, i.e., weak but non-uniform stability, corresponds to existence of a one-parameter family of (nondecaying) transverse traveling-wave solutions $e^{i(\kappa\tau + \kappa\tilde{\xi}\cdot\tilde{x})} w(\kappa z)$ of the linearized equations, $\kappa \in \mathbb{R}$, moving with speed $\sigma = \tau/|\tilde{\xi}|$ in direction $\tilde{\xi}/|\tilde{\xi}|$, and is in general associated with loss of smoothness in the front $X(\cdot,\cdot)$ and loss of derivatives in the iteration used to prove non-linear stability.[12] When w has finite energy, $w \in L^2$, such solutions are known as *surface waves*; however, this is not always the case. In particular, $w \in L^2$ only if

$$a^{\pm}_{j_\pm}(\xi_1, \tilde{\xi}) < \tau < a^{\pm}_{k_\pm}(\xi_1, \tilde{\xi}) \tag{1.34}$$

for some $j_\pm$, $k_\pm$ (without loss of generality, $a_{k_\pm}(\xi) = -a_{j_\pm}(-\xi)$): see Proposition 3.1 of [BRSZ]. For $\xi_1 = 0$, this implies that the traveling wave is "subsonic" in the sense that σ lies between the $j_\pm$th and $k_\pm$th hyperbolic characteristic speeds in direction $\tilde{\xi}/|\tilde{\xi}|$ at endstates $U_\pm$. It is easily seen that the set determined by (1.34) is bounded by the glancing set defined in Definition 1.3; indeed, the complement of the closure of this set is the union $\mathcal{G}$ of the components of $(\tilde{\xi}, \tau) = (0, \pm 1)$ in the complement of the glancing set. A more fundamental division is between the larger *hyperbolic domain* $\mathcal{R}$, defined as the set of $(\tilde{\xi}, \tau)$ for which the stable and unstable subspaces of $A(\tilde{\xi}, i\tau)$ in (1.30) admit real bases, and its complement. (Note: $\mathcal{G} \subset \mathcal{R}$.) On $\mathcal{R}$, the eigenvalues of $A(\tilde{\xi}, i\tau)$ in (1.30) determining exponential rates of spatial decay in z are pure imaginary, and so surface waves cannot occur. The set $\mathcal{R}$ like $\mathcal{G}$ is bounded by the glancing set, at which certain eigenvalues of $A(\tilde{\xi}, i\tau)$ bifurcate away from the imaginary axis.

As discussed in [BRSZ], these various cases are in fact quite different. In particular, (i) in the surface wave case, the linearized equations can sometimes be "weakly well posed" in L^2, permitting a degraded version of the energy estimates in the strongly stable case, whereas in the complementary case they are not even weakly well posed, and (ii) zeroes $(\tilde{\xi}, \tau)$ lying in the closure of $\mathcal{R}^c$ generically mark a boundary between regions of strong stability and strong instability in parameter space (U_-, U_+, s), whereas zeroes lying in the interior of $\mathcal{R}$ generically persist, marking an open set in parameter space. The explanation for the latter, at first surprising fact is that for real $(\tilde{\xi}, \tau) \in \mathcal{R}$, it may be arranged by choosing $r_j^\pm$ real that $\Delta(\tilde{\xi}, \cdot)$ take the imaginary axis to itself; thus, zeroes may leave the imaginary axis only by coalescing at a double (or higher multiplicity) root, either finite,

$$\Delta(\tilde{\xi}, i\tau) = 0, \qquad (\partial/\partial\lambda)\Delta(\tilde{\xi}, i\tau) = 0, \tag{1.35}$$

or at infinity (recall, by complex symmetry, that roots escape to infinity in complex conjugate pairs). The latter possibility corresponds to one-dimensional instability $\Delta(0, 1) = 0$,

[12] An exception is the scalar case; see [M.1–M.4].

as discussed in [Se.2,BRSZ,Z.3].[13] The important category of weakly stable shocks possessing a zero $\Delta(\tilde{\xi}, \mathrm{i}\tau) = 0$ with $(\tilde{\xi}, \tau)$ in $\mathcal{R}$ was denoted in [BRSZ] as "weakly stable of real type", and forms a third open (i.e., generic) type besides the strongly stable and strongly unstable ones.

Evaluation in one dimension. In one dimension, $\tilde{\xi} = 0$, the basis vectors $r_j^{\pm}$ may be fixed as the outgoing characteristic directions for $A^1_{\pm} = \mathrm{d}F^1(U_{\pm})$, i.e., eigenvectors with associated eigenvalues $a_j^{\pm} \gtrless 0$, yielding $\Delta(0, \lambda) = \lambda\Delta(0, 1)$. Thus, both weak and uniform stability conditions reduce to

$$\Delta(0, 1) = \det\left(r_1^-, \ldots, r_{p-1}^-, [U], r_{p+1}^+, \ldots, r_n^+\right) \neq 0. \tag{1.36}$$

Condition (1.36) may be regarded as a sharpened form of the Lax characteristic condition (1.23), which asserts only that the number of columns of the matrix on the left-hand side of (1.36) is correct (i.e., equal to n). Condition (1.36) can be recognized also as the condition that the associated Riemann problem (U_-, U_+) be linearly well posed (transverse), since otherwise there is a nearby family of Riemann solutions possessing the same (final) end states $U_{\pm}$, but for which the data at the p-shock is modified by infinitesimal perturbations $\mathrm{d}U_{\pm}$ such that the linearized Rankine–Hugoniot relations remain satisfied:

$$-\mathrm{d}s\,[U] = F^1(U_+)\,\mathrm{d}U_+ - \mathrm{d}F^1(U_-)\,\mathrm{d}U_-. \tag{1.37}$$

That is, failure of one-dimensional inviscid stability is generically associated with the phenomenon of *wave-splitting*; see, e.g., [Er.1,ZH,Z.3], for further details.

As observed by Majda [M.1–M.4], in the case of a simple eigenvalue $a_p(U_*)$, the Lax structure theorem, Proposition 1.10, ensures that *one-dimensional stability holds for sufficiently small amplitude shocks of general systems.* One-dimensional stability holds for arbitrary amplitude shocks of the equations of gas dynamics with a "standard" equation of state (e.g., possessing a global convex entropy and satisfying the "medium condition" of [MeP]), but may fail at large amplitudes for more general equations of state; (see, e.g., [Er.1,Er.2,M.1–M.4,MeP]).

REMARK 1.16. The set of parameters (U_-, U_+, s) on which the one-dimensional dynamical stability condition (1.36) fails is of co-dimension one, hence generically consists of isolated points along a given Hugoniot curve $H_p(U_-, \cdot)$. In other words, the one-dimensional dynamical stability condition *yields essentially no information* further than the Lax structure already imposed on the problem by the condition of structural stability.

---

[13]By positive homogeneity, $0 \equiv \Delta_j(\tilde{\xi}_j, \mathrm{i}\tau_j) = \Delta_j(\tilde{\xi}_j/|\tau_j|, \mathrm{i}\,\mathrm{sgn}\,\tau_j) \to \Delta(0, \pm\mathrm{i})$ as $j \to \infty$, where Δ_j are the Lopatinski conditions for a sequence of converging parameter values indexed by j, with zeroes $(\tilde{\xi}_j, \mathrm{i}\tau_j)$ such that $\sigma_j = \tau_j/|\tilde{\xi}_j| \to \infty$ as $j \to \infty$, and $\Delta = \lim \Delta_j$ is the Lopatinski condition for the limiting parameter values as $j \to \infty$. Hence, Δ vanishes at one of $(0, \pm\mathrm{i})$ (in fact both, by complex symmetry of the spectrum), yielding one-dimensional instability.

Evaluation in multiple dimensions. Evaluation of the stability conditions in multidimensions is significantly more difficult than in the one-dimensional case, even the computation of Δ being in general nontrivial. Nonetheless, we obtain again a small-amplitude result for general systems and an arbitrary-amplitude result for gas dynamics, both with appealingly simple formulations.

Namely, under the assumptions of [Mé.1] – (A1) and (H2) plus the conditions that principal characteristic $a_p(U, \xi)$ be (i) simple and genuinely nonlinear in the normal direction to the shock, without loss of generality $\xi = (1, 0, \dots, 0)$, and (ii) strictly convex in $\tilde{\xi}$ in a vicinity of $\xi = (1, 0, \dots, 0)$ – we have the simple small-amplitude result, generalizing Majda's observation in one dimension, that *sufficiently small-amplitude shocks are strongly ("uniformly") stable.* This is an indirect consequence of the detailed partial differential equation (PDE) estimates carried out by Métivier in the proof of existence and stability of small-amplitude shocks [Mé.1]. We give a direct proof in Appendix B, in the spirit of Métivier's analysis but carried out in the much simpler linear-algebraic setting.

A detailed analysis of the large-amplitude gas-dynamical case recovering the physical stability criteria of Erpenbeck and Majda [Er.1,Er.2,M.1–M.4] is presented in Appendix C.[14] This includes in particular Majda's theorem that for ideal gas dynamics, shocks of arbitrary amplitude are strongly (uniformly) stable. Discussion of more general equations of state may be found, e.g., in [Er.1,Er.2,BE,MeP]. In particular, uniform stability holds for equations of state possessing a global convex entropy and satisfying the "medium condition" of [MeP], but in general may fail, depending on the geometry of the associated Hugoniot curve.

REMARK 1.17. For gas-dynamical shocks, direct calculation shows that weak multidimensional inviscid stability when it occurs is always of real type, with $\mathcal{R} = \mathcal{G}$. Moreover, transition from weak multidimensional inviscid stability of real type to strong multidimensional inviscid instability (failure of (1.28)) can occur only through the infinite speed limit $\sigma = \tau/|\tilde{\xi}| \to \infty$ corresponding to one-dimensional inviscid instability, $\Delta(0, 1) = 0$; see, for example, discussions in [Er.1] (p. 1185), [Fo].[15] Likewise (see Remark 1.15), the transition from weak multidimensional inviscid stability of real type to strong multidimensional inviscid stability can occur only through collision of $(\tilde{\xi}, \tau)$ with the glancing set $\{(\tilde{\xi}, \eta(\tilde{\xi}))\}$ defined in Definition 1.3: easily calculated for gas dynamics; see, e.g., [Z.3], example near (3.12), or Appendix C of this chapter. By rotational invariance and homogeneity, this transition may be detected by the single condition $\Delta(\tilde{\xi}_0, \eta(\tilde{\xi}_0)) \neq 0$, $\tilde{\xi}_0 = (1, 0, \dots, 0)$, similarly as in the previous case. These two observations greatly simplify the determination of the various inviscid shock stability regions once Δ has been determined; see, e.g., [BRSZ] or Section 6.5 of [Z.3].[16] From our point of view, however, the main point is simply that, *for gas dynamics, transition from strong inviscid stability to instability, if it occurs,*

[14] Contributed by K. Jenssen and G. Lyng.

[15] Equivalently, $(\partial/\partial\lambda)\Delta(\tilde{\xi}, i\tau)$ and $\Delta(\tilde{\xi}, i\tau)$ do not simultaneously vanish for $\tilde{\xi}, \tau$ real (indeed, $(\partial/\partial\lambda)\Delta(\tilde{\xi}, i\tau)$ does not vanish for $(\tilde{\xi}, \tau)$ in $\mathcal{R}$); see Appendix C and Remark 1.15.

[16] Precisely, together with Métivier's theorem guaranteeing uniform stability of small-amplitude shocks, we recover by these observations the results of the full Nyquist diagram analysis carried out in Appendix C. (In fact, these two approaches are closely related; see discussion of [Z.3], Section 6.5.)

occurs via passage through the open region of weak inviscid stability of real type, on which (inviscid) stability is indeterminate.

REMARK 1.18 (Nonclassical case). Freistühler [Fre.1,Fre.2] has extended the Majda analysis to the nonclassical, undercompressive case, obtaining in place of (1.28), (1.29) and (1.32) the extended, $((n+q)\times(n+q))$-dimensional determinant conditions:

$$\Delta(\tilde{\xi},\lambda)\neq 0 \quad \text{for all } \tilde{\xi}\in\mathbb{R}^{d-1}, \operatorname{Re}\lambda>0 \text{ (resp. } \operatorname{Re}\lambda\geqslant 0 \text{ and } \lambda\neq 0), \tag{1.38}$$

$$\Delta(\tilde{\xi},\lambda):=\det\begin{pmatrix} \frac{\partial g}{\partial U_-}{A_-^1}^{-1}r_1^-,\dots,\frac{\partial g}{\partial U_-}{A_-^1}^{-1}r_{n-i_-}^-, & \frac{\partial g}{\partial U_+}{A_+^1}^{-1}r_{i_++1}^+,\dots,\frac{\partial g}{\partial U_+}{A_+^1}^{-1}r_n^+, & \mathrm{i}\frac{\partial g}{\partial\omega}\tilde{\xi}+\frac{\partial g}{\partial s}\lambda \\ r_1^-,\dots,r_{n-i_-}^-, & r_{i_++1},\dots,r_n^+, & \mathrm{i}[F^{\tilde{\xi}}(U)]+\lambda[U] \end{pmatrix},$$

where

$$g(U_-,U_+,\omega,s)=0,\qquad g\in\mathbb{R}^q,$$

encodes the q kinetic conditions needed along with (RH) to determine existence of an undercompressive connection in direction $\omega\in S^{d-1}$ (i.e., a traveling-wave solution $U=\overline{U}(x\cdot\omega-st)$), and all partial derivatives are taken at the base parameters $(U_-,U_+,e_1,0)$ corresponding to the one-dimensional stationary profile $U=\overline{U}(x_1)$. This condition has been evaluated in [B-G.1,B-G.2] for the physically interesting case of phase boundaries, with the result of strong (i.e., uniform) stability for sufficiently small but positive viscosities.[17] *Applied to the overcompressive case, on the other hand, the analysis of Majda yields automatic instability.* For, (1.31) is in this case ill posed for all $\tilde{\xi}$, λ, since the matrix on the left-hand side has fewer than n columns.

1.4. *Viscous stability analysis*

The formal, matched asymptotic expansion arguments underlying the classical, inviscid stability analysis leave unclear its precise relation to the original, physical stability problem (1.1). In particular, rigorous validation of matched asymptotic expansion generally proceeds through stability estimates on the inner solution (viscous shock profile), which in the planar case is somewhat circular (recall: the inner problem *is* the planar evolution problem, with additional terms comprising the formal truncation error); hence, we cannot rigorously conclude even necessity of these conditions for viscous stability. This is more than a question of mathematical nicety: in certain physically interesting cases arising in connection with nonclassical under- and overcompressive shocks, the classical stability analysis is known to give wrong results both positive and negative; see Remarks 1.18,

[17] Recall that, in the undercompressive case, higher-order derivative terms influence the inviscid problem through their effect on the connection conditions g.

1.20.1 and 1.22.5, or Section 1.3 of [Z.3]. Moreover, even in the classical Lax case, it is not sufficiently sharp to determine the transition from stability to instability, either in one or multiple dimensions; see Remarks 1.16 and 1.17.

These and related issues may be resolved by a direct stability analysis at the level of the viscous equations (1.1), as we now informally describe. Rigorous statements and proofs will be established throughout the rest of the article. In particular, the viscous theory not only *refines* (and in some cases corrects) the inviscid theory but also *completes* it by the addition of a third, "strong spectral stability" condition augmenting the classical conditions of structural and dynamical stability. Strong spectral stability is somewhat weaker than linearized stability – in the one-dimensional case, for example, it is more or less equivalent to stability with respect to zero-mass perturbations [ZH] – and may be recognized as the missing condition needed to validate the formal asymptotic expansion; see, for example, the related investigations [GX,GR,R,MéZ.1,GMWZ.2–GMWZ.4].

1.4.1. *Connection with classical theory.* The starting point for the viscous analysis is a linearized, Laplace–Fourier transform analysis very similar in spirit to that of the inviscid case, again taking without loss of generality $s = 0$. This yields the Evans function condition

$$D(\tilde{\xi}, \lambda) \neq 0, \quad \tilde{\xi} \in \mathbb{R}^{d-1}, \operatorname{Re} \lambda > 0, \tag{1.39}$$

as a necessary condition for linearized viscous stability, where the Evans function D (defined in Section 2.2) is a determinant analogous to the Lopatinski determinant $\Delta(\cdot, \cdot)$ of the inviscid theory. Similarly as in the inviscid case, zeroes of $D(\cdot, \cdot)$ correspond to growing modes $v = e^{\lambda t} e^{i\tilde{\xi}\cdot\tilde{x}} w(x_1)$ of the linearized equations about the background profile $\overline{U}(\cdot)$.

DEFINITION 1.19. We define *weak spectral stability* (necessary but not sufficient for linearized viscous stability) as (1.39). We define *strong spectral stability* (by itself, neither necessary nor sufficient for linearized stability) as

$$D(\tilde{\xi}, \lambda) \neq 0 \quad \text{for } (\tilde{\xi}, \lambda) \in \mathbb{R}^{d-1} \times \{\operatorname{Re} \lambda \geqslant 0\} \setminus \{0, 0\}. \tag{1.40}$$

The fundamental relation between viscous and inviscid theories, quantifying the formal rescaling argument alluded to in (1.22), Section 1.3, is given by

RESULT 1 ([ZS,Z.3], Theorem 3.5). *$D(\tilde{\xi}, \lambda) \sim \gamma \Delta(\tilde{\xi}, \lambda)$ as $(\tilde{\xi}, \lambda) \to (0, 0)$, where γ is a constant measuring transversality of the connection $\overline{U}(\cdot)$ in* (1.1), *i.e., D is tangent to Δ in the low frequency limit. Here, Δ in the Lax or undercompressive case is given by* (1.29) *or* (1.38), *respectively, and in the overcompressive case by a modified determinant correcting the inviscid theory, as described in* (3.9) *and* (3.10), *Theorem* 3.5.

That is, in the low-frequency limit ($\sim$ large space–time, by the standard duality between spatial and frequency variables), higher-derivative terms become negligible in (1.1), and behavior is governed by its homogeneous first-order (inviscid) part (1.4). The departure from inviscid theory in the overcompressive case reflects a subtle but important distinction between low frequency/large space–time and vanishing-viscosity limits, as discussed in Section 1.3 of [Z.3].[18]

1.4.2. *One-dimensional stability.* The most dramatic improvement in the viscous theory over the inviscid theory comes in the one-dimensional case, for which the inviscid dynamical stability condition yields essentially no information, Remark 1.16. From Result 1, we find that vanishing of the one-dimensional Lopatinski determinant $\Delta(0, 1)$ generically corresponds to crossing from the stable to the unstable complex half-plane $\operatorname{Re}\lambda > 0$ of a root of the Evans function D. Indeed, we have

RESULT 2 ([Z.3,Z.4], Propositions 4.2, 4.4 and 4.5). *With appropriate normalization, the condition $\gamma\Delta(0, 1) > 0$ is necessary for one-dimensional linearized viscous stability.*

More precisely, we have the following sharp one-dimensional stability criterion.

RESULT 3 ([MaZ.3,MaZ.4], Theorems 4.8 and 4.9). *Necessary and sufficient conditions for linearized and nonlinear one-dimensional viscous stability are one-dimensional structural and dynamical stability, $\gamma\Delta(0, 1) \neq 0$, plus strong spectral stability, $D(0, \lambda) \neq 0$ on $\{\operatorname{Re}\lambda \geqslant 0\} \setminus \{0\}$.*

REMARKS 1.20. 1. Result 2 replaces the co-dimension one inviscid instability condition $\gamma\Delta(0, 1) = 0$ with an open condition $\gamma\Delta(0, 1) > 0$ of which it is the boundary. That is, one-dimensional inviscid instability theory marks the boundary between one-dimensional viscous stability and instability. The strengthened, viscous condition $\gamma\Delta(0, 1) > 0$ in particular resolves the problem of physical selection discussed in [MeP], p. 110, for which multiple Riemann solutions exist, each satisfying the classical conditions of dynamical and structural stability, but only one – the one satisfying the one-dimensional viscous stability condition – is selected by numerical experiment; see related discussion, Section 6.3 of [Z.3]. Likewise, it yields instability of undercompressive, pulse-type (i.e., homoclinic) profiles arising in phase-transitional flow [GZ,Z.5], which, again, satisfy the classical inviscid stability criteria but numerically appear to be unstable [AMPZ.1]. Result 3 as claimed completes the inviscid stability conditions with the addition of the third condition of strong spectral stability. Spectral stability has been verified for small amplitude, Lax-type profiles in [HuZ,PZ], and for special large amplitude profiles in [MN.1] and [Z.5]. Efficient algorithms for numerical testing of spectral stability are described, e.g., in [Br.1,Br.2,BrZ,KL,BDG].

2. It is an interesting and physically important open question *which* of the three criteria of structural, dynamical, or spectral stability is the more restrictive in different regimes; in particular, it would be very interesting to know whether, or under what circumstances, spectral stability can be the determining factor in transition to instability as $|\theta|$ is gradually increased along the Hugoniot curve $U_+ = H_p(U_-, \theta)$ defined in Section 1.3.1. The latter

[18] Briefly, initial data scales as ε in the low-frequency limit, but in the vanishing-viscosity limit is fixed, order one. For ε-order data, underdetermination of the inner problem associated with nonuniqueness of overcompressive orbits can balance with overdetermination of the outer problem; for general data, the effects occur on different scales. For related discussion outside the matched asymptotic setting, see [Fre.3,L.2] or [FreL].

question is closely related to the following generalized Sturm–Liouville question: *Do the global symmetry assumptions* (4.2)–(1.9), *or perhaps the stronger condition that the system* (1.1) *possess a global convex entropy, as discussed in Section* 1.1, *preclude the existence of nonzero pure imaginary eigenvalues*? For, as pointed out in [ZH,Z.3], the same calculations used to establish the signed stability condition show that eigenvalues can cross through the origin from stable to unstable complex half-plane only if $\gamma\Delta = 0$, i.e., Δ vanishes or transversality is lost. Thus, beginning with small amplitude profiles (stable, by [HuZ,PZ]) and continuously varying the endstate $\widetilde{U}_+$, we find that loss of stability, if it occurs before structural or dynamical stability fail, must correspond to a pair of complex-conjugate eigenvalues crossing the imaginary axis, presumably signaling a Poincaré–Hopf-type bifurcation[19] to a family of time-periodic solutions.

Recall that such "galloping" instabilities *do* in fact occur in the closely related situation of detonation (see, e.g., [Ly,LyZ.1,LyZ.2]); thus, the question of whether or not or under what circumstances they can occur for shock waves seems physically quite interesting. If they cannot occur for the class of systems we consider, then the relatively simple conditions of structural and dynamical stability – i.e., the classical conditions of inviscid stability – completely determine the transition from viscous stability to instability, a very satisfactory conclusion. If, on the other hand, they can occur, then this implies that heretofore neglected viscous effects play a far stronger role in stability than we have imagined, and this result would be philosophically still more significant.

1.4.3. *Multidimensional stability.* A deficiency of the inviscid multidimensional stability theory is that transition from strong inviscid stability to strong inviscid instability typically occurs through the open region of weak inviscid stability of real type, on which stability is indeterminate; see Remarks 1.15 and 1.17, or Section 1.4 of [Z.3]. This indeterminacy may be remedied by a next-order correction in the low-frequency limit including second-order, viscous effects. There is also some question [BE] whether the physical transition to instability might sometimes occur *before* the passage to weak inviscid instability, i.e., for parameters within the region of strong inviscid stability. This likewise can be determined, at least in principle, by the inclusion of neglected viscous effects, this time in the (global) form of the strong spectral stability condition.

DEFINITION 1.21. We define *weak refined dynamical stability* as inviscid weak stability (1.28) augmented with the second-order condition that $\operatorname{Re}\beta(\tilde\xi, i\tau) \geqslant 0$ for any real $\tilde\xi$, τ such that Δ (hence also $D^{\tilde\xi, i\tau}(\rho) := D(\rho\tilde\xi, \rho i\tau)$, by Result 1) is analytic at $(\tilde\xi, i\tau)$, $\Delta(\tilde\xi, i\tau) = 0$, and $(\partial/\partial\lambda)\Delta(\tilde\xi, i\tau) \neq 0$, where

$$\begin{aligned}\beta(\tilde\xi, i\tau) &:= \frac{(\partial/\partial\rho)^{\ell+1} D(\rho\tilde\xi, \rho i\tau)|_{\rho=0}}{(\partial/\partial\lambda)\overline{\Delta}(\tilde\xi, i\tau)} \\ &= \frac{(\partial/\partial\rho)^{\ell+1} D(\rho\tilde\xi, \rho\lambda)}{(\partial/\partial\lambda)(\partial/\partial\rho)^{\ell} D(\rho\tilde\xi, \rho\lambda)|_{\rho=0,\lambda=i\tau}}.\end{aligned} \tag{1.41}$$

[19]Degenerate, because of the well-known absence [Sat] of a spectral gap between $\lambda = 0$ and the essential spectrum of the linearized operator about the wave.

We define *strong refined dynamical stability* as inviscid weak stability (1.28) augmented with the conditions that Δ be analytic at $(\tilde{\xi}, i\tau)$, $(\partial/\partial\lambda)\Delta(\tilde{\xi}, i\tau) \neq 0$,

$$\{r_1^-, \ldots, r_{p-1}^-, r_{p+1}^+, \ldots, r_n^+\}(\tilde{\xi}, i\tau) \tag{1.42}$$

be independent for $r_j^\pm$ defined as in (1.29) (automatic for "extreme" shocks $p = 1$ or n),[20] and $\operatorname{Re}\beta(\tilde{\xi}, i\tau) > 0$ for any real $\tilde{\xi}$, τ such that $\Delta(\tilde{\xi}, i\tau) = 0$. Condition $\Delta_\lambda(\tilde{\xi}, i\tau) \neq 0$ implies that such imaginary zeroes are confined to a finite union of smooth curves

$$\tau = \tau_j(\tilde{\xi}). \tag{1.43}$$

RESULT 4 ([Z.3]; Corollary 5.2). *Given one-dimensional inviscid and structural stability, $\gamma\Delta(0, 1) \neq 0$, weak refined dynamical stability is a necessary condition for multidimensional linearized viscous stability (indeed, for weak spectral stability as well).*

RESULT 5 (New; Theorems 5.4–5.5). *Structural and strong refined dynamical stability together with strong spectral stability are sufficient for linearized viscous stability in dimensions $d \geqslant 2$ and for nonlinear viscous stability in dimensions $d \geqslant 3$ (Lax or overcompressive case) or $d \geqslant 4$ (undercompressive case). In the strongly inviscid stable Lax or overcompressive case, we obtain nonlinear viscous stability in all dimensions $d \geqslant 2$.*

REMARKS 1.22. 1. By the parenthetical comment in Result 4, we may substitute in Result 5 for strong refined dynamical stability the condition of weak inviscid stability together with the nondegeneracy conditions that for any real $\tilde{\xi}$, τ for which $\Delta(\tilde{\xi}, i\tau) = 0$, $\tilde{\xi}$, τ lie off of the glancing set ($\tau \neq \eta_q(\tilde{\xi})$ in the notation of Definition 1.3), $(\partial/\partial\lambda)\Delta(\tilde{\xi}, i\tau) \neq 0$, and $\beta(\tilde{\xi}, i\tau) \neq 0$. This gives a formulation more analogous to that of Result 3 in the one-dimensional case.

2. Results 4 and 5 together resolve the indeterminacy associated with transition to instability through the open region of weak inviscid stability of real type, at least in dimensions $d \geqslant 3$.[21] For, Δ is analytic at $(\tilde{\xi}, i\tau)$ for any real $(\tilde{\xi}, \tau)$ not lying in the glancing set described in Definition 1.3; see discussion below Remark 1.4. Thus, referring to Remark 1.15, we find that the transition from refined dynamical stability to instability must occur at one of three co-dimension one sets: the boundary between strong inviscid stability and weak inviscid stability, marked either by one-dimensional inviscid instability $\Delta(0, 1) = 0$ or coalescence of two roots $(\tilde{\xi}, i\tau)$ at a double zero of Δ; the boundary between weak and strong inviscid stability, at which $(\tilde{\xi}, \tau)$ associated with a root $(\tilde{\xi}, i\tau)$ of Δ strikes the glancing set; or else the (newly defined) boundary between refined dynamical stability and instability, which $(\tilde{\xi}, \tau)$ associated with a root $(\tilde{\xi}, i\tau)$ of Δ strikes the set $\beta(\tilde{\xi}, i\tau) = 0$ within the interior of $\mathcal{R}$. In the scalar case, $\beta(\tilde{\xi}, i\tau)$ is uniformly positive and, moreover, Δ is linear, hence globally analytic with simple roots; indeed, viscous stability holds for profiles

[20] This condition is for Lax or overcompressive shocks, and must be replaced by a modified version in the undercompressive case [Z.3].

[21] Stability is more delicate in the critical dimension $d = 2$ or the undercompressive case, and requires a refined analysis as discussed respectively in Remark 2, end of Section 5.2.3, and Section 5.6 of [Z.3].

of arbitrary strength [HoZ.3,HoZ.4,Z.3]. More generally, β can be expressed in terms of a generalized Melnikov integral, as described in [BSZ.2,B-G.4]; evaluation (either numerical or analytical) of this integral in physically interesting cases is an important area of current investigation.

3. The condition in the definition of strong refined stability that Δ be analytic at roots on the imaginary boundary is generically satisfied in dimension $d = 2$, or for rotationally invariant systems such as gas dynamics. However, it typically fails for general systems in dimensions $d \geqslant 3$. A weaker condition appropriate to this general case is to require that the curve of imaginary zeroes intersect the glancing set transversely. At the expense of further complications, all of our analysis goes through with this relaxed assumption [Z.4]. Moreover, it fails again on a co-dimension one boundary.

4. Multidimensional spectral stability has been verified for small amplitude Lax-type profiles in [FreS.2,PZ], completing the verification of viscous stability by Result 5 and the small amplitude structural and dynamical stability theorems of Pego and Métivier. Classification (either numerical or analytical) of large amplitude stability, and the associated question of whether refined dynamical stability or multidimensional spectral stability in practice is the determining factor, remain outstanding open problems. In particular, large amplitude multidimensional spectral stability has so far not been verified for a single physical example. (It is automatic in the scalar case, by the maximum principle [HoZ.3,HoZ.4]. It has been verified in [PZ] for a class of artificial systems constructed for this purpose.)

5. Freistühler and Zumbrun [FreZ.2] have recently established for the overcompressive version of Δ defined in (3.9) and (3.10), Theorem 3.5, a weak version of Métivier's theorem analogous to that presented for the Lax case in Appendix B, asserting uniform low-frequency stability under the assumptions of Métivier plus the additional nondegeneracy condition of invertibility of an associated one-dimensional "mass map" introduced in [FreZ.1] taking potential time-asymptotic states to corresponding initial perturbation mass. This agrees with numerical and analytical evidence given in [Fre.3,FreL, L.2,FreP.1,FreP.2,DSL], in contrast to the result of automatic inviscid instability described in Remark 1.18. The corresponding investigation of small amplitude spectral stability remains an important open problem in both the over- and undercompressive case.

Plan of the paper. In Section 2 we recall the general Evans function and inverse Laplace transform (C^0-semigroup) machinery needed for our analysis. In Section 3 we construct the Evans function and establish the key low-frequency limit described in Result 1. In Section 4 we carry out the one-dimensional viscous theory described in Results 2 and 3, and in Section 5 the multidimensional viscous theory described in Results 4 and 5.

2. Preliminaries

Consider a profile $\overline{U}$ of a hyperbolic–parabolic system (1.1), satisfying assumptions (A1)–(A3) and (H0)–(H3). By the change of coordinate frame $x_1 \to x_1 - st$ if necessary, we may arrange that $U \equiv \overline{U}(x_1)$ be a *standing-wave solution*, i.e., an equilibrium of (1.1). We make this normalization throughout the rest of the paper, fixing $s = 0$ once and for all.

Linearizing (1.1) about the stationary solution $\overline{U}(\cdot)$ gives

$$U_t = LU := \sum_{j,k} (B^{jk} U_{x_k})_{x_j} - \sum_j (A^j U)_{x_j}, \tag{2.1}$$

where

$$B^{jk} := B^{jk}(\overline{U}(x_1)) \tag{2.2}$$

and

$$A^j U := \mathrm{d}F^j(\overline{U}(x_1))U - \mathrm{d}B^{j1}(\bar{u}(x_1))(U, \overline{U}_{x_1}) \tag{2.3}$$

are C^{s-1} functions of x_1 alone. In particular,

$$(A^j_{11}, A^j_{12}) = (\mathrm{d}F^j_{11}, \mathrm{d}F^j_{12}). \tag{2.4}$$

Taking now the Fourier transform in transverse coordinates $\tilde{x} = (x_2, \ldots, x_d)$, we obtain

$$\begin{aligned}\widehat{U}_t = L_{\tilde{\xi}}\widehat{U} := {} & \overbrace{(B^{11}\widehat{U}')' - (A^1\widehat{U})'}^{L_0\widehat{U}} - \mathrm{i}\sum_{j\neq 1} A^j \xi_j \widehat{U} \\ & + \mathrm{i}\sum_{j\neq 1} B^{j1}\xi_j \widehat{U}' + \mathrm{i}\sum_{k\neq 1} (B^{1k}\xi_k \widehat{U})' - \sum_{j,k\neq 1} B^{jk}\xi_j\xi_k\widehat{U},\end{aligned} \tag{2.5}$$

where "$'$" denotes $\partial/\partial x_1$, and $\widehat{U} = \widehat{U}(x_1, \tilde{\xi}, t)$ denotes the Fourier transform of $U = U(x_1, \tilde{x}, t)$. Hereafter, where there is no danger of confusion we shall drop the hats and write U for $\widehat{U}$. Operator $L_{\tilde{\xi}}$ reduces at $\tilde{\xi} = 0$ to the linearized operator L_0 associated with the one-dimensional stability problem.

A necessary condition for stability is that the family of linear operators $L_{\tilde{\xi}}$ have no unstable point spectrum, i.e., the eigenvalue equations

$$(L_{\tilde{\xi}} - \lambda)U = 0 \tag{2.6}$$

have no solution $U \in L^2(x_1)$ for $\tilde{\xi} \in \mathbb{R}^{d-1}$, $\mathrm{Re}\,\lambda > 0$. For, unstable (L^p) point spectrum of $L_{\tilde{\xi}}$ corresponds to unstable (L^p) essential spectrum of the operator L for $p < \infty$, by a standard limiting argument (see, e.g., [He,Z.1]), and unstable point spectrum for $p = \infty$. This precludes $L^p \to L^p$ stability, by the Hille–Yosida theorem (see, e.g., [He,Fr,Pa,Z.1]). Moreover, standard spectral continuity results [Kat,He,Z.1] yield that instability, if it occurs, occurs for a band of $\tilde{\xi}$ values, from which we may deduce by inverse Fourier transform the exponential instability of (2.1) for test function initial data $U_0 \in C_0^\infty$, with respect to any L^p, $1 \leqslant p \leqslant \infty$. We shall see later that linearized stability of constant solutions $U \equiv U_\pm$ (embodied in (A1)–(A3)) together with convergence at $x_1 \to \pm\infty$ of the coefficients of L, implies that all spectrum of $L_{\tilde{\xi}}$ is point spectrum on the unstable (open) complex half-plane $\operatorname{Re}\lambda > 0$;[22] thus, we are not losing any information by restricting to point spectrum in (2.6). Of course, condition (2.6) is not sufficient for even neutral, or bounded, linearized stability of (2.1).

2.1. *Spectral resolution formulae*

We begin by deriving spectral resolution, or inverse Laplace transform formulae

$$e^{L_{\tilde\xi}t}\phi = \frac{1}{2\pi i}\text{P.V.}\int_{\eta-i\infty}^{\eta+i\infty} e^{\lambda t}(L_{\tilde\xi}-\lambda)^{-1}\phi\,d\lambda, \tag{2.7}$$

and

$$G_{\tilde\xi}(x_1,t;y_1) = \frac{1}{2\pi i}\text{P.V.}\int_{\eta-i\infty}^{\eta+i\infty} e^{ikt}G_{\tilde\xi,\lambda}(x_1,y_1)\,d\lambda, \tag{2.8}$$

for the solution operator $e^{L_{\tilde\xi}t}$ and Green distribution $G_{\tilde\xi}(x_1,t;y_1) := e^{L_{\tilde\xi}t}\delta_{y_1}(x_1)$ associated with the Fourier-transformed linearized evolution equations (2.5), where $G_{\tilde\xi,\lambda}(x_1,y_1)$ denotes the resolvent kernel $(L_{\tilde\xi}-\lambda)^{-1}\delta_{y_1}(x_1)$, and η is any real constant greater than or equal to some $\eta_0(\tilde\xi)$.

LEMMA 2.1. *Operator $L_{\tilde\xi}$ satisfies*

$$\begin{aligned} &|U|_{H^1} + \left|B^{11}U\right|_{H^2} \leqslant C(\tilde\xi)\big(|L_{\tilde\xi}U|_{L^2} + |U|_{L^2}\big),\\ &|SU|_{L^2} \leqslant |\lambda-\lambda_*|^{-1}\left|S(L_{\tilde\xi}-\lambda)U\right|_{L^2}, \end{aligned} \tag{2.9}$$

for all $U \in \mathcal{H} := \{f\colon f\in H^1, B^{11}f\in H^2\}$, for some well-conditioned coordinate transformation $U \to S(x_1)U$, $\sup_{x_1}|S||S^{-1}| \leqslant C(\tilde\xi)$, and all real λ greater than some value $\lambda_* = \lambda_*(\tilde\xi)$.

PROOF. Consider the resolvent equation $(L_{\tilde\xi}-\lambda)U = f$. By the lower triangular coordinate transformation $U \to S_1U$,

$$S_1 := \begin{pmatrix} I & 0\\ -b_1^{11}(b_2^{11})^{-1} & I\end{pmatrix}, \tag{2.10}$$

we may convert to the case that $b_1^{11} \equiv 0$, with conjugation error $EU = E_1U_x + E_0U$,

$$E_1 = \begin{pmatrix} 0 & 0\\ e_1 & e_2\end{pmatrix},$$

[22] In the strictly parabolic case $\operatorname{Re}\sigma\sum\xi_j\xi_kB^{jk} > 0$, or for any asymptotically constant operator L of nondegerate type, this follows by a standard argument of Henry; see [He,Z.1]. More generally, it holds whenever the eigenvalue equation may be expressed as a nondegenerate first-order ODE [MaZ.3].

where b_2^{11} and $A_*^1 := A_{11}^1 - A_{12}^1 b_1^{11}(b_2^{11})^{-1} = A_{11}^1$ retain their former values, and thus their former properties of strict parabolicity and hyperbolicity with constant multiplicity, respectively. By a further, block-diagonal transformation, we may arrange also that b_2^{11} be strictly positive definite and A_{11}^1 symmetric; moreover, this may be chosen so that the combined transformation is uniformly well conditioned.

Setting $\lambda = 0$ and taking the L^2 inner product of $(A_*^1 u_x^I, b_2^{11} u_{xx}^{II})$ against the resulting equation $(L_{\tilde\xi} + E)U = f$, we obtain after some rearrangement the bound

$$C^{-1}\left(|u_x^I|_{L^2} + |u_{xx}^{II}|_{L^2}\right) \leqslant |A_*^1 u_x^I|_{L^2} + |b_2^{11} u_{xx}^{II}|_{L^2} \leqslant C\left(|U|_{L^2} + |u^{II}|_{H^1}\right),$$

which in the original coordinates implies (2.9)(i). Note that we have so far only used nondegeneracy of the respective leading order coefficients A_*^1 and b_2^{11} in the u^I and u^{II} equations.

Taking the real part of the L^2 inner product of U against the full resolvent equation $(L_{\tilde\xi} + E - \lambda)U = f$, integrating by parts, using the assumptions on A_{11}^1, b_2^{11}, and bounding terms of order $|U|_{L^2}|u_x^{II}|_{L^2}$ by Youngs inequality as $\mathrm{O}(|U|_{L^2}^2) + \epsilon|u_x^{II}|_{L^2}^2$ with ϵ as small as needed, we obtain after some rearrangement the estimate

$$\operatorname{Re}\lambda|U|_{L^2}^2 + \mathrm{O}\left(|U|_{L^2}^2\right) + \theta|u_x^{II}|_{L^2}^2 \leqslant |f|_{L^2}|U|_{L^2}, \tag{2.11}$$

whence we obtain the second claimed bound

$$|U|_{L^2} \leqslant |\lambda - \lambda_*|^{-1}|f|_{L^2} = |\lambda - \lambda_*|^{-1}\left|(L_{\tilde\xi} - \lambda)U\right|_{L^2} \tag{2.12}$$

for all $\lambda > \lambda_*$ on the real axis, and λ_* sufficiently large. □

COROLLARY 2.2. *Operator $L_{\tilde\xi}$ is closed and densely defined on L^2 with domain $\mathcal{H}$, generating a C^0 semigroup $\mathrm{e}^{L_{\tilde\xi}t}$ satisfying $|\mathrm{e}^{L_{\tilde\xi}t}|_{L^2} \leqslant C\mathrm{e}^{\omega t}$ for some real $\omega = \omega(\tilde\xi)$.*

PROOF. The first assertion follows in standard fashion from bound (2.9)(i); see, e.g., [Pa] or [Z.1]. The bound (2.9)(ii) applies also to the limiting, constant-coefficient operators $L_{\tilde\xi\pm}$ as $x \to \pm\infty$, whence the spectra of these operators is confined to $\operatorname{Re}\lambda \leqslant \lambda_*$. Because the eigenvalue equation $(L_{\tilde\xi} - \lambda)U = 0$, after the coordinate transformation described in the proof of Lemma 2.1, may evidently be written as a nondegenerate ODE in $(U, u^{II'})$, we may conclude by the general theory described in Section 2.4 that the essential spectrum of $L_{\tilde\xi}$ is also confined to the set $\operatorname{Re}\lambda \leqslant \lambda_*$; see also Proposition 4.4. of [MaZ.3]. (This observation generalizes a standard result of Henry ([He], Lemma 2, pp. 138–139) in the case of a nondegenerate operator L.) Since (2.9) precludes point spectrum for real $\lambda > \lambda_*$, we thus find that all such λ belong to the resolvent set $\rho(L_{\tilde\xi})$, with the resolvent bound (after the coordinate transformation described above)

$$\left|(L_{\tilde\xi} - \lambda)^{-1}\right|_{L^2} \leqslant |\lambda - \lambda_*|^{-1}. \tag{2.13}$$

But, this is a standard sufficient condition that a closed, densely defined operator $L_{\tilde\xi}$ generate a C^0 semigroup, with $|\mathrm{e}^{L_{\tilde\xi}t}|_{L^2} \leqslant C\mathrm{e}^{\omega t}$ for all real $\omega > \lambda_*$ (see, e.g., [Pa], Theorem 5.3, or [Fr,Y]). □

COROLLARY 2.3. *For* $L_{\tilde\xi}$, $e^{L_{\tilde\xi}t}$ *as in Corollary* 2.2, *the inverse Laplace transform* (*spectral resolution*) *formula* (2.7) *holds for any real* η *greater than some* η_0, *for all* $\phi \in D(L_{\tilde\xi} \cdot L_{\tilde\xi})$, *where domain* $D(L_{\tilde\xi} \cdot L_{\tilde\xi})$ *is defined as in* [Pa], *p.* 1. *Likewise*, (2.8) *holds in the distributional sense.*[23]

PROOF. The first claim is a general property of C^0 semigroups (see, e.g., [Pa], Corollary 7.5). The second follows from the first upon pairing with test functions ϕ (see, e.g., [MaZ.3], Section 2). □

REMARKS 2.4. 1. By standard semigroup properties, the Green distribution $G_{\tilde\xi}(x, t; y)$ satisfies

(i) $(\partial_t - L_{\tilde\xi})G_{\tilde\xi}(\cdot, \cdot; y_1) = 0$ in the distributional sense, for all $t > 0$, and

(ii) $G_{\tilde\xi}(x_1, t; y_1) \rightharpoonup \delta(x - y)$ as $t \to 0$.

Here, $G_{\tilde\xi}$ is to be interpreted in (i) as a distribution in the joint variables (x_1, y_1, t), and in (ii) as a distribution in (x_1, y_1), continuously parametrized by t. We shall see by explicit calculation that $G_{\tilde\xi}$ so defined (uniquely, by uniqueness of weak solutions of (2.5) within the class of test function initial data), is a measure but not a function. Note that, on the resolvent set $\rho(L)$, the resolvent kernel $G_{\tilde\xi,\lambda}$ likewise has an alternative, intrinsic characterization as the unique distribution satisfying

$$(L_{\tilde\xi} - \lambda)G_{\tilde\xi,\lambda}(x, y) = \delta(x, y) \tag{2.14}$$

and taking $f \in L^2$ to $\langle G_{\tilde\xi,\lambda}(x, y), f(y)\rangle_y \in L^2$. This is the characterization that we shall use in constructing $G_{\tilde\xi,\lambda}$ in Section 2.4.

2. A semigroup on even a still more restricted function class than L^2 would have sufficed in this construction, since we are constructing objects in the very weak class of distributions. Later, by explicit computation, we will verify that $L_{\tilde\xi}$ in fact generates a C^0 semigroup in any L^p. Note also that (2.8) implies the expected, standard solution formula $e^{L_{\tilde\xi}t} f = \langle G_{\tilde\xi}(x, t; y), f(y)\rangle_y$, or, formally:

$$e^{L_{\tilde\xi}t} f = \int G_{\tilde\xi}(x, t; y) f(y)\, dy \tag{2.15}$$

for f in any underlying Banach space on which $e^{L_{\tilde\xi}t}$ is defined, in this case on L^p, $1 \leqslant p \leqslant \infty$.

3. As noted in [MaZ.1,MaZ.3], there is another, more concrete route to (2.7), (2.8) and (2.15), generalizing the approach followed in [ZH] for the parabolic case. Namely, we may observe that, in each finite integral in the approximating sequence defined by the principal value integral (2.8), we may exchange orders of integration and distributional differentiation, using Fubini's theorem together with the fact, established in

[23]Note: From the definition in [Pa], we find that the domain $D(L_{\tilde\xi})$ of $L_{\tilde\xi}$ satisfies $H^2 \subset \mathcal{H} \subset D(L_{\tilde\xi})$. Likewise, the domain $D(L_{\tilde\xi} \cdot L_{\tilde\xi})$, consisting of those functions V such that $L_{\tilde\xi} V \in D(L_{\tilde\xi})$, satisfies $H^4 \subset D(L_{\tilde\xi} \cdot L_{\tilde\xi})$.

the course of our analysis, that $G_{\xi,\lambda}$ is uniformly bounded on the contours under consideration (in fact, we establish the much stronger result that $G_{\xi,\lambda}$ decays exponentially in $|x-y|$, with uniform rate), to obtain

$$(\partial_t - L_{\xi}) \int_{\eta-\mathrm{i}K}^{\eta+\mathrm{i}K} \mathrm{e}^{\lambda t} G_{\xi,\lambda}(x,y)\,\mathrm{d}\lambda = \delta(x-y) \int_{\eta-\mathrm{i}K}^{\eta+\mathrm{i}K} \mathrm{e}^{\lambda t}\,\mathrm{d}\lambda \rightharpoonup \delta(x-y) \otimes \delta(t) \tag{2.16}$$

as $K \to \infty$, for all $t \geqslant 0$. So, all that we must verify is: (i′) the principal value integral (2.8) in fact converges to some distribution $G_{\xi}(x,t;y)$ as $K \to \infty$; and (ii′) $G_{\xi}(x,t;y)$ has a limit as $t \to 0$. For, distributional limits and derivatives commute (essentially by definition), so that (i′) and (ii′) together with (2.16) imply (i) and (ii) above. Facts (i′) and (ii′) will be established by direct calculation in the course of our analysis, thus verifying formula (2.8) at the same time that we use it to obtain estimates on G_{ξ}. Note that we have made no reference in this argument to the semigroup machinery cited above.

2.2. *Evans function framework*

We next recall the basic Evans function/asymptotic ODE tools that we will need.

2.2.1. *The gap/conjugation lemma.* Consider a family of first-order systems

$$W' = \mathbb{A}(x,\lambda)W, \tag{2.17}$$

where λ varies within some domain $\Omega \subset \mathbb{C}^k$ and x varies within $\mathbb{R}^1$. Equations (2.17) are to be thought of as generalized eigenvalue equations, with parameter λ representing a suite of frequencies. We make the basic assumption:

(h0) Coefficient $\mathbb{A}(\cdot,\lambda)$, considered as a function from Ω into $L^\infty(x)$ is analytic in λ. Moreover, $\mathbb{A}(\cdot,\lambda)$ approaches exponentially to limits $\mathbb{A}_\pm(\lambda)$ as $x \to \pm\infty$, with uniform exponential decay estimates

$$\left|(\partial/\partial x)^k(\mathbb{A} - \mathbb{A}_\pm)\right| \leqslant C_1 \mathrm{e}^{-\theta|x|/C_2} \quad \text{for } x \gtrless 0,\ 0 \leqslant k \leqslant K, \tag{2.18}$$

$C_j, \theta > 0$, on compact subsets of Ω.

Then, there holds the following *conjugation lemma* of [MéZ.1], a refinement of the "gap lemma" of [GZ,KS], relating solutions of the variable-coefficient ODE (2.17) to solutions of its constant-coefficient limiting equations as

$$Z' = \mathbb{A}_\pm(\lambda)Z \tag{2.19}$$

as $x \to \pm\infty$.

LEMMA 2.5 [MéZ.1]. *Under assumption* (h0) *alone, there exists locally to any given* $\lambda_0 \in \Omega$ *a pair of linear transformations* $P_+(x,\lambda) = I + \Theta_+(x,\lambda)$ *and* $P_-(x,\lambda) =$

$I+\Theta_-(x,\lambda)$ on $x \geqslant 0$ and $x \leqslant 0$, respectively, $\Theta_\pm$ analytic in λ as functions from Ω to $L^\infty[0,\pm\infty)$, such that:

(i) *$|P_\pm|$ and their inverses are uniformly bounded, with*

$$\begin{aligned}&\left|(\partial/\partial\lambda)^j(\partial/\partial x)^k\Theta_\pm\right|\\&\quad\leqslant C(j)C_1C_2\mathrm{e}^{-\theta|x|/C_2}\quad \text{for } x \gtrless 0,\ 0\leqslant k\leqslant K+1,\end{aligned}\tag{2.20}$$

$j \geqslant 0$, where $0<\theta<1$ is an arbitrary fixed parameter, and $C>0$ and the size of the neighborhood of definition depend only on θ, j, the modulus of the entries of $\mathbb{A}$ at λ_0, and the modulus of continuity of $\mathbb{A}$ on some neighborhood of $\lambda_0 \in \Omega$.

(ii) *The change of coordinates $W := P_\pm Z$ reduces* (2.17) *on $x \geqslant 0$ and $x \leqslant 0$, respectively, to the asymptotic constant-coefficient equations* (2.19).

Equivalently, solutions of (2.17) may be conveniently factorized as

$$W = (I+\Theta_\pm)Z_\pm,\tag{2.21}$$

where $Z_\pm$ are solutions of the constant-coefficient equations (2.19), and $\Theta_\pm$ satisfy bounds (2.20).

PROOF OF LEMMA 2.5. As described in [MéZ.1], for $j=k=0$ this is a straightforward corollary of the gap lemma as stated in [Z.3], applied to the "lifted" matrix-valued ODE for the conjugating matrices $P_\pm$; see also Appendix A.3. The x-derivative bounds $0<k\leqslant K+1$ then follow from the ODE and its first K derivatives. Finally, the λ-derivative bounds follow from standard interior estimates for analytic functions. □

DEFINITION 2.6. Following [AGJ], we define the *domain of consistent splitting* for problem (2.17) as the (open) set of λ such that the limiting matrices $\mathbb{A}_+$ and $\mathbb{A}_-$ are hyperbolic (have no center subspace), and the dimensions of their stable (resp. unstable) subspaces S_+ and S_- (resp. unstable subspaces U_+ and U_-) agree.

LEMMA 2.7. *On any simply connected subset of the domain of consistent splitting, there exist analytic bases $\{V_1,\dots,V_k\}^\pm$ and $\{V_{k+1},\dots,V_N\}^\pm$ for the subspaces $S_\pm$ and $U_\pm$ defined in Definition* 2.6.

PROOF. By spectral separation of $U_\pm$, $S_\pm$, the associated (group) eigenprojections are analytic. The existence of analytic bases then follows by a standard result of Kato (see [Kat], pp. 99–102). □

By Lemma 2.5, on the domain of consistent splitting, the subspaces

$$\mathcal{S}^+ = \operatorname{Span}\{W_1^+,\dots,W_k^+\} := \operatorname{Span}\{P_+V_1^+,\dots,P_+V_k^+\}\tag{2.22}$$

and

$$\mathcal{U}^- := \operatorname{Span}\{W_{k+1}^-,\dots,W_N^-\} := \operatorname{Span}\{P_-V_{k+1}^-,\dots,P_-V_N^-\}\tag{2.23}$$

uniquely determine the stable manifold as $x \to +\infty$ and the unstable manifold as $x \to -\infty$ of (2.17), defined as the manifolds of solutions decaying as $x \to \pm\infty$, respectively, independent of the choice of $P_\pm$.

DEFINITION 2.8. On any simply connected subset of the domain of consistent splitting, let $V_1^+, \ldots, V_k^+$ and $V_{k+1}^-, \ldots, V_N^-$ be analytic bases for S_+ and U_-, as described in Lemma 2.7. Then, the *Evans function* for (2.17) associated with this choice of limiting bases is defined as

$$\begin{aligned} D(\lambda) &:= \det\big(W_1^+, \ldots, W_k^+, W_{k+1}^-, \ldots, W_N^-\big)_{|x=0,\lambda} \\ &= \det\big(P_+V_1^+, \ldots, P_+V_k^+, P_-V_{k+1}^-, \ldots, P_-V_N^-\big)_{|x=0,\lambda}, \end{aligned} \tag{2.24}$$

where $P_\pm$ are the transformations described in Lemma 2.5.

REMARK 2.9. Note that D is independent of the choice of $P_\pm$; for, by uniqueness of stable/unstable manifolds, the exterior products (minors) $P_+V_1^+ \wedge \cdots \wedge P_+V_k^+$ and $P_-V_{k+1}^- \wedge \cdots \wedge P_-V_N^-$ are uniquely determined by their behavior as $x \to +\infty, -\infty$, respectively.

PROPOSITION 2.10. *Both the Evans function and the stable/unstable subspaces $\mathcal{S}^+$ and $\mathcal{U}^-$ are analytic on the entire simply connected subset of the domain of consistent splitting on which they are defined. Moreover, for λ within this region,* (2.17) *admits a nontrivial solution $W \in L^2(x)$ if and only if $D(\lambda) = 0$.*

PROOF. Analyticity follows by uniqueness, and local analyticity of $P_\pm$, $V_k^\pm$. Noting that the first k columns of the matrix on the right-hand side of (2.24) are a basis for the stable manifold of (2.17) at $x \to +\infty$, while the final $N-k$ columns are a basis for the unstable manifold at $x \to -\infty$, we find that its determinant vanishes if and only if these manifolds have nontrivial intersection, and the second assertion follows. □

REMARKS 2.11. 1. In the case that (2.17) describes an eigenvalue equation associated with an ordinary differential operator L, $\lambda \in \mathbb{C}^1$, Proposition 2.10 implies that eigenvalues of L agree in location with zeroes of D. In [GJ.1,GJ.2], Gardner and Jones have shown that they agree also in multiplicity (see also Lemma 6.1 of [ZH], or Proposition 6.15 of [MaZ.3].[24]

2. If, further, L is a real-valued operator (i.e., has real coefficients), or, more generally, $\mathbb{A}$ has complex symmetry $\mathbb{A}(x, \bar\lambda) = \overline{\mathbb{A}}(x, \lambda)$, where bar denotes complex conjugate, then D may be chosen with the same symmetry

$$\overline{D(\lambda)} = D(\bar\lambda). \tag{2.25}$$

[24]The latter result applies specifically to ordinary differential operators of degenerate type.

For, all steps in the construction of the Evans function respect complex symmetry: projection onto stable/unstable subspaces by the spectral resolution formula

$$P_\pm(\lambda) = \frac{1}{2\pi \mathrm{i}} \oint_{\Gamma_\pm} \left(\mathbb{A}_\pm(\lambda) - \mu\right)^{-1} \mathrm{d}\mu$$

($\Gamma_\pm$ denoting fixed contours enclosing stable/unstable eigenvalues of $\mathbb{A}_\pm$), choice of asymptotic bases $V_\pm(\lambda)$ by solution of the analytic ODE prescribed by Kato [Kat], and finally conjugation to variable-coefficient solutions by a contraction mapping that likewise respects complex symmetry.

2.2.2. *The tracking/reduction lemma.* Next, consider the complementary situation of a family of equations of form

$$W' = \mathbb{A}^\epsilon(x, \lambda) W, \tag{2.26}$$

on an (ϵ, λ)-neighborhood for which the coefficient $\mathbb{A}^\epsilon$ does not exhibit uniform exponential decay to its asymptotic limits, but instead is *slowly varying*. This occurs quite generally for rescaled eigenvalue equations arising in the study of the large frequency regime (see, e.g., [GZ,ZH,MaZ.1,Z.3]).

In this situation, it frequently occurs that not only $\mathbb{A}^\epsilon$ but also certain of its invariant (group) eigenspaces are slowly varying with x, i.e., there exist matrices

$$L^\epsilon = \begin{pmatrix} L_1^\epsilon \\ L_2^\epsilon \end{pmatrix}(x), \qquad R^\epsilon = \begin{pmatrix} R_1^\epsilon & R_2^\epsilon \end{pmatrix}(x) \tag{2.27}$$

for which $L^\epsilon R^\epsilon(x) \equiv I$ and $|LR'| = |L'R|$ is small relative to

$$\mathbb{M}^\epsilon := L^\epsilon \mathbb{A}^\epsilon R^\epsilon(x) = \begin{pmatrix} M_1^\epsilon & 0 \\ 0 & M_2^\epsilon \end{pmatrix}(x), \tag{2.28}$$

where "$'$" as usual denotes $\partial/\partial x$. In this case, making the change of coordinates $W^\epsilon = R^\epsilon Z$, we may reduce (2.26) to the approximately block-diagonal equation

$$Z^{\epsilon\prime} = \mathbb{M}^\epsilon Z^\epsilon + \delta^\epsilon \Theta^\epsilon Z^\epsilon, \tag{2.29}$$

where $\mathbb{M}^\epsilon$ is as in (2.28), $\Theta^\epsilon(x)$ are uniformly bounded matrices, and $\delta^\epsilon(x) \leqslant \delta(\epsilon)$ is a (relatively) small scalar.

Let us assume that such a procedure has been successfully carried out, and, moreover, that the approximate flows (i.e., solution operators) $\overline{\mathcal{F}}_j^{y\to x}$ generated by the decoupled equations

$$Z' = M_j^\epsilon Z \tag{2.30}$$

are *uniformly exponentially separated* to order $2\eta(\epsilon)$, in the sense that

$$\left|\overline{\mathcal{F}}_1^{y\to x}\right| \left|\left(\overline{\mathcal{F}}_2^{y\to x}\right)^{-1}\right| \leqslant C\mathrm{e}^{-2\eta(\epsilon)|x-y|} \quad \text{for } x \geqslant y. \tag{2.31}$$

REMARK 2.12. A sufficient condition for and the usual means of verification of (2.31) is existence of an approximate *uniform spectral gap*:

$$\max \sigma\big(\mathrm{Re}\big(M_1^\epsilon\big)\big) - \min \sigma\big(\mathrm{Re}\big(M_2^\epsilon\big)\big) \leqslant -2\eta(\epsilon) + \alpha^\epsilon(x) \tag{2.32}$$

for all x, $\eta(\epsilon) > 0$, where α^ϵ is uniformly integrable, $\int |\alpha^\epsilon(x)|\,\mathrm{d}x \leqslant C$, or, alternatively, existence of flows conjugate to $\mathcal{F}_j^{y\to x}$ and satisfying this condition. Here, $\mathrm{Re}(M) := (1/2)(M^* + M)$ denotes the symmetric part of a matrix or linear operator M. Note that an exact uniform spectral gap (2.32) with $\alpha^\epsilon \equiv 0$ may be achieved from (2.32) by the coordinate change $Z_1^\epsilon \to \omega Z_1^\epsilon$, where ω is a well-conditioned (scalar) exponential weight defined by $\omega' = -\alpha^\epsilon \omega$, $\omega(0) = 1$. From the exact version, we may obtain (2.31) by standard energy estimates, with $C = 1$.

Then, there holds the following *reduction lemma*, a refinement established in [MaZ.1] of the "tracking lemma" given in varying degrees of generality in [GZ,ZH,Z.3].

PROPOSITION 2.13 [MaZ.1]. *Consider a system* (2.29) *satisfying the exponential separation assumption* (2.31), *with* Θ^ϵ *uniformly bounded for* ϵ *sufficiently small. If, for* $0 < \epsilon < \epsilon_0$, *the ratio* $\delta(\epsilon)/\eta(\epsilon)$ *of formal error vs. spectral gap is sufficiently small relative to the bounds on* Θ, *in particular if* $\delta(\epsilon)/\eta(\epsilon) \to 0$, *then, for all* $0 < \epsilon \leqslant \epsilon_0$, *there exist (unique) linear transformations* $\Phi_1^\epsilon(x,\lambda)$ *and* $\Phi_2^\epsilon(x,\lambda)$, *possessing the same regularity with respect to the parameters* ϵ, λ *as do coefficients* $\mathbb{M}^\epsilon$ *and* $\delta(\epsilon)\Theta^\epsilon$, *for which the graphs* $\{(Z_1, \Phi_2^\epsilon Z_1)\}$ *and* $\{(\Phi_1^\epsilon Z_2, Z_2)\}$ *are invariant under the flow of* (2.29) *and satisfying*

$$\big|\Phi_j^\epsilon\big|,\ \big|\partial_x \Phi_j^\epsilon\big| \leqslant C\delta(\epsilon)/\eta(\epsilon) \quad \textit{for all } x. \tag{2.33}$$

PROOF. As described in Appendix C of [MaZ.1], this may be established by a contraction mapping argument carried out for the projectivized "lifted" equations governing the flow of exterior forms; see, e.g., [Sat] for a corresponding argument in the case that M_1, M_2 are scalar. We give a more direct, "matrix-valued" version of this argument in Appendix A.3, from which one may obtain further an explicit (but rather complicated) Neumann series for Φ_j^ϵ in powers of δ/η. □

From Proposition 2.13, we obtain reduced flows

$$Z_1^{\epsilon\prime} = M_1^\epsilon Z_1^\epsilon + \delta^\epsilon\big(\Theta_{11} + \Theta_{12}^\epsilon \Phi_2^\epsilon\big) Z_1^\epsilon = M_1^\epsilon Z_1^\epsilon + \mathcal{O}\big(\delta^\epsilon(x)\big) Z_1^\epsilon \tag{2.34}$$

and

$$Z_2^{\epsilon\prime} = M_2^\epsilon Z_2^\epsilon + \delta^\epsilon\big(\Theta_{22} + \Theta_{21}^\epsilon \Phi_1^\epsilon\big) Z_2^\epsilon = M_2^\epsilon Z_2^\epsilon + \mathcal{O}\big(\delta^\epsilon(x)\big) Z_2^\epsilon \tag{2.35}$$

on the two invariant manifolds described. Let us focus on the flow (2.34), assuming without loss of generality (by a centering exponential weighting if necessary) that

$$\big|\mathcal{F}_1^{y\to x}\big| \leqslant C e^{-\eta(\epsilon)|x-y|} \quad \text{for } x \leqslant y. \tag{2.36}$$

COROLLARY 2.14 [MaZ.1]. *Assuming* (2.36), *the flow* $\mathcal{F}_1^{y\to x}$ *of the reduced equation* (2.34) *satisfies*

$$\mathcal{F}_1^{y\to x} = \overline{\mathcal{F}}_1^{y\to x} + \sum_{j=1}^{J} (\delta/\eta)^j \mathcal{E}_j(x,y,\epsilon) + \mathcal{O}(\delta/\eta)^{J+1} \mathrm{e}^{-\tilde{\theta}\eta|x-y|} \tag{2.37}$$

for $x \leqslant y$, *for any* $0 < \tilde{\theta} < 1$, *where* $\overline{\mathcal{F}}_1^{y\to x}$ *is the flow of the associated decoupled system* (2.30); *the iterated integrals*

$$\begin{aligned} \mathcal{E}_{j+1} &:= \eta \int_y^x \overline{\mathcal{F}}_1^{z\to x} (\Theta_{11} + \Theta_{12}\Phi_2)(z,\epsilon) \mathcal{E}_j(z,y,\epsilon)\,\mathrm{d}z, \\ \mathcal{E}_0(x,y,\epsilon) &:= \overline{\mathcal{F}}_1^{y\to x}, \end{aligned} \tag{2.38}$$

satisfy the uniform exponential decay estimates

$$\left|\mathcal{E}_j(x,y,\epsilon)\right| \leqslant C_j \mathrm{e}^{-\tilde{\theta}\eta|x-y|}, \tag{2.39}$$

and are as smooth in x *and* y *and as smooth in* ϵ *as is* $(\Theta_{11} + \Theta_{12}\Phi_2)$; *and* $C_j > 0$ *and* $\mathcal{O}(\cdot)$ *are uniform in* ϵ, *depending only on* $\tilde{\theta}$, M_j, *and the constant* C *in* (2.33). *Symmetric bounds hold for the flow* $\mathcal{F}_2^{y\to x}$ *of* (2.35).

PROOF. This follows by a contraction mapping argument similar to (but simpler than) that used in the proof of Proposition 2.13, where expansion (2.37) is just the Jth-order Neumann expansion with remainder associated with the contraction mapping; see Appendix C of [MaZ.1], or Appendix A.3 of this chapter. □

2.2.3. *Formal block diagonalization.* To complete our discussion of ODE with slowly varying coefficients, we now supply a formal block-diagonalization procedure to be used in tandem with the rigorous error bounds of Corollary 2.14. Consider (2.17) with coefficients of the form

$$\mathbb{A}^\epsilon(x,\lambda) = A\big(\epsilon x, \epsilon\lambda/|\lambda|\big), \tag{2.40}$$

ϵ small, $\lambda/|\lambda|$ (and possibly other indexing parameters) restricted to a compact set depending on ϵ, where A has formal Taylor expansion

$$A(y,\epsilon) = \sum_{k=0}^{p} \epsilon^k A_k(y) + \mathrm{O}\big(\epsilon^{p+1}\big). \tag{2.41}$$

Assume:

(h0) A decays uniformly exponentially as $y \to \pm\infty$ to limits $A_\pm$.

(h1) $A_j \in C^{p+1-j}(y)$ for $0 \leqslant j \leqslant p$, with derivatives uniformly bounded for all $y \in \mathbb{R}$, and $\mathcal{O}(\cdot) \in C^0$ and uniformly bounded for $y \in \mathbb{R}$, $\epsilon \leqslant \epsilon_0$, some $\epsilon_0 > 0$.

(h2) $A_0(y)$ is block-diagonalizable to form

$$D_0(y) = T_0 A_0 T_0^{-1}(y) = \operatorname{diag}\{d_{0,1}, \ldots, d_{0,s}\}, \quad d_j \in \mathbb{C}^{n_j \times n_j}, \tag{2.42}$$

with spectral separation between blocks (i.e., complex distance between their eigenvalues) uniformly bounded below by some $\gamma > 0$, for $\epsilon \leqslant \epsilon_0$, $y \in \mathbb{R}$.

Then, we have:

PROPOSITION 2.15 [MaZ.1]. *Given* (h1) *and* (h2), *there exists a uniformly well-conditioned change of coordinates* $W = T\widetilde{W}$ *such that*

$$\widetilde{W}' = D(\epsilon x, \epsilon)\widetilde{W} + O(\epsilon^{p+1})\widetilde{W} \tag{2.43}$$

uniformly for $x \in \mathbb{R}$, $\epsilon \leqslant \epsilon_0$, *with*

$$D(y, \epsilon) = \sum_{k=0}^{p} \epsilon^k D_k(y) + O(\epsilon^{p+1}) \tag{2.44}$$

and

$$T(y, \epsilon) = \sum_{k=0}^{p} \epsilon^k T_k(y), \tag{2.45}$$

where D_j, $T_j \in C^{p+1-j}(y)$ *with uniformly bounded derivatives for all* $0 \leqslant j \leqslant p$, $\mathcal{O}(\epsilon^{p+1}) \in C^0(y)$ *uniformly bounded, and each* D_j *has the same block-diagonal form*

$$D_j(y) = \operatorname{diag}\{d_{j,1}, \ldots, d_{j,s}\}, \tag{2.46}$$

$d_{j,k} \in \mathbb{C}^{n_k \times n_k}$, *as does* D_0.

If there holds (h0) *as well, then it is possible to choose* $T_0(\cdot)$ *in such a way that, also,* $D_0 + \epsilon D_1$ *is determined simply by the block-diagonal part of* $T_0^{-1}(A_0 + \epsilon A_1)T_0$, *or, equivalently,*

$$T_0^{-1}(\partial/\partial y)T_0 \equiv 0. \tag{2.47}$$

PROOF. See Propositions 4.4 and 4.8 of [MaZ.1], or Appendix A of this chapter. Note that our slightly weakened version of (h1) is what is actually used in the proof, and not the stronger version given in [MaZ.1]. This distinction was unimportant in the relaxation case, but will be important here. □

REMARKS 2.16. 1. An intuitive way to see Proposition 2.15 is to express T as a product $T_0 T_1 \cdots T_p$ and solve the resulting succession of nearby diagonalization problems; a more efficient procedure is given in [MaZ.1]. From this point of view, it is clear that one derivative in y must be given up for each power of ϵ, since each transformation T_j introduces a diagonalization error of form $T_j(\partial/\partial y)T_j^{-1}$; this explains also the origin of condition (2.47).

The class of perturbation series (h1) is preserved under slowly varying changes of coordinates, hence quite natural for the problem at hand. In particular, the series D obtained by diagonalization is in the same class, and so the procedure is convenient for iteration. If desired, $W^{j,\infty}(y)$ may be substituted for $C^j(y)$ everywhere.

2. We call attention to the important statement in Proposition 2.15 that any additional regularity with respect to ϵ in the error term $\mathcal{O}(\epsilon^{p+1})$ in (2.41) translates to the same regularity $C^q(\epsilon \to C^0(y))$ for the error terms $\mathcal{O}(\epsilon^{p+1})$ in (2.44) and (2.45). In particular, if the remainder term is $C^\omega(\epsilon \to C^0(y))$, as often happens (for example, for both relaxation and real viscous profiles), then the error term has regularity $C^\omega(\epsilon \to C^0(y))$. This observation is used in conjunction with Proposition 2.13 and Corollary 2.14; specifically, it gives regularity with respect to ϵ in error terms Θ_{jk}, and the resulting graphs Φ_j, and thus in terms $\mathcal{F}^{y\to x}$, $\mathcal{E}_j$, and $\mathcal{O}(\delta/\eta)^{J+1}e^{-\tilde{\theta}|x-y|}$ of expansion (2.37). Likewise, we point out, in the typical case that (2.40) is obtained by rescaling, that analyticity in λ in the *original* (unrescaled) coordinates, if it holds, may also preserved by appropriate choice of T_0, even though we have expressed A in terms of the nonanalytic parameter $\lambda/|\lambda|$, and this property also is inherited in the fixed-point constructions of Proposition 2.13 and Corollary 2.14; for details, see Remarks 4.11 and 4.12 of [MaZ.1].

2.2.4. *Splitting of a block-Jordan block.* Augmenting the above discussion, we briefly consider the complementary case of an expansion (2.41) for which (h0)–(h2) hold, but (h3) is replaced by

(h3′)(i) A_0 is a standard block-Jordan block of order s,

$$A_0(y) = J_{s,q} + \mu(y) I_{sq}, \qquad J_{s,q} := \begin{pmatrix} 0 & I_q & \cdots & 0 \\ 0 & 0 & I_q \cdots & 0 \\ \vdots & \vdots & \cdots & \vdots \\ 0 & \cdots & 0 & I_q \\ 0 & \cdots & \cdots & 0 \end{pmatrix}; \tag{2.48}$$

and (ii) for each y, the spectrum of the lower left-hand block $M := (A_1)_{s1}$ of A_1 lies in a compact set on which $z \to z^{1/s}$ is analytic; in particular, $\det(M) \neq 0$.

Such an expansion typically arises as a single block $d_j(y,\epsilon) := \sum_{k=0}^{p} \epsilon^k d_{k,j}(y)$ of the matrix $D(y,\epsilon)$ obtained by the block-diagonalization procedure just described. Indeed, the situation may be expected to occur in general for operators of degenerate (mixed) type, indicating the simultaneous presence of multiple scales. (Clearly, still more degenerate situations may occur, and could be treated on a case-by-case basis.)

REMARK 2.17. We have assumed that A_0 has already been put in form (2.48) because (h3′)(ii) is not preserved under general coordinate changes, due to dynamical effects $T_0^{-1}(\partial/\partial y)T_0$. On the other hand a calculation similar to that of Proposition 4.8 of [MaZ.1] shows that $(T_0^{-1}(\partial/\partial y)T_0)_{s1} = 0$ for transformations preserving the block form (2.48) (namely, block upper-triangular matrices with diagonal blocks equal to some common invertible α), and so (h3′)(ii) *is* independent of the manner in which (2.48) is achieved from (2.41).

Performing the "balancing" transformation $A \to \mathcal{B}^{-1} A \mathcal{B}$,

$$\mathcal{B} := \operatorname{diag}\{1, \epsilon^{1/s}, \ldots, \epsilon^{(s-1)/s}\}, \tag{2.49}$$

we may convert this to an expansion

$$A(y, \epsilon) = \mu I + \sum_{k=1}^{p} \epsilon^{k/s} A_k(y) + \mathcal{O}(\epsilon^{(p+1)/s}) \tag{2.50}$$

in powers of $\epsilon^{1/s}$, $A_j \in C^{p+1-j}$, $\mathcal{O}(\epsilon^{(p+1)/s}) \in C^0$, where

$$A_1 = \begin{pmatrix} 0 & I_q & \cdots & 0 \\ 0 & 0 & I_q \cdots & 0 \\ \vdots & \vdots & \cdots & \vdots \\ 0 & \cdots & 0 & I_q \\ M & \cdots & \cdots & 0 \end{pmatrix} \tag{2.51}$$

and M satisfies the spectral criterion (h3′)(ii).

Then, the spectrum $(\sigma M)^{1/s}$ of A_1 separates into s groups of q, corresponding to the s different analytic representatives of $z^{1/s}$, and so A_1 is $(q \times q)$-block-diagonalizable, reducing to the previous case with ϵ replaced by $\epsilon^{1/s}$ and A_0 shifted to A_1. This corresponds to the identification of a new, slow scale $\epsilon^{1/s}$ in the problem. Note that differentiation still brings down a full factor of $\epsilon \ll \epsilon^{1/s}$, so that the first step in the diagonalization process can still be carried out, despite the shift in the series. Alternatively, we could remove the μI term by an exponential weighting, and rescale y to remove the initial factor $\epsilon^{1/s}$, thus converting to the standard situation described in Section 2.2.3.

2.2.5. *Approximation of stable/unstable manifolds.* Proposition 2.15 and Corollary 2.14 together give a recipe for the estimation of the stable/unstable manifolds of ODE with slowly varying coefficients, given (h0)–(h2): namely, expand formally to order $\delta = \epsilon^p$ such that $\delta/\eta \to 0$, then apply Corollary 2.14 to obtain a rigorous approximation in terms of the resulting block-decoupled system. In the case relevant to traveling waves that all A_j (not just A_0) converge exponentially in y to constants $A_j^{\pm}$ as $y \to \pm\infty$, we have, provided that (i) $d_{0,j}$ are scalar multiples of the identity; and (ii) there holds the necessary condition of neutral separation to zeroth order, $\operatorname{Re}\sigma(d_{0,j}) \leqslant 0 \leqslant \operatorname{Re}\sigma(d_{0,k})$, $j \leqslant S < k$, for all y; that this procedure is possible if and only if the corresponding eigenspaces of the original matrix $\mathbb{A}^{\epsilon}$, or equivalently of $\epsilon^p D_p$, have a uniform spectral gap of order ϵ^p at the limits $x = \pm\infty$, i.e., there holds the easily checkable condition

$$\operatorname{Re}\sigma(d_{p,j})^{\pm} \leqslant -\theta < 0 < \theta \leqslant \operatorname{Re}\sigma(d_{p,k})^{\pm}, \quad j \leqslant S < k, \tag{2.52}$$

for some uniform $\theta > 0$. For, this condition is clearly necessary by Lemma 2.5. On the other hand, suppose that the condition holds. The diagonalizing construction described in Proposition 2.15 preserves the property of exponential decay to constant states. Thus, making a change of coordinates such that

$$\operatorname{Re} d^{\pm}_{p,j} \leqslant -\theta/2 < 0 < \theta/2 \leqslant \operatorname{Re} d^{\pm}_{p,k}, \quad j \leqslant S < k,$$

we obtain by zeroth-order neutrality (preserved, by assumption (i)) plus exponential decay, that (2.32) holds with integrable error $\alpha(x) = \mathcal{O}(\epsilon \mathrm{e}^{-\theta\epsilon|x|})$, verifying (2.31).

We remark that $d_{0,j}$ scalar is equivalent to constant multiplicity of the associated eigenvalue of A_0. For neutrally zeroth-order stable (resp. unstable) blocks, this is a standard structural assumption, corresponding in the critical regime $\lambda/|\lambda|$ pure imaginary to local well-posedness of the underlying evolution equations. For strictly zeroth-order stable (resp. unstable) blocks, diagonalizability is not necessary either for our construction or for local well-posedness.

Likewise, uniform spectral gap at $\pm\infty$ for a set of problems ϵ, $(\lambda/|\lambda|)(\epsilon)$ is roughly equivalent to asymptotic direction $\lim(\lambda/|\lambda|)$ pointing into the domain of consistent splitting as $|\lambda| \to \infty$, or equivalently $\epsilon \to 0$. *Thus, the requirement of exponential separation is in practice no restriction for applications to stability of traveling waves*, the domain of feasibility lying in general asymptotic to the domain of consistent splitting. The only real requirement is sufficient regularity in the coefficients of the linearized equations to carry out the formal diagonalization procedure to the necessary order p. Whether or not our regularity requirement is sharp is an interesting technical question, to which we do not know the answer. We point out only that the regularity required in the sectorial case is just $C^{0+\alpha}$, as is correct; see [ZH], or Remark 5.4 of [MaZ.3].

2.3. *Hyperbolic–parabolic smoothing*

We now recall some important ideas of Kawashima et al. concerning the smoothing effects of hyperbolic–parabolic coupling. The following results assert that hyperbolic effects can compensate for degenerate viscosity B, as revealed by the existence of a *compensating matrix* K.

LEMMA 2.18 [KSh]. *Assuming A^0, A, B symmetric, $A^0 > 0$ and $B \geqslant 0$, the genuine coupling condition*

(GC) *No eigenvector of A lies in* $\ker B$ *is equivalent to either of*:

(K1) *There exists a smooth skew-symmetric matrix function $K(A^0, A, B)$ such that*

$$\operatorname{Re}\big(K(A^0)^{-1}A + B\big)(U) > 0. \tag{2.53}$$

(K2) *For some $\theta > 0$, there holds*

$$\operatorname{Re}\sigma\big(-\mathrm{i}\xi(A^0)^{-1}A - |\xi|^2(A^0)^{-1}B\big) \leqslant -\frac{\theta|\xi|^2}{1+|\xi|^2} \tag{2.54}$$

for all $\xi \in \mathbb{R}$.

PROOF. These and other useful equivalent formulations are established in [KSh] (see also [MaZ.4,Z.4]). □

COROLLARY 2.19. *Under* (A1)–(A3), *there holds the uniform dissipativity condition*

$$\operatorname{Re}\sigma\left(\sum_j i\xi_j A^j - \sum_{j,k}\xi_j\xi_k B^{jk}\right)_\pm \leqslant -\frac{\theta|\xi|^2}{1+|\xi|^2}, \qquad \theta > 0. \tag{2.55}$$

Moreover, there exist smooth skew-symmetric "compensating matrices" $K_\pm(\xi)$, *homogeneous degree one in* ξ, *such that*

$$\operatorname{Re}\left(\sum_{j,k}\xi_j\xi_k\widetilde{B}^{jk} - K(\xi)\big(\widetilde{A}^0\big)^{-1}\sum_k \xi_k\widetilde{A}^k\right)_\pm \geqslant \theta > 0 \tag{2.56}$$

for all $\xi \in \mathbb{R}^d \setminus \{0\}$.

PROOF. By the block–diagonal structure of $\widetilde{B}^{jk}$ (GC) holds also for $A^j_\pm$ and $\widehat{B}^{jk} := (\widetilde{A}^0)^{-1}\operatorname{Re}\widetilde{B}^{jk}$, since

$$\ker\sum_{j,k}\xi_j\xi_k\widehat{B}^{jk} = \ker\sum_{j,k}\xi_j\xi_k\operatorname{Re}\widetilde{B}^{jk} = \ker\sum\xi_j\xi_k\widetilde{B}^{jk} = \ker\sum\xi_j\xi_k B^{jk}.$$

Applying Lemma 2.18 to

$$\widetilde{A}^0 := \widetilde{A}^0_\pm, \qquad A := \left(\big(\widetilde{A}^0\big)^{-1}\sum_k\xi_k\widetilde{A}^k\right)_\pm,$$

$$B := \left(\big(\widetilde{A}^0\big)^{-1}\sum_{j,k}\xi_j\xi_k\operatorname{Re}\widetilde{B}^{jk}\right)_\pm,$$

we thus obtain (2.56) and

$$\operatorname{Re}\sigma\left[\big(\widetilde{A}^0\big)^{-1}\left(-\sum_j i\xi_j\widetilde{A}^j - \sum_{j,k}\xi_j\xi_k\operatorname{Re}\widetilde{B}^{jk}\right)\right]_\pm \leqslant -\frac{\theta_1|\xi|^2}{1+|\xi|^2}, \tag{2.57}$$

$\theta_1 > 0$, from which we readily obtain

$$\left(-\sum_j i\xi_j\widetilde{A}^j - \sum_{j,k}\xi_j\xi_k\widetilde{B}^{jk}\right)_\pm \leqslant -\frac{\theta_2|\xi|^2}{1+|\xi|^2}. \tag{2.58}$$

Observing that $M > \theta_1 \Leftrightarrow (\widetilde{A}^0)_\pm^{-1/2} M (\widetilde{A}^0)_\pm^{-1/2} > \theta$ and $\sigma(\widetilde{A}^0)_\pm^{-1/2} M (\widetilde{A}^0)_\pm^{-1/2} > \theta \Leftrightarrow \sigma(\widetilde{A}^0)_\pm^{-1} M > \theta$, together with $S > \theta \Leftrightarrow \sigma S > \theta$ for S symmetric, we obtain (2.55) from (2.58). Because all terms other than K in the left-hand side of (2.56) are homogeneous, it is evident that we may choose $K(\cdot)$ homogeneous as well (restrict to the unit sphere, then take homogeneous extension). □

REMARK 2.20 [GMWZ.4]. In the special case $A^0 = I$, $B = \text{block-diag}\{0, b\}$, and $\operatorname{Re} b > 0$, (GC) is equivalent to the condition that no eigenvector of A_{11} lie in the kernel of A_{21}. If, also, A_{11} is a scalar multiple of the identity, therefore, $\operatorname{Re} A_{12} A_{21} > 0$. Thus, on any compact set of such A, B satisfying (GC), $K(I, A, B)$ may be taken as a linear function

$$K(A) = \theta \begin{pmatrix} 0 & A_{12} \\ -A_{21} & 0 \end{pmatrix} \tag{2.59}$$

of A alone, for $\theta > 0$ sufficiently small. This implies, in particular, for systems (1.1) that can be simultaneously symmetrized, and for which each A_*^j is a scalar multiple of the identity, there is a *linear*, or "differential", choice $K(\sum_j \xi_j A^j) = \sum_j \xi_j K^j$ such that $\sum_{jk} (K^j A^k + B^{jk}) \xi_j \xi_k \geqslant 0$ on any compact set of U. In particular, this holds for the standard Navier–Stokes equations of gas dynamics and MHD, as pointed out in [Kaw]; more generally, it holds for all of the variants described in Appendix A.1: that is, for all examples considered in this chapter.

2.4. *Construction of the resolvent kernel*

We conclude these preliminaries by deriving explicit representation formulae for the resolvent kernel $G_{\tilde{\xi},\lambda}(x, y)$ using a variant of the classical construction (see, e.g., [Ch], pp. 371–376) of the Green distribution of an ordinary differential operator in terms of decaying solutions of the homogeneous eigenvalue equation $(L_{\tilde{\xi}} - \lambda) U = 0$, or

$$\begin{aligned} &\overbrace{(B^{11} U')' - (A^1 U)'}^{L_0 U} - \mathrm{i} \sum_{j \neq 1} A^j \xi_j U + \mathrm{i} \sum_{j \neq 1} B^{j1} \xi_j U' \\ &\quad + \mathrm{i} \sum_{k \neq 1} (B^{1k} \xi_k U)' - \sum_{j,k \neq 1} B^{jk} \xi_j \xi_k U - \lambda U = 0, \end{aligned} \tag{2.60}$$

matched across the singularity $x = y$ by appropriate jump conditions, in the process obtaining standard decay estimates on the resolvent kernel (see (2.85), Proposition 2.23) suitable for analysis of intermediate frequencies λ. Here, A^j and B^{jk} are as defined as in (2.1)–(2.4), $U \in \mathbb{R}^n$, and "$'$" as usual denotes $\mathrm{d}/\mathrm{d}x_1$. Our treatment follows the general approach introduced in [MaZ.3] to treat the case of an ordinary differential operator, such as L above, that is of degenerate type, i.e., the coefficient of the highest-order derivative is singular, but for which the eigenvalue equation can nonetheless be written as a nondegenerate first-order ODE (2.17) in an appropriate reduced phase space: in this case,

$$W' = \mathbb{A}(x_1, \tilde{\xi}, \lambda)W \tag{2.61}$$

with $W = (U, b_1^{11}u^{I'} + b_2^{11}u^{II'}) \in \mathbb{C}^{n+r}$. For related analyses in the nondegenerate case, see, e.g., [AGJ,K.1,K.2,ZH,MaZ.1].

2.4.1. *Domain of consistent splitting.* Define

$$\Lambda^{\tilde{\xi}} := \bigcap_{j=1}^{n} \Lambda_j^{\pm}(\tilde{\xi}), \tag{2.62}$$

where $\Lambda_j^{\pm}(\tilde{\xi})$ denote the open sets bounded on the left by the algebraic curves $\lambda_j^{\pm}(\xi_1, \tilde{\xi})$ determined by the eigenvalues of the symbols $-\xi^2 B_{\pm} - \mathrm{i}\xi A_{\pm}$ of the limiting constant–coefficient operators

$$L_{\pm}w := B_{\pm}w'' - A_{\pm}w' \tag{2.63}$$

as ξ_1 is varied along the real axis, with $\tilde{\xi}$ held fixed. The curves $\lambda_j^{\pm}(\cdot, \tilde{\xi})$ comprise the essential spectrum of operators $L_{\tilde{\xi}\pm}$.

LEMMA 2.21. *Assuming* (A1)–(A3) *and* (H0)–(H1), *the eigenvalue equation may be written as first-order equation* (2.61), *for which the set $\Lambda^{\tilde{\xi}}$ is equal to the component containing real $+\infty$ of the domain of consistent splitting. Moreover,*

$$\Lambda^{\tilde{\xi}} \subset \{\lambda\colon \operatorname{Re}\lambda > -\eta(|\operatorname{Im}\lambda|^2 + |\tilde{\xi}|^2)/(1 + |\operatorname{Im}\lambda|^2 + |\tilde{\xi}|^2)\}, \quad \eta > 0. \tag{2.64}$$

PROOF. We do not need to explicitly calculate the matrix $\mathbb{A} \in \mathbb{C}^{(n+r)\times(n+r)}$ that results, but only to point out that its existence follows by invertibility of A_*^1 and b_2^{11}, (H1), once we rewrite (2.60) as the lower triangular system of equations

$$\begin{aligned}
u' &= (A_*^1)^{-1}\big(-A_{12}^1(b_2^{11})^{-1}z - (A_{11}^{1'} + \lambda)u - A_{12}^{1'}v\big) - \mathrm{i}\sum_{j\neq 1}(A_{11}^j u + A_{12}^j v),\\
v' &= (b_2^{11})^{-1}z - (b_2^{11})^{-1}b_1^{11}u',\\
z' &= \big(A_{21}^1 - A_{22}^1(b_2^{11})^{-1}b_1^{11}\big)u' + A_{22}^1(b_2^{11})^{-1}z + A_{21}^{1'}u + (A_{22}^{1'} + \lambda)v + \overbrace{\cdots}^{\tilde{\xi}\ \text{terms}},
\end{aligned} \tag{2.65}$$

with $(u, v, z) := (u^I, u^{II}, b_1^{11}u' + b_2^{11}v')$. (We will need to carry out explicit computations only for our treatment in Section 4.3.2 of one-dimensional high-frequency bounds.) Note, as follows from the second equation, the important fact that every solution of (2.65) indeed corresponds to a solution $U = (u, v)$ of (2.60), with $z = b_1^{11}u' + b_2^{11}v'$.

The final assertion, bound (2.64), follows easily from the bound (2.55) on the dispersion curves $\lambda_j^{\pm}(\tilde{\xi})$. To establish the second assertion, we must show that, on the set Λ, the limiting eigenvalue equations

$$(L_{\tilde{\xi}\pm} - \lambda)w = 0 \tag{2.66}$$

have no center manifold and the stable manifold at $+\infty$ and the unstable manifold at $-\infty$ have dimensions summing to the full dimension n; moreover, that these properties do not hold together on the boundary of $\Lambda^{\tilde{\xi}}$.

The fundamental modes of (2.66) are of form $e^{\mu x}V$, where μ, V satisfy the *characteristic equation*

$$\begin{aligned}\Bigg[\mu^2 B_{\pm}^{11} + \mu\Big(-A_{\pm}^1 + \mathrm{i}\sum_{j\neq 1} B_{\pm}^{j1}\xi_j + \mathrm{i}\sum_{k\neq 1} B^{1k}\xi_k\Big)\\ - \Big(\mathrm{i}\sum_{j\neq 1} A^j\xi_j + \sum_{jk\neq 1} B^{jk}\xi_j\xi_k + \lambda I\Big)\Bigg]V = 0.\end{aligned} \tag{2.67}$$

The existence of a center manifold thus corresponds with existence of solutions $\mu = \mathrm{i}\xi_1$, V of (2.67), ξ_1 real, i.e., solutions of the *dispersion relation*

$$\Big(-\sum_{j,k} B_{\pm}^{jk}\xi_j\xi_k - \mathrm{i}\sum_{j} A_{\pm}^{j}\xi_j - \lambda I\Big)V = 0. \tag{2.68}$$

But, $\lambda \in \sigma(-B^{\xi\xi} - \mathrm{i}A_{\pm}^{\xi})$ implies, by definition (2.62), that λ lies outside of $\Lambda^{\tilde{\xi}}$, establishing nonexistence of a center manifold. Moreover, it is clear from the same argument that a center manifold does exist on the boundary of Λ, since this corresponds to existence of pure imaginary eigenmodes.

Finally, nonexistence of a center manifold, together with connectivity of Λ, implies that the dimensions of stable/unstable manifolds at $+\infty/-\infty$ are constant on Λ. Taking $\lambda \to +\infty$ along the real axis, with $\tilde{\xi} \equiv 0$, we find that these dimensions sum to the full dimension $n + r$ as claimed. For, Fourier expansion about $\xi_1 = \infty$ of the one-dimensional ($\tilde{\xi} = 0$) dispersion relation (see Appendix A.4) yields $n - r$ "hyperbolic" modes

$$\lambda_j = -\mathrm{i}\xi_1 a_j^* + \cdots, \quad j = 1, \ldots, n - r,$$

where a_j^* denote the eigenvalues of A_*^1 and r "parabolic" modes

$$\lambda_{n-r+j} = -b_j\xi_1^2 + \cdots, \quad j = 1, \ldots r,$$

where b_j denote the eigenvalues of b_2^{11}; here, we have suppressed the $\pm$ indices for readability. Inverting these relationships to solve for $\mu := \mathrm{i}\xi_1$, we find, for $\lambda \to \infty$, that there are $n - r$ hyperbolic roots $\mu_j \sim -\lambda/a_j^*$, and $2r$ parabolic roots $\mu_{n-r+j}^{\pm} \sim \pm\sqrt{\lambda/b_j}$. By assumption (H1)(iii), the former yield a fixed number $k/(n-r-k)$ of stable/unstable roots, independent of x_1, and thus of $\pm$. Likewise, (H1)(i) implies that the latter yields r stable, r unstable roots. Combining, we find the desired consistent splitting, with $(k+r)/(n-k)$ stable/unstable roots at both $\pm\infty$. □

A standard fact, for asymptotically constant-coefficient ordinary differential operators (see, e.g., [He], Lemma 2, pp. 138–139) is that components of the domain of consistent splitting agree with components of the complement of the essential spectrum of the variable-coefficient operator L. Thus, Λ is a maximal domain in the essential spectrum complement, consisting of the component containing real, plus infinity. Moreover, provided that the coefficients of L approach their limits at integrable rate, it can be shown (see, e.g., [AGJ,GZ,ZH,MaZ.1]) that each connected component consists either entirely of eigenvalues, or else entirely of *normal points*, defined as resolvent points or isolated eigenvalues of finite multiplicity. The latter fact will be seen directly in the course of our construction.

2.4.2. *Basic construction.* We now carry out the resolvent construction at points λ in Λ, more generally, at any point in the domain of consistent splitting, by the method of [MaZ.3]. Our starting point is a duality relation between solutions U of the eigenvalue equation and solutions $\widetilde{U}$ of the adjoint eigenvalue equation, which may be obtained by the observation that E defined by

$$\frac{\mathrm{d}}{\mathrm{d}x_1}E := \langle \widetilde{U}, (L_{\tilde{\xi}} - \lambda)U\rangle - \langle (L_{\tilde{\xi}}^* - \bar{\lambda})\widetilde{U}, U\rangle = \langle \widetilde{U}, L_{\tilde{\xi}}U\rangle - \langle L_{\tilde{\xi}}^*\widetilde{U}, U\rangle$$

is a conserved quantity whenever U and $\widetilde{U}$ are solutions of the eigenvalue and adjoint eigenvalue equations, respectively.

Since $\langle \widetilde{U}, L_{\tilde{\xi}}U\rangle - \langle L_{\tilde{\xi}}^*\widetilde{U}, U\rangle$ is a perfect derivative involving derivatives of U, $\widetilde{U}$ of order strictly less than the maximum order of $L_{\tilde{\xi}}$, the quantity E may always be expressed as a quadratic form in phase space, in this case

$$E(x_1) = \langle \widetilde{U}, B^{11}U'\rangle + \langle \widetilde{U}, (-A^1 + \mathrm{i}B^{1\tilde{\xi}} + \mathrm{i}B^{\tilde{\xi}1})U\rangle + \langle \widetilde{U}', -B^{11}U\rangle, \tag{2.69}$$

or

$$E(x_1) = \left\langle \begin{pmatrix} \widetilde{U} \\ \widetilde{U}' \end{pmatrix}, \mathcal{S}^{\tilde{\xi}} \begin{pmatrix} U \\ U' \end{pmatrix} \right\rangle (x_1), \tag{2.70}$$

where

$$\mathcal{S}^{\tilde{\xi}} := \begin{pmatrix} -A^1 + \mathrm{i}B^{1\tilde{\xi}} + \mathrm{i}B^{\tilde{\xi}1} & B^{11} \\ -B^{11} & 0 \end{pmatrix}. \tag{2.71}$$

In general, we obtain a cross-banded matrix S, with elements of the jth band consisting of alternating $\pm$ multiples of the jth coefficient of the operator $L_{\tilde{\xi}}$, the anti-diagonal being filled with multiples of the principal coefficient, and the bands below the anti-diagonal being filled with zero blocks. Thus, for nondegenerate operators, it is immediate that $\mathcal{S}$ is invertible, and we obtain the characterization [ZH,MaZ.1] of solutions of the eigenvalue equation as those functions U with the property that $E(x_1)$ is preserved for any solutions $\widetilde{U}$ of the adjoint equation.

For degenerate operators, $\mathcal{S}$ is of course not invertible, and we must work instead in a reduced phase space. In the present case, we may rewrite (2.69) as

$$E(x_1) = \left\langle \begin{pmatrix} \widetilde{U} \\ \widetilde{Z} \end{pmatrix}, \overline{\mathcal{S}} \begin{pmatrix} U \\ Z \end{pmatrix} \right\rangle, \tag{2.72}$$

where

$$\overline{\mathcal{S}}^{\tilde{\xi}} := \begin{pmatrix} -A^1 + \mathrm{i}B^{1\tilde{\xi}} + \mathrm{i}B^{\tilde{\xi}1} & \begin{pmatrix} 0 \\ I_r \end{pmatrix} \\ \begin{pmatrix} -(b_2^{11})^{-1} b_1^{11} & -I_r \end{pmatrix} & 0 \end{pmatrix} \tag{2.73}$$

and

$$Z := (b_1^{11}, b_2^{11}) U', \qquad \widetilde{Z} := (0, b_2^{11}*) \widetilde{U}'. \tag{2.74}$$

Applying an elementary column operation (subtracting from the first column the second column times $(b_2^{11})^{-1} b_1^{11}$), we find that

$$\det \overline{\mathcal{S}}^{\tilde{\xi}} \equiv \det A_*^1 \neq 0, \tag{2.75}$$

by (H1)(i). Thus, we may conclude, similarly as in the nondegenerate case:

LEMMA 2.22. *$W = (U, Z)$ satisfies eigenvalue equation* (2.60) *if and only if*

$$\widetilde{W}^* \overline{\mathcal{S}} W \equiv \text{constant} \tag{2.76}$$

for all $\widetilde{W} = (\widetilde{U}, \widetilde{Z})$ satisfying the adjoint eigenvalue equation, and vice versa.

For future reference, we note the representation

$$(\overline{\mathcal{S}}^{\tilde{\xi}})^{-1} = \begin{pmatrix} -(A_*^1)^{-1} & 0 & (A_*^1)^{-1} \mathcal{A}(\tilde{\xi})_{12} \\ (b_2^{11})^{-1} b_1^{11} (A_*^1)^{-1} & 0 & -(b_2^{11})^{-1} b_1^{11} (A_*^1)^{-1} \mathcal{A}(\tilde{\xi})_{12} - I \\ -\widetilde{\mathcal{A}}(\tilde{\xi})(A_*^1)^{-1} & I & -\mathcal{A}(\tilde{\xi})_{22} + \widetilde{\mathcal{A}}(\tilde{\xi})(A_*^1)^{-1} A_{12} \end{pmatrix}, \tag{2.77}$$

$$A_*^1 := A_{11}^1 - A_{12}^1 (b_2^{11})^{-1} b_1^{11},$$

$$\widetilde{\mathcal{A}}(\tilde{\xi}) := \alpha_{21}(\tilde{\xi}) - \alpha_{22}(\tilde{\xi}) (b_2^{11})^{-1} b_1^{11},$$

$$\alpha(\tilde{\xi}) := A^1 - \mathrm{i}B^{\tilde{\xi}1} - \mathrm{i}B^{1\tilde{\xi}},$$

which may be obtained by the above-mentioned column operation, followed by straightforward row reduction.

Let

$$\phi_j^+ := P_+ v_j^+, \quad j = 1\dots, k, \tag{2.78}$$

and

$$\phi_j^-, \quad j = k+1, \dots, n+r, \tag{2.79}$$

denote the locally analytic bases of the stable manifold at $+\infty$ and the unstable manifold at $-\infty$ of solutions of the variable-coefficient equation (2.60) that are found in Section 3.1, and set

$$\Phi^+ := (\phi_1^+, \dots, \phi_k^+), \qquad \Phi^- := (\phi_{k+1}^-, \dots, \phi_{n+r}^-), \tag{2.80}$$

and

$$\Phi := (\Phi^+, \Phi^-). \tag{2.81}$$

Define the solution operator from y_1 to x_1 of (2.60), denoted by $\mathcal{F}^{y_1 \to x_1}$, as

$$\mathcal{F}^{y_1 \to x_1} = \Phi(x_1; \lambda)\, \Phi^{-1}(y_1; \lambda) \tag{2.82}$$

and the projections $\Pi_{y_1}^{\pm}$ on the stable manifolds at $\pm\infty$ as

$$\begin{aligned} \Pi_{y_1}^+ &= \left(\Phi^+(y_1; \lambda) \quad 0\right)\Phi^{-1}(y_1; \lambda) \quad \text{and} \\ \Pi_{y_1}^- &= \left(0 \quad \Phi^-(y_1; \lambda)\right)\Phi^{-1}(y_1; \lambda). \end{aligned} \tag{2.83}$$

Then, we have the following universal result (see, e.g., [ZH,MaZ.1] in the nondegenerate case, or [MaZ.3] for the one-dimensional hyperbolic–parabolic case).

PROPOSITION 2.23. *With respect to any L^p, $1 \leqslant p \leqslant \infty$, the domain of consistent splitting consists entirely of normal points of $L_{\tilde\xi}$, i.e., resolvent points, or isolated eigenvalues of constant multiplicity. On this domain, the resolvent kernel $G_{\tilde\xi,\lambda}$ is meromorphic, with representation*

$$G_{\tilde\xi,\lambda}(x_1, y_1) = \begin{cases} (I_n, 0)\mathcal{F}^{y_1 \to x_1}\Pi_{y_1}^+ \left(\overline{\mathcal{S}}^{\tilde\xi}\right)^{-1}(y_1)(I_n, 0)^{\mathrm{tr}}, & x_1 > y_1, \\ -(I_n, 0)\mathcal{F}^{y_1 \to x_1}\Pi_{y_1}^- \left(\overline{\mathcal{S}}^{\tilde\xi}\right)^{-1}(y_1)(I_n, 0)^{\mathrm{tr}}, & x_1 < y_1, \end{cases} \tag{2.84}$$

$(\overline{\mathcal{S}}^{\tilde\xi})^{-1}$ as described in (2.77). Moreover, on any compact subset K of $\rho(L_{\tilde\xi}) \cap \Lambda$ ($\rho(L_{\tilde\xi})$ denoting resolvent set), there holds the uniform decay estimate

$$\left|G_{\tilde\xi,\lambda}(x_1, y_1)\right| \leqslant C\mathrm{e}^{-\eta|x_1 - y_1|}, \tag{2.85}$$

where $C > 0$ *and* $\eta > 0$ *depend only on* K, $L_{\tilde{\xi}}$.

PROOF. As discussed further in [Z.1,MaZ.1], to show that λ is in the resolvent set of $L_{\tilde{\xi}}$, with respect to any L^p, $1 \leqslant p \leqslant \infty$, it is enough to: (i) construct a resolvent kernel $G_{\tilde{\xi},\lambda}(x_1, y_1)$ satisfying

$$(L_{\tilde{\xi}} - \lambda)G_{\tilde{\xi},\lambda} = \delta_{y_1}(x_1) \tag{2.86}$$

and obeying a uniform decay estimate

$$\left|G_{\tilde{\xi},\lambda}(x_1, y_1)\right| \leqslant C\mathrm{e}^{-\eta|x_1-y_1|}; \tag{2.87}$$

and, (ii) show that there are no L^p solutions of $(L_{\tilde{\xi}} - \lambda)W = 0$, i.e., λ is not an eigenvalue of $L_{\tilde{\xi}}$, or, equivalently, a zero of the Evans function $D(\cdot)$ (for $p < \infty$, these are necessary as well as sufficient [Z.1]). For, then, the Hausdorff–Young inequality yields that the distributional solution formula $(L_{\tilde{\xi}} - \lambda)^{-1} f := \int G_\lambda(x_1, y_1) f(y_1)\,\mathrm{d}y$ yields a bounded right inverse of $(L_{\tilde{\xi}} - \lambda)$ taking L^p to L^p, while nonexistence of eigenvalues implies that this is also a left inverse.

On the domain of consistent splitting, we shall show that (ii) implies (i) (they are in fact equivalent [Z.1,MaZ.1]), which immediately implies the first assertion by the properties of analytic functions and the correlation between eigenvalues and zeroes of D, Remark 2.11. It remains, then, to establish existence of a solution of (2.86) satisfying the bound (2.87), under the assumption that $D(\lambda) \neq 0$.

Note, by Lemma 2.5, that the stable manifold Φ^+ decays $x_1 \geqslant 0$ with uniform rate $C\mathrm{e}^{\eta|x_1|}$, $\eta > 0$. Moreover, if $D \neq 0$, then Φ^+ does not decay at $-\infty$, whereupon we may conclude from Lemma 2.5 that it grows exponentially as $x_1 \to -\infty$, so that it in fact decays uniformly exponentially in x_1 on the whole line. Likewise, Φ^- decays exponentially as x_1 decreases, uniformly on the whole line. Thus, we find that (2.84) satisfies (2.85) as claimed. Clearly, it also satisfies (2.86) away from the singular point $x_1 = y_1$. Finally, we note that, by the construction of $\overline{\mathcal{S}^{\tilde{\xi}}}$, we have the general fact that

$$\int_{y_1^-}^{y_1^+} L_{\tilde{\xi}} G_{\tilde{\xi},\lambda} = (I_n, 0)\overline{\mathcal{S}^{\tilde{\xi}}}.$$

Substituting from (2.84), we obtain

$$\begin{pmatrix} [G_{\tilde{\xi},\lambda}] \\ [(b_1^{11}, b_2^{11})G'_{\tilde{\xi},\lambda}] \end{pmatrix} = (I_n, 0)\overline{\mathcal{S}^{\tilde{\xi}}}\mathcal{F}^{y_1 \to y_1}\left(\overline{\mathcal{S}^{\tilde{\xi}}}\right)^{-1}(I_n, 0)^{\mathrm{tr}} = I_n,$$

validating the jump condition across $x_1 = y_1$, and completing the proof. □

Proposition 2.23 includes a satisfactory intermediate-frequency bound (2.85) on the resolvent kernel. More careful analyses will be required in the large- and small-frequency limits.

REMARK 2.24. Similarly as in the strictly parabolic case [ZH], formula (2.84) extends to the full phase-variable representation

$$\begin{pmatrix} G_{\tilde{\xi},\lambda}(x_1,y_1) & (\partial/\partial y_1)G_{\tilde{\xi},\lambda}(x_1,y_1)(0,(b_2^{11})^t)^t \\ (b_1^{11},b_2^{11})(\partial/\partial x_1)G_{\tilde{\xi},\lambda}(x_1,y_1) & (b_1^{11},b_2^{11})(\partial/\partial x_1)(\partial/\partial y_1)G_{\tilde{\xi},\lambda}(x_1,y_1)(0,(b_2^{11})^t)^t \end{pmatrix}$$

$$= \begin{cases} \mathcal{F}^{y_1\to x_1}\Pi_{y_1}^+\left(\overline{\mathcal{S}^{\tilde{\xi}}}\right)^{-1}(y_1), & x_1 > y_1, \\ -\mathcal{F}^{y_1\to x_1}\Pi_{y_1}^-\left(\overline{\mathcal{S}^{\tilde{\xi}}}\right)^{-1}(y_1), & x_1 < y_1. \end{cases} \tag{2.88}$$

This formula (though not the specific phase variable) is also universal, and follows from (2.84), which states that the upper left-hand corner is correct, together with the fact that the columns of the right-hand side satisfy the forward eigenvalue equation (written as a first-order system) with respect to x_1, by inspection, while the rows satisfy the adjoint eigenvalue equation (again, written as a first-order system) with respect to y_1, by duality relation (2.76), a simple consequence of which is that the columns and rows do respect the phase-variable formulation and so we may deduce from correctness of the upper left-hand corner that all entries are correct. Formula (2.88) is important in the development of the effective spectral theory given in Section 4.3.3; see discussion, proof of Proposition 4.38.

2.4.3. *Generalized spectral decomposition.* For the treatment of low-frequency behavior, we develop a modified representation of the resolvent kernel consisting of a *scattering decomposition* in solutions of the forward and adjoint eigenvalue equations.

From (2.76), it follows that if there are k independent solutions $\phi_1^+,\dots,\phi_k^+$ of $(L_{\tilde{\xi}} - \lambda I)W = 0$ decaying at $+\infty$, and $n-k$ independent solutions $\phi_{k+1}^-,\dots,\phi_n^-$ of the same equations decaying at $-\infty$, then there exist $n-k$ independent solutions $\tilde{\psi}_{k+1}^+,\dots,\tilde{\psi}_n^+$ of $(L_{\tilde{\xi}}^* - \bar{\lambda}I)\widetilde{W} = 0$ decaying at $+\infty$, and k independent solutions $\tilde{\psi}_1^-,\dots,\tilde{\psi}_k^-$ decaying at $-\infty$. More precisely, setting

$$\Psi^+(x_1;\lambda) = \left(\psi_{k+1}^+(x_1;\lambda) \ \cdots \ \psi_n^+(x_1;\lambda)\right) \in \mathbb{R}^{n\times(n-k)}, \tag{2.89}$$

$$\Psi^-(x_1;\lambda) = \left(\psi_1^-(x_1;\lambda) \ \cdots \ \psi_k^-(x_1;\lambda)\right) \in \mathbb{R}^{n\times k} \tag{2.90}$$

and

$$\Psi(x_1;\lambda) = \left(\Psi^-(x_1;\lambda) \quad \Psi^+(x_1;\lambda)\right) \in \mathbb{R}^{n\times n}, \tag{2.91}$$

where $\psi_j^\pm$ are exponentially growing solutions obtained through Lemma 2.5, we may define dual exponentially decaying and growing solutions $\tilde{\psi}_j^\mp$ and $\tilde{\phi}_j^\mp$ via

$$(\widetilde{\Psi} \quad \widetilde{\Phi})_\pm^* \mathcal{S}^{\tilde{\xi}} (\Psi \quad \Phi)_\pm \equiv I. \tag{2.92}$$

Then, we have:

PROPOSITION 2.25. *The resolvent kernel may alternatively be expressed as*

$$G_{\xi,\lambda}(x_1,y_1) = \begin{cases} (I_n,0)\Phi^+(x_1;\lambda)M^+(\lambda)\widetilde{\Psi}^{-*}(y_1;\lambda)(I_n,0)^{\mathrm{tr}}, & x_1>y_1,\\ -(I_n,0)\Phi^-(x_1;\lambda)M^-(\lambda)\widetilde{\Psi}^{+*}(y_1;\lambda)(I_n,0)^{\mathrm{tr}}, & x_1<y_1, \end{cases} \tag{2.93}$$

where

$$M(\lambda) := \operatorname{diag}\bigl(M^+(\lambda), M^-(\lambda)\bigr) = \Phi^{-1}(z;\lambda)\bigl(\mathcal{S}^{\xi}\bigr)^{-1}(z)\widetilde{\Psi}^{-1*}(z;\lambda), \tag{2.94}$$

$\widetilde{\Psi} := (\widetilde{\Psi}^- \quad \widetilde{\Psi}^+)$. (Note: the right-hand side of (2.94) is independent with respect to z, as a consequence of Lemma 2.22.)

PROOF. Immediate, by rearrangement of (2.86). □

REMARKS. 1. Representation (2.84) reflects the classical duality principle (see, e.g., [ZH], Lemma 4.2) that the transposition $G^*_{\xi,\lambda}(y_1,x_1)$ of the Green distribution $G_{\xi,\lambda}(x_1,y_1)$ associated with operator $(L_\xi-\lambda)$ should be the Green distribution for the adjoint operator $(L^*_\xi-\bar{\lambda})$.

2. As before, it is possible to represent the matrix $G_{\xi,\lambda}$ by means of intrinsic objects such as solution operators and projections on stable manifolds

$$G_{\xi,\lambda}(x_1,y_1) = \begin{cases} (I_n,0)\mathcal{F}^{z\to x_1}\Pi_z^+\bigl(\mathcal{S}^{\xi}\bigr)^{-1}(z)\widetilde{\Pi}_z^-\widetilde{\mathcal{F}}^{z\to y_1}(I_n,0)^{\mathrm{tr}}, & x_1>y_1,\\ -(I_n,0)\mathcal{F}^{z\to x_1}\Pi_z^-\bigl(\mathcal{S}^{\xi}\bigr)^{-1}(z)\widetilde{\Pi}_z^+\widetilde{\mathcal{F}}^{z\to y_1}(I_n,0)^{\mathrm{tr}}, & x_1<y_1, \end{cases} \tag{2.95}$$

$$\widetilde{\mathcal{F}}^{z\to y_1} := \widetilde{\Psi}^{-1}(z;\lambda)\widetilde{\Psi}(y_1;\lambda),$$

$$\widetilde{\Pi}_z^+ := \widetilde{\Psi}^{-1}\begin{pmatrix}0\\ \widetilde{\Psi}^+\end{pmatrix}(z;\lambda),$$

$$\widetilde{\Pi}_z^- := \widetilde{\Psi}^{-1}\begin{pmatrix}\widetilde{\Psi}^-\\ 0\end{pmatrix}(z;\lambda).$$

The exponential decay asserted in Proposition 2.23 is somewhat more straightforward to see in this dual formulation, by judiciously choosing z in (2.95) and using the exponential decay of forward and adjoint flows on $x_1 \gtrless 0$ alone.

From Proposition 2.25, we obtain the following scattering decomposition, generalizing the Fourier transform representation in the constant-coefficient case.

COROLLARY 2.26 [Z.3]. *On $\Lambda\cap\rho(L_\xi)$, there hold*

$$G_{\xi,\lambda}(x_1,y_1) = \sum_{j,k} M^+_{jk}(\lambda)\phi_j^+(x_1;\lambda)\tilde{\psi}_k^-(y_1;\lambda)^* \tag{2.96}$$

for $y_1 \leqslant 0 \leqslant x_1$,

$$G_{\tilde{\xi},\lambda}(x_1, y_1) = \sum_{j,k} d^+_{jk}(\lambda)\phi^-_j(x_1;\lambda)\tilde{\psi}^-_k(y_1;\lambda)^* - \sum_k \psi^-_k(x_1;\lambda)\tilde{\psi}^-_k(y_1;\lambda)^* \tag{2.97}$$

for $y_1 \leqslant x_1 \leqslant 0$ *and*

$$G_{\tilde{\xi},\lambda}(x_1, y_1) = \sum_{j,k} d^-_{jk}(\lambda)\phi^-_j(x_1;\lambda)\tilde{\psi}^-_k(y_1;\lambda)^* + \sum_k \phi^-_k(x_1;\lambda)\tilde{\phi}^-_k(y_1;\lambda)^* \tag{2.98}$$

for $x_1 \leqslant y_1 \leqslant 0$, *with*

$$M^+ = (-I, 0)\begin{pmatrix}\Phi^+ & \Phi^-\end{pmatrix}^{-1}\Psi^- \tag{2.99}$$

and

$$d^\pm = (0, I)\begin{pmatrix}\Phi^+ & \Phi^-\end{pmatrix}^{-1}\Psi^-. \tag{2.100}$$

Symmetric representations hold for $y_1 \geqslant 0$.

PROOF. The matrix M^+ in (2.94) may be expanded using duality relation (2.76) as

$$\begin{aligned} M^+ &= (-I, 0)\begin{pmatrix}\Phi^+ & \Phi^-\end{pmatrix}^{-1}A^{-1}\begin{pmatrix}\tilde{\Psi}^- & \tilde{\Phi}^-\end{pmatrix}^{*-1}\begin{pmatrix}I\\0\end{pmatrix}_{|z} \\ &= (-I, 0)\begin{pmatrix}\Phi^+ & \Phi^-\end{pmatrix}^{-1}\begin{pmatrix}\Psi^- & \Phi^-\end{pmatrix}\begin{pmatrix}I\\0\end{pmatrix}_{|z} \\ &= (-I, 0)\begin{pmatrix}\Phi^+ & \Phi^-\end{pmatrix}^{-1}\Psi^-_{|z}, \end{aligned} \tag{2.101}$$

yielding (2.99) for $x_1 \geqslant y_1$, in particular for $y_1 \leqslant 0 \leqslant x_1$.

Next, expressing $\phi^\pm_j(x_1;\lambda)$ as a linear combination of basis elements at $-\infty$, we obtain the preliminary representation

$$G_\lambda(x_1, y_1) = \sum_{j,k} d^+_{jk}(\lambda)\phi^-_j(x_1;\lambda)\tilde{\psi}^-_k(y_1;\lambda)^* + \sum_{j,k} e^+_{jk}\psi^-_j(x_1;\lambda)\tilde{\psi}^-_k(y_1;\lambda)^* \tag{2.102}$$

valid for $y_1 \leqslant x_1 \leqslant 0$. Duality, (2.76), with (2.84), and the fact that $\Pi_+ = I - \Pi_-$, gives

$$
\begin{aligned}
-\begin{pmatrix} d^+ \\ e^+ \end{pmatrix} &= (\tilde{\Phi}^- \quad \tilde{\Psi}^-)^* A\Pi_+\Psi^-|_{x_1} \\
&= -(\Phi^- \quad \Psi^-)^{-1}\bigl[I - (0 \quad \Phi^-)(\Phi^+ \quad \Phi^-)^{-1}\bigr]\Psi^- \\
&= \begin{pmatrix} 0 \\ -I_k \end{pmatrix} - \begin{pmatrix} 0 & I_{n+r-k} \\ 0 & 0 \end{pmatrix}(\Phi^+ \quad \Phi^-)^{-1}\Psi^-,
\end{aligned}
\tag{2.103}
$$

yielding (2.97) and (2.100) for $y_1 \leqslant x_1 \leqslant 0$. Relations (2.98) and (2.100) follow for $x_1 \leqslant y_1 \leqslant 0$ in similar, but more straightforward fashion from (2.76) and (2.84). □

REMARK 2.27. In the constant-coefficient case, with a choice of common bases $\Psi^\pm = \Phi^\mp$ at $\pm\infty$, (2.96)–(2.100) reduce to the simple formula

$$
G_{\xi,\lambda}(x_1, y_1) = \begin{cases} -\sum_{j=k+1}^{n+r} \phi_j^+(x_1;\lambda)\tilde{\phi}_j^{+*}(y_1;\lambda), & x_1 > y_1, \\ \sum_{j=1}^{k} \phi_j^-(x_1;\lambda)\tilde{\phi}_j^{-*}(y_1;\lambda), & x_1 < y_1, \end{cases}
\tag{2.104}
$$

where, generically, $\phi_j^\pm$, $\tilde{\phi}_j^\pm$ may be taken as pure exponentials

$$
\phi_j^\pm(x_1)\tilde{\phi}_j^{\pm*}(y_1) = \mathrm{e}^{\mu_j^\pm(\lambda)(x_1-y_1)} V_j^\pm(\lambda)\tilde{V}_j^{\pm*}(\lambda).
\tag{2.105}
$$

Note that, moving individual modes $\phi_j^\pm\tilde{\phi}_j^\pm$ in the spectral resolution formula (2.8), using Cauchy theorem, to contours $\mu_j(\lambda) \equiv \mathrm{i}\xi_1$ lying along corresponding dispersion curves $\lambda = \lambda_j(\xi)$, we obtain the standard decomposition of $\mathrm{e}^{L_{\tilde{\xi}} t}$ into eigenmodes of continuous spectrum: in this (constant-coefficient) case, just the usual representation obtained by Fourier transform solution.

3. The Evans function, and its low-frequency limit

We next explore the key link between the Evans function D and the Lopatinski determinant Δ in the limit as frequency goes to zero, establishing Result 1 of the Introduction.

3.1. *The Evans function*

We first recall the following important result, established by Kreiss in the strictly hyperbolic case (see [Kr], Lemma 3.2, and discussion just above; see also [CP], Theorem 3.5, p. 431, and [ZS], Lemma 3.5) and by Métivier [Mé.4] in the constant multiplicity case. The result holds, more generally, for systems satisfying the block structure condition of Kreiss and Majda [Kr,M.1–M.4].

LEMMA 3.1. *Let there hold* (A1) *and* (H1)–(H2), *and* (H4), *or, more generally,* $\sigma(\mathrm{d}F^{\xi}(U_{\pm}))$ *real, semisimple, and of constant multiplicity for* $\xi \in \mathbb{R}^d \setminus \{0\}$ *and* $\det \mathrm{d}F^1 \neq 0$. *Then, the vectors* $\{r_1^-, \dots, r_{n-i_-}^-\}$, $\{r_{i_+ +1}^+, \dots, r_n^+\}$ *(defined as in* (1.30), *Section* 1.3), *and thus* $\Delta(\tilde{\xi}, \lambda)$ *may be chosen to be homogeneous (degree zero for* $r_j^{\pm}$, *degree one for* Δ), *analytic on* $\tilde{\xi} \in \mathbb{R}^{d-1}$, $\operatorname{Re}\lambda > 0$ *and continuous at the boundary* $\tilde{\xi} \in \mathbb{R}^{d-1} \setminus \{0\}$, $\operatorname{Re}\lambda = 0$.

PROOF. See Exercises 4.23 and 4.24 and Remark 4.25, Section 4.5.2 of [Z.3] for a proof in the general case (three alternative proofs, based, respectively, on [Kr], [CP], and [ZS]). In the case of main interest, when (H5) holds as well, we will see this later in the course of the explicit computations of Section 5.3; see Remark 5.20. □

REMARK 3.2. Typically, Δ has a conical singularity at $(\tilde{\xi}, \lambda) = (0, 0)$, in the sense that the level set through the origin is a cone and not a plane. That is, it is degree one homogeneous but not linear, with a gradient discontinuity at the origin. This reflects the fact that $r_j^{\pm}$ are degree zero homogeneous but are not constant unless all $\mathrm{d}F^j$ commute.

Following the construction of Section 2.2, we may define an Evans function

$$\begin{aligned} D(\tilde{\xi}, \lambda) &:= \det\big(W_1^+, \dots, W_k^+, W_{k+1}^-, \dots, W_{n+r}^-\big)\big|_{x=0,\lambda} \\ &= \det\big(P_+ V_1^+, \dots, P_+ V_k^+, P_- V_{k+1}^-, \dots, P_- V_{n+r}^-\big)\big|_{x=0,\lambda}, \end{aligned} \tag{3.1}$$

associated with the linearized operator $L_{\tilde{\xi}}$ about the viscous shock profile, on its domain of consistent splitting, in particular on the set

$$\begin{aligned} \Lambda &:= \big\{(\tilde{\xi}, \lambda) : \lambda \in \Lambda^{\tilde{\xi}}\big\} \\ &= \left\{(\tilde{\xi}, \lambda) : \operatorname{Re}\lambda > -\frac{\eta(|\operatorname{Im}\lambda|^2 + |\tilde{\xi}|^2)}{1 + |\operatorname{Im}\lambda|^2 + |\tilde{\xi}|^2}\right\}, \end{aligned}$$

$\Lambda^{\tilde{\xi}}$ as defined as in (2.62), where $P_{\pm}$ are the transformations described in Lemma 2.5, $W_j^{\pm}$ denote solutions of variable-coefficient eigenvalue problem $(L_{\tilde{\xi}} - \lambda)U = 0$ written as a first-order system in phase-variables $W = (U, z_2')$, $z_2 = b_1^{11} u^I + b_2^{11} u^{II}$, and $V_j^{\pm}$ solutions of the associated limiting, constant-coefficient systems as $x_1 \to \pm\infty$. By construction, D is analytic on Λ.

For comparison with the inviscid case, it is convenient to introduce polar coordinates

$$(\tilde{\xi}, \lambda) =: (\rho\tilde{\xi}_0, \rho\lambda_0), \tag{3.2}$$

$\rho \in \mathbb{R}^1$, $(\tilde{\xi}_0, \lambda_0) \in \mathbb{R}^{d-1} \times \{\operatorname{Re}\lambda \geqslant 0\} \setminus \{(0, 0)\}$, and consider $W_j^{\pm}$, $V_j^{\pm}$ as functions of $(\rho, \tilde{\xi}_0, \lambda_0)$.

LEMMA 3.3 [MéZ.2]. *Under assumptions* (A1)–(A3) *and* (H0)–(H3), *the functions* $V_j^{\pm}$ *may be chosen within groups of r "fast modes" bounded away from the center subspace of coefficient* $\mathbb{A}_{\pm}$, *analytic in* $(\rho, \tilde{\xi}_0, \lambda_0)$ *for* $\rho \geqslant 0$, $\tilde{\xi}_0 \in \mathbb{R}^{d-1}$, $\operatorname{Re}\lambda_0 \geqslant 0$, *and n "slow modes" approaching the center subspace as* $\rho \to 0$, *analytic in* $(\rho, \tilde{\xi}_0, \lambda_0)$ *for* $\rho > 0$, $\tilde{\xi}_0 \in \mathbb{R}^{d-1}$, $\operatorname{Re}\lambda_0 \geqslant 0$ *and continuous at the boundary* $\rho = 0$, *with limits*

$$V_j^{\pm}(0, \tilde{\xi}_0, \lambda_0) = \begin{pmatrix} (A_{\pm}^1)^{-1} r_j^{\pm}(\tilde{\xi}_0, \lambda_0) \\ 0 \end{pmatrix}, \tag{3.3}$$

$r_j^{\pm}$ *defined as in Lemma* 3.1.

PROOF. We here carry out the case $\operatorname{Re}\lambda_0 > 0$. In the case of interest, that (H5) also holds, we shall later obtain the case $\operatorname{Re}\lambda = 0$ as a straightforward consequence of the more detailed computations in Section 5.3; see Remark 5.20. The general case (without (H5)) follows by the analysis of [MéZ.2].

Substituting $v = e^{\mu x_1}\mathbf{v}$ into the limiting eigenvalue equations $(L_{\tilde{\xi}} - \lambda)v = 0$ written in polar coordinates, we obtain the polar characteristic equation,

$$\begin{aligned} \Bigg[\mu^2 B_{\pm}^{11} + \mu\Big(-A_{\pm}^1 + \mathrm{i}\rho \sum_{j\neq 1} B_{\pm}^{j1}\xi_j + \mathrm{i}\rho \sum_{k\neq 1} B^{1k}\xi_k\Big) \\ - \Big(\mathrm{i}\rho \sum_{j\neq 1} A^j\xi_j + \rho^2 \sum_{jk\neq 1} B^{jk}\xi_j\xi_k + \rho\lambda I\Big)\Bigg]\mathbf{v} = 0, \end{aligned} \tag{3.4}$$

where for notational convenience we have dropped subscripts from the fixed parameters $\tilde{\xi}_0, \lambda_0$. At $\rho = 0$, this simplifies to

$$(\mu^2 B_{\pm}^{11} - \mu A_{\pm}^1)\mathbf{v} = 0,$$

which, by the analysis in Appendix A.2 of the linearized traveling-wave ordinary differential equation $(\mu B_{\pm}^{11} - A_{\pm}^1)\mathbf{v} = 0$, has n roots $\mu = 0$, and r roots $\operatorname{Re}\mu \neq 0$. The latter, "fast" roots correspond to stable and unstable subspaces, which extend analytically as claimed by their spectral separation from other modes; thus, we need only focus on the bifurcation as ρ varies near zero of the n-dimensional center manifold associated with "slow" roots $\mu = 0$.

Positing a first-order Taylor expansion

$$\begin{cases} \mu = 0 + \mu^1\rho + o(\rho), \\ \mathbf{v} = \mathbf{v}^0 + \mathbf{v}^1\rho + o(\rho), \end{cases} \tag{3.5}$$

and matching terms of order ρ in (3.4), we obtain:

$$\Big(-\mu^1 A_{\pm}^1 - \mathrm{i}\sum_{j\neq 1} A^j\xi_j - \lambda I\Big)\mathbf{v}^0 = 0, \tag{3.6}$$

or $-\mu^1$ is an eigenvalue of $(A^1)^{-1}(\lambda + \mathrm{i}A^{\tilde{\xi}})$ with associated eigenvector $\mathbf{v}^0$.

For $\operatorname{Re}\lambda > 0$, $(A^1)^{-1}(\lambda + \mathrm{i}A^{\tilde{\xi}})$ has no center subspace. For, substituting $\mu^1 = \mathrm{i}\xi_1$ in (3.6), we obtain $\lambda \in \sigma(\mathrm{i}A^{\xi})$, pure imaginary, a contradiction. Thus, the stable/unstable spectrum *splits* to first order, and we obtain the desired analytic extension by standard matrix perturbation theory, though not in fact the analyticity of individual eigenvalues μ. □

COROLLARY 3.4. *Denoting* $W = (U, b_1^{11}u^I + b_2^{11}u^{II})$ *as above, we may arrange at* $\rho = 0$ *that all* $W_j^{\pm}$ *satisfy the linearized traveling-wave ODE* $(B^{11}U')' - (A^1U)' = 0$, *with constant of integration*

$$B^{11}U_j^{\pm'} - A^1U_j^{\pm} \equiv \begin{cases} 0 & \text{for fast modes,} \\ r_j^{\pm} & \text{for slow modes,} \end{cases} \tag{3.7}$$

$r_j^{\pm}$ *as above, with fast modes analytic at the* $\rho = 0$ *boundary and independent of* $(\tilde{\xi}_0, \lambda_0)$ *for* $\rho = 0$, *and slow modes continuous at* $\rho = 0$.

PROOF. Immediate. □

3.2. *The low-frequency limit*

We are now ready to establish the main result of this section, validating Result 1 of the Introduction. This may be recognized as a generalization of the basic Evans function calculation pioneered by Evans [E.4], relating behavior near the origin to geometry of the phase space of the traveling-wave ODE and thus giving an explicit link between PDE and ODE dynamics. The corresponding one-dimensional result was established in [GZ]; for related calculations, see, e.g., [J,AGJ,PW].

THEOREM 3.5 [ZS,Z.3]. *Under assumptions* (A1)–(A3) *and* (H0)–(H4), *there holds*

$$D(\tilde{\xi}, \lambda) = \gamma\Delta(\tilde{\xi}, \lambda) + \mathcal{O}(|\tilde{\xi}| + |\lambda|)^{\ell+1}, \tag{3.8}$$

where Δ *is given in the Lax case by the inviscid stability function described in* (1.29), *in the undercompressive case by the analogous undercompressive inviscid stability function* (1.38) *with g appropriately chosen,*[25] *and in the overcompressive case by the special, low-frequency stability function*

$$\Delta(\tilde{\xi}, \lambda) := \det\big(r_1^-, \ldots, r_{n-i_-+1}^-, m_{\delta_1}, \ldots, m_{\delta_\ell}, r_{i_++1}^+, \ldots, r_n^+\big), \tag{3.9}$$

[25] Namely, as a Melnikov separation function associated with the undercompressive connection; see [ZS] for details.

where

$$m(\tilde{\xi}, \lambda, \delta) := \lambda \int_{-\infty}^{\infty} \big(\overline{U}^{\delta}(x) - \overline{U}(x)\big)\,\mathrm{d}x + \mathrm{i} \int_{-\infty}^{\infty} \big(F^{\tilde{\xi}}(\overline{U}^{\delta}(x)) - F^{\tilde{\xi}}(\overline{U}(x))\big)\,\mathrm{d}x. \tag{3.10}$$

In each case the factor γ is a constant measuring transversality of the intersection of unstable/resp. stable manifolds of U_-/U_+, in the phase space of the traveling-wave ODE ($\gamma \neq 0 \Leftrightarrow$ transversality), while the constant ℓ as usual denotes the dimension of the manifold $\{\overline{U}^{\delta}\}$, $\delta \in \mathcal{U} \subset \mathbb{R}^{\ell}$, of connections between $U_{\pm}$ (see (H3), *Section* 1.1, *and discussion, Section* 1.2). *In the Lax or undercompressive case, $\ell = 1$ and $\{\overline{U}^{\delta}\} = \{\overline{U}(\cdot - \delta)\}$ is simply the manifold of translates of $\overline{U}$.*

That is, $D(\cdot,\cdot)$ is *tangent* to $\Delta(\cdot,\cdot)$ at $(0,0)$; equivalently, $\Delta(\cdot,\cdot)$ *describes the low-frequency behavior of* $D(\cdot,\cdot)$. Note that $\Delta(\cdot,\cdot)$ is evidently homogeneous degree ℓ in each case (recall, $\ell = 1$ for Lax, undercompressive cases), hence $\mathcal{O}(|\tilde{\xi}| + |\lambda|)^{\ell+1}$ is indeed a higher-order term.

REMARK 3.6. In the one-dimensional setting, $m(0, 1, \delta)$ has an interpretation as "mass-map", see [FreZ.1]; likewise, $\Delta(0, 1)$ (Δ as in (3.10)) arises naturally in determining shock shift/distribution of mass resulting from a given perturbation mass.

PROOF OF THEOREM 3.5. We will carry out the proof in the Lax case only. The proofs for the under- and overcompressive cases are quite similar, and may be found in [ZS]. We are free to make any analytic choice of bases, and any nonsingular choice of coordinates, since these affect the Evans function only up to a nonvanishing analytic multiplier which does not affect the result. Choose bases $W_j^{\pm}$ as in Lemma 3.4, $W = (U, z_2')$, $z_2 = b_1^{11} u^I + b_2^{11} u^{II}$. Noting that $L_0 \overline{U}' = 0$, by translation invariance, we have that $\overline{U}'$ lies in both $\mathrm{Span}\{U_1^+, \ldots, U_K^+\}$ and $\mathrm{Span}(U_{K+1}^-, \ldots, U_{n+r}^-)$ for $\rho = 0$, hence without loss of generality

$$U_1^+ = U_{n+r}^- = \overline{U}', \tag{3.11}$$

independent of $\tilde{\xi}$, λ. (Here, as usual, "$'$" denotes $\partial/\partial x_1$).

More generally, we order the bases so that $W_1^+, \ldots, W_k^+$ and $W_{n+r-k-1}^-, \ldots, W_{n+r}^-$ are fast modes (decaying for $\rho = 0$) and $W_{k+1}^+, \ldots, W_K^+$ and $W_{K+1}^-, \ldots, W_{n+r-k-2}^-$ are slow modes (asymptotically constant for $\rho = 0$), fast modes analytic and slow modes continuous at $\rho = 0$ (Corollary 3.4).

Using the fact that U_1^+ and U_{n+r}^- are analytic, we may express

$$U_1^+(\rho) = U_1^+(0) + U_{1,\rho}^+ \rho + \mathrm{o}(\rho),$$

$$U_{n+r}^-(\rho) = U_{n+r}^-(0) + U_{n+r,\rho}^- \rho + \mathrm{o}(\rho).$$

Writing out the eigenvalue equation in polar coordinates,

$$\begin{aligned}(B^{11}U')' &= (A^1U)' - \mathrm{i}\rho \sum_{j\neq 1} B^{j1}\xi_j U' - \mathrm{i}\rho\Big(\sum_{k\neq 1} B^{1k}\xi_k U\Big)' \\ &\quad + \mathrm{i}\rho\sum_{j\neq 1} A^j\xi_j U + \rho\lambda U - \rho^2 \sum_{j,k\neq 1} B^{jk}\xi_j\xi_k U,\end{aligned} \tag{3.12}$$

we find that $Y^+ := U^+_{1,\rho}(0)$ and $Y^- := U^-_{n+r,\rho}(0)$ satisfy the variational equations

$$\begin{aligned}(B^{11}Y')' &= (A^1Y)' - \mathrm{i}\sum_{j\neq 1} B^{j1}\xi_j \overline{U}'' \\ &\quad - \mathrm{i}\Big(\sum_{k\neq 1} B^{1k}\xi_k \overline{U}'\Big)' + \mathrm{i}\sum_{j\neq 1} A^j\xi_j\overline{U}' + \lambda\overline{U}',\end{aligned} \tag{3.13}$$

with boundary conditions $Y^+(+\infty) = Y^-(-\infty) = 0$. Integrating from $+\infty$, $-\infty$ respectively, we obtain therefore

$$\begin{aligned}&B^{11}Y^{\pm'} - A^1Y^{\pm} \\ &\quad = \mathrm{i}F^{\tilde\xi}(\overline{U}) - \mathrm{i}B^{1\tilde\xi}(\overline{U})\overline{U}' - \mathrm{i}B^{\tilde\xi 1}(\overline{U})\overline{U}' + \lambda\overline{U} - \big[\mathrm{i}F^{\tilde\xi}(U_\pm) + \lambda U_\pm\big],\end{aligned} \tag{3.14}$$

hence $\widetilde{Y} := (Y^- - Y^+)$ satisfies

$$B^{11}\widetilde{Y}' - A^1\widetilde{Y} = \mathrm{i}\big[F^{\tilde\xi}\big] + \lambda[u]. \tag{3.15}$$

By hypothesis (H1), $\begin{pmatrix} A^1_{11} & A^1_{12} \\ b^{11}_1 & b^{11}_2 \end{pmatrix}$ is invertible, hence

$$\begin{aligned}(U, z_2') &\to \big(z_2, -z_1, z_2' - (A^1_{21} + b_1^{11'}, A^1_{22} + b_2^{11'})U\big) \\ &= \big(z_2, B^{11}U' - A^1U\big)\end{aligned} \tag{3.16}$$

is a nonsingular coordinate change, where $\begin{pmatrix} z_1 \\ z_2 \end{pmatrix} := \begin{pmatrix} A^1_{11} & A^1_{12} \\ b^{11}_1 & b^{11}_2 \end{pmatrix} U$.

Fixing $\tilde\xi_0$, λ_0, and using $W_1^+(0) = W_{n+r}^-(0)$, we have

$$\begin{aligned}D(\rho) &= \det\big(W_1^+(0) + \rho W^+_{1_\rho}(0) + \mathrm{o}(\rho), \ldots, W_K^+(0) + \mathrm{o}(1), \\ &\qquad W^-_{K+1}(0) + \mathrm{o}(1), \ldots, W^-_{n+r}(0) + \rho W^-_{n+r_\rho}(0) + \mathrm{o}(\rho)\big) \\ &= \det\big(W_1^+(0) + \rho W^+_{1_\rho}(0) + \mathrm{o}(\rho), \ldots, W_K^+(0) + \mathrm{o}(1), \\ &\qquad W^-_{K+1}(0) + \mathrm{o}(1), \ldots, \rho\widetilde{Y}(0) + \mathrm{o}(\rho)\big) \\ &= \det\big(W_1^+(0), \ldots, W_K^+(0), W^-_{K+1}(0), \ldots, \rho\widetilde{Y}(0)\big) + \mathrm{o}(\rho).\end{aligned} \tag{3.17}$$

Applying now coordinate change (3.16) and using (3.7) and (3.15), we obtain

$$\begin{aligned}
D(\rho) &= C \det \left(\begin{matrix} \overbrace{z_{2,1}^+, \dots, z_{2,k}^+}^{\text{fast}}, & \overbrace{*, \dots, *, *, \dots, *}^{\text{slow}}, \\ 0, \dots, 0, & r_{i_+ +1}^+, \dots, r_n^+, r_1^-, \dots, r_{n-i_-}^- \end{matrix} \right. \\
&\qquad \left. \begin{matrix} \overbrace{z_{2,n+r-k-1}^-, \dots, z_{2,n+r-1}^-}^{\text{fast}}, & * \\ 0, \dots, 0, & \mathrm{i}[F^{\tilde\xi}(U)] + \lambda[U] \end{matrix} \right) \Bigg|_{x_1=0} + \mathrm{o}(\rho) \\
&= \gamma \Delta(\tilde\xi, \lambda) + \mathrm{o}(\rho)
\end{aligned} \tag{3.18}$$

as claimed, where

$$\gamma := C \det\big(z_{2,1}^+, \dots, z_{2,k}^+, z_{2,n+r-k-1}^-, \dots, z_{2,n+r-1}^-\big)\big|_{x_1=0}.$$

Noting that $\{z_{2,1}^+, \dots, z_{2,k}^+\}$ and $\{z_{2,n+r-k-1}^-, \dots, z_{2,n+r}^-\}$ span the tangent manifolds at $\overline{U}(\cdot)$ of the stable/unstable manifolds of traveling wave ODE (1.17) at U_+/U_-, respectively, with $z_{2,1}^+ = z_{2,n+r}^- = (b_1^{11}, b_2^{11})\overline{U}'$ in common, we see that γ indeed measures transversality of their intersection; moreover, γ is constant, by Corollary 3.4. □

4. One-dimensional stability

We now focus attention on the one-dimensional case, establishing Results 2 and 3 of the Introduction. For further discussion/applications, see, e.g., [GS,BSZ.1,ZH,MaZ.4] and [Z.3], Section 6.

4.1. *Necessary conditions: The stability index*

In the one-dimensional case $\tilde\xi \equiv 0$, we obtain by the construction of Section 3 an Evans function $D(\lambda)$ depending only on the temporal frequency, associated with the one-dimensional linearized operator $L := L_0$. Since L is real-valued, we obtain by Remark 2.11.2 that $D(\cdot)$ may be chosen with complex symmetry

$$D(\bar\lambda) = \overline{D}(\lambda),$$

where bar denotes complex conjugate, so that, in particular, $D(\lambda)$ is real-valued for λ real. Further, the energy estimates of Lemma 2.1 show that (since there is no spectrum there) D is nonzero for λ real and sufficiently large, so that

$$\operatorname{sgn} D(+\infty) := \lim_{\lambda \to \mathrm{real}+\infty} \operatorname{sgn} D(\lambda) \tag{4.1}$$

is well defined. Finally, Theorem 3.5 gives $(\mathrm{d}/\mathrm{d}\lambda)^\ell D(0) = \ell! \gamma \Delta(0, 1)$.

DEFINITION 4.1. Combining the above observations, we define the *stability index*

$$\Gamma := \operatorname{sgn}(\mathrm{d}/\mathrm{d}\lambda)^{\ell} D(0)D(+\infty) = \operatorname{sgn} \gamma \Delta(0, 1)D(+\infty), \tag{4.2}$$

where γ, Δ as defined in Section 1.4 denote transversality and inviscid stability coefficients, and ℓ as defined in Section 1.2 the dimension of the manifold $\{\overline{U}^{\delta}\}$ of connections between $U_{\pm}$. In the Lax or undercompressive case, $\ell = 1$.

PROPOSITION 4.2. *The number of unstable eigenvalues $\lambda \in \{\lambda\colon \operatorname{Re}\lambda > 0\}$ has even parity if $\Gamma > 0$ and odd parity if $\Gamma < 0$. In particular, $\Gamma > 0$ is necessary for one-dimensional linearized viscous stability with respect to $L^1 \cap L^{\infty}$ or even test function (C_0^{∞}) initial data.*

PROOF. By complex symmetry (2.25), nonreal eigenvalues occur in conjugate pairs, hence do not affect parity. On the other hand, the number of real roots clearly has the parity claimed. This establishes the first assertion, from which it follows that $\Gamma \geqslant 0$ is necessary for stability with respect to $L^1 \cap L^{\infty}$ data (recall that zeroes of D on $\{\operatorname{Re}\lambda > 0\}$ correspond to exponentially decaying eigenmodes). The assertion that $\gamma \Delta(0, 1) = 0$ implies linearized instability, and instability with respect to test function initial data follow from the more detailed calculations of Section 4.3.4. □

Proposition 4.2 gives a weak version of Result 2 of the Introduction, the deficiency being that the normalizing factor $\operatorname{sgn} D(+\infty)$ is a priori unknown. As described in Section 6 of [Z.3], the basic definition (4.2) may nonetheless yield useful conclusions as a measure of spectral flow/change in stability under homotopy of model or shock parameters. In particular, in the Lax case for which stability of small-amplitude shocks is known, one may conclude that, moving along the Hugoniot curve $H_p(U_-, \theta)$ in direction of increasing $|\theta|$, instability occurs when $\gamma \Delta(0, 1)$ first changes sign.

On the other hand, formula (4.2) is of little use as a measure of absolute stability, e.g., for nonclassical shocks, or Lax-type shocks occurring on disconnected components of the Hugoniot curve. It is useful therefore, as described with increasing levels of generality in [GS,BSZ.1,Z.3], to give an alternative, "absolute" version of the stability index in which $\operatorname{sgn} D(+\infty)$ is explicitly evaluated.

LEMMA 4.3 [Z.3]. *Let matrix A be symmetric, invertible, and matrix B positive semidefinite, $\operatorname{Re}(B) \geqslant 0$. Then, the cones $\mathcal{S}(A^{-1}B) \oplus (N(A) \cap \ker B)$ and $\mathcal{U}(A)$ are transverse, where $\mathcal{S}(M)$, $\mathcal{U}(M)$ refer to stable/unstable subspaces of M and $N(M)$ to the cone $\{v\colon \operatorname{Re}\langle v, Mv\rangle \leqslant 0\}$.*

PROOF. Suppose to the contrary that $x_0 \neq 0$ lies both in the cone $\mathcal{S}(A^{-1}B) \oplus (N(A) \cap \ker B)$ and in $\mathcal{U}(A)$, i.e.,

$$x_0 = x_1 + x_2,$$

where $x_1 \in \mathcal{S}(A^{-1}B)$, $x_2 \in (N(A) \cap \ker B)$, and $x_0 \in \mathcal{U}(A)$. Define $x(t)$ by the ordinary differential equation $x' = A^{-1}Bx, x(0) = x_0$. Then $x(t) \to x_2$ as $t \to +\infty$ and thus $\lim_{t\to+\infty}\langle x(t), Ax(t)\rangle \leqslant 0$. On the other hand,

$$\langle x, Ax\rangle' = 2\langle A^{-1}Bx, Ax\rangle = 2\langle Bx, x\rangle \geqslant 0$$

by assumption, hence $\langle x_0, Ax_0\rangle \leqslant 0$, contradicting the assumption that x_0 belongs to $\mathcal{U}(A)$. □

PROPOSITION 4.4 [Z.3]. *Given* (A1)–(A3) *and* (H0)–(H3), *we have*

$$\operatorname{sgn} D(\lambda) = \operatorname{sgn}\det(\overline{\mathbb{S}}^+, \overline{\mathbb{U}}^+)\det(\pi\mathbb{Z}^+, \varepsilon\overline{\mathbb{S}}^+)\det(\varepsilon\overline{\mathbb{U}}^-, \pi\mathbb{Z}^-)\big|_{\lambda=0} \neq 0, \tag{4.3}$$

for sufficiently large, real λ, *where* π *denotes projection of* $Z = (z_1, z_2, z_2')$ *onto* (z_1, z_2) *with*

$$\begin{pmatrix} z_1 \\ z_2 \end{pmatrix} := \begin{pmatrix} A_{11} & A_{22} \\ b_1 & b_2 \end{pmatrix} U, \tag{4.4}$$

ε *denotes extension of* (z_1) *to* $(z_1, 0)$, *and* $\mathbb{S}(x_1), \mathbb{U}(x_1)$ *are bases of the stable/unstable subspaces of* $A_*^1 = (A_{11}^1 - A_{12}(b_2^{11})^{-1}b_1^{11})$ (*note*: $(n-r)$*-dimensional*).

PROOF. Working in standard coordinates $W = (u^I, u^{II}, z_2')$, we show equivalently that

$$\operatorname{sgn} D(\lambda) = \operatorname{sgn}\det(\overline{\mathbb{S}}^+, \overline{\mathbb{U}}^+)\det(\pi\mathbb{W}^+, \varepsilon\overline{\mathbb{S}}^+)\det(\varepsilon\overline{\mathbb{U}}^-, \pi\mathbb{W}^-)\big|_{\lambda=0} \neq 0, \tag{4.5}$$

where π denotes projection of $W = (u^I, u^{II}, z_2')$ onto (u^I, u^{II}) components, $z_2 := b_1^{11}u^I + b_2^{11}u^{II}$, $\varepsilon u^I := (u^I, -(b_2^{11})^{-1}b_1^{11}u^{II})$ denotes extension, and $\mathbb{S}(x_1)$, $\mathbb{U}(x_1)$ are bases of the stable/unstable subspaces of A_*^1.

Though it is not immediately obvious, we may after a suitable coordinate change arrange that $A_\pm^1$ be symmetric, $B_\pm^{11} = \begin{pmatrix} 0 & 0 \\ 0 & b_2^{11} \end{pmatrix}_\pm$, and $\operatorname{Re}(\tilde{b}_2^{11})_\pm > 0$. For, as pointed out in [KSh], we may without loss of generality take $\widetilde{A}_\pm^0$ block-diagonal in (1.11), by multiplying coefficients on the left by T^t and on the right by T with T a suitably chosen upper block-triangular matrix; for further discussion, see the proof of Proposition 6.3, Appendix A.1. Multiplying on the left and right by $(\widetilde{A}_\pm^0)^{-1/2}$ then reduces the equation to the desired form, in coordinate $V = (\widetilde{A}_\pm^0)^{1/2}W$.

It is sufficient to show that quantity (4.5) does not vanish in the class (A1), (A3). For example, since $D(\lambda)$ does not vanish either, for real λ sufficiently large, we can then establish the result by homotopy of the symmetric matrix $A_\pm^1$ to an invertible real diagonal matrix (straightforward, using the unitary decomposition $A^1 = UDU^*$, $U^*U = I$, and the fact that the unitary group is arcwise connected[26]) and of $B_\pm^{11}$ to $\begin{pmatrix} 0 & 0 \\ 0 & I_r \end{pmatrix}$ (e.g., by linear interpolation of the positive definite $b_{II\pm}^{11}$ to I_r), in which case it can be seen by explicit

[26]See, e.g., [Se.5], Chapter 7.

computation. (Note that the endpoint of this homotopy is on the boundary of but not in the Kawashima class (A2), since eigenvectors of A^1 are in the kernel of B.)

As to behavior at $\lambda = 0$, a bifurcation analysis as in Section 3 of the limiting constant-coefficient equations at $\pm\infty$ shows that the projections Π of slow modes of $\mathbb{W}_+$ may be chosen as the unstable eigenvectors r_j^+ of A_+^1, corresponding to outgoing characteristic modes, and the projections of fast modes as the stable (i.e., $\operatorname{Re}\mu < 0$) solutions of

$$\left(A^1 - \mu B^{11}\right)_+ \begin{pmatrix} u^I \\ u^{II} \end{pmatrix} = \begin{pmatrix} 0 \\ 0 \end{pmatrix}, \tag{4.6}$$

or without loss of generality

$$\begin{pmatrix} A_{11} & A_{12} \\ A_{21} & A_{22} - \mu b_2 \end{pmatrix}_+ \begin{pmatrix} u^I \\ u^{II} \end{pmatrix} = \begin{pmatrix} 0 \\ 0 \end{pmatrix}, \tag{4.7}$$

and thus of form

$$\begin{pmatrix} -(A_{11})^{-1} A_{12} u^{II} \\ u^{II} \end{pmatrix}_+, \tag{4.8}$$

where

$$\left(b_2^{-1}\left(A_{22} - A_{21}(A_{11})^{-1}A_{12}\right) - \mu I\right)_+ u^{II} = 0. \tag{4.9}$$

Likewise, using $b_1^{11+} = 0$, we find from the definitions of $\mathbb{S}$, ε that stable solutions $\mathbb{S}^+$ are in the stable subspace of A_{11+}^1, with $\varepsilon\mathbb{S}^+ = \binom{\mathbb{S}^+}{0}$, hence vectors $\varepsilon\mathbb{S}^+$ lie in the intersection of the stable subspace of A_+^1 and the kernel of B_+^{11}. Our claim is that these three subspaces are independent, spanning $\mathbb{C}^n$. Rewording this assumption, we are claiming that the stable subspace of $(A^1)^{-1}B_+^{11}$, the center subspace $\ker B_+^{11}$ intersected with the stable cone $N(A_+^1)$, and the unstable subspace of A_+^1 are mutually independent. (Note: that dimensions are correct follows by consistent splitting, a consequence of (H1)(i).) However, this follows by Lemma 4.3, with $A := A_+^1$ and $B := B_+^{11}$, hence $\det(\pi\mathbb{Z}^+, \varepsilon\mathbb{S}^+) \neq 0$. A symmetric argument shows that $\det(\varepsilon\mathbb{U}^-, \pi\mathbb{Z}^-)|_{\lambda=0} \neq 0$, completing the result. □

Formula (4.5) gives in principle an explicit evaluation of Γ, involving only information that is linear-algebraic or concerning the dynamics of the traveling-wave ODE; however, it is still quite complicated. In the extreme Lax shock case, the formula simplifies considerably.

PROPOSITION 4.5. *In the case of an extreme right (n-shock) Lax profile,*

$$\Gamma = \operatorname{sgn}\gamma^2 \det\left(r_1^-, \ldots, r_{n-1}^-, [U]\right) \det\left(r_1^-, \ldots, r_{n-1}^-, \bar{U}'/|\bar{U}'|(-\infty)\right) > 0 \tag{4.10}$$

is necessary for one-dimensional stability. A symmetric formula holds for extreme left (1-shock) Lax profiles.

PROOF. From Corollary 1.8, we may deduce that γ for an extreme right (i.e., n-shock) Lax profile consists of a Wronskian involving only modes from the $+\infty$ side, and is therefore explicitly evaluable; in particular, working in (z_1, z_2, z_2') coordinates, we obtain γ as a determinant of z_2 components only. Moreover, the expression (4.5) simplifies greatly. For example, in (z_1, z_2) coordinates, we have that $\mathbb{U} = \emptyset$, $\mathbb{S}$ is full dimension $n - r$, and $\varepsilon\mathbb{S}$ consists of vectors of the simple form $(z, 0)$. This means that $\det(\mathbb{S}, \mathbb{U})$ simplifies to just $\det\mathbb{S}$, while $\det(\mathbb{Z}^-, \varepsilon\mathbb{S}^+)$ simplifies to the product of $\det\mathbb{S}^+$ and γ. Therefore, this term, similarly as in the strictly parabolic case, combines with like term γ in the computation of the stability index. Finally, $\det(\varepsilon\mathbb{U}^0, \Pi\mathbb{Z}^-)$ simplifies to $\det(r_1^-, \dots, r_{n-1}^-, \overline{U}')$, completing the result. □

REMARK 4.6. Condition (4.10), our strongest version of Result 2, is identical with that of the strictly parabolic case. The only very weak information required from the connection problem is the orientation of $\bar{u}'$ as $x \to -\infty$, i.e., the direction in which the profile leaves along the one-dimensional unstable manifold. In the case $r = 1$, for example, for isentropic gas dynamics, the traveling-wave ODE is scalar, and so the orientation of $\overline{U}'$ is determined by the direction of the connection. In this case, (4.10) gives a full evaluation.

4.2. *Sufficient conditions*

By (H3), there exists an ℓ-parameter family of stationary solutions $\{\overline{U}^\delta\}$ of (1.1) near $\overline{U}$, $\ell \geqslant 1$. Indeed, by Lemma 1.6, these lie arbitrarily close to $\overline{U}$ in $L^1 \cap C^4(x_1)$, precluding one-dimensional asymptotic stability of $\overline{U}$ for any reasonable class of initial perturbation (say, test function initial data). The appropriate notion of stability of shock profiles in one dimension is, rather, "orbital" stability, or convergence to the manifold of stationary solutions $\{\overline{U}^\delta\}$.

DEFINITION 4.7. We define *nonlinear orbital stability* as convergence of $\widetilde{U}(\cdot, t)$ as $t \to \infty$ to $\overline{U}^{\delta(t)}$, where $\delta(\cdot)$ is an appropriately chosen function, for any solution $\widetilde{U}$ of (1.1) with initial data sufficiently close in some norm to the original profile $\overline{U}$. Likewise, we define *linearized orbital stability* as convergence of $U(\cdot, t)$ to $(\partial\overline{U}^\delta/\partial\delta) \cdot \delta(t)$ for any solution U of the linearized equations (2.1) with initial data bounded in some chosen norm.

Note that differentiation with respect to δ_j of the standing-wave ODE $F^1(\overline{U}^\delta)_{x_1} = (B^{11}(\overline{U}^\delta)\overline{U}^\delta_{x_1})_{x_1}$ yields $L_0\,\partial\overline{U}^\delta/\partial\delta_j = 0$, so that $(\partial\overline{U}^\delta/\partial\delta_j)$ represent stationary solutions of (2.1), with $\mathrm{Span}\{\partial\overline{U}^\delta/\partial\delta_j)\}$ lying tangent to the stationary manifold $\{\overline{U}^\delta\}$. In the Lax or undercompressive case, $\ell = 1$, and the stationary manifold is just the set of translates $\overline{U}^\delta(x_1) := \overline{U}(x_1 - \delta)$ of $\overline{U}$, with the tangent stationary manifold $\mathrm{Span}\{\overline{U}_{x_1}\}$ corresponding to the usual translational zero eigenvalue arising in the study of stability of traveling waves.

Result 3 of the Introduction is subsumed in the following two results, to be established throughout the rest of the section.

THEOREM 4.8 ([MaZ.4]; Linearized stability). *Let $\overline{U}$ be a shock profile* (1.2) *of* (1.1), *under assumptions* (A1)–(A3) *and* (H0)–(H3). *Then, $\overline{U}$ is $L^1 \cap L^p \to L^p$ linearly orbitally* *stable in dimension $d = 1$, for all $p > 1$, if and only if it satisfies the conditions of structural stability (transversality), dynamical stability (inviscid stability), and strong spectral stability. More precisely, for initial data $U_0 \in L^1 \cap L^p$, $p \geqslant 1$, the solution $U(x,t)$ of Equations* (1.1) *linearized about $\overline{U}$ satisfies*

$$\left|U(\cdot,t) + \delta(t)\overline{U}'(\cdot)\right|_{L^p} \leqslant C(1+t)^{-\frac{1}{2}(1-1/p)}|U_0|_{L^1\cap L^p}. \tag{4.11}$$

THEOREM 4.9 (Nonlinear stability). *Let $\overline{U}$ be a Lax-type profile* (1.2) *of a general real viscosity model* (1.1), *satisfying* (A1)–(A3), (H0)–(H3), *and the necessary conditions of structural, dynamical and strong spectral stability. Then, $\overline{U}$ is $L^1 \cap H^3 \to L^p \cap H^3$ nonlinearly orbitally stable in dimension $d = 1$, for all $p \geqslant 2$. More precisely, the solution $\widetilde{U}(x,t)$ of* (1.1) *with initial data $\widetilde{U}_0$ satisfies*

$$\begin{aligned} \left|\widetilde{U}(x,t) - \overline{U}\big(x-\delta(t)\big)\right|_{L^p} &\leqslant C(1+t)^{-\frac{1}{2}(1-1/p)}|U_0|_{L^1\cap H^3}, \\ \left|\widetilde{U}(x,t) - \overline{U}\big(x-\delta(t)\big)\right|_{H^3} &\leqslant C(1+t)^{-\frac{1}{4}}|U_0|_{L^1\cap H^3} \end{aligned} \tag{4.12}$$

for initial perturbations $U_0 := \widetilde{U}_0 - \overline{U}$ that are sufficiently small in $L^1 \cap H^3$, for all $p \geqslant 2$, for some $\delta(t)$ satisfying $\delta(0) = 0$,

$$\left|\dot{\delta}(t)\right| \leqslant C(1+t)^{-1/2}|U_0|_{L^1\cap H^3} \tag{4.13}$$

and

$$\left|\delta(t)\right| \leqslant C|U_0|_{L^1\cap H^3}. \tag{4.14}$$

REMARKS. 1. For slightly more regular initial perturbation $U_0 \in L^1 \cap H^4$, we may obtain by the same argument, using higher-derivative Green distribution bounds (described, e.g., in [MaZ.2,MaZ.4]), the derivative estimate

$$\left|\widetilde{U}(x,t) - \overline{U}\big(x-\delta(t)\big)\right|_{W^{1,p}} \leqslant C(1+t)^{-\frac{1}{2}(1-1/p)}|U_0|_{L^1\cap H^4}$$

generalizing (4.12), which yields in turn the extension of (4.12) to low norms L^p, $1 \leqslant p \leqslant 2$, by a simple bootstrap argument (described in the remark just below the proof of Theorem 4.9).

2. Theorem 4.8 and the extension of Theorem 4.9 just described contain in particular the results of $L^1 \to L^1$ linear and $L^1 \cap H^4 \to L^1 \cap H^4$ nonlinear *global bounded stability*, indicating that L^1 and $L^1 \cap H^4$, respectively, are natural norms for the linearized and nonlinear viscous shock perturbation problem in one dimension.

4.2.1. *Linearized estimates.* Theorem 4.8 (and to some extent Theorem 4.9) is obtained as a consequence of detailed, pointwise bounds on the Green distribution $G(x,t;y)$ of the linearized evolution equations (2.1), which we now describe. For readability, we defer the rather technical proof to Section 4.3. In the remainder of this section, for ease of notation, we drop superscripts on A^1, F^1 and B^{11}, and subscripts on x_1 and y_1, denoting them simply as A, F, B, x, and y. We also make the simplifying assumption:

(P0) Either $A_\pm$ are strictly hyperbolic, or else $A_\pm$ and $B_\pm$ are simultaneously symmetrizable.

As described in Remark 4.13, the bounds needed for our stability analysis persist in the general case, but the full description of the Green's function does not.

Let $a_j^\pm$, $j = 1, \ldots, n$, denote the eigenvalues of $A(\pm\infty)$, and $l_j^\pm$ and $r_j^\pm$ associated left and right eigenvectors, respectively, normalized so that $l_j^{\pm t} r_k^\pm = \delta_k^j$. In case $A_\pm$ is strictly hyperbolic, these are uniquely defined up to a scalar multiplier. Otherwise, we require further that $l_j^\pm, r_j^\pm$ be left and right eigenvectors also of $P_j^\pm B_\pm P_j^\pm$, $P_j^\pm := R_j^\pm L_j^{\pm t}$, where $L_j^\pm$ and $R_j^\pm$ denote $m_j^\pm \times m_j^\pm$ left and right eigenblocks associated with the $m_j^\pm$-fold eigenvalue $a_j^\pm$, normalized so that $L_j^\pm R_j^\pm = I_{m_j^\pm}$. (Note: The matrix $P_j^\pm B_\pm P_j^\pm \sim L_j^{\pm t} B_\pm R_j^\pm$ is necessarily diagonalizable, by simultaneous symmetrizability of $A_\pm$, $B_\pm$.)

Eigenvalues $a_j^\pm$, and eigenvectors $l_j^\pm, r_j^\pm$ correspond to large-time convection rates and modes of propagation of the degenerate model (1.1). Likewise, let $a_j^*(x)$, $j = 1, \ldots,$ $(n - r)$, denote the eigenvalues of

$$A_* := A_{11} - A_{12} B_{22}^{-1} B_{21}, \tag{4.15}$$

$A := \mathrm{d}F(\overline{U}(x))$ and $l_j^*(x), r_j^*(x) \in \mathbb{R}^{n-r}$ associated left and right eigenvectors, normalized so that $l_j^{*t} r_j \equiv \delta_k^j$. More generally, for an m_j^*-fold eigenvalue, we choose $(n - r) \times m_j^*$ blocks L_j^* and R_j^* of eigenvectors satisfying the *dynamical normalization*

$$L_j^{*t}\, \partial_x R_j^* \equiv 0,$$

along with the usual static normalization $L_j^{*t} R_j \equiv \delta_k^j I_{m_j^*}$; as shown in Lemma 4.9 of [MaZ.1], this may always be achieved with bounded L_j^*, R_j^*. Associated with L_j^*, R_j^*, define extended, $n \times m_j^*$ blocks

$$\mathcal{L}_j^* := \begin{pmatrix} L_j^* \\ 0 \end{pmatrix}, \qquad \mathcal{R}_j^* := \begin{pmatrix} R_j^* \\ -B_{22}^{-1} B_{21} R_j^* \end{pmatrix}. \tag{4.16}$$

Eigenvalues a_j^* and eigenmodes $\mathcal{L}_j^*$, $\mathcal{R}_j^*$ correspond, respectively, to short-time hyperbolic characteristic speeds and modes of propagation for the reduced, hyperbolic part of degenerate system (1.1).

Define time-asymptotic, *effective diffusion coefficients*

$$\beta_j^\pm := \left(l_j^t B r_j\right)_\pm, \quad j = 1, \ldots, n, \tag{4.17}$$

and local, $m_j \times m_j$ *dissipation coefficients*

$$\eta_j^*(x) := -L_j^{*t} D_* R_j^*(x), \quad j = 1, \ldots, J \leqslant n - r, \tag{4.18}$$

where

$$D_*(x) := A_{12}B_{22}^{-1}\big[A_{21} - A_{22}B_{22}^{-1}B_{21} + A_* B_{22}^{-1}B_{21} + B_{22}\,\partial_x\big(B_{22}^{-1}B_{21}\big)\big] \tag{4.19}$$

is an effective dissipation analogous to the effective diffusion predicted by formal, Chapman–Enskog expansion in the (dual) relaxation case.

At $x = \pm\infty$, these reduce to the corresponding quantities identified by Zeng [Ze.1,LZe] in her study by Fourier transform techniques of decay to *constant solutions* $(\bar{u}, \bar{v}) \equiv (u_\pm, v_\pm)$ of hyperbolic–parabolic systems, i.e., of limiting equations

$$U_t = L_\pm U := -A_\pm U_x + B_\pm U_{xx}. \tag{4.20}$$

These arise naturally through Taylor expansion of the one-dimensional (frozen-coefficient) Fourier symbol $(-\mathrm{i}\xi A - \xi^2 B)_\pm$, as described in Appendix A.4; in particular, as a consequence of dissipativity, (A2), we obtain (see, e.g., [Kaw,LZe,MaZ.3], or Lemma 2.18)

$$\beta_j^\pm > 0, \qquad \operatorname{Re}\sigma\big(\eta_j^{*\pm}\big) > 0 \quad \text{for all } j. \tag{4.21}$$

However, note that the dynamical dissipation coefficient $D_*(x)$ *does not* agree with its static counterpart, possessing an additional term $B_{22}\,\partial_x(B_{22}^{-1}B_{21})$, and so we cannot conclude that (4.21) holds everywhere along the profile, but only at the endpoints. This is an important difference in the variable-coefficient case; see Remarks 1.11 and 1.12 of [MaZ.3] for further discussion.

PROPOSITION 4.10 [MaZ.3]. *In spatial dimension $d = 1$, under assumptions* (A1)–(A3), (H0)–(H3) *and* (P0), *for Lax-type shock profiles satisfying the conditions of transversality, hyperbolic stability, and strong spectral stability, the Green distribution $G(x, t; y)$ associated with the linearized evolution equations* (2.1) *may be decomposed as*

$$G(x, t; y) = H + E + S + R, \tag{4.22}$$

where, for $y \leqslant 0$:

$$\begin{aligned} H(x, t; y) &:= \sum_{j=1}^{J} a_j^{*-1}(x) a_j^*(y) \mathcal{R}_j^*(x) \zeta_j^*(y, t) \delta_{x - \bar{a}_j^* t}(-y) \mathcal{L}_j^{*t}(y) \\ &= \sum_{j=1}^{J} \mathcal{R}_j^*(x) \mathcal{O}\big(\mathrm{e}^{-\eta_0 t}\big) \delta_{x - \bar{a}_j^* t}(-y) \mathcal{L}_j^{*t}(y), \end{aligned} \tag{4.23}$$

$$E(x, t; y) := \sum_{a_k^- > 0} \big[c_{k,-}^0\big] \bar{U}'(x) l_k^{-t} \left(\operatorname{errfn}\left(\frac{y + a_k^- t}{\sqrt{4\beta_k^- t}} \right) - \operatorname{errfn}\left(\frac{y - a_k^- t}{\sqrt{4\beta_k^- t}} \right) \right) \tag{4.24}$$

and

$$
\begin{aligned}
S(x,t;y) := {} & \chi_{\{t\geqslant 1\}} \sum_{a_k^-<0} r_k^- l_k^{-t} \left(4\pi\beta_k^- t\right)^{-1/2} \mathrm{e}^{-(x-y-a_k^- t)^2/4\beta_k^- t} \\
& + \chi_{\{t\geqslant 1\}} \sum_{a_k^->0} r_k^- l_k^{-t} \left(4\pi\beta_k^- t\right)^{-1/2} \mathrm{e}^{-(x-y-a_k^- t)^2/4\beta_k^- t} \left(\frac{\mathrm{e}^{-x}}{\mathrm{e}^x+\mathrm{e}^{-x}}\right) \\
& + \chi_{\{t\geqslant 1\}} \sum_{a_k^->0,\, a_j^-<0} \left[c_{k,-}^{j,-}\right] r_j^- l_k^{-t} \left(4\pi\bar\beta_{jk}^- t\right)^{-1/2} \mathrm{e}^{-(x-z_{jk}^-)^2/4\bar\beta_{jk}^- t} \mathrm{O} \\
& \times \left(\frac{\mathrm{e}^{-x}}{\mathrm{e}^x+\mathrm{e}^{-x}}\right) \\
& + \chi_{\{t\geqslant 1\}} \sum_{a_k^->0,\, a_j^+>0} \left[c_{k,-}^{j,+}\right] r_j^+ l_k^{-t} \left(4\pi\bar\beta_{jk}^+ t\right)^{-1/2} \mathrm{e}^{-(x-z_{jk}^+)^2/4\bar\beta_{jk}^+ t} \\
& \times \left(\frac{\mathrm{e}^{x}}{\mathrm{e}^x+\mathrm{e}^{-x}}\right)
\end{aligned} \tag{4.25}
$$

denote hyperbolic, excited, and scattering terms, respectively, $\eta_0 > 0$, *and* R *denotes a faster decaying residual (described in Proposition* 4.39). *Symmetric bounds hold for* $y \geqslant 0$.

Here, $\bar a_j^* = \bar a_j^*(x,t)$ *in* (4.23) *denote the time-averages over* $[0,t]$ *of* $a_j^*(x)$ *along backward characteristic paths* $z_j^* = z_j^*(x,t)$ *defined by*

$$
\frac{\mathrm{d}z_j^*}{\mathrm{d}t} = a_j^*\left(z_j^*\right), \quad z_j^*(t) = x, \tag{4.26}
$$

and the dissipation matrix $\zeta_j^* = \zeta_j^*(x,t) \in \mathbb{R}^{m_j^* \times m_j^*}$ *is defined by the dissipative flow*

$$
\frac{\mathrm{d}\zeta_j^*}{\mathrm{d}t} = -\eta_j^*\left(z_j^*\right)\zeta_j^*, \quad \zeta_j^*(0) = I_{m_j}. \tag{4.27}
$$

Similarly, in (4.25),

$$
z_{jk}^{\pm}(y,t) := a_j^{\pm}\left(t - \frac{|y|}{|a_k^-|}\right) \tag{4.28}
$$

and

$$
\bar\beta_{jk}^{\pm}(x,t;y) := \frac{|x^{\pm}|}{|a_j^{\pm} t|}\beta_j^{\pm} + \frac{|y|}{|a_k^- t|}\left(\frac{a_j^{\pm}}{a_k^-}\right)^2 \beta_k^-, \tag{4.29}
$$

represent, respectively, approximate scattered characteristic paths and the time-averaged diffusion rates along those paths. In all equations, a_j, $a_j^{*\pm}$, l_j, $\mathcal{L}_j^{*\pm}$, r_j, $\mathcal{R}_j^{*\pm}$, $\beta_j^{\pm}$ *and* η_j^* *are as defined just above, and scattering coefficients* $[c_{k,-}^{j,i}]$, $i = -, 0, +$, *are uniquely determined by*

$$\sum_{a_j^-<0}[c_{k,-}^{j,-}]r_j^- + \sum_{a_j^+>0}[c_{k,-}^{j,+}]r_j^+ + [c_{k,-}^0]\big(U(+\infty) - U(-\infty)\big) = r_k^- \tag{4.30}$$

for each $k = 1, \ldots, n$, *and satisfying*

$$\sum_{a_k^->0}[c_{k,-}^0]l_k^- = \sum_{a_k^+<0}[c_{k,+}^0]l_k^+ = \pi, \tag{4.31}$$

where the constant vector π *is the left zero effective eigenfunction of* L *associated with the right eigenfunction* $\overline{U}'$. *Similar decompositions hold in the over- and undercompressive case.*

Proposition 4.10, the variable-coefficient generalization of the constant-coefficient results of [Ze.1,LZe], was established in [MaZ.3] by Laplace transform (i.e., semigroup) techniques generalizing the Fourier transform approach of [Ze.1,Ze.2,LZe]; for discussion/geometric interpretation, see [Z.6,MaZ.1,MaZ.2]. In our stability analysis, we will use only a small part of the detailed information given in the proposition, namely $L^p \to L^q$ estimates on the time-decaying portion $H + S + R$ of the Green distribution G (see Lemma 4.1). However, the stationary portion E of the Green distribution must be estimated accurately for an efficient stability analysis.

REMARK 4.11 [MaZ.2]. The function H described in Proposition 4.10 satisfies

$$H(x, t; y)\Pi_2 \equiv 0 \tag{4.32}$$

and

$$\Pi_2\big(A^0(x)\big)^{-1} H(x, t; y) \equiv 0, \tag{4.33}$$

where $\Pi_2 := \begin{pmatrix} 0 & 0 \\ 0 & I_r \end{pmatrix}$ denotes projection onto final r coordinates, or equivalently $\mathcal{L}_j^{*tr}\Pi_2 \equiv 0$ and $\Pi_2(A^0)^{-1}\mathcal{R}_j^* \equiv 0$, $1 \leqslant j \leqslant J$.

These may be seen by the intrinsic property of $\mathcal{L}_j^*$ and $\mathcal{R}_j^*$ (readily obtained from our formulae) that they lie, respectively, in the left and right kernel of B. This gives the first relation immediately, by the structure of B given in (A3). Likewise, we find that

$$\begin{aligned} 0 \equiv B\mathcal{R}_j^* &:= \begin{pmatrix} 0 & 0 \\ 0 & b \end{pmatrix} (A^0)^{-1}\mathcal{R}_j^* \\ &= \begin{pmatrix} 0 \\ b\Pi_2(A^0)^{-1}\mathcal{R}_j^* \end{pmatrix}, \end{aligned} \tag{4.34}$$

yielding the second result by invertibility of b (see condition (A3)). This quantifies the observation that, in $W=(w^I,w^{II})^t$ coordinates, data in the w^{II} coordinate is smoothed under the evolution of (2.1), whereas data in the w^I coordinate is not; likewise, the "parabolic" variable w^{II} experiences smoothing while the "hyperbolic" coordinate w^I does not. (Recall that $W=(A^0)^{-1}U$.)

REMARK 4.12. Following [Ho], we may recover the hyperbolic evolution equations (4.23), (4.26) and (4.27) by direct calculation, considering the evolution of a jump discontinuity along curve $x=z(t)$ in the original equation $U_t+(AU)_x=(BU_x)_x$; see also the related calculation of Section 1.3, [MaZ.1] in the relaxation case. For simplicity, write $U=(u,v)$. Introducing the parabolic variable $\tilde{v}:=b_1u+b_2v$, and denoting jump across z of a variable f by $[f]$, we may write (2.1) as the pair of equations

$$u_t+(A_*u)_x+\left(A_{12}b_2^{-1}\tilde{v}\right)_x=0 \tag{4.35}$$

and

$$\begin{aligned} v_t &+ \left(\left(A_{21}-A_{22}b_2^{-1}b_1+b_{1,x}-b_{2,x}b_2^{-1}b_1\right)u\right)_x \\ &+ \left(\left(A_{22}b_2^{-1}+b_{2,x}b_2^{-1}\right)\tilde{v}\right)_x-\left(\tilde{v}_x\right)_x=0. \end{aligned} \tag{4.36}$$

Applying the Rankine–Hugoniot conditions, we obtain relations

$$(A_*-\dot{z})[u]=0, \qquad [\tilde{v}]=0$$

and

$$\begin{aligned} [\tilde{v}_x] &= \left(A_{21}-A_{22}b_2^{-1}b_1+\dot{z}b_2^{-1}b_1+b_2\,\partial_x\left(b_2^{-1}b_1\right)\right)[u] \\ &= \left(A_{21}-A_{22}b_2^{-1}b_1+b_2^{-1}b_1A_*+b_2\,\partial_x\left(b_2^{-1}b_1\right)\right)[u] \\ &=: d_*[u], \end{aligned} \tag{4.37}$$

yielding $\dot{z}_j=a_j(z_j(t))$, $[u]=R_j^*(z_j(t))\zeta_j(t)$, $[\tilde{v}]\equiv 0$, and $[\tilde{v}_x]=d_*R_j^*(z_j(t))\zeta_j(t)$, $\zeta_j\in\mathbb{R}$. Substituting these Ansatze into (4.35) yields

$$[u_t]+a_j^*[u_x]+a_{j,x}^*[u]+A_{12}b_2^{-1}[\tilde{v}_x]+\left(A_{12}b_2^{-1}\right)_x[\tilde{v}]=0,$$

and thus

$$[\dot{u}]+a_{j,x}^*[u]+D_*[u]=0,$$

where all coefficients are evaluated at $z_j(t)$ and $D_*=A_{12}b_2^{-1}d_*$ as defined in (4.19). Taking inner product with $L_j^*(z_j(t))$ and rearranging using

$$L_j^{*t}\dot{R}_j^*=-\dot{L}_j^{*t}R_j^*=-a_j^*\left(\partial_x L_j^{*t}\right)R_j^*,$$

we therefore obtain the characteristic evolution equations

$$\dot{\zeta}_j = \big(a_j^*(\partial_x L_j^{*t})R_j^* - \partial_x a_j - \eta_j\big)\zeta_j, \tag{4.38}$$

$\eta_j = L_j^{*t} D_* R_j^*$ as defined in (4.18). Under the normalization $(\partial_x L_j^{*t})R_j^* \equiv 0$ (note: automatic in the gas-dynamical case considered in [Ho], for which L_j^* and R_j^* are scalar), this yields in the original coordinate U precisely the solution operator $a_j^{*-1}(z_j)a_j^*(y)\mathcal{R}_j^*(z_j) \times \zeta_j(y,t)\mathcal{L}_j^*(y)$ described in (4.23).

REMARK 4.13. In the case that $A_\pm$ is nonstrictly hyperbolic and $B_\pm$ are not simultaneously symmetrizable, the simple description of scattering terms breaks down, and much more complicated behavior appears to occur. However, the excited term E remains much the same, with scalar diffusion replaced by a multimode heat kernel and associated multimode error function. Likewise, a review of the pointwise Green's function arguments of Section 4.3.3 shows that they still yield sharp *modulus* bounds on the scattering term even though we do not know the leading-order behavior, and so all bounds relevant to the ultimate stability argument go through.

4.2.2. *Linearized stability.* As described in [MaZ.2,MaZ.3], linearized orbital stability follows immediately from the pointwise bounds of Proposition 4.10. We reproduce the argument here, both for completeness and to motivate the nonlinear argument to follow in Section 4.2.4. Similarly as in [Z.6,MaZ.1–MaZ.3], define the *linear instantaneous projection*:

$$\begin{aligned}\varphi(x,t) &:= \int_{-\infty}^{+\infty} E(x,t;y)U_0(y)\,dy \\ &=: -\delta(t)\overline{U}'(x),\end{aligned} \tag{4.39}$$

where U_0 denotes the initial data for (2.1), and $\overline{U} = \overline{U}(x)$ as usual. The amplitude δ may be expressed, alternatively, as

$$\delta(t) = -\int_{-\infty}^{+\infty} e(y,t)U_0(y)\,dy,$$

where

$$E(x,t;y) =: \overline{U}'(x)e(y,t), \tag{4.40}$$

i.e.,

$$e(y,t) := \sum_{a_k^- > 0} \left(\operatorname{errfn}\left(\frac{y + a_k^- t}{\sqrt{4\beta_k^- t}}\right) - \operatorname{errfn}\left(\frac{y - a_k^- t}{\sqrt{4\beta_k^- t}}\right)\right) \tag{4.41}$$

for $y \leqslant 0$, and symmetrically for $y \geqslant 0$.

Then, the solution U of (2.1) satisfies

$$U(x,t)-\varphi(x,t)=\int_{-\infty}^{+\infty}(H+\widetilde{G})(x,t;0)U_0(y)\,\mathrm{d}y, \tag{4.42}$$

where

$$\widetilde{G}:=S+R \tag{4.43}$$

is the regular part and H the singular part of the time-decaying portion of the Green distribution G.

LEMMA 4.14 [MaZ.2]. *In spatial dimension* $d=1$, *under assumptions* (A1)–(A3) *and* (H0)–(H3), *for Lax-type shock profiles satisfying the conditions of transversality, hyperbolic stability, and strong spectral stability,* $\widetilde{G}$ *and* H *satisfy*

$$\left|\int_{-\infty}^{+\infty}\widetilde{G}(\cdot,t;y)f(y)\,\mathrm{d}y\right|_{L^p}\leqslant C(1+t)^{-\frac{1}{2}(1/q-1/p)}|f|_{L^q}, \tag{4.44}$$

$$\left|\int_{-\infty}^{+\infty}\widetilde{G}_y(\cdot,t;y)f(y)\,\mathrm{d}y\right|_{L^p}\leqslant C(1+t)^{-\frac{1}{2}(1/q-1/p)-1/2}|f|_{L^q}+C\mathrm{e}^{-\eta t}|f|_{L^p} \tag{4.45}$$

and

$$\left|\int_{-\infty}^{+\infty}H(\cdot,t;y)f(y)\,\mathrm{d}y\right|_{L^p}\leqslant C\mathrm{e}^{-\eta t}|f|_{L^p}, \tag{4.46}$$

$$\left|\int_{-\infty}^{+\infty}H_x(\cdot,t;y)f(y)\,\mathrm{d}y\right|_{L^p}\leqslant C\mathrm{e}^{-\eta t}|f|_{W^{1,p}} \tag{4.47}$$

for all $t\geqslant 0$, *some* C, $\eta>0$, *for any* $1\leqslant q\leqslant p$ (*equivalently*, $1\leqslant r\leqslant p$) *and* $f\in L^q\cap W^{1,p}$, *where* $1/r+1/q=1+1/p$.

PROOF. Assuming (P0), bounds (4.44) and (4.45) follow by the Hausdorff–Young inequality together with bounds (4.25) and the more stringent bounds on R and its derivatives given in Proposition 4.39; see [MaZ.2,MaZ.3] for further details. Bound (4.46) follows by direct computation and the fact that particle paths $z_j^*(x,t)$ satisfy uniform bounds

$$\frac{1}{C}\leqslant\left|\frac{\partial}{\partial x}z_j\right|<C,$$

for all x, t, by the fact that characteristic speeds $a_j(z)$ converge exponentially as $z\to\pm\infty$ to constant states. Finally, (4.47) follows from the relations

$$\delta_{x-\bar{a}_j^*t}(-y)=a_j^*(y)^{-1}\delta\left(t-\int_y^x\frac{\mathrm{d}z}{a_j^*(z)}\right)$$

and

$$\left(\frac{\partial}{\partial x}\right)\delta\left(t-\int_y^x \frac{\mathrm{d}z}{a_j^*(z)}\,\mathrm{d}z\right)=a_j^*(y)a_j^*(x)^{-1}\left(\frac{\partial}{\partial y}\right)\delta\left(t-\int_y^x \frac{\mathrm{d}z}{a_j^*(z)}\right),$$

where $\bar{a}_j^* = \bar{a}_j^*(x,t)$ is defined as the average of $a_j^*(z)$ along the backward characteristic $z_j^*(x,t)$ defined by (4.26), which are special cases of the more general relations

$$\delta_{h(x,t)}(y) = f_y(x,y,t)\delta\big(f(x,y,t)\big)$$

and

$$(\partial/\partial x)\delta\big(f(x,y,t)\big) = (f_x/f_y)(\partial/\partial y)\delta\big(f(x,y,t)\big),$$

where $h(x,t)$ is defined by $f(x,h(x,t),t)\equiv 0$. Noting that the same modulus bounds, though not the pointwise description, persist also when (P0) fails, Remark 4.13, we obtain the general case as well. □

PROOF OF THEOREM 4.8. It is equivalent to show that, for initial data $U_0 \in L^1 \cap L^p$, the solution $U(x,t)$ of (2.1) satisfies

$$\big|U(\cdot,t)-\varphi(\cdot,t)\big|_{L^p} \leqslant C(1+t)^{-\frac{1}{2}(1-1/p)}\big(|U_0|_{L^1}+|U_0|_{L^p}\big). \tag{4.48}$$

But, this follows immediately from (4.42) and bounds (4.44) and (4.46), with $q=p$. The estimate (4.11) gives sufficiency of structural, dynamical, and strong spectral stability for viscous linearized orbital stability. Necessity will be established by a separate, and somewhat simpler computation in Section 4.3.5; see Proposition 4.43 and Corollary 4.44. □

4.2.3. *Auxiliary energy estimate.* Along with Proposition 4.10, the proof of Theorem 4.9 relies on the following auxiliary energy estimate, generalizing estimates of Kawashima [Kaw] in the constant-coefficient case. Following [MaZ.2], define nonlinear perturbation

$$U(x,t) := \tilde{U}\big(x+\delta(t),t\big)-\overline{U}(x), \tag{4.49}$$

where the "shock location" δ is to be determined later; for definiteness, fix $\delta(0)=0$. Evidently, decay of U is equivalent to nonlinear orbital stability as described in (4.12).

PROPOSITION 4.15 [MaZ.4]. *Under the hypotheses of Theorem* 4.9 *let* $U_0 \in H^3$, *and suppose that, for* $0 \leqslant t \leqslant T$, *both the supremum of* $|\delta|$ *and the* $W^{2,\infty}$ *norm of the solution* $U=(u^I,u^{II})^t$ *of* (1.1), (4.49) *remain bounded by a sufficiently small constant* $\zeta>0$. *Then, for all* $0\leqslant t\leqslant T$,

$$\big|U(t)\big|_{H^3}^2 \leqslant C\big|U(0)\big|_{H^3}^2 \mathrm{e}^{-\theta t} + C\int_0^t \mathrm{e}^{-\theta_2(t-\tau)}\big(|U|_{L^2}^2+|\dot{\delta}|^2\big)(\tau)\,\mathrm{d}\tau. \tag{4.50}$$

REMARKS 4.16. 1. Estimate (4.50) asserts that the H^3 norm is controlled essentially by the L^2 norm, that is, it reveals strong hyperbolic–parabolic smoothing (more properly speaking, high-frequency damping) of the initial data. Weaker versions of this estimate may be found in [Kaw,MaZ.1,MaZ.2,MaZ.4] (used there, as here, to control higher derivatives in the nonlinear iteration yielding time-asymptotic decay).

2. As described in [MaZ.2], H^s energy estimates may for small-amplitude profiles be obtained essentially as a perturbation of the constant-coefficient case, using the fact that coefficients are in this case slowly varying; in particular, hypothesis (H1) is not needed in this case. In the large-amplitude case, there is a key new ingredient in the analysis beyond that of the constant-coefficient case, connected with the "upwind" hypothesis (H1)(ii) requiring that hyperbolic convection rates be of uniform sign relative to the shock speed s. This allows us to control terms of form $\int \mathcal{O}(|\overline{U}_x|)|v|^2$ that were formerly controlled by the small-amplitude assumption $|\overline{U}_x| = \mathcal{O}(\varepsilon)$, $\varepsilon := |\widetilde{U}_+ - \widetilde{U}_-| \ll 1$, instead by a "Goodman-type" weighted norm estimate in the spirit of [Go.1,Go.2], thus closing the argument.

PROOF OF PROPOSITION 4.15. A straightforward calculation shows that $|U|_{H^r} \sim |W|_{H^r}$,

$$W = \widetilde{W} - \overline{W} := W(\widetilde{U}) - W(\overline{U}), \tag{4.51}$$

for $0 \leqslant r \leqslant 3$ provided $|U|_{W^{2,\infty}}$ remains bounded, hence it is sufficient to prove a corresponding bound in the special variable W. We first carry out a complete proof in the more straightforward case that the equations may be globally symmetrized to exact form (1.7), i.e., with conditions (A1)–(A3) replaced by the following global versions, indicating afterward by a few brief remarks the changes needed to carry out the proof in the general case.

(A1′) $\widetilde{A}^j$, $\widetilde{A}^j_* := \widetilde{A}^j_{11}$, $\widetilde{A}^0$ are symmetric, $\widetilde{A}^0 > 0$.

(A2′) No eigenvector of $\sum \xi_j \, dF^j(U)$ lies in the kernel of $\sum \xi_j \xi_k B^{jk}(U)$, for all nonzero $\xi \in \mathbb{R}^d$. (Equivalently, no eigenvector of $\sum \xi_j \widetilde{A}^j (\widetilde{A}^0)^{-1}(W)$ lies in the kernel of $\sum \xi_j \xi_k \widetilde{B}^{jk}(W)$.)

(A3′) $\widetilde{B}^{jk} = \begin{pmatrix} 0 & 0 \\ 0 & \tilde{b}^{jk} \end{pmatrix}$, with $\operatorname{Re} \sum \xi_j \xi_k \tilde{b}^{jk}(W) \geqslant \theta |\xi|^2$ for some $\theta > 0$, for all W and all $\xi \in \mathbb{R}^d$, and $\widetilde{G} \equiv 0$.

Substituting (4.49) into (1.11), we obtain the quasilinear perturbation equation

$$\bigl(\widetilde{A}^0 \widetilde{W}_t - \overline{A}^0 \overline{W}_t\bigr) + \bigl(\widetilde{A} \widetilde{W}_x - \overline{A} \overline{W}_x\bigr) - \bigl(\widetilde{B} \widetilde{W}_x - \overline{B} \overline{W}_x\bigr)_x = \dot{\delta}(t) \widetilde{A}^0 \widetilde{W}_x, \tag{4.52}$$

where, with slight abuse of notation,

$$\begin{aligned} \widetilde{A}^0 &:= \widetilde{A}^0(\widetilde{W}), & \overline{A}^0 &:= \widetilde{A}^0(\overline{W}); \\ \widetilde{A} &:= \widetilde{A}(\widetilde{W}), & \overline{A} &:= \widetilde{A}(\overline{W}); \\ \widetilde{B} &:= \widetilde{B}(\widetilde{W}), & \overline{B} &:= \widetilde{B}(\overline{W}). \end{aligned} \tag{4.53}$$

Thus, using the quadratic Leibnitz relation

$$A_2 W_2 - A_1 W_1 = A_2(W_2 - W_1) + (A_2 - A_1) W_1, \tag{4.54}$$

and recalling the block structure assumption (A3), we obtain the alternative perturbation equation:

$$\begin{aligned}&\widetilde{A}^0 W_t + \widetilde{A} W_x - (\widetilde{B} W_x)_x \\ &\quad = M_1 \overline{W}_x + (M_2 \overline{W}_x)_x + \dot{\delta}(t)\widetilde{A}^0 W_x + \dot{\delta}(t)\widetilde{A}^0 \overline{W}_x,\end{aligned} \tag{4.55}$$

where

$$M_1 = M_1(W, \overline{W}) := \widetilde{A} - \overline{A} = \left(\int_0^1 \mathrm{d}A\,(\overline{W} + \theta W)\,\mathrm{d}\theta\right) W, \tag{4.56}$$

and

$$M_2 = M_2(W, \overline{W}) := \widetilde{B} - \overline{B} = \begin{pmatrix} 0 & 0 \\ 0 & \left(\int_0^1 \mathrm{d}b\,(\overline{W} + \theta W)\,\mathrm{d}\theta\right) W \end{pmatrix}. \tag{4.57}$$

We now carry out a series of successively higher-order energy estimates of the type formalized by Kawashima [Kaw]. The origin of this approach goes back to [Kan,MNi] in the context of gas dynamics; see, e.g., [HoZ.1] for further discussion/references. The novelty here is that the stationary background state $\overline{W}$ is not constant in x, nor even slowly varying as in [MaZ.2]. We overcome this difficulty by introducing the additional ingredient of a Goodman-type weighted norm in order to control hyperbolic modes, assumed for this purpose (see hypothesis (H1)(ii)) to be uniformly transverse to the background shock profile.

Let $\widetilde{K}$ denote the skew-symmetric matrix described in Lemma 2.18 associated with $\widetilde{A}^0$, $\widetilde{A}$, $\widetilde{B}$, satisfying

$$\widetilde{K}(\widetilde{A}^0)^{-1}\widetilde{A} + \widetilde{B} > 0.$$

Then, regarding $\widetilde{A}^0$, $\widetilde{K}$, we have the bounds

$$\begin{aligned}
\widetilde{A}^0_x &= \mathrm{d}A^0(\widetilde{W})\widetilde{W}_x, & \widetilde{K}_x &= \mathrm{d}K(\widetilde{W})\widetilde{W}_x,\\
\widetilde{A}_x &= \mathrm{d}A(\widetilde{W})\widetilde{W}_x, & \widetilde{B}_x &= \mathrm{d}B(\widetilde{W})\widetilde{W}_x,\\
\widetilde{A}^0_t &= \mathrm{d}A^0(\widetilde{W})\widetilde{W}_t, & \widetilde{K}_t &= \mathrm{d}K(\widetilde{W})\widetilde{W}_t,\\
\widetilde{A}_t &= \mathrm{d}A(\widetilde{W})\widetilde{W}_t, & \widetilde{B}_x &= \mathrm{d}B(\widetilde{W})\widetilde{W}_t,
\end{aligned} \tag{4.58}$$

and (from the defining equations):

$$|\widetilde{W}_x| = |W_x + \overline{W}_x| \leqslant |W_x| + |\overline{W}_x| \tag{4.59}$$

and

$$\begin{aligned}|\widetilde{W}_t| &\leqslant C\big(|\widetilde{W}_x| + |\tilde{w}^{II}_{xx}| + |\dot{\delta}||\widetilde{W}_x|\big)\\ &\leqslant C\big(|W_x| + |\overline{W}_x| + |w^{II}_{xx}| + |\bar{w}^{II}_{xx}| + |\dot{\delta}||W_x| + |\dot{\delta}||\overline{W}_x|\big)\\ &\leqslant C\big(|W_x| + |\overline{W}_x| + |w^{II}_{xx}|\big). \end{aligned} \tag{4.60}$$

Thus, in particular,

$$\begin{aligned}&|\dot{\delta}|, |\widetilde{A}^0_x|, |\widetilde{A}^0_{xx}|, |\widetilde{K}_x|, |\widetilde{K}_{xx}|, |\widetilde{A}_x|, |\widetilde{A}_{xx}|, |\widetilde{B}_x|, |\widetilde{B}_{xx}|, |\widetilde{A}^0_t|, |\widetilde{K}_t|, |\widetilde{A}_t|, |\widetilde{B}_t|\\ &\quad\leqslant C\big(\zeta + |\overline{U}_x|\big). \end{aligned} \tag{4.61}$$

Finally, we introduce the weighted norms and inner product

$$|f|_\alpha := \big|\alpha^{1/2} f\big|_{L^2}, \qquad |f|_{H^s_\alpha} := \sum_{r=0}^{s} \big|\partial_x^r f\big|_\alpha, \qquad \langle f, g\rangle_\alpha := \langle \alpha f, g\rangle_{L^2}, \tag{4.62}$$

$\alpha(x)$ scalar, uniformly positive, and uniformly bounded. For the remainder of this section, we shall for notational convenience omit the subscript α, referring always to α-norms or α-inner products unless otherwise specified. For later reference, we note the commutator relation

$$\langle f, g_x\rangle = -\big\langle f_x + (\alpha_x/\alpha) f, g\big\rangle, \tag{4.63}$$

and the related identities

$$\langle f, S f_x\rangle = -\frac{1}{2}\big\langle f, \big(S_x + (\alpha_x/\alpha) S\big) f\big\rangle, \tag{4.64}$$

$$\big\langle f, (Sf)_x\big\rangle = \frac{1}{2}\big\langle f, \big(S_x - (\alpha_x/\alpha) S\big) f\big\rangle, \tag{4.65}$$

valid for symmetric operators S.

By (H1)(ii), we have that $\overline{A}_{11}(\overline{A}^0_{11})^{-1}$ has real spectrum of uniform sign, without loss of generality negative, so that the similar matrix

$$\big(\overline{A}^0_{11}\big)^{-1/2} \overline{A}_{11} \big(\overline{A}^0_{11}\big)^{-1/2} = \big(\overline{A}^0_{11}\big)^{-1/2} \overline{A}_{11} \big(\overline{A}^0_{11}\big)^{-1} \big(\overline{A}^0_{11}\big)^{1/2}$$

has real, negative spectrum as well. (Recall, $\overline{A}^0_{11}$ is symmetric negative definite as a principal minor of the symmetric negative definite matrix $\overline{A}^0$.) It follows that $\overline{A}_{11}$ itself is uniformly symmetric negative definite, i.e.,

$$\overline{A}_{11} \leqslant -\theta < 0. \tag{4.66}$$

Defining α, following Goodman [Go.2], by the ODE

$$\alpha_x = C_* |\overline{U}_x| \alpha, \quad \alpha(0) = 1, \tag{4.67}$$

where $C_* > 0$ is a large constant to be chosen later, we have by (4.66)

$$(\alpha_x/\alpha)\overline{A}_{11} \leqslant -C_* \theta |\overline{U}_x|. \tag{4.68}$$

Note, because $|\overline{U}_x| \leqslant C\mathrm{e}^{-\theta|x|}$ (see Lemma 1.6), that α is indeed positive and bounded from both zero and infinity, as the solution of the simple scalar exponential growth equation (4.67). In what follows, we shall need to keep careful track of the distinguished constant C_*.

Computing

$$-\langle W, \widetilde{A} W_x \rangle = \frac{1}{2} \langle W, (\widetilde{A}_x + (\alpha_x/\alpha)\widetilde{A}) W \rangle \tag{4.69}$$

and expanding $\widetilde{A} = \overline{A} + \mathcal{O}(\zeta)$, $\widetilde{A}_x = \mathcal{O}(|\overline{U}_x| + \zeta)$, we obtain by (4.68) the key property

$$\begin{aligned} -\langle W, \widetilde{A} W_x \rangle &= \frac{1}{2} \langle w^I, (\alpha_x/\alpha)\overline{A}_{11} w^I \rangle \\ &\quad + \mathcal{O}(\langle |\overline{U}_x| |W|, |W| \rangle + \langle (\alpha_x/\alpha)|W|, \zeta |W| + |w^{II}| \rangle) \\ &\leqslant -(C_*\theta/3) \langle |w^I|, |\overline{W}_x| |w^I| \rangle + C\zeta |w^I|^2 + C(C_*) |w^{II}|^2, \end{aligned} \tag{4.70}$$

by which we shall control transverse modes, provided C_* is chosen sufficiently large, or, more generally,

$$\begin{aligned} &-\langle \partial_x^k W, \widetilde{A} \partial_x^k W_x \rangle \\ &\quad \leqslant -(C_*\theta/3) \langle |\partial_x^k u|, |\overline{U}_x| |\partial_x^k u| \rangle + C\zeta |\partial_x^k u|^2 + C(C_*) |\partial_x^k v|^2. \end{aligned} \tag{4.71}$$

Here and below, $C(C_*)$ denotes a suitably large constant depending on C_*, while C denotes a fixed constant independent of C_*: likewise, $\mathcal{O}(\cdot)$ indicates a bound independent of C_*.

Zeroth-order "Friedrichs-type" estimate. We first perform a standard, zeroth- and first-order "Friedrichs-type" estimate for symmetrizable hyperbolic systems [Fri]. Taking the α-inner product of W against (4.55), we obtain after rearrangement, integration by parts using (4.63) and (4.64), and several applications of Young's inequality, the energy estimate

$$\begin{aligned} \frac{1}{2} \langle W, \widetilde{A}^0 W \rangle_t &= \langle W, \widetilde{A}^0 W_t \rangle + \frac{1}{2} \langle W, \widetilde{A}^0_t W \rangle \\ &= -\langle W, \widetilde{A} W_x \rangle + \langle W, (\widetilde{B} W_x)_x \rangle + \langle W, M_1 \overline{U}_x \rangle + \langle W, (M_2 \overline{W}_x)_x \rangle \\ &\quad + \dot{\delta}(t) \langle W, \widetilde{A}^0 W_x \rangle + \dot{\delta}(t) \langle W, \widetilde{A}^0 \overline{U}_x \rangle + \frac{1}{2} \langle W, \widetilde{A}^0_t W \rangle \end{aligned}$$

$$
\begin{aligned}
&= \frac{1}{2}\langle W, (\widetilde{A}_x + (\alpha_x/\alpha)\widetilde{A})W\rangle \\
&\quad - \langle W_x - (\alpha_x/\alpha)W, \widetilde{B}W_x\rangle + \langle W, M_1\overline{U}_x\rangle \\
&\quad - \langle W_x + (\alpha_x/\alpha)W, M_2\overline{U}_x\rangle - \frac{1}{2}\dot{\delta}(t)\langle W, (\widetilde{A}^0_x + (\alpha_x/\alpha)\widetilde{A}^0)W\rangle \\
&\quad + \dot{\delta}(t)\langle W, \widetilde{A}^0\overline{U}_x\rangle + \frac{1}{2}\langle W, \widetilde{A}^0_t W\rangle \\
&\leqslant -\langle W_x, \widetilde{B}W_x\rangle + C(C_*)\int \alpha\big((|W_x| + |\overline{U}_x|)|W|^2 \\
&\quad + |w^{II}_x||W|(|W_x| + |\overline{U}_x|) + |\dot{\delta}||W||\overline{U}_x|\big) \\
&\leqslant -\theta|w^{II}_x|^2 + C(C_*)\big(|W|^2_{L^2} + |\dot{\delta}|^2\big).
\end{aligned} \tag{4.72}
$$

Here, we have freely used the bounds (2.18), as well as (4.61). We have also used in a crucial way the block-diagonal form of M_2 in estimating $|\langle W_x, M_2\overline{U}_x\rangle| \leqslant C\int |v_x||W||\overline{U}_x|$ in the first inequality.

First-order "Friedrichs-type" estimate. For first and higher derivative estimates, it is crucial to make use of the favorable terms (4.71) afforded by the introduction of α-weighted norms. Differentiating (4.55) with respect to x, taking the α-inner product of W_x against the resulting equation, and substituting the result into the first term on the right-hand side of

$$
\frac{1}{2}\langle W_x, \widetilde{A}^0 W_x\rangle_t = \langle W_x, (\widetilde{A}^0 W_t)_x\rangle - \langle W_x, \widetilde{A}^0_x W_t\rangle + \frac{1}{2}\langle W_x, \widetilde{A}^0_t W_x\rangle, \tag{4.73}
$$

we obtain after various simplifications and integrations by parts:

$$
\begin{aligned}
\frac{1}{2}\langle W_x, \widetilde{A}^0 W_x\rangle_t &= -\langle W_x, (\widetilde{A}W_x)_x\rangle + \langle W_x, (\widetilde{B}W_x)_{xx}\rangle + \langle W_x, (M_1\overline{W}_x)_x\rangle \\
&\quad + \langle W_x, (M_2\overline{W}_x)_{xx}\rangle + \dot{\delta}(t)\langle W_x, (\widetilde{A}^0 W_x)_x\rangle \\
&\quad + \dot{\delta}(t)\langle W_x, (\widetilde{A}^0\overline{W}_x)_x\rangle - \langle W_x, \widetilde{A}^0_x W_t\rangle + \frac{1}{2}\langle W_x, \widetilde{A}^0_t W_x\rangle \\
&= -\langle W_x, \widetilde{A}W_{xx}\rangle - \langle W_x, \widetilde{A}_x W_x\rangle \\
&\quad - \langle W_{xx} + (\alpha_x/\alpha)W_x, \widetilde{B}W_{xx} + \widetilde{B}_x W_x\rangle \\
&\quad + \langle W_x, (M_1\overline{U}_x)_x\rangle - \langle W_{xx} + (\alpha_x/\alpha)W_x, (M_2\overline{U}_x)_x\rangle \\
&\quad + \frac{1}{2}\dot{\delta}(t)\langle W_x, (\widetilde{A}^0_x - (\alpha_x/\alpha)\widetilde{A}^0)W_x\rangle + \dot{\delta}(t)\langle W_x, \widetilde{A}^0_x\overline{W}_x\rangle \\
&\quad + \dot{\delta}(t)\langle W_x, \widetilde{A}^0\overline{W}_{xx}\rangle - \langle W_x, \widetilde{A}^0_x W_t\rangle + \frac{1}{2}\langle W_x, \widetilde{A}^0_t W_x\rangle.
\end{aligned} \tag{4.74}
$$

Estimating the first term on the right-hand side of (4.74) using (4.71), $k=1$, and substituting $(\widetilde{A}^0)^{-1}$ times (4.55) into the second to last term on the right-hand side of (4.74), we obtain by (4.61) plus various applications of Young's inequality the next-order energy estimate:

$$\begin{aligned}
\frac{1}{2}\langle W_x, \widetilde{A}^0 W_x\rangle_t &\leqslant -\langle W_x, \widetilde{A} W_{xx}\rangle - \langle W_{xx}, \widetilde{B} W_{xx}\rangle \\
&\quad + C(C_*)\int \big(|W|^2 + |\dot{\delta}|^2\big)\big(|\overline{U}_{xx}| + |\overline{U}_x|\big) \\
&\quad + C(C_*)\big\langle |W_x^{II}| + \zeta|W_x|, \big(|W| + |W_x|\big)|\overline{U}_x| + |w_{xx}^{II}|\big\rangle \\
&\quad + C\big\langle \big(|W_x| + |w_{xx}^{II}|\big), |\overline{U}_x|\big(|W| + |W_x|\big)\big\rangle \\
&\leqslant -\theta|w_{xx}^{II}|^2 - (C_*\theta/3)\big\langle |\overline{U}_x||w_x^I|, |w_x^I|\big\rangle \\
&\quad + C\zeta|w_x^I|^2 + C(C_*)|w_x^{II}|^2 \\
&\quad + C(C_*)\big((|W|_{L^2}^2 + |\dot{\delta}|^2\big) \\
&\quad + C(C_*)\big\langle |v_x| + \zeta|W_x|, \big(|W| + |W_x|\big)|\overline{U}_x| + |w_{xx}^{II}|\big\rangle \\
&\quad + C\big\langle \big(|W_x| + |w_{xx}^{II}|\big), |\overline{U}_x|\big(|W| + |W_x|\big)\big\rangle \\
&\leqslant -(\theta/2)|w_{xx}^{II}|^2 - (C_*\theta/4)\big\langle |\overline{W}_x||w_x^I|, |w_x^I|\big\rangle \\
&\quad + C(C_*)\zeta|w_x^I|^2 + C(C_*)|w_x^{II}|^2 \\
&\quad + C(C_*)\big((|W|_{L^2}^2 + |\dot{\delta}|^2\big),
\end{aligned} \tag{4.75}$$

provided C_* is sufficiently large and ζ sufficiently small.

First-order "Kawashima-type" estimate. Next, we perform a "Kawashima-type" derivative estimate. Taking the α-inner product of W_x against $\widetilde{K}(\widetilde{A}^0)^{-1}$ times (4.55), and noting that (integrating by parts, and using skew-symmetry of $\widetilde{K}$)

$$\begin{aligned}
\frac{1}{2}\langle W_x, \widetilde{K} W\rangle_t &= \frac{1}{2}\langle W_x, \widetilde{K} W_t\rangle + \frac{1}{2}\langle W_{xt}, \widetilde{K} W\rangle + \frac{1}{2}\langle W_x, \widetilde{K}_t W\rangle \\
&= \frac{1}{2}\langle W_x, \widetilde{K} W_t\rangle - \frac{1}{2}\langle W_t, \widetilde{K} W_x\rangle \\
&\quad - \frac{1}{2}\langle W_t, (\widetilde{K}_x + (\alpha_x/\alpha)) W\rangle + \frac{1}{2}\langle W_x, \widetilde{K}_t W\rangle \\
&= \langle W_x, \widetilde{K} W_t\rangle + \frac{1}{2}\langle W, (\widetilde{K}_x + (\alpha_x/\alpha)) W_t\rangle + \frac{1}{2}\langle W_x, \widetilde{K}_t W\rangle,
\end{aligned} \tag{4.76}$$

we obtain by calculations similar to the above the auxiliary energy estimate:

$$\begin{aligned}\frac{1}{2}\langle W_x, \widetilde{K}W\rangle_t &\leqslant -\langle W_x, \widetilde{K}(\widetilde{A}^0)^{-1}\widetilde{A}W_x\rangle \\ &\quad + C(C_*)|w^{II}_x|^2 + C\langle(|\overline{U}_x| + \bar{\zeta} + \zeta)|w^I_x|, |w^I_x|\rangle \\ &\quad + C\bar{\zeta}^{-1}|w^{II}_{xx}|^2 + C(C_*)(|W|^2_{L^2} + |\dot{\delta}(t)|^2), \end{aligned} \tag{4.77}$$

where $\bar{\zeta} > 0$ is an arbitrary constant arising through Young's inequality. (Here, we have estimated term $\langle \widetilde{A}U_x, (\alpha_x/\alpha)U\rangle$ arising in the middle term of the right-hand side of (4.76) using (4.64) by $C(C_*)\int |\overline{U}_x||U|^2 \leqslant C(C_*)|U|^2_{L^\infty}$.)

Combined, weighted H^1 estimate. Choosing $\zeta \ll \bar{\zeta} \ll 1$, adding (4.77) to the sum of (4.72) and (4.75) times a suitably large positive constant $M(C_*, \bar{\zeta}) \gg \bar{\zeta}^{-1}$, and recalling (2.53), we obtain, finally, the combined first-order estimate

$$\begin{aligned}&\frac{1}{2}\big(M(C_*, \bar{\zeta})\langle W, \widetilde{A}^0 W\rangle + \langle W_x, \widetilde{K}W\rangle + M(C_*, \bar{\zeta})\langle W_x, \widetilde{A}^0 W_x\rangle\big)_t \\ &\quad \leqslant -\theta(|W_x|^2 + |w^{II}_{xx}|^2) + C(C_*)(|W|^2_{L^2} + |\dot{\delta}|^2), \end{aligned} \tag{4.78}$$

$\theta > 0$, for any $\bar{\zeta}$, $\zeta(\bar{\zeta}, C_*)$ sufficiently small, and C_*, $C(C_*)$ sufficiently large.

Higher-order estimates. Performing the same procedure on the twice- and thrice-differentiated versions of equation (4.55), we obtain, likewise, Friedrichs estimates

$$\begin{aligned}&\frac{1}{2}\langle \partial_x^q W, \widetilde{A}^0 \partial_x^q W\rangle_t \\ &\quad \leqslant -(\theta/2)|\partial_x^{q+1} w^{II}|^2 - (C_*\theta/4)\langle|\overline{W}_x||\partial_x^q w^I|, |\partial_x^q w^I|\rangle \\ &\qquad + C(C_*)(\zeta|\partial_x^q w^I|^2 + |\partial_x^q w^{II}|^2 + |W_x|_{H^{q-2}_\alpha} + |W|^2_{L^2} + |\dot{\delta}|^2) \end{aligned} \tag{4.79}$$

and Kawashima estimates

$$\begin{aligned}&\frac{1}{2}\langle \partial_x^q W, \widetilde{K}\partial_x^{q-1} W\rangle_t \\ &\quad \leqslant -\langle \partial_x^q W, \widetilde{K}(\widetilde{A}^0)^{-1}\widetilde{A}\partial_x^q W\rangle \\ &\qquad + C(C_*)|\partial_x^q w^{II}|^2 + C\langle(|\overline{W}_x| + \bar{\zeta} + \zeta)|\partial_x^q w^I|, |\partial_x^q w^I|\rangle \\ &\qquad + C\bar{\zeta}^{-1}|\partial_x^{q+1} w^{II}|^2 + C(C_*)(|W_x|_{H^{q-2}_\alpha} + |W|^2_{L^2} + |\dot{\delta}(t)|^2) \end{aligned} \tag{4.80}$$

for $q = 2, 3$, provided $\bar{\zeta}$, $\zeta(\bar{\zeta}, C_*)$ are sufficiently small, and C_*, $C(C_*)$ are sufficiently large. The calculations are similar to those carried out already; see also the closely related calculations of Appendix A of [MaZ.2].

Final estimate. Adding $M(C_*, \bar{\zeta})^2$ times (4.78), $M(C_*, \bar{\zeta})$ times (4.79), and (4.80), with $q = 2$, where M is chosen still larger if necessary, we obtain

$$\begin{aligned}
&\frac{1}{2}\Big(M(C_*, \bar{\zeta})^3 \langle W, \tilde{A}^0 W\rangle + M(C_*, \bar{\zeta})^2 \langle W_x, \tilde{K} W\rangle + M(C_*, \bar{\zeta})^3 \langle W_x, \tilde{A}^0 W_x\rangle \\
&\quad + \langle \partial_x^2 W, \tilde{K}\partial_x W\rangle + M(C_*, \bar{\zeta})\langle \partial_x^2 W, \tilde{A}^0 \partial_x^2\rangle\Big)_t \\
&\leqslant -\theta\big(|W_x|^2_{H^1_\alpha} + |w_x^{II}|^2_{H^2_\alpha}\big) + C(C_*)\big(|W|^2_{L^2} + |\dot{\delta}|^2\big).
\end{aligned} \tag{4.81}$$

Adding now $M(C_*, \bar{\zeta})^2$ times (4.81), $M(C_*, \bar{\zeta})$ times (4.79), and (4.80), with $q = 3$, we obtain the final higher-order estimate

$$\begin{aligned}
&\frac{1}{2}\Big(M(C_*, \bar{\zeta})^5 \langle W, \tilde{A}^0 W\rangle + M(C_*, \bar{\zeta})^4 \langle W_x, \tilde{K} W\rangle + M(C_*, \bar{\zeta})^5 \langle W_x, \tilde{A}^0 W_x\rangle \\
&\quad + M(C_*, \bar{\zeta})^2 \langle \partial_x^2 W, \tilde{K}\partial_x W\rangle + M(C_*, \bar{\zeta})^3\langle \partial_x^2 W, \tilde{A}^0 \partial_x^2 W\rangle \\
&\quad + \langle \partial_x^3 W, \tilde{K}\partial_x^2 W\rangle + M(C_*, \bar{\zeta})\langle \partial_x^3 W, \tilde{A}^0 \partial_x^3 W\rangle\Big)_t \\
&\leqslant -\theta\big(|W_x|^2_{H^2_\alpha} + |w_x^{II}|^3_{H^2_\alpha}\big) + C(C_*)\big(|W|^2_{L^2} + |\dot{\delta}|^2\big) \\
&\leqslant -\theta|W|^2_{H^3_\alpha} + C(C_*)\big(|W|^2_{L^2} + |\dot{\delta}|^2\big).
\end{aligned} \tag{4.82}$$

Denoting

$$\begin{aligned}
\mathcal{E}(W) := \frac{1}{2}\Big(&M(C_*, \bar{\zeta})^5 \langle W, \tilde{A}^0 W\rangle + M(C_*, \bar{\zeta})^4 \langle W_x, \tilde{K} W\rangle \\
&+ M(C_*, \bar{\zeta})^5 \langle W_x, \tilde{A}^0 W_x\rangle \\
&+ M(C_*, \bar{\zeta})^2 \langle \partial_x^2 W, \tilde{K}\,\partial_x W\rangle + M(C_*, \bar{\zeta})^3\langle \partial_x^2 W, \tilde{A}^0 \partial_x^2 W\rangle \\
&+ \langle \partial_x^3 W, \tilde{K}\,\partial_x^2 W\rangle + M(C_*, \bar{\zeta})\langle \partial_x^3 W, \tilde{A}^0 \partial_x^3 W\rangle\Big),
\end{aligned}$$

we have by Young inequality that $\mathcal{E}^{1/2}$ is equivalent to norms H^3 and H^3_α, hence (4.78) yields

$$\mathcal{E}_t \leqslant -\theta_2 \mathcal{E} + C(C_*)\big(|W|^2_{L^2} + |\dot{\delta}|^2\big).$$

Multiplying by integrating factor $\mathrm{e}^{-\theta_2 t}$, and integrating from 0 to t, we thus obtain

$$\mathcal{E}(t) \leqslant \mathrm{e}^{-\theta_2 t}\mathcal{E}(0) + C(C_*)\int_0^t \mathrm{e}^{-\theta_2 s}\big(|W|^2_{L^2} + |\dot{\delta}|^2\big)(s)\,\mathrm{d}s.$$

Multiplying by $\mathrm{e}^{-\theta_2 t}$, and using again that $\mathcal{E}^{1/2}$ is equivalent to H^3, we obtain the result.

The general case. It remains only to discuss the general case that hypotheses (A1)–(A3) hold as stated and not everywhere along the profile, with $\widetilde{G}$ possibly nonzero. These generalizations requires only a few simple observations. The first is that we may express matrix $\widetilde{A}$ in (4.55) as

$$\widetilde{A} = \widehat{A} + (|\overline{U}_x| + \zeta)\begin{pmatrix} 0 & \mathcal{O}(1) \\ \mathcal{O}(1) & \mathcal{O}(1) \end{pmatrix}, \tag{4.83}$$

where $\widehat{A}$ is a symmetric matrix obeying the same derivative bounds as described for $\widetilde{A}$, identical to $\widetilde{A}$ in the 11 block and obtained in other blocks jk by smoothly interpolating over a bounded interval $[-R, +R]$ between $\overline{A}(\widetilde{W}_{-\infty})_{jk}$ and $\overline{A}(\widetilde{W}_{+\infty})_{jk}$. Replacing $\widetilde{A}$ by $\widehat{A}$ in the qth-order Friedrichs-type bounds above, we find that the resulting error terms may be expressed as (integrating by parts if necessary)

$$\langle \partial_x^q \mathcal{O}(|\overline{U}_x| + \zeta)|W|, |\partial_x^{q+1} w^{II}|\rangle$$

plus lower-order terms, hence absorbed using Young's inequality to recover the same Friedrichs-type estimates obtained in the previous case. Thus, we may relax (A1′) to (A1).

The second observation is that, because of the favorable terms

$$-(C_*\theta/4)\langle |\overline{U}_x||\partial_x^q w^I|, |\partial_x^q w^I|\rangle$$

occurring in the right-hand sides of the Friedrichs-type estimates, we need the Kawashima-type bound only to control the contribution to $|\partial_x^q w^I|^2$ coming from x near $\pm\infty$; more precisely, we require from this estimate only a favorable term

$$-\theta\langle (1 - \mathcal{O}(|\overline{U}_x| + \zeta + \bar{\zeta}))|\partial_x^q w^I|, |\partial_x^q w^I|\rangle \tag{4.84}$$

rather than $-\theta|\partial_x^q w^I|^2$ as in (4.77) and (4.80). But, this may easily be obtained by substituting for $\widetilde{K}$ a skew-symmetric matrix-valued function $\widehat{K}$ defined to be identically equal to $\overline{K}(+\infty)$ and $\overline{K}(-\infty)$ for $|x| > R$, and smoothly interpolating between $\overline{K}(\pm\infty)$ on $[-R, +R]$, and using the fact that

$$(\overline{K}(\overline{A}^0)^{-1}\overline{A} + \overline{B})_\pm \geqslant \theta > 0,$$

hence

$$(\widehat{K}(\widetilde{A}^0)^{-1}\widetilde{A} + \widetilde{B}) \geqslant \theta(1 - \mathcal{O}(|\overline{U}_x| + \zeta)).$$

Thus, we may relax (A2′) to (A2).

Finally, notice that the term $\widetilde{G} - \overline{G}$ in the perturbation equation may be Taylor expanded as

$$\begin{pmatrix} 0 \\ \tilde{g}(\widetilde{W}_x, \overline{U}_x) + g(\overline{U}_x, \widetilde{W}_x) \end{pmatrix} + \begin{pmatrix} 0 \\ \mathcal{O}(|W_x|^2) \end{pmatrix}. \tag{4.85}$$

The first, linear term on the right-hand side may be grouped with term $\widetilde{A}^0 W_x$ and treated in the same way, since it decays at plus and minus spatial infinity and vanishes in the 1–1 block. The $(0, \mathcal{O}(|W_x|^2)$ nonlinear term may be treated as other source terms in the energy estimates Specifically, the worst-case terms $\langle \partial_x^3 W, K\, \partial_x^2 \mathcal{O}(|W_x|^2)\rangle$ and $\langle \partial_x^3 W, \partial_x^3 (0, \mathcal{O}(|W_x|^2))\rangle = \langle \partial_x^4 w^{II}, \partial_x^2 \mathcal{O}(|W_x|^2)\rangle$ may be bounded, respectively, by $|W|_{W^{2,\infty}} |W|_{H^3}^2$ and $|W|_{W^{2,\infty}} |w^{II}|_{H^4} |W|_{H^3}$. Thus, we may relax (A3′) to (A3), completing the proof of the general case (A1)–(A3) and the theorem. □

REMARK 4.17. Given $\widetilde{A}^j$, $\widetilde{B}^{jk}$, $\widetilde{G} \in C^{2r}$, we may obtain by the same argument a corresponding result in H^{2r-1} or H^{2r}, $r \geqslant 1$, assuming that $W^{r,\infty}$ remains sufficiently small.

4.2.4. *Nonlinear stability.* We are now ready to carry out our proof of nonlinear stability. We give a simplified version of the basic iteration scheme of [MaZ.1,MaZ.2,Z.6,Z.4]; for precursors of this scheme, see [Go.2,K.1,K.2,LZ.1,LZ.2,ZH,HZ.1,HZ.2]. For this stage of the argument, it will be convenient to work again with the conservative variable

$$U := \widetilde{U} - \overline{U}, \tag{4.86}$$

writing (4.52) in the more standard form:

$$U_t - LU = Q(U, U_x)_x + \dot{\delta}(t)\big(\overline{U}_x + U_x\big), \tag{4.87}$$

where

$$\begin{aligned} Q(U, U_x) &= \mathcal{O}\big(|U|^2 + |U||U_x|\big), \\ Q(U, U_x)_x &= \mathcal{O}\big(|U|^2 + |U_x|^2 + |U||U_{xx}|\big), \end{aligned} \tag{4.88}$$

so long as $|U|$, $|U_x|$ remain bounded.

By Duhamel's principle, and the fact that

$$\int_{-\infty}^{\infty} G(x, t; y)\overline{U}_x(y)\, dy = e^{Lt}\overline{U}_x(x) = \overline{U}_x(x), \tag{4.89}$$

we have, formally,

$$\begin{aligned} U(x, t) = &\int_{-\infty}^{\infty} G(x, t; y) U_0(y)\, dy \\ &+ \int_0^t \int_{-\infty}^{\infty} G_y(x, t-s; y)\big(Q(U, U_x) + \dot{\delta} U\big)(y, s)\, dy\, ds \\ &+ \delta(t)\overline{U}_x. \end{aligned} \tag{4.90}$$

Defining, by analogy with the linear case, the *nonlinear instantaneous projection*:

$$\begin{aligned}\varphi(x,t) &:= -\delta(t)\overline{U}_x \\ &:= \int_{-\infty}^{\infty} E(x,t;y)U_0(y)\,\mathrm{d}y \\ &\quad - \int_0^t \int_{-\infty}^{\infty} E_y(x,t-s;y)\big(Q(U,U_x)+\dot{\delta}U\big)(y,s)\,\mathrm{d}y,\end{aligned} \tag{4.91}$$

or equivalently, the *instantaneous shock location*:

$$\begin{aligned}\delta(t) &= -\int_{-\infty}^{\infty} e(y,t)U_0(y)\,\mathrm{d}y \\ &\quad + \int_0^t \int_{-\infty}^{+\infty} e_y(y,t-s)\big(Q(U,U_x)+\dot{\delta}U\big)(y,s)\,\mathrm{d}y\,\mathrm{d}s,\end{aligned} \tag{4.92}$$

where E, e are defined as in (4.24), (4.41), and recalling (4.43), we thus obtain the *reduced equations*:

$$\begin{aligned}U(x,t) &= \int_{-\infty}^{\infty} (H+\widetilde{G})(x,t;y)U_0(y)\,\mathrm{d}y \\ &\quad + \int_0^t \int_{-\infty}^{\infty} H(x,t-s;y)\big(Q(U,U_x)_x+\dot{\delta}U_x\big)(y,s)\,\mathrm{d}y\,\mathrm{d}s \\ &\quad - \int_0^t \int_{-\infty}^{\infty} \widetilde{G}_y(x,t-s;y)\big(Q(U,U_x)+\dot{\delta}U\big)(y,s)\,\mathrm{d}y\,\mathrm{d}s,\end{aligned} \tag{4.93}$$

and, differentiating (4.92) with respect to t,

$$\begin{aligned}\dot{\delta}(t) &= -\int_{-\infty}^{\infty} e_t(y,t)U_0(y)\,\mathrm{d}y \\ &\quad + \int_0^t \int_{-\infty}^{+\infty} e_{yt}(y,t-s)\big(Q(U,U_x)+\dot{\delta}U\big)(y,s)\,\mathrm{d}y\,\mathrm{d}s.\end{aligned} \tag{4.94}$$

NOTE. In deriving (4.94), we have used the fact that $e_y(y,s) \rightharpoonup 0$ as $s \to 0$, as the difference of approaching heat kernels, in evaluating the boundary term

$$\int_{-\infty}^{+\infty} e_y(y,0)\big(Q(U,U_x)+\dot{\delta}U\big)(y,t)\,\mathrm{d}y = 0. \tag{4.95}$$

(Indeed, $|e_y(\cdot,s)|_{L^1} \to 0$; see Remark 2.6.)

The defining relation $\delta(t)\bar{u}_x := -\varphi$ in (4.91) can be motivated heuristically by

$$\begin{aligned}&\widetilde{U}(x,t) - \varphi(x,t) \\ &\quad \sim U = U\big(x+\delta(t),t\big) - \overline{U}(x) \sim \widetilde{U}(x,t) + \delta(t)\overline{U}_x(x),\end{aligned}$$

where $\widetilde{U}$ denotes the solution of the linearized perturbation equations, and $\overline{U}$ the background profile. Alternatively, it can be thought of as the requirement that the instantaneous projection of the shifted (nonlinear) perturbation variable U be zero; see [HZ.1,HZ.2].

LEMMA 4.18 [Z.6]. *The kernel e satisfies*

$$\left|e_y(\cdot,t)\right|_{L^p}, \left|e_t(\cdot,t)\right|_{L^p} \leqslant Ct^{-\frac{1}{2}(1-1/p)}, \tag{4.96}$$

$$\left|e_{ty}(\cdot,t)\right|_{L^p} \leqslant Ct^{-\frac{1}{2}(1-1/p)-1/2} \tag{4.97}$$

for all $t > 0$. Moreover, for $y \leqslant 0$, we have the pointwise bounds

$$\left|e_y(y,t)\right|, \left|e_t(y,t)\right| \leqslant Ct^{-1/2} \sum_{a_k^- > 0} \left(\mathrm{e}^{-(y+a_k^- t)^2/(Mt)} + \mathrm{e}^{-(y-a_k^- t)^2/(Mt)}\right), \tag{4.98}$$

$$\left|e_{ty}(y,t)\right| \leqslant Ct^{-1} \sum_{a_k^- > 0} \left(\mathrm{e}^{-(y+a_k^- t)^2/(Mt)} + \mathrm{e}^{-(y-a_k^- t)^2/(Mt)}\right) \tag{4.99}$$

for $M > 0$ sufficiently large, and symmetrically for $y \geqslant 0$.

PROOF. For definiteness, take $y \leqslant 0$. Then, (4.41) gives

$$e_y(y,t) := \sum_{a_k^- > 0} \left[c_{k,-}^0\right] l_k^{-t} \big(K(y + a_k^- t, t, \beta_k^-) - K(y - a_k^- t, t, \beta_k^-)\big) \tag{4.100}$$

$$e_t(y,t) := \sum_{a_k^- > 0} \left[c_{k,-}^0\right] l_k^{-t} \big((K + K_y)(y + a_k^- t, t, \beta_k^-) - (K + K_y)(y - a\_k^t, t, \beta_k^-)\big), \tag{4.101}$$

$$e_{ty}(y,t) := \sum_{a_k^- > 0} \left[c_{k,-}^0\right] l_k^{-t} \big((K_y + K_{yy})(y + a_k^- t, t) - (K_y + K_{yy})(y - a_k^- t, t, \beta_k^-)\big), \tag{4.102}$$

where

$$K(y,t,b_-) := \frac{\mathrm{e}^{-y^2/4b_- t}}{\sqrt{4\pi b_- t}} \tag{4.103}$$

denotes a heat kernel with diffusion coefficient β. The pointwise bounds (4.98) and (4.99) follow immediately for $t \geqslant 1$ by properties of the heat kernel, in turn yielding (4.96) and

(4.97) in this case. The bounds for small time $t \leqslant 1$ follow from estimates

$$
\begin{aligned}
\left|K_y(y+at,t,\beta)-K_y(y-at,t,\beta)\right| &= \left|\int_{y+at}^{y-at} K_{yy}(z,t,\beta)\,\mathrm{d}z\right| \\
&\leqslant Ct^{-3/2}\int_{y+at}^{y-at} \mathrm{e}^{-z^2/(Mt)}\,\mathrm{d}z \\
&\leqslant Ct^{-1/2}\mathrm{e}^{-y^2/(Mt)}
\end{aligned}
\tag{4.104}
$$

and, similarly,

$$
\begin{aligned}
\left|K_{yy}(y+at,t,\beta)-K_{yy}(y-a,t,\beta)\right| &= \left|\int_{-at}^{at} K_{yyy}(z,t,\beta)\,\mathrm{d}z\right| \\
&\leqslant Ct^{-2}\int_{y+at}^{y-at} \mathrm{e}^{-z^2/(Mt)}\,\mathrm{d}z \\
&\leqslant Ct^{-1}\mathrm{e}^{-y^2/(Mt)}.
\end{aligned}
\tag{4.105}
$$

The bounds for $|e_y|$ are again immediate. Note that we have taken crucial account of cancellation in the small time estimates of e_t, e_{ty}. □

REMARK 4.19. For $t \leqslant 1$, a calculation analogous to that of (4.104) yields $|e_y(y,t)| \leqslant C\mathrm{e}^{-y^2/(Mt)}$, and thus $|e_y(\cdot,s)|_{L^1} \to 0$ as $s \to 0$.

With these preparations, we are ready to carry out our analysis:

PROOF OF THEOREM 4.9. Define

$$
\begin{aligned}
\zeta(t) := \sup_{0\leqslant s\leqslant t,\,2\leqslant p\leqslant\infty} &\Big[\left|U(\cdot,s)\right|_{L^p}(1+s)^{\frac{1}{2}(1-1/p)} \\
&+\left|\dot\delta(s)\right|(1+s)^{1/2}+\left|\delta(s)\right|\Big].
\end{aligned}
\tag{4.106}
$$

We shall establish:

CLAIM. *For all $t \geqslant 0$ for which a solution exists with ζ uniformly bounded by some fixed, sufficiently small constant, there holds*

$$
\zeta(t) \leqslant C_2\big(|U_0|_{L^1\cap H^3}+\zeta(t)^2\big).
\tag{4.107}
$$

From this result, it follows by continuous induction that, provided

$$
|U_0|_{L^1\cap H^3} < \frac{1}{4}C_2^2,
$$

there holds

$$\zeta(t) \leqslant 2C_2|U_0|_{L^1\cap H^3} \tag{4.108}$$

for all $t \geqslant 0$ such that ζ remains small. For, by standard short-time theory/local well-posedness in H^3, and the standard principle of continuation, there exists a solution $U \in H^3(x)$ on the open time-interval for which $|U|_{H^3}$ remains bounded, and on this interval ζ is well defined and continuous. Now, let $[0, T)$ be the maximal interval on which $|U|_{H^3(x)}$ remains strictly bounded by some fixed, sufficiently small constant $\delta > 0$. By Proposition 4.15, and the one-dimensional Sobolev bound $|U|_{W^{2,\infty}} \leqslant C|U|_{H^3}$, we have

$$\begin{aligned} |U(t)|^2_{H^3} &\leqslant C|U(0)|^2_{H^3}\mathrm{e}^{-\theta t} + C\int_0^t \mathrm{e}^{-\theta_2(t-\tau)}\big(|U|^2_{L^2} + |\dot\delta|^2\big)(\tau)\,\mathrm{d}\tau \\ &\leqslant C_2\big(|U(0)|^2_{H^3} + \zeta(t)^2\big)(1+t)^{-1/2}, \end{aligned} \tag{4.109}$$

and so the solution continues so long as ζ remains small, with bound (4.108), at once yielding existence and the claimed sharp $L^p \cap H^3$ bounds, $2 \leqslant p \leqslant \infty$. □

Thus, it remains only to establish the claim above.

PROOF OF CLAIM. We must show that each of the quantities $|U|_{L^p}(1+s)^{\frac{1}{2}(1-1/p)}$, $|\dot\delta|(1+s)^{1/2}$, and $|\delta|$ is separately bounded by

$$C\big(|U_0|_{L^1\cap H^3} + \zeta(t)^2\big), \tag{4.110}$$

for some $C > 0$, all $0 \leqslant s \leqslant t$, so long as ζ remains sufficiently small. By (4.93) and (4.94), we have

$$\begin{aligned} |U|_{L^p}(t) \leqslant{} & \left|\int_{-\infty}^{\infty} (H+\widetilde{G})(x,t;y)U_0(y)\,\mathrm{d}y\right|_{L^p} \\ &+ \left|\int_0^t\int_{-\infty}^{\infty} H(x,t-s;y)\big(Q(U,U_x)_x + \dot\delta U_x\big)(y,s)\,\mathrm{d}y\,\mathrm{d}s\right|_{L^p} \\ &+ \left|\int_0^t\int_{-\infty}^{\infty} \widetilde{G}_y(x,t-s;y)\big(Q(U,U_x) + \dot\delta U\big)(y,s)\,\mathrm{d}y\,\mathrm{d}s\right|_{L^p} \\ =:{} & I_a + I_b + I_c, \end{aligned} \tag{4.111}$$

$$\begin{aligned} |\dot\delta|(t) \leqslant{} & \left|\int_{\infty}^{\infty} e_t(y,t)U_0(y)\,\mathrm{d}y\right| \\ &+ \left|\int_0^t\int_{-\infty}^{+\infty} e_{yt}(y,t-s)\big(Q(U,U_x) + \dot\delta U\big)(y,s)\,\mathrm{d}y\,\mathrm{d}s\right| \\ =:{} & II_a + II_b \end{aligned} \tag{4.112}$$

and

$$
\begin{aligned}
|\delta|(t) \leqslant{} & \left| \int_{-\infty}^{\infty} e(y,t) U_0(y)\, \mathrm{d}y \right| \\
& + \left| \int_0^t \int_{-\infty}^{+\infty} e_y(y,t-s) \big(Q(U,U_x) + \dot{\delta} U \big)(y,s)\, \mathrm{d}y\, \mathrm{d}s \right| \\
={}: & \, III_a + III_b.
\end{aligned} \tag{4.113}
$$

We estimate each term in turn, following the approach of [Z.6,MaZ.1]. The linear term I_a satisfies bound

$$
I_a \leqslant C |U_0|_{L^1 \cap L^p} (1+t)^{-\frac{1}{2}(1-1/p)}, \tag{4.114}
$$

as already shown in the proof of Theorem 4.8. Likewise, applying the bounds of Lemma 4.14 together with (4.88), (4.109) and definition (4.106), we have

$$
\begin{aligned}
I_b &= \left| \int_0^t \int_{-\infty}^{\infty} H(x,t-s;y) \big(Q(U,U_x)_x + \dot{\delta} U_x \big)(y,s)\, \mathrm{d}y\, \mathrm{d}s \right|_{L^p} \\
&\leqslant C \int_0^t \mathrm{e}^{-\eta(t-s)} \big(|U|_{L^\infty} + |U_x|_{L^\infty} + |\dot{\delta}| \big) |U|_{W^{2,p}}(s)\, \mathrm{d}s \\
&\leqslant C \zeta(t)^2 \int_0^t \mathrm{e}^{-\eta(t-s)} (1+s)^{-1/2}\, \mathrm{d}s \\
&\leqslant C \zeta(t)^2 (1+t)^{-1/2},
\end{aligned} \tag{4.115}
$$

and (taking $q = 2$ in (4.45))

$$
\begin{aligned}
I_c &= \left| \int_0^t \int_{-\infty}^{\infty} \widetilde{G}_y(x,t-s;y) \big(Q(U,U_x) + \dot{\delta} U \big)(y,s)\, \mathrm{d}y\, \mathrm{d}s \right|_{L^p} \\
&\leqslant C \int_0^t \mathrm{e}^{-\eta(t-s)} \big(|U|_{L^\infty} + |U_x|_{L^\infty} + |\dot{\delta}| \big) |U|_{L^p}(s)\, \mathrm{d}s \\
&\quad + C \int_0^t (t-s)^{-3/4+1/2p} \big(|U|_{L^\infty} + |\dot{\delta}| \big) |U|_{H^1}(s)\, \mathrm{d}s \\
&\leqslant C \zeta(t)^2 \int_0^t \mathrm{e}^{-\eta(t-s)} (1+s)^{-\frac{1}{2}(1-1/p)-1/2}\, \mathrm{d}s \\
&\quad + C \zeta(t)^2 \int_0^t (t-s)^{-3/4+1/2p} (1+s)^{-3/4}\, \mathrm{d}s \\
&\leqslant C \zeta(t)^2 (1+t)^{-\frac{1}{2}(1-1/p)}.
\end{aligned} \tag{4.116}
$$

Summing bounds (4.114)–(4.116), we obtain (4.110), as claimed, for $2 \leqslant p \leqslant \infty$.

Similarly, applying the bounds of Lemma 4.18 together with definition (4.106), we find that

$$
\begin{aligned}
II_a &= \left| \int_{-\infty}^{\infty} e_t(y,t) U_0(y)\,\mathrm{d}y \right| \\
&\leqslant \left| e_t(y,t) \right|_{L^\infty}(t) |U_0|_{L^1} \\
&\leqslant C|U_0|_{L^1}(1+t)^{-1/2}
\end{aligned}
\tag{4.117}
$$

and

$$
\begin{aligned}
II_b &= \left| \int_0^t \int_{-\infty}^{+\infty} e_{yt}(y,t-s)\big(Q(U,U_x) + \dot{\delta} U\big)(y,s)\,\mathrm{d}y\,\mathrm{d}s \right| \\
&\leqslant \int_0^t |e_{yt}|_{L^2}(t-s)\big(|U|_{L^\infty} + |\dot{\delta}|\big)|U|_{H^1}(s)\,\mathrm{d}s \\
&\leqslant C\zeta(t)^2 \int_0^t (t-s)^{-3/4}(1+s)^{-3/4}\,\mathrm{d}s \\
&\leqslant C\zeta(t)^2(1+t)^{-1/2},
\end{aligned}
\tag{4.118}
$$

while

$$
\begin{aligned}
III_a &= \left| \int_{-\infty}^{\infty} e(y,t) U_0(y)\,\mathrm{d}y \right| \\
&\leqslant \left| e(y,t) \right|_{L^\infty}(t) |U_0|_{L^1} \\
&\leqslant C|U_0|_{L^1}
\end{aligned}
\tag{4.119}
$$

and

$$
\begin{aligned}
III_b &= \left| \int_0^t \int_{-\infty}^{+\infty} e_y(y,t-s)\dot{\delta} U(y,s)\,\mathrm{d}y\,\mathrm{d}s \right| \\
&\leqslant \int_0^t |e_y|_{L^2}(t-s)\big(|U|_{L^\infty} + |\dot{\delta}|\big)|U|_{H^1}(s)\,\mathrm{d}s \\
&\leqslant C\zeta(t)^2 \int_0^t (t-s)^{-1/4}(1+s)^{-3/4}\,\mathrm{d}s \\
&\leqslant C\zeta(t)^2.
\end{aligned}
\tag{4.120}
$$

Summing (4.117) and (4.118), and (4.119) and (4.120), we obtain (4.110) as claimed.

This completes the proof of the claim, and the result. □

REMARK (Low norm stability analysis, L^p, $1 \leqslant p \leqslant 2$). The source term $\dot{\delta} U_x$ appearing in the reduced equations is convenient for high norm estimates L^p, $p \geqslant 2$, but not for low norm estimates L^p, $1 \leqslant p \leqslant 2$. To treat low norms, we may redefine $U := \widetilde{U}(x,t) - \overline{U}(x - \delta(t))$, which has the effect of replacing $\dot{\delta} U_x$ in the reduced equations with "centering errors"

$$S(U, U_x, \delta)_x := -\big(\big(A(\overline{U}(x-\delta) - A(\overline{U}(x))U\big)_x + \big(\big(B(\overline{U}(x-\delta) - B(\overline{U}(x))U_x\big)_x,$$

satisfying

$$|S|_{L^1} \leqslant C|U|_{W^{1,\infty}}, \qquad |S|_{W^{2,1}} \leqslant C|U|_{W^{3,\infty}} \leqslant C|U|_{H^4}.$$

This does not affect the previously-obtained L^p and H^s estimates, since these norms are invariant under spatial shift.

Repeating our high-norm analysis for more regular initial data $U_0 \in L^1 \cap H^4$, we obtain (using higher-derivative analogs of Green distribution bounds (4.44) and (4.45))

$$|U|_{W^{1,\infty}} \leqslant C(1+t)^{-1/2}|U_0|_{L^1 \cap H^4}, \qquad |U|_{H^4} \leqslant C(1+t)^{-1/4}|U_0|_{L^1 \cap H^4}.$$

Estimating I_a–I_c using this new information, taking $q = 1$ now in (4.45) and estimating $|Q|_{W^{1,p}} \leqslant C|U|_{H^3}^2 \leqslant C_2(1+t)^{-1/2}|U_0|_{L^1 \cap H^4}$, we obtain the sharp rate of decay

$$\big|\widetilde{U}(x,t) - \overline{U}(x)\big|_{L^p} \leqslant C(1+t)^{-(1/2)(1-1/p)}|U_0|_{L^1 \cap H^4}$$

for all $1 \leqslant p < \infty$. (Indeed, we could have carried out the original nonlinear iteration in these "uncentered" coordinates, for all $2 \leqslant p \leqslant \bar{p} < \infty$, using $q = 1$ in (4.45).) We omit the details, as lying outside the scope of our main development.

4.3. *Proof of the linearized bounds*

Finally, we carry out in this subsection the proof of Proposition 4.10, completing the analysis of the one-dimensional case. For simplicity, we restrict to the case that $A_\pm$ have distinct eigenvalues; as discussed in [MaZ.3,MaZ.4], the general case can be treated in the same way, at the expense only of some further bookkeeping. Following the notation of [MaZ.3], we abbreviate by $(\mathcal{D})$ the triple conditions of structural, dynamical, and strong spectral stability.[27]

4.3.1. *Low-frequency bounds on the resolvent kernel.* We begin by estimating the resolvent kernel in the critical regime $|\lambda| \to 0$, corresponding to large time behavior of the Green distribution G, or global behavior in space and time. From a global perspective, the

[27] This refers to an alternate characterization in terms of the Evans function D.

structure of the linearized equations is that of two nearly constant-coefficient regions separated by a thin shock layer near $x = 0$; accordingly, behavior is essentially governed by the two, limiting far-field equations:

$$U_t = L_\pm U, \tag{4.121}$$

coupled by an appropriate transmission relation at $x = 0$. We begin by examining these limiting systems.

LEMMA 4.20 [MaZ.3]. *Let dimension* $d = 1$ *and assume* (A1)–(A3), (H0)–(H3) *and* (P0). *Then, for* $|\lambda|$ *sufficiently small, the eigenvalue equation* $(L_\pm - \lambda)U = 0$ *associated with the limiting, constant-coefficient operator* $L_\pm$ *has an analytic basis of* $n + r$ *solutions* $\overline{U}_j^\pm$ *consisting of r "fast" modes*

$$\begin{aligned} &\left|(\partial/\partial x)^\alpha \overline{U}_j^\pm(x)\right| \leqslant C\mathrm{e}^{-\theta|x|}, \quad x \geqslant 0, j = 1, \dots, k_\pm, \\ &\left|(\partial/\partial x)^\alpha \overline{U}_j^\pm(x)\right| \leqslant C\mathrm{e}^{-\theta|x|}, \quad x \leqslant 0, j = k_\pm + 1, \dots, r, \end{aligned} \tag{4.122}$$

$\theta > 0$, $0 \leqslant |\alpha| \leqslant 1$, *and n "slow" modes* $\overline{U}_{r+j}^\pm = \mathrm{e}^{\mu x} V_j^\pm$, $j = 1, \dots, n$, *where*

$$\begin{aligned} &\mu_{r+j}^\pm(\lambda) := -\lambda/a_j^\pm + \lambda^2 \beta_j^\pm / a_j^{\pm 3} + \mathcal{O}(\lambda^3), \\ &V_j^\pm(\lambda) := r_j^\pm + \mathcal{O}(\lambda), \end{aligned} \tag{4.123}$$

with $a_j^\pm$, $r_j^\pm$ *and* $\beta_j^\pm$ *as in Proposition* 4.10. *Likewise, the adjoint eigenvalue equation* $(L_\pm - \lambda)^* \widetilde{U} = 0$ *has an analytic basis consisting of r fast solutions*

$$\begin{aligned} &\left|(\partial/\partial x)^\alpha \overline{\widetilde{U}}_j^\pm(x)\right| \leqslant C\mathrm{e}^{-\theta|x|}, \quad x \leqslant 0, j = 1, \dots, k_\pm, \\ &\left|(\partial/\partial x)^\alpha \overline{\widetilde{U}}_j^\pm(x)\right| \leqslant C\mathrm{e}^{-\theta|x|}, \quad x \geqslant 0, j = k_\pm + 1, \dots, r, \end{aligned} \tag{4.124}$$

$\theta > 0$, $0 \leqslant |\alpha| \leqslant 1$, *and n slow solutions* $\overline{\widetilde{U}}_{r+j}^\pm = \mathrm{e}^{-\mu_j^\pm(\lambda)x} \widetilde{V}_j(\lambda)$,

$$\widetilde{V}_j^\pm(\lambda) = l_j^\pm + \mathcal{O}(\lambda), \tag{4.125}$$

$j = 1, \dots, n$, *where* $l_j^\pm$ *are as in Proposition* 4.10. *Moreover, for* $\theta > 0$, $0 \leqslant |\alpha| \leqslant 1$,

$$\begin{aligned} &\left| \left(\frac{\partial}{\partial y}\right)^\alpha \sum_{j=1}^k \overline{U}_j(x) \overline{\widetilde{U}}_j(y) \right|_\pm \leqslant C\mathrm{e}^{-\theta|x-y|}, \quad x \geqslant y, \\ &\left| \left(\frac{\partial}{\partial y}\right)^\alpha \sum_{j=k+1}^r \overline{U}_j^+(x) \overline{\widetilde{U}}_j^+(y) \right|_\pm \leqslant C\mathrm{e}^{-\theta|x-y|}, \quad x \leqslant y. \end{aligned} \tag{4.126}$$

PROOF. By Lemma 3.3 and Corollary 3.4, there exist analytic bases of claimed dimension for the "strong" or "fast" unstable and stable subspaces of the coefficient matrix of the eigenvalue ODE written as a first-order system, defined as the total eigenspaces of spectra with real part uniformly bounded away from zero as $|\lambda| \to 0$, and these satisfy the standard constant-coefficient bounds (4.122) and (4.124). The bounds (4.126) may likewise be seen from standard constant-coefficient bounds, together with the fact (see Section 2.4.3, and especially relation (2.92)) that

$$\left(\sum_{j=1}^{k} \overline{W}_j(x)\overline{\widetilde{W}}_j(y)\right)_{\pm} = \left(\overline{\mathcal{F}}^{y\to x}\overline{\Pi}_{\mathrm{fast}}(\overline{\mathcal{S}}^0)^{-1}\right)_{\pm},$$

$\overline{W} := (U, b^{11}U')$, $\widetilde{W} := (\widetilde{U}, (0, b_2^{11\,\mathrm{Tr}})\widetilde{U}')$, where $\overline{\mathcal{F}}^{y\to x}$ is the solution operator of the limiting eigenvalue ODE written as a first-order system $W' = \mathbb{A}_{\pm}W$, $\overline{\Pi}_{\mathrm{fast}\pm}$ is the projection onto the strongly stable subspace of $\mathbb{A}_{\pm}$, well conditioned thanks to spectral separation, and $\overline{\mathcal{S}}_{\pm}^0$ is as defined in (4.25), also well conditioned by (2.75).

There exists also an n-dimensional manifold of slow solutions, of which the stable and unstable manifolds in the general, multidimensional case, are only continuous as $|\lambda| \to 0$. In the one-dimensional case, however, they may be seen to vary analytically, with expansions (4.123) and (4.125), by inversion of the expansions

$$\lambda_j(\mathrm{i}\xi) = -\mathrm{i}a_j^{*\pm}\xi - \beta_j^{*\pm}\xi^2 + \mathcal{O}(\xi^3)$$

carried out in Appendix A.4 for the dispersion curves near $\xi = 0$, together with the fundamental relation $\mu = \mathrm{i}\xi$ between roots of characteristic and dispersion equations (recall (2.67) and (2.68) of Section 3.1). Alternatively, one can carry out directly the associated matrix bifurcation problem as in the proof of Lemma 3.4. A symmetric argument applies to the adjoint problem. □

REMARK 4.21 [MaZ.3]. Under the simplifying assumption that matrices

$$N_{\pm} := (\mathrm{d}f_{21}\ \ \mathrm{d}f_{22})\begin{pmatrix} \mathrm{d}f_{11} & \mathrm{d}f_{12} \\ b_1 & b_2 \end{pmatrix}^{-1}\begin{pmatrix} 0 \\ I_r \end{pmatrix}\Bigg|_{(U_{\pm})}$$

governing fast modes (see discussion of traveling-wave ODE, Appendix A.2) have distinct eigenvalues, then fast modes may be taken as pure exponential solutions $U = \mathrm{e}^{\mu x}V$, $\widetilde{U} = \mathrm{e}^{-\mu x}\widetilde{V}$, and (4.126) follows immediately from (4.122) and (4.124).

Our main result of this section is then:

PROPOSITION 4.22 [MaZ.3]. *Let $d = 1$ and assume* (A1)–(A3), (H0)–(H3), (P0) *and* ($\mathcal{D}$). *Then, for $r > 0$ sufficiently small, the resolvent kernel G_λ has a meromorphic extension onto $B(0, r) \subset \mathbb{C}$, which may in the Lax or overcompressive case be decomposed as*

$$G_\lambda = E_\lambda + S_\lambda + R_\lambda, \tag{4.127}$$

where, for $y \leqslant 0$:

$$E_\lambda(x,t;y) := \lambda^{-1} \sum_{a_k^- > 0} \big[c_{k,-}^{j,0}\big]\big(\partial \bar{U}^\delta / \partial \delta_j\big) l_k^{-t} \mathrm{e}^{(\lambda/a_k^{\pm} - \lambda^2 \beta_k^{\pm}/a_k^{\pm 3})y}, \tag{4.128}$$

$$\begin{aligned} &S_\lambda(x,t;y) \\ &\quad := \sum_{a_k^- > 0, a_j^+ > 0} \big[c_{k,-}^{j,+}\big] r_j^+ l_k^{-t} \mathrm{e}^{(-\lambda/a_j^+ + \lambda^2 \beta_k^+/a_k^{+3})x + (\lambda/a_k^- - \lambda^2 \beta_k^-/a_k^{-3})y} \end{aligned} \tag{4.129}$$

for $y \leqslant 0 \leqslant x$,

$$\begin{aligned} &S_\lambda(x,t;y) \\ &\quad := \sum_{a_k^- > 0} r_k^- l_k^{-t} \mathrm{e}^{(-\lambda/a_k^- + \lambda^2 \beta_k^-/a_k^{-3})(x-y)} \\ &\qquad + \sum_{a_k^- > 0, a_j^- < 0} \big[c_{k,-}^{j,-}\big] r_j^- l_k^{-t} \mathrm{e}^{(-\lambda/a_j^- + \lambda^2 \beta_j^-/a_j^{-3})x + (\lambda/a_k^- - \lambda^2 \beta_k^-/a_k^{-3})y} \end{aligned} \tag{4.130}$$

for $y \leqslant x \leqslant 0$, *and*

$$\begin{aligned} &S_\lambda(x,t;y) \\ &\quad := \sum_{a_k^- < 0} r_k^- l_k^{-t} \mathrm{e}^{(-\lambda/a_k^- + \lambda^2 \beta_k^-/a_k^{-3})(x-y)} \\ &\qquad + \sum_{a_k^- > 0, a_j^- < 0} \big[c_{k,-}^{j,-}\big] r_j^- l_k^{-t} \mathrm{e}^{(-\lambda/a_j^- + \lambda^2 \beta_j^-/a_j^{-3})x + (\lambda/a_k^- - \lambda^2 \beta_k^-/a_k^{-3})y} \end{aligned} \tag{4.131}$$

for $x \leqslant y \leqslant 0$, *and* $R_\lambda = R_\lambda^E + R_\lambda^S$ *is a faster-decaying residual satisfying*

$$\begin{aligned} &(\partial/\partial y)^\alpha R_\lambda^E \\ &\quad = \mathcal{O}\big(\mathrm{e}^{-\theta|x-y|}\big) + \sum_{a_k^- > 0} \mathrm{e}^{-\theta|x|} \mathrm{e}^{(\lambda/a_k^{\pm} - \lambda^2 \beta_k^{\pm}/a_k^{\pm 3})y} \\ &\qquad \times \big(\lambda^{\alpha-1}\mathcal{O}\big(\mathrm{e}^{\mathcal{O}(\lambda^3)y} - 1\big) + \lambda^{\alpha-1}\mathcal{O}\big(\mathrm{e}^{\mathcal{O}(\lambda^3)x} - 1\big) + \mathcal{O}(\lambda^\alpha)\big), \end{aligned} \tag{4.132}$$

$$\begin{aligned} &(\partial/\partial y)^\alpha R_\lambda^S(x,t;y) \\ &\quad = \sum_{a_k^- > 0, a_j^+ > 0} \mathrm{e}^{(-\lambda/a_j^+ + \lambda^2 \beta_k^+/a_k^{+3})x + (\lambda/a_k^- - \lambda^2 \beta_k^-/a_k^{-3})y} \\ &\qquad \times \big(\lambda^\alpha \mathcal{O}\big(\mathrm{e}^{\mathcal{O}(\lambda^3)y} - 1\big) + \lambda^\alpha \mathcal{O}\big(\mathrm{e}^{\mathcal{O}(\lambda^3)x} - 1\big) + \lambda^\alpha \mathcal{O}\big(\mathrm{e}^{-\theta|x|}\big) + \mathcal{O}(\lambda)\big) \end{aligned} \tag{4.133}$$

for $y \leqslant 0 \leqslant x$,

$$
\begin{aligned}
&(\partial/\partial y)^{\alpha} R_{\lambda}^{S}(x,t;y)\\
&\quad= \sum_{a_k^- > 0} e^{(-\lambda/a_k^- + \lambda^2\beta_k^-/a_k^{-3})(x-y)}\\
&\qquad \times \big(\lambda^{\alpha}\mathcal{O}\big(e^{\mathcal{O}(\lambda^3)y} - 1\big) + \lambda^{\alpha}\mathcal{O}\big(e^{\mathcal{O}(\lambda^3)x} - 1\big) + \lambda^{\alpha}\mathcal{O}\big(e^{-\theta|x|}\big) + \mathcal{O}(\lambda)\big)\\
&\qquad + \sum_{a_k^- > 0, a_j^- < 0} e^{(-\lambda/a_j^- + \lambda^2\beta_j^-/a_j^{-3})x + (\lambda/a_k^- - \lambda^2\beta_k^-/a_k^{-3})y}\\
&\qquad \times \big(\lambda^{\alpha}\mathcal{O}\big(e^{\mathcal{O}(\lambda^3)y} - 1\big) + \lambda^{\alpha}\mathcal{O}\big(e^{\mathcal{O}(\lambda^3)x} - 1\big) + \lambda^{\alpha}\mathcal{O}\big(e^{-\theta|x|}\big) + \mathcal{O}(\lambda)\big)
\end{aligned}
\tag{4.134}
$$

for $y \leqslant x \leqslant 0$, *and*

$$
\begin{aligned}
&(\partial/\partial y)^{\alpha} R_{\lambda}^{S}(x,t;y)\\
&\quad= \sum_{a_k^- < 0} e^{(-\lambda/a_k^- + \lambda^2\beta_k^-/a_k^{-3})(x-y)}\\
&\qquad \times \big(\lambda^{\alpha}\mathcal{O}\big(e^{\mathcal{O}(\lambda^3)y} - 1\big) + \lambda^{\alpha}\mathcal{O}\big(e^{\mathcal{O}(\lambda^3)x} - 1\big) + \lambda^{\alpha}\mathcal{O}\big(e^{-\theta|x|}\big) + \mathcal{O}(\lambda)\big)\\
&\qquad + \sum_{a_k^- > 0, a_j^- < 0} e^{(-\lambda/a_j^- + \lambda^2\beta_j^-/a_j^{-3})x + (\lambda/a_k^- - \lambda^2\beta_k^-/a_k^{-3})y}\\
&\qquad \times \big(\lambda^{\alpha}\mathcal{O}\big(e^{\mathcal{O}(\lambda^3)y} - 1\big) + \lambda^{\alpha}\mathcal{O}\big(e^{\mathcal{O}(\lambda^3)x} - 1\big) + \lambda^{\alpha}\mathcal{O}\big(e^{-\theta|x|}\big) + \mathcal{O}(\lambda)\big)
\end{aligned}
\tag{4.135}
$$

for $x \leqslant y \leqslant 0$, $0 \leqslant \alpha \leqslant 1$, *with each* $\mathcal{O}(\cdot)$ *term separately analytic in* λ, *and where* $[c_{k,i}^{j,0}$, $i = -, 0, +$, *are constants to be determined later. Symmetric bounds hold for* $y \geqslant 0$. *Similar, but more complicated formulae hold in the undercompressive case* (*see Remark* 6.9 *of* [MaZ.1] *or Section* 3 *of* [Z.6] *for further discussion*).

Normal modes (Behavior in x). We begin by relating the normal modes of the variable coefficient eigenvalue equation (2.60) to those of the limiting, constant-coefficient equations (4.121).

LEMMA 4.23 [MaZ.3]. *Let* $d = 1$ *and assume* (A1)–(A3), (H0)–(H3) *and* (P0). *Then, for* $\lambda \in B(0,r)$, r *sufficiently small, there exist solutions* $U_j^{\pm}(x;\lambda)$ *of* (2.60), C^1 *in* x *and analytic in* λ, *satisfying, for slow modes, the bounds*

$$
\begin{aligned}
&U_{r+j}^{\pm}(x;\lambda) = V_j^{\pm}(x;\lambda)e^{\mu_j^{\pm}x},\\
&(\partial/\partial\lambda)^k V_j^{\pm}(x;\lambda) = (\partial/\partial\lambda)^k V_j^{\pm}(\lambda) + \mathcal{O}\big(e^{-\tilde{\theta}|x|}\big|V_j^{\pm}(\lambda)\big|\big), \quad x \gtrless 0,
\end{aligned}
\tag{4.136}
$$

for any $k \geqslant 0$ and $0 < \tilde{\theta} < \theta$, where θ is the rate of decay given in (2.18), $\mu_j^{\pm}(\lambda)$, $V_j^{\pm}(\lambda)$ *are as in Lemma* 4.20, *and $\mathcal{O}(\cdot)$ depends only on k, $\tilde{\theta}$, and, for fast modes, the bounds*

$$\begin{aligned} &\left|U_j^+(x)\right| \leqslant Ce^{-\theta|x|}, \quad x \geqslant 0,\ j=1,\ldots,k_+,\\ &\left|\widetilde{U}_j^+(x)\right| \leqslant Ce^{-\theta|x|}, \quad x \geqslant 0,\ j=k_++1,\ldots,r, \end{aligned} \tag{4.137}$$

$$\begin{aligned} &\left|U_j^-(x)\right| \leqslant Ce^{-\theta|x|}, \quad x \leqslant 0,\ j=k_-+1,\ldots,r,\\ &\left|\widetilde{U}_j^-(x)\right| \leqslant Ce^{-\theta|x|}, \quad x \leqslant 0,\ j=1,\ldots,k_-, \end{aligned} \tag{4.138}$$

$$\begin{aligned} &\left|\sum_{j=1}^{k+} U_j^+(x)\widetilde{U}_j^+(y)\right| \leqslant Ce^{-\theta|x-y|}, \qquad 0 \leqslant y \leqslant x,\\ &\left|\sum_{j=k_++1}^{r} U_j^+(x)\widetilde{U}_j^+(y)\right| \leqslant Ce^{-\theta|x-y|}, \quad 0 \leqslant x \leqslant y, \end{aligned} \tag{4.139}$$

and

$$\begin{aligned} &\left|\sum_{j=k_-+1}^{r} U_j^-(x)\widetilde{U}_j^-(y)\right| \leqslant Ce^{-\theta|x-y|}, \quad x \leqslant y \leqslant 0.\\ &\left|\sum_{j=1}^{k-} U_j^-(x)\widetilde{U}_j^-(y)\right| \leqslant Ce^{-\theta|x-y|}, \qquad y \leqslant x \leqslant 0. \end{aligned} \tag{4.140}$$

PROOF. An immediate consequence of Lemma 4.20 together with Lemma 2.5. (Note the helpful factorization

$$\sum W_j(x)\widetilde{W}_j(y) = P(x)\Big(\sum \overline{W}_j(x)\overline{\widetilde{W}}_j(y)\Big)\big(\mathcal{S}^0\big)^{-1}P^{-1}\mathcal{S}^0(y)$$

linking solutions W and $\overline{W}$ of the associated first-order systems, where P is the conjugating matrix of Lemma 2.5 and $\mathcal{S}^0$, $\overline{\mathcal{S}}^0$ are as defined in (4.25), both uniformly well conditioned.) □

The bases $\phi_j^{\pm}$, $\psi_j^{\pm}$ defined in Section 2.4 may evidently be chosen from among $U_j^{\pm}$, yielding an analytic choice of bases in λ, with the detailed description (4.136). It follows that the dual bases $\tilde{\phi}_j^{\pm}$, $\tilde{\psi}_j^{\pm}$ defined in Section 2.4 are also analytic in λ and satisfy corresponding bounds with respect to the dual solutions (4.124) and (4.125). With this observation, we have immediately, from Definition (3.1) and Corollary 2.26:

COROLLARY 4.24 [MaZ.3]. *Let* $d = 1$ *and assume* (A1)–(A3), (H0)–(H3). *Then, the Evans function $D(\lambda)$ admits an analytic extension onto $B(0, r)$, for r sufficiently small, and the resolvent kernel $G_\lambda(x, y)$ admits a meromorphic extension, with an isolated pole of finite multiplicity at $\lambda = 0$.*

REMARK 4.25. Analytic extendability of the Evans function past the origin (more generally, into the essential spectrum of L) is a special feature of the one-dimensional, or scalar multidimensional case, and reflects the simple geometry of hyperbolic propagation in these settings; see [HoZ.3,HoZ.4,Z.3] for discussion of the scalar multidimensional case. The simple hyperbolic propagation makes possible a much more detailed description of the Green's function in these cases, for which analytic extension is a key tool.

For general systems in multidimensions, $D(\tilde{\xi}, \lambda)$ has a conical singularity at the origin; see Proposition 3.5.

Refined derivative bounds. We have close control on the (x, y) behavior of G_λ through the spectral decomposition formulae of Corollary 2.26. For *slow, dual modes*, the bounds (4.136) (in particular, the consequent bounds on first spatial derivatives) can be considerably sharpened, provided that we appropriately initialize our bases at $\lambda = 0$. This observation will be quite significant in the Lax or overcompressive case. Likewise, *fast-decaying forward modes* can be well approximated near $\lambda = 0$ by their representatives at $\lambda = 0$, using only the basic bounds (4.136). These two categories comprise the modes determining behavior of the Green distribution to lowest order.

Specifically, due to the special, conservative structure of the underlying evolution equations, the adjoint eigenvalue equation $L * \widetilde{U} = 0$ at $\lambda = 0$ admits an n-dimensional subspace of constant solutions

$$\widetilde{U} \equiv \text{constant}; \tag{4.141}$$

this is equivalent to the observation that integral quantities in variable U are conserved under time evolution for general conservation laws. Thus, at $\lambda = 0$, we may choose, by appropriate change of coordinates if necessary, that slow-decaying dual modes $\tilde{\phi}_j^{\pm}$ and slow-growing dual modes $\tilde{\psi}_j^{\pm}$ be identically constant. Note that this does not interfere with our previous choice in Lemma 4.23, since that concerned only the choice of limiting solutions $\overline{U}_j^{\pm}$ of the asymptotic, constant coefficient equations at $x \to \pm\infty$, and not the particular representatives $U_j^{\pm}$ that approach them (which, in the case of slow modes, are specified only up to the addition of an arbitrary fast-decaying mode).

REMARK 4.26. The prescription of constant dual bases just described requires that we reverse our previous approach, choosing dual bases first using the conjugation lemma, then defining forward bases using duality. Alternatively, we may rewrite the forward eigenvalue equation at $\lambda = 0$ as

$$\left(\begin{pmatrix} A_{11} & A_{12} \\ b_1 & b_2 \end{pmatrix} U'\right)' = \left(\begin{pmatrix} 0 & 0 \\ A_{12} & A_{22} \end{pmatrix} U\right)' \tag{4.142}$$

to see, after integration from $x = \pm\infty$, that fast-decaying modes satisfy the linearized traveling-wave ODE (cf. Appendix A.2)

$$U' = \begin{pmatrix} A_{11} & A_{12} \\ b_1 & b_2 \end{pmatrix}^{-1} \begin{pmatrix} 0 & 0 \\ A_{12} & A_{22} \end{pmatrix} U. \tag{4.143}$$

By duality relation (2.76), requiring slow dual modes to be constant is equivalent to choosing fast-growing (forward) modes as well as fast-decaying modes from among the solutions of (4.143). (Recall: though fast-decaying modes are uniquely determined as a subspace, fast-growing modes are only determined up to the addition of faster decaying modes.)

LEMMA 4.27 [MaZ.1]. *In dimension* $d = 1$, *assuming* (A1)–(A3), (H0)–(H3) *and* (P0), *with the above choice of bases at* $\lambda = 0$, *and for* $\lambda \in B(0, r)$ *and* r *sufficiently small, slow modes* $\widetilde{U}_j^{\pm}$ *satisfy*

$$\widetilde{U}_j^{\pm}(y;\lambda) = \mathrm{e}^{-\mu(\lambda)y}\overline{\widetilde{V}}_j^{\pm}(0) + \lambda\widetilde{\Theta}_j^{\pm}(y;\lambda), \tag{4.144}$$

where

$$\begin{aligned}&\bigl|\widetilde{\Theta}_j^{\pm}\bigr| \leqslant C\bigl|\mathrm{e}^{-\mu(\lambda)y}\bigr|,\\ &\bigl|(\partial/\partial y)\widetilde{\Theta}_j^{\pm}\bigr| \leqslant C\bigl|\mathrm{e}^{-\mu(\lambda)y}\bigr|\bigl(|\lambda| + \mathrm{e}^{-\theta|y|}\bigr),\end{aligned} \tag{4.145}$$

$\theta > 0$, *as* $y \to \pm\infty$, *and* $\overline{\widetilde{V}}_j^{\pm} \equiv \text{constant}$.[28] *Similarly, fast-decaying (forward) modes* $U_j^{\pm}$ *satisfy*

$$U_j^{\pm}(x;\lambda) = U_j^{\pm}(x;0) + \lambda\Theta_j^{\pm}(x;\lambda), \tag{4.146}$$

where

$$\bigl|\Theta_j^{\pm}\bigr|, \bigl|(\partial/\partial x)\Theta_j^{\pm}\bigr| \leqslant C\mathrm{e}^{-\theta|x|} \tag{4.147}$$

as $x \to \pm\infty$, *for some* $\theta > 0$.

PROOF. Applying Lemma 2.5 to the augmented variables

$$\begin{aligned}\widetilde{\mathbb{U}}_j^{\pm}(y;\lambda) &:= \begin{pmatrix}\widetilde{U}_j^{\pm}\\ \widetilde{U}_j^{\pm\prime}\end{pmatrix}(y;\lambda)\\ &=: \begin{pmatrix}\widetilde{U}_j^{\pm}\\ \widetilde{U}_j^{\pm\prime}\end{pmatrix}(y;\lambda)\\ &= \mathrm{e}^{-\mu(\lambda)y}\begin{pmatrix}\widetilde{V}_j^{\pm}\\ -\mu_j^{\pm}\widetilde{V}_j^{\pm} + \widetilde{V}_j^{\pm\prime}\end{pmatrix}(y;\lambda)\end{aligned}$$

---

[28] In particular, this includes $|(\partial/\partial y)\widetilde{U}_j^{\pm}(y;\lambda)| \leqslant C|\lambda\mathrm{e}^{\mu_j^{\pm}(\lambda)y}|$ in place of the general spatial-derivative bound $|(\partial/\partial y)\widetilde{U}_j^{\pm}(y;\lambda)| \leqslant C(|\mu\widetilde{V}_j^{\pm}| + |(\partial/\partial y)\widetilde{V}_j^{\pm}|)\mathrm{e}^{\mu_j^{\pm}(\lambda)y} \sim C|\mathrm{e}^{\mu_j^{\pm}(\lambda)y}|$.

and

$$\begin{aligned}\mathbb{U}_j^{\pm}(x;\lambda) &:= \begin{pmatrix} U_j^{\pm} \\ U_j^{\pm\prime} \end{pmatrix}(x;\lambda) \\ &=: \mathrm{e}^{\mu(\lambda)x}\mathbb{V}_j^{\pm}(x;\lambda) \\ &= \mathrm{e}^{-\mu(\lambda)y}\begin{pmatrix} \widetilde{V}_j^{\pm} \\ \mu_j^{\pm}V_j^{\pm} + V_j^{\pm\prime} \end{pmatrix}(y;\lambda)\end{aligned}$$

we obtain bounds

$$\begin{aligned}&\widetilde{\mathbb{U}}_j^{\pm}(x;\lambda) = \widetilde{\mathbb{V}}_j^{\pm}(x;\lambda)\mathrm{e}^{-\mu_j^{\pm}(\lambda)x}, \\ &(\partial/\partial\lambda)^k\widetilde{\mathbb{V}}_j^{\pm}(x;\lambda) = (\partial/\partial\lambda)^k\widetilde{\mathbb{V}}_j^{\pm}(\lambda) + \mathcal{O}\big(\mathrm{e}^{-\tilde{\theta}|x|}\big|\widetilde{\mathbb{V}}_j^{\pm}(\lambda)\big|\big), \quad x \gtrless 0,\end{aligned} \tag{4.148}$$

and

$$\begin{aligned}&\mathbb{U}_j^{\pm}(x;\lambda) = \mathbb{V}_j^{\pm}(x;\lambda)\mathrm{e}^{\mu_j^{\pm}(\lambda)x}, \\ &(\partial/\partial\lambda)^k\mathbb{V}_j^{\pm}(x;\lambda) = (\partial/\partial\lambda)^k\mathbb{V}_j^{\pm}(\lambda) + \mathcal{O}\big(\mathrm{e}^{-\tilde{\theta}|x|}\big|\widetilde{\mathbb{V}}_j^{\pm}(\lambda)\big|\big), \quad x \gtrless 0,\end{aligned} \tag{4.149}$$

$\tilde{\theta} > 0$, analogous to (4.136), valid for $\lambda \in B(0,r)$, where

$$\widetilde{\mathbb{V}}_j^{\pm}(\lambda) = \begin{pmatrix} \widetilde{V}(\lambda) \\ -\mu_j^{\pm}\widetilde{V}(\lambda) \end{pmatrix}$$

and

$$\mathbb{V}_j^{\pm}(\lambda) = \begin{pmatrix} V(\lambda) \\ \mu_j^{\pm}V(\lambda) \end{pmatrix}.$$

(Note: an appropriate phase space in which to apply the conjugation lemma is, e.g., $(U, U', (bU')')$ for the forward equations, and similarly for the dual equations.)

By Taylor theorem with differential remainder, applied to $\widetilde{\mathbb{V}}$, we have:

$$\begin{aligned}\widetilde{\mathbb{U}}_j^{\pm}(y,\lambda) = \mathrm{e}^{-\mu_j^{\pm}(\lambda)y}\Big(&\widetilde{\mathbb{V}}_j^{\pm}(y;0) + \lambda(\partial/\partial\lambda)\widetilde{\mathbb{V}}_j^{\pm}(y;0) \\ &+ \frac{1}{2}\lambda^2(\partial/\partial\lambda)^2\widetilde{\mathbb{V}}_j^{\pm}(y;\lambda_*)\Big),\end{aligned} \tag{4.150}$$

for some λ_* on the ray from 0 to λ, where, recall, $(\partial/\partial\lambda)\widetilde{\mathbb{V}}_j^{\pm}(y;\cdot)$ and $(\partial/\partial\lambda)^2\widetilde{\mathbb{V}}_j^{\pm}(y;\cdot)$ are uniformly bounded in $L^\infty[0,\pm\infty]$ for $\lambda \in B(0,r)$. Together with the choice $\widetilde{V}_j^{\pm}(y,0) \equiv$ constant, this immediately gives the first (undifferentiated) bound in (4.145).

Applying now the bound (4.149) with $k = 1$, we may expand the second coordinate of (4.153) as

$$\begin{aligned}(\partial/\partial y)\widetilde{U}_j^{\pm}(y,\lambda) &= \mathrm{e}^{-\mu_j^{\pm}(\lambda)y}\big(-\mu_j^{\pm}\widetilde{V}_j^{\pm}(y;0)+\widetilde{V}_j^{\pm\prime}(y;0)\\ &\quad -\lambda\big((\partial/\partial\lambda)\big(\mu_j^{\pm}\widetilde{V}_j^{\pm}\big)(0)+\mathcal{O}\big(\mathrm{e}^{-\theta|y|}\big)\big)+\mathcal{O}\big(\lambda^2\big)\big)\\ &= \mathrm{e}^{-\mu_j^{\pm}(\lambda)y}\big(-\lambda\big((\partial/\partial\lambda)\mu_j^{\pm}(0)\widetilde{V}_j^{\pm}(0)+\mathcal{O}\big(\mathrm{e}^{-\theta|y|}\big)\big)+\mathcal{O}\big(\lambda^2\big)\big),\end{aligned} \tag{4.151}$$

and subtracting off the corresponding Taylor expansion

$$\begin{aligned}(\partial/\partial y)\big(\mathrm{e}^{-\mu_j^{\pm}(\lambda)y}\widetilde{V}_j^{\pm}(y,0)\big) &= \mu_j^{\pm}(\lambda)\mathrm{e}^{-\mu_j^{\pm}(\lambda)y}\widetilde{V}_j^{\pm}(y,0)\\ &= \mathrm{e}^{-\mu_j^{\pm}(\lambda)y}\big(-\mu_j^{\pm}(0)\widetilde{V}_j^{\pm}(y;0)-\lambda(\partial/\partial\lambda)\mu_j^{\pm}(0)\widetilde{V}_j^{\pm}(y;0)+\mathcal{O}\big(\lambda^2\big)\big)\\ &= \mathrm{e}^{-\mu_j^{\pm}(\lambda)y}\big(-\lambda(\partial/\partial\lambda)\mu_j^{\pm}(0)\widetilde{V}_j^{\pm}(y;0)+\mathcal{O}\big(\lambda^2\big)\big),\end{aligned} \tag{4.152}$$

we obtain

$$(\partial/\partial y)\widetilde{\Theta}_j^{\pm}(y,\lambda) = \mathrm{e}^{-\mu_j^{\pm}(\lambda)y}\big(\lambda\mathcal{O}\big(\mathrm{e}^{-\theta|y|}\big)+\mathcal{O}\big(\lambda^2\big)\big), \tag{4.153}$$

as claimed.

Similarly, we may obtain (4.146) and (4.147) by Taylor theorem with differential remainder applied to $U = \mathrm{e}^{\mu(\lambda)x}$, and the Leibnitz calculation

$$(\partial/\partial\lambda)U = (\mathrm{d}\mu/\mathrm{d}\lambda)x\mathrm{e}^{\mu x}V + \mathrm{e}^{\mu x}(\partial/\partial\lambda)V,$$

together with the observation that $|x|\mathrm{e}^{-\theta|x|} \leqslant C\mathrm{e}^{-\theta|x|/2}$. □

This leaves only the problem of determining behavior in λ through the study of coefficients M, $d^{\pm}$. To this end, we make the following further observations in the Lax or overcompressive case, generalizing the corresponding observation of Lemma 4.30 of [Z.3], in the strictly parabolic case:

LEMMA 4.28 [MaZ.3]. *For transverse ($\gamma \neq 0$) Lax and overcompressive shocks, with the above-specified choice of basis at $\lambda = 0$, fast-growing modes ψ_j^+, ψ_j^- are fast-decaying at $-\infty$, $+\infty$, respectively. Equivalently, fast-decaying modes $\tilde{\psi}_j^+$, $\tilde{\psi}_j^-$ are fast-growing at $-\infty$, $+\infty$: i.e., the only bounded solutions of the adjoint eigenvalue equation are constant solutions.*

PROOF. Noting that the manifold of solutions of the r-dimensional ODE (4.143) that decay at either $x \to +\infty$ or $x \to -\infty$ is by transversality, together with Lemma 1.7, exactly

$d_+ + d_- - \ell = r$, we see that *all* solutions of this ODE in fact decay at least at one infinity. This implies the first assertion, by the alternative characterization of our bases described in Remark 4.26; the second follows by duality, (2.76). Since the manifold of fast-decaying dual modes is uniquely determined, independent of the choice of basis, this implies that nonconstant dual modes that are bounded at one infinity must blow up at the other, hence bounded solutions must be constant as claimed. □

REMARK 4.29. As noted in [LZ.2,ZH,Z.3,Z.6], the property of Lax and overcompressive shocks that the adjoint eigenvalue equation has only constant solutions has the interpretation that the only L^1 time-invariants of the evolution of the one-dimensional linearized equations about $\overline{U}(\cdot)$ are given by conservation of mass, see discussion [LZ.2]. This distinguishes then from undercompressive shocks, which do have additional L^1 time-invariants. The presence of additional time-invariants for undercompressive shocks has significant implications for their behavior (see discussions [LZ.2] Section 3 and [ZH], Section 10): in particular, the time-asymptotic location of a perturbed undercompressive shock (generically) evolves nonlinearly, and is not determinable by any linear functional of the initial perturbation. By contrast, the time-asymptotic location of a perturbed (stable) Lax or overcompressive shock may be determined by the mass of the initial perturbation alone.

Scattering coefficients (Behavior in λ). We next turn to the estimation of scattering coefficients M_{jk}, $d^{\pm}_{jk}$ defined in Corollary 2.26. Consider coefficient M_{jk}. Expanding (2.101) using Cramer's rule, and setting $z=0$, we obtain

$$M^{\pm}_{jk} = D^{-1}C^{\pm}_{jk}, \tag{4.154}$$

where

$$C^{+} := (I, 0)\begin{pmatrix} \Phi^{+} & \Phi^{-} \\ \Phi^{+\prime} & \Phi^{-\prime} \end{pmatrix}^{\mathrm{adj}} \begin{pmatrix} \Psi^{-} \\ \Psi^{-\prime} \end{pmatrix}\Bigg|_{z=0} \tag{4.155}$$

and a symmetric formula holds for C^-. Here, P^{adj} denotes the *adjugate matrix* of a matrix P, i.e., the transposed matrix of minors. As the adjugate is polynomial in the entries of the original matrix, it is evident that $|C^{\pm}|$ is uniformly bounded and therefore

$$|M^{\pm}_{jk}| \leqslant C_1|D^{-1}| \leqslant C_2\lambda^{-\ell} \tag{4.156}$$

by $(\mathcal{D})$, where $C_1, C_2 > 0$ are uniform constants.

However, the crude bound (4.156) hides considerable cancellation, a fact that will be crucial in our analysis. Relabel the $\{\phi^{\pm}\}$ so that, at $\lambda = 0$,

$$\phi^{+}_{n+r-j+1} \equiv \phi^{-}_{j} = (\partial/\partial\delta_j)\overline{U}^{\delta}, \quad j = 1, \ldots, \ell. \tag{4.157}$$

With convention (4.157), we have the sharpened bounds:

LEMMA 4.30 ([MaZ.3]). *In dimension $d = 1$, let there hold* (A1)–(A3), (H0)–(H3) *and* ($\mathcal{D}$), *and let $\phi_j^{\pm}$ be labeled as in* (4.157). *Then, for $|\lambda|$ sufficiently small, there hold*

$$\left|M_{jk}^{\pm}\right|, \left|d_{jk}^{\pm}\right| \leqslant C \begin{cases} \lambda^{-1} & \text{for } j = 1, \ldots, \ell, \\ 1 & \text{otherwise}, \end{cases} \tag{4.158}$$

where $M^{\pm}$, $d^{\pm}$ are as defined in Corollary 2.26. *Moreover,*

$$\operatorname{Res}_{\lambda=0} M_{n+r-j+1,k}^{+} = \operatorname{Res}_{\lambda=0} d_{j,k}^{\pm} \tag{4.159}$$

for $1 \leqslant j \leqslant \ell$, all k.

That is, blowup in M_{jk} occurs to order $\lambda^{\ell-1}|D^{-1}|$ rather than $|D^{-1}|$, and, more importantly, only in (fast-decaying) *stationary modes* $(\partial/\partial\delta_j)(\bar{u}^{\delta}, \bar{v}^{\delta})$. Moreover, each stationary mode $(\partial/\partial\delta_j)(\bar{u}^{\delta}, \bar{v}^{\delta})$ has consistent scattering coefficients with dual mode $\tilde{\psi}_k^-$, to lowest order, across all of the various representations of G_λ given in Corollary 2.26.

PROOF OF LEMMA 4.30. Formula (4.155) may be rewritten as

$$C_{jk}^{+} = \det \begin{pmatrix} \phi_1^{+}, \ldots, \phi_{j-1}^{+}, \psi_k^{-}, \phi_j^{+}, \ldots, \phi_n^{+}, & \Phi^{-} \\ \phi_1^{+\prime}, \ldots, \phi_{j-1}^{+}{}', \psi_k^{-\prime}, \phi_j^{+\prime}, \ldots, \phi_n^{+}, & \Phi^{-\prime} \end{pmatrix}\Bigg|_{z=0}, \tag{4.160}$$

from which we easily obtain the desired cancellation in $M^{+} = C^{+}D^{-1}$. For example, for $j > \ell$, we have

$$\begin{aligned} C_{jk}^{+} &= \det \begin{pmatrix} \phi_1^{+} + \lambda\phi_{1_\lambda}^{+} + \cdots, & \ldots, & \phi_n^{+} + \lambda\phi_{n_\lambda}^{+} + \cdots \\ \ldots, & \ldots, & \phi_n^{+\prime} + \lambda\phi_{n_\lambda}^{+}{}' + \cdots \end{pmatrix} \\ &= \mathcal{O}(\lambda^{\ell}) \leqslant C|D|, \end{aligned} \tag{4.161}$$

yielding $|M_{jk}| = |C_{jk}||D|^{-1} \leqslant C$ as claimed, by elimination of ℓ zero-order terms, using linear dependency among fast modes at $\lambda = 0$. For $1 \leqslant j \leqslant \ell$, there is only an $(\ell - 1)$-order dependency, and we obtain instead the bound $|C_{jk}| \leqslant C|\lambda|^{\ell-1}$, or $|M_{jk}| = |C_{jk}||D|^{-1} \leqslant C|\lambda|^{-1}$. The bounds on $|d_{jk}|$ follow similarly; likewise, (4.159) follows easily from the observation that, since columns ψ_j^{+} and $\psi_{n+r-j+1}^{-}$ agree, and $\tilde{\psi}_j^{-}$ is held fixed, the expansions of the various representations by Cramer's rule yield determinants which at λ^{-1} order differ only by a transposition of columns. □

In the Lax or overcompressive case, we can say a bit more:

LEMMA 4.31 [MaZ.3]. *In dimension $d = 1$, let there hold* (A1)–(A3), (H0)–(H3) *and* ($\mathcal{D}$), *with $|\lambda|$ sufficiently small. Then, for Lax and overcompressive shocks, with appropriate basis at $\lambda = 0$ (i.e., slow dual modes taken identically constant), there hold*

$$|M_{jk}|, |d_{jk}| \leqslant C \tag{4.162}$$

if $\tilde{\psi}_k$ is a fast mode, and

$$|M_{jk}|, |d_{jk}| \leqslant C|\lambda| \tag{4.163}$$

if, additionally, ϕ_j is a slow mode.

As we shall see in the following section, this result has the consequence that only slow, constant dual modes play a role in long-time behavior of G.

PROOF OF LEMMA 4.31. Transversality, $\gamma \neq 0$, follows from (D2), so that the results of Lemma 4.28 hold. We first establish the bound (4.162). By Lemma 4.30, we need only consider $j = 1, \dots, \ell$, for which $\phi_j^- = \phi_j^+$. By Lemma 4.28, all fast-growing modes ψ_k^+, ψ_k^- lie in the fast-decaying manifolds $\mathrm{Span}(\phi_1^-, \dots, \phi_{i_-}^-)$, $\mathrm{Span}(\phi_1^+, \dots, \phi_{i_+}^+)$ at $-\infty$, $+\infty$, respectively. It follows that in the right-hand side of (4.160), there is at $\lambda = 0$ a linear dependency between columns

$$\phi_1^+, \dots, \phi_{j-1}^+, \psi_k^-, \phi_{j+1}^+, \phi_{i_+}^+ \quad \text{and} \quad \phi_j^- = \phi_j^+, \tag{4.164}$$

i.e., an ℓ-fold dependency among columns

$$\phi_1^+, \dots, \psi_k^-, \dots, \phi_{i+}^+ \quad \text{and} \quad \phi_1^-, \dots, \phi_\ell^-. \tag{4.165}$$

It follows as in the proof of Lemma 4.30 that $|C_{jk}| \leqslant C\lambda^\ell$ for λ near zero, giving bound (4.162) for M_{jk}. If φ_j^+ is a slow mode, on the other hand, then the same argument shows that there is a linear dependency in columns

$$\phi_1^+, \dots, \phi_{j-1}^+, \psi_k^-, \phi_{j+1}^+, \phi_{i_+}^+ \tag{4.166}$$

and an $(\ell + 1)$-fold dependency in (4.165), since the omitted slow mode ϕ_j^+ plays no role in either linear dependence; thus, we obtain the bound (4.163), instead. Analogous calculations yield the result for d_{jk}, as well. □

PROOF OF PROPOSITION 4.22. The proof of Proposition 4.22 is now just a matter of collecting the bounds of Lemmas 4.23–4.31, and substituting in the representations of Corollary 2.26. More precisely, approximating fast-decaying dual modes $\phi_1^+, \dots, \phi_\ell^+$ and $\phi_{n+r-\ell+1}^-, \dots, \phi_{n+r}^-$ by stationary modes $(\partial/\partial\delta_j)(\bar{u}^\delta, \bar{v}^\delta)$, and slow dual modes by $\mathrm{e}^{-\mu(\lambda)y}\tilde{\tilde{V}}(\lambda)$ as described in Lemma 4.27, truncating $\mu(\lambda)$ at second order, and keeping only the lowest order terms in the Laurent expansions of scattering coefficients $M^\pm$, $d^\pm$ we obtain E_λ and S_λ, respectively, as order λ^{-1} and order one terms, except for negligible $\mathcal{O}(\mathrm{e}^{-\theta|x-y|})$ terms arising through (4.137) and (4.138), and (4.139) and (4.140), which we have accounted for in error term R_λ^E.

Besides the latter, we have accounted in term R_λ^E for: (i) the truncation errors involved in the aforementioned approximations (difference between stationary modes and actual fast modes, and between constant-coefficient approximant and actual slow dual modes, plus truncation errors in exponential rates); and (ii) neglected fast-decaying forward/slow-decaying dual mode combinations (of order λ, by Lemma 4.30). A delicate point is the fact that term $\mathcal{O}(e^{-\theta|x-y|})\mathcal{O}(e^{-\theta|y|})$ arising in the dual truncation of E through the $\lambda\mathcal{O}(e^{-\theta|y|})$ portion of estimate (4.145) for $(\partial/\partial y)\widetilde{\Theta}_j^{\pm}$, absorbs in the (leading) $\mathcal{O}(e^{-\theta|x-y|})$ term of (4.132). It was to obtain this key reduction that we carried out expansion (4.145) to such high order.

In R_λ^S, we have accounted for truncation errors in the slow/slow pairings approximated by term S_λ (difference between constant-coefficient approximants and actual slow forward modes, corresponding to an additional $e^{-\theta|x|}$ term in the truncation error factor, and between constant-coefficient approximants and actual slow dual modes, plus truncation errors in exponential rates). We have also accounted for the remaining, slow-decaying forward/fast-decaying dual modes, which according to Lemma 4.31 have scattering coefficient of order λ and therefore may be grouped with the difference between dual modes and their constant-coefficient approximants (of smaller order by a factor of $e^{-\theta|y|}$) in the term $\mathcal{O}(\lambda)$. □

4.3.2. *High-frequency bounds on the resolvent kernel.* We now turn to the estimation of the resolvent kernel in the high frequency regime $|\lambda| \to +\infty$. According to the usual duality between spatial and frequency variables, large frequency λ corresponds to small time t, or *local behavior* in space and time. Therefore, we begin by examining the frozen-coefficient equations

$$U_t = L_{x_0} U := B(x_0)U_{xx} - A(x_0)U_x \tag{4.167}$$

at each value x_0 of x, for simplicity taking $b(x_0)$ diagonalizable (for motivation only: we make no such assumption in the variable-coefficient case).

LEMMA 4.32 [MaZ.3]. *In dimension $d = 1$, let $b_2(x_0)$ be diagonalizable, and let $A_*(U(x_0))$ have distinct, real eigenvalues. Then, for $|\lambda|$ sufficiently large, the eigenvalue equation $(L_{x_0} - \lambda)U = 0$ associated with the frozen-coefficient operator L_{x_0} at x_0 has a basis of $n + r$ solutions*

$$\{\bar\phi_1^+, \dots, \bar\phi_k^+, \bar\phi_{k+1}^-, \dots, \bar\phi_{n+r}^-\}(x; \lambda, x_0), \quad \bar\phi_j^{\pm} = e^{\mu_j(\lambda, x_0)x} V_j(\lambda, x_0),$$

of which $n - r$ are "hyperbolic modes", analytic in $1/\lambda$ and satisfying

$$\begin{aligned} \mu_j^{\pm}(\lambda, x_0) &= -\lambda/a_j^*(x_0) - \eta_j^*(x_0)/a_j^*(x_0) + \mathcal{O}(1/\lambda), \\ V_j(\lambda, x_0) &= R_j^*(x_0)/a_j^*(x_0) + \mathcal{O}(1/\lambda), \end{aligned} \tag{4.168}$$

$a_j^*(\cdot)$, $\eta_j^*(\cdot)$ and $R_j^*(\cdot)$ as defined in Section 4.2.1 (now with $\partial_x(b_2^{-1}b_1) \equiv 0$), and $2r$ are

"parabolic modes", analytic in $1/\sqrt{\lambda}$ *and satisfying*

$$\mu_j^{\pm}(\lambda, x_0) = -\sqrt{\lambda/\gamma_j(x_0)} + \mathcal{O}(1/\lambda),$$
$$V_j(\lambda, x_0) = \begin{pmatrix} 0_{n-r} \\ s_j \\ \mu_j s_j \end{pmatrix}(x_0) + \mathcal{O}(1/\lambda), \tag{4.169}$$

$\gamma_j(\cdot)$, $s_j(\cdot)$, *the eigenvalues and right eigenvectors of* $b_2(U(x_0))$. *Likewise, the adjoint eigenvalue equation* $(L_{x_0} - \lambda)^* \widetilde{U} = 0$ *has a basis of solutions*

$$\{\bar{\tilde{\phi}}_1^-, \ldots, \bar{\tilde{\phi}}_k^-, \bar{\tilde{\phi}}_{k+1}^+, \ldots, \bar{\tilde{\phi}}_{n+r}^+\}(x; \lambda, x_0), \quad \bar{\tilde{\phi}}_j^{\pm} = \mathrm{e}^{-\mu_j(\lambda, x_0)x} \widetilde{V}_j(\lambda, x_0),$$

with

$$\widetilde{V}_j(\lambda, x_0) = L_j^*(x_0) + \mathcal{O}(1/\lambda),$$
$$\widetilde{V}_j(\lambda, x_0) = \begin{pmatrix} \gamma_j^{-1} b_1^{\mathrm{tr}} t_j \\ t_j \\ -\mu_j t_j \end{pmatrix}(x_0) + \mathcal{O}(1/\lambda), \tag{4.170}$$

respectively, for hyperbolic and parabolic modes, $L_j^*(\cdot)$ *as defined in Section* 4.2.1 *and* $t_j(\cdot)$ *the left eigenvectors of* $b_2(U(x_0))$.

The expansions (4.168)–(4.170) *hold also in the nonstrictly hyperbolic case, with* $\bar{\phi}_j$, $\bar{\tilde{\phi}}_j$ *and* R_j^*, L_j^* *now denoting* $n \times m_j^*$ *blocks, and* μ_j, η_j *denoting* $m_j^* \times m_j^*$ *matrices,* $j = 1, \ldots, J$, *where* m_j^* *is the multiplicity of eigenvalue* a_j^* *of* $A_*(x_0)$.

PROOF. This follows by inversion of the expansions about $\xi = \infty$ of the eigenvalues $\lambda_j(\xi)$ and eigenvectors $V_j(\xi)$ of the one-dimensional frozen-coefficient Fourier symbol $-\mathrm{i}\xi A(x_0) - \xi^2 B(x_0)$, using the basic relation $\mu_j = \mathrm{i}\xi$ relating characteristic and dispersion equations (2.67) and (2.68). The expansions of the one-dimensional Fourier symbol are carried out in Appendix A.4. □

By (2.104), Lemma 4.32 gives an expression for the constant-coefficient resolvent kernel $G_{\lambda;x_0}$ at x_0 of, to lowest order, a "hyperbolic" part,

$$H_{\lambda;x_0} \sim \begin{cases} \sum_{j=1}^k a_j^*(x_0)^{-1} \mathrm{e}^{(-\lambda/a_j^* - \eta_j^*/a_j^*)(x_0)(x-y)} R_j^*(x_0) L_j^{*t}(x_0), & x < y, \\ -\sum_{j=k+1}^J a_j^*(x_0)^{-1} \mathrm{e}^{(-\lambda/a_j^* - \eta_j^*/a_j^*)(x_0)(x-y)} R_j^*(x_0) L_j^{*t}(x_0), & x > y, \end{cases} \tag{4.171}$$

with next-order contribution a "parabolic" part $P_{\lambda;x_0}$, analytic in $\lambda^{-1/2}$ (i.e., for λ away from the negative real axis) and satisfying

$$P_{\lambda;x_0} = \mathcal{O}\big(|\lambda|^{-1/2} \mathrm{e}^{-\theta|\lambda|^{1/2}|x-y|}\big), \quad \theta > 0, \tag{4.172}$$

on some sector

$$\Omega_P := \{\lambda\colon \operatorname{Re}\lambda \geqslant -\theta_1|\operatorname{Im}\lambda| + \theta_2\}, \quad \theta_j > 0. \tag{4.173}$$

Our main result of this section will be to establish on an appropriate unbounded subset of the resolvent set $\rho(L)$ that the variable coefficient solutions $\phi_j^\pm$, $\tilde{\phi}_j^\pm$, and thus the resolvent kernel, satisfy analogous formulae, with static quantities replaced by the corresponding dynamical ones defined in Section 4.2.1. More precisely, define

$$\Omega := \{\lambda\colon -\eta_1 \leqslant \operatorname{Re}\lambda\}, \tag{4.174}$$

with $\eta_1 > 0$ sufficiently small that $\Omega \setminus B(0, r)$ is compactly contained in the set of consistent splitting Λ (defined in (2.62)), for some small $r > 0$ to be chosen later; this is possible, by Lemma 2.21(i).

PROPOSITION 4.33 [MaZ.3]. *In dimension $d = 1$, assume* (A1)–(A3) *and* (H0)–(H3), *plus strict hyperbolicity of $A_*(\cdot)$. Then, for $R > 0$ sufficiently large, there holds $\Omega \setminus B(0, R) \subset \Lambda \cap \rho(L)$; moreover, there holds on $\Omega \setminus B(0, R)$ the decomposition*

$$G_\lambda(x, y) = H_\lambda(x, y) + P_\lambda(x, y) + \Theta_\lambda(x, y), \tag{4.175}$$

$$H_\lambda(x, y) := \begin{cases} -\sum_{j=k+1}^{J} a_j^*(x)^{-1} e^{\int_y^x (-\lambda/a_j^* - \eta_j^*/a_j^*)(z)\,dz} R_j^*(x) L_j^{*t}(y), & x > y, \\ \sum_{j=1}^{k} a_j^*(x)^{-1} e^{\int_y^x (-\lambda/a_j^* - \eta_j^*/a_j^*)(z)\,dz} R_j^*(x) L_j^{*t}(y), & x < y, \end{cases} \tag{4.176}$$

$$\Theta_\lambda(x, y) = \lambda^{-1} B(x, y; \lambda) + \lambda^{-1}(x - y) C(x, y; \lambda) + \lambda^{-2} D(x, y; \lambda), \tag{4.177}$$

where

$$B(x, y; \lambda) = \begin{cases} \sum_{j=k+1}^{J} e^{-\int_y^x \lambda/a_j^*(z)\,dz} b_j^+(x, y), & x > y, \\ \sum_{j=1}^{k} e^{-\int_y^x \lambda/a_j^*(z)\,dz} b_j^-(x, y), & x < y, \end{cases} \tag{4.178}$$

$$C(x, y; \lambda) = \begin{cases} -\sum_{i,j=k+1}^{J} \operatorname{mean}_{z\in[x,y]} e^{-\int_y^z \lambda/a_i^*(s)\,ds - \int_z^x \lambda/a_j^*(s)\,ds} c_{i,j}^+(x, y; z), & x > y, \\ \sum_{i,j=1}^{k} \operatorname{mean}_{z\in[x,y]} e^{-\int_y^z \lambda/a_i^*(s)\,ds - \int_z^x \lambda/a_j^*(s)\,ds} c_{i,j}^-(x, y; z), & x < y, \end{cases} \tag{4.179}$$

with

$$|b_j^\pm|, |c_{i,j}^\pm| \leqslant C e^{-\theta|x-y|} \tag{4.180}$$

and

$$D(x, y; \lambda) = \mathcal{O}\big(e^{-\theta(1+\operatorname{Re}\lambda)|x-y|} + e^{-\theta|\lambda|^{1/2}|x-y|}\big), \tag{4.181}$$

for some uniform $\theta > 0$ independent of x, y, z, each described term separately analytic in λ, and P_λ is analytic in λ on a (larger) sector Ω_P as in (4.173), *with θ_1 sufficiently small, and θ_2 sufficiently large, satisfying uniform bounds*

$$(\partial/\partial x)^\alpha(\partial/\partial y)^\beta P_\lambda(x,y) = \mathcal{O}\big(|\lambda|^{(|\alpha|+|\beta|-1)/2} \mathrm{e}^{-\theta|\lambda|^{1/2}|x-y|}\big), \quad \theta > 0, \tag{4.182}$$

for $|\alpha| + |\beta \leqslant 2$ and $0 \leqslant |\alpha|, |\beta| \leqslant 1$.

Likewise, there hold derivative bounds

$$\begin{aligned}
&(\partial/\partial x)\Theta_\lambda(x,y)\\
&\quad = \big(B_x^0(x,y;\lambda) + (x-y)C_x^0(x,y;\lambda)\big)\\
&\qquad + \lambda^{-1}\big(B_x^1(x,y;\lambda) + (x-y)C_x^1(x,y;\lambda) + (x-y)^2 D_x^1(x,y;\lambda)\big)\\
&\qquad + \lambda^{-3/2}E_x(x,y;\lambda)
\end{aligned} \tag{4.183}$$

and

$$\begin{aligned}
&(\partial/\partial y)\Theta_\lambda(x,y)\\
&\quad = \big(B_y^0(x,y;\lambda) + (x-y)C_y^0(x,y;\lambda)\big)\\
&\qquad + \lambda^{-1}\big(B_y^1(x,y;\lambda) + (x-y)C_y^1(x,y;\lambda) + (x-y)^2 D_y^1(x,y;\lambda)\big)\\
&\qquad + \lambda^{-3/2}E_y(x,y;\lambda),
\end{aligned} \tag{4.184}$$

where B_β^α and C_β^α satisfy bounds of form (4.178) *and* (4.179), *D_β^1 now denotes the iterated integral*

$$\begin{aligned}
&D_\beta^1(x,y;\lambda)\\
&\quad = \begin{cases}
-\sum_{h,i,j=k+1}^J \operatorname{mean}_{y\leqslant w\leqslant z\leqslant x} \mathrm{e}^{-\int_y^w \lambda/a_h^*(s)\,\mathrm{d}s - \int_w^z \lambda/a_i^*(s)\,\mathrm{d}s - \int_z^x \lambda/a_j^*(s)\,\mathrm{d}s}\\
\qquad \times d_{h,i,j}^{\beta,+}(x,y;z), \quad x > y,\\
\sum_{h,i,j=1}^k \operatorname{mean}_{x\leqslant z\leqslant w\leqslant y} \mathrm{e}^{-\int_y^w \lambda/a_h^*(s)\,\mathrm{d}s - \int_w^z \lambda/a_i^*(s)\,\mathrm{d}s - \int_z^x \lambda/a_j^*(s)\,\mathrm{d}s}\\
\qquad \times d_{h,i,j}^{\beta,-}(x,y;z), \quad x < y,
\end{cases}
\end{aligned} \tag{4.185}$$

with $|d_{h,i,j}^{\beta,-}| \leqslant C\mathrm{e}^{-\theta|x-y|}$, and E_β satisfies a bound of form (4.181).

The bounds (4.177)–(4.184) *hold also in the nonstrictly hyperbolic case, with* (4.176) *replaced by*

$$\begin{aligned}
&H_\lambda(x,y)\\
&\quad = \begin{cases}
-\sum_{j=K+1}^J a_j^*(x)^{-1}\mathrm{e}^{-\int_y^x \lambda/a_j^*(z)\,\mathrm{d}z} R_j^*(x)\tilde\zeta_j(x,y)L_j^{*t}(y), & x > y,\\
\sum_{j=1}^K a_j^*(x)^{-1}\mathrm{e}^{-\int_y^x \lambda/a_j^*(z)\,\mathrm{d}z} R_j^*(x)\tilde\zeta_j(x,y)L_j^{*t}(y), & x < y,
\end{cases}
\end{aligned} \tag{4.186}$$

where $\tilde{\zeta}_j(x, y) \in \mathbb{R}^{m_j^* \times m_j^*}$ *denotes a dissipative flow similar to* ζ_j *in* (4.27), *but with respect to variable* x; *i.e.*,

$$\frac{\mathrm{d}\tilde{\zeta}_j}{\mathrm{d}x} = \frac{\eta_j^*(x)\tilde{\zeta}_j(x, y)}{a_j^*(x)}, \qquad \tilde{\zeta}_j(y, y) = I, \tag{4.187}$$

or, equivalently, $\tilde{\zeta}_j(x, y) = \zeta_j(x, \tau)$ *for* τ *such that* $z_j(x, 0) = y$, *where* $z_j(x, s)$ *as in* (4.26) *denotes the backward characteristic path associated with* a_j^*, $z_j(x, \tau) := x$; R_j^*, L_j^* *now denote* $n \times m_j^*$ *blocks; and* μ_j, η_j^* *denote* $m_j^* \times m_j^*$ *matrices,* $j = 1, \ldots, K$, $j = K + 1, \ldots, J$, *where* m_j^* *is the multiplicity of eigenvalue* a_j^* *of* $A_*(x_0)$, *with*

$$a_1^* \leqslant \cdots \leqslant a_K^* < 0 < a_{K+1}^* \leqslant \cdots \leqslant a_J^*.$$

(Note that we have not here assumed $(\mathcal{D})$.)

PROOF. Our argument is similar in spirit to the proof of Proposition 4.4 of [MaZ.1] in the hyperbolic relaxation case. However, the mixed hyperbolic–parabolic nature of the underlying equations, and the associated presence of multiple scales, makes the analysis more delicate. In accordance with the behavior we seek to identify, it is convenient to express the eigenvalue equation in "local" variables $\tilde{u} := A_* u$, $\tilde{v} := b_1 u + b_2 v$ similar to those used in proving Lemma 2.1, yielding

$$\begin{aligned} \tilde{u}_x &= -\lambda A_*^{-1}\tilde{u} - \left(A_{12} b_2^{-1}\tilde{v}\right)_x, \\ \left(\tilde{v}_x\right)_x &= \left[\left(\left(A_{21} - A_{22} b_2^{-1} b_1 + b_2\, \partial_x\left(b_2^{-1} b_1\right)\right) A_*^{-1}\tilde{u}\right)_x \right. \\ &\quad \left. + \left(A_{22} + \partial_x(b_2)\right) b_2^{-1}\tilde{v}\right)_x + \lambda b_2^{-1} b_1 A_*^{-1}\tilde{u} + \lambda b_2^{-1}\tilde{v}\right]. \end{aligned} \tag{4.188}$$

Similarly as in the relaxation case, the substitution of $A_* u$ for u converts the hyperbolic part of the equation from divergence to nondivergence form, simplifying slightly the flow along characteristics in the hyperbolic term H_λ; for related discussion, see Remark 4.12.

Following standard procedure (e.g., [AGJ,GZ,ZH,Z.3]), we perform the rescaling

$$\tilde{x} := |\lambda| x, \qquad \tilde{\lambda} := \lambda / |\lambda|, \tag{4.189}$$

suggested by the leading order hyperbolic behavior of $\mu(x_0)$, (4.168), to obtain after some rearrangement the perturbation equation

$$Y' = A\left(\tilde{x}, |\lambda|^{-1}\right) Y, \quad Y := \left(\tilde{u}, \tilde{v}, \tilde{v}'\right)^{\mathrm{tr}}, \tag{4.190}$$

where

$$A(\tilde{x}, |\lambda|^{-1}) = A_0(\tilde{x}) + |\lambda|^{-1} A_1(\tilde{x}) + \mathcal{O}(|\lambda|^{-2}), \tag{4.191}$$

$$A_0(\tilde{x}) := \begin{pmatrix} -\tilde{\lambda} A_*^{-1} & 0 & -A_{12} b_2^{-1} \\ 0 & 0 & I_r \\ 0 & 0 & 0 \end{pmatrix},$$
$$A_1(\tilde{x}) := \begin{pmatrix} 0 & \partial_x (A_{12} b_2^{-1}) & 0 \\ 0 & 0 & 0 \\ -\tilde{\lambda} d_* A_*^{-2} & -\tilde{\lambda} b_2^{-1} & e_* b_2^{-1} \end{pmatrix}, \tag{4.192}$$

$$d_* := A_{21} - A_{22} b_2^{-1} b_1 + b_2^{-1} b_1 A_* + b_2 \, \partial_x (b_2^{-1} b_1),$$
$$e_* := A_{22} + d_* A_*^{-1} A_{12} + \partial_x (b_2) \tag{4.193}$$

and "$'$" denotes $\partial/\partial\tilde{x}$. Here, expansion (4.191) is to be considered as a continuous family of one-parameter perturbation equations, indexed by $\tilde{\lambda} \in S^1$. The form of e_* is not important, and is included only for completeness; d_*, on the other hand, may be recognized as the important quantity defined in (4.37).

We seek to develop corresponding expansions of the resolvent kernel in orders of $|\lambda|^{-1}$, or, equivalently, by representation (2.84), expansions of stable/unstable manifolds of (4.190), and the reduced flows therein. This we will accomplish by the methods described in Sections 2.2.2–2.2.5, first reducing to an approximately block-diagonal system segregating spectrally separated (in particular, stable/unstable) modes, with formal error of a suitably small order, then converting the formal error into rigorous error bounds using Proposition 2.13 together with Corollary 2.14. An interesting aspect of this calculation is the block-diagonalization of parabolic modes, which are not uniformly spectrally separated in this (hyperbolic) scaling, but rather comprise a block-Jordan block of order $s = 2$ associated with eigenvalue $\mu = 0$. Accordingly, we treat them separately, by the method sketched in Section 2.2.4, after the initial block-diagonalization has been carried out. This results in a slower decaying error series for these blocks, converging in powers of $|\lambda|^{-1/2}$; parabolic modes are well behaved, however, and so the loss of accuracy is harmless.

We will carry out in detail the basic estimate (4.177), indicating the extension to derivative bounds (4.183) and (4.184) by a few brief remarks. First, observe that in the modified coordinates $Y = \mathcal{Q}W$, (2.84) just becomes

$$G_\lambda(x, y) = \begin{cases} (I_n, 0)\mathcal{Q}^{-1}(x)\mathcal{F}_Y^{y\to x} \Pi_Y^- \mathcal{Q}\overline{\mathcal{S}}^1(y)(I_n, 0)^{\mathrm{tr}}, & x < y, \\ -(I_n, 0)\mathcal{Q}^{-1}(x)\mathcal{F}_Y^{y\to x} \Pi_Y^+ \mathcal{Q}\overline{\mathcal{S}}^{-1}(y)(I_n, 0)^{\mathrm{tr}}, & x > y, \end{cases} \tag{4.194}$$

where $\Pi_Y^\pm$ and $\mathcal{F}_Y^{y\to x}$ denote projections and flows in Y-coordinates, and the lower triangular matrices

$$\mathcal{Q} = \begin{pmatrix} I_{n-r} & 0 & 0 \\ 0 & I_r & 0 \\ 0 & 0 & |\lambda|^{-1} I_r \end{pmatrix} \begin{pmatrix} A_* & 0 & 0 \\ b_1 & b_2 & 0 \\ \partial_x(b_1) & \partial_x(b_2) & I_r \end{pmatrix}, \tag{4.195}$$

$$Q^{-1} = \begin{pmatrix} A_*^{-1} & 0 & 0 \\ -b_2^{-1} b_1 A_*^{-1} & b_2^{-1} & 0 \\ -b_2\,\partial_x(b_2^{-1}b_1)A_*^{-1} & -\partial_x(b_2)b_2^{-1} & I_r \end{pmatrix} \begin{pmatrix} I_{n-r} & 0 & 0 \\ 0 & I_r & 0 \\ 0 & 0 & |\lambda| I_r \end{pmatrix} \tag{4.196}$$

relate Y-coordinates to the original coordinates $W = (u, v, b_1 u_x + b_2 v_x)^{\mathrm{tr}}$ of (2.84):

$$Y = QW, \quad W = Q^{-1}Y. \tag{4.197}$$

It is readily checked that $Q \in C^{p+1}$ for f, $B \in C^{p+2}$, whence $Q \in C^2$ under the regularity hypothesis (H0) ($p = 3$).

Initial diagonalization. For f, $B \in C^{p+2}$, we may check further that $A_j \in C^{p-j+1}$ as required in the theory of Section 2.2.3, suitable for approximate block-diagonalization to formal error $\mathcal{O}(|\lambda|^{-p-1})$: formal error $\mathcal{O}(|\lambda|^{-4})$ in the present case $p = 3$. Applying the formal diagonalization procedure of Proposition 2.15 to (4.190), and choosing the special initialization (2.47), at the same time taking care as described in Remark 2.16.3 to preserve analyticity with respect to λ^{-1} in original coordinates, we obtain the approximately block-diagonalized system

$$Z' = D(\tilde{x}, |\lambda|^{-1})Z, \qquad TZ := Y, \qquad D := TAT^{-1}, \tag{4.198}$$

$$T(\tilde{x}, |\lambda|^{-1}) = T_0(\tilde{x}) + |\lambda|^{-1}T_1(\tilde{x}) + \cdots + |\lambda|^{-3}T_3(\tilde{x}), \tag{4.199}$$

$$D(\tilde{x}, |\lambda|^{-1}) = D_0(\tilde{x}) + |\lambda|^{-1}D_1(\tilde{x}) + \cdots + D_3(\tilde{x})|\lambda|^{-3} + \mathcal{O}(|\lambda|^{-4}), \tag{4.200}$$

where without loss of generality (since T_0 is uniquely determined up to a constant linear coordinate change)

$$T_0^{-1} = \begin{pmatrix} L_*^{\mathrm{tr}} & 0 & \tilde{\lambda}^{-1}L_*^{\mathrm{tr}}A_*A_{12}b_2^{-1} \\ 0 & I_r & 0 \\ 0 & 0 & I_r \end{pmatrix}, \qquad T_0 = \begin{pmatrix} R_* & 0 & -\tilde{\lambda}^{-1}A_*A_{12}b_2^{-1} \\ 0 & I_r & 0 \\ 0 & 0 & I_r \end{pmatrix}, \tag{4.201}$$

$L_* := (L_1^*, \ldots, L_J^*)$, $R_* := (R_1^*, \ldots, R_J^*)$, L_j^*, R_j^* as defined in Section 4.2.1, and so

$$D_0 = \begin{pmatrix} -\operatorname{diag}\{\tilde{\lambda} I_{m_j^*}/a_j^*\} & 0 & 0 \\ 0 & 0 & I_r \\ 0 & 0 & 0 \end{pmatrix}, \qquad D_1 = \begin{pmatrix} -\operatorname{diag}\{\eta_j^* I_{m_j^*}/a_j^*\} & 0 & 0 \\ 0 & 0 & 0 \\ 0 & -\tilde{\lambda} b_2^{-1} & * \end{pmatrix}, \tag{4.202}$$

a_j^* and m_j^* as defined in (4.15), and η_j^* as defined in (4.18) and (4.19). Here, we have used the simple block upper-triangular form of A_0 to deduce the form of (4.201), then calculated (4.202) from the relations $D_0 = T_0^{-1} A_0 T_0$, $D_1 = T_0^{-1} A_1 T_0$, the second a consequence of the special normalization (2.47).

The parabolic block. At this point, we have diagonalized into J $m_j^* \times m_j^*$ hyperbolic blocks corresponding to eigenvalues $\tilde{\mu} = -\tilde{\lambda}/a_j^*$ of A_0, and a $2r \times 2r$ parabolic block-Jordan block (the lower right-hand corner)

$$N := \begin{pmatrix} 0 & I_r \\ 0 & 0 \end{pmatrix} + |\lambda|^{-1} \begin{pmatrix} 0 & 0 \\ -\tilde{\lambda} b_2^{-1} & * \end{pmatrix} + \mathcal{O}(|\lambda|^{-2}), \tag{4.203}$$

corresponding to eigenvalue $\tilde{\mu} = 0$ of A_0. We now focus on the latter block, applying the procedure of Section 2.2.4. "Balancing" by transformation $\mathcal{B} := \{1, |\lambda|^{1/2}\}$ yields

$$\widetilde{M} := \mathcal{B}^{-1} N \mathcal{B} = |\lambda|^{-1/2} \widetilde{M}_1 + \mathcal{O}(|\lambda|^{-1}), \quad \widetilde{M}_1 := \begin{pmatrix} 0 & I_r \\ -\tilde{\lambda} b_2^{-1} & 0 \end{pmatrix}, \tag{4.204}$$

or $|\lambda|^{-1/2}$ times the standard rescaled expansion of the strictly parabolic case [ZH].

Proceeding as in that case, we observe that $\sigma(\widetilde{M}_1) = \pm\sqrt{\sigma(\tilde{\lambda} b_2^{-1})}$ has a uniform spectral gap of order one, by (H1)(i), on the sector Ω_P defined in (4.173), for $\theta_1 > 0$ sufficiently small. Thus (see, e.g. Proposition A.9, p. 361 of [St]), there exists a well-conditioned coordinate change $S = S(\widetilde{M}_1(y))$ depending continuously on $\widetilde{M}_1$, such that

$$\widehat{M}_1 = \operatorname{diag}\{\widehat{M}_1^-, \widehat{M}_1^+\}(\tilde{x}) := S^{-1} \widetilde{M} S(\tilde{x})$$

with $\widehat{M}_1^-$ uniformly negative definite and $\widehat{M}_1^+$ uniformly positive definite. Applying this coordinate change, and noting that the "dynamic error" $S^{-1}(\partial/\partial\tilde{x})S$ is of order $(\partial/\partial\tilde{x})\widetilde{M}_1 = \mathcal{O}(|\lambda|^{-1})$, we obtain, finally, the formal expansion

$$\widehat{M}(\tilde{x}, |\tilde{\lambda}|^{-1}) = |\lambda|^{-1/2} \operatorname{diag}\{\widehat{M}_1^-, \widehat{M}_1^+\} + \mathcal{O}(|\lambda|^{-1}). \tag{4.205}$$

On the sector Ω_P, blocks $\pm$ are exponentially separated to order $|\lambda|^{-1/2}$, by Remark 2.12. Thus, we may apply Proposition 2.13 to find that, on Ω_P, there is a further transformation $\widehat{S} = I_{2r} + \mathcal{O}(|\lambda|^{-1/2})$ converting (4.205) to fully diagonalized form

$$M(\tilde{x}, |\tilde{\lambda}|^{-1}) = |\lambda|^{-1/2} \operatorname{diag}\{M_1^-, M_1^+\}, \tag{4.206}$$

where $M_1^{\pm} = \widehat{M}_1^{\pm} + \mathcal{O}(|\lambda|^{-1/2})$ are still uniformly positive–negative definite. We will define P_λ to be the part of the Green distribution associated to the flow of this diagonal block. Note that it has been obtained independent of any separation (or lack thereof) from hyperbolic modes, so is analytic on all of Ω_p.

More precisely, denote by

$$\mathcal{Z} := \mathcal{T} Y := \begin{pmatrix} I_{n-r} & 0 \\ 0 & \widehat{S} S \mathcal{B} \end{pmatrix} (T_0 + |\lambda|^{-1} T_1) Y \tag{4.207}$$

the concatenation of coordinate changes made in the course of the above reductions, $\mathcal{Z} = (\zeta_1, \ldots, \zeta_J, \rho_-, \rho_+)^{\mathrm{tr}}$. Then, we have, similarly as in (4.194), the representation

$$G_\lambda(x, y) = \begin{cases} (I_n, 0)\mathcal{Q}^{-1}\mathcal{T}^{-1}(x)\mathcal{F}_{\mathcal{Z}}^{y\to x}\Pi_{\mathcal{Z}}^{-}\mathcal{T}\mathcal{Q}\overline{\mathcal{S}}^{1}(y)(I_n, 0)^{\mathrm{tr}}, & x < y, \\ -(I_n, 0)\mathcal{Q}^{-1}\mathcal{T}^{-1}(x)\mathcal{F}_{\mathcal{Z}}^{y\to x}\Pi_{\mathcal{Z}}^{+}\mathcal{T}\mathcal{Q}\overline{\mathcal{S}}^{-1}(y)(I_n, 0)^{\mathrm{tr}}, & x > y, \end{cases} \tag{4.208}$$

where $\Pi_{\mathcal{Z}}^{\pm}$ and $\mathcal{F}_{\mathcal{Z}}^{y\to x}$ denote projections and flows in $\mathcal{Z}$-coordinates, and $\mathcal{Z}$ satisfies the approximately decoupled system of equations

$$\begin{aligned} \zeta_j' &= -\big(\tilde{\lambda}/a_j^* - |\lambda|^{-1}\eta_j^*/a_j^* + c_j|\lambda|^{-2} + d_j|\lambda|^{-3}\big)\zeta_j \\ &\quad + \mathcal{O}\big(|\lambda|^{-4}\big)(\rho, \zeta), \\ \rho_\pm' &= |\lambda|^{-1/2}M_1^{\pm}\rho_\pm + \mathcal{O}\big(|\lambda|^{-7/2}\big)\zeta. \end{aligned} \tag{4.209}$$

(Note: the degraded error estimate in the ρ-equations is due to the ill-conditioning of the balancing transformation $\mathcal{B}$.) Approximating $\Pi_{\mathcal{Z}}^{\pm}$ and $\mathcal{F}_{\mathcal{Z}}^{y\to x}$ by the corresponding projections $\overline{\Pi}_{\mathcal{Z}}^{\pm} = \overline{\Pi}_{\mathcal{Z}}^{H,\pm} + \overline{\Pi}_{\mathcal{Z}}^{P,\pm}$ and flows $\overline{\mathcal{F}}_{\mathcal{Z}}^{y\to x}$ for the associated decoupled equations, where $\overline{\Pi}_{\mathcal{Z}}^{H,\pm}$ and $\overline{\Pi}_{\mathcal{Z}}^{P,\pm}$ denote the subprojections on hyperbolic and parabolic blocks, we obtain for $x < y$ the approximation

$$\begin{aligned} &\overline{G}_\lambda(x, y) \\ &\quad := (I_n, 0)^{\mathrm{tr}}\mathcal{Q}^{-1}\mathcal{T}^{-1}(x)\overline{\mathcal{F}}_{\mathcal{Z}}^{y\to x}\big(\overline{\Pi}_{\mathcal{Z}}^{H,-} + \overline{\Pi}_{\mathcal{Z}}^{H,-}\big)\mathcal{T}\mathcal{Q}\overline{\mathcal{S}}^{1}(y)(I_n, 0), \end{aligned} \tag{4.210}$$

and similarly for $x > y$. Defining

$$\begin{aligned} &P_\lambda(x, y) \\ &\quad := (I_n, 0)\mathcal{Q}^{-1}\mathcal{T}^{-1}(x)\overline{\mathcal{F}}_{\mathcal{Z}}^{y\to x}\overline{\Pi}_{\mathcal{Z}}^{P,-}\mathcal{T}\mathcal{Q}\overline{\mathcal{S}}^{1}(y)(I_n, 0)^{\mathrm{tr}} \\ &\quad = (I_n, 0)\mathcal{Q}^{-1}\mathcal{T}^{-1}(x)\overline{\mathcal{F}}_{\mathcal{Z}}^{y\to x}\operatorname{diag}\{0_{n-r}, I_r, 0_r\}\mathcal{T}\mathcal{Q}\overline{\mathcal{S}}^{1}(y)(I_n, 0)^{\mathrm{tr}}, \end{aligned} \tag{4.211}$$

and untangling the various coordinate changes, we find that P_λ satisfies bounds (4.182), similarly as in the strictly parabolic case [ZH]. Specifically, we have by direct calculation that

$$(I_n, 0)\mathcal{Q}^{-1} = \begin{pmatrix} A_*^{-1} & 0 & 0 \\ -b_2^{-1}b_1A_*^{-1} & b_2^{-1} & 0 \end{pmatrix} = \mathcal{O}(1) \tag{4.212}$$

and

$$\mathcal{Q}\overline{\mathcal{S}}^{-1}(I_n, 0)^{\mathrm{tr}} = \begin{pmatrix} -I_{n-r} & 0 \\ 0 & 0 \\ \mathcal{O}(|\lambda|^{-1}) & |\lambda|^{-1}I_r \end{pmatrix}, \tag{4.213}$$

whence (by the special form of T_0)

$$(T_0 + |\lambda|^{-1} T_1)\mathcal{Q}\overline{\mathcal{S}}^{-1}(I_n, 0)^{\mathrm{tr}} = \begin{pmatrix} \mathcal{O}(1) & \mathcal{O}(1) \\ 0 & 0 \\ \mathcal{O}(|\lambda|^{-1}) & |\lambda|^{-1} I_r \end{pmatrix} \tag{4.214}$$

and therefore

$$\Pi_{\mathcal{Z}}^{P,-} \mathcal{T} \mathcal{Q} \overline{\mathcal{S}}^{-1}(I_n, 0)^{\mathrm{tr}} = \mathcal{O}(|\lambda|^{-1/2}). \tag{4.215}$$

Combining with the evident bound

$$|\overline{\mathcal{F}}_{\mathcal{Z}}^{y \to x} \overline{\Pi}_{\mathcal{Z}}^{P,-}| \leqslant C e^{-\theta|\lambda|^{-1/2}|\tilde{x}-\tilde{y}|} = C e^{-\theta|\lambda|^{1/2}|x-y|} \tag{4.216}$$

for some $\theta > 0$, coming from $M_1^- \leqslant -\theta|\lambda|^{-1/2}$, we obtain (4.182) for $|\alpha| = 0$. Derivative bounds then follow from the fact that x and y derivatives of all transformations are at worst order one, while

$$|(\partial/\partial(x,y))^{\alpha} \overline{\mathcal{F}}_{\mathcal{Z}}^{y \to x} \overline{\Pi}_{\mathcal{Z}}^{P,-}| = |\lambda|^{|\alpha|} |(\partial/\partial(\tilde{x},\tilde{y}))^{\alpha} \overline{\mathcal{F}}_{\mathcal{Z}}^{y \to x} \overline{\Pi}_{\mathcal{Z}}^{P,-}| = \mathcal{O}(|\lambda|^{-|\alpha/2|})$$

by inspection of the decoupled equation $\rho'_- = |\lambda|^{-1/2} M_1^- \rho_-$. For closely related but much simpler calculations in the strictly parabolic case, see the proof of Proposition 7.3 of [ZH] (discussion just below (7.24)).

Approximation error. Let us now return to the approximately decoupled equations (4.209), or, converting back to the original spatial coordinates:

$$\begin{aligned} \zeta_j' &= -(\lambda/a_j^* - \eta_j^*/a_j^* + c_j\lambda^{-1} + d_j\lambda^{-2})\zeta_j + \mathcal{O}(|\lambda|^{-3})(\rho, \zeta), \\ \rho_\pm' &= \lambda^{1/2} M_1^\pm \rho_\pm + \mathcal{O}(|\lambda|^{-5/2})\zeta. \end{aligned} \tag{4.217}$$

By condition (2.52) together with (4.21) to the ζ_j equations and condition (2.32) to the $\rho_\pm$ equations, we find that the stable and unstable manifolds of the decoupled flow, consisting respectively of ρ_- together with ζ_j, $a_j^* > 0$ and ρ_+ together with ζ_j, $a_j^* > 0$, are exponentially separated to order $\theta \min\{|\lambda|^{1/2}, 1 + \operatorname{Re}\lambda\}$, for some $\theta > 0$. Applying Proposition 2.13 together with Corollary 2.14, therefore, yields just the decoupled flows with error terms of order $|\lambda|^{-5/2}(e^{-\theta|\lambda|^{1/2}|x-y|} e^{-\theta(1+\operatorname{Re}\lambda)|x-y|})$, and diagonalization error $\mathcal{O}(|\lambda|^{-5/2})$.

Recalling representation (4.208), we find after some rearrangement that G_λ is given by the decoupled approximant $\overline{G}_\lambda$ of (4.210) plus error terms (coming from flow plus diagonalization error) of form $|\lambda|^{-5/2}(e^{-\theta|\lambda|^{1/2}|x-y|} e^{-\theta(1+\operatorname{Re}\lambda)|x-y|})$, absorbable in term D. The contribution to $\overline{G}_\lambda$ from decoupled parabolic modes, as already discussed, is exactly P_λ. Having separated off the parabolic modes, we have now essentially reduced to the purely hyperbolic case treated in [MaZ.1], and we proceed accordingly. Focusing on the decoupled hyperbolic equations

$$\zeta_j' = -(\lambda/a_j^* - \eta_j^*/a_j^*)\zeta_j + (c_j\lambda^{-1} + d_j\lambda^{-2})\zeta_j, \tag{4.218}$$

and applying Corollary 2.14 once again, considering the second term of the right-hand side as a perturbation, we find, substituting into (4.210), that the contribution to $\overline{G}_\lambda$ from decoupled hyperbolic modes is H_λ plus terms of form Θ_λ as claimed. We omit the details of this latter calculation, which are entirely similar to those of the relaxation case. We note in passing that exponential separation implies that the stable/unstable manifolds of (4.217) are *uniformly transverse* for $\lambda \in \Omega$, $|\lambda|$ sufficiently large, yielding $\Omega \setminus B(0, R) \subset \rho(L) \cap \Lambda$, for R sufficiently large, as claimed.

Derivative estimates. Derivative estimates (4.183) and (4.184) now follow in straightforward fashion, by differentiation of (4.208), noting from the approximately decoupled equations that differentiation of the flow brings down a factor (to absorbable error) of λ in hyperbolic modes, $\lambda^{1/2}$ in parabolic modes.

4.3.3. *Extended spectral theory.* Finally, we cite without proof the extended spectral theory of [ZH,MaZ.3] that we shall need in order to establish necessity of condition $(\mathcal{D})$. In the case that $(\mathcal{D})$ holds, this reduces to the simple relations (4.159). Let C^j_η denote the space of C^j functions $f(x)$ satisfying

$$\left|(\mathrm{d}/\mathrm{d}x)^k f(x)\mathrm{e}^{-\eta|x|}\right| \leqslant C, \quad 0 \leqslant k \leqslant j. \tag{4.219}$$

DEFINITION 4.34. Let L be a (possibly degenerate type) rth-order linear ordinary differential operator for which the associated eigenvalue equation may be written as a first-order system with bounded C^q coefficients (so that L: $C^{q+r}_\eta \to C^q_\eta$), and let G_λ denote the Green distribution of $L - \lambda I$, $G_\lambda \in C^{q+1}_{-\eta/2}(x, y)$ away from $y = x$. Further, let Ω be an open, simply connected domain intersecting the resolvent set of L, on which G_λ has a (necessarily unique) meromorphic extension considered as a function from Ω to L^∞_{loc}. Then, for $\lambda_0 \in \Omega$, we define the *effective eigenprojection* $\mathcal{P}_{\lambda_0}$: $C^q_\eta \to C^q_{-\eta/2}$ by

$$\mathcal{P}_{\lambda_0} f(x) = \int_{-\infty}^{+\infty} P_{\lambda_0}(x, y) f(y)\,\mathrm{d}y,$$

where

$$P_{\lambda_0}(x, y) = \mathrm{Res}_{\lambda_0} G_\lambda(x, y) \tag{4.220}$$

and Res_{λ_0} denotes residue at λ_0. (See [ZH], p. 44, for a proof that $\mathcal{P}_{\lambda_0}$: $C^q_\eta \to C^q_{-\eta/2}$.) We will refer to $P_{\lambda_0}(x, y)$ as the *projection kernel*. Likewise, we define the *effective eigenspace* $\Sigma'_{\lambda_0}(L) \subset C^q_{-\eta}$ by

$$\Sigma'_{\lambda_0}(L) = \mathrm{Range}(\mathcal{P}_{\lambda_0}),$$

and the *effective point spectrum* $\sigma'_p(L)$ of L in Ω to be the set of $\lambda \in \Omega$ such that $\dim \Sigma'_{\lambda_0}(L) \neq 0$.

DEFINITION 4.35. Let L, Ω, λ_0 be as above, and K be the order of the pole of $(L-\lambda I)^{-1}$ at λ_0. For $\lambda_0 \in \Omega$, and k any integer, we define $\mathcal{Q}_{\lambda_0,k}\colon C^q_\eta \to C^q_{-\eta/2}$ by

$$\mathcal{Q}_{\lambda_0,k} f(x) = \int_{-\infty}^{+\infty} Q_{\lambda_0,k}(x,y) f(y)\,\mathrm{d}y,$$

where

$$Q_{\lambda_0,k}(x,y) = \mathrm{Res}_{\lambda_0}(\lambda-\lambda_0)^k G_\lambda(x,y).$$

For $0 \leqslant k \leqslant K$, we define the *effective eigenspace of ascent k* by

$$\Sigma'_{\lambda_0,k}(L) = \mathrm{Range}(\mathcal{Q}_{\lambda_0,K-k}).$$

With the above definitions, we obtain the following, modified Fredholm theory.

PROPOSITION 4.36 [ZH,MaZ.3]. *Let L, λ_0, Ω be as in Definition 4.34, and K be the order of the pole of G_λ at λ_0. Then*:

(i) *The operators $\mathcal{P}_{\lambda_0}$, $\mathcal{Q}_{\lambda_0,k}\colon C^q_\eta \to C^q_{-\eta}$ are L-invariant, with*

$$\mathcal{Q}_{\lambda_0,k+1} = (L-\lambda_0 I)\mathcal{Q}_{\lambda_0,k} = \mathcal{Q}_{\lambda_0,k}(L-\lambda_0 I) \tag{4.221}$$

for all $k \neq -1$, and

$$\mathcal{Q}_{\lambda_0,k} = (L-\lambda_0 I)^k \mathcal{P}_{\lambda_0} \tag{4.222}$$

for $k \geqslant 0$.

(ii) *The effective eigenspace of ascent k satisfies*

$$\Sigma'_{\lambda_0,k}(L) = (L-\lambda_0 I)\Sigma'_{\lambda_0,k+1}(L). \tag{4.223}$$

for all $0 \leqslant k \leqslant K$, with

$$\{0\} = \Sigma'_{\lambda_0,0}(L) \subset \Sigma'_{\lambda_0,1}(L) \subset \cdots \subset \Sigma'_{\lambda_0,K}(L) = \Sigma'_{\lambda_0}(L). \tag{4.224}$$

Moreover, each containment in (4.224) is strict.

(iii) *On $\mathcal{P}^{-1}_{\lambda_0}(C^q_\eta)$, $\mathcal{P}_{\lambda_0}$, $\mathcal{Q}_{\lambda_0,k}$ all commute ($k \geqslant 0$), and $\mathcal{P}_{\lambda_0}$ is a projection. More generally, $\mathcal{P}_{\lambda_0} f = f$ for any $f \in C^{Kr}_\eta$ such that $(L-\lambda_0)^K f = 0$, hence each $\Sigma'_{\lambda_0,k}(L)$ contains all $f \in C^{kr}_\eta$ such that $(L-\lambda_0)^k f = 0$.*

(iv) *The multiplicity of the eigenvalue λ_0, defined as $\dim \Sigma'_{\lambda_0}(L)$, is finite and bounded by Kn, where n is the dimension of the phase space for the first-order eigenvalue ODE. Moreover, for all $0 \leqslant k \leqslant K$,*

$$\dim \Sigma'_{\lambda_0,k}(L) = \dim \Sigma'_{\bar{\lambda}_0,k}(L^*). \tag{4.225}$$

Further, the projection kernel can be expanded as

$$P_{\lambda_0} = \sum_j \varphi_j(x)\pi_j(y), \tag{4.226}$$

where $\{\varphi_j\}$, $\{\pi_j\}$ *are bases for* $\Sigma'_{\lambda_0}(L)$, $\Sigma'_{\lambda_0^*}(L^*)$, *respectively.*

PROOF. Residue calculations similar in spirit to those of [Kat] in the classical case of an isolated eigenvalue of finite multiplicity, but substituting everywhere the resolvent kernel for the resolvent. See [ZH] for details. □

REMARK 4.37. For λ_0 in the resolvent set, $\mathcal{P}_{\lambda_0}$ agrees with the standard definition, hence $\Sigma'_{\lambda_0,k}(L)$ agrees with the usual L^p eigenspace $\Sigma_{\lambda_0,k}(L)$ of generalized eigenfunctions of ascent $\leqslant k$, for all $p < \infty$, since C_η^∞ is dense in L^p, $p > 1$, and $\Sigma'_{\lambda_0,k}(L)$ is closed. In the context of stability of traveling waves (more generally, whenever the coefficients of L exponentially approach constant values at $\pm\infty$), $\Sigma'_{\lambda_0,k}(L)$ lies *between* the L^p subspace $\Sigma_{\lambda_0,k}(L)$ and the corresponding L^p_{Loc} subspace.

Now, suppose further (as holds in the case under consideration) that, on Ω: (i) L is a differential operator for which both the forward and adjoint eigenvalue equations may be expressed in appropriate phase spaces as nondegenerate first-order ODE of the same order, and for which the reduced quadratic form $\mathcal{S}$ of Lemma 2.22 is invertible; and, (ii) there exists an analytic choice of bases for stable/unstable manifolds at $+\infty/-\infty$ of the associated eigenvalue equation. In this case, we can define an analytic Evans function $D(\lambda)$ as in (3.1), which we assume does not vanish identically. It is easily verified that the order to which D vanishes at any λ_0 is independent of the choice of analytic bases $\Phi^\pm$. The following result established in [ZH,MaZ.3] generalizes to the extended spectral framework the standard result of Gardner and Jones [GJ.1,GJ.2] in the classical setting.

PROPOSITION 4.38 ([ZH,MaZ.3]). *Let* L, λ_0 *be as above. Then*:

(i) $\dim \Sigma'_{\lambda_0}(L)$ *is equal to the order* d *to which the Evans function* D_L *vanishes at* λ_0.

(ii) $P_{\lambda_0}(L) = \sum_j \varphi_j(x)\pi_j(y)$, *where* φ_j *and* π_j *are in* $\Sigma'_{\lambda_0}(L)$ *and* $\Sigma'_{\lambda_0}(L^*)$, *respectively, with ascents summing to* $\leqslant K+1$, *where* K *is the order of the pole of* G_λ *at* λ_0.

PROOF. Direct calculation using the augmented resolvent kernel formula (2.88) of Remark 2.24, Section 2.4.2, and working in a Jordan basis similarly as in [GJ.1]. See [ZH] for details. □

4.3.4. *Bounds on the Green distribution.* With these preparations, we may now establish pointwise bounds on the Green distribution G and consequently sufficiency and necessity of condition $(\mathcal{D})$ for linearized orbital stability.

PROPOSITION 4.39. *In dimension* $d = 1$, *assuming* (A1)–(A3), (H0)–(H3) *and* $(\mathcal{D})$, *the Green distribution* $G(x,t;y)$ *associated with* (2.1) *may be decomposed as in Theo-*

rem 4.10, *where, for* $y \leqslant 0$,

$$
\begin{aligned}
&R(x,t;y)\\
&\quad = \mathcal{O}\big(e^{-\eta t}e^{-|x-y|^2/(Mt)}\big)\\
&\qquad + \sum_{k=1}^{n} \mathcal{O}\big((t+1)^{-1/2}e^{-\eta x^+} + e^{-\eta|x|}\big)t^{-1/2}e^{-(x-y-a_k^- t)^2/(Mt)}\\
&\qquad + \sum_{a_k^->0,\, a_j^-<0} \chi_{\{|a_k^- t|\geqslant|y|\}}\mathcal{O}\big((t+1)^{-1/2}t^{-1/2}\big)e^{-(x-a_j^-(t-|y/a_k^-|))^2/(Mt)}e^{-\eta x^+}\\
&\qquad + \sum_{a_k^->0,\, a_j^+>0} \chi_{\{|a_k^- t|\geqslant|y|\}}\mathcal{O}\big((t+1)^{-1/2}t^{-1/2}\big)e^{-(x-a_j^+(t-|y/a_k^-|))^2/(Mt)}e^{-\eta x^-}
\end{aligned}
\tag{4.227}
$$

and

$$
\begin{aligned}
&R_y(x,t;y)\\
&\quad = \sum_{j=1}^{J} \mathcal{O}\big(e^{-\eta t}\big)\delta_{x-\bar a_j^* t}(-y) + \mathcal{O}\big(e^{-\eta t}e^{-|x-y|^2/(Mt)}\big)\\
&\qquad + \sum_{k=1}^{n} \mathcal{O}\big((t+1)^{-1/2}e^{-\eta x^+} + e^{-\eta|x|}\big)t^{-1}e^{-(x-y-a_k^- t)^2/(Mt)}\\
&\qquad + \sum_{a_k^->0,\, a_j^-<0} \chi_{\{|a_k^- t|\geqslant|y|\}}\mathcal{O}\big((t+1)^{-1/2}t^{-1}\big)e^{-(x-a_j^-(t-|y/a_k^-|))^2/(Mt)}e^{-\eta x^+}\\
&\qquad + \sum_{a_k^->0,\, a_j^+>0} \chi_{\{|a_k^- t|\geqslant|y|\}}\mathcal{O}\big((t+1)^{-1/2}t^{-1}\big)e^{-(x-a_j^+(t-|y/a_k^-|))^2/(Mt)}e^{-\eta x^-}
\end{aligned}
\tag{4.228}
$$

for some η, $M > 0$, *where* a_j^*, $\bar a_j^*$ *are as in Theorem* 4.10, $x^\pm$ *denotes the positive/negative part of* x, *and indicator function* $\chi_{\{|a_k^- t|\geqslant|y|\}}$ *is one for* $|a_k^- t| \geqslant |y|$ *and zero otherwise. Symmetric bounds hold for* $y \geqslant 0$.

PROOF. Our starting point is the representation

$$
G(x,t;y) = \frac{1}{2\pi i}\,\mathrm{P.V.}\int_{\eta-i\infty}^{\eta+i\infty} e^{\lambda t}G_\lambda(x,y)\,d\lambda,
\tag{4.229}
$$

valid by Proposition 2.3 for any η sufficiently large.

CASE I ($|x-y|/t$ large). We first treat the simpler case that $|x-y|/t \geqslant S$, S sufficiently large. Fixing x, y, t, set $\lambda = \eta + i\xi$. Applying Proposition 4.33, we obtain, for $\eta > 0$

sufficiently large, the decomposition

$$\begin{aligned} G(x,t;y) &= \frac{1}{2\pi \mathrm{i}} \mathrm{P.V.} \int_{\eta-\mathrm{i}\infty}^{\eta+\mathrm{i}\infty} \mathrm{e}^{\lambda t} H_\lambda(x,y)\,\mathrm{d}\lambda \\ &\quad + \frac{1}{2\pi \mathrm{i}} \mathrm{P.V.} \int_{\eta-\mathrm{i}\infty}^{\eta+\mathrm{i}\infty} \mathrm{e}^{\lambda t} \Theta_\lambda^H(x,y)\,\mathrm{d}\lambda \\ &\quad + \frac{1}{2\pi \mathrm{i}} \mathrm{P.V.} \int_{\eta-\mathrm{i}\infty}^{\eta+\mathrm{i}\infty} \mathrm{e}^{\lambda t} P_\lambda(x,y)\,\mathrm{d}\lambda \\ &\quad + \frac{1}{2\pi \mathrm{i}} \mathrm{P.V.} \int_{\eta-\mathrm{i}\infty}^{\eta+\mathrm{i}\infty} \mathrm{e}^{\lambda t} \Theta_\lambda^P(x,y)\,\mathrm{d}\lambda \\ &=: I + II + III + IV, \end{aligned} \tag{4.230}$$

where

$$\begin{aligned} &H_\lambda(x,y) \\ &= \begin{cases} -\sum_{j=K+1}^{J} a_j^*(y)^{-1} \mathrm{e}^{-\int_y^x \lambda/a_j^*(z)\,\mathrm{d}z} R_j^*(x)\tilde{\zeta}_j^*(x,y) L_j^{*t}(y), & x > y, \\ \sum_{j=1}^{K} a_j^*(y)^{-1} \mathrm{e}^{-\int_y^x \lambda/a_j^*(z)\,\mathrm{d}z} R_j^*(x)\tilde{\zeta}_j^*(x,y) L_j^{*t}(y), & x < y, \end{cases} \end{aligned} \tag{4.231}$$

$$\Theta_\lambda^H(x,y) := \lambda^{-1} B(x,y;\lambda) + \lambda^{-1}(x-y) C(x,y;\lambda) \tag{4.232}$$

and

$$\Theta_\lambda^P(x,y) := \lambda^{-2} D(x,y;\lambda), \tag{4.233}$$

with ζ_j, B, C, D as defined in Proposition 4.33. For definiteness taking $x > y$, we estimate each term in turn.

Term I. Term I of (4.230) contributes to integral (4.229) the explicitly evaluable quantity

$$\begin{aligned} &\frac{1}{2\pi \mathrm{i}} \mathrm{P.V.} \int_{\eta-\mathrm{i}\infty}^{\eta+\mathrm{i}\infty} \mathrm{e}^{\lambda t} H_\lambda(x,y)\,\mathrm{d}\lambda \\ &= \sum_{j=K+1}^{J} \left(a_j^*(x)^{-1} \mathrm{P.V.} \int_{-\infty}^{+\infty} \mathrm{e}^{\mathrm{i}\xi(t - \int_y^x (1/a_j^*(z))\,\mathrm{d}z)}\,\mathrm{d}\xi \right) \\ &\qquad \times \mathrm{e}^{\eta(t - \int_y^x (1/a_j^*(z))\,\mathrm{d}z)} r_j(x)\tilde{\zeta}_j(x,y) l_j^t(y) \\ &= \sum_{j=K+1}^{J} \left(a_j^*(x)^{-1} \delta\left(t - \int_y^x (1/a_j^*(z))\,\mathrm{d}z \right) \right) r_j(x)\tilde{\zeta}_j(x,y) l_j^t(y) \end{aligned}$$

$$= \sum_{j=K+1}^{J} a_j^*(x)^{-1} a_j^*(y) \delta_{x-\bar{a}_j^* t}(-y) r_j(x) \zeta_j^*(y,t) l_j^t(y)$$

$$= H(x,t;y), \tag{4.234}$$

where a_j^*, ζ_j^*, and H are as defined in Theorems 4.10 and 4.39. Note: in the second to last equality of (4.234) we have used the general fact that

$$f_y(x,y,t)\delta\big(f(x,y,t)\big) = \delta_{h(x,t)}(y), \tag{4.235}$$

provided $f_y \neq 0$, where $h(x,t)$ is defined by $f(x,h(x,t),t) \equiv 0$. This term clearly vanishes for x outside $[z_1(y,t), z_J(y,t)]$, hence makes no contribution for S sufficiently large.

Term II. Similar calculations show that the "hyperbolic error term" II in (4.230) also vanishes. For example, the term $\mathrm{e}^{\lambda t}\lambda^{-1}B(x,y;\lambda)$ contributes

$$\frac{1}{2\pi \mathrm{i}} \sum_{j=K+1}^{K} \left(\mathrm{P.V.} \int_{-\infty}^{+\infty} (\eta + \mathrm{i}\xi)^{-1} \mathrm{e}^{\mathrm{i}\xi(t - \int_y^x (1/a_j^*(z))\,\mathrm{d}z)}\,\mathrm{d}\xi \right)$$

$$\times\, \mathrm{e}^{\eta(t-\int_y^x (1/a_j^*(z))\,\mathrm{d}z)} b_j^+(x,y). \tag{4.236}$$

The factor

$$\mathrm{P.V.} \int_{-\infty}^{+\infty} (\eta + \mathrm{i}\xi)^{-1} \mathrm{e}^{\mathrm{i}\xi(t-\int_y^x (1/a_j^*(z))\,\mathrm{d}z)}\,\mathrm{d}\xi, \tag{4.237}$$

though not absolutely convergent, is integrable and uniformly bounded as a principal value integral, for all real η bounded from zero, by explicit computation. On the other hand,

$$\mathrm{e}^{\eta(t-\int_y^x (1/a_j^*(z))\,\mathrm{d}z)} \leqslant C\mathrm{e}^{-\eta d(x,[z_1(y,t),z_J(y,t)])/\min_{j,x}|a_j^*(x)|} \to 0 \tag{4.238}$$

as $\eta \to +\infty$ for each $K+1 \leqslant j \leqslant J$, since $a_j^* > 0$ on this range. Thus, taking $\eta \to +\infty$, we find that the product (4.236) goes to zero, giving the result. Likewise, the contributions of terms $\mathrm{e}^{\lambda t}C(x,y;\lambda)$ and $\mathrm{e}^{\lambda t}D(x,y;\lambda)$ split into the product of a convergent, uniformly bounded integral in ξ, a (constant) factor depending only on (x,y), and a factor $\alpha(x,y,t,\eta)$ going to zero as $\eta \to 0$ at rate (4.238). Thus, each of these terms vanishes also as $\eta \to +\infty$, as claimed.

Term III. The parabolic term III may be treated exactly as in the strictly parabolic case [ZH]. Namely, using analyticity of P_λ within the sector Ω_P defined in (4.173), we may for any fixed η deform the contour in the principal value integral to

$$\int_{\Gamma_1 \cup \Gamma_2} \mathrm{e}^{\lambda t} P_\lambda(x,y)\,\mathrm{d}\lambda, \tag{4.239}$$

where $\Gamma_1 := \partial B(0, R) \cap \overline{\Omega}_P$ and $\Gamma_2 := \partial\Omega_P \setminus B(0, R)$ plus an error term of order not greater than

$$|\xi|^{-1/2} \int_\eta^{-\infty} e^{wt}\, dw \to 0$$

as $|\xi| \to \infty$, where

$$R := \bar{\alpha}^2, \quad \bar{\alpha} := \frac{\theta|x-y|}{2t}, \tag{4.240}$$

θ is as in (4.182).

By estimate (4.182) we have for all $\lambda \in \Gamma_1 \cup \Gamma_2$ that

$$\left|P_\lambda(x,y)\right| \leqslant C\left|\lambda^{-1/2}\right| e^{-\theta|\lambda|^{1/2}|x-y|}. \tag{4.241}$$

Further, we have

$$\begin{aligned} &\operatorname{Re}\lambda \leqslant R\left(1-\eta\omega^2\right), \quad \lambda \in \Gamma_1,\\ &\operatorname{Re}\lambda \leqslant \operatorname{Re}\lambda_0 - \eta\left(|\operatorname{Im}\lambda| - |\operatorname{Im}\lambda_0|\right), \quad \lambda \in \Gamma_2, \end{aligned} \tag{4.242}$$

for R sufficiently large, where ω is the argument of λ and λ_0 and λ_0^* are the two points of intersection of Γ_1 and Γ_2, for some $\eta > 0$ independent of $\bar{\alpha}$.

Combining (4.241), (4.242) and (4.240), we obtain

$$\begin{aligned} \left|\int_{\Gamma_1} e^{\lambda t} P_\lambda\, d\lambda\right| &\leqslant \int_{\Gamma_1} C\left|\lambda^{-1/2}\right| e^{\operatorname{Re}\lambda t - \theta|\lambda|^{1/2}|x-y|}\, d\lambda \\ &\leqslant C e^{-\bar{\alpha}^2 t} \int_{-L}^{+L} R^{-1/2} e^{-R\eta\omega^2 t} R\, d\omega \\ &\leqslant C t^{-1/2} e^{-\bar{\alpha}^2 t}. \end{aligned}$$

Likewise,

$$\begin{aligned} \left|\int_{\Gamma_2} e^{\lambda t} P_\lambda\, d\lambda\right| &\leqslant \int_{\Gamma_2} C\left|\lambda^{-1/2}\right| C e^{\operatorname{Re}\lambda t - \theta|\lambda|^{1/2}|x-y|\, d\lambda} \\ &\leqslant C e^{\operatorname{Re}(\lambda_0)t - \theta|\lambda_0|^{1/2}|x-y|} \int_{\Gamma_2} \left|\lambda^{-1/2}\right| e^{(\operatorname{Re}\lambda - \operatorname{Re}\lambda_0)t}\, |d\lambda| \\ &\leqslant C e^{-\bar{\alpha}^2 t} \int_{\Gamma_2} |\operatorname{Im}\lambda|^{-1/2} e^{-\eta|\operatorname{Im}\lambda - \operatorname{Im}\lambda_0|t} |d \operatorname{Im}\lambda| \\ &\leqslant C t^{-1/2} e^{-\bar{\alpha}^2 t}. \end{aligned} \tag{4.243}$$

Combining these last two estimates, and recalling (4.240), we have

$$III \leqslant Ct^{-1/2}\mathrm{e}^{-\bar{\alpha}^2 t/2}\mathrm{e}^{-\theta^2(x-y)^2/(8t)} \leqslant Ct^{-1/2}\mathrm{e}^{-\eta t}\mathrm{e}^{-(x-y)^2/(Mt)}, \tag{4.244}$$

for $\eta > 0$, $M > 0$ independent of $\bar{\alpha}$. Observing that

$$\frac{|x-at|}{2t} \leqslant \frac{|x-y|}{t} \leqslant \frac{2|x-at|}{t}$$

for any bounded a, for $|x-y|/t$ sufficiently large, we find that III can be absorbed in the residual term $\mathcal{O}(\mathrm{e}^{-\eta t}\mathrm{e}^{-|x-y|^2/(Mt)})$ for $t \geqslant \epsilon$, any $\epsilon > 0$, and by any summand $\mathcal{O}(t^{-1/2}(t+1)^{-1/2}\mathrm{e}^{-(x-y-a_k^{\pm}t)^2/(Mt)})\mathrm{e}^{-\eta x^{\pm}}\mathrm{e}^{-\eta y^{\pm}}$ for t small.

Likewise, differentiating within the absolutely convergent integral (4.239), we find that $|III_y|$ can be absorbed in the corresponding summand of (4.228).

NOTE. A delicate point about the derivative bounds to first move the contour to (4.239) and then differentiate. Differentiating within the original principal value integral yields an integral that is not obviously convergent, and for which it is not clear that the contour may be so deformed.

Term IV. Similarly as in the treatment of term III, the principal value integral for the "parabolic error term" IV may be shifted to $\eta = R = \bar{\alpha}^2$, $\bar{\alpha}$ as in (4.240), at the cost of an error term that vanishes as $|\xi| \to \infty$. But, this yields an estimate

$$|IV| \leqslant C\mathrm{e}^{-\bar{\alpha}^2 t}\int_{-\infty}^{+\infty} |\eta_0 + \mathrm{i}\xi|^{-2}\,\mathrm{d}\xi \leqslant \mathrm{e}^{-\bar{\alpha}^2 t}$$

that by (4.244) may be absorbed in $\mathcal{O}(\mathrm{e}^{-\eta t}\mathrm{e}^{-|x-y|^2/(Mt)})$ for all t. Derivatives IV_x and IV_y may be similarly absorbed, while IV_{xy} may be absorbed in

$$(\partial/\partial y)\mathcal{O}\big(\mathrm{e}^{-\eta t}\mathrm{e}^{-|x-y|^2/(Mt)}\big).$$

CASE II ($|x-y|/t$ bounded). We now turn to the critical case that $|x-y|/t \leqslant S$ for some fixed S. In this regime, note that any contribution of order $\mathrm{e}^{\theta t}$, $\theta > 0$, may be absorbed in the residual (error) term R (resp. R_x, R_y); we shall use this observation repeatedly.

Decomposition of the contour. We begin by converting contour integral (4.229) into a more convenient form decomposing high, intermediate, and low frequency contributions.

OBSERVATION 4.40. *In dimension $d = 1$, assuming* (A1)–(A3) *and* (H0)–(H3), *there holds the representation*

$$
\begin{aligned}
G(x,t;y) &= I_a + I_b + I_c + II_a + II_b \\
&:= \frac{1}{2\pi\mathrm{i}} P.V. \int_{\eta-\mathrm{i}\infty}^{\eta+\mathrm{i}\infty} \mathrm{e}^{\lambda t} H_\lambda(x,y)\,\mathrm{d}\lambda \\
&\quad + \frac{1}{2\pi\mathrm{i}} P.V. \left(\int_{-\eta_1-\mathrm{i}\infty}^{-\eta_1-\mathrm{i}R} + \int_{-\eta_1+\mathrm{i}R}^{-\eta_1+\mathrm{i}\infty} \right) \mathrm{e}^{\lambda t} (G_\lambda - H_\lambda - P_\lambda)(x,y)\,\mathrm{d}\lambda \\
&\quad + \frac{1}{2\pi\mathrm{i}} \oint_{\Gamma_2} \mathrm{e}^{\lambda t} P_\lambda(x,y)\,\mathrm{d}\lambda \\
&\quad + \frac{1}{2\pi\mathrm{i}} \oint_{\Gamma} \mathrm{e}^{\lambda t} G_\lambda(x,y)\,\mathrm{d}\lambda \\
&\quad - \frac{1}{2\pi\mathrm{i}} \int_{-\eta_1-\mathrm{i}R}^{-\eta_1+\mathrm{i}R} \mathrm{e}^{\lambda t} H_\lambda(x,y)\,\mathrm{d}\lambda,
\end{aligned}
\tag{4.245}
$$

$$
\begin{aligned}
\Gamma &:= [-\eta_1 - \mathrm{i}R, \eta - \mathrm{i}R] \cup [\eta - \mathrm{i}R, \eta + \mathrm{i}R] \\
&\qquad \cup [\eta + \mathrm{i}R, -\eta_1 + \mathrm{i}R],
\end{aligned}
\tag{4.246}
$$

$$
\Gamma_2 := \partial\Omega_P \setminus \Omega,
\tag{4.247}
$$

Ω_P *as defined in* (4.173), *for any $\eta > 0$ such that* (4.229) *holds, R sufficiently large, and $-\eta_1 < 0$ as in* (4.174). (*Note that we have not here assumed* ($\mathcal{D}$).)

PROOF. We first observe that, by Proposition 4.33, L has no spectrum on the portion of Ω lying outside of the rectangle

$$
\mathcal{R} := \{\lambda \colon -\eta_1 \leqslant \operatorname{Re}\lambda \leqslant \eta, -R \leqslant \operatorname{Im}\lambda \leqslant R\}
\tag{4.248}
$$

for $\eta > 0$, $R > 0$ sufficiently large, hence G_λ is analytic on this region. Since, also, H_λ is analytic on the whole complex plane, contours involving either one of these contributions may be arbitrarily deformed within $\Omega \setminus \mathcal{R}$ without affecting the result, by Cauchy's theorem. Likewise, P_λ is analytic on $\Omega_P \setminus \mathcal{R}$, and so contours involving this contribution may be arbitrarily deformed within this region.

Recalling, further, that

$$
|G_\lambda - H_\lambda - P_\lambda| = \mathcal{O}(\lambda^{-1})
$$

by Proposition 4.33, we find that

$$
\frac{1}{2\pi\mathrm{i}} \mathrm{P.V.} \int_{\eta-\mathrm{i}\infty}^{\eta+\mathrm{i}\infty} \mathrm{e}^{\lambda t} (G_\lambda - H_\lambda - P_\lambda)(x,y)\,\mathrm{d}\lambda
$$

may be deformed to

$$\frac{1}{2\pi \mathrm{i}}\mathrm{P.V.}\oint_{\partial(\Omega\setminus\mathcal{R})} \mathrm{e}^{\lambda t}(G_\lambda - H_\lambda - P_\lambda)(x,y)\,\mathrm{d}\lambda.$$

And, similarly as in the treatment of term *III* of the large $|x-y|/t$ case, we find using bounds (4.182) that

$$\frac{1}{2\pi \mathrm{i}}\mathrm{P.V.}\int_{\eta-\mathrm{i}\infty}^{\eta+\mathrm{i}\infty} \mathrm{e}^{\lambda t} P_\lambda(x,y)\,\mathrm{d}\lambda$$

may be deformed to the absolutely convergent integral

$$\frac{1}{2\pi \mathrm{i}}\oint_{\partial(\Omega_P\setminus\mathcal{R})} \mathrm{e}^{\lambda t} P_\lambda(x,y)\,\mathrm{d}\lambda.$$

Noting, finally, that

$$+\frac{1}{2\pi \mathrm{i}}\oint_{\partial\mathcal{R}} \mathrm{e}^{\lambda t} H_\lambda(x,y)\,\mathrm{d}\lambda = 0,$$

by Cauchy's theorem, we obtain, finally,

$$\begin{aligned}
&\frac{1}{2\pi \mathrm{i}}\mathrm{P.V.}\int_{\eta-\mathrm{i}\infty}^{\eta+\mathrm{i}\infty} \mathrm{e}^{\lambda t} G_\lambda(x,y)\,\mathrm{d}\lambda\\
&\quad= \frac{1}{2\pi \mathrm{i}}\int_{\eta-\mathrm{i}\infty}^{\eta+\mathrm{i}\infty} \mathrm{e}^{\lambda t} H_\lambda(x,y)\,\mathrm{d}\lambda\\
&\qquad+ \frac{1}{2\pi \mathrm{i}}\mathrm{P.V.}\int_{\eta-\mathrm{i}\infty}^{\eta+\mathrm{i}\infty} \mathrm{e}^{\lambda t}(G_\lambda - H_\lambda - P_\lambda)(x,y)\,\mathrm{d}\lambda\\
&\qquad+ \frac{1}{2\pi \mathrm{i}}\oint_{\partial(\Omega_P\setminus\mathcal{R})} \mathrm{e}^{\lambda t} P_\lambda(x,y)\,\mathrm{d}\lambda\\
&\quad= \frac{1}{2\pi \mathrm{i}}\mathrm{P.V.}\int_{\eta-\mathrm{i}\infty}^{\eta+\mathrm{i}\infty} \mathrm{e}^{\lambda t} H_\lambda(x,y)\,\mathrm{d}\lambda\\
&\qquad+ \frac{1}{2\pi \mathrm{i}}\mathrm{P.V.}\oint_{\partial(\Lambda\setminus\mathcal{R})} \mathrm{e}^{\lambda t}(G_\lambda - H_\lambda - P_\lambda)(x,y)\,\mathrm{d}\lambda\\
&\qquad+ \frac{1}{2\pi \mathrm{i}}\oint_{\partial\mathcal{R}} \mathrm{e}^{\lambda t} H_\lambda(x,y)\,\mathrm{d}\lambda + \frac{1}{2\pi \mathrm{i}}\oint_{\partial(\Omega_P\setminus\mathcal{R})} \mathrm{e}^{\lambda t} P_\lambda(x,y)\,\mathrm{d}\lambda,
\end{aligned}\tag{4.249}$$

from which the result follows by combining the second third and fourth contour integrals along their common edges Γ. □

OBSERVATION 4.41. *In dimension $d = 1$, assuming* (A1)–(A3), (H0)–(H3) *and* ($\mathcal{D}$), *we may replace* (4.245) *by*

$$G(x,t;y) = I_a + I_b + I_c + II_{\tilde{a}} + II_b + III,$$

where I_a, I_b, I_c and II_b are as in (4.245), *and*

$$II_{\tilde{a}} := \frac{1}{2\pi}\left(\int_{-\eta_1-\mathrm{i}R}^{-\eta_1-\mathrm{i}r/2} + \int_{-\eta_1+\mathrm{i}r/2}^{-\eta_1+\mathrm{i}R}\right) \mathrm{e}^{\lambda t} G\lambda(x,y)\,\mathrm{d}x\,\mathrm{i},$$

$$III := \frac{1}{2\pi\mathrm{i}} \oint_{\widetilde{\Gamma}} \mathrm{e}^{\lambda t} G_\lambda(x,y)\,\mathrm{d}\lambda, \tag{4.250}$$

and

$$\begin{aligned}\widetilde{\Gamma} &:= [-\eta_1 - \mathrm{i}r/2, \eta - \mathrm{i}r/2] \cup [\eta - \mathrm{i}r/2, \eta + \mathrm{i}r/2] \\ &\quad \cup [\eta + \mathrm{i}r/2, -\eta_1 + \mathrm{i}r/2],\end{aligned} \tag{4.251}$$

for any $\eta, r > 0$, and η_1 sufficiently small with respect to r.

PROOF. By assumption ($\mathcal{D}$), L has no spectrum on the region between contour Γ and the union of contour $\widetilde{\Gamma}$ and the contour of term $II_{\tilde{a}}$, hence G_λ is analytic on that region, and

$$II_a = II_{\tilde{a}} + III$$

by Cauchy's theorem, giving the result. □

Using the final decomposition (4.250), we shall estimate in turn the high-frequency contributions I_a, I_b and I_c, the intermediate frequency contributions $II_{\tilde{a}}$ and II_b, and the low-frequency contributions III.

High-frequency contribution. We first carry out the straightforward estimation of the high-frequency terms I_a, I_b and I_c. The principal term I_a has already been computed in (4.234) to be $H(x,t;y)$. Likewise, calculations similar to those of (4.236)–(4.238) show that the error term

$$\begin{aligned}I_b &= \frac{1}{2\pi\mathrm{i}}\mathrm{P.V.}\left(\int_{-\eta_1-\mathrm{i}\infty}^{-\eta_1-R} \int_{-\eta_1+\mathrm{i}R}^{-\eta_1+\mathrm{i}\infty}\right) \mathrm{e}^{\lambda t} \Theta_\lambda(x,y)\,\mathrm{d}\lambda \\ &= \frac{1}{2\pi\mathrm{i}}\mathrm{P.V.}\left(\int_{-\eta_1-\mathrm{i}\infty}^{-\eta_1-R} \int_{-\eta_1+\mathrm{i}R}^{-\eta_1+\mathrm{i}\infty}\right) \mathrm{e}^{\lambda t} \\ &\quad \times \left(\lambda^{-1} B(x,y;\lambda) + \lambda^{-1}(x-y)C(x,y;\lambda) + \lambda^{-2} D(x,y;\lambda)\right)\mathrm{d}\lambda\end{aligned}$$

is time-exponentially small.

For example, for $x > y$ the term $e^{\lambda t}\lambda^{-1}B(x, y; \lambda)$ contributes

$$\sum_{j=K+1}^{K} \frac{1}{2\pi i} \mathrm{P.V.}\left(\int_{-\infty}^{-R} \int_{R}^{+\infty}\right)(-\eta_1 + i\xi)^{-1} \times e^{i\xi(t-\int_y^x (1/a_j^*(z))\,dz)}\,d\xi\, e^{-\eta_1(t-\int_y^x (1/a_j^*(z))\,dz)} b_j^+(x, y), \tag{4.252}$$

where

$$\frac{1}{2\pi i} \mathrm{P.V.}\left(\int_{-\infty}^{-R} \int_{R}^{+\infty}\right)(\eta + i\xi)^{-1} e^{i\xi(t-\int_y^x (1/a_j^*(z))\,dz)}\,d\xi < \infty \tag{4.253}$$

and

$$e^{\eta_1 \int_y^x (1/a_j^*(z))\,dz} b(x, y) \leqslant C_1 e^{\eta_1 \int_y^x (1/a_j^*(z))\,dz - \theta|x-y|} \leqslant C_2 \tag{4.254}$$

for η_1 sufficiently small. This may be absorbed in the first term of R, (4.227). Likewise, the contributions of terms $e^{\lambda t}C(x, y; \lambda)$ and $e^{\lambda t}D(x, y; \lambda)$ split into the product of a convergent, uniformly bounded integral in ξ, a bounded factor analogous to (4.254), and the factor $e^{-\eta_1 t}$, giving the result.

The term I_c may be estimated exactly as was term *III* in the large $|x - y|/t$ case, to obtain contribution $\mathcal{O}(t^{-1/2}e^{-\eta_1 t})$ absorbable again in the residual term $\mathcal{O}(e^{-\eta t}e^{-|x-y|^2/(Mt)})$ for $t \geqslant \epsilon$, any $\epsilon > 0$, and by any summand $\mathcal{O}(t^{-1/2}(t+1)^{-1/2}e^{-(x-y-a_k^{\pm}t)^2/(Mt)})e^{-\eta x \pm}e^{-\eta y \pm}$ for t small.

Derivative bounds. Derivatives $(\partial/\partial y)I_b$ may be treated in identical fashion using (4.184) to show that they are absorbable in the estimates given for R_y. We point out that the integral arising from term B_y^0 of (4.184) corresponds to the first, delta-function term in (4.228), while the integral arising from term $(x - y)C_y^0$ vanishes, as the product of $(x - y)$ and a delta function $\delta_z(y)$ with $z \neq x$. Derivatives of term I_c may be treated exactly as were derivatives of term *III* in the large $|x - y|/t$ case.

Intermediate frequency contribution. Error term II_b is time-exponentially small for η_1 sufficiently small, by the same calculation as in (4.252)–(4.254), hence negligible. Likewise, term $II_{\tilde{a}}$ by the basic estimate (2.85) is seen to be time-exponentially small of order $e^{-\eta_1 t}$ for any $\eta_1 > 0$ sufficiently small that the associated contour lies in the resolvent set of L.

Low-frequency contribution. It remains to estimate the low-frequency term *III*, which is of essentially the same form as the low-frequency contribution analyzed in [ZH,Z.3,Z.6] in the strictly parabolic case, in that the contour is the same and the resolvent kernel G_λ satisfies identical bounds in this regime. Thus, we may conclude from these previous analyses that *III* gives contribution $E + S + R$, as claimed, exactly as in the strictly parabolic case. For completeness, we indicate the main features of the argument here.

Case $t \leqslant 1$. First observe that estimates in the short-time regime $t \leqslant 1$ are trivial, since then $|e^{\lambda t}G_\lambda(x, y)|$ is uniformly bounded on the compact set $\widetilde{\Gamma}$, and we have $|G(x, t; y)| \leqslant$

$C \leqslant \mathrm{e}^{-\theta t}$ for $\theta > 0$ sufficiently small. But, likewise, E and S are uniformly bounded in this regime, hence time-exponentially decaying. As observed previously, all such terms are negligible, being absorbable in the error term R. Thus, we may add $E + S$ and subtract G to obtain the result.

Case $t \geqslant 1$. Next, consider the critical (long-time) regime $t \geqslant 1$. For definiteness, take $y \leqslant x \leqslant 0$; the other two cases are similar. Decomposing

$$G(x,t;y) = \frac{1}{2\pi \mathrm{i}} \oint_{\widetilde{\Gamma}} \mathrm{e}^{\lambda t} E_\lambda(x,y)\,\mathrm{d}\lambda + \frac{1}{2\pi \mathrm{i}} \oint_{\widetilde{\Gamma}} \mathrm{e}^{\lambda t} S_\lambda(x,y)\,\mathrm{d}\lambda + \frac{1}{2\pi \mathrm{i}} \oint_{\widetilde{\Gamma}} \mathrm{e}^{\lambda t} R_\lambda(x,y)\,\mathrm{d}\lambda, \tag{4.255}$$

with E_λ, S_λ and R_λ as defined in Proposition 4.22, we consider in turn each of the three terms on the right-hand side.

E_λ term. Let us first consider the dominant term

$$\frac{1}{2\pi \mathrm{i}} \oint_{\widetilde{\Gamma}} \mathrm{e}^{\lambda t} E_\lambda(x,y)\,\mathrm{d}\lambda, \tag{4.256}$$

which, by (4.128), is given by

$$\sum_{a_k^- > 0} \left[c_{k,-}^{j,0}\right] (\partial \bar{U}^\delta / \partial \delta)(x) l_k^{-t} \alpha_k(x,t;y), \tag{4.257}$$

where

$$\alpha_k(x,t;y) := \frac{1}{2\pi \mathrm{i}} \oint_{\widetilde{\Gamma}} \lambda^{-1} \mathrm{e}^{\lambda t} \mathrm{e}^{(\lambda/a_k^- - \lambda^2 \beta_k^- / a_k^{-3}) y}\,\mathrm{d}\lambda. \tag{4.258}$$

Using Cauchy's theorem, we may move the contour $\widetilde{\Gamma}$ to obtain

$$\begin{aligned}
\alpha_k(x,t;y) = {} & \frac{1}{2\pi} \mathrm{P.V.} \int_{-r/2}^{+r/2} (\mathrm{i}\xi)^{-1} \mathrm{e}^{\mathrm{i}\xi t} \mathrm{e}^{(\mathrm{i}\xi/a_k^- + \xi^2 \beta_k^- / a_k^{-3}) y}\,\mathrm{d}\xi \\
& + \frac{1}{2\pi \mathrm{i}} \left(\int_{-\eta_1 - \mathrm{i}r/2}^{-\mathrm{i}r/2} + \int_{\mathrm{i}r/2}^{-\eta_1 + \mathrm{i}r/2} \right) \lambda^{-1} \mathrm{e}^{\lambda t} \mathrm{e}^{(\lambda/a_k^- - \lambda^2 \beta_k^- / a_k^{-3}) y}\,\mathrm{d}\lambda \\
& + \frac{1}{2} \mathrm{Res}_{\lambda=0}\, \mathrm{e}^{\lambda t} \mathrm{e}^{(\lambda/a_k^- - \lambda^2 \beta_k^- / a_k^{-3}) y},
\end{aligned} \tag{4.259}$$

or, rearranging and evaluating the final, residue term:

$$\begin{aligned}
\alpha_k(x,t;y) = {} & \left(\frac{1}{2\pi} \mathrm{P.V.} \int_{-\infty}^{+\infty} (\mathrm{i}\xi)^{-1} \mathrm{e}^{\mathrm{i}\xi(t + y/a_k^-)} \mathrm{e}^{\xi^2 (\beta_k^- / a_k^{-3}) y}\,\mathrm{d}\xi + \frac{1}{2} \right) \\
& - \frac{1}{2\pi} \left(\int_{-\infty}^{-r/2} + \int_{r/2}^{+\infty} \right) (\mathrm{i}\xi)^{-1} \mathrm{e}^{\mathrm{i}\xi(t + y/a_k^-)} \mathrm{e}^{\xi^2 (\beta_k^- / a_k^{-3}) y}\,\mathrm{d}\xi
\end{aligned}$$

$$+\frac{1}{2\pi\mathrm{i}}\left(\int_{-\eta_1-\mathrm{i}r/2}^{-\mathrm{i}r/2}+\int_{\mathrm{i}r/2}^{-\eta_1+\mathrm{i}r/2}\right)$$
$$\times\,\lambda^{-1}\mathrm{e}^{\lambda t}\mathrm{e}^{(\lambda/a_k^- -\lambda^2\beta_k^-/a_k^{-3})y}\,\mathrm{d}\lambda. \tag{4.260}$$

The first term in (4.260) may be explicitly evaluated to give

$$\mathrm{errfn}\left(\frac{y+a_k^-t}{\sqrt{4\beta_k^-|y/a_k^-|}}\right), \tag{4.261}$$

where

$$\mathrm{errfn}(z):=\frac{1}{2\pi}\int_{-\infty}^{z}\mathrm{e}^{-y^2}\,\mathrm{d}y, \tag{4.262}$$

whereas the second and third terms are clearly time-exponentially small for $t\leqslant C|y|$ and η_1 sufficiently small relative to r (see detailed discussion of similar calculations below, under R_λ term). In the trivial case $t\geqslant C|y|$, $C>0$ sufficiently large, we can simply move the contour to $[-\eta_1-\mathrm{i}r/2, -\eta_1+\mathrm{i}r/2]$ to obtain (complete) residue 1 plus a time-exponentially small error corresponding to the shifted contour integral, which result again agrees with (4.261) up to a time-exponentially small error.

Expression (4.261) may be rewritten as

$$\mathrm{errfn}\left(\frac{y+a_k^-t}{\sqrt{4\beta_k^-|y/a_k^-|}}\right) \tag{4.263}$$

plus error

$$\begin{aligned}&\mathrm{errfn}\left(\frac{y+a_k^-t}{\sqrt{4\beta_k^-|y/a_k^-|}}\right)-\mathrm{errfn}\left(\frac{y+a_k^-t}{\sqrt{4\beta_k^-t}}\right)\\&\quad\sim\mathrm{errfn}'\left(\frac{y+a_k^-t}{\sqrt{4\beta_k^-t}}\right)\left(-\frac{4\beta_k^-}{2}(y+a_k^-t)^2(4\beta_k^-t)^{-3/2}\right)\\&\quad=\mathcal{O}\left(t^{-1}\mathrm{e}^{(y+a_k^-t)^2/(Mt)}\right),\end{aligned} \tag{4.264}$$

for $M>0$ sufficiently large, and similarly for x- and y-derivatives. Multiplying by

$$\left[c_{k,-}^{j,0}\right]\left(\partial\overline{U}^\delta/\partial\delta\right)(x)l_k^{-t}=\mathcal{O}\left(\mathrm{e}^{-\theta|x|}\right),$$

we find that term (4.263) gives contribution

$$\left[c_{k,-}^{j,0}\right]\left(\partial\overline{U}^\delta/\partial\delta\right)(x)l_k^{-t}\,\mathrm{errfn}\left(\frac{y+a_k^-t}{\sqrt{4\beta_k^-|y/a_k^-|}}\right), \tag{4.265}$$

whereas term (4.264) gives a contribution absorbable in R (resp. R_x, R_y).

Finally, observing that

$$[c_{k,-}^{j,0}]l_k^{-t}\,\mathrm{errfn}\left(\frac{y-a_k^- t}{\sqrt{4\beta_k^- t}}\right) \tag{4.266}$$

is time-exponentially small for $t \geqslant 1$, since $a_k^- > 0$, $y < 0$, and $(\partial/\partial\delta_j)(\bar{u}^\delta, \bar{v}^\delta) \leqslant C\mathrm{e}^{-\theta|x|}$, $\theta > 0$, we may subtract and add this term to (4.265) to obtain a total of $E(x,t;y)$ plus terms absorbable in R (resp. R_x, R_y).

S_λ term. Next, consider the second-order term

$$\frac{1}{2\pi\mathrm{i}}\oint_{\widetilde{\Gamma}} \mathrm{e}^{\lambda t} S_\lambda(x,y)\,\mathrm{d}\lambda, \tag{4.267}$$

which, by (4.130), is given by

$$\sum_{a_k^->0} r_k^- l_k^{-t}\alpha_k(x,t;y) + \sum_{a_k^->0,a_j^-<0} [c_{k,-}^{j,-}] r_j^- l_k^{-t}\alpha_{jk}(x,t;y), \tag{4.268}$$

where

$$\alpha_k(x,t;y) := \frac{1}{2\pi\mathrm{i}}\oint_{\widetilde{\Gamma}} \mathrm{e}^{\lambda t}\mathrm{e}^{(-\lambda/a_k^- + \lambda^2\beta_k^-/a_k^{-3})(x-y)}\,\mathrm{d}\lambda \tag{4.269}$$

and

$$\alpha_{jk}(x,t;y) := \frac{1}{2\pi\mathrm{i}}\oint_{\widetilde{\Gamma}} \mathrm{e}^{\lambda t}\mathrm{e}^{(-\lambda/a_j^- + \lambda^2\beta_j^-/a_j^{-3})x + (\lambda/a_k^- - \lambda^2\beta_k^-/a_k^{-3})y}\,\mathrm{d}\lambda. \tag{4.270}$$

Similarly as in the treatment of the E_λ term, just above, by deforming the contour $\widetilde{\Gamma}$ to

$$\Gamma' := [-\eta_1 - \mathrm{i}r/2, -\mathrm{i}r/2] \cup [-\mathrm{i}r/2, +\mathrm{i}r/2] \cup [+\mathrm{i}r/2, -\eta_1 + \mathrm{i}r/2], \tag{4.271}$$

these may be transformed modulo time-exponentially decaying terms to the elementary Fourier integrals

$$\begin{aligned} &\frac{1}{2\pi}\mathrm{P.V.}\int_{-\infty}^{+\infty} \mathrm{e}^{\mathrm{i}\xi(t-(x-y)/a_k^-)}\mathrm{e}^{\xi^2(-\beta_k^-/a_k^{-3})(x-y)}\,\mathrm{d}\xi \\ &\quad = \left(4\pi\beta_k^- t\right)^{-1/2}\mathrm{e}^{-(x-y-a_k^- t)^2/4\beta_k^- t} \end{aligned} \tag{4.272}$$

and

$$\begin{aligned} &\frac{1}{2\pi}\mathrm{P.V.}\int_{-\infty}^{+\infty} \mathrm{e}^{\mathrm{i}\xi((t-|y/a_k^-|)-x/a_j^-)}\mathrm{e}^{-\xi^2((\beta_k^-/a_k^{-3})|y| + (\beta_j^-/a_j^{-3})|x|)}\,\mathrm{d}\xi \\ &\quad = \left(4\pi\bar{\beta}_{jk}^- t\right)^{-1/2}\mathrm{e}^{-(x-z_{jk}^-)^2/4\bar{\beta}_{jk}^- t}, \end{aligned} \tag{4.273}$$

respectively, where $\bar\beta^{\pm}_{jk}$ and $z^{\pm}_{jk}$ are as defined in (4.29) and (4.28). These correspond to the first and third terms in expansion (4.25), the latter of which has an additional factor $\mathrm{e}^{-x}/(\mathrm{e}^{x}+\mathrm{e}^{-x})$. Noting that the second and fourth terms of (4.25) are time-exponentially small for $t \geqslant 1$, $y \leqslant x \leqslant 0$, and that

$$\begin{aligned}&\left|\left(4\pi\bar\beta^{-}_{jk}t\right)^{-1/2}\mathrm{e}^{-(x-z^{-}_{jk})^2/4\bar\beta^{-}_{jk}t}\left(1-e^{-x}/(e^{x}+e^{-x})\right)\right|\\&\quad\leqslant\left|\left(4\pi\bar\beta^{-}_{jk}t\right)^{-1/2}\mathrm{e}^{-(x-z^{-}_{jk})^2/4\bar\beta^{-}_{jk}t}\mathrm{e}^{-\theta|x|}\right|\end{aligned}$$

for some $\theta > 0$, so is absorbable in error term R, we find that the total contribution of this term, modulo terms absorbable in R, is S.

R_λ term. Finally, we briefly discuss the estimation of error term

$$\frac{1}{2\pi\mathrm{i}}\oint_{\Gamma}\mathrm{e}^{\lambda t}R_\lambda(x,y)\,\mathrm{d}\lambda, \tag{4.274}$$

which decomposes into the sum of integrals involving the various terms of R^E_λ and R^S_λ given in (4.132) and (4.134). Since each of these are separately analytic, they may be split up and estimated sharply via the *Riemann saddlepoint method* (method of steepest descent), as described at great length in [ZH,HoZ.4]. That is, for each summand $\alpha_\lambda(x,y)\sim \mathrm{e}^{\beta(\lambda)x+\gamma(\lambda)y}$ in R_λ we deform the contour $\tilde\Gamma$ to a new, contour in Λ that is a mini–max contour for the modulus

$$m_\alpha(x,y,\lambda):=\left|\mathrm{e}^{\lambda t}\alpha_\lambda(x,y)\right|=\mathrm{e}^{\mathrm{Re}\,\lambda t+\mathrm{Re}\,\beta x+\mathrm{Re}\,\gamma y},$$

passing through an appropriate saddlepoint/critical point of $m_\alpha(x,y,\cdot)$: necessarily lying on the real axis, by the underlying complex symmetry resulting from reality of operator L.

Since terms of each type appearing in R have been sharply estimated in [ZH], we shall omit the details, only describing two sample calculations to illustrate the method:

EXAMPLE 4.42 ($\mathrm{e}^{-\theta|x-y|}$). Contour integrals of form

$$\frac{1}{2\pi\mathrm{i}}\oint_{\tilde\Gamma}\mathrm{e}^{\lambda t}\mathrm{e}^{-\theta|x-y|}\,\mathrm{d}\lambda, \tag{4.275}$$

arising through the pairing of fast forward and fast dual modes, may be deformed to

$$\frac{1}{2\pi\mathrm{i}}\int_{-\eta_1-\mathrm{i}r/2}^{-\eta_1+\mathrm{i}r/2}\mathrm{e}^{\lambda t}\mathrm{e}^{-\theta|x-y|}\,\mathrm{d}\lambda \tag{4.276}$$

and estimated as $\mathcal{O}(\mathrm{e}^{-\eta_1 t-\theta|x-y|})$, a negligible, time-exponentially decaying contribution.

EXAMPLE 4.43 ($\lambda^r\mathrm{e}^{(-\lambda/a^-_k+\lambda^2\beta^-_k/a^{-3}_k)(x-y)}$). Contour integrals of form

$$\frac{1}{2\pi\mathrm{i}}\oint_{\tilde\Gamma}\mathrm{e}^{\lambda t}\lambda^r\mathrm{e}^{(-\lambda/a^-_k+\lambda^2\beta^-_k/a^{-3}_k)(x-y)}\,\mathrm{d}\lambda, \tag{4.277}$$

$a_k^- > 0$, arising through the pairing of slow forward and slow dual modes, may be deformed to contour

$$\begin{aligned} \Gamma' &:= [-\eta_1 - \mathrm{i}r/2, \eta_* - \mathrm{i}r/2] \cup [\eta_* - \mathrm{i}r/2, \eta_* + \mathrm{i}r/2] \\ &\quad \cup [\eta_* + \mathrm{i}r/2, -\eta_1 + \mathrm{i}r/2], \end{aligned} \tag{4.278}$$

where saddlepoint η_* is defined as

$$\eta_*(x, y, t) := \begin{cases} \bar{\alpha}/p & \text{if } |\bar{\alpha}/p| \leqslant \epsilon, \\ \pm\epsilon & \text{if } \bar{\alpha}/p \gtrless \epsilon, \end{cases} \tag{4.279}$$

with

$$\bar{\alpha} := \frac{x - y - a_k^- t}{2t}, \qquad p := \frac{\beta_k^-(x-y)}{(a_k^-)^2 t} > 0, \tag{4.280}$$

and η_1, $\epsilon > 0$ are chosen sufficiently small with respect to r, to yield, modulo time-exponentially decaying terms, the estimate

$$\begin{aligned} &\mathrm{e}^{-(x-y-a_k^- t)^2/4\beta^- t} \int_{-\infty}^{+\infty} \mathcal{O}\big(|\eta_*|^r + |\xi|^r\big) \mathrm{e}^{-\theta\xi^2 t}\, \mathrm{d}\xi \\ &\quad = \mathcal{O}\big(t^{-(r+1)/2} \mathrm{e}^{-(x-y-a_k^- t)^2/(Mt)}\big) \end{aligned} \tag{4.281}$$

if $|\bar{\alpha}/p| \leqslant \epsilon$, and

$$\mathrm{e}^{-\epsilon t/M} \int_{-\infty}^{+\infty} \mathcal{O}\big(|\eta_*|^r + |\xi|^r\big) \mathrm{e}^{-\theta\xi^2 t}\, \mathrm{d}\xi = \mathcal{O}\big(t^{-(r+1)/2} \mathrm{e}^{-\eta t}\big) \tag{4.282}$$

if $|\bar{\alpha}/p| \geqslant \epsilon$, $\theta > 0$. In either case, the main contribution lies along the central portion $[\eta_* - \mathrm{i}r/2, \eta_* + \mathrm{i}r/2]$ of contour Γ'.

To see this, note that $|x - y|$ and t are comparable when $|\bar{\alpha}/p|$ is bounded, whence the evident spatial decay $\mathrm{e}^{-\theta\xi^2|x-y|}$ of the integrand along contour $\lambda = \eta_* + \mathrm{i}\xi$ may be converted to the $\mathrm{e}^{-\xi^2 t}$ decay displayed in (4.281) and (4.282); likewise, temporal growth in $\mathrm{e}^{\lambda t}$ on the horizontal portions of the contour, of order $\leqslant \mathrm{e}^{(|\eta_1|+|\varepsilon|)t}$, is dominated by the factor $\mathrm{e}^{-\theta\xi^2} = \mathrm{e}^{-\theta r^2/4}$, provided $|\eta_1| + |\varepsilon|$ is sufficiently small with respect to r^2. For $|\bar{\alpha}/p|$ large, on the other hand, η_* is uniformly negative, and also $|x - y|$ is negligible with respect to t, whence the estimate (4.282) holds trivially.

We point out that η_* is easily determined, as the minimal point on the real axis of the quadratic function

$$f_{x,y,t}(\lambda) := \lambda t\big(-\lambda/a_k^- + \lambda^2\beta_k^-/a_k^{-3}\big)(x - y), \tag{4.283}$$

the argument of the integrand of (4.277).

Other terms may be treated similarly: All "constant–coefficient" terms $\phi_k\tilde{\phi}_k^*$ or $\psi_k\tilde{\psi}_k^*$ are of either the form treated in Example 4.42 (fast modes) or in Example 4.43 (slow modes). Scattering pairs involving slow forward and slow dual modes from different families (i.e., terms with coefficients M_{jk}, $d_{jk}^{\pm}$) may be treated similarly as in Example 4.43. Scattering pairs involving two fast modes are of the form already treated in Example 4.42, since both modes of a scattering pair are decaying. Scattering pairs involving one fast and one slow mode may be factored as the product of a term of the form treated in Example 4.43 and a term that is uniformly exponentially decaying in either x or y; factoring out the exponentially decay, we may treat such terms as in Example 4.43.

Scattering coefficients. The relation (4.30) may now be deduced, a posteriori, from conservation of mass in the linearized flow (2.1). For, by inspection, all terms save E and S in the decomposition (4.22) of G decay in L^1, whence these terms must, time-asymptotically, carry exactly the mass in U of the initial perturbation $\delta_y(x)I_n$, i.e.,

$$\lim_{t\to\infty}\int_{-\infty}^{+\infty}\big(E(x,t;y)+S(x,t;y)\big)\,\mathrm{d}x = I_n. \tag{4.284}$$

Taking $y \leqslant 0$ for definiteness, and right-multiplying both sides of (4.284) by r_k^-, we thus obtain the result from definitions (4.24) and (4.25).

Moreover, rewriting (4.30) as

$$\left(r_1^-,\dots,r_{n-i_-}^-,r_{i_+}^+,\dots,r_n^+,m_1,\dots,m_\ell\right)\begin{pmatrix}[c_{k,-}^{1,-}]\\ \vdots\\ [c_{k,-}^{n-i_-,-}]\\ [c_{k,-}^{i_+,+}]\\ \vdots\\ [c_{k,-}^{n,+}]\\ [c_{k,-}^{1,0}]\\ \vdots\\ [c_{k,-}^{\ell,0}]\end{pmatrix} = r_k^-, \tag{4.285}$$

where

$$m_j := \int_{-\infty}^{+\infty}(\partial/\partial\delta_j)\overline{U}^\delta(x)\,\mathrm{d}x, \tag{4.286}$$

we find that, under assumption $(\mathcal{D})$, it uniquely determines the scattering coefficients $[c_{k,-}^{j,\pm}]$. For, the determinant

$$\det\left(r_1^-,\dots,r_{n-i_-}^-,r_{i_+}^+,\dots,r_n^+,m_1,\dots,m_\ell\right) \tag{4.287}$$

of the matrix on the right-hand side of (4.285) is exactly the inviscid stability coefficient Δ defined in [ZS,Z.3], hence is nonvanishing by the equivalence of ($\mathcal{D}2$) and (D2) (recall discussion of Section 1.2, just below Theorem 4.8). Note that (4.287) reduces to the inviscid stability determinant (1.36) in the Lax case, $\ell = 1$, with $\overline{U}^{\delta}(x)$ parametrized as $\overline{U}^{\delta}(x) := \overline{U}(x - \delta_1 x)$, for which m_1 reduces to $[U]$.

Relation (4.31) follows from the observation that

$$P_\lambda(x, y) := \mathrm{Res}_{\lambda=0}\, G_\lambda(x, y) = \sum_{j=1}^{\ell} (\partial/\partial\delta_j)\overline{U}^{\delta}(x)\pi_j,$$

with π_j defined by the first, or second expressions appearing in (4.31), according as y is less than or equal to/greater than or equal to zero. For example, the extended spectral theory of Section 4.3.3 then implies that π_j are the left effective eigenfunctions associated with right eigenfunctions $(\partial/\partial\delta_j)\overline{U}^{\delta}$. Alternatively, (4.31) may be deduced by linear-algebraic manipulation directly from (4.285) and its counterpart for $y \geqslant 0$ (the same identity with $k, -$ everywhere replaced by $k, +$). □

REMARK 4.44 (The undercompressive case). In the undercompressive case, the result of Lemma 4.28 is false, and consequently the estimates of Lemma 4.31 do not hold. This fact has the implication that shock dynamics are not governed solely by conservation of mass, as in the Lax or overcompressive case, but by more complicated dynamics of front interaction; for related discussion, see [LZ.2,Z.6]. At the level of Proposition 4.22, it means that the simple representations of E_λ and S_λ in terms of slow dual modes alone (corresponding to characteristics that are incoming to the shock) are no longer valid in the undercompressive case, and there appear new terms involving rapidly decaying dual modes $\sim \mathrm{e}^{-\theta|y|}$ related to inner layer dynamics. Though precise estimates can nonetheless be carried out, we have not found a similarly compact representation of the resulting bounds as that of the Lax/overcompressive case, and so we shall not state them here. We mention only that this rapid variation in the y-coordinate precludes the L^p stability arguments used here and in [Z.6], requiring instead detailed pointwise bounds as in [HZ.2]. See [LZ.1,LZ.2,ZH,Z.6] for further discussion of this interesting case.

4.3.5. *Inner layer dynamics.* Similarly as in [ZH], we now investigate dynamics of the inner shock layer, in the case that ($\mathcal{D}$) does not necessarily hold, in the process establishing the necessity of ($\mathcal{D}$) for linearized orbital stability.

PROPOSITION 4.45. *In dimension $d = 1$, given* (A1)–(A3) *and* (H0)–(H3), *there exists $\eta > 0$ such that, for x, y restricted to any bounded set, and t sufficiently large,*

$$G(x, y; t) = \sum_{\lambda\in\sigma'_p(L)\cap\{\mathrm{Re}(\lambda)\geqslant 0\}} \mathrm{e}^{\lambda t} \sum_{k\geqslant 0} t^k (L - \lambda I)^k P_\lambda(x, y) + \mathcal{O}(\mathrm{e}^{-\eta t}), \qquad (4.288)$$

where $P_\lambda(x, y)$ is the effective projection kernel described in Definition 4.34, *and $\sigma'_-(L)$ the effective point spectrum.*

PROOF. Similarly as in the proof of Proposition 4.39, decompose G into terms I_a, I_b, I_c, II_a and II_b of (4.245). Then, the same argument (Case II: $|x-y|/t$ bounded) yields that $I_a = H(x,t;y)$, while I_b and II_b are time-exponentially small. Observing that $H = 0$ for x, y bounded, and t sufficiently large, we have reduced the problem to the study of

$$II_a := \frac{1}{2\pi \mathrm{i}} \oint_\Gamma \mathrm{e}^{\lambda t} G_\lambda(x, y)\, \mathrm{d}\lambda, \tag{4.289}$$

where, recall,

$$\Gamma := [-\eta_1 - \mathrm{i}R, \eta - \mathrm{i}R] \cup [\eta - \mathrm{i}R, \eta + \mathrm{i}R] \cup [\eta + \mathrm{i}R, -\eta_1 + \mathrm{i}R]. \tag{4.290}$$

Note that the high-frequency bounds of Proposition 4.33 imply that L has no point spectrum in Ω outside of the rectangle $\mathcal{R}$ enclosed by $\Gamma \cup [-\eta_1 - \mathrm{i}R, -\eta + \mathrm{i}R]$. Choosing η_1 sufficiently small, therefore, we may ensure that no effective point spectrum lies within the strip $-\eta_1 \leqslant \operatorname{Re}\lambda < 0$, by compactness of $\mathcal{R}$ together with the fact that effective eigenvalues are isolated from one another, as zeroes of the analytic Evans function.

Recall that G_λ is meromorphic on Ω. Thus, we may express (4.289), using Cauchy's theorem, as

$$\frac{1}{2\pi \mathrm{i}} \int_{-\eta_1 - \mathrm{i}R}^{-\eta_1 + \mathrm{i}R} \mathrm{e}^{\lambda t} G_\lambda(x, y)\, \mathrm{d}\lambda + \operatorname{Res}_{\lambda \in \mathcal{R}} \mathrm{e}^{\lambda t} G_\lambda(x, t; y). \tag{4.291}$$

By Definition 4.34 and Proposition 4.36,

$$\begin{aligned}
&\operatorname{Res}_{\lambda \in \{\operatorname{Re}\lambda \geqslant 0\}} \mathrm{e}^{\lambda t} G_\lambda(x, y) \\
&\quad = \sum_{\lambda_0 \in \sigma'_p(L) \cap \{\operatorname{Re}\lambda \geqslant 0\}} \mathrm{e}^{\lambda_0 t} \operatorname{Res}_{\lambda_0} \mathrm{e}^{(\lambda - \lambda_0) t} G_\lambda(x, y) \\
&\quad = \sum_{\lambda_0 \in \sigma'_p(L) \cap \{\operatorname{Re}\lambda \geqslant 0\}} \mathrm{e}^{\lambda_0 t} \sum_{k \geqslant 0} \big(t^k / k!\big) \operatorname{Res}_{\lambda_0} (\lambda - \lambda_0)^k G_\lambda(x, y) \\
&\quad = \sum_{\lambda_0 \in \sigma'_p(L) \cap \{\operatorname{Re}\lambda \geqslant 0\}} \mathrm{e}^{\lambda_0 t} \sum_{k \geqslant 0} \big(t^k / k!\big) Q_{\lambda_0, k}(x, y) \\
&\quad = \sum_{\lambda_0 \in \sigma'_p(L) \cap \{\operatorname{Re}\lambda \geqslant 0\}} \mathrm{e}^{\lambda_0 t} \sum_{k \geqslant 0} \big(t^k / k!\big) (L - \lambda_0 I)^k P_{\lambda_0}(x, y).
\end{aligned} \tag{4.292}$$

On the other hand, for x, y bounded and t sufficiently large, t dominates $|x|$ and $|y|$ and we obtain from Propositions 2.23 and 4.22 that

$$\big|G_\lambda(x, t; y)\big| \leqslant C$$

for all $\lambda \in \Omega$, for η_1 sufficiently small, hence

$$\frac{1}{2\pi \mathrm{i}} \int_{-\eta_1 - \mathrm{i}R}^{-\eta_1 + \mathrm{i}R} \mathrm{e}^{\lambda t} G_\lambda(x, y)\, \mathrm{d}\lambda \leqslant 2CR\mathrm{e}^{-\eta_1 t}. \tag{4.293}$$

Combining (4.291), (4.292), and (4.293), we obtain the result. □

COROLLARY 4.46. *Let dimension* $d = 1$, *and assume* (A1)–(A3) *and* (H0)–(H3). *Then* $(\mathcal{D})$ *is necessary for linearized orbital stability with respect to compactly supported initial data, as measured in any* L^p *norm*, $1 \leqslant p \leqslant \infty$.

PROOF. From (4.288), we find that $\overline{U}(\cdot)$ is linearly orbitally stable only if $P_\lambda = 0$ for all $\lambda \in \{\operatorname{Re}\lambda \geqslant 0\} \setminus \{0\}$ and $\operatorname{Range} \mathcal{P}_0 = \operatorname{Span}\{(\partial/\partial\delta_j)\overline{U}^\delta\}$. By Propositions 4.36 and 4.38, this is equivalent to $(\mathcal{D})$. □

This completes our treatment of the one-dimensional case.

5. Multidimensional stability

We now focus on the multidimensional case, establishing Results 4 and 5 of the Introduction.

5.1. *Necessary conditions*

Define the *reduced Evans function* as

$$\overline{\Delta}(\tilde{\xi}, \lambda) := \lim_{\rho \to 0} \rho^{-\ell} D(\rho\tilde{\xi}, \rho\lambda). \tag{5.1}$$

By the results of the previous section, the limit $\overline{\Delta}$ exists and is analytic, with

$$\overline{\Delta} = \gamma \Delta, \tag{5.2}$$

for shocks of pure type (indeed, such a limit exists for all types). Evidently, $\overline{\Delta}(\cdot, \cdot)$ is homogeneous, degree ℓ.[29]

Recall that the restriction

$$D(\lambda) := D(0, \lambda) \tag{5.3}$$

is the one-dimensional Evans function considered in [GZ,ZH].

[29] Here, and elsewhere, homogeneity is with respect to the positive reals, as in most cases should be clear from the context. Recall that Δ (and thus $\overline{\Delta}$) is only defined for real $\tilde{\xi}$, and $\operatorname{Re}\lambda \geqslant 0$.

LEMMA 5.1 ([ZS]). *Let* $\overline{\Delta}(0,1) \neq 0$. *Then, near any root* $(\tilde{\xi}_0, \lambda_0)$ *of* $\overline{\Delta}(\cdot,\cdot)$, *there exists a continuous branch* $\lambda(\tilde{\xi})$, *homogeneous degree one, of solutions of*

$$\overline{\Delta}\big(\tilde{\xi}, \lambda(\tilde{\xi})\big) \equiv 0 \tag{5.4}$$

defined in a neighborhood V *of* $\tilde{\xi}_0$, *with* $\lambda(\tilde{\xi}_0) = \lambda_0$. *Likewise, there exists a continuous branch* $\lambda_*(\tilde{\xi})$ *of roots of*

$$D\big(\tilde{\xi}, \lambda_*(\tilde{\xi})\big) \equiv 0, \tag{5.5}$$

defined on a conical neighborhood $V_{\rho_0} := \{\tilde{\xi} = \rho\tilde{\eta}\colon\ \tilde{\eta} \in V,\ 0 < \rho < \rho_0\}$, $\rho_0 > 0$ *sufficiently small, "tangent" to* $\lambda(\cdot)$ *in the sense that*

$$\big|\lambda_*(\tilde{\xi}) - \lambda(\tilde{\xi})\big| = o\big(|\tilde{\xi}|\big) \tag{5.6}$$

as $|\tilde{\xi}| \to 0$, *for* $\tilde{\xi} \in V_{\rho_0}$.

PROOF. The statement (5.4) follows by Rouche's theorem, provided $\overline{\Delta}(\tilde{\xi}_0, \cdot) \not\equiv 0$, since $\overline{\Delta}(\tilde{\xi}_0, \cdot)$ are a continuous family of analytic functions. For, otherwise, restricting λ to the positive real axis, we have by homogeneity that

$$0 = \lim_{\lambda\to+\infty} \overline{\Delta}\big(\tilde{\xi}_0, \lambda\big) = \lim_{\lambda\to+\infty} \overline{\Delta}\big(\tilde{\xi}_0/\lambda, 1\big) = \overline{\Delta}(0,1),$$

in contradiction with the hypothesis. Clearly we can further choose $\lambda(\cdot)$ homogeneous degree one, by homogeneity of $\overline{\Delta}$. Similar considerations yield existence of a branch of roots $\bar{\lambda}(\tilde{\xi}, \rho)$ of the family of analytic functions

$$g^{\tilde{\xi},\rho}(\lambda) := \rho^{-\ell} D\big(\rho\tilde{\xi}, \rho\lambda\big) \tag{5.7}$$

for ρ sufficiently small, since $g^{\tilde{\xi},0} = \overline{\Delta}(\tilde{\xi}, \cdot)$. Setting $\lambda_*(\tilde{\xi}) := |\tilde{\xi}|\bar{\lambda}(\tilde{\xi}/|\tilde{\xi}|, |\tilde{\xi}|)$, we have

$$D\big(\tilde{\xi}, \lambda_*(\tilde{\xi})\big) = \big|\tilde{\xi}\big|^{\ell} g^{\tilde{\xi}/|\tilde{\xi}|,|\tilde{\xi}|}(\lambda_*) \equiv 0,$$

as claimed. "Tangency", in the sense of (5.6), follows by continuity of $\bar{\lambda}$ at $\rho = 0$, the definition of λ_*, and the fact that $\lambda(\tilde{\xi}) = |\tilde{\xi}|\lambda(\tilde{\xi}/|\tilde{\xi}|)$, by homogeneity of $\overline{\Delta}$. □

COROLLARY 5.2 [ZS]. *Given one-dimensional inviscid stability,* $\overline{\Delta}(0,1) \neq 0$, *weak refined dynamical stability (Definition* 1.21) *is necessary for viscous multidimensional weak spectral stability, and thus for multidimensional linearized viscous stability with respect to test function* (C_0^∞) *initial data.*

PROOF. By the discussion just preceding Section 2.1, if is sufficient to prove the first assertion, i.e., that failure of weak refined dynamical stability implies existence of a zero

$D(\xi, \lambda) = 0$ for $\xi \in \mathbb{R}^{d-1}$, $\operatorname{Re} \lambda > 0$. Failure of weak inviscid stability, or $\Delta(\tilde{\xi}, \lambda) = 0$ for $\tilde{\xi} \in \mathbb{R}^{d-1}$, $\operatorname{Re} \lambda > 0$, implies immediately the existence of such a root, by tangency of the zero-sets of D and Δ at the origin, Lemma 5.1. Thus, it remains to consider the case that weak inviscid stability holds, but there exists a root $D(\xi, i\tau)$ for ξ, τ real, at which Δ is analytic, $\Delta_\lambda \neq 0$, and $\beta(\xi, i\tau) < 0$, where β is defined as in (4.17).

Recalling that $D(\rho\tilde{\xi}, \rho\lambda)$ vanishes to order ℓ in ρ at $\rho = 0$, we find by L'Hopital's rule that

$$(\partial/\partial\rho)^{\ell+1} D(\rho\tilde{\xi}, \rho\lambda)\big|_{\rho=0,\lambda=i\tau} = (1/\ell!)(\partial/\partial\rho) g^{\tilde{\xi}, i\tau}(0)$$

and

$$(\partial/\partial\lambda) D(\rho\tilde{\xi}, \rho\lambda)\big|_{\rho=0,\lambda=i\tau} = (1/\ell!)(\partial/\partial\lambda) g^{\tilde{\xi}, i\tau}(0),$$

where $g^{\tilde{\xi},\rho}(\lambda) := \rho^{-\ell} D(\rho\tilde{\xi}, \rho\lambda)$ as in (5.7), whence

$$\beta = \frac{(\partial/\partial\rho) g^{\tilde{\xi},\lambda}(0)}{(\partial/\partial\lambda) g^{\tilde{\xi},\lambda}(0)}, \tag{5.8}$$

with $(\partial/\partial\lambda\,) g^{\tilde{\xi},\lambda}(0) \neq 0$.

By the (analytic) Implicit Function Theorem, therefore, $\lambda(\tilde{\xi}, \rho)$ is analytic in $\tilde{\xi}, \rho$ at $\rho = 0$, with

$$(\partial/\partial\rho)\lambda(\tilde{\xi}, 0) = -\beta, \tag{5.9}$$

where $\lambda(\tilde{\xi}, \rho)$ as in the proof of Lemma 5.1 is defined implicitly by $g^{\tilde{\xi},\lambda}(\rho) = 0$, $\lambda(\tilde{\xi}, 0) := i\tau$. We thus have, to first order,

$$\lambda(\tilde{\xi}, \rho) = i\tau - \beta\rho + \mathcal{O}(\rho^2). \tag{5.10}$$

Recalling the definition $\lambda_*(\tilde{\xi}) := |\tilde{\xi}|\bar{\lambda}(\tilde{\xi}/|\tilde{\xi}|, |\tilde{\xi}|)$, we have then, to second order, the series expansion

$$\lambda_*(\rho\tilde{\xi}) = i\rho\tau - \beta\rho^2 + \mathcal{O}(\rho^3), \tag{5.11}$$

where $\lambda_*(\tilde{\xi})$ is the root of $D(\tilde{\xi}, \lambda) = 0$ defined in Lemma 5.1. It follows that there exist unstable roots of D for small $\rho > 0$ unless $\operatorname{Re} \beta \geqslant 0$. □

REMARK 5.3. When $\Delta_\lambda \neq 0$, $\lambda = i\tau$ is a simple root of $\Delta(\tilde{\xi}, \cdot)$ and, likewise, $\lambda_*(\tilde{\xi})$ defined in (5.11) represents an isolated branch of the zeroes of $D(\tilde{\xi}, \cdot)$. The quantity $-\beta$ gives the curvature of the zero level set of D tangent to the level set $\{\rho\tilde{\xi}, \rho i\tau\}$ for Δ. The value β also represents the effective diffusion coefficient for the transverse traveling waves associated with this frequency (see [Z.3], Section 3.3).

5.2. *Sufficient conditions for stability*

Result 5 of the Introduction is subsumed in the following two theorems, to be established throughout the remainder of the section.

THEOREM 5.4 (Linearized stability). *Under assumptions* (A1)–(A3) *and* (H0)–(H5), *structural and strong refined dynamical stability together with strong spectral stability are sufficient for linearized viscous stability in* L^2 *with respect to initial perturbations* $U_0 \in L^1 \cap L^2$, *or in* L^p, $p \geqslant 2$, *with respect to initial perturbations* $U_0 \in L^1 \cap H^{[(d-1)/2]+2}$, *for all dimensions* $d \geqslant 2$, *with rate of decay*

$$\begin{aligned} \|U(t)\|_{L^2} &\leqslant C(1+t)^{-\frac{d-1}{4}+\beta\epsilon} |U_0|_{L^1 \cap L^2}, \\ \|U(t)\|_{L^p} &\leqslant C(1+t)^{-\frac{d-1}{2}(1-1/p)+\beta\epsilon} |U_0|_{L^1 \cap H^{[(d-1)/2]+2}} \end{aligned} \tag{5.12}$$

for all $t \geqslant 0$, *where*

$$\beta = \begin{cases} 0 & \text{for uniformly inviscid stable (UIS) shocks,} \\ 1 & \text{for weakly inviscid stable (WIS) shocks,} \end{cases} \tag{5.13}$$

$\epsilon > 0$ *is arbitrary, and* $C = C(\epsilon)$ *is independent of* p.

THEOREM 5.5 (Nonlinear stability). *Under assumptions* (A1)–(A3) *and* (H0)–(H5), *structural and strong refined dynamical stability together with strong spectral stability are sufficient for nonlinear viscous stability in* $L^p \cap H^{s-1}$, $p \geqslant 2$ *with respect to initial perturbations* $U_0 := \widetilde{U}_0 - \overline{U}$ *that are sufficiently small in* $L^1 \cap H^{s-1}$, *where* $s \geqslant s(d)$ *is as defined in* (H0), *Section* 1.1, *for dimensions* $d \geqslant 2$ *in the case of a uniformly inviscid stable Lax or overcompressive shock,* $d \geqslant 3$ *in the case of a weakly inviscid stable Lax or overcompressive shock, and* $d \geqslant 4$ *in the case of an undercompressive shock, with rate of decay*

$$\begin{aligned} \|\widetilde{U}(\cdot,t) - \overline{U}\|_{L^p} &\leqslant C(1+t)^{-\frac{d-1}{2}(1-1/p)+\beta\epsilon} \|U_0\|_{L^1 \cap H^{s-1}}, \\ \|\widetilde{U}(\cdot,t) - \overline{U}\|_{H^{s-1}} &\leqslant C(1+t)^{-\frac{d-1}{4}+\beta\epsilon} \|U_0\|_{L^1 \cap H^{s-1}} \end{aligned} \tag{5.14}$$

for all $t \geqslant 0$, *where* $\epsilon > 0$ *is arbitrary,* $C = C(\epsilon)$ *and* $\beta = 1$ *if dimension* $d = 2$ *and* $p > 2$, *or in the weakly inviscid stable (WIS) case, and otherwise* $\beta = 0$. *(In particular,* $\beta = 0$ *for uniformly inviscid stable Lax or overcompressive shocks, in dimension* $d \geqslant 3$ *or in dimension* $d = 2$ *with* $p = 2$.*)*

REMARKS 5.6. 1. Stability here is in the usual sense of asymptotic stability, and not only bounded or orbital stability as in the one-dimensional case, with rate of decay in dimension d equal to that of a $(d-1)$-dimensional heat kernel, corresponding to transverse diffusion along the front. This reflects the fact that the class of L^1 perturbations is much more restrictive in multiple than in single dimensions; in particular, conservation of mass precludes convergence of an L^1 perturbation of a planar shock front to any translate of the shock front other than the initial front itself [Go.3]. Recall that our necessary stability results concerned the still more restrictive class of test function initial data, so that the problems are consistent.

2. The diffusive decay rate given here is sharp, neglecting error term β, for the weakly inviscid stable case; indeed, the basic iteration scheme [Z.3] was motivated by the scalar analysis of [Go.3] (recall: scalar shocks are always weakly inviscid stable). However, uniformly inviscid stable shocks, for which transverse front disturbances are strongly damped, are expected to decay at the faster rate governing far-field behavior, of a d-dimensional heat kernel. That is, our analysis here as in [Z.3] is focused on the WIS regime, and the transition from viscous stability to instability; see again Remark 1.22.2.

5.2.1. *Linearized estimates.* Theorems 5.4 and 5.5 are obtained using the following $L^q \to L^p$ bounds on the linearized solution operator, analogous to the one-dimensional bounds described in Lemma 4.14 (proof deferred to Section 5.3). Note that we no longer require detailed Green distribution bounds as described in Proposition 4.10, nor, because of the more complicated geometry of characteristic surfaces in multidimensions, does there exist such a simple description of the propagation of solutions; see, e.g., [HoZ.1,HoZ.2] for further discussion in the constant-coefficient case.

PROPOSITION 5.7. *Let* (A1)–(A3) *and* (H0)–(H5) *hold, together with structural, strong refined dynamical, and strong spectral stability. Then, solution operator* $\mathcal{S}(t) := \mathrm{e}^{Lt}$ *of the linearized equations* (2.1) *may be decomposed into* $\mathcal{S} = \mathcal{S}_1 + \mathcal{S}_2$ *satisfying*

$$\left\| \mathcal{S}_1(t) \partial_x^\sigma F \right\|_{L^p(x)} \leqslant C(1+t)^{-\frac{d-1}{2}(1-1/p)-(1-\alpha)\frac{|\delta|}{2}+\beta\epsilon} \|F\|_{L^1(x)} \tag{5.15}$$

for $0 \leqslant |\delta| \leqslant 1$, $2 \leqslant p \leqslant \infty$ *and* $t \geqslant 0$, *where* $\alpha = 0$ *for Lax and overcompressive shocks and* $\alpha = 1$ *for undercompressive shocks,* $\beta = 0$ *for uniformly inviscid stable* (*UIS*) *shocks and* $\beta = 1$ *for weakly inviscid stable* (*WIS*) *shocks,* $\epsilon > 0$ *is arbitrary and* $C = C(d, \epsilon)$ *is independent of* p, *and, for* $0 \leqslant |\gamma_1| \leqslant 1$, $0 \leqslant |\tilde{\gamma}|$,

$$\left\| \partial_{x_1}^{\gamma_1} \partial_{\tilde{x}}^{\tilde{\gamma}} \mathcal{S}_2(t) F \right\|_{L^2} \leqslant C \mathrm{e}^{-\eta t} \|F\|_{H^{|\gamma_1|+|\tilde{\gamma}|}}. \tag{5.16}$$

REMARKS 5.8. 1. Operators $\mathcal{S}_1$ and $\mathcal{S}_2$ represent low- and high-frequency parts of $\mathcal{S}$. Though we do not state them, $\mathcal{S}_2$ satisfies refined bounds analogous to (4.32) and (4.33) in the one-dimensional case.

2. The time-asymptotic decay rate of solutions is determined by the bounds on $\mathcal{S}_1$, which may be recognized as those for a $(d-1)$-dimensional heat equation. In contrast to the situation for the heat equation, x-differentiation of solution $U = \mathcal{S}F$ does not improve the rate of decay for the variable-coefficient shock problem, due to commutator terms arising in the differentiated equations (that is, terms for which derivatives fall on coefficients), nor in general does differentiation with respect to source F. A single derivative of the source does improve the rate of decay in the Lax or overcompressive case, but further derivatives would not (and are not needed in our later argument). This is a slight difference from the constant-coefficient case considered in, e.g., [Kaw,HoZ.1,HoZ.2], and requires

some modification in the arguments. However, in spirit our nonlinear stability analysis in dimension d follows that of [Kaw,HoZ.1,HoZ.2] in one lower dimension, $d-1$.

PROOF OF THEOREM 5.4. Applying (5.15) and (5.16), we have

$$\begin{aligned}|U(t)|_{L^2} &\leqslant |S_1(t)U_0|_{L^2} + |S_2(t)U_0|_{L^2}\\ &\leqslant C(1+t)^{-\frac{d-1}{4}+\beta\epsilon}|U_0|_1 + Ce^{-\eta t}|U_0|_{L^2}\\ &\leqslant C(1+t)^{-\frac{d-1}{4}+\beta\epsilon}|U_0|_{L^1\cap L^2}.\end{aligned}$$

Likewise, we have, using the Sobolev embedding

$$|f|_{L^p} \leqslant |f|_{L^2\cap L^\infty} \leqslant |f|_{H^{[(d-1)/2]+1}(\tilde{x};H^1(x_1))}$$

for $2 \leqslant p \leqslant \infty$, that

$$\begin{aligned}|U(t)|_{L^p} &\leqslant |S_1(t)U_0|_{L^p} + |S_2(t)U_0|_{H^{[(d-1)/2]+1}(\tilde{x};H^1(x_1))}\\ &\leqslant C(1+t)^{-\frac{d-1}{2}(1-1/p)+\beta\epsilon}|U_0|_1 + Ce^{-\eta t}|U_0|_{H^{[d/2]+2}}\\ &\leqslant C(1+t)^{-\frac{d-1}{2}(1-1/p)+\beta\epsilon}|U_0|_{L^1\cap H^{[d/2]+2}}.\end{aligned}$$

□

5.2.2. *Auxiliary energy estimate.* In our analysis of nonlinear stability, we shall make use also of the following auxiliary energy estimate, analogous to that of the one-dimensional case.

PROPOSITION 5.9. *Under the hypotheses of Theorem* 5.5, *suppose that, for* $0 \leqslant t \leqslant T$, *the* H^{s-1} *norm of the perturbation* $U := \widetilde{U} - \overline{U} = (u^I, u^{II})^t$ *remains bounded by a sufficiently small constant, where* $\widetilde{U} = \overline{U} + U$ *denotes a solution of* (1.1). *Then, for all* $0 \leqslant t \leqslant T$,

$$|U(t)|^2_{H^{s-1}} \leqslant C|U(0)|^2_{H^{s-1}}e^{-\theta t} + C\int_0^t e^{-\theta_2(t-\tau)}|U|^2_{L^2}(\tau)\,d\tau. \tag{5.17}$$

PROOF. The proof follows almost exactly the argument of the one-dimensional case; see Remark 4.17. Indeed, it is somewhat simpler, since there are no front location terms $\dot{\delta}\widetilde{U}_x$, We therefore omit most details, pointing out only two small points in which the argument must be modified for the multidimensional case.

The first is the standard issue that "good" terms

$$-\sum_{jk}\langle\partial_x^\ell W_{x_j}, \widetilde{B}^{jk}\,\partial_x^\ell W_{x_k}\rangle_\alpha = -\sum_{jk}\langle\partial_x^\ell w^{II}_{x_j}, \tilde{b}^{jk}\,\partial_x^\ell w^{II}_{x_k}\rangle_\alpha$$

arising in the ℓth-order Friedrichs estimate must now be estimated indirectly, using Gärding inequality

$$\sum_{jk}\langle\partial_x^\ell w^{II}_{x_j}, \tilde{b}^{jk}\,\partial_x^\ell w^{II}_{x_k}\rangle_\alpha \geqslant \theta|\partial_x^{\ell+1}w^{II}|^2_\alpha - C|\partial_x^\ell w^{II}|_{L^2} \tag{5.18}$$

(a consequence of uniform ellipticity, (1.6)). Recall that $\langle f, g\rangle_\alpha := \langle f, \alpha g\rangle$, where α is a scalar weight function. Recall that $\mathcal{O}(|\partial_x^\ell|^2_{L^2})$ terms already appear in the ℓth-order estimate with arbitrarily large constant, and so the second, lower-order error term in (5.18) is harmless.

A second, related issue, is the treatment of Kawashima estimates. In the case (as in gas dynamics and MHD) that there exists a linear (i.e., differential) compensating matrix $\widetilde{K}(\xi) := \sum_j \widetilde{K}^j\xi_j$ for $\widetilde{A}(\xi) := \sum_j \widetilde{A}^j\xi_j$ and $\widetilde{B}(\xi) := \sum_{jk} \widetilde{B}^{jk}\xi_j\xi_k$ in the sense that

$$\big(\widetilde{K}(\xi)\big(\widetilde{A}^0\big)^{-1}\widetilde{A}(\xi) + \widetilde{B}(\xi)\big) \geqslant \theta > 0, \tag{5.19}$$

the ℓth-order Kawashima estimate takes the form

$$(\mathrm{d}/\mathrm{d}t)\frac{1}{2}\Big\langle \partial_x^{\ell-1}W, -\sum_j \widetilde{K}\partial_{x_j}\partial_x^{\ell-1}W\Big\rangle_\alpha, \tag{5.20}$$

and yields a "good" term

$$\Big\langle \partial_x^{\ell-1}W, \sum_{jk} \widetilde{K}\big(\widetilde{A}^0\big)^{-1}\widetilde{A}^k\partial_{x_j}\partial_{x_k}\partial_x^{\ell-1}W\Big\rangle_\alpha,$$

which may be again estimated by an appropriate Gärding inequality as

$$\leqslant -\theta\big|\partial_x^\ell w^I\big|_\alpha^2 + C\big|\partial_x^\ell w^{II}\big|_\alpha^2 + C(C_*)\big|\partial_x^{\ell-1}W\big|_{L^2}^2$$

for some uniform θ, $C > 0$ and $C(C_*)$ depending on the details of weight α, and used as before to compensate the Friedrichs estimate near plus and minus spatial infinity. Recall that lower-order derivative terms are harmless for the Kawashima estimate, whether in $|w^I|_{H^{\ell-1}}$ or $|w^{II}|_{H^{\ell-1}}$.

Of course, the compensating matrix $\widetilde{K}(\xi)$ in general depends on ξ in pseudodifferential rather than differential fashion (that is, it is homogeneous degree one and not linear in ξ), and in this case the argument must be modified. A convenient method is to consider the variables $W_\alpha^\pm := \chi_\pm\alpha^{1/2}W$, where $\chi_\pm = \chi_\pm(x_1)$ are smooth, scalar cutoff functions supported on $x_1 \geqslant L$ and $x_1 \leqslant -L$, respectively. For L sufficiently large, the equations for $W_\alpha^\pm$ have coefficients that are constant up to an arbitrarily small error. Taking the Fourier transform now in x, we may perform an energy estimate in the frequency domain, treating coefficient errors and nonlinear terms alike as a small source, to obtain a similar α-weighted Kawashima estimate, this time explicitly localized near plus and minus spatial infinity (essentially, the pseudodifferential Gärding inequality in a simple case). We omit the details, which may be found in [Z.4].

Finally, we estimate nonlinear source terms more systematically using the fact that H^r is an algebra, $|fg|_{H^r} \leqslant |f|_{H^r}|g|_{H^r}$ for $r > d/2$, in particular for $r = s-1$ (Moser's inequality and Sobolev embedding; see, e.g., [M.1,M.4,Kaw,GMWZ.2,Z.4]. With these changes, the argument goes through as before, yielding the result. See [Z.4] for further details. □

5.2.3. *Nonlinear stability.* Using the linearized bounds of Section 5.2.1 together with the energy estimate of Section 5.2.2, it is now straightforward to establish $L^1 \cap H^{s-1} \to L^p \cap H^{s-1}$ asymptotic nonlinear stability, $p \geqslant 2$.

PROOF OF THEOREM 5.5. We present in detail the case of a uniformly inviscid stable Lax or overcompressive shock in dimension $d \geqslant 3$. Other cases follow similarly; see, e.g., [Z.3,Z.4].

H^{s-1} stability. We first carry out a self-contained stability analysis in H^{s-1}. Afterwards, we shall establish sharp L^∞ decay rates (and thus, by interpolation, L^p rates for $2 \leqslant p \leqslant \infty$) by a bootstrap argument. Defining

$$U := \widetilde{U} - \overline{U}, \tag{5.21}$$

and Taylor expanding as usual, we obtain the nonlinear perturbation equation

$$U_t - LU = \sum_j Q^j(U, \partial_x U)_{x_j}, \tag{5.22}$$

where

$$\begin{aligned} &Q^j(U, \partial_x U) = \mathcal{O}\big(|U|^2 + |U||\partial_x U|\big)\\ &\partial_{x_j} Q^j(U, \partial_x U) = \mathcal{O}\big(|U||\partial_x U| + |U|\big|\partial_x^2 U\big| + |\partial_x U|^2\big) \end{aligned} \tag{5.23}$$

so long as $|U|$ remains bounded by some fixed constant. Applying Duhamel's principle, and integrating by parts, we can thus express

$$U(x,t) = \mathcal{S}(t)U(0) + \int_0^t \mathcal{S}(t-s)\sum_j \partial_{x_j} Q^j(s)\,\mathrm{d}s. \tag{5.24}$$

Define now

$$\zeta(t) := \sup_{0 \leqslant s \leqslant t} \big\|U(s)\big\|_{L^2}(1+s)^{(d-1)/4}. \tag{5.25}$$

By standard, short-time existence theory (see, e.g., [Kaw] or [Z.4]) and the principle of continuation, there exists a solution $U \in H^{s-1}(x)$, $U_t \in H^{s-3}(x) \subset L^2(x)$ on the open time-interval for which $|U|_{H^{s-1}}$ remains bounded. On this interval, ζ is well defined and continuous.

Now, let $[0, T)$ be the maximal interval on which $|U|_{H^{s-1}(x)}$ remains strictly bounded by some fixed, sufficiently small constant $\delta > 0$. By Proposition 5.9,

$$\begin{aligned} \big|U(t)\big|_{H^{s-1}}^2 &\leqslant C\big|U(0)\big|_{H^{s-1}}^2 \mathrm{e}^{-\theta t} + C\int_0^t \mathrm{e}^{-\theta_2(t-\tau)}\big|U(\tau)\big|_{L^2}^2\,\mathrm{d}\tau\\ &\leqslant C_2\big(\big|U(0)\big|_{H^{s-1}}^2 + \zeta(t)^2\big)(1+\tau)^{-(d-1)/2}. \end{aligned} \tag{5.26}$$

Combining this with the bounds of Proposition 5.7 and $|Q|_{L^1\cap H^1} \leqslant |U|^2_{H^{s-1}}$, and recalling that $d \geqslant 3$ and $|U(0)|_{H^{s-1}}$ is small, we thus obtain

$$\begin{aligned}
\big|U(t)\big|_{L^2} &\leqslant \big|\mathcal{S}(t)U(0)\big|_{L^2} + \int_0^t \sum_j \big(\mathcal{S}_1(t-s)\partial_{x_j}\big) Q^j(s)\,\mathrm{d}s \\
&\quad + \int_0^t \mathcal{S}_2(t-s)\Big(\sum_j \partial_{x_j} Q^j(s)\Big)\,\mathrm{d}s \\
&\leqslant C(1+t)^{-d-1/4}\big|U(0)\big|_{L^1\cap L^2} \\
&\quad + \int_0^t (1+t-s)^{-(d-1)/4-1/2}\big|Q^j(s)\big|_{L^1}\,\mathrm{d}s \\
&\quad + \int_0^t \mathrm{e}^{-\theta(t-s)}\Big|\sum_j \partial_{x_j} Q^j(s)\Big|_{L^2}\,\mathrm{d}s \\
&\leqslant C(1+t)^{-(d-1)/4}\big|U(0)\big|_{L^1\cap L^2} \\
&\quad + \int_0^t (1+t-s)^{-(d-1)/4-1/2}\big|Q^j(s)\big|_{L^1\cap H^1}\,\mathrm{d}s \\
&\leqslant C(1+t)^{-(d-1)/4}\big|U(0)\big|_{L^1\cap L^2} \\
&\quad + C_2\big(\big|U(0)\big|^2_{H^{s-1}} + \zeta(t)^2\big) \\
&\quad \times \int_0^t (1+t-s)^{-(d-1)/4-1/2}(1+s)^{-(d-1)/2}\,\mathrm{d}s \\
&\leqslant C_3(1+t)^{-(d-1)/4}\big(\big|U(0)\big|_{L^1\cap H^{s-1}} + \zeta(t)^2\big),
\end{aligned} \tag{5.27}$$

or, dividing by $(1+t)^{-(d-1)/4}$,

$$\zeta(t) \leqslant C_2\big(\big|U(0)\big|_{L^1\cap H^{s-1}} + \zeta(t)^2\big). \tag{5.28}$$

Bound (5.28) together with continuity of ζ implies that

$$\zeta(t) \leqslant 2C_2\big|U(0)\big|_{L^1\cap H^{s-1}} \tag{5.29}$$

for $t \geqslant 0$, provided $|U(0)|_{L^1\cap H^{s-1}} < 1/4C_2^2$. With (5.26), this gives

$$\big|U(t)\big|_{H^{s-1}} \leqslant 4C_2\big|U(0)\big|_{L^1\cap H^{s-1}}$$

and so we find that the maximum time of existence T is in fact $+\infty$, and (5.29) holds globally in time. Definition (5.25) then yields

$$\begin{aligned}\|U(t)\|_{L^2} &\leqslant 2C_2\zeta(1+t)^{-(d-1)/4} \leqslant C_3\zeta(t)\\ &\leqslant 2C_2|U(0)|_{L^1\cap H^{s-1}}(1+t)^{-(d-1)/4}\end{aligned} \tag{5.30}$$

as claimed.

L^∞ *stability.* We now carry out the L^∞ estimate. Together with the L^2 result already obtained, this yields rates (5.14) by interpolation, completing the proof.

Denoting by U_1 and U_2 the low- and high-frequency parts

$$U_j(t) := \mathcal{S}_1(t)U_0 + \int_0^t S_j(t-s)\Big(\sum_j \partial_{x_j} Q^j(s)\Big)\,\mathrm{d}s, \tag{5.31}$$

$j = 1, 2$ of U, we have, using the bounds of Proposition 5.7 together with the previously obtained H^{s-1} bounds, Sobolev bound $|f|_{L^\infty} \leqslant C|\partial_{x_1}^{\gamma_1}\partial_{\tilde{x}}^{\tilde{\gamma}} f|_{H^\gamma}$ for $\gamma_1 = 1$, $\tilde{\gamma} = [(d-1)/2)]+1 \leqslant s-4$, and the fact that H^r is an algebra, $|fg|_{H^r} \leqslant |f|_{H^r}|g|_{H^r}$ for $r > d/2$ (Moser's inequality and Sobolev embedding; see, e.g., [M.1–M.4,Kaw,GMWZ.2,Z.4]), that

$$\begin{aligned}|U_1(t)|_{L^\infty} &\leqslant |\mathcal{S}_1(t)U_0|_{L^1} + \left|\int_0^t \big(\mathcal{S}_1(t)\partial_x\big)Q(s)\,\mathrm{d}s\right|_{L^\infty}\\ &\leqslant C(1+t)^{(d-1)/2}|U_0|_{L^1}\\ &\quad + C\int_0^t (1+t-s)^{-(d-1)/2-1/2}|U(s)|_{H^1}^2\,\mathrm{d}s\\ &\leqslant C\bigg((1+t)^{(d-1)/2}\\ &\quad + \int_0^t (1+t-s)^{-(d-1)/2-1/2}(1+s)^{-(d-1)/2}\,\mathrm{d}s\bigg)|U_0|_{L^1\cap H^{s-1}}\\ &\leqslant C(1+t)^{(d-1)/2}|U_0|_{L^1\cap H^{s-1}}\end{aligned} \tag{5.32}$$

and

$$\begin{aligned}|U_2(t)|_{L^\infty} &\leqslant C\big|\partial_{x_1}^{\gamma_1}\partial_{\tilde{x}}^{\gamma_2}U_2(t)\big|_{L^2}\\ &\leqslant \big|\partial_x^\gamma \mathcal{S}_2(t)U_0\big|_{L^2} + \left|\int_0^t \partial_x^\gamma S_2(t)\big(\partial_x Q(s)\big)\,\mathrm{d}s\right|_{L^2}\\ &\leqslant C\mathrm{e}^{-\theta t}|U_0|_{H^\gamma} + C\int_0^t \mathrm{e}^{-\theta(t-s)}|\partial_x Q(s)|_{H^{|\gamma|}}\,\mathrm{d}s\end{aligned}$$

$$\leqslant C\left(e^{-\theta t} + \left|\int_0^t e^{-\theta(t-s)}(1+s)^{-(d-1)/2}\,ds\right|\right)|U(0)|_{H^{|\gamma|+2}}$$

$$\leqslant C(1+t)^{-(d-1)/2}|U(0)|_{H^{s-1}}, \tag{5.33}$$

giving the claimed bound. □

REMARKS. 1. Results of [HoZ.1] show that $\|G\|_{L^p}$ for $p \leqslant 2$ degrade for systems in multi-dimensions, with blow up at $p = 1$. So, $p \geqslant 2$ is the optimal result.

2. The restriction to $d \geqslant 3$ in the case of weakly inviscid stable Lax or overcompressive shocks results from the degraded Green distribution bounds (5.15) obtained in dimension $d = 2$, as indicated by the presence of error terms β. These, in turn, result from our nonsharp estimation of "pole" terms of order $\int_{\Gamma^{\tilde\xi}} (\lambda - \lambda_*(\tilde\xi))^{-1}$ occurring in the inverse Laplace transform of the resolvent via modulus estimates $\int_{\Gamma^{\tilde\xi}} |\lambda - \lambda_*(\tilde\xi)|^{-1}$; see (5.154)–(5.158), Section 5.3.3. To obtain optimal estimates, one should rather treat this as a residue term similarly as in the one-dimensional case, identifying cancellation in the nonintegrable $1/\lambda$ singularity; see, e.g., the treatment of the (always weakly inviscid stable) scalar case in [HoZ.3,HoZ.4,Z.3].

5.3. *Proof of the linearized estimates*

Finally, we carry out in this section the proof of Proposition 5.7, completing the analysis of the multidimensional case.

5.3.1. *Low-frequency bounds on the resolvent kernel.* Consider the family of elliptic Green distributions $G_{\tilde\xi,\lambda}(x_1, y_1)$,

$$G_{\tilde\xi,\lambda}(\cdot, y_1) := (L_{\tilde\xi} - \lambda I)^{-1}\delta_{y_1}(\cdot), \tag{5.34}$$

associated with the ordinary differential operators $(L_{\tilde\xi} - \lambda I)$, i.e., the *resolvent kernel* of the Fourier transformed operator $L_{\tilde\xi}$. The function $G_{\tilde\xi,\lambda}(x_1, y_1)$ is the Laplace–Fourier transform in variables $\tilde x = (x_2, \ldots, x_d)$ and t, respectively, of the time-evolutionary Green distribution

$$G(x, t; y) = G(x_1, \tilde x, t; y_1, \tilde y) := e^{Lt}\delta_y(x) \tag{5.35}$$

associated with the linearized evolution operator $(\partial/\partial t - L)$. Restrict attention to the surface

$$\Gamma^{\tilde\xi} := \partial\Lambda^{\tilde\xi} = \{\lambda\colon \operatorname{Re}\lambda = -\theta_1(|\operatorname{Im}\lambda|^2 + |\tilde\xi|^2)\} \tag{5.36}$$

$\theta_1 > 0$ sufficiently small.

Then, our main result, to be proved in the remainder of the section, is:

PROPOSITION 5.10. *Under the hypotheses of Theorem* 5.5, *for* $\lambda \in \Gamma^{\tilde\xi}$ (*defined in* (5.36)) *and* $\rho := |(\tilde\xi, \lambda)|$, $\theta_1 > 0$ *and* $\theta > 0$ *sufficiently small, there hold*:

$$\left|G_{\tilde\xi,\lambda}(x_1, y_1)\right| \leqslant C\gamma_2\gamma_1\left(\rho^{-1}\mathrm{e}^{-\theta|x_1|}\mathrm{e}^{-\theta\rho^2|y_1|} + \mathrm{e}^{-\theta\rho^2|x_1-y_1|}\right), \tag{5.37}$$

and

$$\begin{aligned}&\left|(\partial/\partial y_1)G_{\tilde\xi,\lambda}(x_1, y_1)\right| \\ &\quad\leqslant C\gamma_2\gamma_1\left[\rho^{-1}\mathrm{e}^{-\theta|x_1|}\left(\rho\mathrm{e}^{-\theta\rho^2|y_1|} + \alpha\mathrm{e}^{-\theta|y_1|}\right) + \mathrm{e}^{-\theta\rho^2|x_1-y_1|}\left(\rho + \alpha\mathrm{e}^{-\theta|y_1|}\right)\right],\end{aligned} \tag{5.38}$$

where

$$\gamma_1(\tilde\xi, \lambda) := \begin{cases} 1 & \text{in UIS case}, \\ 1 + \sum_j \left[\rho^{-1}\left|\operatorname{Im}\lambda - \mathrm{i}\tau_j(\tilde\xi)\right| + \rho\right]^{-1} & \text{in WIS case}, \end{cases} \tag{5.39}$$

$$\gamma_2(\tilde\xi, \lambda) := 1 + \sum_{j,\pm}\left[\rho^{-1}\left|\operatorname{Im}\lambda - \eta_j^{\pm}(\tilde\xi)\right| + \rho\right]^{1/s_j-1}, \tag{5.40}$$

$\eta_j(\cdot)$, $s_j(\cdot) \geqslant 1$ *as in* (H5) *and* $\tau_j(\cdot)$ *as in* (1.43), *and*

$$\alpha := \begin{cases} 0 & \text{for Lax or overcompressive case}, \\ 1 & \text{for undercompressive case}. \end{cases} \tag{5.41}$$

(Here, as above, UIS and WIS denote "uniform inviscid stable" and "weak inviscid stable", as defined, respectively, in (1.32), Section 1.4, and (1.28), Section 1.3; the classification of Lax, overcompressive, and undercompressive types is given in Section 1.2.) *More precisely*, $t := 1 - 1/K_{\max}$, *where* $K_{\max} := \max K_j^{\pm} = \max s_j^{\pm}$ *is the maximum among the orders of all branch singularities* $\eta_j^{\pm}(\cdot)$, $s_j^{\pm}$ *and* $\eta_j^{\pm}$ *defined as in* (H5); *in particular*, $t = 1/2$ *in the* (*generic*) *case that only square-root singularities occur.*[30]

COROLLARY 5.11. *Under the hypotheses of Theorem* 5.5, *for* $\lambda \in \Gamma^{\tilde\xi}$ (*defined in* (5.36)) *and* $\rho := |(\tilde\xi, \lambda)|$, $\theta_1 > 0$, *and* $\theta > 0$ *sufficiently small, there holds the resolvent bound*

$$\left|(L_{\tilde\xi} - \lambda)^{-1}\partial_{x_1}^{\beta} f\right|_{L^p(x_1)} \leqslant C\gamma_1\gamma_2\rho^{(1-\alpha)|\beta|-1}|f|_{L^1(x_1)} \tag{5.42}$$

[30] Bound (5.37) corrects a minor error in [Z.3], where it was mistakenly stated as

$$\left|G_{\tilde\xi,\lambda}(x_1, y_1)\right| \leqslant C\gamma_2\left(\gamma_1\rho^{-1}\mathrm{e}^{-\theta|x_1|}\mathrm{e}^{-\theta\rho^2|y_1|} + \mathrm{e}^{-\theta\rho^2|x_1-y_1|}\right).$$

The source of this error was the normalization (4.137) of [Z.3] that $\varphi_1^{\pm}$ agree to first order in ρ at $\rho = 0$, which is incompatible with the assumption (used in the proof of the low-frequency expansion) that fast modes decay at $x_1 = \pm\infty$.

for all $2 \leqslant p \leqslant \infty$, $0 \leqslant |\beta| \leqslant 1$, *where* γ_j, α *are as defined in* (5.39)–(5.41).

PROOF. Direct integration, using the triangle inequality bound

$$\left|(L_{\tilde{\xi}} - \lambda)^{-1}\partial_{x_1}^{\alpha} f\right|_{L^p(x_1)} \leqslant \sup_{y_1}\left|\partial_{y_1}^{\alpha} G_{\tilde{\xi},\lambda}(\cdot, y_1)\right|_{L^p(x_1)} |f|_{L^1(x_1)}.$$

□

REMARK 5.12. By standard considerations (see, e.g., [He,ZH,Z.1]), condition (H3) implies that surface $\Gamma^{\tilde{\xi}}$ strictly bounds the essential spectrum of operator $L_{\tilde{\xi}}$ to the right. That is, in this analysis, we shall work entirely in the *resolvent set* of $L_{\tilde{\xi}}$. This is in marked contrast to our approach in the one-dimensional case, wherein bounds on the resolvent kernel G_λ *within* the essential spectrum were used to obtain sharp pointwise estimates on G in the one-dimensional case. The nonoptimal bounds we obtain here could perhaps be sharpened by a similar analysis; for further discussion, see [Z.3,HoZ.3,HoZ.4].

REMARK. Terms $\rho^{-1}e^{-\theta|x_1|}e^{-\theta\rho^2|y_1|}$ and $e^{-\theta\rho^2|x_1-y_1|}$ in (5.37), respectively encoding near and far-field behavior with respect to the front location $x_1 = 0$, are, roughly speaking, associated respectively with point and essential spectrum of the operator L. In particular, the simple pole ρ^{-1} in the first term corresponds to a semisimple eigenvalue at $\rho = 0$, and the factors $e^{-\theta|x_1|}$ and $e^{-\theta\rho^2|y_1|}$ to associated right and left eigenfunctions. However, this is clearly not the usual point spectrum encountered in the case of an operator with positive spectral gap. For example, note the lack of spatial decay/localization in the "left eigenfunction" term $e^{-\theta\rho^2|y_1|}$ at $\rho = 0$. Also, though it is not apparent from the simple bound (5.37), it is a fact that the "left eigenspace" at $\rho = 0$ (defined by continuation along rays from $\rho > 0$ following the approach of Section 4.3.3) depends on direction $(\tilde{\xi}_0, \lambda_0)$; that is, there is a conical singularity at the origin in the spectral projection, reflecting that of the Evans function.

REMARK 5.13. The α-terms appearing in (5.37) and (5.38) are not technical artifacts, but in fact reflect genuinely new effects arising in the undercompressive case, related to L^1 time-invariants and conservation of mass. See [Z.6] for a detailed discussion in the one-dimensional case.

Normal modes. Paralleling our one-dimensional analysis, we begin by identifying normal modes of the limiting, constant-coefficient equations $(L_{\tilde{\xi}\pm} - \lambda)U = 0$. The key estimates of slow and super-slow mode decay will be seen to reduce to a matrix perturbation problem closely related to the one arising in the corresponding inviscid theory; the crucial issue here, as there, is the careful treatment of branch singularities, corresponding to inviscid glancing modes.

Introduce the curves

$$(\tilde{\xi}, \lambda)(\rho, \tilde{\xi}_0, \tau_0) := (\rho\tilde{\xi}_0, \rho i\tau_0 - \theta_1\rho^2), \tag{5.43}$$

where $\tilde{\xi}_0 \in \mathbb{R}^{d-1}$ and $\tau_0 \in \mathbb{R}$ are restricted to the unit sphere S^d: $|\tilde{\xi}_0|^2 + |\tau_0|^2 = 1$. Evidently, as $(\tilde{\xi}_0, \tau_0, \rho)$ range in the compact set $S^d \times [0, \delta]$, $(\tilde{\xi}, \lambda)$ traces out the portion of the surface $\Gamma^{\tilde{\xi}}$ contained in the set $|\tilde{\xi}|^2 + |\lambda|^2 \leqslant \delta$ of interest. As before, fixing $\tilde{\xi}_0$, τ_0, we denote by $v_j^{\pm}(\rho)$, the solutions of the limiting, constant coefficient equations at $(\tilde{\xi}, \lambda)(\rho)$ from which $\phi_j^{\pm}(\rho)$, $\psi_j^{\pm}(\rho)$, etc. are constructed; it is these solutions that we wish to estimate.

Making as usual the Ansatz

$$v^{\pm} =: e^{\mu x_1}\mathbf{v}, \tag{5.44}$$

and substituting $\lambda = i\rho\tau_0 - \theta_1\rho^2$ into (3.4), we obtain the characteristic equation

$$\begin{aligned}\Big[\mu^2 B_{\pm}^{11} + \mu\Big(-A_{\pm}^1 + i\rho\sum_{j\neq 1} B_{\pm}^{j1}\xi_{0_j} + i\rho\sum_{k\neq 1} B^{1k}\xi_{0_k}\Big)\\ - \Big(i\rho\sum_{j\neq 1} A^j\xi_{0_j} + \rho^2\sum_{jk\neq 1} B^{jk}\xi_{0_j}\xi_{0_k} + \big(\rho i\tau_0 - \theta_1\rho^2\big)I\Big)\Big]\mathbf{v} = 0.\end{aligned} \tag{5.45}$$

Note that this agrees with (3.4) up to first order in ρ, hence the (first-order) matrix bifurcation analyses of Lemma 3.3 applies for any θ.

As in past sections, we first separate normal modes (5.44) into fast and slow modes, the former having exponential growth/decay rate with real part bounded away from zero, and the latter having growth/decay rate close to zero (indeed, vanishing for $\rho = 0$). The fast modes have $\mathcal{O}(1)$ decay rate, and are spectrally separated by assumption (A1), hence admit a straightforward treatment. We now focus on the slow modes, which we will further subdivide into *intermediate-* and *super-slow* types, corresponding respectively to elliptic and hyperbolic/glancing modes in the inviscid terminology.

Positing the Taylor expansion

$$\begin{cases}\mu = 0 + \mu^1\rho + \cdots,\\ \mathbf{v} = \mathbf{v}^0 + \cdots\end{cases} \tag{5.46}$$

(or Puiseux expansion, in the case of a branch singularity), and matching terms of order ρ in (5.45), we obtain

$$\Big(-\mu^1 A_{\pm}^1 - i\sum_{j\neq 1} A^j\xi_{0_j} - i\tau_0 I\Big)\mathbf{v}^0 = 0,$$

just as in (3.6), or equivalently

$$\big[(A^1)^{-1}\big(i\tau_0 + iA^{\tilde{\xi}_0}\big) - \alpha_0 I\big]\mathbf{v}^0 = 0, \tag{5.47}$$

with $\mu^1 =: -\alpha_0$, which can be recognized as the equation occurring in the inviscid theory on the imaginary boundary $\lambda = i\tau_0$. For eigenvalues α_0 of nonzero real part, denoted as *intermediate-slow* modes, we have growth or decay at rate $\mathcal{O}(\rho)$, and spectral separation by assumption (A2); thus, they again admit straightforward treatment.

It remains to study *super-slow* modes, corresponding to pure imaginary eigenvalues $\alpha_0 =: \mathrm{i}\xi_{0_1}$; here, we must consider quadratic order terms in ρ, and the viscous and inviscid theory part ways. Using $\mu = -\mathrm{i}\rho\xi_{0_1} + \mathrm{o}(\rho)$, and dividing (5.45) by ρ, we obtain the modified equation

$$\big[(A^1)^{-1}\big(\mathrm{i}\tau_0 + \mathrm{i}A^{\tilde{\xi}_0} + \rho\big(B^{\xi_0\xi_0} - \theta_1\big) + \mathrm{o}(\rho)\big) - \alpha I\big]\mathbf{v} = 0, \tag{5.48}$$

where $\alpha := -\mu/\rho$ and $\mathbf{v}$ denote exact solutions; that is, we expand the equations rather than the solutions to second order. (Note that this derivation remains valid near branch singularities, since we have only assumed continuity of μ/ρ and not analyticity at $\rho = 0$.) Here, $B^{\xi_0\xi_0}$ as usual denotes $\sum B^{jk}\xi_{0_j}\xi_{0_k}$, where $\xi_0 := (\xi_{0_1}, \tilde{\xi}_0)$. Equation (5.48) generalizes the (all-orders) perturbation equation

$$\big[(A^1)^{-1}\big(\mathrm{i}\tau_0 + \mathrm{i}A^{\tilde{\xi}_0} + \rho I\big) - \alpha I\big]\mathbf{v} = 0, \tag{5.49}$$

$\rho \to 0^+$, arising in the inviscid theory near the imaginary boundary $\lambda = \mathrm{i}\tau_0$ [Kr,Mé.5].

Note that τ_0 is an eigenvalue of A^{ξ_0}, as can be seen by substituting $\alpha_0 = \mathrm{i}\xi_{0_1}$ in (5.47), hence $|\tau_0| \leqslant C|\xi_0|$ and therefore (since clearly also $|\xi_0| \geqslant |\tilde{\xi}_0|$)

$$|\xi_0| \geqslant \frac{1}{C}\big|(\tilde{\xi}_0, \tau_0)\big| = \frac{1}{C}.$$

Thus, for B positive definite, and θ_1 sufficiently small, perturbation $\rho(B^{\xi_0\xi_0} - \theta_1)$, roughly speaking, enters (5.48) with the same sign as does ρI in (5.49). Indeed, for identity viscosity $B^{jk} := \delta_k^j I$, (5.48) reduces for fixed $(\tilde{\xi}_0, \tau_0)$ exactly to (5.49), by the rescaling $\rho \to \rho/(|\xi_0|^2 - \theta_1)$.

For $(\tilde{\xi}_0, \tau_0)$ bounded away from the set of branch singularities $\bigcup(\tilde{\xi}, \eta_j(\tilde{\xi}))$, we may treat (5.48) as a continuous family of single-variable matrix perturbation problem in ρ, indexed by $(\tilde{\xi}_0, \tau_0)$; the resulting continuous family of analytic perturbation series will then yield uniform bounds by compactness. For $(\tilde{\xi}_0, \tau_0)$ near a branch singularity, on the other hand, we must vary both ρ and $(\tilde{\xi}_0, \tau_0)$, in general a complicated multi-variable perturbation problem. Using homogeneity, however, and the uniform structure assumed in (H6), this can be reduced to a two-variable perturbation problem that again yields uniform bounds. For, noting that $\eta_j(\tilde{\xi}) \equiv 0$, we find that $\tilde{\xi}_0$ must be bounded away from the origin at branch singularities; thus, we may treat the direction $\tilde{\xi}_0/|\tilde{\xi}_0|$ as a fixed parameter and vary only ρ and the ratio $|\tau_0|/|\tilde{\xi}_0|$. Alternatively, relaxing the restriction of $(\tilde{\xi}_0, \tau_0)$ to the unit sphere, we may fix $\tilde{\xi}_0$ and vary ρ and τ_0, obtaining after some rearrangement the rescaled equation

$$\begin{aligned}\big[(A^1)^{-1}\big(&\mathrm{i}\tau_0 + \mathrm{i}A^{\tilde{\xi}_0} \\ &+ \big[\mathrm{i}\sigma + \rho\big(B^{\xi_0\xi_0} - \theta_1\big(|\tilde{\xi}_0|^2 + |\tau_0|^2\big) + \mathrm{o}(|\rho, \sigma|)\big)\big]\big) - \alpha I\big]\mathbf{v} = 0,\end{aligned} \tag{5.50}$$

where σ denotes variation in τ_0.

We shall find it useful to introduce at this point the wedge, or exterior algebraic product of vectors, determined by the properties of multilinearity and alternation. We fix once and for all a basis, thus determining a norm. (This allows us to conveniently factorize into minors, below.)

We make also the provisional assumption (P0) of Section 4, that either the inviscid system is strictly hyperbolic, or else A^j and B^{jk} are simultaneously symmetrizable. This is easily removed by working with vector blocks in place of individual modes.

LEMMA 5.14 [Z.3]. *Under the hypotheses of Theorem* 5.5, *let* $\alpha_0 = \mathrm{i}\xi_{0_1}$ *be a pure imaginary root of the inviscid equation* (5.47) *for some given* $\tilde{\xi}_0$, τ_0, *i.e.*, $\det(A^{\xi_0} + \tau_0) = 0$. *Then, associated with the corresponding root* α *in* (5.48), *we have the following behavior, for some fixed* $\epsilon, \theta > 0$ *independent of* $(\tilde{\xi}_0, \tau_0)$:

(i) *For* $(\tilde{\xi}_0, \tau_0)$ *bounded distance* ϵ *away from any branch singularity* $(\tilde{\xi}, \eta_j(\tilde{\xi}))$ *involving* α, *the root* $\alpha(\rho)$ *in* (5.48) *such that* $\alpha(0) := \alpha_0$ *bifurcates smoothly into* m *roots* $\alpha_1, \dots, \alpha_m$ *with associated vectors* $\mathbf{v}_1, \dots, \mathbf{v}_m$, *where* m *is the dimension of* $\ker(A^{\xi_0} + \tau_0)$, *satisfying*

$$|\operatorname{Re} \alpha_j| \geqslant \theta\rho \tag{5.51}$$

and

$$|\mathbf{v}_1 \wedge \cdots \wedge \mathbf{v}_m| \geqslant \theta > 0 \tag{5.52}$$

for $0 < \rho \leqslant \epsilon$, *where the modulus of the wedge product is evaluated with respect to its coordinatization in some fixed reference basis.*

(ii) *For* $(\tilde{\xi}_0, \tau_0)$ *lying at a branch singularity* $(\tilde{\xi}, \eta_j(\tilde{\xi}))$ *involving* α, *the root* $\alpha(\rho, \sigma)$ *in* (5.50) *such that* $\alpha(0,0) = \alpha_0$ *bifurcates* (*nonsmoothly*) *into* m *groups of* s *roots each*:

$$\{\alpha_1^1, \dots, \alpha_s^1\}, \dots, \{\alpha_1^m, \dots, \alpha_s^m\}, \tag{5.53}$$

with associated vectors $\mathbf{v}_k^j$, *where* m *is the dimension of* $\ker(A^{\xi_0} + \tau_0)$ *and* s *is some positive integer, such that, for* $0 \leqslant \rho \leqslant \epsilon$ *and* $|\sigma| \leqslant \epsilon$,

$$\alpha_k^j = \alpha + \pi_k^j + \mathrm{o}\big(|\sigma| + |\rho|\big)^{1/s}, \tag{5.54}$$

and, in appropriately chosen coordinate system,

$$\mathbf{v}_k^j = \begin{pmatrix} 0 \\ \vdots \\ 0 \\ 1 \\ \pi_k^j \\ (\pi_k^j)^2 \\ \vdots \\ (\pi_k^j)^{m-1} \\ 0 \\ \vdots \\ 0 \end{pmatrix} + \mathrm{o}\big(|\sigma| + |\rho|\big)^{1/s}, \tag{5.55}$$

where

$$\pi_k^j := \varepsilon^k \mathrm{i}\big(p\sigma - \mathrm{i}q_j\rho\big)^{1/s}, \tag{5.56}$$

$\varepsilon := 1^{1/s}$, *and the functions* $p(\tilde{\xi}_0)$ *and* $q_j(\tilde{\xi}_0)$ *are real-valued and uniformly bounded both above and away from zero, with* $\operatorname{sgn} p = \operatorname{sgn} q$. *Moreover,*

$$\big|\mathrm{Re}\,\alpha_k^j\big| \geqslant \theta\rho \tag{5.57}$$

and the "group vectors" $\mathbf{v}^j$ *satisfy*

$$\big|\mathbf{v}^1 \wedge \cdots \wedge \mathbf{v}^m\big| \geqslant \theta > 0 \tag{5.58}$$

with respect to some fixed reference basis.

PROOF.

CASE $(\widetilde{\mathrm{H}2})$(i). We first treat the considerably simpler *strictly hyperbolic* case, which permits a direct and relatively straightforward treatment. In this case, the dimension of $\ker(A^{\xi_0} + \tau_0)$ is one, hence m is simply one and (5.52) and (5.58) are irrelevant. Let $l(\tilde{\xi}_0, \tau_0)$ and $r(\tilde{\xi}_0, \tau_0)$ denote left and right zero eigenvectors of $(A^{\xi_0} + \tau_0) = 1$, spanning co-kernel and kernel, respectively; these are necessarily real, since $(A^{\xi_0} + \tau_0)$ is real. Clearly r is also a right (null) eigenvector of $(A^1)^{-1}(\mathrm{i}\tau_0 + \mathrm{i}A^{\xi_0})$, and lA^1 a left eigenvector.

Branch singularities are signaled by the relation

$$lA^1 r = 0, \tag{5.59}$$

which indicates the presence of a single Jordan chain of generalized eigenvectors of $(A^1)^{-1}(\mathrm{i}\tau_0 + \mathrm{i}A^{\xi_0})$ extending up from the genuine eigenvector r; we denote the length of this chain by s.

OBSERVATION 5.15 [Z.3]. *Bound* (2.55) *implies that*

$$lB^{\xi_0\xi_0}r \geqslant \theta > 0, \tag{5.60}$$

uniformly in $\tilde{\xi}$.

PROOF. In our present notation, (2.55) can be written as

$$\operatorname{Re}\sigma\big(-\mathrm{i}A^{\xi_0} - \rho B^{\xi_0\xi_0}\big) \leqslant -\theta_1\rho,$$

for all $\rho > 0$, some $\theta_1 > 0$. (Recall: $|\xi_0| \geqslant \theta_2 > 0$, by previous discussion.) By standard matrix perturbation theory [Kat], the simple eigenvalue $\gamma = \mathrm{i}\tau_0$ of $-\mathrm{i}A^{\xi_0}$ perturbs analytically as ρ is varied around $\rho = 0$, with perturbation series

$$\gamma(\rho) = \mathrm{i}\tau_0 - \rho lB^{\xi_0\xi_0}r + \mathrm{o}(\rho).$$

Thus,

$$\operatorname{Re}\gamma(\rho) = -\rho lB^{\xi_0\xi_0}r + \mathrm{o}(\rho) \leqslant -\theta_1\rho,$$

yielding the result. □

In case (i), $\alpha(0) = \alpha_0$ is a simple eigenvalue of $(A^1)^{-1}(\mathrm{i}\tau_0 + \mathrm{i}A\tilde{\xi}_0)$, and so perturbs analytically in (5.48) as ρ is varied around zero, with perturbation series

$$\alpha(\rho) = \alpha_0 + \rho\alpha^1 + \mathrm{o}(\rho), \tag{5.61}$$

where $\alpha^1 = \tilde{l}(A^1)^{-1}\tilde{r}$, $\tilde{l}$, $\tilde{r}$ denoting left and right eigenvectors of $(A^1)^{-1}(\mathrm{i}\tau_0 + A\tilde{\xi}_0)$. Observing by direct calculation that $\tilde{r} = r$, $\tilde{l} = lA^1/lA^1r$, we find that

$$\alpha^1 = lB^{\xi_0\xi_0}r/lA^1r$$

is real and bounded uniformly away from zero, by Observation 5.15, yielding the result (5.51) for any fixed $(\tilde{\xi}_0, \tau_0)$, on some interval $0 \leqslant \rho \leqslant \epsilon$, where ϵ depends only on a lower bound for α^1 and the maximum of $\gamma''(\rho)$ on the interval $0 \leqslant \rho \leqslant \epsilon$. By compactness, we can therefore make a uniform choice of ϵ for which (5.51) is valid on the entire set of $(\tilde{\xi}_0, \tau_0)$ under consideration.

In case (ii), $\alpha(0, 0) = \alpha_0$ is an s-fold eigenvalue of $(A^1)^{-1}(\mathrm{i}\tau_0 + \mathrm{i}A\tilde{\xi}_0)$, corresponding to a single $s \times s$ Jordan block. By standard matrix perturbation theory, the corresponding s-dimensional invariant subspace (or "total eigenspace") varies analytically with ρ and σ, and admits an analytic choice of basis with arbitrary initialization at $\rho, \sigma = 0$ [Kat]. Thus, by restricting attention to this subspace we can reduce to an s-dimensional perturbation problem; moreover, up to linear order in ρ, σ, the perturbation may be calculated with respect to the fixed, initial coordinization at $\rho, \sigma = 0$.

Choosing the initial basis as a real, Jordan chain reducing the restriction (to the subspace of interest) of $(A^1)^{-1}(\mathrm{i}\tau_0 + \mathrm{i}A\tilde{\xi}_0)$ to i times a standard Jordan block, we thus reduce (5.50) to the canonical problem

$$\big(\mathrm{i}J + \mathrm{i}\sigma M + \rho N + \mathrm{o}\big(\big|(\rho,\sigma)\big|\big) - (\alpha - \alpha_0)\big)\mathbf{v}_I = 0, \tag{5.62}$$

where

$$J := \begin{pmatrix} 0 & 1 & 0 & \cdots & 0 \\ 0 & 0 & 1 & 0 & \cdots \\ \vdots & \vdots & \vdots & \vdots & \vdots \\ 0 & 0 & 0 & \cdots & 0 \end{pmatrix}, \tag{5.63}$$

$\mathbf{v}_I$ is the coordinate representation of $\mathbf{v}$ in the s-dimensional total eigenspace, and M and N are given by

$$M := \widetilde{L}\big(A^1\big)^{-1}\widetilde{R} \tag{5.64}$$

and

$$N := \widetilde{L}\big(A^1\big)^{-1}\big(B^{\xi_0\xi_0} - \theta_1\big)\widetilde{R}, \tag{5.65}$$

respectively, where $\widetilde{R}$ and $\widetilde{L}$ are the initializing (right) basis, and its corresponding (left) dual.

Now, we have only to recall that, as may be readily seen by the defining relation

$$\widetilde{L}\big(A^1\big)^{-1}\big(\mathrm{i}\tau_0 + \mathrm{i}A\tilde{\xi}_0\big)\widetilde{R} = J,$$

or equivalently $(A^1)^{-1}(\mathrm{i}\tau_0 + \mathrm{i}A\tilde{\xi}_0)\widetilde{R} = \widetilde{R}J$ and $\widetilde{L}(A^1)^{-1}(\mathrm{i}\tau_0 + \mathrm{i}A\tilde{\xi}_0) = J\widetilde{L}$, the first column of $\widetilde{R}$ and the last row of $\widetilde{L}$ are genuine left and right eigenvectors $\tilde{r}$ and $\tilde{l}$ of $(A^1)^{-1}(\mathrm{i}\tau_0 + \mathrm{i}A\tilde{\xi}_0)$, hence without loss of generality

$$\tilde{r} = r, \qquad \tilde{l} = plA^1$$

as in the previous (simple eigenvalue) case, where p is an appropriate nonzero real constant. Applying again Observation 5.15, we thus find that the crucial s, 1 entries of the perturbations M, N, namely p and $pl(B^{\xi_0\xi_0} - \theta_1)r =: q$, respectively, are real, nonzero and of the same sign. Recalling, by standard matrix perturbation theory, that this entry when nonzero is the only significant one, we have reduced finally (modulo $\mathrm{o}(|\sigma| + |\rho|)^{1/s}$) errors) to the computation of the eigenvalues/eigenvectors of

$$\mathrm{i}\begin{pmatrix} 0 & 1 & 0 & \cdots & 0 \\ 0 & 0 & 1 & 0 & \cdots \\ \vdots & \vdots & \vdots & \vdots & \vdots \\ p\sigma - \mathrm{i}q\rho & 0 & 0 & \cdots & 0 \end{pmatrix}, \tag{5.66}$$

from which results (5.54)–(5.56) follow by an elementary calculation, for any fixed $(\tilde{\xi}_0, \tau_0)$, and some choice of $\epsilon > 0$; as in the previous case, the corresponding global results then follow by compactness. Finally, bound (5.57) follows from (5.54) and (5.56) by direct calculation. (Note that the addition of further $\mathcal{O}(|\sigma| + |\rho|)$ perturbation terms in entries other than the lower left-hand corner of (5.66) does not affect the result.) This completes the proof in the strictly hyperbolic case.

CASE $(\widetilde{\text{H2}})$(ii). We next turn to the more complicated *symmetrizable, constant-multiplicity* case; here, we make essential use of recent results of Métivier [Mé.4] concerning the spectral structure of matrix $(A^1)^{-1}(\mathrm{i}\tau + \mathrm{i}A^{\tilde{\xi}})$. Without loss of generality, take A^{ξ}, $B^{\xi,\xi}$ (but not necessarily A^1) *symmetric*; this may be achieved by the change of coordinates $A^{\xi} \to A_0^{1/2} A^{\xi} A_0^{-1/2}$, $B^{\xi,\xi} \to A_0^{1/2} B^{\xi,\xi} A_0^{-1/2}$. From (H3), we find, additionally, that $B^{\xi,\xi} \geqslant \theta|\xi|^2$ for all $\xi \in \mathbb{R}^d$.

With these assumptions, the kernel and co-kernel of $(A^{\xi_0} + \tau_0)$ are of fixed dimension m, not necessarily equal to one, and are spanned by a common set of zero-eigenvectors $r_1, \dots, r_m$. Vectors $r_1, \dots, r_m$ are necessarily right zero-eigenvectors of $(A^1)^{-1}(\mathrm{i}\tau_0 + \mathrm{i}A^{\xi_0})$ as well. Branch singularities correspond to the existence of one or more Jordan chains of generalized zero-eigenvectors extending up from genuine eigenvectors in their span, which by the argument of Lemma 3.4 is equivalent to

$$\det\bigl(r_j^t A^1 r_k\bigr) = 0. \tag{5.67}$$

In fact, as pointed out by Métivier [Mé.4], the assumption of constant multiplicity implies considerable additional structure.

OBSERVATION 5.16 [Mé.4]. *Let $(\tilde{\xi}_0, \tau_0)$ lie at a branch singularity involving root $\alpha_0 = \mathrm{i}\xi_{0_1}$ in (5.47), with τ_0 an m-fold eigenvalue of A^{ξ_0}. Then, for $(\tilde{\xi}, \tau)$ in the vicinity of $(\tilde{\xi}_0, \tau_0)$, the roots α bifurcating from α_0 in (5.47) consist of m copies of s roots $\alpha_1, \dots, \alpha_s$, where s is some fixed positive integer.*

PROOF. Let $a(\tilde{\xi}, \alpha)$ denote the unique eigenvalue of A^{ξ} lying near $-\tau_0$, where, as usual, $-\mathrm{i}\xi_1 := \alpha$; by the constant multiplicity assumption, $a(\cdot,\cdot)$ is an analytic function of its arguments. Observing that

$$\begin{aligned}\det\bigl[(A^1)^{-1}\bigl(\mathrm{i}\tau + \mathrm{i}A^{\tilde{\xi}}\bigr) - \alpha\bigr] &= \det \mathrm{i}\bigl(A^1\bigr)^{-1} \det\bigl(\tau + A^{\xi}\bigr) \\ &= e\bigl(\tilde{\xi}, \tau, \alpha\bigr)\bigl(\tau + a\bigl(\tilde{\xi}, \alpha\bigr)\bigr)^m,\end{aligned}$$

where $e(\cdot,\cdot,\cdot)$ does not vanish for $(\tilde{\xi}, \tau, \alpha)$ sufficiently close to $(\tilde{\xi}_0, \tau_0, \alpha_0)$, we see that the roots in question occur as m-fold copies of the roots of

$$\tau + a\bigl(\tilde{\xi}, \alpha\bigr) = 0. \tag{5.68}$$

But, the left-hand side of (5.68) is a family of analytic functions in α, continuously varying in the parameters $(\tilde{\xi}, \tau)$, whence the number of zeroes is constant near $(\tilde{\xi}_0, \tau_0)$. □

OBSERVATION 5.17 [Z.3]. *The matrix* $(r_j^t A^1 r_k)$, $j, k = 1, \dots, m$, *is a real multiple of the identity,*

$$(r_j^t A^1 r_k) = (\partial a/\partial \xi_1) I_m, \tag{5.69}$$

where $a(\xi)$ *denotes the (unique, analytic)* m*-fold eigenvalue of* A^ξ *perturbing from* $-\tau_0$.
More generally, if

$$(\partial a/\partial \xi_1) = \cdots = (\partial^{s-1} a/\partial \xi_1^{s-1}) = 0, \quad (\partial^s a/\partial \xi_1^s) \neq 0 \tag{5.70}$$

at ξ_0, *then, letting* $r_1(\tilde{\xi}), \dots, r_m(\tilde{\xi})$ *denote an analytic choice of basis for the eigenspace corresponding to* $a(\tilde{\xi})$, *orthonormal at* $(\tilde{\xi}_0, \tau_0)$, *we have the relations*

$$(A^1)^{-1}(\tau_0 + A^{\xi_0}) r_{j,p} = r_{j,p-1} \tag{5.71}$$

for $1 \leqslant p \leqslant s-1$ *and*

$$(r_{j,0}^t A^1 r_{k,p-1}) = p!(\partial^p a/\partial \xi_1^p) I_m \tag{5.72}$$

for $1 \leqslant p \leqslant s$, *where*

$$r_{j,p} := (-1)^p p(\partial^p r_j/\partial \xi_1^p).$$

In particular,

$$\{r_{j,0}, \dots, r_{j,s-1}\}, \quad j = 1, \dots, m,$$

is a right Jordan basis for the total zero eigenspace of $(A^1)^{-1}(\tau_0 + A^{\xi_0})$, *for which the genuine zero-eigenvectors* $\tilde{l}_j$ *of the dual, left basis are given by*

$$\tilde{l}_j = (1/s!(\partial^s a/\partial \xi_1^s)) A^1 r_j. \tag{5.73}$$

PROOF. Considering A^ξ as a matrix perturbation in ξ_1, we find by standard spectral perturbation theory that the bifurcation of the m-fold eigenvalue τ_0 as ξ_1 is varied is governed to first order by the spectrum of $(r_j^t A^1 r_k)$. Since these eigenvalues in fact do not split, it follows that $(r_j^t A^1 r_k)$ has a single eigenvalue. But, also, $(r_j^t A^1 r_k)$ is symmetric, hence diagonalizable, whence we obtain result (5.69).

Result (5.72) may be obtained by a more systematic version of the same argument. Let $R(\xi_1)$ denote the matrix of right eigenvectors

$$R(\xi_1) := (r_1, \dots, r_m)(\xi_1).$$

Denoting by

$$a(\xi_1 + h) =: a^0 + a^1 h + \cdots + a^p h^p + \cdots \tag{5.74}$$

and

$$R(\xi_1 + h) =: R^0 + R^1 h + \cdots + R^p h^p + \cdots, \tag{5.75}$$

the Taylor expansions of functions $a(\cdot)$ and $R(\cdot)$ around ξ_0 as ξ_1 is varied, and recalling that

$$A^{\xi} = A^{\xi_0} + hA^1,$$

we obtain in the usual way, matching terms of common order in the expansion of the defining relation $(A - a)R = 0$, the hierarchy of relations:

$$\begin{aligned}
&(A^{\xi_0} - a^0)R^0 = 0,\\
&(A^{\xi_0} - a^0)R^1 = -(A^1 - a^1)R^0,\\
&(A^{\xi_0} - a^0)R^2 = -(A^1 - a^1)R^1 + a^2 R^0,\\
&\qquad\vdots\\
&(A^{\xi_0} - a^0)R^p = -(A^1 - a^1)R^{p-1} + a^2 R^{p-2} + \cdots + a^p R^0.
\end{aligned} \tag{5.76}$$

Using $a^0 = \cdots = a^{s-1} = 0$, we obtain (5.71) immediately, from equations $p = 1, \ldots, s-1$, and $R^p = (1/p!)(r_{1,p}, \ldots, r_{m,p})$. Likewise, (5.72) follows from equations $p = 1, \ldots, s$, upon left multiplication by $L^0 := (R^0)^{-1} = (R^0)^t$, using relations $L^0(A^{\xi_0} - a^0) = 0$ and $a^p = (\partial^p a/\partial \xi_1^p)/p!$.

From (5.71), we have the claimed right Jordan basis. But, defining $\tilde{l}_j$ as in (5.73), we can rewrite (5.72) as

$$(\tilde{l}_j^t r_{k,p-1}) = \begin{cases} 0, & 1 \leqslant p \leqslant s-1,\\ I_m, & p = s, \end{cases} \tag{5.77}$$

these ms criteria uniquely define $\tilde{l}_j$ (within the ms-dimensional total left eigenspace) as the genuine left eigenvectors dual to the right basis formed by vectors $r_{j,p}$ (see also exercise just below). □

Observation 5.17 implies, in particular, that Jordan chains extend from *all* or *none* of the genuine eigenvectors $r_1, \ldots, r_m$, with common height s. As suggested by Observation 5.16 (but not directly shown here), this uniform structure in fact persists under variations in $\tilde{\xi}$, τ, see [Mé.4]. Observation 5.17 is a slightly more concrete version of Lemma 2.5 in [Mé.4]; note the close similarity between the argument used here, based on successive variations in basis r_j, and the argument of [Mé.4], based on variations in the associated total projection.

With these preparations, the result goes through essentially as in the strictly hyperbolic case. Set

$$p := 1/\bigl(s!(\partial^s a/\partial \xi_1^s)\bigr). \tag{5.78}$$

In the result of Observation 5.17, we are free to choose any orthonormal basis r_j at $(\tilde{\xi}_0, \tau_0)$; choosing a basis diagonalizing the symmetric matrix $B^{\tilde{\xi}_0,\tilde{\xi}_0}$, we define

$$pR^t B^{\tilde{\xi}_0,\tilde{\xi}_0} R =: \operatorname{diag}\{q_1, \dots, q_m\}. \tag{5.79}$$

Note, as claimed, that $p \neq 0$ by assumption $(\partial^s a/\partial \xi_1^s) \neq 0$ in Observation 5.17, and $\operatorname{sgn} q_j = \operatorname{sgn} p$ by positivity of $B^{\tilde{\xi}_0,\tilde{\xi}_0}$ (an easy consequence of (2.55)).

Then, working in the Jordan basis defined in Observation 5.17, we find similarly as in the strictly hyperbolic case that the matrix perturbation problem (5.50) reduces to an $ms \times ms$ system consisting of m $s \times s$ equations

$$\begin{aligned}&\big(\mathrm{i}J + \mathrm{i}\sigma M_j + \rho N_j + \mathrm{o}\big(|(\rho,\sigma)|\big) - (\alpha - \alpha_0)\big)\mathbf{v}_{I_j}\\&\quad = \sum_{k \neq j} (\mathrm{i}\sigma M_{jk} + \rho N_{jk})\mathbf{v}_{I_k}\end{aligned} \tag{5.80}$$

(J denoting the standard Jordan block (5.63)), weakly coupled in the sense that lower left-hand corner elements vanish in the coupling blocks M_{jk}, N_{jk}, but in the diagonal blocks $\sigma \mathrm{i} M_j + \rho N_j$ are $p\sigma - \mathrm{i}q_j\rho \sim |\sigma| + |\rho|$. To order $(|\sigma| + |\rho|)^{1/s}$, therefore, the problem decouples, reducing to the computation of eigenvectors and eigenvalues of the block diagonal matrix

$$\operatorname{diag}\left\{\mathrm{i}\begin{pmatrix} 0 & 1 & 0 & \cdots & 0 \\ 0 & 0 & 1 & 0 & \cdots \\ \vdots & \vdots & \vdots & \vdots & \vdots \\ p\sigma - \mathrm{i}q_j\rho & 0 & 0 & \cdots & 0 \end{pmatrix}\right\}, \tag{5.81}$$

from which results (5.54)–(5.57) follow as before; see Section 2.2.4, *Splitting of a block-Jordan block*. (Note that the simple eigenvalue case $s = 1$ follows as a special case of the Jordan block computation.) Finally, the uniform condition (5.58) is a consequence of (5.55), orthonormality of bases $\{r_j\}$, and continuity of the Taylor expansions (5.75). □

REMARK 5.18 [Z.3]. Fixing j, consider a single cycle of roots $\pi_k^j := \varepsilon^k \mathrm{i}(p\sigma - \mathrm{i}q_j\rho)^{1/s}$ in (5.56), $j = 1, \dots, s$, $\varepsilon := 1^{1/s}$, $\operatorname{sgn} p = \operatorname{sgn} q_j$, on the set $\rho \geqslant 0$, $\rho, \sigma \in \mathbb{R}$ of interest. By direct evaluation at $\sigma = 0$, $\rho = 1/|q_j|$, for $\rho > 0$, the roots π_k^j split into s_+ unstable modes ($\operatorname{Re}\pi > 0$) and s_- stable modes ($\operatorname{Re}\pi < 0$), where

$$(s_+, s_-) = \begin{cases} (r, r) & \text{for } s = 2r, \\ (r+1, r) & \text{for } s = 2r+1 \text{ and } p > 0, \\ (r, r+1) & \text{for } s = 2r+1 \text{ and } p < 0. \end{cases} \tag{5.82}$$

(At $\sigma = 0$, $\rho = 1/|q_j|$, we have simply $\pi_k^j = \varepsilon^k (-\mathrm{i}^{s+1} \operatorname{sgn} q_k)^{1/s}$ which in the critical case $s = 2r + 1$ becomes $\varepsilon^k (-1)^{r+2} \operatorname{sgn} q_k = \varepsilon^k (-1)^r \operatorname{sgn} p$.)

REMARK 5.19 [Z.3]. Defining $w_j \in \mathbb{R}^s$ by $w_j := (1, \alpha_j, \alpha_j^2, \dots, \alpha_j^{s-1})^t$, $1 \leqslant j \leqslant p \leqslant s$, with $|\alpha_j| \leqslant C_1$, we find that $|w_1 \wedge \cdots \wedge w_p| \sim \Pi_{j<k}|\alpha_j - \alpha_k|$, with constants depending only on C_1, or equivalently

$$\left|\det\begin{pmatrix} \alpha_1^{q_1} & \cdots & \alpha_p^{q_1} \\ \vdots & \vdots & \vdots \\ \alpha_1^{q_p} & \cdots & \alpha_p^{q_p} \end{pmatrix}\right| \leqslant C\Pi_{j<k}|\alpha_j - \alpha_k| \tag{5.83}$$

for each minor $0 \leqslant q_1 < \cdots < q_p \leqslant s-1$ of matrix $(w_1, \dots, w_p) \in \mathbb{R}^{s\times p}$, with equality for the Vandermonde minor $q_r := r-1$. For, considering the determinants in (5.83) as complex polynomials $P(\alpha_1, \dots, \alpha_p)$, and noting that $P = 0$ whenever $\alpha_j = \alpha_k$, we may use successive applications of the remainder theorem to conclude that $\Pi_{j<k}(\alpha_j - \alpha_k)$ divides P.

REMARK 5.20. The direct calculations of Lemma 5.14 verifies Lemma 3.3 in the case that (H5) holds.

REMARK 5.21. With Lemma 5.14, Corollary 5.11 follows, also, by the Kreiss symmetrizer argument of [GMWZ.2].

LEMMA 5.22 [Z.3]. *Under the hypotheses of Theorem 5.5, for $\lambda \in \Gamma^{\tilde\xi}$ (defined in (5.36)) and $\rho := |(\tilde\xi, \lambda)|$, $\theta_1 > 0$, and $\theta > 0$ sufficiently small, there exists a choice of bases $\{\phi_j^{\pm}\}$, $\{\tilde\phi_j^{\pm}\}$, $\{\psi_j^{\pm}\}$, $\{\tilde\psi_j^{\pm}\}$ of solutions of the variable-coefficient eigenvalue problem such that, at $z = 0$, the wedge products*

$$\begin{pmatrix} \phi_1 \\ \phi_1{}' \end{pmatrix} \wedge \cdots \wedge \begin{pmatrix} \phi_n \\ \phi_n{}' \end{pmatrix}_{\pm} (z) \tag{5.84}$$

and determinants

$$\det\begin{pmatrix} \Phi & \Psi \\ \Phi' & \Psi' \end{pmatrix}_{\pm} (z) \tag{5.85}$$

and

$$\det\begin{pmatrix} \tilde\Phi & \tilde\Psi \\ \tilde\Phi' & \tilde\Psi' \end{pmatrix}_{\pm} (z) \tag{5.86}$$

are bounded in modulus (measured, in the case of wedge products, with respect to coordinates in some fixed basis) uniformly above and below, and, moreover, there hold bounds:

$$\begin{pmatrix} \phi_j^{\pm} \\ \phi_j^{\pm\prime} \end{pmatrix} = \gamma_{21,\phi_j^{\pm}}\left[e^{\mu_j^{\pm} x}\begin{pmatrix} \mathbf{v}_j^{\pm} \\ \mu_j \mathbf{v}_j^{\pm} \end{pmatrix} + \mathcal{O}(e^{-\theta|x|})\right], \qquad x_1 \gtrless 0, \tag{5.87}$$

$$\begin{pmatrix} \tilde\phi_j^{\pm} \\ \tilde\phi_j^{\pm\prime} \end{pmatrix} = \gamma_{21,\phi_j^{\pm}}\left[e^{-\mu_j^{\pm} x}\begin{pmatrix} \tilde{\mathbf{v}}_j^{\pm} \\ -\mu_j \tilde{\mathbf{v}}_j^{\pm} \end{pmatrix} + \mathcal{O}(e^{-\theta|x|})\right], \quad x_1 \gtrless 0, \tag{5.88}$$

and

$$\begin{pmatrix}\psi_j^{\pm}\\ \psi_j^{\pm\prime}\end{pmatrix} = \gamma_{21,\psi_j^{\pm}}\left[e^{\nu_j^{\pm}x}\begin{pmatrix}\mathbf{v}_j^{\pm}\\ \nu_j\mathbf{v}_j^{\pm}\end{pmatrix} + \mathcal{O}(e^{-\theta|x|})\right], \qquad x_1 \gtrless 0, \tag{5.89}$$

$$\begin{pmatrix}\tilde\psi_j^{\pm}\\ \tilde\psi_j^{\pm\prime}\end{pmatrix} = \gamma_{21,\tilde\psi_j^{\pm}}\left[e^{-\nu_j^{\pm}x}\begin{pmatrix}\mathbf{v}_j^{\pm}\\ -\nu_j\mathbf{v}_j^{\pm}\end{pmatrix} + \mathcal{O}(e^{-\theta|x|})\right], \qquad x_1 \gtrless 0, \tag{5.90}$$

where $|\mathbf{v}_j^{\pm}|$, $|\tilde{\mathbf{v}}_j^{\pm}|$ *are uniformly bounded above and below, and*:

(i) *For fast and intermediate-slow modes, or super-slow modes for which* $\operatorname{Im}\lambda$ *is bounded distance* $\theta_1\epsilon$ *away from any associated branch singularities* $\eta_h(\tilde\xi)$, $\mathbf{v}_j^{\pm}$ *are uniformly transverse to all other* $\mathbf{v}_k^{\pm}$, *while, for super-slow modes for which* $\operatorname{Im}\lambda$ *is within* $\theta_1\epsilon$ *of an (necessarily unique) associated branch singularity* $\eta_j(\tilde\xi)$, $\mathbf{v}_j^{\pm}$ *are as described in Lemma* 5.14, *with* $\sigma := \rho^{-1}(\operatorname{Im}\lambda - \eta_h(\tilde\xi))$; *moreover, the "group vectors" associated with each cluster* $\mathbf{v}_k^{j}$ *are uniformly transverse to each other and to all other modes. The dual vectors* $\tilde{\mathbf{v}}_j^{\pm}$ *satisfy identical bounds.*

(ii) *The decay/growth rates* $\mu_j^{\pm}/\nu_j^{\pm}$ *satisfy*

$$|\operatorname{Re}\mu_j^{\pm}|, |\operatorname{Re}\nu_j^{\pm}| \sim 1, \tag{5.91}$$

for fast modes,

$$|\operatorname{Re}\mu_j^{\pm}|, |\operatorname{Re}\nu_j^{\pm}| \sim \rho \tag{5.92}$$

for intermediate-slow modes and

$$|\operatorname{Re}\mu_j^{\pm}|, |\operatorname{Re}\nu_j^{\pm}| \sim \rho^2 \tag{5.93}$$

for super-slow modes; moreover,

$$|\mu_j^{\pm}|, |\nu_j^{\pm}| = \mathcal{O}(\rho) \tag{5.94}$$

for both intermediate- and super-slow modes.

(iii) *The factors* $\gamma_{21,\phi_j^{\pm}}$, $\gamma_{21,\tilde\phi_j^{\pm}}$, $\gamma_{21,\psi_j^{\pm}}$ *and* $\gamma_{21,\psi_j^{\pm}}$ *satisfy*

$$\gamma_{21,\beta} \sim 1, \quad \beta = \phi_j^{\pm}, \psi_j^{\pm}, \tilde\phi_j^{\pm}, \tilde\psi_j^{\pm}, \tag{5.95}$$

for fast and intermediate-slow modes, and for super-slow modes for which $\operatorname{Im}\lambda$ *is bounded distance* $\theta_1\epsilon$ *away from any associated branch singularities* $\eta_h(\tilde\xi)$, *and*

$$\gamma_{21,\beta} \sim \big(|\sigma| + |\rho|\big)^{-t_\beta}, \quad \beta = \phi_j^{\pm}, \psi_j^{\pm}, \tilde\phi_j^{\pm}, \tilde\psi_j^{\pm}, \tag{5.96}$$

for super-slow modes for which $\operatorname{Im}\lambda$ *is within* $\theta_1\epsilon$ *of an (necessarily unique) associated branch singularity* $\eta_h(\tilde\xi)$, *with*

$$\begin{aligned}&\left(t_{\phi_j^\pm}, t_{\psi_j^\pm}, t_{\tilde\phi_j^\pm}, t_{\tilde\psi_j^\pm}\right)\\ &:= \begin{cases} \left(\frac{r-1}{4r}, \frac{3r-1}{4r}, \frac{3r-1}{4r}, \frac{r-1}{4r}\right) & \textit{for } s = 2r,\\ \left(\frac{r}{2(2r+1)}, \frac{3r+1}{2(2r+1)}, \frac{3r}{2(2r+1)}, \frac{r-1}{2(2r+1)}\right) & \textit{for } s = 2r+1 \textit{ and } p \gtrless 0,\\ \left(\frac{r-1}{2(2r+1)}, \frac{3r}{2(2r+1)}, \frac{3r+1}{2(2r+1)}, \frac{r}{2(2r+1)}\right) & \textit{for } s = 2r+1 \textit{ and } p \lessgtr 0; \end{cases}\end{aligned} \tag{5.97}$$

here, $s := K_h^\pm$ *denotes the order of the associated branch singularity* $\eta_h^\pm(\tilde\xi)$ *and* $\sigma := \rho^{-1}(\operatorname{Im}\lambda - \eta_h(\tilde\xi))$, *as above.*

PROOF. Rescaling $\mathbf{v}_j^\pm \to \gamma_{21,\mathbf{v}_j^\pm}\mathbf{v}_j^\pm$ in (5.44), for any choice of $\gamma_{21,\mathbf{v}_j^\pm}$, we obtain immediately relations (5.87)–(5.93) and statements (i) and (ii) from the bounds (2.20)–(2.21) guaranteed by the conjugation lemma together with Lemma 5.14 and discussion above; recall the relation $\mu \sim -\rho\tilde\alpha$ in (5.48).

Using the results of Remarks 5.18 and 5.19, we readily obtain bounds (5.96) and (5.97) from the bounds on wedge product (5.84) and determinants (5.85) and (5.86), or equivalently (bounds (2.20)–(2.21); see also Exercise 4.20 of [Z.3]) the limiting, constant-coefficient versions

$$\begin{pmatrix}\bar{\bar\phi}_1\\ \bar{\bar\phi}'_1\end{pmatrix} \wedge\cdots\wedge \begin{pmatrix}\bar{\bar\phi}_n\\ \bar{\bar\phi}'_n\end{pmatrix}_\pm (z), \tag{5.98}$$

$$\det\begin{pmatrix}\bar{\bar\Phi} & \bar{\bar\Psi}\\ \bar{\bar\Phi}' & \bar{\bar\Psi}'\end{pmatrix}_\pm (z), \tag{5.99}$$

and

$$\det\begin{pmatrix}\bar{\bar{\tilde\Phi}} & \bar{\bar{\tilde\Psi}}\\ \bar{\bar{\tilde\Phi}}' & \bar{\bar{\tilde\Psi}}'\end{pmatrix}_\pm (z), \tag{5.100}$$

be bounded above and below at $z = 0$: more precisely, from the corresponding requirements on the restriction to each $s := K_h^\pm$-dimensional subspace of solutions corresponding to a single branch singularity $\eta_h^\pm(\cdot)$ (clearly sufficient, by group transversality, statement (i)).

For example, in the representative case $s = 2r$, we find from Remark 5.18(c) that there are r decaying modes $\bar{\bar\phi}_j^\pm$ and r growing modes $\bar{\bar\psi}_j^\pm$, which at $z = 0$ are given by rescalings $\gamma_{21,\mathbf{v}_{j,k}^\pm}\mathbf{v}_{j,k}^\pm$ of the vectors $\mathbf{v}_{j,k}^\pm$ associated with the branch cycle in question. Choosing common scalings $\gamma_{21,\mathbf{v}_{j,k}^\pm} :\equiv \gamma_\phi$ for all decaying modes $\bar{\bar\phi}_j^\pm$ and $\gamma_{21,\mathbf{v}_{j,k}^\pm} :\equiv \gamma_\psi$ for all growing modes $\bar{\bar\psi}_j^\pm$, and applying the result of Remark 5.19, we find that $(\bar{\bar\phi}_1 \wedge\cdots\wedge \bar{\bar\phi}_n)_\pm(0) \sim 1$

and $\det(\overline{\overline{\Phi}}, \overline{\overline{\Psi}})_\pm(0) \sim 1$ are equivalent to

$$\gamma_\phi^r \big((|\sigma| + |\rho|)^{1/s} \big)^{r(r-1)/2} \sim 1$$

and

$$\gamma_\phi^r \gamma_\psi^r \big((|\sigma| + |\rho|)^{1/s} \big)^{2r(2r-1)/2} \sim 1,$$

respectively, from which we immediately obtain the stated bounds on $t_{\phi_j^\pm}$ and $t_{\tilde{\phi}_j^\pm}$. The bounds on $t_{\tilde{\phi}_j^\pm}$ and $t_{\tilde{\psi}_j^\pm}$ then follow by duality,

$$\begin{aligned}(\widetilde{\Phi}, \widetilde{\Psi}) &= \big[(\Phi, \Psi)^{-1} \mathcal{S}^{\tilde{\xi}-1} \big]^* \\ &= \mathcal{S}^{\tilde{\xi}-1*} (\Phi, \Psi)^{-1*},\end{aligned} \tag{5.101}$$

$\mathcal{S}^{\tilde{\xi}} \sim I$ as defined in (4.25), which, restricted to the $s \times s$ cycle in question, yields $\gamma_{\tilde{\phi}} \sim \gamma_\phi^{-1} (|\sigma| + |\rho|)^{(s-1)/s}$ and $\gamma_{\tilde{\psi}} \sim \gamma_\psi^{-1} (|\sigma| + |\rho|)^{(s-1)/s}$, where $\gamma_{\tilde{\phi}}, \gamma_{\tilde{\psi}}$ denote common scalings for the dual growing, decaying modes associated with the branch cycle in question. The computations in other cases are analogous. □

EXAMPLE. In the "generic" case of a square-root singularity $s = 2$, calculation (5.101) becomes simply

$$(\tilde{\phi}, \tilde{\psi}) := \widetilde{\mathcal{S}}^{-1*} (\phi, \psi)^{-1*} = \frac{1}{2} \widetilde{\mathcal{S}}^{-1*} \begin{pmatrix} 1 & \pi^{1/2} \\ \pi^{-1/2} & -1 \end{pmatrix},$$

where

$$(\phi, \psi) := \begin{pmatrix} 1 & \pi^{-1/2} \\ \pi^{1/2} & -1 \end{pmatrix}$$

(recall: chosen so that $\det(\phi, \psi) = 2 \sim 1$) and $\widetilde{\mathcal{S}}$ denotes the (undetermined) restriction of $\mathcal{S}^{\tilde{\xi}}$ to the two-dimensional total eigenspace associated with the branching eigenvalue. Note that we do not require knowledge of $\widetilde{\mathcal{S}}$ in order to determine the orders of $\tilde{\phi}$ and $\tilde{\psi}$. This simple case is the one relevant to gas dynamics, for which glancing involves only the pair of strictly hyperbolic acoustic modes; see Remark 4.14 of [Z.3].

REMARK. The requirements $|\phi_1 \wedge \cdots \wedge \phi_n|_\pm(0) \sim 1$ are needed in order that the bases chosen in Lemma 5.14 be consistent with those defined in previous sections, in particular that they induce the same Evans function. (Note: since we do not require individual analyticity of our basis elements, we are free to choose C^∞ rescalings $\gamma_{21,\beta}$ so as to exactly match the Evans function defined in Section 3.) The requirements $\det(\Phi, \Psi)_\pm(0), \det(\widetilde{\Phi}, \widetilde{\Psi})_\pm(0) \sim 1$ are only a convenient normalization, to simplify certain calculations later on.

From now on, we will assume that the bases $\{\phi_j^\pm\}$, $\{\tilde\phi_j^\pm\}$, $\{\psi_j^\pm\}$, $\{\tilde\psi_j^\pm\}$ have been chosen as described in Lemma 5.22. For later use, we record the useful relations:

$$\begin{aligned}
&(t_{\phi_j}+t_{\tilde\phi_j})^\pm, (t_{\psi_j}+t_{\tilde\psi_j})^\pm \equiv 1-1/s,\\
&(t_{\psi_j}-t_{\phi_j})^\pm, (t_{\tilde\phi_j}-t_{\tilde\psi_j})^\pm \equiv 1/2,\\
&(t_{\phi_j}+t_{\tilde\psi_j})^\pm, (t_{\psi_j}+t_{\tilde\phi_j})^\pm \equiv 1/2-1/s,\\
&t_{\phi_j^\pm}, t_{\tilde\psi_j^\pm} \leqslant \frac{1}{4}(1-1/s)
\end{aligned} \tag{5.102}$$

between exponents t_β in (5.97); these may be verified case-by-case.

Refined derivative bounds. Similarly as in the one-dimensional case, for *slow, dual modes* the derivative bounds (5.88), (5.90) can be considerably sharpened, provided that we appropriately initialize our bases at $\rho = 0$, and this observation is significant in the Lax and overcompressive case.

Due to the special, conservative structure of the underlying evolution equations, the adjoint eigenvalue equation

$$\big(B^{11*}\tilde w'\big)' = -A^{1*}\tilde w' \tag{5.103}$$

at $\rho = 0$ admits an n-dimensional subspace of constant solutions; this is equivalent to the familiar fact that integral quantities are conserved under time evolution for systems of conservation laws. Thus, at $\rho = 0$, we may choose, by appropriate change of coordinates if necessary, that slow-decaying dual modes $\tilde\phi_j^\pm$ and slow-growing dual modes $\tilde\psi_j^\pm$ be identically constant, or equivalently that not only fast-decaying modes $\phi_j^\pm$, but also fast-growing modes $\psi_j^\pm$ be solutions of the linearized first-order traveling-wave ODE

$$B^{11}w' = A^1 w. \tag{5.104}$$

Note that this does not interfere with our previous choice in Lemma 3.4, nor does it interfere with the specifications of Lemma 5.22, since these concern only the choice of limiting solutions $\bar w_j^\pm$ of the asymptotic, constant coefficient equations at $x_1 \to \pm\infty$, and not the particular representatives $w_j^\pm$ that approach them (which, in the case of slow modes, are specified only up to the addition of an arbitrary fast-decaying mode).

LEMMA 5.23 [Z.3]. *With the above choice of bases at $\rho = 0$, slow modes $\tilde\phi_k^\pm$, $\tilde\psi_k^\pm$, for $\lambda \in \Gamma^{\tilde\xi}$ and $\rho := |(\tilde\xi, \lambda)|$ sufficiently small, satisfy*

$$\begin{aligned}
&\big|(\partial/\partial_{y_1})\tilde\psi_k^\pm\big| \leqslant C\rho\big|\tilde\psi_k^\pm\big|,\\
&\big|(\partial/\partial_{y_1})\tilde\phi_k^\pm\big| \leqslant C\rho\big|\tilde\phi_k^\pm\big|.
\end{aligned} \tag{5.105}$$

PROOF. Recall that slow dual modes $\bar{\tilde{\phi}}_k$ or $\bar{\tilde{\phi}}_k$, by our choice of basis, are identically constant for $\rho = 0$, (i.e, $(\partial/\partial_{y_1})\tilde{\phi}_{k|_{\rho=0}}$, $(\partial/\partial_{y_1})\tilde{\phi}_{k|_{\rho=0}} \equiv 0$, making (5.105) plausible). Indeed, the manifold of *all* slow dual modes, both decaying and growing, is exactly the manifold of constant functions. Now, consider the evolution of this manifold along the curve $(\tilde{\xi}, \lambda)(\tilde{\xi}_0, \tau_0, \rho)$ defined in (5.43), $\lambda_0 =: \mathrm{i}\tau_0$. The conglomerate slow manifold is C^1 in ρ, even though the separate decaying and growing manifolds may be only C^0; for, it is spectrally separated from all fast modes. Thus, the subspace

$$\operatorname{Span}\{(w, w')^t\colon\ w \in \text{slow manifold}\} \tag{5.106}$$

lies for each y_1 within angle $C\rho$ of the space $\{(w, 0)^t\}$ of constant functions, where "$'$" denotes ∂/∂_{y_1}. But, this is exactly the statement (5.105). □

Bounds (5.105) are to be compared with the bounds

$$\begin{aligned}
&\bigl|(\partial/\partial_{y_1})\tilde{\psi}_k^{\pm}\bigr| \leqslant C\rho\bigl|\tilde{\psi}_k^{\pm}\bigr| + C\gamma_{21,\tilde{\psi}_k^{\pm}}\mathrm{e}^{-\theta|y_1|},\\
&\bigl|(\partial/\partial_{y_1})\tilde{\phi}_k^{\pm}\bigr| \leqslant C\rho\bigl|\tilde{\phi}_k^{\pm}\bigr| + C\gamma_{21,\tilde{\psi}_k^{\pm}}\mathrm{e}^{-\theta|y_1|},
\end{aligned}$$

resulting from (5.88), (5.90), and (5.94).

Low-frequency Evans function bounds. The results of Section 5.1 suggest that the zeroes $\lambda_*(\tilde{\xi})$ of the Evans function satisfy

$$\operatorname{Re}\lambda_*(\tilde{\xi}) \leqslant -\theta|\tilde{\xi}|, \qquad \bigl|\operatorname{Im}\lambda_*(\tilde{\xi})\bigr| \leqslant C|\tilde{\xi}|, \quad \theta, C > 0, \tag{5.107}$$

for uniformly inviscid stable shocks, i.e. they move *linearly* into the stable complex half-plane with respect to $|\tilde{\xi}|$, and

$$\lambda_*(\tilde{\xi}) = \lambda_{*j} := \mathrm{i}\tau_j(\tilde{\xi}) - \beta_j(\tilde{\xi})|\tilde{\xi}|^2 + \mathrm{o}(|\tilde{\xi}|^2), \quad j = 1, \ldots, \ell, \tag{5.108}$$

τ_j real, distinct, β_j real, greater than 0, for weakly inviscid stable shocks satisfying the refined dynamical stability condition, where $\mathrm{i}\tau_j = \lambda_j(\tilde{\xi})$ are the roots of $\Delta(\tilde{\xi}_j, \cdot)$ (recall, strong refined dynamical stability requires that they be simple, Definition 1.21). The following result quantifies these observations via appropriate polar coordinate computations centered around the refined dynamical stability condition.

LEMMA 5.24 [Z.3]. *Under the hypotheses of Theorem 5.5, for $\lambda \in \partial\Lambda^{\tilde{\xi}}$ and $\rho := |(\tilde{\xi}, \lambda)|$ sufficiently small, there holds*

$$\bigl|D(\tilde{\xi}, \lambda)\bigr|^{-1} \leqslant C\rho^{-\ell} \tag{5.109}$$

for uniformly inviscid stable (UIS) shocks and

$$
\begin{aligned}
&|D(\tilde\xi,\lambda)|^{-1}\\
&\quad\leqslant C\rho^{-\ell+1}\Big[\min_j\big(|\mathrm{Im}(\lambda)-i\tau_j(\tilde\xi)|+|\mathrm{Re}(\lambda)+\beta_j(\tilde\xi)|\tilde\xi|^2|+o(|\tilde\xi|^2)\big)\Big]^{-1}
\end{aligned}
\tag{5.110}
$$

otherwise, where τ_j, β_j *are as in* (5.108) *and* $\Lambda^{\tilde\xi}$ *is as in* (2.62).

PROOF. Define the curves

$$
(\tilde\xi,\lambda)(\rho,\tilde\xi_0,\lambda_0):=\big(\rho\tilde\xi_0,\rho\lambda_0-\theta(\rho|\lambda_0|)^2-\theta(\rho|\tilde\xi_0|)^2\big),
\tag{5.111}
$$

for $\tilde\xi_0\in\mathbb{R}^{d-1}$, $\mathrm{Re}\,\lambda_0=0$, $|\tilde\xi_0|+|\lambda_0|=1$, and set

$$
\overline{\overline{D}}(\rho,\tilde\xi_0,\lambda_0):=D\big(\tilde\xi(\rho,\tilde\xi_0,\lambda_0),\lambda(\rho,\tilde\xi_0,\lambda_0)\big).
\tag{5.112}
$$

Inspection of the proof of Lemma 3.3 (either in [MéZ.2] or the later calculations in Section 5.3.2) show that continuity of slow decaying modes holds not only along rays, but also along the curved paths $(\tilde\xi,\lambda)(\rho,\tilde\xi_0,\lambda_0)$ defined in (5.111), provided $\theta>0$ is taken sufficiently small. Thus, we obtain by a calculation identical to that in the proof of Theorem 3.5, for $\theta>0$ sufficiently small, that

$$
\overline{\overline{D}}(\tilde\xi_0,\lambda_0,0)=\rho^\ell\overline{\Delta}(\tilde\xi_0,\lambda_0)+o(\rho^\ell).
\tag{5.113}
$$

In the UIS case, $\overline{\Delta}(\tilde\xi_0,\lambda_0)\neq0$, hence, for ρ sufficiently small,

$$
\begin{aligned}
\big|\overline{\overline{D}}(\tilde\xi_0,\lambda_0,\rho)\big|^{-1}&\leqslant C\big|\overline{\Delta}(\tilde\xi_0,\lambda_0)\big|^{-1}\rho^{-\ell}\\
&\leqslant C_2\big(|\tilde\xi|+|\lambda|\big)^{-\ell}
\end{aligned}
\tag{5.114}
$$

as claimed.

It remains to verify (5.110) for λ_0 in the vicinity of a point $i\tau_j(\tilde\xi_0)$ such that $\Delta(\tilde\xi_0, i\tau_j(\tilde\xi_0))=0$, where, by assumption, $\overline{\Delta}$ and D are analytic, with $\overline{\Delta}_\lambda\neq0$. Introducing again the function

$$
g^{\tilde\xi,\rho}(\lambda):=\rho^{-\ell}D(\rho\tilde\xi,\rho\lambda)
\tag{5.115}
$$

used in Section 5.1, we thus have

$$
\begin{aligned}
\overline{\overline{D}}(\tilde\xi_0,\lambda_0,\rho)&=\rho^\ell g^{\tilde\xi_0,\rho}(\lambda_0-\theta\rho)\\
&=\rho^\ell\big[g^{\xi_0,0}(i\tau_j)+g_\rho^{\xi_0,0}(i\tau_j)\rho
\end{aligned}
$$

$$+ g_\lambda^{\xi_0,0}(\mathrm{i}\tau_j)(\lambda_0 - \theta\rho_0 - \mathrm{i}\tau_j)$$
$$+ \mathcal{O}\big(\rho^2 + |\lambda_0 - \mathrm{i}\tau_j|^2 + \rho^2\big)\big]. \tag{5.116}$$

Using $g^{\tilde{\xi}_0,0}(\mathrm{i}\tau_j) = \overline{\Delta}(\tilde{\xi}_0, \mathrm{i}\tau_j) = 0$, $g_\lambda^{\tilde{\xi}_0,0}(\mathrm{i}\tau_j) = \overline{\Delta}_\lambda(\tilde{\xi}_0, \mathrm{i}\tau_j) \neq 0$, and $\overline{\Delta}_\lambda \neq 0$, together with definitions $\beta_j := (g_\rho/g_\lambda)|\tilde{\xi}_0|^{-2}$ and $\lambda(\tilde{\xi}_0, \lambda_0, \rho) := (\lambda_0 - \theta\rho)\rho$, we obtain (factoring out the term g_λ)

$$\begin{aligned}\big|\overline{\overline{D}}(\tilde{\xi}_0, \lambda_0, \rho)\big| &\geqslant C^{-1}|\rho^\ell||\overline{\Delta}_\lambda||\lambda_0 - \theta\rho - \mathrm{i}\tau_j + \beta_j|\tilde{\xi}_0|^2\rho + \Theta| \\ &= C^{-1}|\rho^{\ell-1}||\overline{\Delta}_\lambda||\lambda - \mathrm{i}\tau_j(\tilde{\xi}) + \beta_j|\tilde{\xi}|^2 + \rho\Theta|,\end{aligned} \tag{5.117}$$

where

$$\Theta = \mathcal{O}\big(|\lambda_0 - \mathrm{i}\tau_j|^2 + \rho^2\big). \tag{5.118}$$

The result then follows by the observation that $|\tilde{\xi}_0|/|\tau_j(\tilde{\xi}_0)|$ is bounded from zero (recall, $\Delta(0,1) \neq 0$), whence $\rho = \mathcal{O}(\tilde{\xi})$. □

Scattering coefficients. It remains to estimate the behavior of coefficients $M_{jk}^\pm$, $d_{jk}^\pm$ defined in Corollary 2.26, as $\rho \to 0$, $\rho := |(\tilde{\xi}, \lambda)|$. Consider coefficients $M_{jk}^\pm$. Expanding (2.101) using Cramer's rule, and setting $z = 0$, we obtain

$$M_{jk}^\pm = D^{-1} C_{jk}^\pm, \tag{5.119}$$

where

$$C^+ := (I, 0) \begin{pmatrix} \Phi^+ & \Phi^- \\ \Phi^{+\prime} & \Phi^{-\prime} \end{pmatrix}^{\mathrm{adj}} \begin{pmatrix} \Psi^- \\ \Psi^{-\prime} \end{pmatrix}\Bigg|_{z=0} \tag{5.120}$$

and a symmetric formula holds for C^-. Here, P^{adj} denotes the *adjugate matrix* of a matrix P, i.e. the transposed matrix of minors. As the adjugate is polynomial in the entries of the original matrix, it is evident that, at least away from branch singularities $\eta_j^\pm(\tilde{\xi})$, $|C^\pm|$ is uniformly bounded and therefore

$$\begin{aligned}|M_{jk}^\pm| &\leqslant C_1|D^{-1}| \\ &\leqslant C_2\rho^{-\ell}\begin{cases} 1, & \text{UIS} \\ \big[\min_j \rho^{-1}\big(|\mathrm{Im}(\lambda) - \tau_j(\tilde{\xi})| + \rho\big)\big]^{-1}, & \text{otherwise,}\end{cases}\end{aligned} \tag{5.121}$$

by Lemma 5.24, where $C_1, C_2 > 0$ are uniform constants.

However, the crude bound (5.121) hides considerable cancellation, a fact that will be crucial in our analysis. Again focusing on the curve $(\tilde{\xi}, \lambda)(\tilde{\xi}_0, \lambda_0, \rho)$ defined in (5.43), let us relabel the $\{\phi^\pm\}$ so that, at $\rho = 0$,

$$\phi_j^+ \equiv \phi_j^- = \partial\overline{U}^\delta/\partial\delta_j, \quad j = 1, \ldots, \ell. \tag{5.122}$$

(Note: as observed in Chapter 3, the fast decaying modes, among them $\{\partial\overline{U}^\delta/\partial\delta_j\}$, can be chosen independent of $(\tilde{\xi}_0, \lambda_0)$ at $\rho = 0$.)

In the case that $\overline{\Delta}(\tilde{\xi}_0, \lambda_0) = 0$ or equivalently $(\partial/\partial\rho)^\ell \overline{\overline{D}}(\xi_0, \lambda_0, \rho) = 0$, there is an additional dependency among

$$(\partial/\partial\rho)\phi_1^\pm, \ldots, (\partial/\partial\rho)\phi_\ell^\pm \quad \text{and} \quad \phi_1^\pm, \ldots, \phi_n^\pm,$$

so that we can arrange either

$$\phi_{\ell+1}^+ = \phi_{\ell+1}^- \tag{5.123}$$

or else (after C^1 change in coordinates)

$$(\partial/\partial\rho)\phi_1^+ = (\partial/\partial\rho)\phi_1^- \tag{5.124}$$

modulo slow decaying modes. The corresponding functions $\phi_{\ell+1}$, $(\partial/\partial\rho)\phi_1$, respectively, have an interpretation in the one-dimensional case as "effective eigenfunctions", see [ZH], Sections 5 and 6, for further discussion.

By assumption (1.42), in fact only case (5.124) can occur. For, review of the calculation giving $(\partial/\partial\rho)^\ell \overline{\overline{D}}(\tilde{\xi}_0, \lambda, \rho) = \Delta(\tilde{\xi}_0, \lambda)$ reveals that a dependency of form (5.123) implies dependence of rows involving $r_j^\pm$ in Δ. With convention (5.124), we have the sharpened bounds:

LEMMA 5.25 [Z.3]. *Under the hypotheses of Theorem* 5.5, *for* $\lambda \in \Gamma^{\tilde{\xi}}$ *and* $\rho := |(\tilde{\xi}, \lambda)|$ *sufficiently small, there holds*

$$|M_{jk}^\pm|, |d_{jk}^\pm| \leqslant C\gamma_{22,\beta}\tilde{\gamma}_1, \quad \beta = M_{jk}^\pm, d_{jk}^\pm, \tag{5.125}$$

where

$$\gamma_{22,M_{jk}^\pm}\gamma_{21,\phi_j^\pm}\gamma_{21,\tilde{\psi}_k^\mp} \leqslant \Big[1 + \sum_j \big(|\sigma_j^+| + |\rho|\big)^{-1/2(1-1/K_j^+)}\Big] \times \Big[1 + \sum_j \big(|\sigma_j^-| + |\rho|\big)^{-1/2(1-1/K_j^-)}\Big], \tag{5.126}$$

$$(\gamma_{22,d_{jk}}\gamma_{21,\phi_j}\gamma_{21,\tilde{\psi}_k})_\pm \leqslant 1 + \sum_j \big(|\sigma_j^\pm| + |\rho|\big)^{-(1-1/K_j^\pm)}, \tag{5.127}$$

with $\gamma_{21,\beta}$ *as defined in* (5.96)–(5.97), $\sigma_j^\pm := \rho^{-1}(\operatorname{Im}\lambda - \eta_j^\pm(\tilde{\xi}))$, $\eta_j(\cdot)$ *and* $K_j^\pm$ *as in* (H6), *and, for uniformly inviscid stable* (*UIS*) *shocks,*

$$\tilde{\gamma}_1 := \begin{cases} \rho^{-1} & \text{for } j = 1, \ldots, \ell, \\ 1 & \text{otherwise}, \end{cases} \tag{5.128}$$

while, for weakly inviscid stable (WIS) shocks,

$$\tilde{\gamma}_1 := \begin{cases} \left(\min_j \left|\mathrm{Im}(\lambda) - \mathrm{i}\tau_j(\tilde{\xi})\right| + \rho^2\right)^{-1} & \textit{for } j = 1, \\ \rho^{-1} & \textit{for } j = 2, \ldots, \ell, \\ 1 & \textit{otherwise}, \end{cases} \tag{5.129}$$

$\tau_j(\cdot)$ *as in* (5.108). (Here, as above, UIS and WIS denote "uniform inviscid stable" and "weak inviscid stable", as defined, respectively, in (1.32), Section 1.4, and (1.28), Section 1.3; the classification of Lax, overcompressive, and undercompressive types is given in Section 1.2.)

That is, apart from the factor induced by branch singularities, blow up in M_{jk} occurs to order $\rho^{\ell-1}|D^{-1}|$ rather than $|D^{-1}|$, and, more importantly only in *fast-decaying* modes.

PROOF OF LEMMA 5.25. Formula (5.120) may be rewritten as

$$C_{jk}^{+} = \det\begin{pmatrix} \phi_1^+, \ldots, \phi_{j-1}^+, \psi_k^-, \phi_j^+, \ldots, \phi_n^+, & \Phi^- \\ \phi_1^{+\prime}, \ldots, \phi_{j-1}^{+\prime}, \psi_k^{-\prime}, \phi_j^{+\prime}, \ldots, \phi_n^+, & \Phi^{-\prime} \end{pmatrix}\Bigg|_{z=0}, \tag{5.130}$$

from which we easily obtain the desired cancellation in $M^+ = C^+ D^{-1}$. For example, in the UIS case, we obtain for $j > \ell$ that, along path $(\tilde{\xi}, \lambda)(\tilde{\xi}_0, \lambda_0, \rho)$, we have, away from branch singularities $\eta_j^{\pm}(\tilde{\xi})$:

$$\begin{aligned} C_{jk}^{+} &= \det\begin{pmatrix} \phi_1^+ + \rho\phi_{1_\rho}^+ + \cdots, & \ldots, & \phi_n^+ + \rho\phi_{n_\rho}^+ + \cdots \\ \ldots, & \ldots, & \phi_n^{+\prime} + \rho\phi_{n_\rho}^{+\prime} + \cdots \end{pmatrix} \\ &= \mathcal{O}(\rho^\ell) \leqslant C|D|, \end{aligned} \tag{5.131}$$

yielding $|M_{jk}| = |C_{jk}||D|^{-1} \leqslant C$ as claimed, by elimination of ℓ zero-order terms, using linear dependency among fast modes at $\rho = 0$. Note, similarly as in the proof of Lemma 5.24, that we require in this calculation only that fast modes be C^1 in ρ, while slow modes may be only continuous. For this situation, $\gamma_{22} \sim 1$ suffices.

In the vicinity of one or more branch singularities, we must also take account of blowup in associated slow modes, as quantified by the factors $\gamma_{21,\beta}$ in Lemma 5.22. Branch singularities not involving ϕ_j^+ or $\tilde{\psi}_k^-$ clearly do not affect the calculation, since we have chosen a normalization for which complete cycles of branching (super-slow) modes $\phi_j^{\pm}$ have wedge products of order one; thus, $\gamma_{22} \sim 1$ again suffices for this case. In the vicinity of a branch singularity involving either or both of the modes ϕ_j^+ or $\tilde{\psi}_k^-$, however, there arises an additional blowup that must be accounted for.

Since this blowup involves only slow modes, it enters (5.131) simply as an additional multiplicative factor corresponding to the modulus of their wedge products at $z = 0$; as in the proof of Lemma 5.22, these can be estimated by the wedge products of the corresponding modes for the limiting, constant coefficient equations at $x_1 \to \pm\infty$, which at $z = 0$ are exactly the vectors $\mathbf{v}_j$, $\mathbf{v}_k^j$ described in Lemma 5.14. Since complete cycles have order one wedge product, as noted above, we need only compute the blowup in branch the cycles directly involving $\phi_j^+/\tilde{\psi}_k^-$, since these are modified in expression (5.131), respectively by deleting one decaying mode, ϕ_j^+, and augmenting one dual, growing mode, ψ_k^-.

Using the result of Remark 5.19, and the scaling $\gamma_{21,\beta}$ given in Lemma 5.22 for the modulus of individual modes, we find that the deleted cycle has wedge product of order

$$\pi^{(\tilde{r}-1)(\tilde{r}-2)/2}\pi^{-(\tilde{r}-1)(\tilde{r}-1)/2} = \pi^{-(\tilde{r}-1)/2} = \gamma_{21,\phi_j^+}, \tag{5.132}$$

where $\tilde{r}$ denotes the number of decaying modes in the original cycle: $\tilde{r} = r$ for $s := K_h^+ = 2r$ or $s = 2r+1$ and $p > 0$, otherwise $\tilde{r} = r+1$, and $\pi := (|\sigma| + |\rho|)^{1/s}$ as in the notation of Lemma 5.22. Here, $s := K_h^+$ is the multiplicity of the nearby branch singularity $\eta_h^+(\xi)$ associated with mode ϕ_j^+, and $\gamma = 21, \phi_j^+$ is as defined in (5.97).

Similarly, the augmented cycle involving ψ_k^- yields a wedge product of order

$$\pi^{(\tilde{r}+1)(\tilde{r})/2}\pi^{-(\tilde{r})(\tilde{r}-1)/2}\pi^{-(2s-\tilde{r}-2)/2} = \pi^{-(3\tilde{r}-2s+1)/2} = \gamma_{21,\tilde{\psi}_k^-}, \tag{5.133}$$

where $\gamma_{21,\tilde{\psi}_k^-}$ is as defined in (5.96) and (5.97). The asserted bound on γ_{22} then follows by the observation (see (5.102)) that

$$t_{\phi_j^\pm}, t_{\tilde{\psi}_j^\pm} \leqslant \frac{1}{4}\left(1 - 1/K_h^\pm\right)$$

in (5.97) ($K_h^\pm =: s =: 2r$ in the notation of formulae (5.97)). This completes the estimation of $|M_{jk}|$ in the case $x_1 \geqslant y_1$; the case $x_1 \leqslant y_1$ may be treated by a symmetric argument.

The bounds on $|d_{jk}^\pm|$ follow similarly, using (2.100) in place of (2.99). In these cases, however, there is the new possibility that the augmented mode ψ_k and the deleted mode ϕ_j may belong to the same cycle, since both are associated with the same spatial infinity. In this case, the total number of modes remains fixed, with a growth mode replacing a decay mode, and the resulting blowup in the associated wedge product is simply

$$(\gamma_{21,\psi_j}/\gamma_{21,\phi_j})_\pm. \tag{5.134}$$

The asserted bound on $(\gamma_{22}\gamma_{21,\phi_j}\gamma_{21,\tilde{\psi}_k})_\pm$ then follows from relation

$$\begin{aligned}(t_{\psi_j} - t_{\phi_j} + t_{\phi_j} + t_{\tilde{\psi}_k})_\pm &= (t_{\psi_j} - t_{\phi_j} + t_{\phi_j} + t_{\tilde{\psi}_j})_\pm \\ &\equiv 1 - 1/K_h^\pm,\end{aligned}$$

again obtained by inspection of formulae (5.97) (see (5.102)), where $K_h^\pm$ as usual denotes the order of the associated branch singularity. If on the other hand the augmented and deleted modes belong to different cycles, then the computation reduces essentially to that of the previous case, yielding

$$\gamma_{22} = (\gamma_{21,\phi_j}\gamma_{21,\tilde{\psi}_k})_\pm,$$

from which the result follows as before (indeed, we obtain a slightly better bound, see (5.102)). □

REMARK 5.26. Note that frequency blowup in term $M_{jk}\phi_j^+\tilde{\psi}_k^-$ splits evenly between coefficient M_{jk} and modes $\phi_j^+\tilde{\psi}_k^-$, see (5.132) and (5.133).

Lax and overcompressive case. For Lax and overcompressive shocks, we can say a bit more. First, observe that duality relation

$$\begin{aligned}
&\det\begin{pmatrix} \widetilde{\Psi}^- & \widetilde{\Phi}^- \\ \widetilde{\Psi}^{-\prime} & \widetilde{\Phi}^{-\prime} \end{pmatrix}^* \mathcal{S} \begin{pmatrix} \Phi^+ & \Phi^- \\ \Phi^{+\prime} & \Phi^{-\prime} \end{pmatrix} \\
&\quad = \det\begin{pmatrix} \begin{pmatrix} \widetilde{\Psi}^- \\ \widetilde{\Psi}^{-\prime} \end{pmatrix}^* \mathcal{S} \begin{pmatrix} \Phi^+ \\ \Phi^{+\prime} \end{pmatrix} & 0 \\ * & I \end{pmatrix} \\
&\quad = \det\begin{pmatrix} \begin{pmatrix} \widetilde{\Psi}^- \\ \widetilde{\Psi}^{-\prime} \end{pmatrix}^* \mathcal{S} \begin{pmatrix} \Phi^+ \\ \Phi^{+\prime} \end{pmatrix} & * \\ 0 & I \end{pmatrix} \\
&\quad = \det\begin{pmatrix} \widetilde{\Psi}^- & \widetilde{\Psi}^+ \\ \widetilde{\Psi}^{-\prime} & \widetilde{\Psi}^{+\prime} \end{pmatrix}^* \mathcal{S} \begin{pmatrix} \Phi^+ & \Psi^+ \\ \Phi^{+\prime} & \Psi^{+\prime} \end{pmatrix}
\end{aligned} \tag{5.135}$$

yields that $D(\tilde{\xi}, \lambda)$ and the *dual Evans function*

$$\widetilde{D} := \det\begin{pmatrix} \widetilde{\Psi}^- & \widetilde{\Psi}^+ \\ \widetilde{\Psi}^{-\prime} & \widetilde{\Psi}^{+\prime} \end{pmatrix}\bigg|_{x_1=0} \tag{5.136}$$

vanish to the same order as $(\tilde{\xi}, \lambda) \to (0, 0)$, since the determinants of $\begin{pmatrix} \widetilde{\Psi}^- & \widetilde{\Phi}^- \\ \widetilde{\Psi}^{-\prime} & \widetilde{\Phi}^{-\prime} \end{pmatrix}$, $\mathcal{S}$, and $\begin{pmatrix} \Phi^+ & \Psi^+ \\ \Phi^{+\prime} & \Psi^{+\prime} \end{pmatrix}$ are by construction bounded above and below. Indeed, subspaces $\ker\begin{pmatrix} \Phi^+ & \Phi^- \\ \Phi^{+\prime} & \Phi^{-\prime} \end{pmatrix}$ and $\ker\begin{pmatrix} \widetilde{\Psi}^+ & \widetilde{\Psi}^+ \\ \widetilde{\Psi}^{-\prime} & \widetilde{\Psi}^{+\prime} \end{pmatrix}$ can be seen to have the same dimension, $\dim\ker\begin{pmatrix} \widetilde{\Psi}^- \\ \widetilde{\Psi}^{-\prime} \end{pmatrix}^* \mathcal{S}\begin{pmatrix} \Phi^+ \\ \Phi^{+\prime} \end{pmatrix}$, which at $\rho = 0$ is equal to ℓ. This computation partially recovers the results of [ZH], section 6, without the assumption of smoothness in $\{\phi_j^{\pm}\}$, $\{\psi_j^{\pm}\}$. The "contracted" Evans function

$$\det\begin{pmatrix} \widetilde{\Psi}^- \\ \widetilde{\Psi}^{-\prime} \end{pmatrix}^* \mathcal{S} \begin{pmatrix} \Phi^+ \\ \Phi^{+\prime} \end{pmatrix}$$

is discussed further in [BSZ.1].

It follows that there is an ℓ-fold intersection between

$$\operatorname{Span}\begin{pmatrix} \widetilde{\Psi}_j^- \\ \widetilde{\Psi}_j^{-\prime} \end{pmatrix} \quad \text{and} \quad \operatorname{Span}\begin{pmatrix} \widetilde{\Psi}_j^+ \\ \widetilde{\Psi}_j^{+\prime} \end{pmatrix}.$$

We now recall the important observation of Lemma 4.28. Recall, Section 4.5.2, that we have fixed a choice of bases such that, at $\rho = 0$, both slow-decaying dual modes $\tilde{\phi}_j^{\pm}$

and slow-growing dual modes $\tilde{\psi}_j^{\pm}$ are identically constant, or equivalently that not only fast-decaying modes $\phi_j^{\pm}$, but also fast-growing modes $\psi_j^{\pm}$ are solutions of the linearized traveling wave ODE (5.104). With this choice of basis, we have for Lax and overcompressive shocks that the only bounded solutions of the adjoint eigenvalue equation at $\rho = 0$ are constant solutions, or, equivalently, fast-decaying modes ψ_j at one infinity are fast-growing at the other, and fast-decaying $\tilde{\psi}_j$ at one infinity are fast-growing at the other. Combining the above two results, we have:

LEMMA 5.27. *Under the hypotheses of Theorem 5.5, with $|(\tilde{\xi}, \lambda)|$ sufficiently small, for Lax and overcompressive shocks, with appropriate basis at $\rho = 0$ (i.e., slow dual modes taken identically constant), there holds*

$$|M_{jk}|, |d_{jk}| \leqslant C\gamma_{22} \tag{5.137}$$

if $\tilde{\psi}_k$ is a fast mode, and

$$|M_{jk}|, |d_{jk}| \leqslant C\gamma_{22}\rho \tag{5.138}$$

if, additionally, ϕ_j is a slow mode, where γ_{22} is as defined in (5.126) *and* (5.127).

PROOF. Transversality, $\gamma \neq 0$, follows from (D2), so that the results of Lemma 4.28 hold. Consider first, the more difficult case of uniform inviscid stability, $\Delta(\tilde{\xi}_0, \lambda) \neq 0$.

We first establish the bound (5.137). By Lemma 5.25, we need only consider $j = 1, \dots, \ell$, for which $\phi_j^- = \phi_j^+$. By Lemma 4.28, all fast-growing modes ψ_k^+, ψ_k^- lie in the fast-decaying manifolds $\mathrm{Span}(\phi_1^-, \dots, \phi_{i_-}^-)$, $\mathrm{Span}(\phi_1^+, \dots, \phi_{i_+}^+)$ at $-\infty$, $+\infty$, respectively. It follows that in the right-hand side of (5.130), there is at $\rho = 0$ a linear dependency between columns

$$\phi_1^+, \dots, \phi_{j-1}^+, \psi_k^-, \phi_{j+1}^+, \phi_{i_+}^+ \quad \text{and} \quad \phi_j^- = \phi_j^+, \tag{5.139}$$

i.e., an ℓ-fold dependency among columns

$$\phi_1^+, \dots, \psi_k^-, \dots, \phi_{i+}^+ \quad \text{and} \quad \phi_1^-, \dots, \phi_\ell^-, \tag{5.140}$$

all of which, as fast modes, are C^1 in ρ. It follows as in the proof of Lemma 5.25 that $|C_{jk}| \leqslant C\gamma_{22}\rho^\ell$ for ρ near zero, giving bound (5.137) for M_{jk}. If ϕ_j^+ is a slow mode, on the other hand, then the same argument shows that there is a linear dependency in columns

$$\phi_1^+, \dots, \phi_{j-1}^+, \psi_k^-, \phi_{j+1}^+, \phi_{i_+}^+ \tag{5.141}$$

and an $(\ell + 1)$-fold dependency in (5.140), since the omitted slow mode ϕ_j^+ plays no role in either linear dependence; thus, we obtain the bound (5.138), instead. Analogous calculations yield the result for $d_{jk}^{\pm}$ as well.

In the case of weak inviscid stable shocks, the above argument suffices near $(\tilde{\xi}, \lambda)$ such that $\Delta(\tilde{\xi}, \lambda) \neq 0$ (again, computing as in the proof of Lemma 5.25 along the curves

$(\tilde{\xi}, \rho)(\tilde{\xi}_0, \lambda, \rho))$. Near points such that $\Delta(\tilde{\xi}_0, \lambda) = 0$, all modes are C^1 in ρ, and the result can be obtained more simply by a dual version of the calculation in the proof of Lemma 5.25, together with the observation that the "dual eigenfunctions" $\tilde{\psi}_k^{\pm}$ lying in $\mathrm{Span}(\widetilde{\Psi}^-) \cap \mathrm{Span}(\widetilde{\Psi}^+)$ are by Lemma 4.28 restricted to the *slow modes*. □

PROOF OF PROPOSITION 5.10. The proof of Proposition 5.10 is now straightforward. Collecting the results of Lemmas 5.22 and 5.23, we have the spatial decay bounds

$$\begin{aligned} |\phi_j(x_1)| &\leqslant C\gamma_{21,\phi_j} e^{-\theta|x_1|}, \\ |\phi_j'(x_1)| &\leqslant C\gamma_{21,\phi_j} e^{-\theta|x_1|} \end{aligned} \tag{5.142}$$

for fast forward modes,

$$\begin{aligned} |\phi_j(x_1)| &\leqslant C\gamma_{21,\phi_j} e^{-\theta\rho^2|x_1|}, \\ |\phi_j'(x_1)| &\leqslant C\gamma_{21,\phi_j} \left(\rho e^{-\theta\rho^2|x_1|} + e^{-\theta|x_1|}\right) \end{aligned} \tag{5.143}$$

for slow forward modes,

$$\begin{aligned} |\tilde{\psi}_k(y_1)| &\leqslant C\gamma_{21,\tilde{\psi}_k} e^{-\theta|y_1|}, \\ |\tilde{\psi}_k'(y_1)| &\leqslant C\gamma_{21,\tilde{\psi}_k} e^{-\theta|y_1|} \end{aligned} \tag{5.144}$$

for fast dual modes, and

$$\begin{aligned} |\tilde{\psi}_k(y_1)| &\leqslant C\gamma_{21,\tilde{\psi}_k} e^{-\rho^2\theta|y_1|}, \\ |\tilde{\psi}_k'(y_1)| &\leqslant C\gamma_{21,\tilde{\psi}_k} \left(\rho e^{-\theta\rho^2|y_1|} + \alpha e^{-\theta|y_1|}\right) \end{aligned} \tag{5.145}$$

for slow dual modes, where $\gamma_{21,\beta}$ are as defined in (5.96) and (5.97) and

$$\alpha = \begin{cases} 0 & \text{for Lax and overcompressive shocks,} \\ 1 & \text{for undercompressive shocks.} \end{cases}$$

Likewise, the results of Lemmas 5.25 and 5.27 give the frequency growth bounds

$$|M_{jk}^{\pm}|, |d_{jk}^{\pm}| \leqslant C\gamma_{22} \tag{5.146}$$

for slow modes $\phi_j^{\pm}$, and

$$|M_{jk}^{\pm}|, |d_{jk}^{\pm}| \leqslant C\rho^{-1}\gamma_{22}\gamma_1 \tag{5.147}$$

for fast modes $\phi_j^{\pm}$, where γ_1 is as defined in (5.37) and γ_{22} as in (5.127) and (5.126). In the critical case that $\phi_j^{\pm}$ is slow and $\tilde{\psi}_k^{\pm}$ is fast, (5.146) can be sharpened to

$$|M_{jk}^{\pm}|, |d_{jk}^{\pm}| \leqslant C\gamma_{22}(\rho + \alpha), \tag{5.148}$$

α as above.

Combining these bounds, we may estimate the various terms arising in the decompositions of Corollary 2.26 as:

$$|\phi_j M_{jk}\tilde{\psi}_k|, |\phi_j d_{jk}\tilde{\psi}_k| \leqslant C\gamma_{21,\phi_j}\gamma_{21,\tilde{\psi}_k}\gamma_{22}\gamma_1\rho^{-1}\mathrm{e}^{-\theta|x_1|}\mathrm{e}^{-\theta\rho^2|y_1|}$$

and

$$\begin{aligned}&|(\partial/\partial y_1)\phi_j M_{jk}\tilde{\psi}_k|, |(\partial/\partial y_1)\phi_j d_{jk}\tilde{\psi}_k|\\&\quad\leqslant C\gamma_{21,\phi_j}\gamma_{21,\tilde{\psi}_k}\gamma_{22}\gamma_1\rho^{-1}\mathrm{e}^{-\theta|x_1|}\big(\rho\mathrm{e}^{-\theta\rho^2|y_1|}+\alpha\mathrm{e}^{-\theta|y_1|}\big)\end{aligned}$$

for fast modes ϕ_j,

$$|\phi_j M_{jk}\tilde{\psi}_k|, |\phi_j d_{jk}\tilde{\psi}_k| \leqslant C\gamma_{21,\phi_j}\gamma_{21,\tilde{\psi}_k}\gamma_{22}\mathrm{e}^{-\theta\rho^2|x_1|}\mathrm{e}^{-\theta\rho^2|y_1|}$$

and

$$\begin{aligned}&|(\partial/\partial y_1)\phi_j M_{jk}\tilde{\psi}_k|, |(\partial/\partial y_1)\phi_j d_{jk}\tilde{\psi}_k|\\&\quad\leqslant C\gamma_{21,\phi_j}\gamma_{21,\tilde{\psi}_k}\gamma_{22}\mathrm{e}^{-\theta\rho^2|x_1|}\big(\rho\mathrm{e}^{-\theta\rho^2|y_1|}+\alpha\mathrm{e}^{-\theta|y_1|}\big)\end{aligned}$$

for slow modes ϕ_j,

$$|\phi_j\tilde{\phi}_j|_\pm \leqslant C(\gamma_{21,\phi_j}\gamma_{21,\tilde{\phi}_j})_\pm\mathrm{e}^{-\theta|x_1-y_1|}$$

and

$$|(\partial/\partial y_1)\phi_j\tilde{\phi}_j|_\pm \leqslant C(\gamma_{21,\phi_j}\gamma_{21,\tilde{\phi}_j})_\pm\mathrm{e}^{-\theta|x_1-y_1|}$$

for fast modes ϕ_j,

$$|\phi_j\tilde{\phi}_j|_\pm \leqslant C(\gamma_{21,\phi_j}\gamma_{21,\tilde{\phi}_j})_\pm\mathrm{e}^{-\theta\rho^2|x_1-y_1|}$$

and

$$|(\partial/\partial y_1)\phi_j\tilde{\phi}_j|_\pm \leqslant C(\gamma_{21,\phi_j}\gamma_{21,\tilde{\phi}_j})_\pm\rho\mathrm{e}^{-\theta\rho^2|x_1-y_1|}$$

for slow modes ϕ_j,

$$|\psi_j\tilde{\psi}_j|_\pm \leqslant C(\gamma_{21,\psi_j}\gamma_{21,\tilde{\psi}_j})_\pm\mathrm{e}^{-\theta|x_1-y_1|}$$

and

$$|(\partial/\partial y_1)\psi_j\tilde{\psi}_j|_\pm \leqslant C(\gamma_{21,\psi_j}\gamma_{21,\tilde{\psi}_j})_\pm\mathrm{e}^{-\theta|x_1-y_1|}$$

for fast modes ψ_j, and

$$|\psi_j \tilde{\psi}_j|_{\pm} \leqslant C(\gamma_{21,\psi_j}\gamma_{21,\tilde{\psi}_j})_{\pm} e^{-\theta\rho^2|x_1-y_1|}$$

and

$$|(\partial/\partial y_1)\psi_j \tilde{\psi}_j|_{\pm} \leqslant C(\gamma_{21,\psi_j}\gamma_{21,\tilde{\psi}_j})_{\pm} \rho e^{-\theta\rho^2|x_1-y_1|}$$

for slow modes ψ_j. Using bounds (5.127) and (5.126), and recalling (see (5.102)) that

$$(t_{\phi_j} + t_{\tilde{\phi}_j})_{\pm}, (t_{\psi_j} + t_{\tilde{\psi}_j})_{\pm} \equiv 1 - 1/K_j^{\pm},$$

we thus obtain the result by checking term by term. □

High-frequency bounds. Modifying the auxiliary energy estimate and one-dimensional resolvent estimates already derived, we obtain in straightforward fashion the following high-frequency resolvent bounds.

PROPOSITION 5.28. *Under the hypotheses of Theorem 5.5, for $|\tilde{\xi}|$ sufficiently large, and some $C, \theta > 0$,*

$$|e^{L_{\tilde{\xi}}t} f|_{\hat{H}^1(x_1,\tilde{\xi})} \leqslant Ce^{-\theta t}|f|_{\hat{H}^1(x_1,\tilde{\xi})} \tag{5.149}$$

for $|f|_{\hat{H}^1(x_1,\tilde{\xi})} := (1 + |\tilde{\xi}|)|f|_{H^1(x_1)}$.

PROOF. Carrying out a Fourier-transformed version of the auxiliary energy estimate of Propositions 4.15 and 5.9 in the simpler, linearized setting, we obtain

$$\frac{d}{dt}\mathcal{E}(W(t)) \leqslant -\theta\mathcal{E}(t) + C|W|^2_{L^2(x_1)},$$

$\theta > 0$, where

$$\begin{aligned}\mathcal{E}(W) := {} & M(1 + |\tilde{\xi}|^2)\langle W, \alpha\widetilde{A}^0 W\rangle + \langle \alpha K(\partial x_1, i\tilde{\xi})W, W\rangle \\ & + M\langle \partial_{x_1} W, \alpha\widetilde{A}^0 \partial_{x_1} W\rangle,\end{aligned} \tag{5.150}$$

M, $C > 0$ sufficiently large constants, $K(\partial_x)$ a first-order differential operator as defined in (5.19) and (5.20), or, more generally, an analogous first-order pseudodifferential operator as discussed in the proof of Proposition 5.9 (see also [Z.4]). Observing that $|W|^2_{L^2} \leqslant C|\tilde{\xi}|^{-2}\mathcal{E}(W)$ can be absorbed in $\theta\mathcal{E}(W)$, we obtain

$$\frac{d}{dt}\mathcal{E}(W(t)) \leqslant -\frac{\theta}{2}\mathcal{E}(t),$$

from which the result follows by the fact that $\mathcal{E}(W)^{1/2}$ is equivalent to the $\widehat{H}^1(x_1,\tilde{\xi})$ norm of W. □

PROPOSITION 5.29. *Under the hypotheses of Theorem 5.5, for $|\tilde{\xi}| \leqslant R$ and $|\lambda|$ sufficiently large,* $\operatorname{Re}\lambda \geqslant -\eta$ *is contained in the resolvent set $\rho(L_{\tilde{\xi}})$ for some uniform $\eta > 0$ sufficiently small, with resolvent kernel $G_{\tilde{\xi},\lambda}(x,y)$ satisfying the same decomposition* (4.175)–(4.187) *as in the one-dimensional case, but with dissipation coefficients η_j^* replaced by multidimensional versions $\eta_j(\tilde{\xi},x)^*$, exponentially decaying to asymptotic values*

$$\operatorname{Re}\sigma\eta_j(\tilde{\xi},\pm\infty)^* > 0. \tag{5.151}$$

PROOF. For $|\tilde{\xi}|$ bounded and $|\lambda|$ sufficiently large, we may treat $\tilde{\xi}$ terms as first-order perturbations in the one-dimensional analysis in the proof of Proposition 4.33, order $|\lambda|^{-1}$ after the rescaling of (4.189)–(4.193). These affect only the bounds on the hyperbolic block, modifying the form of η_j^* in (4.202). We do not require the precise dependence of the resulting $\eta_j(\tilde{\xi},x)^*$ on $\tilde{\xi}$, but only the bound (5.151), which follows as in the one-dimensional case by (2.55) together with the fact that $-\operatorname{Re}\sigma\eta_j(\tilde{\xi},\pm\infty)$ corresponds to the zero-order term in the expansion of hyperbolic modes of symbol $-A^{\tilde{\xi}} - B^{\tilde{\xi},\tilde{\xi}}$. □

5.3.2. *Bounds on the solution operator.* We are now ready to establish the claimed bounds on the linearized solution operator. Define "low-" and "high-frequency" parts of the linearized solution operator $S(t)$

$$S_1(t) := \frac{1}{(2\pi)^d \mathrm{i}} \int_{|\tilde{\xi}|\leqslant r} \oint_{\Gamma^{\tilde{\xi}}} \mathrm{e}^{\lambda t + \mathrm{i}\tilde{\xi}\cdot\tilde{x}} (L_{\tilde{\xi}} - \lambda)^{-1} \,\mathrm{d}\lambda\,\mathrm{d}\tilde{\xi} \tag{5.152}$$

and

$$S_2(t) := \mathrm{e}^{Lt} - S_1(t), \tag{5.153}$$

$\Gamma^{\tilde{\xi}}$ as in (5.36) and $r > 0$ sufficiently small.

PROOF OF PROPOSITION 5.7. We estimate S_1 and S_2 in turn, for simplicity restricting to the case of a uniformly stable Lax or overcompressive shock, $\gamma_1 = 1$.

S_1 bounds. Let $\hat{u}(x_1,\tilde{\xi},\lambda)$ denote the solution of $(L_{\tilde{\xi}} - \lambda)\hat{u} = \hat{f}$, where $\hat{f}(x_1,\tilde{\xi})$ denotes Fourier transform of f, and

$$u(x,t) := S_1(t) f = \frac{1}{(2\pi)^d \mathrm{i}} \int_{|\tilde{\xi}|\leqslant r} \oint_{\Gamma^{\tilde{\xi}}} \mathrm{e}^{\lambda t + \mathrm{i}\tilde{\xi}\cdot\tilde{x}} (L_{\tilde{\xi}} - \lambda)^{-1} \hat{f}(x_1,\tilde{\xi}) \,\mathrm{d}\lambda\,\mathrm{d}\tilde{\xi}.$$

Bounding $|\tilde{f}|_{L^\infty(\tilde{\xi},L^1(x_1))} \leqslant |f|_{L^1(x_1,\tilde{x})} = |f|_1$ using Hausdorff–Young inequality and appealing to the $L^1 \to L^p$ resolvent estimates of Corollary 5.11, we may thus bound

$$\big|\hat{u}(x_1,\tilde{\xi},\lambda)\big|_{L^p(x_1)} \leqslant |f|_1 b(\tilde{\xi},\lambda),$$

where $b := C\gamma_2\rho^{-1}$.

L^2 bounds. Using in turn Parseval's identity, Fubini's theorem, the triangle inequality, and our $L^1 \to L^2$ resolvent bounds, we may estimate

$$\begin{aligned}
&|u|_{L^2(x_1,\tilde{x})}(t)\\
&= \left(\frac{1}{(2\pi)^d}\int_{x_1}\int_{\tilde{\xi}\in\mathbb{R}^{d-1}}\left|\oint_{\lambda\in\tilde{\Gamma}(\tilde{\xi})} e^{\lambda t}\hat{u}(x_1,\tilde{\xi},\lambda)\,d\lambda\right|^2 d\tilde{\xi}\,dx_1\right)^{1/2}\\
&= \left(\frac{1}{(2\pi)^d}\int_{\tilde{\xi}\in\mathbb{R}^{d-1}}\left|\oint_{\lambda\in\tilde{\Gamma}(\tilde{\xi})} e^{\lambda t}\hat{u}(x_1,\tilde{\xi},\lambda)\,d\lambda\right|^2_{L^2(x_1)} d\tilde{\xi}\right)^{1/2}\\
&\leqslant \left(\frac{1}{(2\pi)^d}\int_{\tilde{\xi}\in\mathbb{R}^{d-1}}\left|\oint_{\lambda\in\tilde{\Gamma}(\tilde{\xi})} |e^{\lambda t}||\hat{u}(x_1,\tilde{\xi},\lambda)|_{L^2(x_1)}\,d\lambda\right|^2 d\tilde{\xi}\right)^{1/2}\\
&\leqslant |f|_1\left(\frac{1}{(2\pi)^d}\int_{\tilde{\xi}\in\mathbb{R}^{d-1}}\left|\oint_{\lambda\in\tilde{\Gamma}(\tilde{\xi})} e^{\mathrm{Re}\,\lambda t}b(\tilde{\xi},\lambda)\,d\lambda\right|^2 d\tilde{\xi}\right)^{1/2},
\end{aligned}\tag{5.154}$$

from which we readily obtain the claimed bound on $|S_1(t)f|_{L^2(x)}$ using the bounds on b. Specifically, parametrizing $\Gamma(\tilde{\xi})$ by

$$\lambda(\tilde{\xi},k) = ik - \theta_1\big(k^2 + |\tilde{\xi}|^2\big), \quad k\in\mathbb{R},$$

and observing that in nonpolar coordinates

$$\begin{aligned}
\rho^{-1}\gamma_2 &\leqslant \left[\big(|k|+|\tilde{\xi}|\big)^{-1}\left(1+\sum_{j\geqslant 1}\left(\frac{|k-\tau_j(\tilde{\xi})|}{\rho}\right)^{1/s_j-1}\right)\right]\\
&\leqslant \left[\big(|k|+|\tilde{\xi}|\big)^{-1}\left(1+\sum_{j\geqslant 1}\left(\frac{|k-\tau_j(\tilde{\xi})|}{\rho}\right)^{\epsilon-1}\right)\right],
\end{aligned}\tag{5.155}$$

where $\epsilon := \frac{1}{\max_j s_j}$ ($0<\epsilon<1$ chosen arbitrarily if there are no singularities), we obtain a contribution bounded by

$$\begin{aligned}
&C|f|_1\left(\int_{\tilde{\xi}\in\mathbb{R}^{d-1}}\left|\int_{-\infty}^{+\infty} e^{-\theta(k^2+|\tilde{\xi}|^2)t}(\rho)^{-1}\gamma_2\,dk\right|^2 d\tilde{\xi}\right)^{1/2}\\
&\quad\leqslant C|f|_1\int_{\tilde{\xi}\in\mathbb{R}^{d-1}}\left(e^{-2\theta|\tilde{\xi}|^2 t}|\tilde{\xi}|^{-2\epsilon}\left|\int_{-\infty}^{+\infty} e^{-\theta|k|^2 t}|k|^{\epsilon-1}\,dk\right|^2 d\tilde{\xi}\right)^{1/2}\\
&\qquad + C\sum_{j\geqslant 1}|f|_1\\
&\qquad \times\int_{\tilde{\xi}\in\mathbb{R}^{d-1}}\left(e^{-2\theta|\tilde{\xi}|^2 t}|\tilde{\xi}|^{-2\epsilon}\left|\int_{-\infty}^{+\infty} e^{-\theta|k|^2 t}|k-\tau_j(\tilde{\xi})|^{\epsilon-1}\,dk\right|^2 d\tilde{\xi}\right)^{1/2}
\end{aligned}$$

$$\leqslant C|f|_1 \int_{\tilde\xi\in\mathbb{R}^{d-1}} \left(\mathrm{e}^{-2\theta|\tilde\xi|^2 t}|\tilde\xi|^{-2\epsilon} \left| \int_{-\infty}^{+\infty} \mathrm{e}^{-\theta|k|^2 t}|k|^{\epsilon-1}\,\mathrm{d}k\right|^2 \mathrm{d}\tilde\xi \right)^{1/2}$$
$$\leqslant C|f|_1 t^{-(d-1)/4} \tag{5.156}$$

as claimed. Derivative bounds follow similarly.

L^∞ bounds. Similarly, using Hausdorff–Young's inequality, we may estimate

$$\begin{aligned}
|u|_{L^\infty(x_1,\tilde x)}(t) &\leqslant \sup_{x_1} \frac{1}{(2\pi)^d} \int_{\tilde\xi\in\mathbb{R}^{d-1}} \left| \oint_{\lambda\in\tilde\Gamma(\tilde\xi)} \mathrm{e}^{\lambda t}\hat u(x_1,\tilde\xi,\lambda)\,\mathrm{d}\lambda \right| \mathrm{d}\tilde\xi \\
&\leqslant \frac{1}{(2\pi)^d} \int_{\tilde\xi\in\mathbb{R}^{d-1}} \oint_{\lambda\in\tilde\Gamma(\tilde\xi)} |\mathrm{e}^{\lambda t}| \left|\hat u(x_1\tilde\xi,\lambda)\right|_{L^\infty(x_1)} \mathrm{d}\lambda\,\mathrm{d}\tilde\xi \\
&\leqslant |f|_1 \frac{1}{(2\pi)^d} \int_{\tilde\xi\in\mathbb{R}^{d-1}} \oint_{\lambda\in\tilde\Gamma(\tilde\xi)} \mathrm{e}^{\operatorname{Re}\lambda t} b(\tilde\xi,\lambda)\,\mathrm{d}\lambda\,\mathrm{d}\tilde\xi,
\end{aligned} \tag{5.157}$$

to obtain the claimed bound on $|S_1(t)f|_\infty$. Parametrizing $\Gamma(\tilde\xi)$ again by

$$\lambda(\tilde\xi,k) = \mathrm{i}k - \theta_1(k^2 + |\tilde\xi|^2), \qquad k\in\mathbb{R},$$

we obtain a contribution bounded by

$$\begin{aligned}
&C|f|_1 \int_{\tilde\xi\in\mathbb{R}^{d-1}} \int_{-\infty}^{+\infty} \mathrm{e}^{-\theta(k^2+|\tilde\xi|^2)t} \rho^{-1}\gamma_2\,\mathrm{d}k\,\mathrm{d}\tilde\xi \\
&\quad\leqslant C|f|_1 \int_{\tilde\xi\in\mathbb{R}^{d-1}} \mathrm{e}^{-\theta|\tilde\xi|^2 t}|\tilde\xi|^{-\epsilon} \int_{-\infty}^{+\infty} \mathrm{e}^{-\theta|k|^2 t}|k|^{\epsilon-1}\,\mathrm{d}k\,\mathrm{d}\tilde\xi \\
&\quad\leqslant C|f|_1 t^{-(d-1)/2}
\end{aligned} \tag{5.158}$$

as claimed. Derivative bounds follow similarly.

General $2\leqslant p\leqslant\infty$. Finally, the general case follows by interpolation between L^2 and L^∞ norms.

S_2 bounds. Using

$$S(t) = \frac{1}{(2\pi)^{d-1}} \int \mathrm{e}^{\tilde\xi\cdot\tilde x}\mathrm{e}^{L_{\tilde\xi}t}\,\mathrm{d}\tilde\xi,$$

we may split $S_2(t)$ into two parts,

$$S_2^I(t) := \frac{1}{(2\pi)^{d-1}} \int_{|\tilde\xi|\geqslant R} \mathrm{e}^{\tilde\xi\cdot\tilde x}\mathrm{e}^{L_{\tilde\xi}t}\,\mathrm{d}\tilde\xi,$$

$R > 0$ sufficiently large, and

$$S_2^{II}(t) := \frac{1}{(2\pi)^{d-1}} \int_{|\tilde{\xi}| \leqslant R} e^{\tilde{\xi}\cdot\tilde{x}} e^{L_{\tilde{\xi}} t} \, d\tilde{\xi} - S_1(t)$$

$$= \frac{1}{(2\pi)^{d-1}} \int_{|\tilde{\xi}| \leqslant R} \text{P.V.} \int_{\gamma_0 - i\infty}^{\gamma_0 + i\infty} e^{i\tilde{\xi}\cdot\tilde{x}+\lambda t} (L_{\tilde{\xi}} - \lambda)^{-1} \, d\lambda \, d\tilde{\xi}$$

$$- \frac{1}{(2\pi)^{d-1}} \int_{|\tilde{\xi}| \leqslant r} \oint_{\Gamma^{\tilde{\xi}}} e^{i\tilde{\xi}\cdot\tilde{x}+\lambda t} (L_{\tilde{\xi}} - \lambda)^{-1} \, d\lambda \, d\tilde{\xi}.$$

The first part, $S_2^I(t)$ satisfies the claimed bounds by Parseval's identity together with (5.149), together with the observation that $\partial_{\tilde{x}}$ commutes with $L_{\tilde{\xi}}$ and thus all S_j^k, by invariance of coefficients with respect to $\tilde{x}$. The second part, $S_2^{II}(t)$, satisfies the claimed bounds by Parseval's identity together with calculations like those of the corresponding one-dimensional high- and intermediate-frequency estimates in the proof of Proposition 4.39 and the $H^1(x_1) \to H^1(x_1)$ estimates of Lemma 4.14, together with the same observation that $\partial_{\tilde{x}}$ commutes all S_j^k.

This completes the proof and the article. □

Appendix A: Auxiliary calculations

A.1. *Applications to example systems*

In this appendix, we present simple conditions for (A1)–(A3) and (H1)–(H2), and use them to verify the hypotheses for various interesting examples from gas dynamics and MHD. Consider a system (1.1) satisfying conditions (1.5) and (1.6). (Note: (1.6) is implied by (A3) under the class of transformations considered, since $\operatorname{Re} \tilde{b} > 0 \Rightarrow \operatorname{Re} \sigma(\widetilde{A}^0)^{-1}\tilde{b} = (\partial W/\partial U) b (\partial U/\partial W) > 0$.)

DEFINITION A.1. Following [Kaw], we define functions $\eta(G), q^j(G)\colon \mathbb{R}^n \to \mathbb{R}^1$ to be a convex entropy, entropy flux ensemble for (1.1), (1.6) if: (i) $d^2\eta > 0$ (η convex), (ii) $d\eta \, d\widetilde{F}^j = dq^j$, and (iii) $d^2\eta \widetilde{B}^{jk}$ is symmetric (hence positive semidefinite, by (1.6)).

PROPOSITION A.2 [KSh]. *A necessary and sufficient condition that a system* (1.1), (1.6) *can be put in symmetric form* (1.7) *by a change of coordinates $U \to W(U)$ for U in a convex set $\mathcal{U}$ is existence of a convex entropy, entropy flux ensemble η, q^j defined on $\mathcal{U}$, with $W = d\eta(U)$ (known as an "entropy variable").*

PROOF. ($\Leftarrow$) The change of coordinates $U \to W := d\eta(U)$ is invertible on any convex domain, since its Jacobian $d^2\eta$ is symmetric positive definite, by Definition A.1(i), and thus we can rewrite (1.1) as (1.7), as claimed, where $\widetilde{F}^j(W) := F^j(U(W))$ and $\widetilde{B}^{jk}(W) := B^{jk}(U(W))(\partial U/\partial W) = B^{jk} d^2\eta^{-1}$. Symmetry of $d\widetilde{F}^j = dF^j d^2\eta^{-1}$ and $\widetilde{B}^{jk} = B^{jk} d^2\eta^{-1}$ follow from symmetry of $d^2\eta \, dF^j$ and $d^2\eta B^{jk}$, the first a well-known consequence of Definition A.1(ii) [G,Bo,Se.3], the second just Definition A.1(iii). Finally, block structure and uniform ellipticity of $\widetilde{B}^{jk}$ follow by symmetry of $\widetilde{B}^{jk}$, the fact that the w^I rows must vanish, by the corresponding property of left factor B^{jk}, and the fact that $\operatorname{Re}\sigma \sum_{jk} b^{jk}\xi_j\xi_k \geqslant \theta|\xi|^2$, since $\widetilde{B}^{jk}$ is similar to B^{jk}.

($\Rightarrow$) Symmetry of $\widetilde{A}^0 = \mathrm{d}W(U)$ implies that $\mathrm{d}\eta(U) := W(u)$ is exact on any simply connected domain, while positive definiteness implies that η is convex. Likewise, symmetry of $\mathrm{d}(\mathrm{d}\eta\,\mathrm{d}F^j)$ (following by symmetry of $\mathrm{d}^2\eta\,\mathrm{d}F^j$, a consequence of symmetry of $\mathrm{d}F^j\,\mathrm{d}^2\eta^{-1}$, plus the reverse calculation to that cited above) implies that $\mathrm{d}q^j(U) := \mathrm{d}\eta\,\mathrm{d}F^j$ is exact. Finally, symmetry of $\mathrm{d}^2\eta B^{jk}$ follows by symmetry of $B^{jk}\mathrm{d}^2\eta^{-1}$. □

PROPOSITION A.3 [GMWZ.4]. *Necessary and sufficient conditions that there exist coordinate change $W(U)$ and matrix multiplier $S(W)$ taking* (1.1) *to form* (1.11) *satisfying* (A1)–(A3) *are*: (i) *there exist symmetric, positive definite (right) symmetrizers $Q(U)$ such that $(A^jQ)_\pm$ and $(A^jQ)_{11}$ are symmetric, and $B^{jk}Q = \text{block-diag}\{0, \beta^{jk}\}$ with β^{jk} satisfying uniform ellipticity condition* (1.6), *and* (ii) *there exists $w^{II} : U \to \mathbb{R}^r$ such that $B^{jk}(U)U_{x_k} = (0, \hat{b}^{jk}(U)w^{II}(U)_{x_k})$.*

If there exist such transformations, then S must be lower block-triangular and $B^{jk}(\partial U/\partial W)$ block-diagonal, with $S(U) = (\partial U/\partial W)^{\mathrm{tr}}Q^{-1}$, and

$$W(U) := \left(w^I\left(u^I\right), w^{II}(U)\right)$$

for w^I invertible and w^{II} satisfying (ii). *Moreover, we may always choose W with $(\partial W/\partial U)$ lower block-triangular, in which case $\widetilde{A}^0 = S(\partial U/\partial W)$ is block-diagonal as well. (Indeed, this structure follows already from block structure of $\widetilde{B}^{jk}$ and positive symmetric definiteness of $\widetilde{A}^0$ alone, without any further considerations; conversely, $W(U) := (w^I(u^I), w^{II}(U))$ and $S = \widetilde{A}^0(\partial U/\partial W)$ for symmetric positive definite block-diagonal $\widetilde{A}^0$ is sufficient to imply block structure of $\widetilde{B}^{jk}$.)*

PROOF. ($\Rightarrow$) Define $W(U) = (w^I, w^{II})(U)$, where w^I, w^{II} are as described in the hypotheses. Then, $\partial W/\partial U$ is lower block triangular, of form

$$\begin{pmatrix} \mathrm{d}w^I & 0 \\ \partial w^{II}/\partial u^I & \partial w^{II}/\partial u^{II} \end{pmatrix},$$

with $B^{jk}_{22} = \hat{b}^{jk}(\partial w^{II}/\partial u^{II})$ by assumption (ii), hence (taking, e.g., $j = k = 1$) is invertible, by assumption (1.6). Moreover, $\widehat{B}^{jk} := B^{jk}(\partial U/\partial W)$ by assumption (ii) is block-diagonal of form block-diag$\{0, \hat{b}^{jk}\}$. Thus, choosing a lower block-triangular multiplier

$$S(W) := P(W)(\partial W/\partial U), \qquad P = \text{block-diag}\{P_1, P_2\},$$

for any symmetric positive definite P_j, we obtain form (1.11), with $\widetilde{A}^0 = P$ symmetric positive definite and $\widetilde{B}^{jk} = \text{block-diag}\{0, \tilde{b}^{jk}\}$ in the required block-diagonal form. Indeed, only multipliers of this form will suffice, since $(\partial W/\partial U)B^{jk}(\partial U/\partial W)$ is also of block-diagonal form, by lower triangularity of $(\partial W/\partial U)$, so that P must be lower triangular as well to preserve this property, but also $P = \widetilde{A}^0$ is required to be symmetric. Writing the block-diagonal $\widehat{B}^{jk}$ as $B^{jk}Q(Q^{-1}(\partial U/\partial W))$, where $B^{jk}Q$ is block-diagonal, we find similarly that $Q^{-1}(\partial U/\partial W)$ must be upper block-triangular to preserve block structure. Therefore, $S(U) := (\partial U/\partial W)^{\mathrm{tr}} Q^{-1}$ is lower block-triangular, and so is its product with the lower block-triangular $\partial U/\partial W$, so that $\widetilde{A}^0 = S(\partial U/\partial W) = (\partial U/\partial W)^{\mathrm{tr}} Q^{-1}(\partial U/\partial W)$ is both lower block-triangular and symmetric positive definite, hence block-diagonal as well. Further, expanding

$$\begin{aligned} S(U)B^{jk}(\partial U/\partial W) &= (\partial U/\partial W)^{\mathrm{tr}} Q^{-1}\big(B^{jk}Q\big)Q^{-1}(\partial U/\partial W) \\ &= S\big(B^{jk}Q\big)S^{\mathrm{tr}} = \text{block-diag}\big\{0, S_{22}\big(B^{jk}Q\big)_{22}S_{22}^{\mathrm{tr}}\big\}, \end{aligned}$$

we find that $\widetilde{B}^{jk}$ satisfies ellipticity condition (1.6) as well. Likewise, lower block-triangularity of S yields $SA^j(\partial U/\partial W)_{11} = S_{11}(A^jQ)_{11}S_{11}^{\mathrm{tr}}$, verifying symmetry of $SA^j(\partial U/\partial W)_{11}$, and $SA^j(\partial U/\partial W)_{\pm} = (S(A^jQ)S^{\mathrm{tr}})_{\pm}$ and $SB^{jk}(\partial U/\partial W)_{\pm} = (S(B^{jk}Q)S^{\mathrm{tr}})_{\pm}$, verifying symmetry of $SA^j(\partial U/\partial W)_{\pm}$ and $SB^{jk}(\partial U/\partial W)_{\pm}$.

($\Leftarrow$) By nonsingularity of $(\partial U/\partial W)$, and vanishing of the first block-row of $SB^{jk}(\partial U/\partial W)$, we find that the first block-row of SB^{jk} must vanish, hence S must be lower block-triangular by vanishing of the first block-row of B^{jk} together with nonsingularity of B_{22}^{jj}. It follows by nonsingularity of S that S_{11} and S_{22} are nonsingular, whence the product of lower block-triangular S with vanishing first block-row $B^{jk}(\partial U/\partial W)$ can be block-diagonal only if $B^{jk}(\partial U/\partial W)$ is already block-diagonal, implying condition (ii). From the assumption that $\widetilde{A}^0 = S(\partial U/\partial W)$ is symmetric positive definite, we find that

$$Q := S^{-1}(\partial U/\partial W)^{\mathrm{tr}} = (\partial U/\partial W)\big(\widetilde{A}^0\big)^{-1}(\partial U/\partial W)^{\mathrm{tr}} = (\partial U/\partial W)S^{-1,\mathrm{tr}}$$

is symmetric positive definite as well. Factoring $A^jQ = S^{-1}(SA^j(\partial U/\partial W))S^{-1,\mathrm{tr}}$, $B^{jk}Q = S^{-1}(SB^{jk}(\partial U/\partial W))S^{-1,\mathrm{tr}}$, we find, again by lower block-triangularity of S, that right symmetrizer Q preserves all properties of $\widetilde{A}^j$ and $\widetilde{B}^{jk}$, thus verifying (i). Once existence of such Q is verified, it follows from the analysis in the first part that we can take W of form $(w^I(u^I), w^{II}(U))$ as claimed. □

REMARK A.4. The above construction may be recognized as a generalization of the Kawashima normal form (N), described, e.g., in [GMWZ.4]; indeed, at the endstates (more generally, wherever symmetry of A^j, B^{jk} are enforced), the two constructions coincide. From the proof of Proposition A.3, we find, more generally, that, assuming any subset of the properties asserted on A^jQ, $B^{jk}Q$, these properties will be inherited by the system (1.11), independent of the others.

COROLLARY A.5. *Suppose it is known that system* (1.1) *admits right symmetrizers* $Q_{\pm}$ *at* $U_{\pm}$ *as described above, for example, if there exist local convex entropies* $\eta_{\pm}(U)$ *near these states. Then, necessary and sufficient conditions for existence of a transformed system* (1.11) *satisfying* (A1)–(A3) *are* (ii) *of Proposition* A.3, *together with the existence of symmetric positive definite matrices* P_1 *and* P_2 *such that* P_1 *simultaneously symmetrizes* $\alpha_*^j := (\partial w^I/\partial u^I)A_*^j(\partial w^I/\partial u^I)^{-1}$, $P_1\alpha_*^j$ *symmetric, and* P_2 *simultaneously sta-*

bilizes $\beta^{jk} := (\partial w^{II}/\partial u^{II})\hat{b}^{jk} = (\partial w^{II}/\partial u^{II})B_{22}^{jk} = (\partial w^{II}/\partial u^{II})^{-1}$, W *as defined in* (ii) *above, in the sense that* β^{jk} *satisfy a uniform ellipticity condition* (1.6).

PROOF. Necessity follows from Proposition A.3, block-diag$\{P_1, P_2\} = \widetilde{A}^0$. Sufficiency follows by a partition of unity argument, interpolating the symmetric positive definite matrices P_j guaranteed at each point U, taking care at the endstates $U_\pm$ to choose the special P_j (guaranteed by the existence of symmetrizers $Q_\pm$) that also symmetrize A^j. □

Corollary A.5 suggests a strategy for verifying (A1)–(A3), namely to make the complete coordinate change $U \to W(U)$, $S(U) = (\partial W/\partial U)$, with $W = (u^I, w^{II}(U))$ for simplicity, then look directly for simultaneous symmetrizer/stabilizers P_1/P_2. This works extremely well for the Navier–Stokes equations of compressible gas dynamics, which appear as

$$\rho_t + \operatorname{div}(\rho u) = 0,$$

$$(\rho u)_t + \operatorname{div}(\rho u \otimes u) + \nabla p = \overbrace{\mu \Delta u + (\lambda + \mu)\nabla \operatorname{div} u}^{\operatorname{div} \tau}, \tag{A.1}$$

$$\left(\rho\left(e + \frac{1}{2}u^2\right)\right)_t + \operatorname{div}\left(\rho\left(e + \frac{1}{2}u^2\right)u + pu\right) = \operatorname{div}(\tau \cdot u) + \kappa \Delta T,$$

where $\rho > 0$ denotes density, $u \in \mathbb{R}^d$ fluid velocity, $T > 0$ temperature, $e = e(\rho, T)$ internal energy, and $p = p(\rho, T)$ pressure. Here, $\tau := \lambda \operatorname{div}(u) I + 2\mu D u$, where $Du_{jk} = \frac{1}{2}(u^j_{x_j} + u^j_{x_k})$ is the deformation tensor, $\lambda(\rho, T) > 0$ and $\mu(\rho, T) > 0$ are viscosity coefficients, and $\kappa(\rho, T) > 0$ is the coefficient of thermal conductivity.

For these equations, $U = (\rho, \rho u, E = e + |u|^2/2)$ and $W = (\rho, u, T)$, and α^j, β^{jk} are symmetric, with $\sum_j \alpha^j \rho_{x_j} = u \cdot \nabla \rho$ (scalar convection) and

$$\sum_{jk} \left(\beta^{jk}(u, T)_{x_k}\right)_{x_j} = \rho^{-1}\left(\mu \Delta u + (\mu + \lambda)\nabla \operatorname{div} u, \kappa T_e \Delta T\right)$$

(see, e.g., [GMWZ.4]), whence suitable P_j exist if and only if $T_e > 0$, in which case they may be simply taken as the identity. Here, E represents total energy density. Likewise, MHD may be treated similarly, with $W = (\rho, u, T, B)$, B the magnetic field. Note that this indeed allows the case of van der Waals equation of state.

Alternatively, we may start with the Kawashima form (N) at the endstates, and try to deduce directly an appropriate modification preserving block structure along the profile. For example, examining form (N) for Navier–Stokes equations, for which (see, e.g., [KS], or equation (2.13) of [GMWZ.4])

$$\widetilde{A}^0 = \begin{pmatrix} p_\rho/\rho & 0 & 0 \\ 0 & \rho I_3 & 0 \\ 0 & 0 & \rho e_T/T \end{pmatrix},$$

$$\sum_j \xi_j \widetilde{A}^j = \begin{pmatrix} (p_\rho/\rho)(u \cdot \xi) & p_\rho \xi & 0 \\ p_\rho \xi^{\mathrm{tr}} & \rho(u \cdot \xi) I_3 & p_T \xi^{\mathrm{tr}} \\ 0 & p_\rho \xi & (\rho e_T/T)(u \cdot \xi) \end{pmatrix}$$

and

$$\sum_{j,k} \widetilde{B}^{jk} = \begin{pmatrix} 0 & 0 & 0 \\ 0 & \mu|\xi|^2 I_3 + (\mu+\lambda)\xi^{tr}\xi & 0 \\ 0 & 0 & T^{-1}|\xi|^2 \end{pmatrix},$$

we find that we may simply multiply the first row by $\chi(U)p_\rho^{-1} + (1-\chi(U))$ to obtain form (1.11) satisfying (A1)–(A3), where χ is a smooth cutoff function supported on the set where the Navier–Stokes equations support a convex entropy and equal to unity at $U_\pm$. A similar procedure applies to MHD.

By direct calculation, we find that the eigenvalues of $(\widetilde{A}^0)^{-1}\sum_j \widetilde{A}^j\xi_j = \sum_j A^j\xi_j$ for gas dynamics are of constant multiplicity with respect to both U and ξ; see Appendix C. For MHD on the other hand, they are of constant multiplicity (under appropriate nondegeneracy conditions) only with respect to U (see [Je], pp. 122–127), and this limits the applications for the moment to one dimension. In both cases, $A_* = (u\cdot\xi)$ is scalar, hence constant multiplicity. Thus, to complete the verification of (A1)–(A3), (H0)–(H2), it remains only to check that $\det(A^1 - s)$ does not vanish at the endstates $U_\pm$, and that $A^1_* - s = u^1 - s$ does not vanish along the profile, both purely one-dimensional consideration associated with the traveling-wave ODE.

We conclude by examining further various example systems in one dimension, expressed for simplicity in Lagrangian coordinates. For ease of notation, we drop the superscripts j and jk from A^j, A^j_* and B^{jk}.

EXAMPLE A.6 (Standard gas dynamics). The general Navier–Stokes equations of compressible gas dynamics, written in Lagrangian coordinates, appear as

$$\begin{aligned} &v_t - u_x = 0,\\ &u_t + p_x = \big((\nu/v)u_x\big)_x,\\ &\big(e+u^2/2\big)_t + (pu)_x = \big((\kappa/v)T_x + (\mu/v)uu_x\big)_x, \end{aligned} \tag{A.2}$$

where $v > 0$ denotes specific volume, u velocity, e internal energy, $T = T(v,e)$ temperature, $p = p(v,e)$ pressure, and $\mu > 0$ and $\kappa > 0$ are coefficients of viscosity and heat conduction, respectively.

Note, as described in [MeP], that the "incomplete equation of state" $p = p(v,e)$ is sufficient to close the compressible Euler equations, consisting of the associated hyperbolic system (1.4), whereas the compressible Navier–Stokes equations require also the relation of temperature to v and e. Accordingly, following [MeP], consider the "complete equation of state"

$$e = e(v,s) \tag{A.3}$$

expressing internal energy in terms of v and entropy s, defining temperature and pressure through the fundamental thermodynamic relation (combining first and second laws)

$$\mathrm{d}e = T\,\mathrm{d}s - p\,\mathrm{d}v, \tag{A.4}$$

under the standard constraints of positive temperature,

$$T = e_s > 0, \tag{A.5}$$

and global thermodynamic stability (precluding van der Waals type equation of state),

$$e(v, s) \text{ convex.} \tag{A.6}$$

Note that we do not require positivity of $p = e_v$ or p_e, which may fail under certain interesting circumstances (for example, in water, near the transition from liquid to ice).

Under assumption (A.5), we may invert relation (A.3) to obtain $s = s(v, e)$ as a function of the primary variables v and e. A straightforward calculation then gives:

LEMMA A.7. *For $T = e_s$ positive, $e(v, s)$ is convex if and only if $-s(v, e)$ is convex, if and only if $-s(v, E - u^2/2)$ is convex with respect to (v, u, E), where $E := e + u^2/2$.*

An immediate consequence of Lemma A.7 is that

$$(\partial/\partial E)T\big(v, E - u^2/2\big) = T_e = (e_s)_e = (1/s_e)_e = -s_{ee}/s_e^2 > 0, \tag{A.7}$$

whence (A.2) has the structural form (1.1)–(1.5), where the lower triangular matrix

$$b_2 = \begin{pmatrix} 1/v & 0 \\ * & (\partial/\partial E)T \end{pmatrix}$$

evidently has real, positive spectrum $1/v$, $(\partial/\partial E)T$. Likewise, it is easily checked that genuine coupling, (A2), holds.

Appealing to relation (A.4), we find that

$$\eta(v, u, E) := -s\big(v, u, E - u^2/2\big), \qquad q(v, u, E) := 0 \tag{A.8}$$

satisfy conditions (ii) and (iii) of Definition A.1. (Here, it is helpful to take $u = 0$ by Galilean invariance; for details, see, e.g., Section 9 of [LZe].) By Lemma A.7, condition (i) is satisfied as well, whence η, q is a convex entropy, entropy flux pair for (A.2). Applying Proposition A.2, we thus find that (A1)–(A3) are satisfied for (A.2) written in terms of the entropy variable

$$\mathrm{d}\eta = (-s_v, us_e, -s_e) = (e_v/e_s, u/e_s, -1/e_s) = T^{-1}(-p, u, -1). \tag{A.9}$$

Straightforward calculation then gives $A_* = \mathrm{d}\widetilde{F}_{11} = \mathrm{d}\widetilde{F}_{12} b_2^{-1} b_1 \equiv 0$ as in the isentropic case, verifying (H1) and completing the verification of the hypotheses under assumptions

(A.5) and (A.6). Note: zero-speed profiles, $s = 0$ do not exist. For, stationary solutions of (A.2) are easily seen to be constant in u, p, and T. But, $(T, -p) = (e_s, e_v)$ is the (necessarily invertible) Legendre transform of (s, v) with respect to the convex function $e(s, v)$, whence s, v, and therefore e and E must be constant as well.

As discussed above, the assumption of global thermodynamic stability may be relaxed to $T_e > 0$ together with thermodynamic stability at endstates $U_\pm$, thus accommodating models for van der Waals gas dynamics. Again, stationary profiles cannot occur, since $U = (v, u, e)$ must remain constant so long as it stays sufficiently near $U_\pm$, and thus remains so forever. This completes the verification of (A1)–(A3) and (H0)–(H2) in the general case; condition (H3) holds for extreme, Lax-type profiles, but not necessarily in general.

EXAMPLE A.8 (MHD). Next, consider the equations of MHD:

$$\begin{aligned}
&v_t - u_{1x} = 0,\\
&u_{1t} + \big(p + (1/2\mu_0)\big(B_2^2 + B_3^2\big)\big)_x = \big((\nu/v)u_{1x}\big)_x,\\
&u_{2t} - \big((1/\mu_0)B_1^* B_2\big)_x = \big((\nu/v)u_{2x}\big)_x,\\
&u_{3t} - \big((1/\mu_0)B_1^* B_3\big)_x = \big((\nu/v)u_{3x}\big)_x,\\
&(vB_2)_t - \big(B_1^* u_2\big)_x = \big((1/\sigma\mu_0 v)B_{2x}\big)_x,\\
&(vB_3)_t - \big(B_1^* u_3\big)_x = \big((1/\sigma\mu_0 v)B_{3x}\big)_x,\\
&\big(e + (1/2)\big(u_1^2 + u_2^2 + u_3^2\big) + (1/2\mu_0)v\big(B_2^2 + B_3^2\big)\big)_t\\
&\qquad + \big[\big(p + (1/2\mu_0)\big(B_2^2 + B_3^2\big)\big)u_1 - (1/\mu_0)B_1^*(B_2u_2 + B_3u_3)\big]_x\\
&\quad = \big[(\nu/v)u_1u_{1x} + (\mu/v)(u_2u_{2x} + u_3u_{3x})\\
&\qquad + (\kappa/v)T_x + \big(1/\sigma\mu_0^2 v\big)(B_2B_{2x} + B_3B_{3x})\big]_x,
\end{aligned} \tag{A.10}$$

where v denotes specific volume, $u = (u_1, u_2, u_3)$ velocity, $p = P(v, e)$ pressure, $B = (B_1^*, B_2, B_3)$ magnetic induction, B_1^* constant, e internal energy, $T = T(v, e)$ temperature, and $\mu > 0$ and $\nu > 0$ the two coefficients of viscosity, $\kappa > 0$ the coefficient of heat conduction, $\mu_0 > 0$ the magnetic permeability, and $\sigma > 0$ the electrical resistivity.

Calculating similarly as in the Navier–Stokes case, under the same assumptions (A.5) and (A.6), we find again that $\eta := -s$, $q := 0$ is an entropy, entropy flux pair for system (A.10), η, q now considered as functions of the conserved quantities

$$(v, u_1, u_2, u_3, vB_2, vB_3, E),$$

where $E := e + (1/2)(u_1^2 + u_2^2 + u_3^2) + (1/2\mu_0)v(B_2^2 + B_3^2)$ denotes total energy; for details, see again Section 9 of [LZe]. Likewise, we obtain by straightforward calculation that $A_* = \mathrm{d}\widetilde{F}_{11} = \mathrm{d}\widetilde{F}_{12} b_2^{-1} b_1 \equiv 0$, verifying (H1) and completing the verification of the hypotheses under assumptions (A.5) and (A.6), plus the additional assumption (possibly superfluous, as in the Navier–Stokes case) that speed s be nonzero.

EXAMPLE A.9 (MHD with infinite resistivity/permeability). An interesting variation of (A.10) that is of interest in certain astrophysical parameter regimes is the limit in which either electrical resistivity σ, magnetic permeability μ_0, or both, go to infinity, in which case the right-hand sides of the fifth and sixth equations of (A.10) go to zero and there is a three-dimensional set of hyperbolic modes (v, vB_2, vB_3) instead of the usual one. Nonetheless, $A_* = \mathrm{d}\widetilde{F}_{11} = \mathrm{d}\widetilde{F}_{12} b_2^{-1} b_1 \equiv 0$ in this case as well; that is, though now vectorial, hyperbolic modes still experience passive, scalar convection, and so (H1) remains valid whenever $s \neq 0$. (Note: in Lagrangian coordinates, zero speed corresponds to "particle", or fluid velocity.) Reduction to symmetric form may be achieved by the same entropy, entropy flux pair as in the standard case of Example A.10, under assumptions (A.5) and (A.6).

EXAMPLE A.10 (Multispecies gas dynamics or MHD). Another simple example for which the hyperbolic modes are vectorial is the case of miscible, multispecies flow, neglecting species diffusion, in either gas- or magnetohydrodynamics. In this case, the hyperbolic modes consist of k copies of the hyperbolic modes for a single species, where k is the number of total species, with a single, scalar convection rate $A_* \equiv 0$, corresponding to passive convection by the common fluid velocity.

A.2. *Structure of viscous profiles*

In this appendix, we prove the results cited in the introduction concerning structure of viscous profiles.

PROOF OF LEMMA 1.6. Differentiating (1.16) and rearranging, we may write (1.16) and (1.17) in the alternative form

$$\begin{pmatrix} u^I \\ u^{II} \end{pmatrix}' = \begin{pmatrix} \mathrm{d}F_{11}^1 & \mathrm{d}F_{12}^1 \\ b_1^{11} & b_2^{11} \end{pmatrix}^{-1} \begin{pmatrix} 0 \\ f^{II} - f_-^{II} \end{pmatrix}. \tag{A.11}$$

Linearizing (A.11) about $U_\pm$, we obtain

$$\begin{pmatrix} u^I \\ u^{II} \end{pmatrix}' = \begin{pmatrix} \mathrm{d}F_{11}^1 & \mathrm{d}F_{12}^1 \\ b_1^{11} & b_2^{11} \end{pmatrix}^{-1} \begin{pmatrix} 0 & 0 \\ \mathrm{d}F_{21}^1 & \mathrm{d}F_{22}^1 \end{pmatrix}\Bigg|_{(U_\pm)} \begin{pmatrix} u^I \\ u^{II} \end{pmatrix}, \tag{A.12}$$

or, setting

$$\begin{pmatrix} z_1 \\ z_2 \end{pmatrix} := \begin{pmatrix} \mathrm{d}F_{11}^1 & \mathrm{d}F_{12}^1 \\ b_1^{11} & b_2^{11} \end{pmatrix}\Bigg|_{(U_\pm)} \begin{pmatrix} u^I \\ u^{II} \end{pmatrix},$$

the pair of equations

$$z_1' = 0$$

and

$$z_2' = \begin{pmatrix} dF_{21}^1 & dF_{22}^1 \end{pmatrix} \begin{pmatrix} dF_{11}^1 & dF_{12}^1 \\ b_1^{11} & b_2^{11} \end{pmatrix}^{-1} \begin{pmatrix} 0 \\ I_r \end{pmatrix} \bigg|_{(U_\pm)} z_2, \tag{A.13}$$

the latter of which evidently describes the linearized ODE on manifold (1.16).

Observing that

$$\begin{aligned} &\det\begin{pmatrix} dF_{21}^1 & dF_{22}^1 \end{pmatrix} \begin{pmatrix} dF_{11}^1 & dF_{12}^1 \\ b_1^{11} & b_2^{11} \end{pmatrix}^{-1} \begin{pmatrix} 0 \\ I_r \end{pmatrix} \bigg|_{(U_\pm)} \\ &\quad = \det \begin{pmatrix} dF_{11}^1 & dF_{12}^1 \\ dF_{21}^1 & dF_{22}^1 \end{pmatrix} \begin{pmatrix} dF_{11}^1 & dF_{12}^1 \\ b_1^{11} & b_2^{11} \end{pmatrix}^{-1} \bigg|_{(U_\pm)} \\ &\quad \neq 0 \end{aligned}$$

by (H2) and (H1)(i), we find that the coefficient matrix of (A.13) has no zero eigenvalues. On the other hand, it can have no nonzero purely imaginarily eigenvalues $i\xi$, since otherwise

$$\begin{pmatrix} F_{11}^1 & F_{12}^1 \\ dF_{21}^1 & dF_{22}^1 \end{pmatrix} \begin{pmatrix} F_{11}^1 & F_{12}^1 \\ b_1^{11} & b_2^{11} \end{pmatrix}^{-1} \begin{pmatrix} 0 \\ u^{II} \end{pmatrix} = i\xi \begin{pmatrix} 0 \\ u^{II} \end{pmatrix},$$

and thus

$$\left[-i\xi \begin{pmatrix} dF_{11}^1 & dF_{12}^1 \\ dF_{21}^1 & dF_{22}^1 \end{pmatrix} - \xi^2 \begin{pmatrix} 0 & 0 \\ b_1^{11} & b_2^{11} \end{pmatrix} \right] \left[\begin{pmatrix} dF_{11}^1 & dF_{12}^1 \\ b_1^{11} & b_2^{11} \end{pmatrix}^{-1} \begin{pmatrix} 0 \\ u^{II} \end{pmatrix} \right] = \begin{pmatrix} 0 \\ 0 \end{pmatrix}$$

for $\xi \neq 0 \in \mathbb{R}$, in violation of (2.55), Corollary 2.19. Thus, we find that $U_\pm$ are hyperbolic rest points, from which (2.18) follows. □

PROOF OF LEMMA 1.7. We follow the notation of Sections 1.2 and 3.1. Equating by consistent splitting/continuous extension to $\lambda = 0$ the dimensions of the stable subspace S^+ of the limiting coefficient matrix $\mathbb{A}_+(\lambda)$ appearing in eigenvalue equation (2.17) at $\lambda = 0$ and at $\lambda \to +\infty$, we obtain

$$\dim \mathcal{U}(A_+) + d_+ = \dim \mathcal{U}(A_{*+}) + r$$

or

$$(n - i_+) + d_+ = \dim \mathcal{U}(A_{*+}) + r,$$

and similarly at $x \to -\infty$, where the left-hand side follows from Lemma 4.20 and the right-hand side from the proof of Lemma 2.21. Rearranging, we obtain (1.19). If there exists a connecting profile, then we have by (H1)(i) that $\dim \mathcal{S}(A_*)$ and $\dim \mathcal{U}(A_*)$ are constant along the profile and sum to $n - r$. Thus, $\dim \mathcal{S}(A_{*-}) + \dim \mathcal{U}(A_{*+}) = n - r$, and we obtain $n - i = r - d$ by summing relations (1.19). □

A.3. *Asymptotic ODE estimates*

For completeness, we prove in this appendix the asymptotic ODE estimates cited in Section 2.2. We first recall the gap lemma of [GZ,KS,ZH].

LEMMA A.11 (The gap lemma). *Consider* (2.17), *under assumption* (h0), *with* $\bar\theta < \theta$. *If* $V^-(\lambda)$ *is an eigenvector of* $\mathbb{A}_-$ *with eigenvalue* $\mu(\lambda)$, *both analytic in* λ, *then there exists a solution of* (2.17) *of form*

$$W(\lambda, x) = V(x, \lambda)\mathrm{e}^{\mu(\lambda)x},$$

where V *is* C^1 *in* x *and locally analytic in* λ *and for each* $j = 0, 1, \ldots$ *satisfies*

$$V(x,\lambda) = V^-(\lambda) + \mathcal{O}\big(\mathrm{e}^{-\bar\theta|x|}\big|V^-(\lambda)\big|\big), \quad x < 0, \tag{A.14}$$

for all $\bar\theta < \theta$. *Moreover, if* $\operatorname{Re}\mu(\lambda) > \operatorname{Re}\tilde\mu(\lambda) - \theta$ *for all (other) eigenvalues* $\tilde\mu$ *of* $\mathbb{A}_-$, *then also* W *is uniquely determined by* (A.14), *and* (A.14) *holds for* $\bar\theta = \theta$.

PROOF. Setting $\mathbb{W}(x) = \mathrm{e}^{\mu x}V(x)$, we may rewrite $\mathbb{W}' = \mathbb{A}\mathbb{W}$ as

$$V' = (\mathbb{A}_- - \mu I)V + \theta V, \quad \theta := (\mathbb{A} - \mathbb{A}_-) = \mathcal{O}\big(\mathrm{e}^{-\theta|x|}\big), \tag{A.15}$$

and seek a solution $V(x,\lambda) \to V^-(x)$ as $x \to \infty$. Choose $\bar\theta < \theta_1 < \theta$ such that there is a spectral gap $|\operatorname{Re}(\sigma\mathbb{A}_- - (\mu + \theta_1))| > 0$ between $\sigma\mathbb{A}_-$ and $\mu + \theta_1$. Then, fixing a base point λ_0, we can define on some neighborhood of λ_0 to the complementary $\mathbb{A}_-$-invariant projections $P(\lambda)$ and $Q(\lambda)$ where P projects onto the direct sum of all eigenspaces of $\mathbb{A}_-$ with eigenvalues $\tilde\mu$ satisfying $\operatorname{Re}(\tilde\mu) < \operatorname{Re}(\mu) + \theta_1$, and Q projects onto the direct sum of the remaining eigenspaces, with eigenvalues satisfying $\operatorname{Re}(\tilde\mu) > \operatorname{Re}(\mu) + \theta_1$. By basic matrix perturbation theory (eg. [Kat]) it follows that P and Q are analytic in a neighborhood of λ_0, with

$$\begin{aligned} \big|\mathrm{e}^{(\mathbb{A}_- - \mu I)x}P\big| &\leqslant C\big(\mathrm{e}^{\theta_1 x}\big), \quad x > 0, \\ \big|\mathrm{e}^{(\mathbb{A}_- - \mu I)x}Q\big| &\leqslant C\big(\mathrm{e}^{\theta_1 x}\big), \quad x < 0. \end{aligned} \tag{A.16}$$

It follows that, for $M > 0$ sufficiently large, the map $\mathcal{T}$ defined by

$$\begin{aligned} \mathcal{T}V(x) = V^- &+ \int_{-\infty}^{x} \mathrm{e}^{(\mathbb{A}_- - \mu I)(x-y)}P\theta(y)V(y)\,\mathrm{d}y \\ &- \int_{x}^{-M} \mathrm{e}^{(\mathbb{A}_- - \mu I)(x-y)}Q\theta(y)V(y)\,\mathrm{d}y \end{aligned} \tag{A.17}$$

is a contraction on $L^\infty(-\infty, -M]$. Applying (A.16), we have

$$\begin{aligned} |\mathcal{T}V_1 - \mathcal{T}V_2|_{(x)} &\leqslant C|V_1 - V_2|_\infty \left(\int_{-\infty}^{x} e^{\theta_1(x-y)} e^{\theta y}\, dy + \int_{x}^{-M} e^{\theta_1(x-y)} e^{\theta y}\, dy \right) \\ &\leqslant C_1|V_1 - V_2|_\infty \left(e^{\theta_1 x} e^{(\theta-\theta_1)y}\big|_{-\infty}^{x} + e^{\theta_1 x} e^{(\theta-\theta_1)y}\big|_{x}^{-M} \right) \\ &\leqslant C_2|V_1 - V_2|_\infty e^{-\bar\theta M} < \frac{1}{2}|V_1 - V_2|_\infty. \end{aligned} \tag{A.18}$$

By iteration, we thus obtain a solution $V \in L^\infty(-\infty, -M]$ of $V = \mathcal{T}V$ with $V \leqslant C_3|V^-|$; since $\mathcal{T}$ clearly preserves analyticity $V(\lambda, x)$ is analytic in λ as the uniform limit of analytic iterates (starting with $V_0 = 0$). Differentiation shows that V is a bounded solution of $V = \mathcal{T}V$ if and only if it is a bounded solution of (A.15) (exercise). Further, taking $V_1 = V$, $V_2 = 0$ in (A.18), we obtain from the second to last inequality that

$$\left|V - V^-\right| = \left|\mathcal{T}(V) - \mathcal{T}(0)\right| \leqslant C_2 e^{\bar\theta x}|V| \leqslant C_4 e^{\bar\theta x}\left|V^-\right|, \tag{A.19}$$

giving (A.14). Analyticity, and the bounds (A.14), extend to $x < 0$ by standard analytic dependence for the initial value problem at $x = -M$. Finally, if $\mathrm{Re}(\mu(\lambda)) > \mathrm{Re}(\tilde\mu(\lambda)) - \frac{\theta}{2}$ for all other eigenvalues, then $P = I$, $Q = 0$, and $V = \mathcal{T}V$ must hold for any V satisfying (A.14), by Duhamel's principle. Further, the only term appearing in (A.18) is the first integral, giving (A.19) with $\bar\theta = \theta$. □

PROOF OF LEMMA 2.5. Substituting $W = P_+ Z$ into (2.17), equating to (2.19), and rearranging, we obtain the defining equation

$$P_+' = \mathbb{A}_+ P_+ - P_+ \mathbb{A}, \qquad P_+ \to I \text{ as } x \to +\infty. \tag{A.20}$$

Viewed as a vector equation, this has the form $P_+' = \mathcal{A}P_+$, where $\mathcal{A}$ approaches exponentially as $x \to +\infty$ to its limit $\mathcal{A}_+$, defined by $\mathcal{A}_+ P := \mathbb{A}_+ P - P\mathbb{A}_+$. The limiting operator $\mathcal{A}_+$ evidently has analytic eigenvalue, eigenvector pair $\mu \equiv 0$, $P_+ \equiv I$, whence the result follows by Lemma A.11 for $j = k = 0$. The x-derivative bounds $0 < k \leqslant K + 1$ then follow from the ODE and its first K derivatives, and the λ-derivative bounds from standard interior estimates for analytic functions. A symmetric argument gives the result for P_-. □

PROOF OF PROPOSITION 2.13. Setting $\Phi_2 := \psi_1 \psi_2^{-1}$, $\psi_1 \in \mathbb{C}^{(N-k)\times k}$, $\psi_2 \in \mathbb{C}^{k\times k}$, where $(\psi_1^t, \psi_2^t)^t \in \mathbb{C}^{N\times k}$ satisfies (2.29), we find after a brief calculation that Φ_2 satisfies

$$\Phi_2' = (M_1\Phi_2 - \Phi_2 M_2) + \delta Q(\Phi_2), \tag{A.21}$$

where Q is the quadratic matrix polynomial $Q(\Phi) := \Theta_{12} + \Theta_{11}\Phi - \Phi\Theta_{22} + \Phi\Theta_{21}\Phi$. Viewed as a vector equation, this has the form

$$\Phi_2' = \mathcal{M}\Phi_2 + \delta Q(\Phi_2), \tag{A.22}$$

with linear operator $\mathcal{M}\Phi := M_1\Phi - \Phi M_2$. Note that a basis of solutions of the decoupled equation $\Phi' = \mathcal{M}\Phi$ may be obtained as the tensor product $\Phi = \phi\tilde{\phi}^*$ of bases of solutions of $\phi' = M_1\phi$ and $\tilde{\phi}' = -M_2^*\tilde{\phi}$, whence we obtain from (4.257) the bound

$$\left|\mathcal{F}^{y\to x}\right| \leqslant \left|\widetilde{\mathcal{F}}_1^{y\to x}\right|\left|\widetilde{\mathcal{F}}_2^{y\to x}\right| = \left|\widetilde{\mathcal{F}}_1^{y\to x}\right|\left|\left(\mathcal{F}_2^{y\to x}\right)^{-1}\right| \leqslant C\mathrm{e}^{-\eta|x-y|} \tag{A.23}$$

for $x > y$, where $\mathcal{F}$ denotes the flow of the full decoupled matrix-valued equation, $\mathcal{F}_j$ the flow of $\phi' = M_j\phi$ and $\widetilde{\mathcal{F}}_j$ the adjoint flow associated with equation $\tilde{\phi}' = -M_j^*\tilde{\phi}$. (Recall the standard duality relation $\widetilde{\mathcal{F}}_j = \mathcal{F}_j^{-1}$.)

That is, $\mathcal{F}$ is uniformly exponentially decaying in the forward direction. Thus, assuming only that Φ_2 is bounded at $-\infty$, we obtain by Duhamel's principle the integral fixed-point equation

$$\Phi_2(x) = \mathcal{T}\Phi_2(x) := \delta\int_{-\infty}^{x} \mathcal{F}^{y\to x} Q(\Phi_2)(y)\,\mathrm{d}y. \tag{A.24}$$

Using (A.23), we find that $\mathcal{T}$ is a contraction of order $\mathcal{O}(\delta/\eta)$, hence (A.24) determines a unique solution for δ/η sufficiently small, which, moreover, is order δ/η as claimed. The bound on $\partial_x\Phi_j^\epsilon$ then follows by differentiation of (A.24), using the fact that $\partial_x\mathcal{F}^{y\to x} \leqslant C\mathcal{F}^{y\to x}$, since the coefficients of the decoupled (linear) flow are bounded. A symmetric argument establishes existence of Φ_1. □

REMARKS A.12. 1. Note that the above stable/unstable manifold construction requires only boundedness of coefficients at infinity, and not any special (e.g., asymptotically constant or periodic) structure. This reflects the local nature of large frequency/short time estimates in the applications.

2. Equation (A.21) could alternatively be derived from the point of view of Lemma 2.5, as defining a change of coordinates

$$P := \begin{pmatrix} I & \Phi_2 \\ \Phi_1 & I \end{pmatrix}$$

conjugating (2.29) to diagonal form. The result of Proposition 2.13 shows that there is a unique such transformation that is bounded on the whole line.

3. If desired, one can obtain a full, and explicit Neumann series expansion in powers of δ/η from the fixed-point equation (A.24), using our explicit description of the flow $\mathcal{F}$ as the tensor product $\Phi = \phi\tilde{\phi}^*$ of bases of solutions of $\phi' = M_1\phi$ and $\tilde{\phi}' = -M_2^*\tilde{\phi}$. For our purposes, it is only the existence and not the precise form of the expansion that is important.

PROOF OF COROLLARY 2.14. Similarly as in the proof of Proposition 2.13 just above, we may obtain a fixed-point representation

$$\mathcal{F}_j^{y\to x} = \mathcal{T}\mathcal{F}_j^{y\to x} := \overline{\mathcal{F}}_1^{y\to x} + \delta\int_y^x \overline{\mathcal{F}}^{y\to z}(\Theta_{11} + \Theta_{12}\Phi_2)(z,\epsilon)\mathcal{F}_j^{z\to x}\,\mathrm{d}z,$$

where $\mathcal{T}$ is a contraction of order $\mathcal{O}(\delta/\eta)$, yielding the result. □

LEMMA A.13. *Let $d_1 \in \mathbb{C}^{m_1 \times m_1}$ and $d_2 \in \mathbb{C}^{m_2 \times m_2}$ have norm bounded by C_1 and respective spectra separated by $1/C_2 > 0$. Then, the matrix commutator equation*

$$d_1 X - X d_2 = F, \tag{A.25}$$

$X \in \mathbb{C}^{m_1 \times m_2}$, is soluble for all $F \in \mathbb{C}^{m_1 \times m_2}$, with $|X| \leqslant C(C_1, C_2)|F|$.

PROOF. Consider (A.25) as a matrix equation $\mathcal{D}\mathcal{X} = \mathcal{F}$ where $\mathcal{X}$ and $\mathcal{F}$ are vectorial representations of the $m_1 \times m_2$ dimensional quantities X and F, and $\mathcal{D}$ is the matrix representation of the linear operator (commutator) corresponding to the left-hand side. It is readily seen that $\sigma(\mathcal{D})$ is just the difference $\sigma(d_1) - \sigma(d_2)$ between the spectra of d_1 and d_2, with associated eigenvectors of form $r_1 l_2^*$, where r_1 are right eigenvectors associated with d_1 and l_2 are left eigenvectors associated with d_2. (Here as elsewhere, $*$ denotes adjoint, or conjugate transpose of a matrix or vector.) By assumption, therefore, the spectrum of $\mathcal{D}$ has modulus uniformly bounded below by $1/C_2$, whence the result follows. □

PROOF OF PROPOSITION 2.15. Substituting $W = T\widetilde{W}$ into (2.43) and rearranging, we obtain

$$\begin{aligned}\widetilde{W}' &= \left(T^{-1}AT - T^{-1}T'\right)\widetilde{W} \\ &= \left(T^{-1}AT - \varepsilon T^{-1}T_y\right)\widetilde{W},\end{aligned}$$

yielding the defining relation

$$\left(T^{-1}AT - \varepsilon T^{-1}T_y\right) = D \tag{A.26}$$

for T.

By (h2), there exists a uniformly well-conditioned family of matrices $T_0(x)$ such that $T_0^{-1}A_0T_0 = D_0$, D_0 as in (2.42); moreover, these may be chosen with the same regularity in y as A_0 (i.e., the full regularity of A). Expanding

$$T^{-1}(y,\epsilon) = \left(I - \left(\sum_{j=2}^{p} \epsilon^j T_0^{-1}T_j\right) + \left(\sum_{j=2}^{p} \epsilon^j T_0^{-1}T_j\right)^2 - \cdots\right)T_0^{-1} \tag{A.27}$$

by Neumann series, and matching terms of like order ϵ^j, we obtain a hierarchy of systems of linear equations of form:

$$D_0T_0^{-1}T_j - T_0^{-1}T_jD_0 - F_j = D_j, \tag{A.28}$$

where F_j depends only on A_k for $0 \leqslant k \leqslant j$ and T_k, $(\mathrm{d}/\mathrm{d}y)T_k$ for $0 \leqslant k \leqslant j-1$.

On off-diagonal blocks (k, l), (A.28) reduces to

$$d_l\big[T_0^{-1}T_j\big]^{(k,l)} - \big[T_0^{-1}T_j\big]^{(k,l)}d_k = F_j^{(k,l)}, \tag{A.29}$$

uniquely determining $[T_0^{-1}T_j]^{(k,l)}$, by Lemma A.13. On diagonal blocks (k, k), we are free to set $[T_0^{-1}T_j]^{(k,k)} = 0$, whereupon (A.28) reduces to

$$[D_j]^{(k,k)} = -[F_j]^{(k,k)}, \tag{A.30}$$

determining the remaining unknown $[D_j]^{(k,k)}$.

Thus, we may solve for D_j, T_j at each successive stage in a well-conditioned way. Moreover, it is clear that the regularity of D_j, T_j is as claimed, since we lose one order of regularity at each stage, through the dependence of F_j on derivatives of lower order T_k, $k < j$. □

A.4. *Expansion of the one-dimensional Fourier symbol*

In this appendix, we record the Taylor expansions about $\xi = 0$ and $\xi = \infty$ of the Fourier symbol $P(\mathrm{i}\xi) = -\mathrm{i}\xi A^1(x) - \xi^2 B^{11}(x)$ of the one-dimensional frozen, constant-coefficient operator at a given x. Henceforth, we suppress the superscripts for A and B.

Low frequency expansion. Under assumption (P0), straightforward matrix perturbation theory [Kat] yields that the second-order expansion at $\mathrm{i}\xi = 0$ of $P(\mathrm{i}\xi)$ with respect to $\mathrm{i}\xi$ is just

$$\begin{aligned} &P = -\mathrm{i}\xi A - \xi^2 R\,\mathrm{diag}\{\beta_1, \dots, \beta_n\}L, \\ &\mathrm{diag}\{\beta_1, \dots, \beta_n\} := LBR, \end{aligned}$$

where L and R are composed of left and right eigenvectors of A, as described in the Introduction.

High frequency expansion. We next consider the expansion of P at infinity with respect to $(\mathrm{i}\xi)^{-1}$. Recall that

$$B := \begin{pmatrix} 0 & 0 \\ b_1 & b_2 \end{pmatrix}, \qquad A := \begin{pmatrix} A_{11} & A_{12} \\ A_{21} & A_{22} \end{pmatrix}, \qquad A_* = A_{11} - A_{12}b_1b_2^{-1}. \tag{A.31}$$

Claim. *To zeroth order, "hyperbolic" dispersion relations*

$$\lambda(\mathrm{i}\xi) \in \sigma\big(-\xi^2 B - \mathrm{i}\xi A\big), \quad \lambda \sim \xi, \tag{A.32}$$

are given by

$$\lambda_j(\xi) = -\mathrm{i}\xi a_j^* - \eta_j^* + \cdots, \qquad j = 1, \ldots, n - r, \tag{A.33}$$

with corresponding right and left eigenvectors

$$R_j^*(\mathrm{i}\xi) = \begin{pmatrix} r_j^* \\ -b_2^{-1} b_1 r_j^* \end{pmatrix} + \cdots \quad \textit{and} \quad L_j^*(\mathrm{i}\xi) = \begin{pmatrix} l_j^* \\ 0 \end{pmatrix} + \cdots, \tag{A.34}$$

where

$$\eta_j^* := l_j^* D^* r_j^*, \qquad D_* := A_{12} b_2^{-1} \bigl[A_{21} - A_{22} b_2 b_1 + b_2^{-1} b_1 A_*\bigr], \tag{A.35}$$

and a_j^*, r_j^*, l_j^* *are eigenvalues and associated right and left eigenvectors of* A_*. *The remaining* r *"parabolic" relations* $\lambda \sim \xi^2$ *are given to second order by*

$$\lambda_{n-r+j}(\xi) = -\xi^2 \gamma_j + \cdots, \qquad j = 1, \ldots, r, \tag{A.36}$$

with corresponding right and left eigenvectors

$$R_{n-r+j}(\mathrm{i}\xi) = \begin{pmatrix} 0 \\ s_j \end{pmatrix} + \cdots \quad \textit{and} \quad L_{n-r+j}(\mathrm{i}\xi) = \begin{pmatrix} \gamma_j^{-1} b_1^{\mathrm{tr}} t_j \\ t_j \end{pmatrix}, \tag{A.37}$$

where γ_j, s_j, t_j *are the eigenvalues and right and left eigenvectors of* b_2.

PROOF. Expressing $P(\mathrm{i}\xi) = -\xi^2(B + \epsilon A)$, $\epsilon := (\mathrm{i}\xi)^{-1}$ and expanding around $\epsilon = 0$ by standard matrix perturbation theory, we obtain the results after rearrangement; for details, see the dual calculation carried out in Appendix A2 of [MaZ.1] in the relaxation case. □

REMARK A.14. Note that we have without calculating that $\mathrm{Re}\,\sigma \eta_j^* > 0$ as $x \to \pm\infty$, by the spectral bounds of Lemma 2.18.

Appendix B: A weak version of Métivier's theorem

In this appendix, we present a simplified version of the results of Métivier [Mé.1,FM, Mé.5] on inviscid stability of small-amplitude shock waves, establishing sharp uniform lower bounds on the Lopatinski determinant but not the associated uniform L^2 bounds on the solution operator. This gives the flavor of Métivier's analysis, while avoiding many of the technical difficulties; at the same time, it illustrates in the simpler, inviscid setting some of the issues arising in the verification of strong spectral stability in the viscous small-amplitude case [PZ,FreS.2]. This material, presented first in [Z.2], was developed in part together with Freistühler; for an extension to the overcompressive case, see [FreZ.2]. Thanks also to Jenssen for helpful comments and corrections.

Consider an inviscid system

$$U_t + \sum_j F^j(U)_{x_j}, \tag{B.1}$$

and a one-parameter family of stationary Lax p-shocks (defined in (1.23)) $(U_+^\varepsilon, U_-^\varepsilon)$ with normal e_1 and amplitude $|U_+^\varepsilon - U_-^\varepsilon| =: \varepsilon$, satisfying the Rankine–Hugoniot condition

$$\left[F^1(U^\varepsilon)\right] = 0,$$

with $U_\pm$ converging in the small-amplitude limit $\varepsilon \to 0$ to some base point U_0. Setting $A^j := \mathrm{d}F^j(U)$, and denoting by $\bar{a}_j$, $\bar{r}_j$, $\bar{l}_j$ the eigenvalues and associated right and left eigenvectors of A^1, we make the assumptions:

(h0) $F^j \in C^2$ (regularity).

(h1) There exists $A^0(\cdot)$, symmetric positive definite and smoothly depending on U, such that $A^0 A^j(U)$ is symmetric for all $1 \leqslant j \leqslant d$ in a neighborhood of U_0 (simultaneous symmetrizability, $\Rightarrow$ hyperbolicity).

(h2) $\bar{a}_p = 0$ is a simple eigenvalue of $\mathrm{d}F^1(U_0)$ with $\nabla \bar{a}_p \cdot \bar{r}_p(U_0) \neq 0$ (strict hyperbolicity and genuine nonlinearity of the principal characteristic field a_p in the normal spatial direction x_1).

(h3) $\bar{r}_p$ is not an eigenvector of $A^{\tilde{\xi}} := \sum_{j \neq 1} \xi_j A^j$ for $\tilde{\xi} \in \mathbb{R}^{d-1} \neq 0$ (sufficient for extreme shocks, $p = 1$ or n); more generally,

$$\left\langle \bar{r}_p, A^0 A^{\tilde{\xi}} \tilde{\Pi} \left(A^1 - \bar{a}_p\right)^{-1} \tilde{\Pi} A^{\tilde{\xi}} \bar{r}_p \right\rangle \neq 0 \tag{B.2}$$

for all $\tilde{\xi} \in \mathbb{R}^{d-1} \neq 0$, where $\tilde{\Pi}$ denotes the eigenprojection complementary to the eigenprojection of A^1 onto $\mathrm{Range}(\bar{r}_p)$ (genuine hyperbolic coupling [Mé.1–Mé.4,FreZ.2]).

Assumptions (h0)–(h2) are standard conditions satisfied by many physical systems. Assumption (h3) is a technical condition that will emerge through the analysis. It ensures that the system does not decouple in the pth field, in particular excluding the (nonuniformly stable) scalar case.

Propositions 1.10 and 1.12 and Remark 1.14.1, show that the described scenario of a converging family of Lax p-shocks is typical under (h0)–(h2). As pointed out by Métivier, condition (h3) has the following easily checkable characterization (proof deferred to the end of this section), which shows also that (h3) is generically satisfied in two dimensions or for extreme shocks in higher dimensions, in both cases corresponding to nonsingularity of the associated Hessian (1×1 for two dimensions; semidefinite for extreme shocks, as a consequence of hyperbolicity). Conditions (h0)–(h3) are generically satisfied for extreme shocks of gas- and magnetohydrodynamics near points U_0 of thermodynamic stability.

LEMMA B.1 [Mé.1]. *Under assumptions* (h0)–(h2), *condition* (h3) *is equivalent to strict concavity* (*resp. convexity*) *of eigenvalue* $a_p(U, \xi)$ *of* $A^\xi := \sum_j \xi_j A^j$ *with respect to* $\tilde{\xi}$ *at the base point* $U = U_0$, $\xi = (\xi_1, \tilde{\xi}) = (1, 0)$.

Then, we have the following two stability results, respectively concerning the one- and multidimensional case.

PROPOSITION B.2 ([M.4]; one-dimensional stability). *In dimension $d = 1$, hypotheses* (h0)–(h2) *are sufficient to give uniform stability for ε sufficiently small*; *moreover, we have*

$$\left|\Delta(0, \lambda)\right| \geqslant C^{-1}\varepsilon|\lambda|, \tag{B.3}$$

for some $C > 0$, uniformly as $\varepsilon \to 0$.

PROOF. Recalling (1.29), we have

$$\left|\Delta(0, \lambda)\right| := \left|\det\left(\bar{r}_1^-, \ldots, \bar{r}_{p-1}^-, \lambda[U], \bar{r}_{p+1}^+, \ldots, \bar{r}_n^+\right)\right|. \tag{B.4}$$

Without loss of generality choosing $\bar{r}_j(U)$ continuously in U, $|\bar{r}_j| \equiv 1$, we have

$$\bar{r}_j^{\pm} = \bar{r}_j(U_0) + \mathrm{o}(1)$$

as $\varepsilon \to 0$. At the same time,

$$[U] = \varepsilon\left(\bar{r}_p(U_0) + \mathrm{o}(1)\right), \tag{B.5}$$

by the Lax structure theorem, Proposition 1.10. Combining, we have

$$\left|\Delta(0, \lambda)\right| = \varepsilon|\lambda|\left|\det\left(\bar{r}_1, \ldots, \bar{r}_n\right)(U_0) + \mathrm{o}(1)\right|,$$

giving the result. □

PROPOSITION B.3 ([Mé.1]; multidimensional stability). *Assuming* (h0)–(h3), *we have uniform stability for $\varepsilon := |U_+ - U_-|$ sufficiently small*; *moreover*

$$\left|\Delta(\tilde{\xi}, \lambda)\right| \geqslant C^{-1}\varepsilon^{3/2}\left|(\tilde{\xi}, \lambda)\right|, \tag{B.6}$$

for some uniform $C > 0$.

PROOF. Without loss of generality take $d = 2$ and (by the change of coordinates $A^j \to (A^0)^{1/2}A^j(A^0)^{-1/2}$) A^1, A^2 symmetric, $A^0 = I$. For simplicity, restrict to the extreme shock case $p = 1$; the intermediate case goes similarly [Mé.1]. Then, up to a well-conditioned rescaling, we may rewrite (1.29) as

$$\Delta = l_1^+ \cdot \left(\lambda[U] + \mathrm{i}\xi_2\left[F^2(U)\right]\right), \tag{B.7}$$

where $l_1^+(\tilde{\xi}, \lambda)$ denotes the unique stable left eigenvector for $\operatorname{Re}\lambda > 0$ of

$$A_+(\tilde{\xi}, \lambda) := \left(\lambda + \mathrm{i}\xi_2 A_+^2\right)\left(A_+^1\right)^{-1},$$

extended by continuity to $\operatorname{Re}\lambda = 0$. We use form (B.7) to simplify the computations. □

OBSERVATION 1. *By appropriate choice of coordinates, we may arrange that*

$$A^1_+ = \begin{pmatrix} -\hat{\varepsilon} & 0 \\ 0 & \widetilde{A}^1 \end{pmatrix}, \tag{B.8}$$

where $\widetilde{A}^1 \in \mathbb{R}^{(n-1)\times(n-1)}$ *is symmetric, positive definite and bounded uniformly above and below,* $\hat{\varepsilon} \sim \varepsilon$ *by the Lax structure theorem, Proposition* 1.10, *and genuine nonlinearity, condition* (h2);

$$A^2_+ = \begin{pmatrix} 0 & 1 & \dots & 0 \\ 1 & & & \\ \vdots & & \widetilde{A}^2 & \\ 0 & & & \end{pmatrix}, \tag{B.9}$$

where $\widetilde{A}^2 \in \mathbb{R}^{(n-1)\times(n-1)}$ *is symmetric; and*

$$\lambda[U] + \mathrm{i}\xi_2\big[F^2(U)\big] = \varepsilon \begin{pmatrix} \lambda \\ \mathrm{i}\xi_2 \\ 0 \\ \vdots \\ 0 \end{pmatrix} \big(c + \mathcal{O}(\varepsilon)\big), \tag{B.10}$$

$c \neq 0$. *(Here,* $\mathcal{O}(\varepsilon)$ *is matrix-valued, for brevity of exposition.)*

PROOF. First, take coordinates such that A^1 is block-diagonal and A^2 symmetric. Changing to a moving coordinate frame $x_2' = x_2 - ct$ in the transverse direction, we may arrange that $(A^2_+)_{11} = 0$, i.e., $\langle \bar{r}^+_1, A^2_+ \bar{r}^+_1 \rangle = 0$, hence by a change of dependent variables in the second $((n-1) \times (n-1))$ block of A^1 that $A^2_+ \bar{r}^+_1$ lies in the second coordinate direction. Since $A^2_+ \bar{r}^+_1 \neq 0$, by (h3), we may arrange by a rescaling of x_2 that $A^2_+ \bar{r}^+_1 = 1$. This verifies (B.8) and (B.9). Relation (B.10) then follows by the Lax structure theorem, Proposition 1.10, and Taylor expansion about U_+ of $F^2(U)$. □

From here on, we drop the hat, and write ε for $\hat{\varepsilon}$.

OBSERVATION 2. *The matrix* $A_+(\tilde{\xi}, \lambda) := (\lambda + \mathrm{i}\xi_2 A^2_+)(A^1_+)^{-1}$ *then has the form*

$$\left(\begin{pmatrix} -\lambda/\varepsilon & \mathrm{i}\xi_2/a \\ -\mathrm{i}\xi_2/\varepsilon & 0 \end{pmatrix} \begin{matrix} 0 \\ 0 \end{matrix} \atop \begin{matrix} \qquad 0 & & 0 \end{matrix} \right) + \big(|\xi_2| + |\lambda|\big) \begin{pmatrix} 0 & \mathcal{O}(\varepsilon) & \mathcal{O}(1) \\ 0 & \mathcal{O}(1) & \mathcal{O}(1) \\ 0 & \mathcal{O}(1) & \mathcal{O}(1) \end{pmatrix}, \tag{B.11}$$

$a > 0$, *or, "balancing" using transformation*

$$T := \begin{pmatrix} \varepsilon^{-1/2} & 0 & 0 \\ 0 & 1 & 0 \\ 0 & 0 & I_{n-2} \end{pmatrix},$$

$$TA_+T^{-1} = \left(\begin{pmatrix} -\lambda/\varepsilon & \mathrm{i}\xi_2/a\varepsilon^{1/2} \\ -\mathrm{i}\xi_2/\varepsilon^{1/2} & 0 \end{pmatrix} \begin{matrix} 0 \\ \end{matrix} \atop \begin{matrix} 0 & & 0 \end{matrix}\right)$$

$$+ \big(|\xi_2| + |\lambda|\big)\begin{pmatrix} 0 & \mathcal{O}(\varepsilon^{1/2}) & \mathcal{O}(\varepsilon^{-1/2}) \\ 0 & \mathcal{O}(1) & \mathcal{O}(1) \\ 0 & \mathcal{O}(1) & \mathcal{O}(1) \end{pmatrix}$$

$$= \varepsilon^{-1/2}\begin{pmatrix} -\hat{\lambda} & \mathrm{i}\xi_2/a & w^{\mathrm{tr}} \\ -\mathrm{i}\xi_2 & 0 & 0 \\ 0 & 0 & 0 \end{pmatrix}$$

$$+ \big(|\xi_2| + \varepsilon^{1/2}|\hat{\lambda}|\big)\begin{pmatrix} 0 & \mathcal{O}(\varepsilon^{1/2}) & 0 \\ 0 & \mathcal{O}(1) & \mathcal{O}(1) \\ 0 & \mathcal{O}(1) & \mathcal{O}(1) \end{pmatrix}, \tag{B.12}$$

where $\hat{\lambda} := \lambda\varepsilon^{-1/2}$, $|w| = \mathcal{O}(1)$.

PROOF. We have from (B.8) that

$$(A_+^1)^{-1} = \begin{pmatrix} -\varepsilon^{-1} & 0 \\ 0 & (\widetilde{A}^1)^{-1} \end{pmatrix}, \tag{B.13}$$

where

$$(\widetilde{A}^1)^{-1} = \begin{pmatrix} a^{-1} & * \\ * & * \end{pmatrix} \tag{B.14}$$

is symmetric positive definite, whence the diagonal entry a^{-1} is symmetric positive definite as well. The result then follows by direct computations, combining (B.9) with (B.14). □

Without loss of generality fixing $\xi_2 \equiv 1$ (the case $\xi_2 = 0$ may be obtained by continuity, and has anyway been treated already in the one-dimensional case), we consider the resulting matrix perturbation problem (B.12).

Case (i) ($|\lambda| \gg \varepsilon^{1/2}$). First we treat the straightforward case $|\hat{\lambda}| \gg 1$. In this case, $-\hat{\lambda}$ dominates all entries in the rescaled perturbation problem (B.12), from which we find that the associated left eigendirection of TA_+T^{-1} is $(1, 0, \ldots, 0)^{\mathrm{tr}} + \mathrm{o}(1)$, and thus the left eigenvector l_1^+ of A_+ is (rescaled to order one)

$$\varepsilon^{1/2}\big((1, 0, \ldots, 0)^{\mathrm{tr}} + \mathrm{o}(1)\big)T \sim (1, 0, \ldots, 0)^{\mathrm{tr}} + \mathrm{o}\big(\varepsilon^{1/2}\big). \tag{B.15}$$

Computing $\Delta = l_1^+ \cdot (\lambda[U] + \mathrm{i}\xi_2[F^2(U)])$ following (B.7), we obtain from (B.10) and (B.15) that $|\Delta| \sim \varepsilon\lambda \geqslant C^{-1}\varepsilon^{3/2}$, as asserted.

Case (ii) ($|\lambda| \leqslant C\varepsilon^{1/2}$). Next, we treat the critical case that $|\hat{\lambda}|$ is uniformly bounded. Noting that the upper left-hand corner

$$\alpha := \begin{pmatrix} -\hat{\lambda} & \mathrm{i}\xi_2/a \\ -\mathrm{i}\xi_2 & 0 \end{pmatrix} = \begin{pmatrix} -\hat{\lambda} & \mathrm{i}/a \\ -\mathrm{i} & 0 \end{pmatrix} \tag{B.16}$$

of block-upper triangular matrix

$$\mathcal{A} = \begin{pmatrix} -\hat{\lambda} & \mathrm{i}\xi_2/a & w^{\mathrm{tr}} \\ -\mathrm{i}\xi_2 & 0 & 0 \\ 0 & 0 & 0 \end{pmatrix} = \begin{pmatrix} -\hat{\lambda} & \mathrm{i}/a & w^{\mathrm{tr}} \\ -\mathrm{i} & 0 & 0 \\ 0 & 0 & 0 \end{pmatrix}$$

is uniformly nonsingular, with determinant $-\xi_2^2/a = -1/a < 0$, we have by standard matrix perturbation theory that the stable left eigenvector $\mathcal{L}_1^+$ of

$$TA_+T^{-1} = \varepsilon^{-1/2}\mathcal{A} + \mathcal{O}(1)$$

is an $\mathcal{O}(\varepsilon^{1/2})$ perturbation of the left stable eigenvector $\mathcal{L} = (\ell^{\mathrm{tr}}, w^{\mathrm{tr}})^{\mathrm{tr}}$ of $\mathcal{A}$, where

$$\ell := (1, \mathrm{i}/a\mu)^{\mathrm{tr}}$$

is the left stable eigenvector of α and

$$\mu := \frac{-\hat{\lambda} - \sqrt{\hat{\lambda}^2 + 4/a}}{2}$$

is the associated stable eigenvalue. Scaling to order one length, we thus have

$$l_1^+ = \varepsilon^{1/2} T\mathcal{L}_1^+. \tag{B.17}$$

Computing $\Delta = l_1^+ \cdot (\lambda[U] + \mathrm{i}\xi_2[F^2(U)])$ following (B.7), we obtain from (B.10), (B.17), and the above developments that

$$\begin{aligned} \Delta &= \varepsilon^{1/2}\big(\mathcal{L}_1^+ + \mathcal{O}(\varepsilon^{1/2})\big) \cdot T\varepsilon \begin{pmatrix} \lambda \\ \mathrm{i}\xi_2 \\ 0 \\ \vdots \\ 0 \end{pmatrix} \big(c + \mathcal{O}(\varepsilon)\big) \\ &= \varepsilon^{3/2}\ell \cdot \begin{pmatrix} \hat{\lambda} \\ \mathrm{i} \end{pmatrix} \big(1 + \mathcal{O}(\varepsilon^{1/2})\big) \\ &= \varepsilon^{3/2}\mu\big(1 + \mathcal{O}(\varepsilon^{1/2})\big), \end{aligned}$$

yielding $|\Delta| \sim \varepsilon^{3/2}$ as claimed.

This completes the proof. □

REMARKS. 1. The upper bound in case (ii) shows that (B.6) is sharp.

2. [Mé.1] The matrix perturbation problem (B.11) may be recognized as essentially that arising from the canonical 2×2 example

$$F^1(u, v) := \begin{pmatrix} u^2/2 \\ v \end{pmatrix}, \qquad F^2(u, v) := \begin{pmatrix} v \\ u \end{pmatrix},$$

$U_\pm = (\mp\varepsilon, 0)$, for which the system at U_+ is the 2×2 wave-type equation

$$A^1 = \begin{pmatrix} -\varepsilon & 0 \\ 0 & 1 \end{pmatrix}, \qquad A^2 = \begin{pmatrix} 0 & 1 \\ 1 & 0 \end{pmatrix},$$

and the resulting matrix perturbation problem agrees with the determining 2×2 upper-left-hand block of (B.11). That is, whereas one-dimensional behavior is (at least at small amplitudes) essentially scalar, Proposition B.2, multidimensional behavior involves in an essential way the coupling between different hyperbolic modes (precisely two for each fixed transverse direction $\tilde{\xi}$).

3. In the intermediate shock case $1 < p < n$, the strengthened version (B.2) of condition (h3) may be seen to correspond (see development of (B.14)) to nonvanishing of a in the 2×2 blocks analogous to α in (B.16) (one block arising at $+\infty$ and one at $-\infty$).

PROOF OF LEMMA B.1. Without loss of generality take $d = 2$ and (by the change of coordinates $A^j \to (A^0)^{1/2} A^j (A^0)^{-1/2}$) A^1, A^2 symmetric, $A^0 = I$. Let

$$a_p(\tilde{\xi}) = a_p^0 + a_p^1 \tilde{\xi} + a_p^2 \tilde{\xi}^2 \tag{B.18}$$

and

$$r_p(\tilde{\xi}) = r_p^0 + r_p^1 \tilde{\xi} + \cdots, \tag{B.19}$$

denote the eigenvalues and eigenvectors of

$$A(\tilde{\xi}) := (A^1 + \tilde{\xi} A^2), \tag{B.20}$$

analytic by simplicity of a_p, (h2), without loss of generality $|r_p^0| = 1$. Expanding the defining relations

$$\bigl(A(\tilde{\xi}) - a_p(\tilde{\xi})\bigr) r_p(\tilde{\xi}) = 0, \tag{B.21}$$

and matching terms of like order in $\tilde{\xi}$, we obtain:

$$(A^1 - a_p^0) r_p^0 = 0 \qquad \text{(0th order)}, \tag{B.22}$$

$$(A^1 - a_p^0) r_p^1 + (A^2 - a_p^1) r_p^0 = 0 \qquad \text{(1st order)}, \tag{B.23}$$

$$(A^1 - a_p^0) r_p^2 + (A^2 - a_p^1) r_p^1 + (-a_p^2) r_j^0 = 0 \qquad \text{(2nd order)}. \tag{B.24}$$

Taking the inner product of r_p^0 with (B.23), we obtain

$$0 = \langle r_p^0, (A^2 - a_p^1) r_p^0 \rangle, \tag{B.25}$$

and thus

$$a_p^1 = \langle r_p^0, A^2 r_p^0 \rangle. \tag{B.26}$$

Next, we may solve modulo $\operatorname{span}(r_p^0)$ for r_p^1 in (B.23) by taking the inner product with r_k^0, $k \neq p$, to obtain

$$\langle r_k^0, r_p^1\rangle = -\frac{\langle r_k^0, (A^2 - a_p^1) r_p^0\rangle}{a_k^0 - a_p^0}; \tag{B.27}$$

hence

$$r_p^1 = -\sum_{k\neq p} \frac{\langle r_k^0, (A^2 - a_p^1) r_p^0\rangle}{a_k^0 - a_p^0} r_k^0 \tag{B.28}$$

is a solution. Finally, taking the inner product of r_p^0 with (B.24), and using (B.28) plus symmetry of A^j, we obtain expansion

$$\begin{aligned} a_p^2 &= -\sum_{k\neq p} \frac{\langle r_k^0, (A^2 - a_p^1) r_p^0\rangle \langle r_p^0, (A^2 - a_p^1) r_k^0\rangle}{a_k^0 - a_p^0} \\ &= -\sum_{k\neq p} \frac{\langle r_k^0, (A^2 - a_p^1) r_p^0\rangle^2}{a_k^0 - a_p^0}. \end{aligned} \tag{B.29}$$

Observing that r_p^0, by (B.25) is an eigenvector of A^2 if and only if

$$(A^2 - a_1^1) r_p^0 = 0 \quad \operatorname{mod}\big(\operatorname{span}_{k\neq p}(r_k^0)\big), \tag{B.30}$$

we have the result in the extreme case $p = 1$ or n, for which $a_k^0 - a_p^0$ for $k \neq p$ has a definite sign. More generally, observing that $(A^2 - a_p^1) r_p^0 = \tilde{\Pi} A^2 r_p^0$, we may rewrite (B.28) as

$$r_p^1 = (A^1 - a_p^0)^{-1} \tilde{\Pi} A^2 r_p^0$$

to obtain

$$\begin{aligned} a_p^2 &= \langle r_p^0, (A^2 - a_p^1) r_p^1\rangle \\ &= \langle \tilde{\Pi} A^2 \bar{r}_p^0, r_p^1\rangle \\ &= \langle \tilde{\Pi} A^2 \bar{r}_p^0, (A^1 - \bar{a}_p)^{-1} \tilde{\Pi} A^2 r_p^0\rangle \end{aligned}$$

in place of (B.29). Thus, $a_p^2 \neq 0$ is equivalent to

$$\langle \bar{r}_p^0, A^2 \tilde{\Pi} (A^1 - \bar{a}_p)^{-1} \tilde{\Pi} A^2 \bar{r}_p^0\rangle \neq 0,$$

as claimed. □

Appendix C: Evaluation of the Lopatinski condition for gas dynamics

C.1. *The Lopatinski condition for general systems*

There is a large body of literature on the problem of stability for inviscid shock waves and in particular for gas dynamics, see [Be,Ro,D,Free.1,Free.2,Ko.1,Ko.2,Ri,Er.1,Gr,SF,M.1, M.4,BE,Mé.1,Mé.5,Z.3].

The purpose of this appendix is to calculate the Lopatinski determinant, or "stability function", for the Euler equations of compressible gas dynamics. We describe two approaches to this problem. In the first we use a change of variables to simplify the computation. In the second method we take advantage of the Galilean invariance of the Euler equations and exploit a relationship between the eigenvectors of a particular pair of matrices. As a preliminary step for the first technique, we discuss how the Lopatinski determinant behaves under a change of coordinates.

Consider an inviscid system of n hyperbolic conservation laws in d space dimensions,

$$w_t + \sum_{j=1}^{d} f^j(w)_{x_j} = 0, \tag{C.1}$$

where $x = (x_1, \dots, x_d) \in \mathbb{R}^d$, $w \in \mathbb{R}^n$. A planar shock wave solution moving along the x_1-axis is a solution of the form $w(x,t) = W(x_1 - st)$ where W is piecewise constant with a single jump discontinuity at the origin:

$$W(y) = \begin{cases} W^- & \text{for } y < 0, \\ W^+ & \text{for } y > 0. \end{cases} \tag{C.2}$$

The Rankine–Hugoniot condition relating the right and left states is

$$s[w] = \left[f^1(w)\right], \tag{C.3}$$

where $[q] := q^+ - q^-$ for any quantity q. By a linear change of independent variables we can assume that the shock is stationary, that is, $s = 0$.

The linearized stability of planar shock waves for (C.1) can be studied through the Lopatinski determinant which we now define. We begin by fixing the notation

$$A^j = Df^j, \qquad f^\xi := \sum_{j=1}^{d} \xi_j f^j,$$

$$\xi = (\xi_1, \dots, \xi_d)^{\mathrm{tr}}, \qquad \tilde{\xi} := (\xi_2, \dots, \xi_d)^{\mathrm{tr}},$$

$$A^\xi := \sum_{j=1}^{d} \xi_j A^j, \qquad A^{\tilde{\xi}} := A^{(0,\tilde{\xi})},$$

and define

$$\mathcal{A}(\tilde{\xi}, \lambda) := \left(\lambda I + \mathrm{i} A^{\tilde{\xi}}\right)\left(A^1\right)^{-1}. \tag{C.4}$$

Here $\mathcal{A}(\tilde{\xi}, \lambda)$ is evaluated along the solution w and we write $\mathcal{A}_\pm(\tilde{\xi}, \lambda)$ for the evaluation at $W^\pm$. The Lopatinski determinant (stability function) for a Lax p-shock is then defined to be

$$\Delta(\tilde{\xi}, \lambda) := \det\big(r_1^-, \ldots, r_{p-1}^-, \lambda[w] + \mathrm{i}\big[f^{\tilde{\xi}}(w)\big], r_{p+1}^+, \ldots, r_n^+\big), \tag{C.5}$$

where $\tilde{\xi} \in \mathbb{R}^{d-1}$ and $\operatorname{Re}\lambda > 0$. Here $\{r_1^-, \ldots, r_{p-1}^-\}$ is a basis for the stable subspace of $\mathcal{A}_-(\tilde{\xi}, \lambda)$, while $\{r_{p+1}^+, \ldots, r_n^+\}$ is a basis for the unstable subspace of $\mathcal{A}_+(\tilde{\xi}, \lambda)$, and the jumps are given by the Rankine–Hugoniot relation for the unperturbed shock. The expression in (C.5) is obtained through a Fourier–Laplace analysis and a zero $(\tilde{\xi}, \lambda)$ of Δ corresponds to a mode of the form $\sim \mathrm{e}^{\lambda t + \mathrm{i}\tilde{\xi}\cdot\tilde{x}}$ for the equations linearized about $W(x)$. The function Δ is jointly analytic in $(\tilde{\xi}, \lambda)$ for $\tilde{\xi} \in \mathbb{R}^{d-1}$ and $\operatorname{Re}\lambda > 0$ and Δ is homogeneous of degree 1. These features will prove useful below. Also by continuity the Lopatinski determinant may be defined for all points $(\tilde{\xi}, \lambda) \neq 0$ with $\tilde{\xi} \in \mathbb{R}^{d-1}$ and $\operatorname{Re}\lambda \geqslant 0$. See [Z.3, ZS,Se.1,M.4] for details.

DEFINITION 1. Consider a planar shock wave W of the form (C.2), and its corresponding Lopatinski determinant $\Delta(\tilde{\xi}, \lambda)$ defined by (C.5). If Δ has a zero $(\tilde{\xi}, \lambda)$ with $\operatorname{Re}\lambda > 0$, the shock wave is referred to as strongly unstable. If Δ has no zeros λ with $\operatorname{Re}\lambda \geqslant 0$, the shock is called uniformly stable. Finally, in the intermediate case where Δ has some root with $\operatorname{Re}\lambda = 0$ but no root with $\operatorname{Re}\lambda > 0$ the shock is said to be weakly stable.

We observe that in the case of an extreme shock, that is $p = 1$ or $p = n$, the expression for Δ simplifies to

$$\Delta = l \cdot \big(\lambda[w] + \mathrm{i}\big[f^{\tilde{\xi}}(w)\big]\big), \tag{C.6}$$

where l is a left eigenvector of $\mathcal{A}$. More precisely we have that if the shock is of the

- first family, then l is the left eigenvector of $\mathcal{A}_+$ corresponding to the unique stable eigenvalue β, i.e., $\operatorname{Re}\beta < 0$,
- last family, then l is the left eigenvector of $\mathcal{A}_+$ corresponding to the unique unstable eigenvalue β, i.e., $\operatorname{Re}\beta > 0$.

The calculation of these left eigenvectors for specific systems such as gas dynamics may be rather involved, the main problem being to invert the matrix A^1.

We next describe two alternative methods for calculating Δ for extreme shocks, one by a change of variables and the other by utilizing the relationship between left eigenvectors and eigenvalues of $\mathcal{A}$ and A^{ξ}.

METHOD 1. An appropriate change of coordinates may yield a straightforward and explicit calculation of Δ. Indeed Erpenbeck used this approach in [Er.1] for gas dynamics, the first work in which this calculation was completely carried out. We begin by briefly considering how Δ behaves under a change of coordinates. We then apply the analysis to the Euler equations written in terms of specific volume, velocity, and specific entropy in the next section.

The system (C.1) can be written in quasilinear form

$$w_t + \sum_{j=1}^{d} A^j w_{x_j} = 0, \tag{C.7}$$

where $A^j(w) := Df^j(w)$. Consider a change of coordinates in (C.1) given by $w = \varphi(\bar{w})$. The corresponding quasilinear system is thus

$$\bar{w}_t + \sum_{j=1}^{d} M^j \bar{w}_{x_j} = 0, \tag{C.8}$$

where $M^j(\bar{w}) = C^{-1} A^j(\varphi(\bar{w}))C$ and $C = D\varphi$. Using notation as above,

$$\mathcal{A}(\tilde{\xi}, \lambda) = C(\lambda I + \mathrm{i}M^{\tilde{\xi}})(M^1)^{-1} C^{-1}.$$

A proper choice of coordinates simplifies the calculation of the desired eigenvectors. We write $\mathcal{B} = (\lambda I + \mathrm{i}M^{\tilde{\xi}})(M^1)^{-1}$. Since $\bar{l}$ is a left eigenvector of $\mathcal{B}$ with corresponding eigenvalue β if and only if $\bar{l}C^{-1}$ is a left eigenvector of $\mathcal{A}$ with eigenvalue β, we have

$$\Delta = \bar{l}^{\mathrm{tr}} C^{-1}\big(\lambda[w] + \mathrm{i}\big[f^{\tilde{\xi}}(w)\big]\big). \tag{C.9}$$

The point is that by properly choosing coordinates, the matrix M^1 may be easy to invert. In the case of gas dynamics, this technique yields an explicit formulation of Δ in terms of λ and $\tilde{\xi}$.

METHOD 2. An alternative way to deal with the problem of inverting A^1 in (C.4) is to observe that l is a left eigenvector of $\mathcal{A}(\tilde{\xi}, \lambda)$ with eigenvalue β if and only if l is a left eigenvector of $A^{(\mathrm{i}\beta, \tilde{\xi})}$ with eigenvalue $\mathrm{i}\lambda$. Thus it is not necessary to invert A^1, provided one can compute the left eigenvectors and eigenvalues of A^{ξ}. For gas dynamics these are straightforward (but tedious) to compute explicitly for any ξ. Furthermore, the relevant left eigenvector l for a 1-shock, say, is the left eigenvector of $\mathcal{A}_+$ corresponding to the unique stable eigenvalue β, $\mathrm{Re}\beta < 0$. Now, a calculation shows that

$$\mathrm{i}\lambda = \mathrm{i}\beta u_1 \pm c\sqrt{|\tilde{\xi}|^2 - \beta^2}, \tag{C.10}$$

where c denotes the sound speed and u_1 denotes the particle velocity in the x_1 direction. It follows that the eigenvector l is expressed in terms of β. To obtain Δ explicitly in terms of λ and $\tilde{\xi}$, we solve (C.10) for β, substitute into the expression for l and then calculate Δ. This approach works particularly well for isentropic gas dynamics, see [Se.1] and [Z.2]. The calculations for the full Euler equations are given below.

C.2. *Calculation for full Euler equations*

The Euler equations in d space dimensions takes the form (C.1) with $w = (\rho, m_1, \ldots, m_d, \mathcal{E})^{\mathrm{tr}}$ as follows

$$\rho_t + \sum_{j=1}^{d} (m_j)_{x_j} = 0, \tag{C.11}$$

$$m_{i,t} + \sum_{j=1}^{d} \left(\frac{m_i m_j}{\rho} + \delta_{ij} p \right)_{x_j} = 0, \quad i = 1, 2, \ldots, d, \tag{C.12}$$

$$\mathcal{E}_t + \sum_{j=1}^{d} \left(\frac{m_j \mathcal{E}}{\rho} + \frac{m_j p}{\rho} \right)_{x_j} = 0. \tag{C.13}$$

Here ρ, m_j, p and $\mathcal{E}$ denote density, momentum in the jth coordinate direction, pressure, and total energy, respectively. We have $m_j = \rho u_j$, where u_j is velocity in the jth coordinate direction, p is a given function of the thermodynamic variables, and the total energy is given by $\mathcal{E} = \rho(e + (u_1^2 + \cdots + u_d^2)/2)$, where e denotes the internal energy, $\mathrm{d}e = T\,\mathrm{d}S - p\,\mathrm{d}v$, $v = 1/\rho$ is specific volume, S is specific entropy, and T is temperature.

C.2.1. *Method 1.* In our first calculation we use the relation (C.9) to calculate Δ. It is computationally more convenient to write the Euler equations in terms of the (nonconservative) variables $\bar{w} = (v, u_1, \ldots, u_d, S)^{\mathrm{tr}}$:

$$v_t + \sum_{j=1}^{d} (v_{x_j} u_j - v u_{j,x_j}) = 0, \tag{C.14}$$

$$u_{i,t} + \sum_{j=1}^{d} u_j u_{i,x_j} + v p_{x_i} = 0, \quad i = 1, 2, \ldots, d, \tag{C.15}$$

$$S_t + \sum_{j=1}^{d} u_j S_{x_j} = 0. \tag{C.16}$$

(For a calculation in the one-dimensional case using (ρ, u, e) as dependent variables, see [Se.2].) Letting $u^{\mathrm{tr}} = (u_1, \ldots, u_d)$, the change of coordinates $\phi : \mathbb{R}^{1+d+1} \to \mathbb{R}^{1+d+1}$ is given by

$$\phi \begin{pmatrix} v \\ u \\ S \end{pmatrix} = \begin{pmatrix} v^{-1} \\ v^{-1} u \\ \mathcal{E} \end{pmatrix},$$

where

$$\mathcal{E} = v^{-1}\left(e(v, S) + \frac{|u|^2}{2}\right).$$

The matrix $C = D\phi$ is thus given as,

$$C = \begin{pmatrix} -v^{-2} & 0 & 0 \\ -v^{-2}u & v^{-1}I_d & 0 \\ -v^{-1}(\mathcal{E} + p) & v^{-1}u^{\mathrm{tr}} & v^{-1}T \end{pmatrix},$$

with inverse

$$C^{-1} = \begin{pmatrix} -v^{2} & 0 & 0 \\ -vu & vI_d & 0 \\ vT^{-1}(|u|^2 - v(\mathcal{E} + p)) & -vu^{\mathrm{tr}}T^{-1} & vT^{-1} \end{pmatrix},$$

where I_d denotes the $d \times d$ identity matrix. Note that both C and C^{-1} above as well as the matrices M^j below have the block structure

$$\left(\begin{array}{c|c|c} 1 \times 1 & 1 \times d & 1 \times 1 \\ \hline d \times 1 & d \times d & d \times 1 \\ \hline 1 \times 1 & 1 \times d & 1 \times 1 \end{array}\right).$$

The matrices M^j are given by (C.14)–(C.16) as

$$M^j = \begin{pmatrix} u_j & -v\mathbf{e}_j^{\mathrm{tr}} & 0 \\ vp_v\mathbf{e}_j & u_jI_d & vp_S\mathbf{e}_j \\ 0 & 0 & u_j \end{pmatrix},$$

where $\mathbf{e}_j$ denotes the jth standard unit vector. The sound speed c is defined as

$$c := \sqrt{-v^2 p_v},$$

and we let

$$\mu := u_1^2 - c^2.$$

Taking advantage the simple structure of M^1, we can easily find that its inverse is

$$(M^1)^{-1} = \begin{pmatrix} \frac{u_1}{\mu} & \frac{v}{\mu}\mathbf{e}_1^{\mathrm{tr}} & \frac{-v^2 p_S}{u_1\mu} \\ \frac{c^2 v^{-1}}{\mu}\mathbf{e}_1 & \frac{1}{u_1}I_d + \frac{c^2}{u_1\mu}\mathbf{e}_1\mathbf{e}_1^{\mathrm{tr}} & \frac{-vp_S}{\mu}\mathbf{e}_1 \\ 0 & 0 & \frac{1}{u_1} \end{pmatrix}.$$

To compute Δ as given by (C.9) we need the eigenvectors and eigenvalues of the matrix $\mathcal{B}$. To find these it is convenient to write M^j in the form

$$M^j = u_j I_{d+2} + \overline{M}^j,$$

where

$$\overline{M}^j = \begin{pmatrix} 0 & -v\mathbf{e}_j^{\mathrm{tr}} & 0 \\ vp_v\mathbf{e}_j & 0 & vp_S\mathbf{e}_j \\ 0 & 0 & 0 \end{pmatrix}.$$

Then

$$M^{\tilde{\xi}} := \sum_{j=2}^{d} \xi_j M^j = (\tilde{u}\cdot\tilde{\xi}) I_{d+2} + \overline{M}^{\tilde{\xi}},$$

where the matrix $\overline{M}^{\tilde{\xi}}$ is

$$\overline{M}^{\tilde{\xi}} = \begin{pmatrix} 0 & 0 & -v\tilde{\xi}^{\mathrm{tr}} & 0 \\ 0 & 0 & 0 & 0 \\ vp_v\tilde{\xi} & 0 & 0 & vp_S\tilde{\xi} \\ 0 & 0 & 0 & 0 \end{pmatrix}.$$

It follows that the matrix $\mathcal{B}$ is given as

$$\mathcal{B} = \left(\alpha I_{d+2} + \mathrm{i}\overline{M}^{\tilde{\xi}}\right)\left(M^1\right)^{-1},$$

where $\alpha = \lambda + \mathrm{i}(\tilde{u}\cdot\tilde{\xi})$. We thus have

$$\mathcal{B} = \begin{pmatrix} \frac{\alpha u_1}{\mu} & \frac{\alpha v}{\mu} & -\mathrm{i}\frac{v}{u_1}\tilde{\xi}^{\mathrm{tr}} & \frac{-\alpha v^2 p_S}{u_1\mu} \\ \frac{\alpha c^2 v^{-1}}{\mu} & \frac{\alpha u_1}{\mu} & 0 & \frac{-\alpha v p_S}{\mu} \\ \mathrm{i}p_v\frac{vu_1}{\mu}\tilde{\xi} & \mathrm{i}p_v\frac{v^2}{\mu}\tilde{\xi} & \frac{\alpha}{u_1}I_{d-1} & \mathrm{i}p_S\frac{vu_1}{\mu}\tilde{\xi} \\ 0 & 0 & 0 & \frac{\alpha}{u_1} \end{pmatrix}$$

which has the block structure

$$\left(\begin{array}{c|c|c|c} 1\times 1 & 1\times 1 & 1\times(d-1) & 1\times 1 \\ \hline 1\times 1 & 1\times 1 & 1\times(d-1) & 1\times 1 \\ \hline (d-1)\times 1 & (d-1)\times 1 & (d-1)\times(d-1) & (d-1)\times 1 \\ \hline 1\times 1 & 1\times 1 & 1\times(d-1) & 1\times 1 \end{array}\right).$$

We proceed to find the eigenvalues of $\mathcal{B}$. The determinant $\det(\mathcal{B} - \beta I_{d+2})$ has the form

$$\det(\mathcal{B} - \beta I_{d+2}) = \begin{vmatrix} a_1 & a_2 & \delta\tilde{\xi}^{\mathrm{tr}} & * \\ a_3 & a_1 & 0 & * \\ a_4\tilde{\xi} & a_5\tilde{\xi} & a_6 I_{d-1} & * \\ 0 & 0 & 0 & a_6 \end{vmatrix}.$$

Using Schur's complement formula (see [HJ], p. 22) this determinant is easily reduced to

$$\det(\mathcal{B} - \beta I_{d+2}) = a_6^{d-1}\big[a_6\big(a_1^2 - a_2a_3\big) + \delta|\tilde{\xi}|^2(a_3a_5 - a_1a_4)\big].$$

Substituting the appropriate values for $a_1, \ldots, a_6$ and simplifying, one obtains

$$\det(\mathcal{B} - \beta I_{d+2}) = \left(\frac{\alpha}{u_1} - \beta\right)^d \left[\left(\frac{\alpha u_1}{\mu} - \beta\right)^2 + \frac{c^2}{\mu^2}\left(\mu|\tilde{\xi}|^2 - \alpha^2\right)\right],$$

which yields three different eigenvalues $\beta = \frac{\alpha}{u_1}$ (a d-fold eigenvalue), and

$$\beta_\pm = \frac{\alpha u_1}{\mu} \pm \frac{c}{\mu}\sqrt{\alpha^2 - \mu|\tilde{\xi}|^2}. \tag{C.17}$$

The complex square root in (C.17) is chosen to have its branch cut along the negative real axis. Recalling that $\alpha = \lambda + \mathrm{i}(\tilde{u} \cdot \tilde{\xi})$ this ensures that β varies analytically as λ varies over the right complex half plane. The eigenvalues of interest for the stability analysis of (extreme) gas dynamical shocks are given by (C.17). We next find the eigenvectors $\bar{l}_\pm$ of $\mathcal{B}$ corresponding to $\beta\pm$. Letting $\bar{l}_\pm^{\mathrm{tr}} = (l_{1\pm}, l_{2\pm}, \tilde{l}_\pm, l_{*\pm})$ with $l_{1\pm} = \alpha - u_1\beta_\pm$, we obtain

$$\bar{l}_\pm = \begin{pmatrix} \alpha - u_1\beta_\pm \\ -v\beta_\pm \\ \mathrm{i}v\tilde{\xi} \\ \frac{v^2 p_S}{\mu}\left(2\alpha + \frac{u_1^2|\tilde{\xi}|^2 - \alpha^2}{\alpha - u_1\beta_\pm}\right) \end{pmatrix}.$$

Multiplying on the right by C^{-1} then yields the desired left eigenvector $l_\pm$ of $\mathcal{A}_\pm$. To compute (C.9), we now need only the jumps which are easily calculated from the Rankine–Hugoniot conditions.

For simplicity and concreteness we carry out the remaining calculations for a Lax 1-shock in dimension $d = 2$. Due to Galilean invariance we may without loss of generality assume that the transversal velocity is zero, that is $\tilde{u} = 0$. As explained in the preceding introductory discussion, β, the eigenvalue relevant for a 1-shock, is a stable eigenvalue of $\mathcal{A}_+$, i.e., β should have *negative* real part. By considering the one-dimensional case where $\xi_2 = 0$ and letting λ be a positive real number, it follows that we must choose the $+$ sign in (C.17).

In the following calculation of Δ all quantities without a subscript are evaluated on the right of (i.e., behind) the shock ($v = v_+$, $p = p_+$, etc.), and u will denote u_1^+. A subscript denotes evaluation on the left of (i.e., in front of) the shock. We note at this point that, as a consequence of the entropy inequalities for Lax shocks, the flow is supersonic in front of the shock and subsonic behind the shock, see [Sm].

In this setting the Rankine–Hugoniot conditions for a stationary gas dynamical shock take the form

$$[m_1] = 0, \tag{C.18}$$

$$\left[\frac{m_1^2}{\rho} + p\right] = 0, \tag{C.19}$$

$$\left[\frac{m_1 m_2}{\rho}\right] = 0, \tag{C.20}$$

$$\left[\frac{m_1(\mathcal{E} + p)}{\rho}\right] = 0, \tag{C.21}$$

so that

$$\lambda[w] + \mathrm{i}\left[f^{\tilde{\xi}}(w)\right] = \begin{pmatrix} \lambda[\rho] \\ 0 \\ \mathrm{i}\xi_2[p] \\ \lambda[\mathcal{E}] \end{pmatrix}. \tag{C.22}$$

Substituting into (C.9) we finally get the desired expression for the stability function Δ,

$$\Delta = v[\rho]\left\{\lambda v(\beta u - l_1) - v\xi_2^2\frac{[p]}{[\rho]} + \frac{\lambda}{T}l_*\left(\frac{[\mathcal{E}]}{[\rho]} + u^2 - v(\mathcal{E} + p)\right)\right\}. \tag{C.23}$$

To simplify this expression we first observe that (C.21) implies

$$\left[\frac{\mathcal{E}}{\rho}\right] = -\left[\frac{p}{\rho}\right],$$

so that

$$[e] + \left[\frac{p}{\rho}\right] + \frac{m^2}{2}\left[\frac{1}{\rho^2}\right] = 0, \tag{C.24}$$

where $m = m_1$. From (C.19) it follows that

$$m^2 = -\frac{[p]}{[1/\rho]},$$

whence

$$\left[\frac{p}{\rho}\right] + \frac{m^2}{2}\left[\frac{1}{\rho^2}\right] = \frac{p_+ + p_-}{2}\left[\frac{1}{\rho}\right] =: \langle p\rangle\left[\frac{1}{\rho}\right].$$

Thus (C.24) takes the form

$$[e] + \langle p \rangle \left[\frac{1}{\rho} \right] = 0,$$

and we obtain

$$\frac{[\mathcal{E}]}{[\rho]} + u^2 - v(\mathcal{E} + p) = \frac{m^2}{\rho} \left[\frac{1}{\rho} \right].$$

Before using this relation to simplify (C.23), we slightly change notation to emphasize the important physical quantities:

$$\text{the Mach number} \quad M = \frac{|u|}{c}$$

and

$$\text{the compression ratio} \quad r = \frac{u}{u_-} = \frac{v}{v_-} = \frac{\rho_-}{\rho}.$$

We also let

$$\omega = \frac{\lambda}{u}$$

and introduce

$$s(\omega) = \omega + \sqrt{\xi_2^2(1 - M^2) + M^2\omega^2},$$

where the square root has positive real part. With this notation we have

$$l_1 = \frac{u}{1 - M^2} s(\omega)$$

and

$$l_* = \frac{uv^2 p_S}{c^2(1 - M^2)} \left(\frac{2\lambda}{u} + \frac{(1 - M^2)(\xi_2^2 - \omega^2)}{s(\omega)} \right) = \frac{uv^2 p_S}{c^2(1 - M^2)} s(\omega),$$

where we have used that $(1 - M^2)\xi_2^2 = (s(\omega) - \omega)^2 - \omega^2 M^2$. Substituting into the expression for Δ, including the compression ratio, and using Rankine–Hugoniot conditions to simplify the resulting expression we find

$$\Delta(\xi_2, \omega) = -\frac{\lambda v^2 u[\rho]}{1 - M^2} \left\{ rk\omega s(\omega) + \left(1 - M^2\right)\left(\xi_2^2 - r\omega^2\right) \right\}, \tag{C.25}$$

where

$$k := 2 - \frac{p_S}{T}(v_- - v)M^2. \tag{C.26}$$

We summarize our findings:

THEOREM B.4. *Consider the Euler equations* (C.11)–(C.13) *in two space dimensions. The Lopatinski determinant* (C.5) *for a* 1*-shock is given by* (C.25) *and* (C.26), *where all unmarked quantities are evaluated to the right of the shock.*

C.2.2. *Method 2.* In this section we give an alternative computation of Δ in (C.6) for a gas dynamical shock of the first family. As explained above the key observation is that l is a left eigenvector of $\mathcal{A}(\tilde{\xi}, \lambda)$ with eigenvalue β if and only if l is a left eigenvector of $A^{(\mathrm{i}\beta,\tilde{\xi})}$ with eigenvalue $\mathrm{i}\lambda$. We begin by outlining how to effectively compute the eigenvalues and left eigenvectors of A^{ξ}, for any ξ. We then use the result to express Δ as given by (C.6).

Writing the Euler equations (C.11)–(C.13) in quasilinear form (C.7) we obtain that

$$A^{\xi} = \begin{pmatrix} 0 & \xi^{\mathrm{tr}} & 0 \\ -(u\cdot\xi)u & (u\cdot\xi)I_d & \frac{P_e}{\rho}\xi \\ +\xi(P_\rho + P_e e_\rho) & +u\xi^{\mathrm{tr}} - \frac{P_e}{\rho}\xi u^{\mathrm{tr}} & \\ (u\cdot\xi)\big(-E - \frac{P}{\rho} + P_\rho + P_e e_\rho\big) & -(u\cdot\xi)\frac{P_e}{\rho}u^{\mathrm{tr}} + \big(E + \frac{P}{\rho}\big)\xi^{\mathrm{tr}} & u\cdot\xi\big(1 + \frac{P_e}{\rho}\big) \end{pmatrix}.$$

Here we consider the pressure P to be a given function of ρ and e, so P is related to p above by

$$p(v, S) = P\big(v^{-1}, e(v, S)\big).$$

In computing the eigenvalues and eigenvectors of A^{ξ} it is convenient, taking advantage of Galilean invariance of the equations, to conjugate with the "shift" matrix S given by

$$S = \begin{pmatrix} 1 & 0 & 0 \\ u & I_d & 0 \\ \frac{|u|^2}{2} & u^{\mathrm{tr}} & 1 \end{pmatrix},$$

whose inverse is

$$S^{-1} = \begin{pmatrix} 1 & 0 & 0 \\ -u & I_d & 0 \\ \frac{|u|^2}{2} & -u^{\mathrm{tr}} & 1 \end{pmatrix}.$$

This yields

$$S^{-1}A^{\xi}S = (u\cdot\xi)I_{d+2} + \overline{A}^{\xi},$$

where

$$\overline{A}^{\xi} = \begin{pmatrix} 0 & \xi^{\mathrm{tr}} & 0 \\ \big(P_\rho - \frac{P_e}{\rho}e\big)\xi & 0 & \frac{P_e}{\rho}\xi \\ 0 & \big(e + \frac{P}{\rho}\big) & 0 \end{pmatrix}.$$

We denote the eigenvalues of $\overline{A}^{\xi}$ by $\bar{a}_j$, the right eigenvectors by $\bar{r}_j$ and the left eigenvectors by $\bar{l}_j$. If a_j, r_j and l_j denote the corresponding quantities for A^{ξ}, then

$$a_j = u \cdot \xi + \bar{a}_j, \qquad r_j = S\bar{r}_j, \qquad l_j = \bar{l}_j S^{-1}.$$

Now the eigenvalues and eigenvectors of $\bar{A}^{\xi}$ are straightforward to compute, and we get that the eigenvalues of A^{ξ} are

$$a_1 = u \cdot \xi - c|\xi|, \tag{C.27}$$

$$a_j = u \cdot \xi, \quad j = 2, \ldots, d+1, \tag{C.28}$$

$$a_{d+2} = u \cdot \xi + c|\xi|, \tag{C.29}$$

where $c^2 = P_\rho + \rho^{-2} P P_e$ is the sound speed. To write down the eigenvectors we introduce the coefficients

$$\theta = P_\rho - \frac{P_e}{\rho} e, \qquad \eta = \frac{P_e}{\rho}, \qquad \zeta = e + \frac{P}{\rho}.$$

The left eigenvectors of A^{ξ} are found to be

$$l_1 = \left(\theta + \frac{c(u \cdot \xi)}{|\xi|} + \frac{\eta |u|^2}{2}, -\frac{c\xi}{|\xi|} - \eta u, \eta\right)^{\mathrm{tr}}, \tag{C.30}$$

$$l_2 = \left(-\zeta + \frac{|u|^2}{2}, -u, 1\right)^{\mathrm{tr}}, \tag{C.31}$$

$$l_j = \left(-u \cdot \xi^{\perp}, \xi^{\perp}, 0\right)^{\mathrm{tr}}, \quad j = 3, \ldots, d+1, \tag{C.32}$$

$$l_{d+2} = \left(\theta - \frac{c(u \cdot \xi)}{|\xi|} + \frac{\eta |u|^2}{2}, \frac{c\xi}{|\xi|} - \eta u, \eta\right)^{\mathrm{tr}}, \tag{C.33}$$

where $\xi^{\perp}$ is orthogonal to ξ. We note here that the right eigenvectors r_j are also easily calculated from $\bar{r}_j$, the corresponding right eigenvectors of $\overline{A}^{\xi}$. As we have no use for these eigenvectors in what follows, we omit them.

We next have to decide which eigenvector to use in the computation of Δ. From the discussion above we know that the left eigenvector of $\mathcal{A}(\tilde{\xi}, \lambda)$ to be used in (C.6) is the same as the left eigenvector of $A^{(\mathrm{i}\beta, \tilde{\xi})}$ with eigenvalue $\mathrm{i}\lambda$, with $\operatorname{Re}\beta < 0$ and $\operatorname{Re}\lambda > 0$. Thus $\mathrm{i}\lambda$ must be equal to either $u \cdot \xi \pm c|\xi|$ or $u \cdot \xi$, where $\xi = (\mathrm{i}\beta, \tilde{\xi})$. In the special case $\tilde{u} = 0$ the last choice yields $\lambda = \beta u_1$. Since $u_1 > 0$ for a 1-shock while $\operatorname{Re}\beta < 0$ and $\operatorname{Re}\lambda > 0$, this is not the right choice. It remains to decide which sign and which branch of the square root is to be used in the relation

$$\mathrm{i}\lambda = \mathrm{i}\beta u_1 \pm c\sqrt{|\tilde{\xi}|^2 - \beta^2}. \tag{C.34}$$

To do so we use the fact that (C.34) must give λ as a continuous function of β as β varies over the left complex half plane and $\tilde{\xi}$ varies over $\mathbb{R}^{d-1}$. In particular, for $\tilde{\xi} = 0$, $\sqrt{-\beta^2}$ must be continuous in $\{\operatorname{Re}\beta < 0\}$. It follows that the square root in (C.34) is the branch with positive imaginary part. For $\tilde{\xi} = 0$ we thus have $\lambda = \beta u_1 \pm c|\beta|$. Taking real parts and using the fact that $c > u$, $\operatorname{Re}\lambda > 0$ and $\operatorname{Re}\beta < 0$, we conclude that the correct sign in (C.34) is the $+$, i.e., we must use l_{d+2} (and not l_1) in the computation of Δ for shock of the first family.

For the actual computation of Δ we restrict ourselves as before to the two-dimensional case and we assume without loss of generality that $\tilde{u} = 0$. From the preceding discussion and the expression (C.33) we thus have that the left eigenvector in (C.6) is given as

$$l = \left(\theta - \frac{\mathrm{i}c\beta u}{\sqrt{\xi_2^2 - \beta^2}} + \frac{\eta u^2}{2}, \frac{\mathrm{i}c\beta}{\sqrt{\xi_2^2 - \beta^2}} - \eta u, \frac{c\xi_2}{\sqrt{\xi_2^2 - \beta^2}}, \eta\right)^{\mathrm{tr}},$$

where $u = u_1$, and β with $\operatorname{Re}\beta < 0$ is related to λ through

$$c\sqrt{\xi_2^2 - \beta^2} = \mathrm{i}(\lambda - \beta u). \tag{C.35}$$

Taking the inner product with $\lambda[w] + \mathrm{i}[f^{\tilde{\xi}}(w)]$ as given in (C.22), and using (C.35), we see that Δ is given by

$$\begin{aligned}\Delta(\xi_2, \lambda) &= \lambda[\rho]\left(\theta - \frac{\mathrm{i}c\beta u}{\sqrt{\xi_2^2 - \beta^2}} + \frac{\eta u^2}{2}\right) + \frac{\mathrm{i}c\xi_2^2}{\sqrt{\xi_2^2 - \beta^2}}[P] + \eta\lambda[\mathcal{E}] \\ &= \frac{\mathrm{i}}{c\sqrt{\xi_2^2 - \beta^2}}\left\{\lambda(\lambda - \beta u)\left([\rho]\left(\theta + \frac{\eta u^2}{2}\right) + \eta[\mathcal{E}]\right)\right. \\ &\qquad \left. + c^2\xi_2^2[P] - \lambda[\rho]c^2\beta u\right\},\end{aligned}$$

where all unmarked quantities are evaluated on the right of the shock. Making use of the identities that follow from the Rankine–Hugoniot relation in the same way as above, we see that

$$[\rho]\left(\theta + \frac{\eta u^2}{2}\right) + \eta[\mathcal{E}] = [\rho]\left(c^2 + \eta u^2\left(1 - \frac{1}{r}\right)\right),$$

r being the compression ratio as defined above. Also, $\eta = P_e/\rho = v p_S/T$, and $[P]/[\rho] = u u_-$ so that the expression for Δ reduces to

$$\begin{aligned}\Delta(\xi_2, \lambda) &= \frac{\mathrm{i}c[\rho]}{\sqrt{\xi_2^2 - \beta^2}}\left\{(\lambda^2 - \beta u\lambda)(k-1) - \beta\lambda u + \xi_2^2\frac{u^2}{r}\right\} \\ &= \frac{\mathrm{i}cu^2[\rho]}{\sqrt{\xi_2^2 - \beta^2}}\left\{(k-1)\omega^2 - k\beta\omega + \xi_2^2\frac{1}{r}\right\},\end{aligned}$$

where $\omega = \lambda/u$ and k is given by (C.26) above. Now, solving (C.35) for β shows that

$$\beta = \frac{-u\lambda - \sqrt{\lambda^2 + \xi_2^2(c^2 - u^2)}}{c^2 - u^2} = -\frac{M^2\omega + \sqrt{\xi_2^2(1 - M^2) + M^2\omega^2}}{1 - M^2}. \tag{C.36}$$

The branch and sign of the square root are determined as in (C.17), i.e. the square root in (C.36) is the standard one with positive real part. Using the notation from above we have that

$$\beta = \omega - \frac{s(\omega)}{1 - M^2}.$$

Substituting back into the expression for Δ and simplifying finally gives

$$\Delta(\xi_2, \lambda) = \frac{\mathrm{i}cu^2[\rho]}{r(1 - M^2)\sqrt{\xi_2^2 - \beta^2}} \{rk\omega s(\omega) + (1 - M^2)(\xi_2^2 - r\omega^2)\},$$

which agrees with the expression in (C.25) up to a multiplicative factor, which is due to a different normalization of the left eigenvector l in (C.6).

C.3. *The stability conditions*

We now derive conditions under which a planar 1-shock is uniformly stable, weakly stable, or strongly unstable. According to Definition 1 this amounts to determining the sign of $\operatorname{Re}\lambda$ whenever $\Delta(\xi_2, \lambda) = 0$.

Here we exploit the analyticity of Δ with respect to λ, and as in [Er.1], we apply the argument principle to count the number of roots of Δ in the right complex half-plane, see [Lan].

We note that $\xi_2 = 0$ corresponds to the one-dimensional situation. Restricting ourselves to the multidimensional setting we may assume without loss of generality that $\xi_2 = 1$ by the homogeneity of Δ. For a thorough discussion of the one-dimensional analysis and its relation to the multidimensional situation, see [Se.2].

We also introduce polar coordinates (R, θ) and define the function

$$G(R, \theta) = G(\omega) := -\frac{1 - M^2}{\lambda v^2 u[\rho]}\Delta(1, \omega), \quad \omega = R\mathrm{e}^{\mathrm{i}\theta}.$$

Thus the zeros of Δ and G are the same. We consider $G(\omega)$ as ω varies counterclockwise along the closed contour C consisting of a semicircle together with a vertical segment joining the ends, see Figure 1. Special care must be taken during the mapping of the imaginary axis to avoid the branch cuts arising from the square root in $s(\omega)$. We first obtain the limiting behavior as $R \to +\infty$. For large R, we find

$$G \approx (1 + M)r(k - 1 + M)R^2\mathrm{e}^{\mathrm{i}2\theta}. \tag{C.37}$$

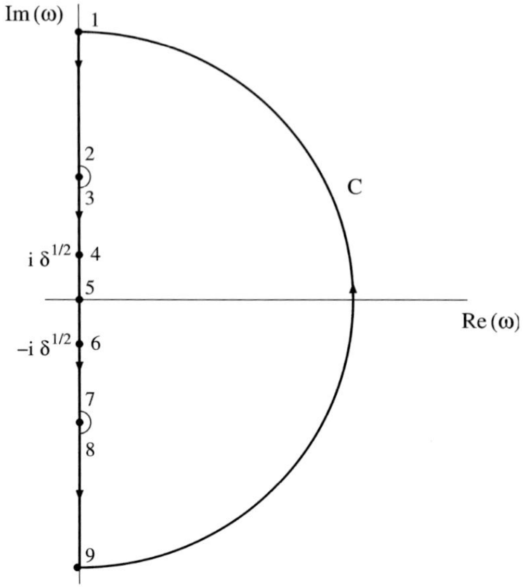

Fig. 1. The path C in the ω plane.

Next consider the part of C on the imaginary axis where $\theta = \pm\pi/2$. Substitution in (C.25) yields

$$G\left(R, \pm\frac{\pi}{2}\right) = \left(1 - M^2\right) + rR^2\left(1 - M^2 - k\right)$$
$$+ rkMR \begin{cases} -\sqrt{R^2 - \delta} & \text{for } R^2 > \delta, \\ \pm \mathrm{i}\sqrt{\delta - R^2} & \text{for } R^2 < \delta, \end{cases}$$

where we have put $\delta = (1 - M^2)/M^2 > 0$. We observe from this that G restricted to the imaginary axis has nonzero imaginary part only on that interval of the axis lying between $-\sqrt{\delta}$ and $+\sqrt{\delta}$. Thus any zero of G on the imaginary axis must be outside this interval. We also note that $G(0, 0) = 1 - M^2$ is real and positive, while for large R it's clear from (C.37) that the sign of the real part of $G(R, \pm\frac{\pi}{2})$ depends on the sign of the quantity $k - 1 + M$. There are thus two cases to consider:

(I) $1 - M - k < 0$ (in which case $\operatorname{Re} G(R, \pm\frac{\pi}{2}) < 0$ for large R)
(II) $1 - M - k > 0$ (in which case $\operatorname{Re} G(R, \pm\frac{\pi}{2}) > 0$ for large R).

It is clear from (C.37) that the change in argument of $G(\omega)$ along the semi circular part of C is 2π in either case. Now, a straightforward computation shows that in case (I),

$$\frac{\partial G(R, \pm\frac{\pi}{2})}{\partial R} < 0 \quad \text{for } R^2 > \delta.$$

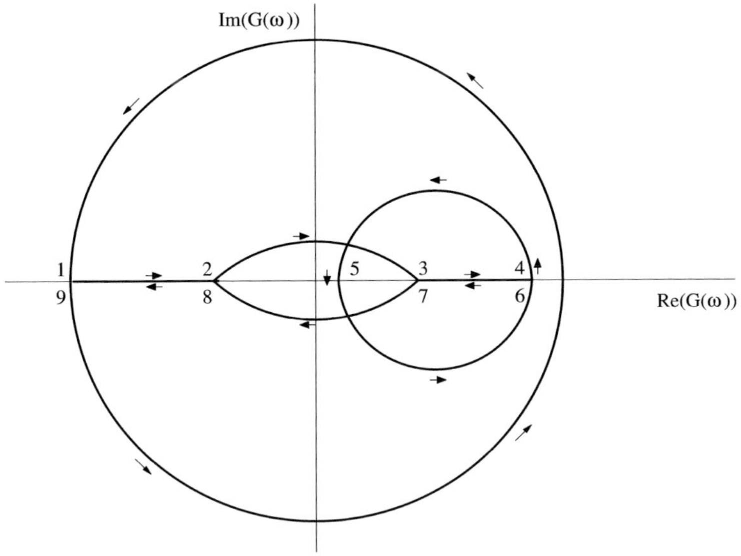

Fig. 2. Case (Ia), the change in $\mathrm{Arg}(G(\omega))$ is 0.

Consequently G in case (I) has a single root on each half of the imaginary axis if and only if

$$G\left(\sqrt{\delta}, \pm\frac{\pi}{2}\right) = \left(1 - M^2\right) + \left(1 - M^2 - k\right)r\frac{1 - M^2}{M^2} > 0. \tag{C.38}$$

We thus further divide case (I) into two subcases depending on whether or not (C.38) holds. We label them as (Ia) if (C.38) is valid and (Ib) otherwise; thus in subcase (Ib) G has no zeros on the imaginary axis. Also it is easily seen that

$$G\left(R, \pm\frac{\pi}{2}\right) > 0 \quad \text{for all } R > \sqrt{\delta}$$

in case (II), whence G has no root on the imaginary axis in that case either.

Finally, a careful analysis of G as ω moves down the vertical part of C shows that its argument changes by -2π in case (I), while it is unchanged in case (II). The mapping in case (Ia) is shown in Figure 2. We note that in this case the cut-outs between points 2 and 3 and points 7 and 8 in Figure 1 must be followed to avoid the roots of G on the imaginary axis. The change in the argument for the mapping of the imaginary axis is thus -2π and in combination with the 2π change in argument along the circular portion of the contour, we find that in case (Ia), there are no roots of G, and hence of Δ, for $\mathrm{Re}\,\lambda > 0$. By Definition 1 the pure imaginary roots indicate that the shock is *weakly stable* in this case. Combining

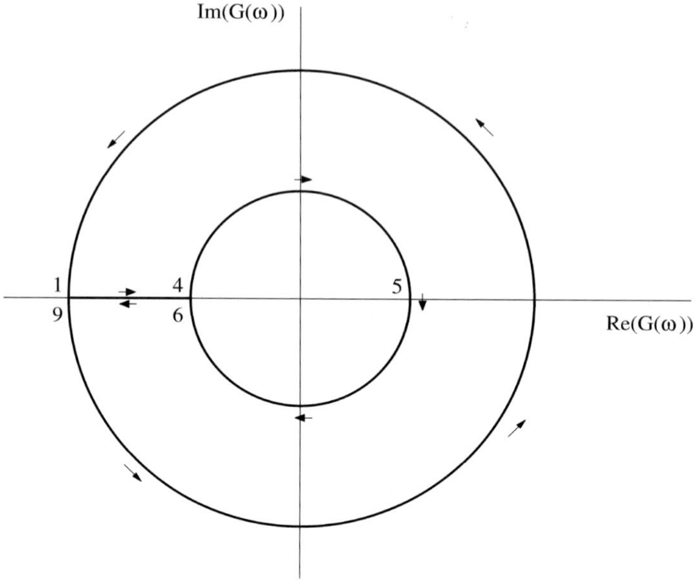

Fig. 3. Case (Ib), the change in $\mathrm{Arg}(G(\omega))$ is 0.

the condition defining case (I) and (C.38), we find that a shock is weakly stable if and only if

$$1 - M < k < 1 - M^2 + M^2/r.$$

The mapping in case (Ib) is simpler and is shown in Figure 3. In this case there are no pure imaginary roots with which to contend, and the change in argument as the imaginary axis is mapped is again -2π. In combination with the 2π change in argument along the circular portion of the contour, it follows then that there are no roots of G, and again therefore of Δ, for $\mathrm{Re}\,\lambda > 0$. It follows then from Definition 1, that the shock is *uniformly stable*. In case (Ib) the inequality (C.38) is assumed not to hold in which case it follows that the shock under consideration is uniformly stable if and only if

$$k > 1 - M^2 + M^2/2.$$

Finally we examine the mapping of the imaginary axis in case (II). The mapping in this case is shown in Figure 4. There is no change in the argument from the mapping of the imaginary axis. It follows then from the 2π change in argument due to the mapping of the circular portion of the contour that there is a zero with positive real part and thus that the shock is *strongly unstable*. The condition for strong instability is thus the condition that characterizes case (II). A shock is strongly unstable if and only if

$$k < 1 - M.$$

Summing up the discussion above we have the following result.

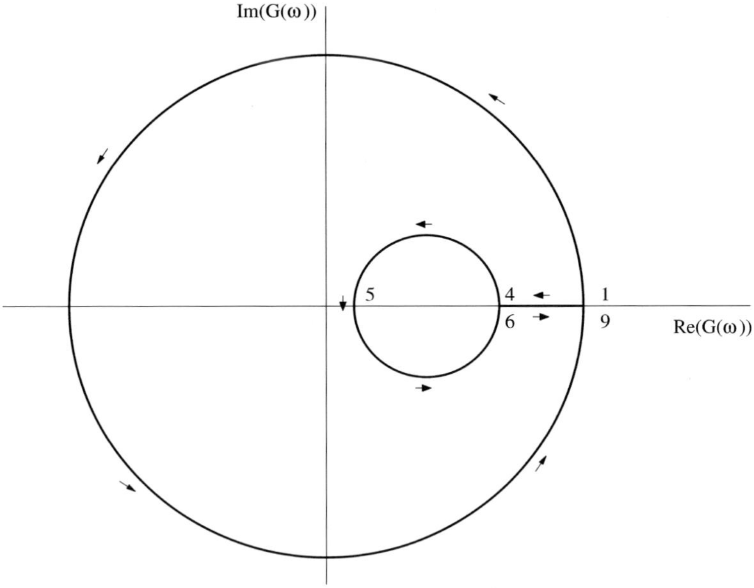

Fig. 4. Case (II), the change in Arg$(G(\omega))$ is 2π.

THEOREM B.5. *With the same assumptions as in Theorem* B.4 *a shock of the first family is*

- *strongly unstable if and only if* $k < 1 - M$,
- *weakly (but not uniformly) stable if and only if* $1 - M < k < 1 - M^2 + M^2/r$,
- *uniformly stable if and only if* $k > 1 - M^2 + M^2/r$,

where $r = \frac{u}{u_-}$ *is the compression ratio, k is given by* (C.26), *and* $M = \frac{|u|}{c}$ *is the Mach number, all of which are evaluated on the right of the* 1*-shock.*

We point out that these results were (apart from the terminology) first proved by Erpenbeck [Er.1], and later derived in the more comprehensive work of Majda [M.1,M.4]. Serre [Se.2] has shown (under certain conditions) that the transition from weak stability to strong instability in several space dimensions is signaled by instability with respect to one dimensional perturbations.

REMARK B.6. We note that the stability conditions above are stated in a slightly different form in [M.4]. Defining μ as the inverse of the compression ratio r, i.e., $\mu := v_-/v$, and introducing the Gruneisen coefficient

$$\Gamma := \frac{v}{T} p_S = -\frac{v}{T} \frac{\partial^2 e}{\partial S \, \partial v},$$

we have that a 1-shock is

- strongly unstable if and only if $\frac{1+M}{\Gamma} < (\mu - 1)M^2$,
- weakly (but not uniformly) stable if and only if $\frac{1}{\Gamma+1} < (\mu - 1)M^2 < \frac{1+M}{\Gamma}$,

- uniformly stable if and only if $(\mu - 1)M^2 < \frac{1}{\Gamma+1}$.

C.3.1. *The Case of an Ideal Gas.* We end by considering the special case of an ideal, polytropic gas; so the equation of state is given by

$$p = R\rho T,$$

and $e = c_v T$, and the pressure can also be written in terms of density and entropy as

$$p = K_1 \rho^\gamma \exp(S/c_v),$$

where c_v is the specific heat at constant volume, K_1 and R are positive constants, and $\gamma > 1$. The Gruneisen coefficient is now given as $\Gamma = \frac{R}{c_v}$. Using the fact that $1 + \gamma = \Gamma$, the condition for uniform stability can be rewritten as

$$\left(\frac{1}{r} - 1\right)M^2 < \frac{1}{\Gamma + 1} = \frac{1}{\gamma}.$$

A straightforward manipulation of the Rankine–Hugoniot conditions yields the relationship

$$\left(\frac{1}{r} - 1\right)M^2 = \frac{1}{\gamma}\left(1 - \frac{p}{p_-}\right).$$

Then combining the two expressions above, we see that for an ideal gas the uniform stability condition is

$$\frac{1}{\gamma}\left(1 - \frac{p}{p_-}\right) < \frac{1}{\gamma},$$

which is clearly satisfied since $p > 0$. It follows then that the uniform stability condition is satisfied, and all ideal gas shocks are uniformly stable.

Acknowledgment

The authors of Appendix C would like to thank Kevin Zumbrun and Denis Serre for numerous helpful conversations and suggestions.

References

[AGJ] J. Alexander, R. Gardner and C.K.R.T. Jones, *A topological invariant arising in the analysis of traveling waves*, J. Reine Angew. Math. **410** (1990), 167–212.

[AMPZ.1] A. Azevedo, D. Marchesin, B. Plohr and K. Zumbrun, *Nonuniqueness of solutions of Riemann problems*, Z. Angew. Math. Phys. **47** (6) (1996), 977–998.

[AMPZ.2] A. Azevedo, D. Marchesin, B. Plohr and K. Zumbrun, *Bifurcation of nonclassical viscous shock profiles from the constant state*, Comm. Math. Phys. **202** (2) (1999), 267–290.

[AMPZ.3] A. Azevedo, D. Marchesin, B. Plohr and K. Zumbrun, *Capillary instability in multiphase flow models*, Z. Angew. Math. Phys. **53** (2002), 713–746.

[AMPZ.4] A. Azevedo, D. Marchesin, B. Plohr and K. Zumbrun, *Long-lasting diffusive solutions of systems of conservation laws*, VI Workshop on Partial Differential Equations, Part I (Rio de Janeiro, 1999), Mat. Contemp. **18** (2000), 1–29.

[BDG] T. Bridges, G. Derks and G. Gottwald, *Stability and instability of solitary waves of the fifth-order KdV equation: A numerical framework*, Phys. D **172** (1–4) (2002), 190–216.

[BE] A.A. Barmin and S.A. Egorushkin, *Stability of shock waves*, Adv. Mech. **15** (1/2) (1992), 3–37.

[B-G.1] S. Benzoni-Gavage, *Stability of multi-dimensional phase transitions in a van der Waals fluid*, Nonlinear Anal. **31** (1/2) (1998), 243–263.

[B-G.2] S. Benzoni-Gavage, *Stability of subsonic planar phase boundaries in a van der Waals fluid*, Arch. Ration. Mech. Anal. **150** (1) (1999), 23–55.

[B-G.3] S. Benzoni-Gavage, *Sur la stabilité de profils de choc semi-discrets au moyen d'une fonction d'Evans en dimension infinie* (in French) [On the stability of semidiscrete shock profiles by means of an Evans function in infinite dimension], C. R. Acad. Sci. Paris Sér. I Math. **329** (5) (1999), 377–382.

[B-G.4] S. Benzoni-Gavage, *Linear stability of propagating phase boundaries in capillary fluids*, Phys. D **155** (2001), 235–273.

[BHR] S. Benzoni-Gavage, P. Huot and F. Rousset, *Nonlinear stability of semidiscrete shock waves*, Preprint (2003).

[BRSZ] S. Benzoni-Gavage, F. Rousset, D. Serre and K. Zumbrun, *Generic types and transitions in hyperbolic initial-boundary-value problems*, Proc. Roy. Soc. Edinburgh Sect. A **132** (2002), 1073–1104.

[BSZ.1] S. Benzoni-Gavage, D. Serre and K. Zumbrun, *Alternate Evans functions and viscous shock waves*, SIAM J. Math. Anal. **32** (2001), 929–962.

[BSZ.2] S. Benzoni-Gavage, D. Serre and K. Zumbrun, *Transition to instability of planar viscous shock fronts*, in preparation.

[BMFC] A. Bertozzi, A. Münch, X. Fanton and A.M. Cazabat, *Contact line stability and 'undercompressive shocks' in driven thin film flow*, Phys. Rev. Lett. **81** (23) (1998), 5169–5172.

[BMS.1] A. Bertozzi, A. Münch and M. Shearer, *Undercompressive shocks in waves in thin film flow*, Phys. D **134** (1999), 431–464.

[BMS.2] A. Bertozzi, A. Münch and M. Shearer, *Stability of capillary shocks in thin film flow*, Preprint (1999).

[BMSZ] A. Bertozzi, A. Münch, M. Shearer and K. Zumbrun, *Stability of compressive and undercompressive thin film travelling waves, The dynamics of thin fluid films*, European J. Appl. Math. **12** (2001), 253–291.

[BS] A. Bertozzi and M. Shearer, *Existence of undercompressive traveling waves in thin film equations*, SIAM J. Math. Anal. **32** (1) (2000), 194–213 (electronic).

[Be] H.A. Bethe, *On the theory of shock waves for an arbitrary equation of state*, Classic Papers in Shock Compression Science, Office of Scientific Research and Development, Report 545, Serial No. NDRC-B-237, Office Sci. Res. Develop., U.S. Army Ballistic Research Laboratory, Aberdeen Proving Ground, MD (1942), Springer-Verlag, New York (1982), 421–492.

[Bo] G. Boillat, *On symmetrization of partial differential systems*, Appl. Anal. **57** (1995), 17–21.

[BoR] G. Boillat and T. Ruggeri, *On the shock structure problem for hyperbolic system of balance laws and convex entropy*, Contin. Mech. Thermodyn. **10** (5) (1998), 285–292.

[B] F. Bouchut, *Construction of BGK models with a family of kinetic entropies for a given system of conservation laws*, J. Statist. Phys. **95** (1/2) (1999), 113–170.

[Br.1] L. Brin, *Numerical Testing of the Stability of Viscous Shock Waves*, Doctoral Thesis, Indiana Univ. (1998).

[Br.2] L. Brin, *Numerical testing of the stability of viscous shock waves*, Math. Comp. **70** (235) (2001), 1071–1088.

[BrZ] L. Brin and K. Zumbrun, *Analytically Varying Eigenvectors and the Stability of Viscous Shock Waves*, Mat. Contemp. (2003).

[BGS] M. Bultelle, M. Grassin and D. Serre, *Unstable Godunov discrete profiles for steady shock waves*, SIAM J. Num. Anal. **35** (1998), 2272–2297.

[Ch] S. Chandrasekhar, *Hydrodynamics and Hydromagnetic Stability*, *The International Series of Monographs on Physics*, Clarendon Press, Oxford (1961).

[CP] J. Chazarain and A. Piriou, *Introduction to the Theory of Linear Partial Differential Equations*, Transl. from French, Stud. Math. Appl., Vol. 14, North-Holland, Amsterdam–New York (1982).

[CL] E.A. Coddington and N. Levinson, *Theory of Ordinary Differential Equations*, McGraw-Hill, New York–Toronto–London (1955).

[CS] C. Conley and J. Smoller, *On the structure of magnetohydrodynamic shock waves*, Comm. Pure Appl. Math. **27** (1974), 367–375.

[Co] W.A. Coppel, *Stability and Asymptotic Behavior of Differential Equations*, Heath, Boston, MA (1965).

[CF] R. Courant and K.O. Friedrichs, *Supersonic Flow and Shock Waves*, Springer-Verlag, New York (1976).

[CH.1] R. Courant and D. Hilbert, *Methods of Mathematical Physics, Vol. I*, Interscience, New York (1953).

[CH.2] R. Courant and D. Hilbert, *Methods of Mathematical Physics, Vol. II: Partial Differential Equations* (Vol. II by R. Courant), Interscience, New York–London (1962).

[DSL] H. De Sterck and B.C. Low, *Complex magnetohydrodynamic bow shock topology in field-aligned low-β flow around a perfectly conducting cylinder*, Phys. Plasmas **5** (11) (1998), 4015–4027.

[Dr] P.G. Drazin, *Discontinuous velocity profiles for the Orr–Sommerfeld equation*, J. Fluid Mech. **10** (1961), 571–583.

[DR] P.G. Drazin and W.H. Reid, *Hydrodynamic Stability*, Cambridge Monographs Mech. Appl. Math., Cambridge Univ. Press, Cambridge (1981).

[D] P. D'yakov, Zh. Eksper. Teoret. Fiz. **27** (288) (1954); English transl.: Atomic Energy Research Establishment AERE Lib./Trans. **648** (1956).

[Er.2] J.J. Erpenbeck, *Stability of steady-state equilibrium detonations*, Phys. Fluids **5** (5) (1962), 604–614.

[Er.1] J.J. Erpenbeck, *Stability of step shocks*, Phys. Fluids **5** (10) (1962), 1181–1187.

[Er.4] J.J. Erpenbeck, *Stability of idealized one-reaction detonations*, Phys. Fluids **7** (5) (1964), 684–696.

[Er.5] J.J. Erpenbeck, *Steady detonations in idealized two-reaction systems*, Phys. Fluids **7** (9) (1964), 1424–1432.

[Er.3] J.J. Erpenbeck, *Detonation stability for disturbances of small transverse wavelength*, Phys. Fluids **9** (7) (1966), 1293–1306.

[Er.6] J.J. Erpenbeck, *Nonlinear theory of unstable one-dimensional detonations*, Phys. Fluids **10** (2) (1967), 274–289.

[E.1] J.W. Evans, *Nerve axon equations: I. Linear approximations*, Indiana Univ. Math. J. **21** (1972), 877–885.

[E.2] J.W. Evans, *Nerve axon equations: II. Stability at rest*, Indiana Univ. Math. J. **22** (1972), 75–90.

[E.3] J.W. Evans, *Nerve axon equations: III. Stability of the nerve impulse*, Indiana Univ. Math. J. **22** (1972), 577–593.

[E.4] J.W. Evans, *Nerve axon equations: IV. The stable and the unstable impulse*, Indiana Univ. Math. J. **24** (1975), 1169–1190.

[FD] W. Fickett and W.C. Davis, *Detonation*, Univ. California Press, Berkeley, CA (1979).

[Fo] G.R. Fowles, *On the evolutionary condition for stationary plane waves in inert and reactive substances*, Shock Induced Transitions and Phase Structures in General Media, IMA Vol. Math. Appl., Vol. 52, Springer-Verlag, New York (1993), 93–110.

[FM] J. Francheteau and G. Métivier, *Existence de chocs faibles pour des systèmes quasi-linéaires hyperboliques multidimensionnels*, C. R. Acad. Sci. Paris Sér. I **327** (1998), 725–728.

[Free.1] N.C. Freeman, *A theory of the stability of plane shock waves*, Proc. Roy. Soc. London Ser. A. **228** (1955), 341–362.

[Free.2] N.C. Freeman, *On the stability of plane shock waves*, J. Fluid Mech. **2** (1957), 397–411.

[Fre.3] H. Freistühler, *Dynamical stability and vanishing viscosity: A case study of a nonstrictly hyperbolic system of conservation laws*, Comm. Pure Appl. Math. **45** (1992), 561–582.

[Fre.2] H. Freistühler, *A short note on the persistence of ideal shock waves*, Arch. Math. (Basel) **64** (4) (1995), 344–352.

[Fre.1] H. Freistühler, *Some results on the stability of non-classical shock waves*, J. Partial Differential Equations **11** (1998), 23–38.

[Fre.4] H. Freistühler, *Profiles for small Laxian shock waves in Kawashima type systems of conservation laws*, Preprint (2000).

[FreL] H. Freistühler and T.-P. Liu, *Nonlinear stability of overcompressive shock waves in a rotationally invariant system of viscous conservation laws*, Comm. Math. Phys. **153** (1993), 147–158.

[FreP.1] H. Freistühler and E.B. Pittman, *A numerical study of a rotationally degenerate hyperbolic system, I. The Riemann problem*, J. Comput. Phys. **100** (2) (1992), 306–321.

[FreP.2] H. Freistühler and E.B. Pittman, *A numerical study of a rotationally degenerate hyperbolic system, II. The Cauchy problem*, SIAM J. Numer. Anal. **32** (3) (1995), 741–753.

[FreS.1] H. Freistühler and P. Szmolyan, *Existence and bifurcation of viscous profiles for all intermediate magnetohydrodynamic shock waves*, SIAM J. Math. Anal. **26** (1) (1995), 112–128.

[FreS.2] H. Freistühler and P. Szmolyan, *Spectral stability of small shock waves*, Arch. Ration. Mech. Anal. **164** (2002), 287–309.

[FreZe] H. Freistühler and Y. Zeng, *Shock profiles for systems of balance laws with relaxation*, Preprint (1998).

[FreZ.1] H. Freistühler and K. Zumbrun, *Examples of Unstable Viscous Shock Waves*, unpublished note, Institut für Mathematik, RWTH Aachen, February (1998).

[FreZ.2] H. Freistühler and K. Zumbrun, *Long-wave stability of small-amplitude overcompressive viscous shock fronts, with applications to MHD*, in preparation.

[Fr] A. Friedman, *Partial Differential Equations of Parabolic Type*, Prentice-Hall, Englewood Cliffs, NY (1964); Reprinted (1983).

[Fri] K.O. Friedrichs, *Symmetric positive linear differential equations,* Comm. Pure Appl. Math. **11** (1958), 333–418.

[FriL] K.O. Friedrichs and P. Lax, *Boundary value problems for first order operators*, Comm. Pure Appl. Math. **18** (1965), 355–388.

[Gr] C.S. Gardner, *Comment on "Stability of step shocks"*, Phys. Fluids **6** (1962), 1366–1367.

[GJ.2] R. Gardner and C.K.R.T. Jones, *Traveling waves of a perturbed diffusion equation arising in a phase field model*, Indiana Univ. Math. J. **38** (4) (1989), 1197–1222.

[GJ.1] R. Gardner and C.K.R.T. Jones, *A stability index for steady state solutions of boundary value problems for parabolic systems*, J. Differential Equations **91** (2) (1991), 181–203.

[GZ] R. Gardner and K. Zumbrun, *The Gap lemma and geometric criteria for instability of viscous shock profiles*, Comm. Pure Appl. Math. **51** (7) (1998), 797–855.

[GS] I. Gasser and P. Szmolyan, *A geometric singular perturbation analysis of detonation and deflagration waves*, SIAM J. Math. Anal. **24** (4) (1993), 968–986.

[Ge] I.M. Gel'fand, *Some problems in theory of quasilinear equations*, Amer. Mat. Soc. Transl. Ser. 2 **29** (1963), 295–381.

[Gi] D. Gilbarg, *The existence and limit behavior of the one-dimensional shock layer*, Amer. J. Math. **73** (1951), 256–274.

[God.1] P. Godillon, *Stabilité linéaire des profils pour les systèmes avec relaxation semi-linéaire*, Phys. D **148** (2001), 289–316.

[God.2] P. Godillon, *Stabilité de profils de chocs dans les systèmes de conservation*, Doctoral Thesis, ENS Lyon (2003).

[G] S.K. Godunov, *An interesting class of quasilinear systems*, Sov. Math. **2** (1961), 947–948.

[Go.1] J. Goodman, *Nonlinear asymptotic stability of viscous shock profiles for conservation laws*, Arch. Ration. Mech. Anal. **95** (4) (1986), 325–344.

[Go.3] J. Goodman, *Stability of viscous scalar shock fronts in several dimensions*, Trans. Amer. Math. Soc. **311** (2) (1989), 683–695.

[Go.2] J. Goodman, *Remarks on the Stability of Viscous Shock Waves*, Viscous Profiles and Numerical Methods for Shock Waves (Raleigh, NC, 1990), SIAM, Philadelphia, PA (1991), 66–72.

[GM] J. Goodman and J.R. Miller, *Long-time behavior of scalar viscous shock fronts in two dimensions*, J. Dynam. Differential Equations **11** (2) (1999), 255–277.

[GX] J. Goodman and Z. Xin, *Viscous limits for piecewise smooth solutions to systems of conservation laws*, Arch. Ration. Mech. Anal. **121** (3) (1992), 235–265.

[GG] E. Grenier and O. Guès, *Boundary layers for viscous perturbations of noncharacteristic quasilinear hyperbolic problems*, J. Differential Equations **143** (1998), 110–146.

[GR] E. Grenier and F. Rousset, *Stability of one-dimensional boundary layers by using Green's functions*, Comm. Pure Appl. Math. **54** (2001), 1343–1385.

[GMWZ.1] O. Guès, G. Métivier, M. Williams and K. Zumbrun, *Multidimensional viscous shocks, I: Degenerate symmetrizers and long time stability*, Preprint (2002).

[GMWZ.2] O. Guès, G. Métivier, M. Williams and K. Zumbrun, *Multidimensional viscous shocks, II: The small viscosity limit*, Comm. Pure Appl. Math. (2004), to appear.

[GMWZ.3] O. Guès, G. Métivier, M. Williams and K. Zumbrun, *A new approach to stability of multidimensional viscous shocks*, Preprint (2003).

[GMWZ.4] O. Guès, G. Métivier, M. Williams and K. Zumbrun, *Navier–Stokes Regularization of Multidimensional Euler Shocks*, in preparation.

[GW] O. Guès and M. Williams, *Curved shocks as viscous limits: a boundary problem approach*, Indiana Univ. Math. J. **51** (2002), 421–450.

[He] D. Henry, *Geometric Theory of Semilinear Parabolic Equations*, Lecture Notes in Math., Springer-Verlag, Berlin (1981).

[Ho] D. Hoff, *Global solutions of the Navier–Stokes equations for multidimensional, compressible flow with discontinuous initial data*, J. Differential Equations **120** (1995), 215–254.

[HoZi.1] D. Hoff and M. Ziane, *Compact attractors for the Navier–Stokes equations of one-dimensional, compressible flow*, C. R. Acad. Sci. Paris Sér. I Math. **328** (1999), 239–244.

[HoZi.2] D. Hoff and M. Ziane, *The global attractor and finite determining nodes for the Navier–Stokes equations of compressible flow with singular initial data*, Indiana Univ. Math. J. **49** (2000), 843–889.

[HoZ.1] D. Hoff and K. Zumbrun, *Multi-dimensional diffusion waves for the Navier–Stokes equations of compressible flow*, Indiana Univ. Math. J. **44** (2) (1995), 603–676.

[HoZ.2] D. Hoff and K. Zumbrun, *Pointwise decay estimates for multidimensional Navier–Stokes diffusion waves*, Z. Angew. Math. Phys. **48** (4) (1997), 597–614.

[HoZ.3] D. Hoff and K. Zumbrun, *Asymptotic behavior of multi-dimensional scalar viscous shock fronts*, Indiana Univ. Math. J. **49** (2000), 427–474.

[HoZ.4] D. Hoff and K. Zumbrun, *Pointwise Green's function bounds for multi-dimensional scalar viscous shock fronts*, J. Differential Equations **183** (2002), 368–408.

[HJ] R.A. Horn and C.R. Johnson, *Matrix Analysis*, Cambridge Univ. Press, Cambridge (1985).

[H.1] P. Howard, *Pointwise methods for stability of a scalar conservation law*, Doctoral Thesis (1998).

[H.2] P. Howard, *Pointwise estimates on the Green's function for a scalar linear convection–diffusion equation*, J. Differential Equations **155** (1999), 327–367.

[H.3] P. Howard, *Pointwise Green's function approach to stability for scalar conservation laws*, Comm. Pure Appl. Math. **52** (1999), 1295–1313.

[H.4] P. Howard, *Pointwise estimates and stability for degenerate viscous shock waves*, J. Reine Angew. Math. **545** (2002), 19–65.

[H.5] P. Howard, *Local tracking and stability for degenerate viscous shock waves*, J. Differential Equations **186** (2002), 440–469.

[HZ.1] P. Howard and K. Zumbrun, *A tracking mechanism for one-dimensional viscous shock waves*, Preprint (1999).

[HZ.2] P. Howard and K. Zumbrun, *Pointwise estimates and stability for dispersive–diffusive shock waves*, Arch. Ration. Mech. Anal. **155** (2000), 85–169.

[HZ.3] P. Howard and K. Zumbrun, *The Evans function and stability criteria for degenerate viscous shock waves*, Discrete Contin. Dyn. Syst. (2004).

[Hu.2] J. Humpherys, *Stability of Jin–Xin relaxation shocks*, Quart. Appl. Math. **61** (2003), 251–263.

[Hu.3] J. Humpherys, *Viscous-dispersive traveling waves, I: Admissibility and existence*, Preprint (2003).

[Hu.1] J. Humpherys, *On spectral stability of strong shocks for isentropic gas dynamics*, in preparation.

[HuZ] J. Humpherys and K. Zumbrun, *Spectral stability of small-amplitude shock profiles for dissipative symmetric hyperbolic–parabolic systems*, Z. Angew. Math. Phys. **53** (2002), 20–34.

[Je] A. Jeffrey, *Magnetohydrodynamics*, University Mathematical Texts (33) Oliver & Boyd, Edinburgh-London; Interscience, New York (1966).

[JeT] A. Jeffrey and T. Taniuti, *Nonlinear Wave Propagation. With Applications to Physics and Magnetohydrodynamics*, Academic Press, New York (1964).

[JLy.2] K. Jenssen and G. Lyng, *Low frequency stability of multi-dimensional inviscid planar detonation waves*, Preprint (2003).

[JLy.1] K. Jenssen and G. Lyng, *Low frequency multi-dimensional stability of viscous detonation waves*, in preparation.

[J] C.K.R.T. Jones, *Stability of the travelling wave solution of the Fitz–Hugh–Nagumo system*, Trans. Amer. Math. Soc. **286** (2) (1984), 431–469.

[JGK] C.K.R.T. Jones, R.A. Gardner and T. Kapitula, *Stability of travelling waves for non-convex scalar viscous conservation laws*, Comm. Pure Appl. Math. **46** (1993), 505–526.

[Kan] Y. Kanel, *On a model system of equations of one-dimensional gas motion*, Differential Equations **4** (1968), 374–380.

[K.1] T. Kapitula, *Stability of weak shocks in λ–ω systems*, Indiana Univ. Math. J. **40** (4) (1991), 1193–1200.

[K.2] T. Kapitula, *On the stability of travelling waves in weighted L^∞ spaces*, J. Differential Equations **112** (1) (1994), 179–215.

[K.3] T. Kapitula, *Multidimensional stability of planar travelling waves*, Trans. Amer. Math. Soc. **349** (1) (1997), 257–269.

[K.4] T. Kapitula, *The Evans function and generalized Melnikov integrals*, SIAM J. Math. Anal. **30** (2) (1999), 273–297 (electronic).

[KR] T. Kapitula and J. Rubin, *Existence and stability of standing hole solutions to complex Ginzburg–Landau equations*, Nonlinearity **13** (1) (2000), 77–112.

[KS] T. Kapitula and B. Sandstede, *Stability of bright solitary-wave solutions to perturbed nonlinear Schrodinger equations*, Phys. D **124** (1–3) (1998), 58–103.

[Kat] T. Kato, *Perturbation Theory for Linear Operators*, Springer-Verlag, Berlin–Heidelberg (1985).

[Kaw] S. Kawashima, *Systems of a hyperbolic–parabolic composite type, with applications to the equations of magnetohydrodynamics*, Thesis, Kyoto University (1983).

[KM] S. Kawashima and A. Matsumura, *Asymptotic stability of traveling wave solutions of systems for one-dimensional gas motion*, Comm. Math. Phys. **101** (1) (1985), 97–127.

[KMN] S. Kawashima, A. Matsumura and K. Nishihara, *Asymptotic behavior of solutions for the equations of a viscous heat-conductive gas*, Proc. Japan Acad. Ser. A Math. Sci. **62** (7) (1986), 249–252.

[KSh] S. Kawashima and Y. Shizuta, *On the normal form of the symmetric hyperbolic–parabolic systems associated with the conservation laws*, Tohoku Math. J. **40** (1988), 449–464.

[Ke.1] L. Kelvin, *On a disturbing infinity in Lord Rayleigh's solution for waves in a plane vortex stratum*, Mathematical and Physical Papers, Vol. 4, Cambridge Univ. Press (1880), 186–187.

[Ke.2] L. Kelvin, *Rectilinear motion of a viscous fluid between parallel planes*, Mathematical and Physical Papers, Vol. 4, Cambridge Univ. Press (1880), 330–337.

[Ko.1] V.M. Kontorovich, *Stability of shock waves in relativistic hydrodynamics*, Soviet Phys. JETP **34** (7) (1958), 127–132.

[Ko.2] V.M. Kontorovich, *On the interaction between small disturbances and discontinuities in magnetohydrodynamics and on the stability of shock waves*, Soviet Phys. JETP **35** (8) (1959), 851–858.

[Kr] H.O. Kreiss, *Initial boundary value problems for hyperbolic systems*, Comm. Pure Appl. Math. **23** (1970), 277–298.

[KL] G. Kreiss and M. Liefvendahl, *Numerical investigation of examples of unstable viscous shock waves*, Hyperbolic Problems: Theory, Numerics, Applications, Vols I, II (Magdeburg, 2000), Internat. Ser. Numer. Math., Vols 140, 141, Birkhäuser, Basel (2001), 613–621.

[KK.1] G. Kreiss and H.O. Kreiss, *Stability of systems of viscous conservation laws*, Comm. Pure Appl. Math. **51** (11/12) (1998), 1397–1424.

[LL] L.D. Landau and E.M. Lifshitz, *Course of Theoretical Physics, Vol. 7. Theory of Elasticity*, Transl. from Russian by J.B. Sykes and W.H. Reid, Third edn, Pergamon Press, Oxford–Elmsford, NY (1986).

[Lan] S. Lang, *Complex Analysis*, Springer-Verlag, New York (1999).

[La] P.D. Lax, *Hyperbolic Systems of Conservation Laws and the Mathematical Theory of Shock Waves*, CBMS-NSF Regional Conf. Ser. in Appl. Math., Vol. 11, Society for Industrial and Applied Mathematics, Philadelphia, PA (1973).

[LaP.2] P.D. Lax and R.S. Phillips, *Local boundary conditions for dissipative symmetric linear differential operators*, Comm. Pure Appl. Math. **13** (1960), 427–456.

[LaP.1] P.D. Lax and R.S. Phillips, *S. Scattering Theory*, Second edn, with appendices by C.S. Morawetz and G. Schmidt, Pure Appl. Math., Vol. 26, Academic Press, Boston, MA (1989).

[LS] H.I. Lee and D.S. Stewart, *Calculation of linear detonation instability: One-dimensional instability of plane detonation*, J. Fluid Mech. **216** (1990), 103–132.

[Lin] C.C. Lin, *The Theory of Hydrodynamic Stability*, Cambridge Univ. Press, Cambridge (1966).

[Li] H. Liu, *Asymptotic stability of relaxation shock profiles for hyperbolic conservation laws*, J. Differential Equations **192** (2003), 285–307.

[L.4] T.-P. Liu, *The entropy condition and the admissibility of shocks*, J. Math. Anal. Appl. **53** (1976), 78–88.

[L.1] T.-P. Liu, *Nonlinear stability of shock waves for viscous conservation laws*, Mem. Amer. Math. Soc. **56** (328) (1985).

[L.5] T.-P. Liu, *Hyperbolic conservation laws with relaxation*, Comm. Math. Phys. **108** (1) (1987), 153–175.

[L.2] T.-P. Liu, *Nonlinear stability and instability of overcompressive shock waves*, Shock Induced Transitions and Phase Structures in General Media, IMA Vol. Math. Appl., Vol. 52, Springer-Verlag, New York (1993), 159–167.

[L.3] T.-P. Liu, *Pointwise convergence to shock waves for viscous conservation laws*, Comm. Pure Appl. Math. **50** (11) (1997), 1113–1182.

[LZe] T.-P. Liu and Y. Zeng, *Large Time Behavior of Solutions for General Quasilinear Hyperbolic–Parabolic Systems of Conservation Laws*, Mem. Amer. Math. Soc., Vol. 599, Amer. Math. Soc., Providence, RI (1997).

[LZ.1] T.-P. Liu and K. Zumbrun, *Nonlinear stability of an undercompressive shock for complex Burgers equation*, Comm. Math. Phys. **168** (1) (1995), 163–186.

[LZ.2] T.-P. Liu and K. Zumbrun, *On nonlinear stability of general undercompressive viscous shock waves*, Comm. Math. Phys. **174** (2) (1995), 319–345.

[Ly] G. Lyng, *One-Dimensional Stability of Combustion Waves*, Thesis, Indiana Univ. (2003).

[LyZ.1] G. Lyng and K. Zumbrun, *On the one-dimensional stability of viscous strong detonation waves*, Arch. Ration. Mech. Anal. (2004), to appear.

[LyZ.2] G. Lyng and K. Zumbrun, *A stability index for detonation waves in Majda's model for reacting flow*, Phys. D (2004), to appear.

[M.5] A. Majda, *A qualitative model for dynamic combustion*, SIAM J. Appl. Math. **41** (1) (1981), 70–93.

[M.3] A. Majda, *The existence and stability of multidimensional shock fronts*, Bull. Amer. Math. Soc. (N.S.) **4** (3) (1981), 342–344.

[M.1] A. Majda, *The Stability of Multi-Dimensional Shock Fronts*, Mem. Amer. Math. Soc., Vol. 275, Amer. Math. Soc., Providence, RI (1983).

[M.2] A. Majda, *The Existence of Multi-Dimensional Shock Fronts*, Mem. Amer. Math. Soc., Vol. 281, Amer. Math. Soc., Providence, RI (1983).

[M.4] A. Majda, *Compressible Fluid Flow and Systems of Conservation Laws in Several Space Variables*, Springer-Verlag, New York (1984).

[MO] A. Majda and S. Osher, *Initial-boundary value problems for hyperbolic equations with uniformly characteristic boundary*, Comm. Pure Appl. Math. **28** (1975), 607–676.

[MP] A. Majda and R. Pego, *Stable viscosity matrices for systems of conservation laws*, J. Differential Equations **56** (1985), 229–262.

[MaZ.1] C. Mascia and K. Zumbrun, *Pointwise Green's function bounds and stability of relaxation shocks*, Indiana Univ. Math. J. **51** (4) (2002), 773–904.

[MaZ.5] C. Mascia and K. Zumbrun, *Stability of large-amplitude shock profiles for general relaxation systems*, Preprint (2003).

[MaZ.3] C. Mascia and K. Zumbrun, *Pointwise Green's function bounds for shock profiles with degenerate viscosity*, Arch. Ration. Mech. Anal. **169** (2003), 177–263.

[MaZ.2] C. Mascia and K. Zumbrun, *Stability of small-amplitude shock profiles of dissipative symmetric hyperbolic–parabolic systems*, Comm. Pure Appl. Math. **57** (2004), 141–218.

[MaZ.4] C. Mascia and K. Zumbrun, *Stability of large-amplitude viscous shock profiles of hyperbolic–parabolic systems*, Arch. Ration. Mech. Anal. **172** (2004), 93–131.

[MNi] A. Matsumura and T. Nishida, *Initial boundary value problems for the equations of motion of compressible viscous and heat conductive fluids*, Comm. Math. Phys. **89** (1983).

[MN.1] A. Matsumura and K. Nishihara, *On the stability of travelling wave solutions of a one-dimensional model system for compressible viscous gas*, Japan J. Appl. Math. **2** (1) (1985), 17–25.

[MN.2] A. Matsumura and K. Nishihara, *Asymptotic stability of traveling waves with shock profile for non-convex viscous scalar conservation laws*, Mathematical Analysis of Phenomena in Fluid and Plasma Dynamics, Kyoto (1993) (in Japanese).

[MN.3] A. Matsumura and K. Nishihara, *Asymptotic stability of traveling waves for scalar viscous conservation laws with non-convex nonlinearity*, Comm. Math. Phys. **165** (1) (1994), 83–96.

[MeP] R. Menikoff and B. Plohr, *The Riemann problem for fluid flow of real materials*, Rev. Modern Phys. **61** (1) (1989), 75–130.

[Mé.2] G. Métivier, *Interaction de deux chocs pour un système de deux lois de conservation, en dimension deux d'espace*, Trans. Amer. Math. Soc. **296** (1986), 431–479.

[Mé.1] G. Métivier, *Stability of multidimensional weak shocks*, Comm. Partial Differential Equations **15** (1990), 983–1028.

[Mé.3] G. Métivier, *Ondes soniques*, J. Math. Pures Appl. **70** (1991), 197–268.

[Mé.4] G. Métivier, *The block structure condition for symmetric hyperbolic systems*, Bull. London Math. Soc. **32** (2000), 689–702.

[Mé.5] G. Métivier, *Stability of multidimensional shocks*, Advances in the Theory of Shock Waves, Progr. Nonlinear Differential Equations Appl., Vol. 47, Birkhäuser, Boston, MA (2001), 25–103.

[MéZ.1] G. Métivier and K. Zumbrun, *Large viscous boundary layers for noncharacteristic nonlinear hyperbolic problems*, Preprint (2002); Mem. Amer. Math. Soc. (2004), to appear.

[MéZ.2] G. Métivier and K. Zumbrun, *Symmetrizers and continuity of stable subspaces for parabolic–hyperbolic boundary value problems*, Discrete Contin. Dyn. Syst., to appear.

[MéZ.3] G.Métivier and K.Zumbrun, *Hyperbolic boundary value problems for symmetric systems with variable multiplicities*, J. Differential Equations (2004), to appear.

[OZ.1] M. Oh and K. Zumbrun, *Stability of periodic solutions of conservation laws with viscosity: Analysis of the Evans function*, Arch. Ration. Mech. Anal. **166** (2) (2003), 99–166.

[OZ.2] M. Oh and K. Zumbrun, *Stability of periodic solutions of conservation laws with viscosity: Pointwise bounds on the Green function*, Arch. Ration. Mech. Anal. **166** (2) (2003), 167–196.

[Pa] A. Pazy, *Semigroups of Linear Operators and Applications to Partial Differential Equations*, Appl. Math. Sci., Vol. 44, Springer-Verlag, New York–Berlin (1983).

[P.1] R.L. Pego, *Stable viscosities and shock profiles for systems of conservation laws*, Trans. Amer. Math. Soc. **282** (1984), 749–763.

[P.2] R.L. Pego, *Phase transitions in one-dimensional nonlinear viscoelasticity: Admissibility and stability*, Arch. Ration. Mech. Anal. **97** (1987), 353–394.

[PW] R.L. Pego and M.I. Weinstein, *Eigenvalues, and instabilities of solitary waves*, Philos. Trans. Roy. Soc. London Ser. A **340** (1656) (1992), 47–94.

[PZ] R. Plaza and K. Zumbrun, *An Evans function approach to spectral stability of small-amplitude shock profiles*, Discrete Contin. Dyn. Syst. Ser. B **10** (2004), 885–924.

[Ra.1] Lord Rayleigh (J.W. Strutt), *On the instability of jets*, Scientific Papers, 1, Cambridge Univ. Press (1880), 361–371.

[Ra.2] Lord Rayleigh (J.W. Strutt), *On the stability, or instability, of certain fluid motions, II*, Scientific Papers, 3, Cambridge Univ. Press (1887), 2–23.

[Ra.4] Lord Rayleigh (J.W. Strutt), *On the question of the stability of the flow of fluids*, Scientific Papers, 3, Cambridge Univ. Press (1892), 575–584.

[Ra.3] Lord Rayleigh (J.W. Strutt), *On the stability, or instability, of certain fluid motions, III*, Scientific Papers, 4, Cambridge Univ. Press (1895), 203–209.

[Ra.5] Lord Rayleigh (J.W. Strutt), *Further remarks on the stability of viscous fluid motion*, Scientific Papers, 6, Cambridge Univ. Press (1914), 266–275.

[Ra.6] Lord Rayleigh (J.W. Strutt), *On the stability of the simple shearing motion of a viscous incompressible fluid*, Scientific Papers, 6, Cambridge Univ. Press (1915), 341–349.

[Ri] R.D. Richtmyer, *Taylor instability in shock acceleration of compressible fluids*, Comm. Pure Appl. Math. **13** (1960), 297–319.

[Ro] A.E. Roberts, *Stability of a Steady Plane Shock*, Los Alamos Scientific Laboratory Report LA-299 (1945).

[R] F. Rousset, *Inviscid boundary conditions and stability of viscous boundary layers*, Asymptot. Anal. **26** (3/4), (2001), 285–306.

[Sa.1] R. Sakamoto, *Mixed problems for hyperbolic equations, I, II*, J. Math. Kyoto Univ. **10** (1970), 349–373 and 403–417.

[Sa.2] R. Sakamoto, *Hyperbolic Boundary Value Problems*, Cambridge Univ. Press (1982).

[Sat] D. Sattinger, *On the stability of waves of nonlinear parabolic systems*, Adv. Math. **22** (1976), 312–355.

[Se.3] D. Serre, *Systems of Conservation Laws, 1. Hyperbolicity, Entropies, Shock Waves*, Transl. from the (1996) French original by I.N. Sneddon, Cambridge Univ. Press, Cambridge (1999).

[Se.1] D. Serre, *Discrete shock profiles and their stability*, Hyperbolic Problems: Theory, Numerics, Applications, 7th International Conference, Zurich, 1998, M. Fey and R. Jeltsch, eds, Internat. Ser. Numer. Math., Vol. 130, Birkhäuser, Basel (1999), 843–854.

[Se.4] D. Serre, *Systems of Conservation Laws, 2. Geometric Structures, Oscillations, and Initial-Boundary Value Problems*, Transl. from the (1996) French original by I.N. Sneddon, Cambridge Univ. Press, Cambridge (2000).

[Se.2] D. Serre, *La transition vers l'instabilité pour les ondes de chocs multi-dimensionnelles*, Trans. Amer. Math. Soc. **353** (2001), 5071–5093.

[Se.5] D. Serre, *Matrices. Theory and Applications*, Transl. from the (2001) French original, Grad. Texts in Math., Vol. 216, Springer-Verlag, New York (2002).

[Sl.5] M. Slemrod, *Admissibility criteria for propagating phase boundaries in a van der Waals fluid*, Arch. Ration. Mech. Anal. **81** (4) (1983), 301–315.

[Sl.4] M. Slemrod, *Dynamic phase transitions in a van der Waals fluid*, J. Differential Equations **52** (1) (1984), 1–23.

[Sl.1] M. Slemrod, *The viscosity-capillarity approach to phase transitions*, PDEs and Continuum Models of Phase Transitions, Nice, 1988, Lecture Notes in Phys., Vol. 344, Springer-Verlag, Berlin–New York (1989), 201–206.

[Sl.2] M. Slemrod, *The vanishing viscosity-capillarity approach to the Riemann problem for a van der Waals fluid*, Nonclassical Continuum Mechanics, Durham, 1986, London Math. Soc. Lecture Note Ser., Vol. 122, Cambridge Univ. Press, Cambridge, NY (1987), 325–335.

[Sl.3] M. Slemrod, *A limiting "viscosity" approach to the Riemann problem for materials exhibiting change of phase*, Arch. Ration. Mech. Anal. **105** (4) (1989), 327–365.

[Sm] J. Smoller, *Shock Waves and Reaction–Diffusion Equations*, Second edn, Grundlehren Math. Wiss. [Fundamental Principles of Mathematical Sciences], Vol. 258, Springer-Verlag, New York (1994).

[St] J.C. Strikwerda, *Finite Difference Schemes and Partial Differential Equations*, Chapman and Hall, New York (1989).

[SF] G.W. Swan and G.R. Fowles, *Shock wave stability*, Phys. Fluids **18** (1975), 28–35.

[SzX] A. Szepessy and Z. Xin, *Nonlinear stability of viscous shock waves*, Arch. Ration. Mech. Anal. **122** (1993), 53–103.

[T] G.I. Taylor, *The air wave surrounding an expanding sphere,* Proc. Roy. Soc. A **218** (1946), 273–292.

[vN] J. von Neumann, *Collected Works, Vol. VI: Theory of Games, Astrophysics, Hydrodynamics and Meteorology*, General ed. A.H. Taub, Pergamon Press, New York (1963).

[Wen.1] B. Wendroff, *The Riemann problem for materials with nonconvex equations of state, I. Isentropic flow*, J. Math. Anal. Appl. **38** (1972), 454–466.

[Wen.2] B. Wendroff, *The Riemann problem for materials with nonconvex equations of state, II. General flow*, J. Math. Anal. Appl. **38** (1972), 640–658.

[We] H. Weyl, *Shock waves in arbitrary fluids*, Comm. Pure Appl. Math. **2** (1949), 103–122.

[Wh] G.B. Whitham, *Linear and Nonlinear Waves*, reprint of the (1974) original, Pure Appl. Math., Wiley, New York (1999).

[W] M. Williams, *Highy oscillatory multidimensional shocks*, Comm. Pure Appl. Math. **52** (2) (1999), 129–192.

[Yo] W.-A. Yong, *Basic properties of hyperbolic relaxation systems*, Advances in the Theory of Shock Waves, Progr. Nonlinear Differential Equations Appl., Vol. 47, Birkhäuser, Boston, MA (2001), 307–516.

[Y] K. Yosida, *Functional Analysis*, reprint of the sixth (1980) edition, Classics in Mathematics, Springer-Verlag, Berlin (1995).

[Ze.1] Y. Zeng, *L^1 asymptotic behavior of compressible, isentropic, viscous 1-d flow*, Comm. Pure Appl. Math. **47** (1994), 1053–1092.

[Ze.2] Y. Zeng, *Gas dynamics in thermal nonequilibrium and general hyperbolic systems with relaxation*, Arch. Ration. Mech. Anal. **150** (3) (1999), 225–279.

[Z.1] K. Zumbrun, *Stability of viscous shock waves*, Lecture Notes, Indiana Univ. (1998).

[Z.2] K. Zumbrun, *Multidimensional stability of viscous shock fronts,* Lecture Notes, Indiana Univ. (2000).

[Z.5] K. Zumbrun, *Dynamical stability of phase transitions in the p-system with viscosity–capillarity*, SIAM J. Appl. Math. **60** (6) (2000), 1913–1924 (electronic).

[Z.6] K. Zumbrun, *Refined wave–tracking and nonlinear stability of viscous Lax shocks*, Meth. Appl. Anal. **7** (2000), 747–768.

[Z.3] K. Zumbrun, *Multidimensional stability of planar viscous shock waves*, Advances in the Theory of Shock Waves, Progr. Nonlinear Differential Equations Appl., Vol. 47, Birkhäuser, Boston, MA (2001), 307–516.

[Z.4] K. Zumbrun, *Planar stability criteria for viscous shock waves of systems with real viscosity*, CIME Summer School Notes, Preprint (2004).

[ZH] K. Zumbrun and P. Howard, *Pointwise semigroup methods and stability of viscous shock waves*, Indiana Univ. Math. J. **47** (4) (1998), 741–871.

[ZH.e] K. Zumbrun and P. Howard, *Errata to*: *"Pointwise semigroup methods, and stability of viscous shock waves"* [Indiana Univ. Math. J. **47** (3) (1998), 741–871], Indiana Univ. Math. J. **51** (4) (2002), 1017–1021.

[ZPM] K. Zumbrun, B. Plohr and D. Marchesin, *Scattering behavior of transitional shock waves*, Mat. Contemp. **3** (1992), 191–209.

[ZS] K. Zumbrun and D. Serre, *Viscous and inviscid stability of multidimensional planar shock fronts*, Indiana Univ. Math. J. **48** (3) (1999), 937–992.

CHAPTER 6

Some Mathematical Problems in Geophysical Fluid Dynamics

Roger Temam

The Institute for Scientific Computing and Applied Mathematics, Indiana University, Bloomington, IN 47405, USA, and
Laboratoire d'Analyse Numérique, Université Paris-Sud, Bâtiment 425, 91405 Orsay, France
E-mail: temam@indiana.edu

Mohammed Ziane

University of Southern California, DRB 155, 1042 W. 36 Place, Los Angeles, CA 90049, USA
E-mail: ziane@math.usc.edu

Contents

HANDBOOK OF MATHEMATICAL FLUID DYNAMICS, VOLUME III
Edited by S.J. Friedlander and D. Serre

Abstract

This chapter addresses some mathematical aspects of the equations of geophysical fluid dynamics namely, existence, uniqueness, and regularity of solutions of the primitive equations (PEs) of the ocean, the atmosphere and the coupled atmosphere–ocean. The emphasis is on the case of the ocean which encompasses most of the mathematical difficulties.

A guide and summary of results for the physics oriented reader is provided at the end of the Introduction (Section 1).

1. Introduction

The aim of this chapter is to address some mathematical aspects of the equations of geophysical fluid dynamics, namely existence, uniqueness and regularity of solutions.

The equations of geophysical fluid dynamics are the equations governing the motion of the atmosphere and the ocean, and are derived from the conservation equations from physics, namely conservation of mass, momentum, energy and some other components such as salt for the ocean, humidity (or chemical pollutants) for the atmosphere. The basic equations of conservation of mass and momentum, that is the three-dimensional compressible Navier–Stokes equations contain however too much information and we cannot hope to numerically solve these equations with enough accuracy in a foreseeable future. Owing to the difference of sizes of the vertical and horizontal dimensions, both in the atmosphere and in the ocean (10–20 km versus several thousands of kilometers), the most natural simplification leads to the so-called *primitive equations* (PEs) which we study in this chapter.

We continue this Introduction by briefly describing the physical and mathematical backgrounds of the PEs.

1.1. *Physical background*

The primitive equations are based on the so-called hydrostatic approximation, in which the conservation of momentum in the vertical direction is replaced by the simpler, hydrostatic equation (see, e.g., (2.25)).

As far as we know, the primitive equations were essentially introduced by Richardson in 1922; when it appeared that they were still too complicated, they were left out and, instead, attention was focused on even simpler models, the geostrophic and quasi-geostrophic models, considered in the late 1940's by von Neumann and his collaborators, in particular Charney. With the increase of computing power, interest eventually returned to the PEs, which are now the core of many Global Circulation Models (GCM) or Ocean Global Circulation Models (OGCM), available at the National Center for Atmospheric Research (NCAR) and elsewhere. GCMs and OGCMs are very complex models which contain many components, but still, the PEs are the central component for the dynamics of the air or the water. For some phenomena there is need to give up the hydrostatic hypothesis and then nonhydrostatic models are considered, such as in Laprise [18] or Smolarkiewicz, Margolin and Wyszogrodzki [34]; these models stand at an intermediate level of physical complexity between the full Navier–Stokes equations and the PEs-hydrostatic equations. Research on nonhydrostatic models is ongoing and, at this time, there is no agreement, in the physical community, for a specific model.

In this hierarchy of models for geophysical fluid dynamics, let us add also the Shallow Water equation corresponding essentially to a vertically integrated form of the Navier–Stokes equations; from the physical point of view they stand as an intermediate model between the primitive and the quasi-geostrophic equations.

In summary, in term of physical relevance and the level of complexity of the physical phenomena they can account for, the hierarchy of models in geophysical fluid dynamics is as follows:

$$\begin{array}{c}\text{Three-dimensional Navier–Stokes equations}\\ \Downarrow\\ \text{Nonhydrostatic models}\\ \Downarrow\\ \text{Primitive equations}\\ \text{(hydrostatic equations)}\\ \Downarrow\\ \text{Shallow water equations}\\ \Downarrow\\ \text{Quasi-geostrophic models}\\ \Downarrow\\ \text{Two-dimensional barotropic equations.}\end{array}$$

We remark here also that much study is needed for the boundary conditions from both the physical and mathematical points of views. As we said, our aim in this chapter is the study of mathematical properties of the PEs.

1.2. *Mathematical background*

The level of mathematical complexity of the equations above is not the same as the level of physical complexity: at both ends, the quasi-geostrophic models and barotropic equations are mathematically well understood (at least in the presence of viscosity; see Wang [40,41]), and we know the level of complexity of the Navier–Stokes equations to which this handbook is devoted. On the other hand, nonhydrostatic models are mathematically out of reach, and there are much less mathematical results available for the shallow water equations than for the Navier–Stokes equations, even in space dimension two (see however Orenga [31]). As we indicate hereafter, the primitive equations although physically simpler are, in fact, slightly more complicated than the incompressible Navier–Stokes equations.

Indeed this is due to the fact that the nonlinear term in the Navier–Stokes equations, also called inertial term, is of the form

velocity × *first-order derivatives of velocity*,

whereas, the nonlinear term for the primitive equations, is of the form

first-order derivatives of horizontal velocity
× *first-order derivatives of horizontal velocity*.

The mathematical study of the primitive equations was initiated by Lions, Temam and Wang in the early 1990s. They produced a mathematical formulation of the PEs which resembles that of the Navier–Stokes due to Leray, and obtained the existence for all time of weak solutions; see Section 2, and the original articles [21,22,24] in the list of references.

Further works, conducted during the 1990s and more especially during the past few years, have improved and supplemented the early results of [21,22,24] by a set of results which, essentially, brings the mathematical theory of the PEs to that of the three-dimensional incompressible Navier–Stokes equations, despite the added complexity mentioned above; this added complexity is overcome by a nonisotropic treatment of the equations (of certain nonlinear terms), in which the horizontal and vertical directions are treated differently.

In summary the following results are now available which will be presented *in details* in this chapter:

(i) Existence of weak solutions for all time (dimension two and three).
(ii) In space dimension three, existence of a strong solution for a limited time (local in time existence).
(iii) In space dimension two, existence and uniqueness for all time of a strong solution.
(iv) Uniqueness of a weak solutionin space dimension two.

In the above, the terminology is that normally used for Navier–Stokes equations: the weak solutions are those with finite (fluid) kinematic energy ($L^\infty(L^2)$ and $L^2(H^1)$), and the strong solutions are those with finite (fluid) enstrophy ($L^\infty(H^1)$ and $L^2(H^2)$). Essential in the most recent developments (ii)–(iv) above is the H^2 regularity result for a Stokes type problem appearing in the PEs, the analog of the H^2 regularity in the Cattabriga–Solonnikov results on the usual Stokes problem; the whole Section 4 is devoted to this problem.

1.3. *Content of this chapter*

Because of space limitation it was not possible to consider all relevant cases here. Relevant cases include:

The Ocean, The Atmosphere and The Coupled Ocean and Atmosphere,

on the one hand, and, on the other hand, the study of global phenomena on the sphere (involving the writing of the equations in spherical coordinates), and the study of mid-latitude regional models in which the equations are projected on a space tangent to the sphere (the Earth), corresponding to the so-called β-plane approximation: here $0x$ is the west–east axis, $0y$ the south–north axis, and $0z$ the ascending vertical.

For this more mathematically oriented chapter we have chosen to concentrate on the cases mathematically most significant. Hence for each case, after a brief description of the equations on the sphere (in spherical coordinates), we concentrate our efforts on the corresponding β-plane case (in Cartesian coordinates). Indeed, in general, going from the β-plane case in Cartesian coordinates to the spherical case necessitates only the proper handling of terms involving lower order derivatives; full details concerning the spherical case can be found also in the original articles [21,22,24], ...

In the Cartesian case of emphasis, generally we first concentrate our attention on the ocean. Indeed, as we will see in Section 2, the domain occupied by the ocean contains corners (in dimension two) or wedges (in dimension three); some regularity issues occur in this case which must be handled using the theory of regularity of elliptic problems in nonsmooth domains (Grisvard [11], Kozlov, Maz'ya and Rossmann [17], Maz'ya and Rossmann [25]). For the atmosphere or the coupled atmosphere–ocean, the difficulties are similar or easier to handle – hence most of the mathematical efforts will be devoted to the ocean in Cartesian coordinates.

In Section 2 we describe the governing equations and derive the result of existence of weak solutions with a different method than in the original articles [21,22,24], thus allowing more generality (for the ocean, the atmosphere and the coupled atmosphere–ocean).

In Section 3 we study the existence of strong solutions in space dimension three and two (solutions local in time in dimension three, and for all time in dimension two). We establish in dimension three the existence and uniqueness of strong solutions on a limited interval of time (Section 3.2); in dimension two we prove the existence and uniqueness, for all time, of such strong solutions. Finally, in Section 3.3 we consider the two-dimensional space-periodic case and prove the existence of solutions for all time, in all H^m, $m \geqslant 2$.

Section 4 is technically very important, and many results of Sections 2 and 3 rely on it: this section contains the proof of the H^2 regularity of elliptic problems which arise in the primitive equations. This proof relies, as we said, on the theory of regularity of solutions of elliptic problems in nonsmooth domains.

More explanations and references will be given in the Introduction of or within each section.

As mentioned earlier, the mathematical formulation of the equations of the atmosphere, of the ocean and of the coupled atmosphere ocean were derived in the articles by Lions, Temam and Wang [21,22,24]. For each of these problems, these articles also contain the proof of existence of weak solutions for all time (in dimension three with a proof which easily extends to dimension two). An alternative slightly more general proof of this result, is given in Section 2. Concerning the strong solutions, the proof given here of the local existence in dimension three is based on the article by Hu, Temam and Ziane [16]. An alternate proof of this result is due to [12]. In dimension two, the proof of existence and uniqueness of strong solutions, for all time, for the considered system of equations and boundary conditions is new, and based on an unpublished manuscript of Ziane [46]. This result is also established, for a simpler system (without temperature and salinity), by Bresch, Kazhikov and Lemoine [6]. In the space periodic case, existence and uniqueness of solutions in all the spaces H^m is proved by Petcu, Temam and Wirosoetisno [29].

1.4. *Summary of results for the physics oriented reader*

The physics oriented reader will recognize in (2.1)–(2.5) the basic conservation laws: conservations of momentum, mass, energy and salt for the ocean, equation of state. In (2.6) and (2.7) appears the simplification due to the Boussinesq approximation, and in (2.11)–(2.16) the simplifications resulting from the hydrostatic balance assumption. Hence (2.11)–(2.16) are the PEs of the ocean. The PEs of the atmosphere appear in (2.116)–(2.121), and those of the coupled atmosphere and ocean are described in Section 2.5. Concerning, to begin, the ocean, the first task is to write these equations, supplemented by the initial and boundary conditions, as an initial value problem in a phase space H of the form

$$\frac{\mathrm{d}U}{\mathrm{d}t} + AU + B(U,U) + E(U) = \ell, \tag{1.1}$$

$$U(0) = U_0, \tag{1.2}$$

where U is the set of prognostic variables of the problem, that is the horizontal velocity $\mathbf{v} = (u, v)$, the temperature T and the salinity S, $U = (\mathbf{v}, T, S)$; see (2.66). The phase space H consists, for its fluid mechanics part, of (horizontal) vector fields with finite kinetic energy. We then study the stationary solutions of (1.1) in Section 2.2.2 and, in Theorem 2.2, we prove the existence for all times of weak solutions of (1.1) and (1.2), which are solutions in $L^\infty(0, t_1; L^2)$ and $L^2(0, t_1; H^1)$ (bounded kinetic energy and square integrable enstrophy for the fluid mechanics part). A parallel study is conducted for the atmosphere and the coupled atmosphere–ocean in Sections 2.4 and 2.5.

In Section 3 we consider in dimension three and two the strong solutions, which are solutions bounded for all times in the Sobolev space H^1 ("finite enstrophy" space). The main results are Theorems 3.1 and 3.2. Section 4 is mathematically very important although technical. It is shown there, that the solutions to certain elliptic problems enjoy certain regularity properties (H^2 regularity, that is the function and their first and second derivatives are square integrable); the problems corresponding to the (horizontal) velocity, the temperature and the salinity are successively considered. The study in Section 4 contains many specific aspects which are explained in details in the long introduction to that section.

The study presented in this chapter is only a small part of the mathematical problems on geophysical flows, but we believe it is an important part. We did not try to produce here an exhaustive bibliography. Further mathematical references on geophysical flows will be given in the text; see also the bibliography of the articles and books that we quote. There is also of course a very large literature in the physical context; we only mentioned some of the books which were very useful to us such as [13,28,39,42,43].

Beside the efforts of the authors, we mention in several places that this study is based on joint works with Lions, Wang, Hu and others. Their help is gratefully acknowledged and we pay tribute to the memory of Jacques-Louis Lions. The authors wish to thank Denis Serre and Shouhong Wang for their careful reading of the whole manuscript and for their numerous comments which significantly improved the manuscript. They extend also their gratitude to Daniele Le Meur and Teresa Bunge who typed significant parts of the manuscript.

2. The primitive equations: Weak formulation, existence of weak solutions

As explained in the Introduction to this chapter, our aim in this section is first to present the derivation of the PEs from the basic physical conservation laws. We then describe the natural boundary conditions. Then, on the mathematical side, we introduce the function spaces and derive the mathematical formulation of the PEs. Finally we derive the existence for all time of weak solutions.

We successively consider the ocean, the atmosphere and the coupled atmosphere–ocean.

2.1. *The primitive equations of the ocean*

Our aim in this section is to describe the PEs of the ocean (see Section 2.1.1), we then describe the corresponding boundary conditions and the associated initial and boundary value problems (Section 2.1.2).

2.1.1. *The primitive equations.* Generally speaking, it is considered that the ocean is made up of a slightly compressible fluid with Coriolis force. The full set of equations of the large-scale ocean are the following: the conservation of momentum equation, the continuity equation (conservation of mass), the thermodynamics equation (that is the conservation of energy equation), the equation of state and the equation of diffusion for the salinity S:

$$\rho \frac{d\mathbf{V}_3}{dt} + 2\rho\Omega \times \mathbf{V}_3 + \nabla_3 p + \rho \mathbf{g} = D, \tag{2.1}$$

$$\frac{d\rho}{dt} + \rho \operatorname{div}_3 \mathbf{V}_3 = 0, \tag{2.2}$$

$$\frac{dT}{dt} = Q_T, \tag{2.3}$$

$$\frac{dS}{dt} = Q_S, \tag{2.4}$$

$$\rho = f(T, S, p). \tag{2.5}$$

Here $\mathbf{V}_3$ is the three-dimensional velocity vector, $\mathbf{V}_3 = (u, v, w)$, ρ, p, T are the density, pressure and temperature, and S is the concentration of salinity; $\mathbf{g} = (0, 0, g)$ is the gravity vector, D is the molecular dissipation, Q_T and Q_S are the heat and salinity diffusions. The analytic expressions of D, Q_T and Q_S will be given below.

The Boussinesq approximation. From both the theoretical and the computational points of view, the above systems of equations of the ocean seem to be too complicated to study. So it is necessary to simplify them according to some physical and mathematical considerations. The Mach number for the flow in the ocean is not large and therefore, as a starting point, we can make the so-called *Boussinesq approximation* in which the density is assumed constant, $\rho = \rho_0$, except in the buoyancy term and in the equation of state.

This amounts to replacing (2.1) and (2.2) by

$$\rho_0 \frac{d\mathbf{V}_3}{dt} + 2\rho_0\Omega \times \mathbf{V}_3 + \nabla_3 \mathbf{p} + \rho \mathbf{g} = \mathbf{D}, \tag{2.6}$$

$$\operatorname{div}_3 \mathbf{V}_3 = 0. \tag{2.7}$$

Consider the spherical coordinate system (θ, ϕ, r), where θ $(-\pi/2 < \theta < \pi/2)$ stands for the latitude on the Earth, ϕ $(0 \leqslant \phi \leqslant 2\pi)$ on the longitude of the Earth, r for the radial distance, and $z = r - a$ for the vertical coordinate with respect to the sea level, and let

$\mathbf{e}_\theta, \mathbf{e}_\phi, \mathbf{e}_r$ be the unit vectors in the θ-, ϕ- and z-directions, respectively. Then we write the velocity of the ocean in the form

$$\mathbf{V}_3 = v_\theta \mathbf{e}_\theta + v_\phi \mathbf{e}_\phi + v_r \mathbf{e}_r = \mathbf{v} + w, \tag{2.8}$$

where $\mathbf{v} = v_\theta \mathbf{e}_\theta + v_\phi \mathbf{e}_\phi$ is the horizontal velocity field and w is the vertical velocity.

Another common simplification is to replace, to first order, r by the radius a of the Earth. This is based on the fact that the depth of the ocean is small compared with the radius of the Earth. In particular,

$$\frac{\mathrm{d}}{\mathrm{d}t} = \frac{\partial}{\partial t} + \frac{v_\theta}{r}\frac{\partial}{\partial \theta} + \frac{v_\phi}{r\cos\theta}\frac{\partial}{\partial \phi} + v_r\frac{\partial}{\partial r} \tag{2.9}$$

becomes

$$\frac{\mathrm{d}}{\mathrm{d}t} = \frac{\partial}{\partial t} + \frac{v_\theta}{a}\frac{\partial}{\partial \theta} + \frac{v_\phi}{a\cos\theta}\frac{\partial}{\partial \phi} + v_z\frac{\partial}{\partial z}, \tag{2.10}$$

and, taking the viscosity into consideration, we obtain the equations of the large-scale ocean with Boussinesq approximation, which are simply called *Boussinesq equations of the ocean* (BEs), i.e., equations (2.11)–(2.16) hereafter (for the equation of state (2.16), see Remark 2.1):

$$\frac{\partial \mathbf{v}}{\partial t} + \nabla_{\mathbf{v}}\mathbf{v} + w\frac{\partial \mathbf{v}}{\partial z} + \frac{1}{\rho_0}\nabla p + 2\Omega\sin\theta \times \mathbf{v} - \mu_{\mathbf{v}}\Delta\mathbf{v} - \nu_{\mathbf{v}}\frac{\partial^2 \mathbf{v}}{\partial \mathbf{z}^2} = \mathbf{0}, \tag{2.11}$$

$$\frac{\partial w}{\partial t} + \nabla_{\mathbf{v}} w + w\frac{\partial w}{\partial z} + \frac{1}{\rho_0}\frac{\partial p}{\partial z} + \frac{\rho}{\rho_0}g - \mu_{\mathbf{v}}\Delta w - \nu_{\mathbf{v}}\frac{\partial^2 w}{\partial z^2} = 0, \tag{2.12}$$

$$\operatorname{div}\mathbf{v} + \frac{\partial w}{\partial z} = 0, \tag{2.13}$$

$$\frac{\partial T}{\partial t} + \nabla_{\mathbf{v}} T + w\frac{\partial T}{\partial z} - \mu_T\Delta T - \nu_T\frac{\partial^2 T}{\partial z^2} = 0, \tag{2.14}$$

$$\frac{\partial S}{\partial t} + \nabla_{\mathbf{v}} S + w\frac{\partial S}{\partial z} - \mu_S\Delta S - \nu_S\frac{\partial^2 S}{\partial z^2} = 0, \tag{2.15}$$

$$\rho = \rho_0\bigl(1 - \beta_T(T - T_\mathrm{r}) + \beta_S(S - S_\mathrm{r})\bigr), \tag{2.16}$$

where $\mathbf{v}$ is the horizontal velocity of the water, w is the vertical velocity, and, T_r, S_r are averaged (or reference) values of T and S. The diffusion coefficients $\mu_{\mathbf{v}}, \mu_T, \mu_S$ and $\nu_{\mathbf{v}}, \nu_T, \nu_S$ are different in the horizontal and vertical directions, accounting for some eddy diffusions in the sense of Smagorinsky [33].

The differential operators are defined as follows. The (horizontal) gradient operator grad $=\nabla$ is defined by

$$\operatorname{grad} p=\nabla p=\frac{1}{a}\frac{\partial p}{\partial\theta}\mathbf{e}_\theta+\frac{1}{a\cos\theta}\frac{\partial p}{\partial\phi}\mathbf{e}_\phi. \tag{2.17}$$

The (horizontal) divergence operator div $=\nabla\cdot$ is defined by

$$\operatorname{div}(v_\theta\mathbf{e}_\theta+v_\phi\mathbf{e}_\phi)=\nabla\cdot\mathbf{v}=\frac{1}{a\cos\theta}\left(\frac{\partial(v_\theta\cos\theta)}{\partial\theta}+\frac{\partial v_\phi}{\partial\phi}\right). \tag{2.18}$$

The derivatives $\nabla_{\mathbf{v}}\tilde{\mathbf{v}}$ and $\nabla_{\mathbf{v}}\widetilde{T}$ of a vector function $\tilde{\mathbf{v}}$ and a scalar function $\widetilde{T}$ (covariant derivatives with respect to $\mathbf{v}$) are

$$\begin{aligned}\nabla_{\mathbf{v}}\tilde{\mathbf{v}}=&\left\{\frac{v_\theta}{a}\frac{\partial\tilde{v}_\theta}{\partial\theta}+\frac{v_\phi}{a\cos\theta}\frac{\partial\tilde{v}_\theta}{\partial\phi}-\frac{v_\phi\tilde{v}_\phi}{a}\cot\theta\right\}\mathbf{e}_\theta\\&+\left\{\frac{v_\theta}{a}\frac{\partial\tilde{v}_\phi}{\partial\theta}+\frac{v_\phi}{a\cos\theta}\frac{\partial\tilde{v}_\phi}{\partial\phi}-\frac{\tilde{v}_\theta v_\phi}{a}\tan\theta\right\}\mathbf{e}_\phi,\end{aligned} \tag{2.19}$$

$$\nabla_{\mathbf{v}}\widetilde{T}=\frac{v_\theta}{a}\frac{\partial\widetilde{T}}{\partial\theta}+\frac{v_\phi}{a\cos\theta}\frac{\partial\widetilde{T}}{\partial\phi}. \tag{2.20}$$

Moreover, we have used the same notation Δ to denote the Laplace–Beltrami operators for both scalar functions and vector fields on S_a^2, the two-dimensional sphere of radius a centrered at 0. More precisely, we have

$$\Delta T=\frac{1}{a^2\cos\theta}\left\{\frac{\partial}{\partial\theta}\left(\cos\theta\frac{\partial T}{\partial\theta}\right)+\frac{1}{\cos\theta}\frac{\partial^2 T}{\partial\phi^2}\right\}, \tag{2.21}$$

$$\begin{aligned}\Delta\mathbf{v}&=\Delta(v_\theta\mathbf{e}_\theta+v_\phi\mathbf{e}_\phi)\\&=\left\{\Delta v_\theta-\frac{2\sin\theta}{a^2\cos^2\theta}\frac{\partial v_\phi}{\partial\phi}-\frac{v_\theta}{a^2\cos^2\theta}\right\}\mathbf{e}_\theta\\&\quad+\left\{\Delta v_\phi-\frac{2\sin\theta}{a^2\cos^2\theta}\frac{\partial v_\theta}{\partial\phi}-\frac{v_\phi}{a^2\cos^2\theta}\right\}\mathbf{e}_\phi,\end{aligned} \tag{2.22}$$

where in (2.22), Δv_θ, Δv_ϕ are defined by (2.21), and in (2.21), T is any given (smooth) function on S_a^2 the two-dimensional sphere of radius a.

REMARK 2.1. Generally speaking, the equation of state for the ocean is given by (2.5). Only empirical forms of the function $\rho=f(T,S,p)$ are known (see Washington and Parkinson [42], pp. 131–132). This equation of state is generally derived on a phenomenological basis. It is natural to expect that ρ decreases if T increases and that ρ increases if S increases.

The simplest law is (2.16) corresponding to a linearization around average (or reference) values ρ_0, T_{r}, S_{r} of the density, the temperature and the salinity, β_T and β_S are positive constant expansion coefficients. Much of what follows extends to more general nonlinear equations of state.

REMARK 2.2. The replacement of r by (2.10) in the differential operators implies a change of metric in $\mathbb{R}^3$, where the usual metric is replaced by that of $\mathbb{S}_a^2 \times \mathbb{R}$, $\mathbb{S}_a^2$ the two-dimensional sphere of radius a centered at O.

REMARK 2.3. In a classical manner, the Coriolis force $2\rho\Omega \times \mathbf{V}_3$ produces the term $2\Omega \sin\theta \mathbf{k} \times \mathbf{v}$ and a horizontal gradient term which is combined with the pressure, so that p in (2.11) is the so-called *augmented pressure*.

The hydrostatic approximation. It is known that for large-scale ocean, the horizontal scale is much bigger than the vertical one (5–10 km versus a few thousands km's). Therefore, the scale analysis (see Pedlosky [28]) shows that $\partial p/\partial z$ and ρg are the dominant terms in (2.12), leading to the hydrostatic approximation

$$\frac{\partial p}{\partial z} = -\rho g, \tag{2.23}$$

which then replaces (2.12). The approximate relation is highly accurate for the large-scale ocean and it is considered as a fundamental equation in oceanography. From the mathematical point of view, its justification relies on tools similar to those used in Section 4.1. It will not be discussed in this chapter; see however Remark 4.1 in Section 4.1. Using the hydrostatic approximation, we obtain the following equations called the *primitive equations of the large-scale ocean* (*PEs*):

$$\frac{\partial \mathbf{v}}{\partial t} + \nabla_{\mathbf{v}}\mathbf{v} + w\frac{\partial \mathbf{v}}{\partial z} + \frac{1}{\rho_0} p + 2\Omega \sin\theta \mathbf{k} \times \mathbf{v} - \mu_{\mathbf{v}}\Delta\mathbf{v} - \nu_{\mathbf{v}}\frac{\partial^2 \mathbf{v}}{\partial z^2} = F_{\mathbf{v}}, \tag{2.24}$$

$$\frac{\partial p}{\partial z} = -\rho g, \tag{2.25}$$

$$\operatorname{div} \mathbf{v} + \frac{\partial w}{\partial z} = 0, \tag{2.26}$$

$$\frac{\partial T}{\partial t} + \nabla_{\mathbf{v}} T + w\frac{\partial T}{\partial z} - \mu_T \Delta T - \nu_T \frac{\partial^2 T}{\partial z^2} = F_T, \tag{2.27}$$

$$\frac{\partial S}{\partial t} + \nabla_{\mathbf{v}} S + w\frac{\partial S}{\partial z} - \mu_S \Delta S - \nu_S \frac{\partial^2 S}{\partial z^2} = F_S, \tag{2.28}$$

$$\rho = \rho_0\big(1 - \beta_T(T - T_{\mathrm{r}}) + \beta_S(S - S_{\mathrm{r}})\big). \tag{2.29}$$

Note that $F_{\mathbf{v}}$, F_T and F_S corresponding to volumic sources (of horizontal momentum, heat and salt), vanish in reality; they are introduced here for mathematical generality. We also set $\Omega = \Omega\mathbf{k}$, where $\mathbf{k}$ is the unit vector in the direction of the poles (from south to north).

REMARK 2.4. At this stage the unknown functions can be divided into two sets. The first one, called the *prognostic variables*, $\mathbf{v}, T, S$ (4 scalar functions); we aim to write the PEs as an initial (boundary value problem) for these unknowns, and we set $U = (\mathbf{v}, T, S)$. The second set of variables comprises p, ρ, w; they are called the *diagnostic variables*. In Section 2.1.2, we will see how, using the boundary condition, one can, at each instant of time, express the diagnostic variables in terms of the prognostic variables (a fact which is already transparent for ρ in 2.29).

REMARK 2.5. We integrate (2.28) over the domain $\mathcal{M}$ occupied by the fluid which is described in Section 2.1.2. Using then the Stokes formula, and taking into account (2.26) and the boundary conditions (also described in Section 2.1.2) we arrive at

$$\frac{\mathrm{d}}{\mathrm{d}t}\int_{\mathcal{M}} S\,\mathrm{d}\mathcal{M} = \int_{\mathcal{M}} F_S\,\mathrm{d}\mathcal{M}; \tag{2.30}$$

hence

$$\int_{\mathcal{M}} S\,\mathrm{d}\mathcal{M}\Big|_t = \int_{\mathcal{M}} S\,\mathrm{d}\mathcal{M}\Big|_0 + \int_0^t \int_{\mathcal{M}} F_S\,\mathrm{d}\mathcal{M}\,\mathrm{d}t'.$$

In practical applications, $F_S = 0$ as we said, and the total amount of salt $\int_{\mathcal{M}} S\,\mathrm{d}\mathcal{M}$ is conserved. In all cases we write

$$S' = S - \frac{1}{|\mathcal{M}|}\int_{\mathcal{M}} S\,\mathrm{d}\mathcal{M}, \qquad F_S' = F_S - \frac{1}{|\mathcal{M}|}\int_{\mathcal{M}} F_S\,\mathrm{d}\mathcal{M}, \tag{2.31}$$

where $|\mathcal{M}|$ is the volume of $\mathcal{M}$, and we see that S' satisfies the same equation (2.15), with F_S replaced by F_S'. From now on, dropping the primes, we consider (2.15) as the equation for S' and we thus have

$$\int_{\mathcal{M}} S\,\mathrm{d}\mathcal{M} = 0, \qquad \int_{\mathcal{M}} F_S\,\mathrm{d}\mathcal{M} = 0. \tag{2.32}$$

2.1.2. *The initial and boundary value problems.* We assume that the ocean fills a domain $\mathcal{M}$ of $\mathbb{R}^3$ which we describe as follows (see Figure 2.1):

The top of the ocean is a domain Γ_{i} included in the surface of the Earth S_a (sphere centered at 0 of radius a). The bottom Γ_{b} of the ocean is defined by $(z = x_3 = r - a)$

$$z = -h(\theta, \varphi),$$

where h is a function of class $\mathcal{C}^2$ at least on $\overline{\Gamma_{\mathrm{i}}}$; it is assumed also that h is bounded from below,

$$0 < \underline{h} \leqslant h(\theta, \varphi) \leqslant \bar{h}, (\theta, \varphi) \in \Gamma_{\mathrm{i}}. \tag{2.33}$$

The lateral surface Γ_ℓ consists of the part of cylinder

$$(\theta, \varphi) \in \partial\Gamma_{\mathrm{i}}, -h(\theta, \varphi) \leqslant z \leqslant 0. \tag{2.34}$$

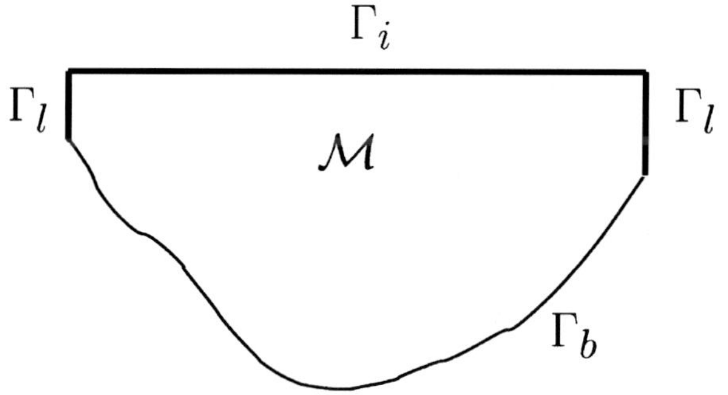

Fig. 1. The ocean $\mathcal{M}$.

REMARK 2.6. Let us make two remarks concerning the geometry of the ocean; the first one is that, for mathematical reasons, the depth is not allowed to be 0 ($h \geqslant \underline{h} > 0$), and thus "beaches" are excluded. The second one is that the top of the ocean is flat (spherical), not allowing waves; this corresponds to the so-called *rigid lid assumption* in oceanography. The assumption $h > 0$ can be relaxed for some of the following results, but this will not be discussed here. The rigid lid assumption can be also relaxed by the introduction of an additional equation for the free surface but this also will not be considered.

Boundary conditions. There are several sets of natural boundary conditions that one can associate to the primitive equations; for instance the following:
On the top of the ocean Γ_{i} ($z = 0$):

$$\begin{aligned} &\nu_{\mathbf{v}} \frac{\partial \mathbf{v}}{\partial z} + \alpha_{\mathbf{v}}\big(\mathbf{v} - \mathbf{v}^{\mathrm{a}}\big) = \tau_{\mathbf{v}}, \quad w = 0, \\ &\nu_T \frac{\partial T}{\partial z} + \alpha_T\big(T - T^{\mathrm{a}}\big) = 0, \\ &\frac{\partial S}{\partial z} = 0. \end{aligned} \tag{2.35}$$

At the bottom of the ocean Γ_{b} ($z = -h(\theta, \varphi)$):

$$\begin{aligned} &\mathbf{v} = 0, \qquad w = 0, \\ &\frac{\partial T}{\partial \mathbf{n}_T} = 0, \qquad \frac{\partial S}{\partial \mathbf{n}_S} = 0. \end{aligned} \tag{2.36}$$

On the lateral boundary Γ_ℓ ($-h(\theta, \varphi) < z < 0$, $(\theta, \varphi) \in \partial\Gamma_{\mathrm{i}}$):

$$\mathbf{v} = 0, \qquad w = 0, \qquad \frac{\partial T}{\partial \mathbf{n}_T} = 0, \qquad \frac{\partial S}{\partial \mathbf{n}_S} = 0. \tag{2.37}$$

Here $\mathbf{n} = (n_H, n_z)$ is the unit outward normal on $\partial\mathcal{M}$ decomposed into its horizontal and vertical components; the co-normal derivatives $\partial/\partial \mathbf{n}_T$ and $\partial/\partial \mathbf{n}_S$ are those associated with the linear (temperature and salinity) operators,

$$\begin{aligned} \frac{\partial}{\partial \mathbf{n}_T} &= \mu_T n_H \cdot \nabla + \nu_T n_z \frac{\partial}{\partial z}, \\ \frac{\partial}{\partial \mathbf{n}_S} &= \mu_S n_H \cdot \nabla + \nu_S n_z \frac{\partial}{\partial z}. \end{aligned} \tag{2.38}$$

REMARK 2.7. (i) The boundary conditions (which are the same) on Γ_{b} and Γ_ℓ express the no-slip boundary conditions for the water and the absence of fluxes of heat or salt. For Γ_{i}, $w = 0$ is the geometrical (kinematical) boundary condition required by the rigid lid assumption; the Neumann boundary condition on S expresses the absence of salt flux.

(ii) In general, the boundary conditions on $\mathbf{v}$ and T on Γ_{i} are not fully settled from the physical point of view. These above correspond to some resolution of the viscous boundary layers on the top of the ocean. Here $\alpha_{\mathbf{v}}$ and α_T are given $\geqslant 0$, $\mathbf{v}^{\mathrm{a}}$ and T^{a} correspond to the values in the atmosphere and $\tau_{\mathbf{v}}$ corresponds to the shear of the wind.

(iii) The first boundary condition (2.35) could be replaced by $\mathbf{v} = \mathbf{v}^{\mathrm{a}}$ expressing a no-slip condition between air and sea. However such a boundary condition necessitating an exact resolution of the boundary layer would not be practically (computationally) realistic, and as indicated in (ii) we use instead some classical resolution of the boundary layer ([32]).

(iv) As we said the boundary condition of Γ_{i} are standard unless more involved interactions are taken into consideration. However for Γ_{b} and Γ_ℓ different combinations of the Dirichlet and Neumann boundary conditions can be (have been) considered; [22].

Beta-plane approximation. For mid-latitude regional studies it is usual to consider the beta-plane approximation of the equations in which $\mathcal{M}$ is a domain in the space $\mathbb{R}^3$ with Cartesian coordinates denoted x, y, z or x_1, x_2, x_3. In the beta-plane approximation, $\Omega = 2f\mathbf{k}$, $f = f_0 + \beta y$, $\mathbf{k}$ the unit vector along the south to north poles, $\Delta = \partial^2/\partial x^2 + \partial^2/\partial y^2$, ∇ is the usual nabla vector $(\partial/\partial x, \partial/\partial y)$ and $\nabla_{\mathbf{v}} = u\partial/\partial x + v\partial/\partial y$ ($\mathbf{v} = (u, v)$). With these notations, the equations (2.24)–(2.29) and the boundary conditions (2.35)–(2.38) keep the same form; here the depth $h = h(x, y)$ satisfies, like (2.33),

$$0 < \underline{h} \leqslant h(x, y) \leqslant \bar{h}, \tag{2.39}$$

and the boundary of $\mathcal{M}$ consists of Γ_{i}, Γ_{b}, Γ_ℓ, defined as before.

As indicated in the Introduction, we will emphasize in this chapter the regional model which is slightly simpler, in particular because of the use of Cartesian coordinates. Usually the general model in spherical coordinates simply requires the treatment of lower-order terms.

From now on we consider the regional (Cartesian coordinate) case.

The diagnostic variables. The first step in the mathematical formulation of the PEs consists in showing how to express the diagnostic variables in terms of the prognostic variables, thanks to the equations and boundary conditions.

Since $w = 0$ on $\Gamma_{\rm i}$ and $\Gamma_{\rm b}$, integration of (2.26) in z gives

$$w = w(\mathbf{v}) = \int_z^0 \operatorname{div} \mathbf{v} \, dz' \tag{2.40}$$

and

$$\int_{-h}^0 \operatorname{div} \mathbf{v} \, dz = 0. \tag{2.41}$$

Note that

$$\operatorname{div} \int_{-h}^0 \mathbf{v} \, dz = \int_{-h}^0 \operatorname{div} \mathbf{v} \, dz + \nabla h \cdot \mathbf{v}\Big|_{z=-h},$$

and since $\mathbf{v}$ vanishes on $\Gamma_{\rm b}$, condition (2.41) is the same as

$$\operatorname{div} \int_{-h}^0 \mathbf{v} \, dz = 0. \tag{2.42}$$

Similarly, integration of (2.26) in z gives

$$p = p_{\rm s} + P, \qquad P = P(T, S) = g \int_z^0 \rho \, dz'. \tag{2.43}$$

Here ρ is expressed in terms of T and S through (2.29) and $p_{\rm s} = p_{\rm s}(x, y, t) = p(x, y, 0, t)$ is the pressure at the surface of the ocean.

Hence (2.40) and (2.43) provide an expression of the diagnostic variables in terms of the prognostic variables (and the surface pressure), and (2.42) is an additional equation which, we will see, is mathematically related to the surface pressure.

REMARK 2.8. The introduction of the nonlocal constraint (2.41) and of the surface pressure $p_{\rm s}$ was first carried out in [21,22]. This new formulation has played a crucial role in much of the mathematical analysis of the PEs in various cases.

2.2. *Weak formulation of the PEs of the ocean. The stationary PEs*

We denote by U the triplet $(\mathbf{v}, T, S)$ (four scalar functions). In summary the equations that we consider for the subsequent mathematical theory (the PEs) are (2.24), (2.27) and (2.28), with $w = w(\mathbf{v})$ given by (2.40), and p given by (2.43) (ρ given by (2.29)); furthermore $\mathbf{v}$ satisfies (2.41); hence

$$\frac{\partial \mathbf{v}}{\partial t} + \nabla_{\mathbf{v}} \mathbf{v} + w \frac{\partial \mathbf{v}}{\partial z} + \frac{1}{\rho_0} \nabla p + 2f \mathbf{k} \times \mathbf{v} - \mu_{\mathbf{v}} \Delta \mathbf{v} - \nu_{\mathbf{v}} \frac{\partial^2 \mathbf{v}}{\partial z^2} = F_{\mathbf{v}}, \tag{2.44}$$

$$\frac{\partial T}{\partial t} + \nabla_{\mathbf{v}} T + w \frac{\partial T}{\partial z} - \mu_T \Delta T - \nu_T \frac{\partial^2 T}{\partial z^2} = F_T, \tag{2.45}$$

$$\frac{\partial S}{\partial t} + \nabla_{\mathbf{v}} S + w \frac{\partial S}{\partial z} - \mu_S \Delta S - \nu_S \frac{\partial^2 S}{\partial z^2} = F_S, \tag{2.46}$$

$$w = w(\mathbf{v}) = \int_z^0 \operatorname{div} \mathbf{v} \, \mathrm{d}z', \tag{2.47}$$

$$\operatorname{div} \int_{-h}^0 \mathbf{v} \, \mathrm{d}z = 0, \tag{2.48}$$

$$p = p_{\mathrm{s}} + P, \qquad P = P(T, S) = g \int_z^0 \rho \, \mathrm{d}z', \tag{2.49}$$

$$\rho = \rho_0 \big(1 - \beta_T (T - T_{\mathrm{r}}) + \beta_S (S - S_{\mathrm{r}})\big), \tag{2.50}$$

$$\int_{\mathcal{M}} S \, \mathrm{d}\mathcal{M} = 0. \tag{2.51}$$

The boundary conditions are (2.35)–(2.38).

2.2.1. *Weak formulation and functional setting.* For the weak formulation of this problem, we introduce the following function spaces V and H:

$$V = V_1 \times V_2 \times V_3, \qquad H = H_1 \times H_2 \times H_3,$$

$$V_1 = \left\{ \mathbf{v} \in H^1(\mathcal{M})^2, \operatorname{div} \int_{-h}^0 \mathbf{v} \, \mathrm{d}z = 0, \mathbf{v} = 0 \text{ on } \Gamma_{\mathrm{b}} \cup \Gamma_\ell \right\},$$

$$V_2 = H^1(\mathcal{M}),$$

$$V_3 = \dot{H}^1(\mathcal{M}) = \left\{ S \in H^1(\mathcal{M}), \int_{\mathcal{M}} S \, \mathrm{d}\mathcal{M} = 0 \right\},$$

$$H_1 = \left\{ \mathbf{v} \in L^2(\mathcal{M})^2, \operatorname{div} \int_{-h}^0 \mathbf{v} \, \mathrm{d}z = 0, n_H \cdot \int_{-h}^0 \mathbf{v} \, \mathrm{d}z = 0 \text{ on } \partial \Gamma_{\mathrm{i}} \text{ (i.e., on } \Gamma_\ell) \right\},$$

$$H_2 = L^2(\mathcal{M}),$$

$$H_3 = \dot{L}^2(\mathcal{M}) = \left\{ S \in L^2(\mathcal{M}), \int_{\mathcal{M}} S \, \mathrm{d}\mathcal{M} = 0 \right\}.$$

These spaces are endowed with the following scalar products and norms:

$$((U, \widetilde{U})) = ((\mathbf{v}, \tilde{\mathbf{v}}))_1 + K_T ((T, \widetilde{T}))_2 + K_S ((S, \widetilde{S}))_3,$$

$$((\mathbf{v}, \tilde{\mathbf{v}}))_1 = \int_{\mathcal{M}} \left(\mu_{\mathbf{v}} \nabla \mathbf{v} \cdot \nabla \tilde{\mathbf{v}} + \nu_{\mathbf{v}} \frac{\partial \mathbf{v}}{\partial z} \frac{\partial \tilde{\mathbf{v}}}{\partial z} \right) \mathrm{d}\mathcal{M},$$

$$((T,\widetilde{T}))_2 = \int_{\mathcal{M}} \left(\mu_T \nabla T \cdot \nabla \widetilde{T} + \nu_T \frac{\partial T}{\partial z} \frac{\partial \widetilde{T}}{\partial z} \right) \mathrm{d}\mathcal{M} + \int_{\Gamma_\mathrm{i}} \alpha_T T \widetilde{T} \, \mathrm{d}\Gamma_\mathrm{i},$$

$$((S,\widetilde{S}))_3 = \int_{\mathcal{M}} \left(\mu_S \nabla S \cdot \nabla \widetilde{S} + \nu_S \frac{\partial S}{\partial z} \frac{\partial \widetilde{S}}{\partial z} \right) \mathrm{d}\mathcal{M},$$

$$(U,\widetilde{U})_H = \int_{\mathcal{M}} (\mathbf{v} \cdot \tilde{\mathbf{v}} + K_T T \widetilde{T} + K_S S \widetilde{S}) \, \mathrm{d}\mathcal{M},$$

$$\|U\| = \big((U,U)\big)^{1/2}, \qquad |U|_H = (U,U)_H^{1/2}.$$

Here K_T and K_S are suitable positive constants chosen below. The norm on H is of course equivalent to the L^2-norm and because of the Poincaré inequality, $\mathbf{v}$ vanishing on $\Gamma_\mathrm{b} \cup \Gamma_\ell$, and (2.51), $\|\cdot\|_i = ((\cdot,\cdot))_i^{1/2}$ is a Hilbert norm on V_i, and $\|\cdot\|$ is a Hilbert norm on V; more precisely we have, with $c_0 > 0$ a suitable constant depending on $\mathcal{M}$:

$$|U|_H \leqslant c_0 \|U\| \quad \forall U \in V. \tag{2.52}$$

Let $\mathcal{V}_1$ be the space of $\mathcal{C}^\infty$ (two-dimensional) vector functions $\mathbf{v}$ which vanish in a neighborhood of $\Gamma_\mathrm{b} \cup \Gamma_\ell$ and such that

$$\operatorname{div} \int_{-h}^{0} \mathbf{v} \, \mathrm{d}z = 0.$$

Then $\mathcal{V}_1 \subset V_1$ and it has been shown in [21] that

$$\mathcal{V}_1 \text{ is dense in } V_1. \tag{2.53}$$

We also denote by $\mathcal{V}_2 \subset V_2$ the set of $\mathcal{C}^\infty$ functions on $\overline{\mathcal{M}}$ and by $\mathcal{V}_3 \subset V_3$ the set of $\mathcal{C}^\infty$ functions on $\overline{\mathcal{M}}$ with zero average; $\mathcal{V} = \mathcal{V}_1 \times \mathcal{V}_2 \times \mathcal{V}_3$ is dense in V.

To derive the weak formulation of this problem we consider a sufficiently regular test function $\widetilde{U} = (\tilde{\mathbf{v}}, \widetilde{T}, \widetilde{S})$ in V. We multiply (2.44) by $\tilde{\mathbf{v}}$, the second one by $K_T \widetilde{T}$, the third one by $K_S \widetilde{S}$, integrate over $\mathcal{M}$ and add the resulting equations; $K_T, K_S > 0$ are two constants to be chosen later on.

The term involving $\operatorname{grad} p_S$ vanishes; indeed, by the Stokes formula,

$$\int_{\mathcal{M}} \nabla p_S \cdot \mathbf{v} \, \mathrm{d}\mathcal{M} = \int_{\partial\mathcal{M}} p_S n_H \cdot \mathbf{v} \, \mathrm{d}(\partial\mathcal{M}) - \int_{\mathcal{M}} p_S \nabla \mathbf{v} \, \mathrm{d}\mathcal{M},$$

where $\mathbf{n} = (n_H, n_z)$ is the unit outward normal on $\partial\mathcal{M}$, and n_H its horizontal component. The integral on $\partial\mathcal{M}$ vanishes because $n_H \cdot \mathbf{v}$ vanishes on $\partial\mathcal{M}$; the remaining integral on $\mathcal{M}$ vanishes too since by Fubini's theorem, (2.48), and $\mathbf{v} = 0$ on Γ_b:

$$\int_{\mathcal{M}} p_S \nabla \mathbf{v} \, \mathrm{d}\mathcal{M} = \int_{\Gamma_\mathrm{i}} p_S \int_{-h}^{0} \nabla \tilde{\mathbf{v}} \, \mathrm{d}z \, \mathrm{d}\Gamma_\mathrm{i} = \int_{\Gamma_\mathrm{i}} p_S \left(\nabla \cdot \int_{-h}^{0} \tilde{\mathbf{v}} \, \mathrm{d}z \right) \mathrm{d}\Gamma_\mathrm{i} = 0.$$

Using Stokes' formula and the boundary conditions (2.35)–(2.38) we arrive after some easy calculations at

$$\left(\frac{\mathrm{d}}{\mathrm{d}t}U, \widetilde{U}\right)_H + a(U, \widetilde{U}) + b(U, U, \widetilde{U}) + e(U, \widetilde{U}) = \ell(\widetilde{U}). \tag{2.54}$$

The notations are as follows:

$$(U, \widetilde{U})_H = \int_{\mathcal{M}} (\mathbf{v} \cdot \tilde{\mathbf{v}} + K_T T\widetilde{T} + K_S S\widetilde{S})\,\mathrm{d}\mathcal{M},$$

$$a(U, \widetilde{U}) = a_1(U, \widetilde{U}) + K_T a_2(U, \widetilde{U}) + K_S a_3(U, \widetilde{U}),$$

$$a_1(U, \widetilde{U}) = \int_{\mathcal{M}} \left(\mu_{\mathbf{v}} \nabla\mathbf{v} \cdot \nabla\tilde{\mathbf{v}} + \nu_{\mathbf{v}} \frac{\partial \mathbf{v}}{\partial z}\frac{\partial \tilde{\mathbf{v}}}{\partial z}\right) \mathrm{d}\mathcal{M}$$

$$\qquad - \int_{\mathcal{M}} \overline{P}(T, S)\nabla\tilde{\mathbf{v}}\,\mathrm{d}\mathcal{M} + \int_{\Gamma_{\mathrm{i}}} \alpha_{\mathbf{v}} \mathbf{v}\tilde{\mathbf{v}}\,\mathrm{d}\Gamma_{\mathrm{i}},$$

$$\overline{P}(T, S) = \rho_0(-\beta_T T + \beta_S S) \quad \big(\text{see (2.49) and (2.50)}\big),$$

$$a_2(U, \widetilde{U}) = \int_{\mathcal{M}} \left(\mu_T \nabla T \cdot \nabla\widetilde{T} + \nu_T \frac{\partial T}{\partial z}\frac{\partial \widetilde{T}}{\partial z}\right) \mathrm{d}\mathcal{M} + \int_{\Gamma_{\mathrm{i}}} \alpha_T T\widetilde{T}\,\mathrm{d}\Gamma_{\mathrm{i}},$$

$$a_3(U, \widetilde{U}) = \int_{\mathcal{M}} \left(\mu_S \nabla S \cdot \nabla\widetilde{S} + \nu_S \frac{\partial S}{\partial z}\frac{\partial \widetilde{S}}{\partial z}\right) \mathrm{d}\mathcal{M},$$

$$b = b_1 + K_T b_2 + K_S b_3,$$

$$b_1(U, \widetilde{U}, U^\sharp) = \int_{\mathcal{M}} \left(\mathbf{v} \cdot \nabla\tilde{\mathbf{v}} + w(\mathbf{v})\frac{\partial \tilde{\mathbf{v}}}{\partial z}\right) \mathbf{v}^\sharp\,\mathrm{d}\mathcal{M},$$

$$b_2(U, \widetilde{U}, U^\sharp) = \int_{\mathcal{M}} \left(\mathbf{v} \cdot \nabla\widetilde{T} + w(\mathbf{v})\frac{\partial \widetilde{T}}{\partial z}\right) T^\sharp\,\mathrm{d}\mathcal{M},$$

$$b_3(U, \widetilde{U}, U^\sharp) = \int_{\mathcal{M}} \left(\mathbf{v} \cdot \nabla\widetilde{S} + w(\mathbf{v})\frac{\partial \widetilde{S}}{\partial z}\right) S^\sharp\,\mathrm{d}\mathcal{M},$$

$$e(U, \widetilde{U}) = 2\int_{\mathcal{M}} (f\mathbf{k} \times \mathbf{v}) \cdot \tilde{\mathbf{v}}\,\mathrm{d}\mathcal{M}$$

and

$$\ell(\widetilde{U}) = \int_{\mathcal{M}} (F_{\mathbf{v}}\tilde{\mathbf{v}} + K_T F_T \widetilde{T} + K_S F_S \widetilde{S})\,\mathrm{d}\mathcal{M}$$

$$\qquad + \int_{\mathcal{M}} (\beta_T T_r - \beta_S S_{\mathrm{r}})\nabla \cdot \tilde{\mathbf{v}}\,\mathrm{d}\mathcal{M} + \int_{\Gamma_{\mathrm{i}}} \big[(g_{\mathbf{v}}) \cdot \tilde{\mathbf{v}} + g_T \widetilde{T}\big]\,\mathrm{d}\Gamma_{\mathrm{i}}, \tag{2.55}$$

where (see (2.35))

$$g_{\mathbf{v}} = \tau_{\mathbf{v}} + \alpha_{\mathbf{v}} \mathbf{v}^{\mathrm{a}}, \qquad g_T = \alpha_T T^{\mathrm{a}}.$$

For ℓ we observe that, if T_{r} and S_{r} are constant, then

$$\begin{aligned}
&\int_{\mathcal{M}} (1 + \beta_T T_{\mathrm{r}} - \beta_S S_{\mathrm{r}}) \nabla \cdot \mathbf{v} \, \mathrm{d}\mathcal{M} \\
&\quad = \int_{\partial\mathcal{M}} (1 + \beta_T T_{\mathrm{r}} - \beta_S S_{\mathrm{r}}) n_H \cdot \mathbf{v} \, \mathrm{d}(\partial\mathcal{M}) \\
&\quad = 0.
\end{aligned} \tag{2.56}$$

It is clear that each a_i, and thus a, is a trilinear continuous form on V; furthermore if K_T and K_S are sufficiently large, a is coercive (a_2, a_3 are automatically coercive on V_2, V_3):

$$a(U, U) \geqslant c_1 \|U\|^2 \quad \forall U \in V \ (c_1 > 0). \tag{2.57}$$

Similarly e is bilinear continuous on V_1 and even H_1, and

$$e(U, U) = 0 \quad \forall U \in H. \tag{2.58}$$

Before studying the properties of the form b, we introduce the space $V_{(2)}$:

$$V_{(2)} \text{ is the closure of } \mathcal{V} \text{ in } \left(H^2(\mathcal{M})\right)^4. \tag{2.59}$$

Then we have the following

LEMMA 2.1. *The form b is trilinear continuous on $V \times V \times V_{(2)}$ and $V \times V_{(2)} \times V$,*[1]

$$\begin{aligned}
&\left|b\big(U, \widetilde{U}, U^\sharp\big)\right| \\
&\quad \leqslant \begin{cases} c_2 \|U\| \|\widetilde{U}\| \|U^\sharp\|_{V_{(2)}}, & \forall U, \widetilde{U} \in V, U^\sharp \in V_{(2)}, \\ c_2 \|U\| \|\widetilde{U}\|_{V_{(2)}} \|U^\sharp\|, & \forall U, U^\sharp \in V, \widetilde{U} \in V_{(2)}, \end{cases}
\end{aligned} \tag{2.60}$$

or

$$\begin{aligned}
&\left|b\big(U, \widetilde{U}, U^\sharp\big)\right| \\
&\quad \leqslant c_2 \|U\| |\widetilde{U}|_H^{1/2} \|\widetilde{U}\|^{1/2} \|U^\sharp\|_{V_{(2)}}, \quad \forall U, \widetilde{U} \in V, U^\sharp \in V_{(2)}.
\end{aligned} \tag{2.61}$$

[1] For (2.60) and (2.61), the specific form of V and $V_{(2)}$ is not important: b is as well trilinear continuous on $H^1(\mathcal{M})^4 \times H^2(\mathcal{M})^4 \times H^1(\mathcal{M})^4$ and $H^1(\mathcal{M})^4 \times H^1(\mathcal{M})^4 \times H^2(\mathcal{M})$, and the estimates are similar, the H^1 and H^2 norms replacing the V and $V_{(2)}$ norms.

Furthermore,

$$b(U, \widetilde{U}, \widetilde{U}) = 0 \quad \text{for } U \in V, \widetilde{U} \in V_{(2)}, \tag{2.62}$$

and

$$b(U, \widetilde{U}, U^\sharp) = -b(U, U^\sharp, \widetilde{U}) \tag{2.63}$$

for $U, \widetilde{U}, U^\# \in V$, *and* $\widetilde{U}$ *or* $U^\sharp$ *in* $V_{(2)}$.

PROOF. To show first that b is defined on $V \times V \times V_{(2)}$ let us consider the typical and most problematic term

$$\int_{\mathcal{M}} w(\mathbf{v}) \frac{\partial \widetilde{T}}{\partial z} T^\sharp \, \mathrm{d}\mathcal{M}. \tag{2.64}$$

We have

$$\int_{\mathcal{M}} \left| w(\mathbf{v}) \frac{\partial \widetilde{T}}{\partial z} T^\sharp \right| \mathrm{d}\mathcal{M} \leqslant \left| w(\mathbf{v}) \right|_{L^2(\mathcal{M})} \left| \frac{\partial \widetilde{T}}{\partial z} \right|_{L^2(\mathcal{M})} \left| T^\sharp \right|_{L^\infty(\mathcal{M})}.$$

The first two terms in the right-hand side of this inequality are bounded by $const \cdot \|\mathbf{v}\|_1$ (using (2.40)) and $\|\widetilde{T}\|_2$. In dimension three, $H^2(\mathcal{M}) \subset L^\infty(\mathcal{M})$ so that the third term is bounded by $const \cdot \|T^\sharp\|_{V_{(2)}}$, and hence the right-hand side of the last inequality is bounded by

$$c\|U\| \|\widetilde{U}\| \|U^\sharp\|_{V_{(2)}}.$$

With similar (and easier) inequalities for the other integral, we conclude that b is defined and trilinear continuous on $V \times V \times V_{(2)}$.

For the continuity on $V \times V_{(2)} \times V$, the typical term above is bounded by

$$\left| w(\mathbf{v}) \right|_{L^2(\mathcal{M})} \left| \frac{\partial \widetilde{T}}{\partial z} \right|_{L^4(\mathcal{M})} \left| T^\sharp \right|_{L^4(\mathcal{M})},$$

which is bounded by

$$c\|\mathbf{v}\|_1 \|\widetilde{T}\|_{H^2} \|T^\sharp\|_{H^1} \leqslant c\|U\| \|\widetilde{U}\|_{V_{(2)}} \|U^\sharp\|,$$

since $H^1(\mathcal{M}) \subset L^6(\mathcal{M})$; hence the second bound (2.60).

We easily prove (2.62) and (2.63) by integration by parts for $U, \widetilde{U}, U^\sharp \in \mathcal{V}$; the relations are then extended by continuity to the other cases, using (2.60).

To establish the improvement (2.61) of the first inequality (2.60), we observe that $b(U, \widetilde{U}, U^\sharp) = -b(U, U^\sharp, \widetilde{U})$ and consider again the most typical term $\int_{\mathcal{M}} w(\mathbf{v})(\partial U^\sharp/\partial z) \times \widetilde{U}\, \mathrm{d}\mathcal{M}$, that we bound by

$$\left|w(\mathbf{v})\right|_{L^2} \left|\frac{\partial U^\sharp}{\partial z}\right|_{L^6} \left|\widetilde{U}\right|_{L^3}.$$

Remembering that $H^1 \subset L^6$ and $H^{1/2} \subset L^3$ in space dimension three, we bound this term by

$$c\|\mathbf{v}\| \left\|U^\sharp\right\|_{V_{(2)}} \left|\widetilde{U}\right|^{1/2} \left\|\widetilde{U}\right\|^{1/2},$$

and (2.61) follows.

The operator form of the equation. We can write equation (2.54) in the form of an evolution equation in the Hilbert space $V'_{(2)}$. For that purpose we observe that we can associate to the forms a, b, e above, the following operators:

- A linear continuous from V into V', defined by

$$\langle AU, \widetilde{U}\rangle = a(U, \widetilde{U}) \quad \forall U, \widetilde{U} \in V,$$

- B bilinear continuous from $V \times V$ into $V'_{(2)}$ defined by

$$\langle B(U, \widetilde{U}), U^\sharp\rangle = b(U, \widetilde{U}, U^\sharp) \quad \forall U, \widetilde{U} \in V,\ \forall U^\sharp \in V_{(2)},$$

- E linear continuous from H into itself, defined by

$$\langle E(U), \widetilde{U}\rangle = e(U, \widetilde{U}) \quad \forall U, \widetilde{U} \in H.$$

Since $V_{(2)} \subset V \subset H$, with continuous injections, each space being dense in the next one, we also have the Gelfand–Lions inclusions

$$V_{(2)} \subset V \subset H \subset V' \subset V'_{(2)}. \tag{2.65}$$

With this we see that (2.54) is equivalent to the following operator evolution equation

$$\frac{\mathrm{d}U}{\mathrm{d}t} + AU + B(U, U) + E(U) = \ell, \tag{2.66}$$

understood in $V'_{(2)}$ and with ℓ defined in (2.55). To this equation we will naturally add an initial condition:

$$U(0) = U_0. \tag{2.67}$$

□

2.2.2. *The stationary PEs.* We now establish the existence of solutions of the stationary PEs. Beside its intrinsic interest, this result will be needed in the next section for the study of the time dependent case.

The equations to be considered are the same as (2.54), with the only difference that the derivatives $\partial \mathbf{v}/\partial t$, $\partial T/\partial t$ and $\partial S/\partial t$ are removed, and that the source terms $F_{\mathbf{v}}$, F_T, F_S are given independent of time t.

The weak formulation proceeds as before:

$$\begin{aligned}&\textit{Given } F=(F_{\mathbf{v}},F_T,F_S) \textit{ in } H \bigl(\text{or } L^2(\mathcal{M})^4\bigr), \textit{ and}\\ &g=(g_{\mathbf{v}},g_T) \textit{ in } L^2(\Gamma_{\mathrm{i}})^2, \textit{ find } U=(\mathbf{v},T,S)\in V, \textit{ such that}\\ &a\bigl(U,\widetilde{U}\bigr)+b\bigl(U,U,\widetilde{U}\bigr)+e\bigl(U,\widetilde{U}\bigr)=\ell\bigl(\widetilde{U}\bigr), \textit{ for every } \widetilde{U}\in V_{(2)};\end{aligned} \tag{2.68}$$

a, b, e and ℓ are the same as above.

We have the following result.

THEOREM 2.1. *We are given $F=(F_{\mathbf{v}},F_T,F_S)$ in $L^2(\mathcal{M})^4$ (or in H), and $g=(g_{\mathbf{v}},g_T)$ in $L^2(\Gamma_{\mathrm{i}})^3$; then problem* (2.68) *possesses at least one solution $U\in V$ such that*

$$\|U\|\leqslant \frac{1}{c_1}\|\ell\|_{V'}. \tag{2.69}$$

PROOF. The proof of existence is done by Galerkin method, a priori estimates and passage to the limit. The proof is essentially standard, but we give the details because of some specificities in this case.

We consider a family of elements $\{\Phi_j\}_j$ of $V_{(2)}$ which is free and total in V ($V_{(2)}$ is dense in V); and for each $m\in\mathbb{N}$, we look for an approximate solution of (2.68), $U_m=\sum_{j=1}^m \xi_{jm}\Phi_j$, such that

$$a(U_m,\Phi_k)+b(U_m,U_m,\Phi_k)=\ell(\Phi_k),\quad k=1,\dots,m. \tag{2.70}$$

The existence of U_m is shown below. An a priori estimate on U_m is obtained by multiplying each equation (2.68) by ξ_{km} and summing for $k=1,\dots,m$. This amounts to replacing Φ_k by U_m in (2.70); since $b(U_m,U_m,U_m)=0$ by Lemma 2.1, we obtain

$$a(U_m,U_m)=\ell(U_m)$$

and, with (2.57),

$$c_1\|U_m\|^2\leqslant \|\ell\|_{V'}\|U_m\|,$$

$$\|U_m\|\leqslant \frac{1}{c_1}\|\ell\|_{V'}. \tag{2.71}$$

From (2.71) we see that there exists $U \in V$ and a subsequence $U_{m'}$, such that $U_{m'}$, converges weakly to U as $m' \to \infty$. Since we cannot replace $\widetilde{U}$ by U in (2.68) it is useful to notice that

$$\|U\| \leqslant \liminf_{m' \to \infty} \|U'_m\| \leqslant \frac{1}{c_1}\|\ell\|_{V'},$$

so that (2.69) is satisfied. Then we pass to the limit in (2.70) written with m', and k fixed less than or equal to m'. We observe below that

$$b(U_{m'}, U_{m'}, \Phi_k) \to b(U, U, \Phi_k), \tag{2.72}$$

so that, at the limit, U satisfies (2.68) for $\widetilde{U} = \Phi_k$, k fixed arbitrary; hence (2.68) is valid for any $\widetilde{U}$ linear combination of Φ_k and, by continuity (Lemma 2.1), for $\widetilde{U} \in V_{(2)}$.

The proof is complete after we prove the results used above. □

Convergence of the b term. To prove (2.72) we first observe, with (2.63), that $b(U_{m'}, U_{m'}, \Phi_k) = -b(U_{m'}, \Phi_k, U_{m'})$. We also observe that each component of $U_{m'}$ converges weakly in $H^1(\mathcal{M})$ to the corresponding component of Φ_k. Therefore, by compactness, the convergence takes place in $H^{3/4}(\mathcal{M})$ strongly; by Sobolev embedding, $H^{3/4}(\mathcal{M}) \subset L^4(\mathcal{M})$ in dimension three, and the convergence holds in $L^4(\mathcal{M})$ strongly. Writing $\Phi_k = \Phi = (\mathbf{v}_\Phi, T_\Phi, S_\Phi)$, the typical most problematic term is

$$\int_{\mathcal{M}} w(\mathbf{v}_{m'}) \frac{\partial \mathbf{v}_\Phi}{\partial z} \mathbf{v}_{m'}\, \mathrm{d}\mathcal{M}. \tag{2.73}$$

Since $\operatorname{div} \mathbf{v}_{m'}$ converges weakly to $\operatorname{div} \mathbf{v}$ in $L^2(\mathcal{M})$, $w(\mathbf{v}_{m'})$ converges weakly in $L^2(\mathcal{M})$ to $w(\mathbf{v})$; $\mathbf{v}_{m'}$ converges strongly to $\mathbf{v}$ in $L^4(\mathcal{M})$ as observed before, and since $\partial \mathbf{v}_\Phi/\partial z$ belongs to $L^4(\mathcal{M})$, the term above converges to the corresponding term where $\mathbf{v}_{m'}$ is replaced by $\mathbf{v}(U = (\mathbf{v}, T, S))$. Hence (2.69).

Existence of U_m. Equations (2.70) amount to a system of m nonlinear equations for the components of the vector $\xi = (\xi_1, \ldots, \xi_m)$, where we have written $\xi_{jm} = \xi_j$ for simplicity. Existence follows from the following consequence of the Brouwer fixed point theorem. (See Lions [19].)

LEMMA 2.2. *Let $\mathcal{F}$ be a continuous mapping of $\mathbb{R}^m$ into itself such that*

$$\big[\mathcal{F}(\xi), \xi\big] > 0 \quad \text{for } [\xi] = k, \text{ for some } k > 0, \tag{2.74}$$

where $[\cdot, \cdot]$ and $[\cdot]$ are the scalar product and norm in $\mathbb{R}^m$.

Then there exists $\xi \in \mathbb{R}^m$ with $[\xi] < k$, such that $\mathcal{F}(\xi) = 0$.

PROOF. If $\mathcal{F}$ never vanishes, then $\mathcal{G} = -k\mathcal{F}(\xi)/[\mathcal{F}(\xi)]$ is continuous on $\mathbb{R}^m$, and we can apply the Brouwer fixed point theorem to $\mathcal{G}$ which maps the ball $\mathcal{C}$ centered at 0 of radius k

into itself. Then $\mathcal{G}$ has a fixed point ξ_0 in $\mathcal{C}$ and we have

$$\big[\mathcal{G}(\xi_0)\big] = [\xi_0] = k,$$

$$\big[\mathcal{G}(\xi_0), \xi_0\big] = -k\frac{[\mathcal{F}(\xi_0), \xi_0]}{[\mathcal{F}(\xi_0)]} = [\xi_0]^2.$$

This contradicts the hypothesis (2.74) on $\mathcal{F}$; the lemma is proven. □

We apply this lemma to (2.70) as follows: $\mathcal{F} = (\mathcal{F}_1, \dots, \mathcal{F}_m)$, with

$$\mathcal{F}_k(\xi) = a(U_m, \Phi_k) + b(U_m, V_m, \Phi_k) + e(U_m, V_m) - l(\Phi_k). \tag{2.75}$$

The space $\mathbb{R}^m$ is equipped with the usual Euclidean scalar product, so that

$$\begin{aligned}\big[\mathcal{F}(\xi), \xi\big] &= \sum_{k=1}^{m} \mathcal{F}_k(\xi)\xi_k \\ &= a(U_m, U_m) + b(U_m, U_m, V_m) - l(U) \\ &\geqslant \big(\text{with (2.57), (2.58), (2.62) and Schwarz' inequality}\big) \\ &\geqslant c_1\|U_m\|^2 - \|\ell\|_{V'}\|U_m\|. \end{aligned} \tag{2.76}$$

Since the last expression converges to $+\infty$ as $\|U_m\| \sim [\xi]$ converges to $+\infty$, there exists $k > 0$ such that (2.74) holds. The existence of U_m follows.

REMARK 2.9. A perusal of the proof of Theorem 2.1 shows that we proved the following more general result.

LEMMA 2.3. *Let V, W be two Hilbert spaces with $W \subset V$, the injection being continuous. Assume that $\bar{a}$ is bilinear continuous coercive on V, and that $\bar{b}$ is trilinear continuous on $V \times W \times V$, $V \times V \times W$, and continuous on $V_w \times V_w \times W$, where V_w is V equipped with the weak topology. Furthermore*

$$b\big(U, \widetilde{U}, U^\sharp\big) = -b\big(U, U^\sharp, \widetilde{U}\big) \quad \text{if } U, \widetilde{U}, U^\sharp \in V \text{ and } \widetilde{U} \text{ or } U^\sharp \in W.$$

Then, for $\bar{l}$ given in V', there exists at least one solution U of

$$\bar{a}\big(U, \widetilde{U}\big) + \bar{b}\big(U, U, \widetilde{U}\big) = \bar{l}\big(\widetilde{U}\big) \quad \forall \widetilde{U} \in W, \tag{2.77}$$

which satisfies

$$\bar{a}(U, U) \leqslant \bar{\ell}(U). \tag{2.78}$$

Lemma 2.3 will be useful in the next section.

2.3. *Existence of weak solutions for the PEs of the ocean*

In this section we establish the existence, for all time, of weak solutions for the equations of the ocean. The main result is Theorem 2.2 given at the end of the section.

We consider the primitive equations in their formulation (2.54), that is, with the notations of Section 2.2:

$$\begin{aligned}&\textit{Given } t_1 > 0,\ U_0 \textit{ in } H,\ F = (F_{\mathbf{v}}, F_T, F_S) \textit{ in } L^2(0, t_1; H), \textit{ and}\\ &g = (g_{\mathbf{v}}, g_T) \textit{ in } L^2\big(0, t_1; L^2(\Gamma_i)\big)^3, \textit{ to find}\\ &U \in L^\infty(0, t_1; H) \cap L^2(0, t_1; V), \textit{ such that}\\ &\big(\tfrac{\mathrm{d}}{\mathrm{d}t}U, \widetilde{U}\big) + a\big(U, \widetilde{U}\big) + b\big(U, U, \widetilde{U}\big) + e\big(U, \widetilde{U}\big) = \ell\big(\widetilde{U}\big) \quad \forall \widetilde{U} \in V_{(2)},\end{aligned} \tag{2.79}$$

$$U(0) = U_0. \tag{2.80}$$

Alternatively, and as explained in the previous section (see (2.66) and (2.67)), we can write (2.79) and (2.80) in the form of an operator evolution equation

$$\frac{\mathrm{d}U}{\mathrm{d}t} + AU + B(U, U) + E(U) = \ell, \tag{2.81}$$

$$U(0) = U_0. \tag{2.82}$$

To establish the existence of weak solutions of this problem we proceed by finite differences in time.[2]

Finite differences in time. Given $t_1 > 0$ which is arbitrary, we consider N an arbitrary integer and introduce the time step $k = \Delta t = t_1/N$. By time discretization of (2.79) and (2.80), we are naturally led to define a sequence of elements of V, U^n, $0 \leqslant n \leqslant N$, defined by

$$U^0 = U_0, \tag{2.83}$$

and then, recursively for $n = 1, \dots, N$, by

$$\begin{aligned}&\frac{1}{\Delta t}\big(U^n - U^{n-1}, \widetilde{U}\big)_H + a\big(U^n, \widetilde{U}\big) + b\big(U^n, U^n, \widetilde{U}\big) + e\big(U^n, \widetilde{U}\big)\\ &\quad = \ell^n\big(\widetilde{U}\big) \quad \forall \widetilde{U} \in V_{(2)}.\end{aligned} \tag{2.84}$$

Here $\ell^n \in V_{(2)}$ is given by

$$\ell^n\big(\widetilde{U}\big) = \frac{1}{\Delta t}\int_{(n-1)\Delta t}^{n\Delta t} \ell\big(t; \widetilde{U}\big)\,\mathrm{d}t, \tag{2.85}$$

[2]At this level of generality it has not been possible to prove the existence of weak solutions to the PEs by any other classical method for parabolic equations. In particular, the proofs in the articles [21,22,24] based on the Galerkin method, assume the H^2 regularity of the solutions of the GFD–Stokes problem, and this result is not available at this level of generality. We recall that the whole Section 4 is devoted to this regularity question.

where $\ell(t;\widetilde{U})$ is defined exactly as in (2.54), the dependence of ℓ on t reflecting now the dependence on t of F, g_{v} and g_T.

The existence, for all n, of $U^n \in V$ solution of (2.84) follows from Lemma 2.3, equation (2.84) being the same as (2.77); the notations are obvious and the verification of the hypotheses of the lemma is easy; furthermore by (2.78), and after multiplication by $2\Delta t$:

$$\begin{aligned}&|U^n|_H^2 - |U^{n-1}|_H^2 + |U^n - U^{n-1}|_H^2 + 2\Delta t a(U^n, U^n)\\&\quad\leqslant 2\Delta t \ell^n(U^n).\end{aligned} \tag{2.86}$$

For (2.86), we also used (2.58) and the elementary relation

$$2(\widetilde{U} - U^\sharp, \widetilde{U})_H = |\widetilde{U}|_H^2 - |U^\sharp|_H^2 + |\widetilde{U} - U^\sharp|_H^2 \quad \forall \widetilde{U}, U^\sharp \in H. \tag{2.87}$$

A priori estimates. We now proceed and derive a priori estimates for the U^n and then for some associated approximate functions.

Using (2.85), (2.54) and Schwarz' inequality, we bound $\Delta t\ \ell^n(U^n)$ by $\Delta t^{1/2}(\xi^n)^{1/2} \times \|U^n\|$ with

$$\begin{aligned}&\xi^n = \int_{(n-1)\Delta t}^{n\Delta t} \xi(t)\,\mathrm{d}t,\\&\xi(t) = c_1'\biggl[|F(t)|_H^2 + \int_{\mathcal{M}} |1 + \beta_T T_{\mathrm{r}}(t) - \beta_S S_{\mathrm{r}}(t)|^2\,\mathrm{d}t\\&\qquad + \int_{\Gamma_{\mathrm{i}}} \bigl(|g_{\mathrm{v}}(t)|^2 + |g_T(t)|^2\bigr)\,\mathrm{d}t\biggr],\end{aligned} \tag{2.88}$$

where c_1' is an absolute constant related to c_0 (see (2.52)). Hence using also (2.57), we infer from (2.86) that

$$\begin{aligned}&|U^n|_H^2 - |U^{n-1}|_H^2 + |U^n - U^{n-1}|_H^2 + 2\Delta t c_1 \|U^n\|^2\\&\quad\leqslant 2\Delta t^{1/2}(\xi^n)^{1/2}\|U^n\|\\&\quad\leqslant \Delta t c_1\|U^n\|^2 + c_1^{-1}\xi^n.\end{aligned}$$

Hence

$$\begin{aligned}&|U^n|_H^2 - |U^{n-1}|_H^2 + |U^n - U^{n-1}|_H^2 + \Delta t c_1 \|U^n\|^2\\&\quad\leqslant c_1^{-1}\xi^n \quad \text{for } n = 1, \dots, N.\end{aligned} \tag{2.89}$$

Summing all these relations for $n = 1, \dots, N$, we find

$$|U^n|_H^2 + \sum_{n=1}^{N}\bigl[|U^n - U^{n-1}|_H^2 + \Delta t c_1\|U^n\|^2\bigr] \leqslant \kappa_1, \tag{2.90}$$

with

$$\kappa_1 = |U_0|^2 + \frac{1}{c_1} \int_0^T \xi(t)\,\mathrm{d}t.$$

Summing the relations (2.89) for $n = 1, \dots, m$, with m fixed, $1 \leqslant m \leqslant N$, we obtain as well

$$\left|U^m\right|_H^2 \leqslant \kappa_1 \quad \forall m = 0, \dots, N. \tag{2.91}$$

Approximate functions. The subsequent steps follow closely the proof in [36], Chapter 3, Section 4, for the Navier–Stokes equations, and we will skip many details.[3]

We first introduce the approximate functions defined as follows on $(0, t_1)$ $(k = \Delta t)$:

$$\begin{aligned}
&U_k : (0, t_1) \mapsto V, \qquad U_k(t) = U^n, \quad t \in \big((n-1)k, nk\big),\\
&\ell_k : (0, t_1) \mapsto V', \qquad \ell_k(t) = \ell^n, \quad t \in \big((n-1)k, nk\big),\\
&\widetilde{U}_k : (0, t_1) \mapsto V,\\
&\widetilde{U}_k \text{ is continuous, linear on each interval } \big((n-1)k, nk\big) \text{ and}\\
&\widetilde{U}_k(nk) = U^n, \quad n = 0, \dots, N.
\end{aligned}$$

An easy computation (see [36]) shows that:

$$\left|U_k - \widetilde{U}_k\right|_{L^2(0,t_1;H)} \leqslant \left(\frac{k}{3}\right)^{1/2} \left(\sum_{n=1}^{N} \left|U^n - U^{n-1}\right|_H^2\right)^{1/2}, \tag{2.92}$$

and we infer from (2.90) and (2.91) that

$$\begin{aligned}
&U_k \text{ and } \widetilde{U}_k \text{ are bounded independently}\\
&\text{of } \Delta t \text{ in } L^\infty(0, t_1; H) \text{ and } L^2(0, t_1; V).
\end{aligned} \tag{2.93}$$

We infer from (2.92) and (2.93) that there exists $U \in L^\infty(0, t_1; H) \cap L^2(0, t_1; V)$, and a subsequence $k' \to 0$, such that, as $k' \to 0$,

$$\begin{aligned}
&U_{k'} \text{ and } \widetilde{U}_{k'} \rightharpoonup U \text{ in } L^\infty(0, t_1; H)\\
&\text{weak star and in } L^2(0, t_1; V) \text{ weakly,}
\end{aligned} \tag{2.94}$$

$$U_{k'} - \widetilde{U}_{k'} \to 0 \text{ in } L^2(0, t_1; H) \text{ strongly.} \tag{2.95}$$

[3] The proof given here would apply to the Navier–Stokes equations in space dimension $d \geqslant 4$; it extends the proof given in [36] which is only valid for the Navier–Stokes equations in dimension $d = 2$ or 3.

Further a priori estimates and compactness. With the notations above and those used for (2.66) and (2.67) (or (2.81) and (2.82)), we see that the scheme (2.84) can be rewritten as

$$\frac{\mathrm{d}\widetilde{U}_k}{\mathrm{d}t} + AU_k + B(U_k, U_k) + E(U_k) = \ell_k, \quad 0 < t < t_1, \tag{2.96}$$

$$\widetilde{U}_k(0) = U_0. \tag{2.97}$$

From (2.93) and (2.61) we see that $B(U_k, U_k)$ is bounded in $L^{4/3}(0, t_1; V'_{(2)})$; since the other terms in (2.96) are bounded in $L^2(0, t_1; V')$ independently of k, we conclude that

$$\frac{\mathrm{d}U_k}{\mathrm{d}t} \text{ is bounded in } L^{4/3}\big(0, t_1; V'_{(2)}\big). \tag{2.98}$$

We then infer from (2.93) and the Aubin compactness theorem (see, e.g., [36]), that as $k' \to 0$,

$$\widetilde{U}_{k'} \to U \text{ in } L^2(0, t_1; H) \text{ strongly}, \tag{2.99}$$

and the same is true for U_k because of (2.95).

Passage to the limit. The passage to the limit $k' \to 0$ ($k = \Delta t$) follows now closely that of Theorem 4.1, Chapter 2 in [36], we skip the details.

We consider $\widetilde{U} \in \mathcal{V}$ (which is dense in $V_{(2)}$, see (2.59)) and a scalar function ψ in $\mathcal{C}^1([0, t_1])$, such that $\psi(t_1) = 0$. We take the scalar product in H of (2.96) with $\widetilde{U}\psi$, integrate from 0 to t_1 and integrate by parts the first term, we arrive at

$$\begin{aligned} -\int_0^{t_1} \big(\widetilde{U}_k, \widetilde{U}\big)_H \psi' \,\mathrm{d}t + \int_0^{t_1} \big[a\big(\widetilde{U}_k, \widetilde{U}\big) + b\big(U_k, U_k, \widetilde{U}\big) + e\big(U_k, \widetilde{U}\big)\big]\psi \,\mathrm{d}t \\ = \big(U_0, \widetilde{U}\big)_H \psi(0) + \int_0^{t_1} \ell_k\big(\widetilde{U}\big)\psi \,\mathrm{d}t. \end{aligned} \tag{2.100}$$

We can pass to the limit in (2.100) for the sequence k'; for the b term we proceed somehow as for (2.72). For the nonlinear term we write

$$\int_0^{t_1} b\big(U_k, U_k, \widetilde{U}\big)\psi \,\mathrm{d}t = -\int_0^{t_1} b\big(U_k, \widetilde{U}, U_k\big)\psi \,\mathrm{d}t,$$

and, considering the typical most problematic term, we show that, as $k' \to 0$,

$$\int_0^{t_1}\int_{\mathcal{M}} w(\mathbf{v}_k)\frac{\partial \widetilde{T}}{\partial z} T_{k'}\psi \,\mathrm{d}\mathcal{M}\,\mathrm{d}t \to \int_0^{t_1}\int_{\mathcal{M}} w(\mathbf{v})\frac{\partial \widetilde{T}}{\partial z} T\psi \,\mathrm{d}\mathcal{M}\,\mathrm{d}t; \tag{2.101}$$

this follows from the fact that $\operatorname{div} \mathbf{v}_{k'}$ converges to $\operatorname{div} \mathbf{v}$ weakly in $L^2(\mathcal{M} \times (0, t_1))$, that $T_{k'}$ converges to T strongly in $L^2(\mathcal{M} \times (0, t_1))$, and $\psi\, \partial\widetilde{T}/\partial z$ belongs to $L^\infty(\mathcal{M} \times (0, t_1))$.

From this we conclude that U satisfies

$$-\int_0^{t_1} (U, \widetilde{U})_H \psi' \,\mathrm{d}t + \int_0^{t_1} \big[a(U, \widetilde{U}) + b(U, U, \widetilde{U}) + e(U, \widetilde{U})\big] \psi \,\mathrm{d}t$$
$$= (U_0, \widetilde{U})\psi(0) + \int_0^{t_1} \ell(\widetilde{U})\psi \,\mathrm{d}t \tag{2.102}$$

for all $\widetilde{U}$ in $\mathcal{V}$ and all ψ of the indicated type. Also, by continuity (Lemma 2.1), (2.102) is valid as well for all U in $V_{(2)}$ since $\mathcal{V}$ is dense in $V_{(2)}$ by (2.59).

It is then standard to infer from (2.102) that U is solution of (2.79) and (2.80); this leads us to the main result of this section.

THEOREM 2.2. *The domain $\mathcal{M}$ is as before. We are given $t_1 > 0$, U_0 in H and $F = (F_{\mathrm{v}}, F_T, F_S)$ in $L^2(0, t_1; H)$ or $L^2(0, t_1; L^2(\mathcal{M})^4)$; $g = g_{\mathrm{v}}, g_T$ is given in $L^2(0, t_1; L^2(\Gamma_i)^3)$. Then there exists*

$$U \in L^\infty(0, t_1; H) \cap L^2(0, t_1; V), \tag{2.103}$$

which is solution of (2.79) *and* (2.80) (*or* (2.81) *and* (2.82)); *furthermore U is weakly continuous from* $[0, t_1]$ *into H.*

2.4. *The primitive equations of the atmosphere*

In this section we briefly describe the PEs of the atmosphere, introduce their mathematical (weak) formulation and state without proof the existence of weak solutions; the proof is essentially the same as for the ocean.

We start from the conservation equations similar to (2.1)–(2.5). In fact, (2.1) and (2.2) are the same; the equation of energy conservation is slightly different from (2.3) because of the compressibility of the air; the state equation is that of perfect gas instead of (2.5); finally, instead of the concentration of salt in the water, we consider the amount q of water in air. Hence, we have

$$\rho \frac{\mathrm{d}\mathbf{V}_3}{\mathrm{d}t} + 2\rho\Omega \times \mathbf{V}_3 + \nabla_3 p + \rho \mathbf{g} = D, \tag{2.104}$$

$$\frac{\mathrm{d}\rho}{\mathrm{d}t} + \rho \operatorname{div}_3 \mathbf{V}_3 = 0, \tag{2.105}$$

$$c_p \frac{\mathrm{d}T}{\mathrm{d}t} - \frac{RT}{p}\frac{\mathrm{d}p}{\mathrm{d}t} = Q_T, \tag{2.106}$$

$$\frac{\mathrm{d}q}{\mathrm{d}t} = Q_q, \tag{2.107}$$

$$p = R\rho T. \tag{2.108}$$

Here, D, Q_T and Q_q contain the dissipation terms. As we said the difference between (2.106) and (2.3) is due to the compressibility of the air; in (2.106), $c_p > 0$ is the specific heat of the air at constant pressure and R is the specific gas constant for the air; (2.108) is the equation of state for the air.

The hydrostatic approximation. We decompose $\mathbf{V}_3$ into its horizontal and vertical components as in (2.8), $\mathbf{V}_3 = (\mathbf{v}, w)$, and we use the approximation (2.10) of $\mathrm{d}/\mathrm{d}t$. Also, as for the ocean, we use the hydrostatic approximation, replacing the equation of conservation of vertical momentum (third equation (2.104)) by the hydrostatic equation (2.23). We find

$$\frac{\partial \mathbf{v}}{\partial t} + \nabla_{\mathbf{v}}\mathbf{v} + w\frac{\partial \mathbf{v}}{\partial z} + \frac{1}{\rho}\nabla p + 2\Omega \sin\theta\, \mathbf{k} \times \mathbf{v} - \mu_{\mathbf{v}}\Delta \mathbf{v} - \nu_{\mathbf{v}}\frac{\partial^2 \mathbf{v}}{\partial z^2} = 0, \tag{2.109}$$

$$\frac{\partial p}{\partial z} = -\rho g, \tag{2.110}$$

$$\frac{\partial \rho}{\partial t} + \rho\left(\nabla \mathbf{v} + \frac{\partial w}{\partial z}\right) + \mathbf{v}\nabla\rho + w\frac{\partial \rho}{\partial z} = 0, \tag{2.111}$$

$$\frac{\partial T}{\partial t} + \nabla_{\mathbf{v}}T + w\frac{\partial T}{\partial z} - \mu_T\Delta T - \nu_T\frac{\partial^2 T}{\partial z^2} - \frac{RT}{p}\frac{\mathrm{d}p}{\mathrm{d}t} = Q_T, \tag{2.112}$$

$$\frac{\partial q}{\partial t} + \nabla_{\mathbf{v}}q + w\frac{\partial q}{\partial z} - \mu_q\Delta q - \nu_q\frac{\partial^2 q}{\partial z^2} = 0, \tag{2.113}$$

$$p = R\rho T. \tag{2.114}$$

The right-hand side of (2.112), which is different from Q_T in (2.106) now represents the solar heating.

Change of vertical coordinate. Since ρ does not vanish, the hydrostatic equation (2.110) implies that p is a strictly decreasing function of z, and we are thus allowed to use p as the vertical coordinate; hence in spherical geometry the independent variables are now φ, θ, p and t. By an abuse of notation we still denote by $\mathbf{v}, T, q, \rho$ these functions expressed in the φ, θ, p, t variables. We also denote by ω the vertical component of the wind, and one can show (see, e.g., [13]) that

$$\omega = \frac{\mathrm{d}p}{\mathrm{d}t} = \frac{\partial p}{\partial t} + \nabla_{\mathbf{v}}p + w\frac{\partial p}{\partial z}; \tag{2.115}$$

in (2.115), p is a dependent variable expressed as a function of φ, θ, z and t.

In this context, the PEs of the atmosphere become

$$\frac{\partial \mathbf{v}}{\partial t} + \nabla_{\mathbf{v}}\mathbf{v} + \omega\frac{\partial \mathbf{v}}{\partial p} + 2\Omega \sin\theta\, \mathbf{k} \times \mathbf{v} + \nabla\Phi - L_{\mathbf{v}}\mathbf{v} = F_{\mathbf{v}}, \tag{2.116}$$

$$\frac{\partial \Phi}{\partial p} + \frac{RT}{p} = 0, \tag{2.117}$$

$$\operatorname{div} \mathbf{v} + \frac{\partial \omega}{\partial p} = 0, \tag{2.118}$$

$$\frac{\partial T}{\partial t} + \nabla_{\mathbf{v}} T + \omega \frac{\partial T}{\partial p} - \frac{R\overline{T}}{c_p p}\omega - L_T T = F_T, \tag{2.119}$$

$$\frac{\partial q}{\partial t} + \nabla_{\mathbf{v}} q + \omega \frac{\partial q}{\partial p} - L_q q = F_q, \tag{2.120}$$

$$p = R\rho T. \tag{2.121}$$

We have denoted by $\Phi = gz$ the geopotential (z is now a function of φ, θ, p, t); $L_{\mathbf{v}}$, L_T and L_q are the Laplace operators, with suitable eddy viscosity coefficients, expressed in the φ, θ, p variables. Hence, for example,

$$L_{\mathbf{v}}\mathbf{v} = \mu_{\mathbf{v}} \Delta \mathbf{v} + \nu_{\mathbf{v}} \frac{\partial}{\partial p}\left[\left(\frac{gp}{R\overline{T}}\right)^2 \frac{\partial \mathbf{v}}{\partial p}\right], \tag{2.122}$$

with similar expressions for L_T and L_q. Note that F_T corresponds to the heating of the sun, whereas $F_{\mathbf{v}}$ and F_q which vanish in reality, are added here for mathematical generality. In (2.119) T has been replaced by $\overline{T}$ in the term $RT\omega/c_p p$. See in [24] a better approximation of $RT\omega/c_p p$ involving an additional term. With additional precautions, and using the maximum principle for the temperature as in [8], we could keep the exact term $RT\omega/c_p p$.

The change of variable gives for $\partial^2 \mathbf{v}/\partial z^2$ a term different from the coefficient of $\nu_{\mathbf{v}}$. The expression above is a simplified form of this coefficient, the simplification is legitimate because $\nu_{\mathbf{v}}$ is a very small coefficient; in particular T has been replaced by $\overline{T}$ (known) which is an average value of the temperature.

Pseudo-geometrical domain. For physical and mathematical reasons, we do not allow the pressure to go to zero, and assume that $p \geqslant p_0$, with $p_0 > 0$ "small". Physically, in the very high atmosphere (p very small), the air is ionized and the equations above are not valid anymore; mathematically, with $p > p_0$, we avoid the appearance of singular terms as, for example, in the expressions of $L_{\mathbf{v}}$, L_T and L_q. The pressure is then restricted to an interval $p_0 < p < p_1$, where p_1 is a value of the pressure smaller in average than the pressure on Earth, so that the isobar $p = p_1$ is slightly above the Earth and the isobar $p = p_0$ is an isobar high in the sky. We study the motion of the air between these two isobars; as we said, for $p < p_0$ we would need a different set of equations and for the "thin" portion of air between the Earth and the isobar $p = p_1$, another specific simplified model would be necessary.

For the whole atmosphere, the domain is

$$\mathcal{M} = \{(\varphi, \theta, p), p_0 < p < p_1\},$$

and its boundary consists first of an upper part Γ_{u}, $p = p_0$ and a lower part $p = p_1$ which is divided into two parts: Γ_{i} the part of $p = p_1$ at the interface with the ocean, and Γ_{e} the part of $p = p_1$ above the Earth.

Boundary conditions. Typically the boundary conditions are as follows:

On the top of the atmosphere Γ_u ($p = p_0$):

$$\frac{\partial \mathbf{v}}{\partial p} = 0, \qquad \omega = 0, \qquad \frac{\partial T}{\partial p} = 0, \qquad \frac{\partial q}{\partial p} = 0. \tag{2.123}$$

Above the Earth on Γ_e:

$$\begin{aligned} &\mathbf{v} = 0, \qquad \omega = 0, \\ &\nu_T \frac{\partial T}{\partial p} + \alpha_T (T - T_e) = 0, \\ &\frac{\partial q}{\partial p} = g_q. \end{aligned} \tag{2.124}$$

Above the ocean on Γ_i:

$$\begin{aligned} &\nu_{\mathbf{v}} \left(\frac{gp}{R\overline{\overline{T}}} \right)^2 \frac{\partial \mathbf{v}}{\partial p} + \alpha_{\mathbf{v}} \left(\mathbf{v} - \mathbf{v}^s \right) = \tau_{\mathbf{v}}, \qquad \omega = 0, \\ &\nu_T \left(\frac{gp}{R\overline{\overline{T}}} \right)^2 \frac{\partial T}{\partial p} + \alpha_T \left(T - T^s \right) = 0, \\ &\frac{\partial q}{\partial p} = g_q. \end{aligned} \tag{2.125}$$

REMARK 2.10. The equations on Γ_i (2.128) are similar to those on Γ_i for the ocean (2.35), with different values of the coefficients $\nu_{\mathbf{v}}, \nu_T, \ldots$; comparison between these two sets of boundary conditions is made in Section 2.5 devoted to the coupled atmosphere and ocean system. In (2.127) and (2.128) T_e is the (given) temperature on the Earth, and $\mathbf{v}^s$, T^s are the (given) velocity and temperature of the sea. The boundary conditions (2.126) on Γ_u are physically reasonable boundary conditions; they can be replaced by other boundary conditions (e.g., $\mathbf{v} = 0$), which can be treated mathematically in a similar manner.

Regional problems and beta-plane approximation. It is reasonable to study regional problems, in particular at mid-latitudes and, in this case we use the beta-plane approximation. In this case, as for the ocean, we use the Cartesian coordinates denoted x, y, z or x_1, x_2, x_3, and $\Omega = (f_0 + \beta y)\mathbf{k}$. The equations are exactly the same as (2.115)–(2.126), but now $\Delta = \partial^2/\partial x^2 + \partial^2/\partial y^2$, ∇ is the usual nabla vector $(\partial/\partial x, \partial/\partial y)$ and $\nabla_{\mathbf{v}} = u\partial/\partial x + v\partial/\partial y$, $\mathbf{v} = (u, v)$. The domain $\mathcal{M}$ is now some portion of the whole atmosphere:

$$\mathcal{M} = \left\{ (\varphi, \theta, p), (\theta, \varphi) \in \Gamma_i \cup \Gamma_e, \, p_0 < p < p_1 \right\},$$

where $\Gamma_i \cup \Gamma_e$ are only part of the isobar $p = p_1$. The boundary of $\mathcal{M}$ consists of Γ_u, Γ_i, Γ_e defined as before and of a lateral boundary

$$\Gamma_\ell = \left\{ (\varphi, \theta, p), \, p_0 < p < p_1, (\varphi, \theta) \in \partial\Gamma_u \right\}.$$

The boundary conditions are the same as before on Γ_u, Γ_i, Γ_e, and, on Γ_ℓ the conditions would be as follows:

Boundary conditions on Γ_ℓ.

$$\mathbf{v} = 0, \qquad \omega = 0, \qquad \frac{\partial T}{\partial \mathbf{n}_T} = 0, \qquad \frac{\partial q}{\partial \mathbf{n}_q} = 0. \tag{2.126}$$

Here $\partial/\partial\mathbf{n}_T$ and $\partial/\partial\mathbf{n}_q$ are defined as in (2.37). Comparing to (2.37), $\mathbf{v} = 0,\ \omega = 0$, is not a physically satisfactory boundary condition; we would rather assume that $(\mathbf{v}, \omega)$ has a nonzero prescribed value; however in the mathematical treatment of this boundary condition we would then recover $(\mathbf{v}, \omega) = 0$ after removing a background flow; the necessary modifications are minor.

Below we only discuss the regional case.

Prognostic and diagnostic variables. The unknown functions are regrouped in two sets: the prognostic variables $U = (\mathbf{v}, T, q)$ for which and initial value problem will be defined, and the diagnostic variables $\omega, \rho, \Phi\ (= gz)$ which can be defined, at each instant of time as functions (functionals) of the prognostic variables, using the equations and boundary conditions. In fact ω is determined in terms of $\mathbf{v}$ very much as in the case of the ocean:

$$\omega = \omega(\mathbf{v}) = -\int_{p_0}^{p} \operatorname{div}\mathbf{v}\,\mathrm{d}p', \tag{2.127}$$

with

$$\int_{p_0}^{p_1} \operatorname{div}\mathbf{v}\,\mathrm{d}p = 0. \tag{2.128}$$

Then ρ is determined by the equation of state (2.121) and Φ is a function of p and T determined by integration of (2.118):

$$\Phi = \Phi_\mathrm{s} + \int_{p}^{p_1} \frac{RT(p')}{p'}\,\mathrm{d}p'; \tag{2.129}$$

in (2.129), $\Phi_\mathrm{s} = \Phi|_{p=p_1}$ is the geopotential at $p = p_1$, that is g times the height of the isobar $p = p_1$.

Weak formulation of the PEs. For the weak formulation of the PEs, we introduce function spaces similar to those considered for the ocean, namely:

$$V = V_1 \times V_2 \times V_3, \qquad H = H_1 \times H_2 \times H_3,$$

$$V_1 = \left\{\mathbf{v} \in H^1(\mathcal{M})^2, \operatorname{div}\int_{p_0}^{p_1} \mathbf{v}\,\mathrm{d}p = 0, \mathbf{v} = 0 \text{ on } \Gamma_\mathrm{e} \cup \Gamma_\ell\right\},$$

$$V_2 = V_3 = H^1(\mathcal{M}),$$

$$H_1 = \left\{ \mathbf{v} \in L^2(\mathcal{M})^2, \operatorname{div} \int_{p_0}^{p_1} \mathbf{v}\, \mathrm{d}p = 0, \right.$$

$$\left. \mathbf{n}_H \cdot \int_{p_0}^{p_1} \mathbf{v}\, \mathrm{d}p = 0 \text{ on } \partial\Gamma_{\mathrm{u}} \text{ (i.e., on } \Gamma_\ell) \right\},$$

$$H_2 = H_3 = L^2(\mathcal{M}).$$

These spaces are endowed with scalar products similar to those for the ocean:

$$((U, \widetilde{U})) = ((\mathbf{v}, \tilde{\mathbf{v}}))_1 + K_T((T, \widetilde{T}))_2 + K_q((q, \tilde{q}))_3,$$

$$((\tilde{\mathbf{v}}, \tilde{\mathbf{v}}))_1 = \int_{\mathcal{M}} \left(\nabla \mathbf{v} \cdot \nabla \tilde{\mathbf{v}} + \left(\frac{gp}{R\overline{T}} \right)^2 \frac{\partial \mathbf{v}}{\partial p} \frac{\partial \tilde{\mathbf{v}}}{\partial p} \right) \mathrm{d}\mathcal{M},$$

$$((T, \widetilde{T}))_2 = \int_{\mathcal{M}} \left(\nabla T \cdot \nabla \widetilde{T} + \left(\frac{gp}{R\overline{T}} \right)^2 \frac{\partial T}{\partial p} \frac{\partial \widetilde{T}}{\partial p} \right) \mathrm{d}\mathcal{M} + \int_{\Gamma_{\mathrm{i}}} \alpha_T T \widetilde{T}\, \mathrm{d}\Gamma_{\mathrm{i}},$$

$$((q, \tilde{q}))_3 = \int_{\mathcal{M}} \left(\nabla q \cdot \nabla \tilde{q} + \left(\frac{gp}{R\overline{T}} \right)^2 \frac{\partial q}{\partial p} \frac{\partial \tilde{q}}{\partial p} \right) \mathrm{d}\mathcal{M} + K_q' \int_{\mathcal{M}} q\tilde{q}\, \mathrm{d}\mathcal{M},$$

$$(U, \widetilde{U})_H = \int_{\mathcal{M}} (\mathbf{v} \cdot \tilde{\mathbf{v}} + K_T T \widetilde{T} + K_q q \tilde{q})\, \mathrm{d}\mathcal{M},$$

$$\|U\| = ((U, U))^{1/2}, \quad |U|_H = (U, U)_H^{1/2}.$$

Here K_T, K_q are suitable positive constants chosen below, $K_q' > 0$ is a constant of suitable (physical) dimension. The norm on H is of course equivalent to the L^2 norm and, thanks to the Poincaré inequality, $\|\cdot\|$ is a Hilbert norm on V; more precisely, there exists a suitable constant $c_0 > 0$ (different from that in (2.52)) such that

$$|U|_H \leqslant c_0 \|U\| \quad \forall U \in V. \tag{2.130}$$

We denote by $\mathcal{V}_1$ the space of $\mathcal{C}^\infty$ ($\mathbb{R}^2$ valued) vector functions $\mathbf{v}$ which vanish in a neighborhood of $\Gamma_{\mathrm{e}} \cup \Gamma_\ell$ and such that

$$\operatorname{div} \int_{p_0}^{p_1} \mathbf{v}\, \mathrm{d}p = 0.$$

Let $\mathcal{V}_2 = \mathcal{V}_3$ be the space of $\mathcal{C}^\infty$ functions on $\mathcal{M}$, and $\mathcal{V} = \mathcal{V}_1 \times \mathcal{V}_2 \times \mathcal{V}_3$; then, as in (2.53):

$$\mathcal{V}_1 \text{ is dense in } V_1, \qquad \mathcal{V} \text{ is dense in } V. \tag{2.131}$$

We also introduce $V_{(2)}$ the closure of $\mathcal{V}$ in $H^2(\mathcal{M})^4$.

The weak formulation of the PEs of the atmosphere takes the form:

$$\begin{aligned}&\left(\frac{\mathrm{d}U}{\mathrm{d}t},\widetilde{U}\right)_H+a(U,\widetilde{U})+b(U,U,\widetilde{U})+e(U,\widetilde{U})\\&\quad=\ell(\widetilde{U})\quad\forall\widetilde{U}\in V_{(2)}.\end{aligned}\tag{2.132}$$

Here $b=b_1+K_Tb_2+K_qb_3$, and e are essentially as for the ocean, replacing, for b, $\partial/\partial z$ by $\partial/\partial p$ and $w(\mathbf{v})$ by $\omega(\mathbf{v})$. Then $a=a_1+K_Ta_2+K_qa_3$, with

$$\begin{aligned}a_1(U,\widetilde{U})=&\int_{\mathcal{M}}\left(\mu_{\mathbf{v}}\nabla\mathbf{v}\cdot\nabla\tilde{\mathbf{v}}+\nu_{\mathbf{v}}\left(\frac{gp}{R\overline{T}}\right)^2\frac{\partial\mathbf{v}}{\partial p}\frac{\partial\tilde{\mathbf{v}}}{\partial p}\right)\mathrm{d}\mathcal{M}\\&-\int_{\mathcal{M}}\left(\int_p^{p_1}\frac{RT}{p'}\,\mathrm{d}p'\,\nabla\tilde{v}\right)\mathrm{d}\mathcal{M}+\int_{\Gamma_{\mathrm{i}}}\alpha_{\mathbf{v}}\mathbf{v}\tilde{\mathbf{v}}\,\mathrm{d}\Gamma_{\mathrm{i}},\\a_2(U,\widetilde{U})=&\int_{\mathcal{M}}\left(\mu_T\nabla T\cdot\nabla\widetilde{T}+\nu_T\left(\frac{gp}{R\overline{T}}\right)^2\frac{\partial T}{\partial p}\frac{\partial\widetilde{T}}{\partial p}\right)\mathrm{d}\mathcal{M}\\&-\int_{\mathcal{M}}\frac{R\overline{T}(p)}{c_pp}\omega(\mathbf{v})\widetilde{T}\,\mathrm{d}\mathcal{M}+\int_{\Gamma_{\mathrm{i}}}\alpha_TT\widetilde{T}\,\mathrm{d}\Gamma_{\mathrm{i}},\\a_3(U,\widetilde{U})=&\int_{\mathcal{M}}\left(\mu_S\nabla q\cdot\nabla\tilde{q}+\nu_S\left(\frac{gp}{R\overline{T}}\right)^2\frac{\partial q}{\partial p}\frac{\partial\tilde{q}}{\partial p}\right)\mathrm{d}\mathcal{M}.\end{aligned}$$

Finally,

$$\begin{aligned}\ell(\widetilde{U})=&\int_{\mathcal{M}}(F_{\mathbf{v}}\tilde{v}+K_TF_T\widetilde{T}+K_qF_q\tilde{q})\,\mathrm{d}\mathcal{M}\\&+\int_{\Gamma_{\mathrm{i}}}g_{\mathbf{v}}\tilde{v}+g_T\widetilde{T}\,\mathrm{d}\Gamma_{\mathrm{i}}+\int_{\Gamma_{\mathrm{e}}}g_T\widetilde{T}\,\mathrm{d}\Gamma_{\mathrm{e}},\\g_{\mathbf{v}}=&\tau_{\mathbf{v}}+\alpha_{\mathbf{v}}\mathbf{v}^{\mathrm{s}},\qquad g_T=\alpha_TT^{\mathrm{s}}\quad\text{on }\Gamma_{\mathrm{i}},\qquad g_T=\alpha_TT_{\mathrm{e}}\quad\text{on }\Gamma_{\mathrm{e}}.\end{aligned}$$

We find that there exist $c_1,c_2>0$ such that

$$a(U,U)+c_2\int_{\mathcal{M}}q^2\,\mathrm{d}\mathcal{M}\geqslant c_1\|U\|^2\quad\forall U\in V,$$

and that

$$e(U,U)=0\quad\forall U\in V.$$

Properties of b are similar to those in Lemma 2.1, with $V_{(2)}$ defined as the closure of $\mathcal{V}$ in $H^2(\mathcal{M})^4$.

The boundary and initial value problem. As for the ocean, the weak formulation reads:

We are given $t_1 > 0$, U_0 *in* H, $F = (F_{\mathrm{v}}, F_T, F_q)$ *in* $L^2(0, t_1; H)$ (*or* $L^2(0, t_1; L^2(\mathcal{M})^4)$, g_{v} *in* $L^2(0, t_1; L^2(\Gamma_{\mathrm{i}})^2)$, g_T *in* $L^2(0, t_1; \Gamma_{\mathrm{i}} \cup \Gamma_{\mathrm{e}})$). *We look for* U:

$$U \in L^\infty(0, t_1; H) \cap L^2(0, t_1; V), \tag{2.133}$$

such that

$$\begin{aligned} &\left(\frac{\mathrm{d}U}{\mathrm{d}t}, \widetilde{U}\right)_H + a(U, \widetilde{U}) + b(U, U, \widetilde{U}) + e(U, \widetilde{U}) \\ &\quad = \ell(\widetilde{U}) \quad \forall \widetilde{U} \in V_{(2)}, \end{aligned} \tag{2.134}$$

$$U(0) = U_0. \tag{2.135}$$

REMARK 2.11. We can introduce the operators A, B, E and write (2.134) in an operator form, as (2.66).

The analog of Theorem 2.2 can be proved in exactly the same way:

THEOREM 2.3. *The domain* $\mathcal{M}$ *is a before. We are given* $t_1 > 0$, U_0 *in* H, $F = (F_{\mathrm{v}}, F_T, F_q)$ *in* $L^2(0, t_1; H)$ (*or* $L^2(0, t_1; L^2(\mathcal{M})^4)$, g_{v} *in* $L^2(0, t_1; L^2(\Gamma_{\mathrm{i}})^2)$, g_T *in* $L^2(0, t_1; \Gamma_{\mathrm{i}} \cup \Gamma_{\mathrm{e}})$). *Then there exists* U *which satisfies* (2.134) *and* (2.135). *Furthermore* U *is weakly continuous from* $[0, t_1]$ *in* H.

2.5. *The coupled atmosphere and ocean*

After considering the ocean and the atmosphere separately, we consider in this section the coupled atmosphere and ocean (CAO in short). The model presented here was first introduced in [24]; it is amenable to the mathematical and numerical analysis and is physically sound. The model was derived by carefully examining the boundary layer near the interface Γ_{i} between the ocean and the atmosphere. Although some processes are still not fully understood from the physical point of view, the derivation of the boundary condition is based on the work of Gill [10] and Haney [14].

We will present the equations and boundary conditions, the variational formulation and arrive to a point where the mathematical treatment is the same as for the ocean and the atmosphere.

The pseudo-geometrical domain. Let us first introduce the pseudo-geometrical domain. Let h_0 be a typical length (height) for the atmosphere; for harmonization with the ocean we introduce the vertical variable $\eta = z$, for $z < 0$ (in the ocean) and

$$\eta = h_0\left(\frac{p_1 - p}{p_1 - p_0}\right), \quad 0 < \eta < h_0, \tag{2.136}$$

for the atmosphere. The pseudo-geometrical domain is

$$\mathcal{M} = \mathcal{M}^{\mathrm{a}} \cup \mathcal{M}^{\mathrm{s}} \cup \Gamma_{\mathrm{i}},$$

where $\mathcal{M}^{\mathrm{s}}$ is the ocean defined as in Section 2.1.2, $\mathcal{M}^{\mathrm{a}}$ is the atmosphere, $0 < \eta < h_0$, and Γ_{i} is, as before the interface between the ocean and the atmosphere.

All quantities will now be defined as for the atmosphere alone, adding when needed, a superscript "a" or as for the ocean alone, adding a superscript "s".[4] Hence, with obvious notations, the boundary of $\mathcal{M}$ consists of

$$\Gamma_{\ell}^{\mathrm{s}} \cup \Gamma_{\mathrm{b}} \cup \Gamma_{\mathrm{e}} \cup \Gamma_{\mathrm{u}}. \tag{2.137}$$

The governing equations. In $\mathcal{M}^{\mathrm{a}}$, the variable is $U^{\mathrm{a}} = (v^{\mathrm{a}}, T^{\mathrm{a}}, q)$ and in $\mathcal{M}^{\mathrm{s}}$ the variable is $U^{\mathrm{s}} = (v^{\mathrm{s}}, T^{\mathrm{s}}, S)$; we set also $U = \{U^{\mathrm{a}}, U^{\mathrm{s}}\}$, or alternatively $v = \{v^{\mathrm{a}}, v^{\mathrm{s}}\}$, $T = \{T^{\mathrm{a}}, T^{\mathrm{s}}\}$.

The equations for U^{s} are exactly as in (2.24)–(2.29), introducing only a superscript "s" for w, p, ρ_0, F_{v}, F_T, F_S, T_{r}, S_{r}, as well as the eddy viscosity coefficients μ_{v}, ν_{v}, etc.

The equations for U^{a} are exactly as in (2.116)–(2.126), introducing again a superscript "a" for ω, p, ρ, F_{v}, F_T, F_q, as well as the various coefficients.[5] Of course the variable p is replaced by η following (2.136), and the differential operators are changed accordingly.

Boundary conditions. Except for Γ_{i}, the boundary conditions are the same as for the ocean and the atmosphere taken separately. Hence we recover the conditions (2.126) on Γ_{u} and (2.127) on Γ_{e}, adding the superscripts "a".

Let us now describe the boundary conditions on Γ_{i}. These conditions were introduced in [24]; we refer the reader to this monograph for justification and a detailed discussion.

We first have the geometrical (kinematical) condition:

$$w^{\mathrm{s}} = \omega^{\mathrm{a}} = 0 \quad \text{on } \Gamma_{\mathrm{i}}, \tag{2.138}$$

which expresses that $\eta = 0$ ($z = 0$) is indeed the upper limit of the ocean (under the rigid lid hypothesis) and $\eta = 0$ is the lower limit of the atmosphere (the isobar $p = p_1$).

Then for the velocity we express the fact that the tangential shear-stresses exerted by the atmosphere on the ocean have opposite values and vice versa, and this value is expressed as a function of the differences of velocities $\mathbf{v}^{\mathrm{a}} - \mathbf{v}^{\mathrm{s}}$ using a classical empirical model of resolution of boundary layers (see, e.g., [10,14,24]; the boundary layer model is used to model the boundary layer of the atmosphere that is most significant). These conditions read:

$$\rho_0^{\mathrm{s}} \nu_{\mathbf{v}}^{\mathrm{s}} \frac{\partial \mathbf{v}^{\mathrm{s}}}{\partial z} = -\bar{\rho}^{\mathrm{a}} \nu_{\mathbf{v}}^{\mathrm{a}} \frac{\partial \mathbf{v}^{\mathrm{a}}}{\partial z} = \bar{\rho}^{\mathrm{a}} C_D^{\mathrm{a}}(\alpha) \left(\mathbf{v}^{\mathrm{a}} - \mathbf{v}^{\mathrm{s}}\right) \left|\mathbf{v}^{\mathrm{a}} - \mathbf{v}^{\mathrm{s}}\right|^{\alpha}, \tag{2.139}$$ [6]

[4]"s" for sea, rather than "o" for ocean which could be confused with zero.

[5]An additional term linear in ω^{a} with a coefficient depending on p^{a} appears in the equation for T^{a} in [24]; see Equation (1.11), p. 4, and Footnote 2, p.15 of [24]. This term does not affect the discussion hereafter.

[6]The same equation appears in [24] with $\bar{\rho}^{\mathrm{a}}$ replaced by ρ^{a}. Replacing ρ^{a} by $\bar{\rho}^{\mathrm{a}}$ is a necessary simplification for the developments below.

Here $\alpha \geqslant 0$ and $C_D^{\mathrm{a}}(\alpha) \geqslant 0$ are coefficients from boundary layer theory, and $\bar{\rho}^{\mathrm{a}} > 0$ is an averaged value of the atmosphere density. Similar conditions hold for the temperatures, the salinity S in the ocean, and the humidity q in the atmosphere.

For the sake of simplicity, to keep the boundary condition linear, we take $\alpha = 0$; we also need to replace z by $\eta(p)$ in the atmosphere; see [24] for the details. In the end we arrive at the following conditions on Γ_{i}

$$\begin{aligned}
&w^{\mathrm{a}} = \omega^{\mathrm{s}} = 0,\\
&\rho_0^{\mathrm{s}} \nu_{\mathbf{v}}^{\mathrm{s}} \frac{\partial \mathbf{v}^{\mathrm{s}}}{\partial z} = -\bar{\rho}^{\mathrm{a}} \nu_{\mathbf{v}}^{\mathrm{a}} \left(\frac{g p^{\mathrm{a}}}{R\overline{\overline{T}}}\right)^2 \frac{\partial v^{\mathrm{a}}}{\partial p^{\mathrm{a}}} = \alpha_{\mathbf{v}} \left(\mathbf{v}^{\mathrm{a}} - \mathbf{v}^{\mathrm{s}}\right),\\
&c_p^{\mathrm{s}} \rho_0^{\mathrm{s}} \nu_T^{\mathrm{s}} \frac{\partial T^{\mathrm{s}}}{\partial z} = -c_p^{a} \bar{\rho}^{\mathrm{a}} \nu_T^{\mathrm{a}} \left(\frac{g p^{\mathrm{a}}}{R\overline{\overline{T}}}\right)^2 \frac{\partial T^{\mathrm{a}}}{\partial z} = \alpha_T \left(T^{\mathrm{a}} - T^{\mathrm{s}}\right),\\
&\frac{\partial q}{\partial p^{\mathrm{a}}} = \frac{\partial S}{\partial z} = 0.
\end{aligned} \tag{2.140}$$

Weak formulation of the PEs. For the sake of simplicity we restrict ourselves to a regional problem using the beta-plane approximation.

The function spaces that we introduce are similar to those used for the ocean and the atmosphere, hence

$$V = V_1 \times V_2 \times V_3, \qquad H = H_1 \times H_2 \times H_3,$$

where $V_i = V_i^{\mathrm{a}} \times V_i^{\mathrm{s}}$, $H_i = H_i^{\mathrm{a}} \times H_i^{\mathrm{s}}$, the spaces V_i^{a}, H_i^{a}, V_i^{s}, H_i^{s} being exactly like those of the atmosphere and the ocean, respectively. Alternatively we can write, with obvious notations, $V = V^{\mathrm{a}} \times V^{\mathrm{s}}$, $H = H^{\mathrm{a}} \times H^{\mathrm{s}}$.

These spaces are endowed with the following scalar products:

$$\begin{aligned}
&((U, \widetilde{U})) = ((U, \widetilde{U}))_{\mathrm{a}} + ((U, \widetilde{U}))_{\mathrm{s}},\\
&((U, \widetilde{U}))_{\mathrm{a}} = ((\mathbf{v}^{\mathrm{a}}, \tilde{\mathbf{v}}^{\mathrm{a}}))_{\mathrm{a},1} + K_T((T^{\mathrm{a}}, \widetilde{T}^{\mathrm{a}}))_{\mathrm{a},2} + K_q((q, \tilde{q}))_{\mathrm{a},3},\\
&((U, \widetilde{U}))_{\mathrm{s}} = ((\mathbf{v}^{\mathrm{s}}, \tilde{\mathbf{v}}^{\mathrm{s}}))_{\mathrm{s},1} + K_T((T^{\mathrm{s}}, \widetilde{T}^{\mathrm{s}}))_{\mathrm{s},2} + K_S((S, \widetilde{S}))_{\mathrm{s},3},\\
&(U, \widetilde{U})_H = (U, \widetilde{U})_{\mathrm{a}} + (U, \widetilde{U})_{\mathrm{s}},\\
&(U, \widetilde{U})_{\mathrm{a}} = \int_{\mathcal{M}^{\mathrm{a}}} \left(\mathbf{v}^{\mathrm{a}} \cdot \tilde{\mathbf{v}}^{\mathrm{a}} + K_T T^{\mathrm{a}} \widetilde{T}^{\mathrm{a}} + K_q q \tilde{q}\right) \mathrm{d}\mathcal{M},\\
&(U, \widetilde{U})_{\mathrm{s}} = \int_{\mathcal{M}^{\mathrm{s}}} \left(\mathbf{v}^{\mathrm{s}} \cdot \tilde{\mathbf{v}}^{\mathrm{s}} + K_T T^{\mathrm{s}} \widetilde{T}^{\mathrm{s}} + K_S S \widetilde{S}\right) \mathrm{d}\mathcal{M}.
\end{aligned}$$

Note that K_T is chosen the same in the atmosphere and the ocean. By the Poincaré inequality, there exists a constant $c_0 > 0$ (different than those for the ocean and the atmosphere), such that

$$|U|_H \leqslant c_0 \|U\| \quad \forall U \in V. \tag{2.141}$$

From this we conclude that $\|\cdot\|$ is a Hilbert norm on V. We also introduce, in a very similar way the spaces $\mathcal{V}$ and $V_{(2)}$.

With this, the weak formulation of the PEs of the coupled atmosphere and ocean takes the form:

$$\begin{aligned}
&\left(\frac{\mathrm{d}U}{\mathrm{d}t}, \widetilde{U}\right)_H + a(U, \widetilde{U}) + b(U, U, \widetilde{U}) + e(U, \widetilde{U}) \\
&\quad = \ell(\widetilde{U}) \quad \forall \widetilde{U} \in V_{(2)}, \qquad (2.142)\\
&U(0) = U_0. \qquad (2.143)
\end{aligned}$$

Here

$$a = a_1 + a_2 + a_3, \qquad b = b_1 + b_2 + b_3, \qquad e = e^{\mathrm{a}} + e^{\mathrm{s}},$$

where

$$\begin{aligned}
a_1 &= \bar{\rho}^{\mathrm{a}} a_1^{\mathrm{a}} + \rho_0^{\mathrm{s}} a_1^{\mathrm{s}} + \alpha_{\mathbf{v}} \int_{\Gamma_{\mathrm{i}}} \left|\mathbf{v}^{\mathrm{a}} - \mathbf{v}^{\mathrm{s}}\right|^2 \mathrm{d}\Gamma_{\mathrm{i}},\\
a_2 &= K_T c_p^{\mathrm{a}} \bar{\rho}^{\mathrm{a}} a_2^{\mathrm{a}} + K_T c_p^{\mathrm{s}} \rho_0^{\mathrm{s}} a_2^{\mathrm{s}} + \alpha_T \int_{\Gamma_{\mathrm{i}}} \left|T^{\mathrm{a}} - T^{\mathrm{s}}\right|^2 \mathrm{d}\Gamma_{\mathrm{i}},\\
a_3 &= K_q a_3^{\mathrm{a}} + K_s a_3^{\mathrm{s}},\\
b_1 &= \bar{\rho}^{\mathrm{a}} b_1^{\mathrm{a}} + \rho_0^{\mathrm{s}} b_1^{\mathrm{s}},\\
b_2 &= K_T c_p^{\mathrm{a}} \bar{\rho}^{\mathrm{a}} b_2^{\mathrm{a}} + K_T c_p^{\mathrm{s}} \rho_0^{\mathrm{s}} b_2^{\mathrm{s}},\\
b_3 &= K_q b_3^{\mathrm{a}} + K_S b_3^{\mathrm{s}}.
\end{aligned}$$

Here, of course, the forms a_i^{a}, b_i^{a}, e^{a} are those of the atmosphere, and a_i^{s}, b_i^{s}, e^{s} are those of the ocean.

The form ℓ is defined as for the ocean and the atmosphere, the terms concerning Γ_{i} being omitted. Hence

$$\begin{aligned}
\ell &= \ell^{\mathrm{a}} + \ell^{\mathrm{s}},\\
\ell^{\mathrm{a}}(\widetilde{U}) &= \int_{\mathcal{M}^{\mathrm{a}}} \left(\bar{\rho}^{\mathrm{a}} F_{\mathbf{v}}^{\mathrm{a}} \tilde{\mathbf{v}}^{\mathrm{a}} + K_T c_p^{\mathrm{a}} \bar{\rho}^{\mathrm{a}} F_T^{\mathrm{a}} \widetilde{T}^{\mathrm{a}} + K_q F_q \tilde{q}\right) \mathrm{d}\mathcal{M}^{\mathrm{a}} + \int_{\Gamma_{\mathrm{e}}} g_T^{\mathrm{a}} \widetilde{T}^{\mathrm{a}} \,\mathrm{d}\Gamma_{\mathrm{e}},\\
\ell^{\mathrm{s}}(\widetilde{U}) &= \int_{\mathcal{M}^{\mathrm{s}}} \left(\rho_0^{\mathrm{s}} F_{\mathbf{v}}^{\mathrm{s}} \tilde{\mathbf{v}}^{\mathrm{s}} + K_T c_p^{\mathrm{s}} \rho_0^{\mathrm{s}} \widetilde{T}^{\mathrm{s}} + K_S F_S \widetilde{S}\right) \mathrm{d}\mathcal{M}^{\mathrm{s}}\\
&\quad + \int_{\mathcal{M}^{\mathrm{s}}} (\beta_T T_{\mathrm{r}} - \beta_S S_{\mathrm{r}}) \nabla \cdot \tilde{\mathbf{v}}^{\mathrm{s}} \,\mathrm{d}\mathcal{M}^{\mathrm{s}}.
\end{aligned}$$

With these definitions, the properties of a, b, e, ℓ are exactly the same as for the ocean and atmosphere (separately) and we prove, exactly as before, the existence, for all time, of weak solutions:

THEOREM 2.4. *The domain $\mathcal{M}$ is as before. We are given $t_1 > 0$, U_0 in H and F in $L^2(0, t_1; H)$ (or $L^2(0, t_1; L^2(\mathcal{M})^8)$), and g_T^{a} in $L^2(0, t_1; \Gamma_{\mathrm{e}})$.*

Then there exists U which satisfies (2.142) *and* (2.143), *and*

$$U \in L^\infty(0, t_1; H) \cap L^2(0, t_1; V).$$

Furthermore U is weakly continuous from $[0, t_1]$ into H.

3. Strong solutions of the primitive equations in dimension two and three

In this section we first show, in Section 3.1, the existence, local in time, of strong solutions to the PEs in space dimension three, that is solutions whose norm in H^1 remains bounded.

Then, in Section 3.2 we consider the PEs in space dimension two in view of adapting to this case the results of Sections 2 and 3.1. The two-dimensional PEs are presented in Section 3.2.1 as well as their weak formulation (Section 3.2.2). Strong solutions are considered in Section 3.2.3 and we show that the strong solutions already considered in dimension three are now defined for all time $t > 0$.

Essential in all this section is the anisotropic treatment of the equations, the vertical direction $0z$ playing a different role than the horizontal ones ($0x$ in two dimensions, $0x$ and $0y$ in three dimensions).

3.1. *Strong solutions in space dimension three*

In this section we establish the local, in time, existence of strong solutions of the primitive equations of the ocean. The result that we obtain is similar to that for the three-dimensional Navier–Stokes equations. The analysis given in this section also applies to the primitive equations of the atmosphere and the coupled atmosphere–ocean equations using the notations and equations given in Section 2.

We first state the main result of this section (Theorem 3.1). We then prepare its proof in several steps: in Step 1, we consider the linearized primitive equations and establish the global existence of strong solutions. In Step 2 we use the solution of the linearized equations in order to reduce the primitive equations to a nonlinear evolution equation with zero initial data and homogeneous boundary conditions. We also provide the necessary a priori estimates for this new problem with zero initial data and homogeneous boundary conditions. Finally, in the last step, we actually prove Theorem 3.1; in particular, we show how one can establish the existence of solutions for this problem using the Galerkin approximation with basis consisting of the eigenvectors of A (which are in H^2, thanks to the regularity results of Section 4). We use the previous estimates and then pass to the limit.

The main result of this section is as follows:

THEOREM 3.1. *We assume that Γ_{i} is of class $\mathcal{C}^3$ and that $h : \overline{\Gamma_{\mathrm{i}}} \to \mathbb{R}_+$ is of class $\mathcal{C}^3$; we also assume* (*see* (2.92)) *that*

$$\nabla h \cdot n_{\Gamma_{\mathrm{i}}} = 0 \quad \text{on } \partial \Gamma_{\mathrm{i}}, \tag{3.1}$$

where $n_{\Gamma_{\rm i}}$ is the unit outward normal on $\partial\Gamma_{\rm i}$ (in the plane $0xy$). Furthermore, we are given U_0 in V, $F=(F_{\mathbf{v}},F_T,F_S)$ in $L^2(0,t_1;H)$ with $\partial F/\partial t$ in $L^2(0,t_1;L^2(\mathcal{M})^4)$ and $g=(g_{\mathbf{v}},g_T)$ in $L^2(0,t_1;H_0^1(\Gamma_{\rm i})^3)$ with $\partial g/\partial t$ in $L^2(0,t_1;H_0^1(\Gamma_{\rm i})^3)$.[7] *Then there exists $t_*>0$, $t_*=t_*(\|U_0\|)$, and there exists a unique solution $U=U(t)$ of the primitive equations (2.79) such that*

$$U\in C\big([0,t_*];V\big)\cap L^2\big(0,t_*;H^2(\mathcal{M})^4\big). \tag{3.2}$$

STEP 1. The first step in the proof of Theorem 3.1 is the study of the linear primitive equations of the ocean.

Hence we consider the equations (to (2.44)–(2.51)):

$$\frac{\partial\mathbf{v}^*}{\partial t}+\nabla p^*+2f\mathbf{k}\times\mathbf{v}^*-\mu_{\mathbf{v}}\Delta\mathbf{v}^*-\nu_{\mathbf{v}}\frac{\partial^2\mathbf{v}^*}{\partial z^2}=F_{\mathbf{v}}, \tag{3.3}$$

$$\frac{\partial p^*}{\partial z}=-\rho g, \tag{3.4}$$

$$\frac{\partial T^*}{\partial t}-\mu_T\Delta T^*-\nu_T\frac{\partial^2 T^*}{\partial z^2}=F_T, \tag{3.5}$$

$$\frac{\partial S^*}{\partial t}-\mu_S\Delta S^*-\nu_S\frac{\partial^2 S^*}{\partial z^2}=F_S, \tag{3.6}$$

$$\operatorname{div}\int_{-h}^0\mathbf{v}^*\,\mathrm{d}z'=0, \tag{3.7}$$

$$\int_{\mathcal{M}}S^*\,\mathrm{d}\mathcal{M}=0, \tag{3.8}$$

$$p^*=p_{\rm s}^*+g\int_z^0\rho^*\,\mathrm{d}z', \tag{3.9}$$

$$\rho^*=\rho_0\big(1-\beta_T(T^*-T_{\rm r})+\beta_S(S^*-S_{\rm r})\big), \tag{3.10}$$

with the same initial and boundary conditions as for the full nonlinear problem, that is (see (2.35)–(2.38)):

$$\begin{aligned}&\nu_{\mathbf{v}}\frac{\partial\mathbf{v}^*}{\partial z}+\alpha_{\mathbf{v}}\big(\mathbf{v}^*-\mathbf{v}^{\rm a}\big)=0,\\&\nu_T\frac{\partial T^*}{\partial z}+\alpha_T\big(T^*-T^{\rm a}\big)=0,\qquad\frac{\partial S^*}{\partial z}=0\quad\text{on }\Gamma_{\rm i},\end{aligned} \tag{3.11}$$

$$\begin{aligned}&\mathbf{v}^*=0,\\&\frac{\partial T^*}{\partial\mathbf{n}_T}=0,\qquad\frac{\partial S^*}{\partial\mathbf{n}_S}=0\quad\text{on }\Gamma_{\rm b}\cup\Gamma_\ell,\end{aligned} \tag{3.12}$$

[7] The hypotheses on $\partial F/\partial t$ and $\partial g/\partial t$ can be weakened.

$$\mathbf{v}^* = \mathbf{v}_0, \qquad T^* = T_0 \quad \text{and} \quad S^* = S_0 \quad \text{at } t = 0. \tag{3.13}$$

Comparing with the nonlinear problems (2.44)–(2.51), and using the same notations as in (2.66), (2.67), we see that $U^* = (\mathbf{v}^*, T^*, S^*)$ is solution of the following equation written in functional form:

$$\frac{\mathrm{d}U^*}{\mathrm{d}t} + AU^* + E(U^*) = \ell, \tag{3.14}$$

$$U^*(0) = U_0. \tag{3.15}$$

The right-hand side ℓ of (3.14) is exactly the same as in (2.66) (see (2.44)–(2.51) and the expression of ℓ in (2.55)).

We also consider the solution $\overline{U} = (\bar{v}, \overline{T}, \overline{S})$ of the linear stationary problem, namely

$$\begin{aligned}
&-\mu_{\mathbf{v}}\Delta\bar{\mathbf{v}} - \nu_{\mathbf{v}}\frac{\partial^2\bar{\mathbf{v}}}{\partial z^2} + \nabla\bar{p} = F_{\mathbf{v}} - 2f\mathbf{k}\times\bar{\mathbf{v}},\\
&\frac{\partial\bar{p}}{\partial z} = -\bar{\rho}g,\\
&-\mu_T\Delta\overline{T} - \nu_T\frac{\partial\overline{T}}{\partial z^2} = F_T,\\
&-\mu_{TS}\Delta\overline{S} - \nu_S\frac{\partial^2\overline{S}}{\partial z^2} = F_S,\\
&\operatorname{div}\int_{-h}^0 \bar{\mathbf{v}}\,\mathrm{d}z' = 0,\\
&\int_{\mathcal{M}} \overline{S}\,\mathrm{d}\mathcal{M} = 0,\\
&\bar{p} = \bar{p}_{\mathrm{s}} + g\int_z^0 \bar{\rho}\,\mathrm{d}z',\\
&\bar{\rho} = \rho_0\bigl(1 - \beta_T(\overline{T} - T_{\mathrm{r}}) + \beta_S(\overline{S} - S_{\mathrm{r}})\bigr),
\end{aligned} \tag{3.16}$$

with the boundary conditions

$$\begin{aligned}
&\nu_{\mathbf{v}}\frac{\partial\bar{\mathbf{v}}}{\partial z} + \alpha_{\mathbf{v}}(\bar{\mathbf{v}} - \mathbf{v}^{\mathrm{a}}) = \tau_{\mathbf{v}},\\
&\nu_T\frac{\partial\overline{T}}{\partial z} + \alpha_T(\overline{T} - T^{\mathrm{a}}) = 0, \qquad \frac{\partial\overline{S}}{\partial z} = 0 \quad \text{on } \Gamma_{\mathrm{i}},\\
&\bar{\mathbf{v}} = 0, \qquad \frac{\partial\overline{T}}{\partial\mathbf{n}_T} = 0, \qquad \frac{\partial\overline{S}}{\partial\mathbf{n}_S} = 0 \quad \text{on } \Gamma_{\mathrm{b}}\cup\Gamma_\ell.
\end{aligned} \tag{3.17}$$

Note that (as in (3.3)–(3.12)) the left- and right-hand sides in (3.16) and (3.17) depend on the time t. The existence and uniqueness for (almost) every time t for (3.16) and (3.17), follows from the Lax–Milgram theorem as explained, e.g., for the velocity $\bar{\mathbf{v}}$, in Section 4.4.1 and Proposition 4.1. Furthermore, the regularity results of Section 4 (in particular, Theorems 4.1–4.5) show that the solution belong to $H^2(\mathcal{M})$, and that

$$
\begin{aligned}
&\|\bar{\mathbf{v}}\|^2_{\mathbb{H}^2(\mathcal{M})} + \|\overline{T}\|^2_{\mathbb{H}^2(\mathcal{M})} + \|\bar{\mathbf{v}}\|^2_{\mathbb{H}^2(\mathcal{M})} \leqslant C_0\kappa_1, \\
&\kappa_1 = \kappa_1(F, \tau_{\mathrm{v}}, \mathbf{v}_{\mathrm{a}}, T_{\mathrm{a}}) = |F|^2_H + \|\tau_{\mathrm{v}}\|^2_{\mathbb{H}^1(\Gamma_{\mathrm{i}})} + \|\mathbf{v}_{\mathrm{a}}\|^2_{\mathbb{H}^1(\Gamma_{\mathrm{i}})} + \|T_{\mathrm{a}}\|^2_{\mathbb{H}^1(\Gamma_{\mathrm{i}})}.
\end{aligned}
\tag{3.18}
$$

Note that, again, each side of (3.18) depends on t, and (3.18) is valid for (almost) every t. The constant C_0 depends only on $\mathcal{M}$ according to the results of Section 4; in particular C_0 is independent of t. Hypothesis (3.1) is precisely what is needed for the utilization of Theorem 4.5 used for (3.18). It is noteworthy that we have the same estimates as (3.18) for the derivatives $\partial\bar{\mathbf{v}}/\partial t$, $\partial\overline{T}/\partial t$, $\partial\overline{S}/\partial t$, κ_1 being replaced by κ_1' which is defined similarly in terms of the time derivatives $\partial F/\partial t$, etc.

Now let $\tilde{\mathbf{v}} = \mathbf{v}^* - \bar{\mathbf{v}}$, $\tilde{p} = p^* - \bar{p}$, $\tilde{\rho} = \rho^* - \bar{\rho}$, $\widetilde{T} = T^* - \overline{T}$ and $\widetilde{S} = S^* - \overline{S}$. The equations satisfied by $(\tilde{v}, \widetilde{T}, \widetilde{S})$ are the same as (3.3)–(3.13) but with $(F_{\mathrm{v}}, F_T, F_S) = -\mathrm{d}\overline{U}/\mathrm{d}t$, $\mathbf{v}_{\mathrm{a}} = \tau_{\mathrm{v}} = T_{\mathrm{a}} = T_{\mathrm{r}} = T_{\mathrm{s}} = 0$, and with initial data

$$
\tilde{\mathbf{v}}\big|_{t=0} = \mathbf{v}_0 - \bar{\mathbf{v}}(0), \qquad \widetilde{T}\big|_{t=0} = T_0 - \overline{T}(0) \quad \text{and} \quad \widetilde{S}\big|_{t=0} = S_0 - \overline{S}(0). \tag{3.19}
$$

Comparing with the nonlinear problem (2.44)–(2.51), and using the same notation as in (2.66) and (2.67), we see that $\widetilde{U} = (\tilde{\mathbf{v}}, \widetilde{T}, \widetilde{S})$ is solution of the following equation written in functional form:

$$
\frac{\mathrm{d}\widetilde{U}}{\mathrm{d}t} + A\widetilde{U} + E(\widetilde{U}) = -\frac{\mathrm{d}\overline{U}}{\mathrm{d}t}, \tag{3.20}
$$

$$
\widetilde{U}(0) = \widetilde{U}_0 = U_0 - \overline{U}(0). \tag{3.21}
$$

Note that the contribution from ℓ vanishes (see the expression in the equation preceding (2.56)).

The existence for all time of a strong solution $\widetilde{U}$ to (3.20) and (3.21) is classical, and we recall the estimate obtained by multiplying (3.20) by $A\widetilde{U}$ and integrating in time:

$$
\sup_{0\leqslant t\leqslant t_1} \|\widetilde{U}(t)\|^2 + \int_0^{t_1} |A\widetilde{U}(s)|^2_H \,\mathrm{d}s \leqslant c\|\widetilde{U}(0)\|^2 + c\kappa_1'.
$$

Here we have used the analog of (3.18) for $\mathrm{d}\overline{U}/\mathrm{d}t$; see the comments after (3.18). From this we obtain

$$
\begin{aligned}
&\sup_{0\leqslant t\leqslant t_1} \|U^*(t)\|^2 + \int_0^{t_1} |U^*(s)|^2_{H^2(\mathcal{M})^4} \,\mathrm{d}s \\
&\quad \leqslant c\big(\|U(0)\|^2 + \|\overline{U}(0)\|^2\big) + c\int_0^{t_1} |\overline{U}(s)|^2_{H^2(\mathcal{M})^4} \,\mathrm{d}s + c\int_0^{t_1} \kappa_1'(s)\,\mathrm{d}s,
\end{aligned}
\tag{3.22}
$$

and, with (3.18), we bound the right-hand side of (3.22) by an expression κ_2 of the form:

$$\kappa_2 = c\|U_0\|^2 + c_0\left(\kappa_1(0) + \int_0^{t_1} (\kappa_1 + \kappa_1')(s)\,\mathrm{d}s\right). \tag{3.23}$$

STEP 2. We will now use $U^* = (\mathbf{v}^*, T^*, S^*)$ and write the primitive equations of the ocean using the decomposition $U = U^* + U'$; we note that $U'(0) = 0$ and that $U' = (\mathbf{v}', T', S')$ satisfies homogeneous boundary conditions of the same type as U. More precisely, starting from the functional form (2.66) and (2.67) of the equation for U and using (3.14) and (3.15), we see that U' satisfies

$$\begin{aligned}&\frac{\mathrm{d}U'}{\mathrm{d}t} + AU' + B(U', U^*) + B(U^*, U') + B(U', U') + E(U') \\ &\quad = -AU^* - B(U^*, U^*),\end{aligned} \tag{3.24}$$

$$U'(0) = 0. \tag{3.25}$$

The existence of solution in Theorem 3.1 is obtained by proving the existence of solution for this system (on some interval of time $(0, t_*)$). As usual this proof of existence is based on a priori estimates for the solutions U' of (3.24) and (3.25). Some a priori estimates can be obtained by proceeding exactly as for Theorem 2.2, but additional estimates are needed here. Essential for these new estimates is another estimate on the bilinear operator B (or the trilinear form b), which is obtained by an *anisotropic treatment of certain integrals*. We have the following result (cf. to Lemma 2.1):

LEMMA 3.1. *In space dimension three, the form b is trilinear continuous on $H^2(\mathcal{M})^4 \times H^2(\mathcal{M})^4 \times L^2(\mathcal{M})^4$ and we have*

$$\begin{aligned}\left|b(U, U^\flat, U^\sharp)\right| \leqslant c_3\big(&\|U\|_{H^1}\|U^\flat\|_{H^1}^{1/2}\|U^\flat\|_{H^2}^{1/2} \\ &+ \|U\|_{H^1}^{1/2}\|U\|_{H^2}^{1/2}\|U^\flat\|_{H^1}^{1/2}\|U^\flat\|_{H^2}^{1/2}\big)|U^\sharp|_{L^2}\end{aligned} \tag{3.26}$$

for every $(U, U^\flat, U^\sharp)$ in this space.

The proof of this lemma is given below. Using Lemma 3.1, we obtain a priori estimates on the solution U' of (3.24) and (3.25). We denote by $A^{1/2}$ the square root of A so that

$$(A^{1/2}U, A^{1/2}\widetilde{U})_H = a(U, \widetilde{U}) \quad \forall U, \widetilde{U} \in V.$$

Taking the scalar product of (3.24) with AU' in H, we obtain

$$\begin{aligned}&\frac{1}{2}\frac{\mathrm{d}}{\mathrm{d}t}|A^{1/2}U'|_H^2 + |AU'|_H^2 \\ &\quad = -b(U', U^*, AU') - b(U^*, U', AU') - b(U', U', AU') \\ &\qquad - b(U^*, U^*, AU') - (AU^*, AU')_H - (E(U'), AU')_H.\end{aligned} \tag{3.27}$$

We bound each term in the right-hand side of (3.27) as follows, using Lemma 3.1 for the b-terms:

$$
\begin{aligned}
\left|\left(AU^*, AU'\right)_H\right| &\leqslant \frac{1}{12}\left|AU'\right|_H^2 + c\left|U^*\right|_{H^2}^2,^{8} \\
\left|\left(E(U'), AU'\right)_H\right| &\leqslant \frac{1}{12}\left|AU'\right|_H^2 + c\left|U'\right|_{H}^2, \\
\left|b\left(U', U^*, AU'\right)\right| &\leqslant c\left\|U'\right\|\left\|U^*\right\|_{H^2}\left|AU'\right|_H \\
&\qquad + c\left\|U'\right\|^{1/2}\left\|U^*\right\|_{H^1}^{1/2}\left\|U^*\right\|_{H^2}^{1/2}\left|AU'\right|_H^{3/2} \\
&\leqslant \frac{1}{12}\left|AU'\right|_H^2 + c\left\|U'\right\|^2\left\|U^*\right\|_{H^2}^2\left(1 + \left\|U^*\right\|_{H^1}^2\right), \\
\left|b\left(U^*, U', AU'\right)\right| &\leqslant c\left\|U^*\right\|_{H^1}^{1/2}\left\|U^*\right\|_{H^2}^{1/2}\left\|U'\right\|^{1/2}\left|AU'\right|_H^{3/2} \\
&\leqslant \frac{1}{12}\left|AU'\right|_H^2 + c\left\|U'\right\|^2\left\|U^*\right\|_{H^1}^2\left\|U^*\right\|_{H^2}^2, \\
\left|b\left(U', U', AU'\right)\right| &\leqslant c_4\left\|U'\right\|\left|AU'\right|_H^2, \\
\left|b\left(U^*, U^*, AU'\right)\right| &\leqslant c\left\|U^*\right\|_{H^1}\left\|U^*\right\|_{H^2}\left|AU'\right|_H \\
&\leqslant \frac{1}{12}\left|AU'\right|_H^2 + c\left\|U^*\right\|_{H^1}^2\left\|U^*\right\|_{H^2}^2.
\end{aligned}
$$

Here we used the fact that the norm $|AU'|_H$ is equivalent to the norm $|U'|_{H^2}$, thanks to the results of Section 4, and (this is easy), the fact that the norm $|A^{1/2}U'|_H$ is equivalent to the norm $\|U'\| = \|U'\|_{H^1}$.

Taking all these bounds into account, we infer from (3.27) that

$$
\frac{\mathrm{d}}{\mathrm{d}t}\left|A^{1/2}U'\right|_H^2 + \left(1 - c_4\left|A^{1/2}U'\right|_H\right)\left|AU'\right|_H^2 \leqslant \lambda(t)\left|A^{1/2}U'\right|_H^2 + \mu(t), \tag{3.28}
$$

with

$$
\begin{aligned}
\lambda(t) &= c\left\|U^*\right\|_{H^2}^2 + \mu(t), \\
\mu(t) &= c\left\|U^*\right\|_{H^1}^2\left\|U^*\right\|_{H^2}^2.
\end{aligned}
$$

We infer from (3.22) and (3.23) (and from the precise expression of κ_2 in (3.23)), that λ and μ are integrable on $(0, t_1)$ and we set

$$
\kappa_3 = \int_0^{t_1} \lambda(t)\,\mathrm{d}t.
$$

[8] U^* is not in $D(A)$ and $AU' \in V'$, because U^* does not satisfy the homogeneous boundary conditions; however this bound is valid, see the details in [16].

By Gronwall's lemma and since $U'(0) = 0$, we have, on some interval of time $(0, t_*)$, and as long as $1 - c_4|A^{1/2}U'|_H \geqslant 0$,

$$
\begin{aligned}
&\left|A^{1/2}U'(t)\right|_H^2 \leqslant \left(\int_0^t \mu(s)\,\mathrm{d}s\right)\exp\left(\int_0^t \lambda(s)\,\mathrm{d}s\right),\\
&\left|A^{1/2}U'(t)\right|_H^2 \leqslant \exp\left(\int_0^t \lambda(s)\,\mathrm{d}s\right).
\end{aligned}
\tag{3.29}
$$

In fact (3.28) is valid as long as $0 < t < t_*$ where t_* is the smaller of t_1 and t^*, where t^* is either $+\infty$ or the time at which

$$
\int_0^{t^*} \lambda(s)\,\mathrm{d}s = \log\left(\frac{1}{4c_4^2\kappa_3}\right). \tag{3.30}
$$

We then have

$$
\left|A^{1/}U'(t)\right|_H^2 \leqslant \frac{1}{4c_4^2} \quad \text{for } 0 < t < t_*, \tag{3.31}
$$

and returning to (3.29) we find also a bound

$$
\int_0^{t_*} \left|AU'(t)\right|_H^2 \mathrm{d}t \leqslant \mathit{const}. \tag{3.32}
$$

STEP 3 (Proof of the existence in Theorem 3.1). As we said, the existence for Theorem 3.1 is shown by proving the existence of a solution U' of (3.24) and (3.25) in $\mathcal{C}([0, t_*]; V) \cap L^2(0, t_*; D(A))$. For that purpose we implement a Galerkin method using the eigenvectors e_j of A:

$$
Ae_j = \lambda_j e_j, \quad j \geqslant 1,\ 0 < \lambda_1 \leqslant \lambda_2 \leqslant \cdots.
$$

The results of Section 4.1 show that the e_j belong to $H^2(\mathcal{M})^4$ (since $D(A) \subset H^2(\mathcal{M})^4$). We look, for each $m > 0$ fixed, for an approximate solution

$$
U'_m = \sum_{j=1}^m \xi_{jm}(t) e_j,
$$

satisfying (cf. (3.24) and (3.25))

$$
\begin{aligned}
&\left(\frac{\mathrm{d}U'_m}{\mathrm{d}t}, e_k\right)_H + a\left(U'_m, e_k\right) + b\left(U'_m, U^*, e_k\right)\\
&\quad + b\left(U^*, U'_m, e_k\right) + b\left(U'_m, U'_m, e_k\right) + e\left(U'_m, e_k\right)\\
&\quad = -a\left(U^*, e_k\right) - b\left(U^*, U^*, e_k\right), \quad k = 1, \ldots, m,
\end{aligned}
\tag{3.33}
$$

and

$$U'_m(0) = 0.$$

Multiplying (3.33) by $\xi_{km}(t)\lambda_k$, and adding these equations for $k = 1, \dots, m$, we obtain the analogue of (3.27) for U'_m. The same calculations as above show that U'_m satisfies the same estimates independent of m as (3.31) and (3.32), with the same time t_* (also independent of m).

It is then straightforward to pass to the limit $m \to \infty$ and we obtain the existence.

STEP 4 (Proof of uniqueness in Theorem 3.1). The proof of uniqueness is easy. Consider two solutions U_1, U_2 of the primitive equations; let $U = U_1 - U_2$, and consider as above the associated functions $U'_i = U_i - U^*$, $U' = U'_1 - U'_2$. Then U' satisfies

$$\frac{\mathrm{d}U'}{\mathrm{d}t} + AU' + B(U', U_2) + B(U_2, U') + B(U', U') + E(U') = 0,$$
$$U'(0) = 0.$$

Treating this equation exactly as (3.24) we obtain an equation similar to (3.28) but with $\mu = 0$ and a different λ:

$$\frac{\mathrm{d}}{\mathrm{d}t}|A^{1/2}U'|_H^2 + (1 - c_4|A'U|_H)|AU'|_H^2 \leqslant \tilde{\lambda}(t)|A^{1/2}U'|_H^2.$$

The uniqueness follows using Gronwall's lemma.

To conclude this section and the proof of Theorem 3.1, we now prove Lemma 3.1.

PROOF OF LEMMA 3.1. We need only to show how the different integrals in b are bounded by the expressions appearing in the right-hand side of (3.26) and, in fact, we restrict ourselves to two typical terms, the other terms being treated in the same way.

The first one is bounded as follows:

$$\left|\int_{\mathcal{M}} [(\mathbf{v} \cdot \nabla)\mathbf{v}^\flat]\mathbf{v}^\sharp \,\mathrm{d}\mathcal{M}\right| \leqslant |\mathbf{v}|_{L^6}|\nabla\mathbf{v}^\flat|_{L^3}|\mathbf{v}^\sharp|_{L^2}. \tag{3.34}$$

By Sobolev embeddings and interpolation, we bound the right-hand side by

$$c\|\mathbf{v}\|_{H^1}\|\mathbf{v}^\flat\|_{H^2}^{1/2}|\mathbf{v}^\sharp|_{L^2},$$

which corresponds to the first term on the right-hand side of (3.26).

The second typical term is bounded as follows:

$$\begin{aligned}\left|\int_{\mathcal{M}} w(\mathbf{v})\frac{\partial \mathbf{v}^\flat}{\partial z}\mathbf{v}^\sharp\,\mathrm{d}\mathcal{M}\right| &= \left|\int_{\Gamma_{\mathrm{i}}}\int_{-h}^{0} w(\mathbf{v})\frac{\partial \mathbf{v}^\flat}{\partial z}\mathbf{v}^\sharp\,\mathrm{d}z\,\mathrm{d}\Gamma_{\mathrm{i}}\right| \\ &\leqslant \int_{\Gamma_{\mathrm{i}}}\left|w(\mathbf{v})\right|_{L_z^\infty}\left|\frac{\partial \mathbf{v}^\flat}{\partial z}\right|_{L_z^2}\left|\mathbf{v}^\sharp\right|_{L_z^2}\mathrm{d}\Gamma_{\mathrm{i}} \\ &\leqslant \bar{h}^{1/2}\int_{\Gamma_{\mathrm{i}}}|\operatorname{div}\mathbf{v}|_{L_z^2}\left|\frac{\partial \mathbf{v}^\flat}{\partial z}\right|_{L_z^2}\left|\mathbf{v}^\sharp\right|_{L_z^2}\mathrm{d}\Gamma_{\mathrm{i}} \\ &\leqslant \text{(with Holder's inequality)} \\ &\leqslant \bar{h}^{1/2}|\operatorname{div}\mathbf{v}|_{L^4_{\Gamma_{\mathrm{i}}}L^2_z}\left|\frac{\partial \mathbf{v}^\flat}{\partial z}\right|_{L^4_{\Gamma_{\mathrm{i}}}L^2_z}\left|\mathbf{v}^\sharp\right|_{L^2_{\Gamma_{\mathrm{i}}}L^2_z}.\end{aligned} \tag{3.35}$$

For a (scalar or vector) function ξ defined on $\mathcal{M}$, we have written ($1 \leqslant \alpha, \beta \leqslant \infty$):

$$|\xi|_{L_z^\beta} = |\xi|_{L_z^\beta}(x,y) = \left(\int_{-h(x,y)}^{0} |\xi(x,y,z)|^\beta\,\mathrm{d}z\right)^{1/\beta},$$

$$|\xi|_{L^\alpha_{\Gamma_{\mathrm{i}}}L^\beta_z} = \left||\xi|_{L_z^\beta}\right|_{L^\alpha(\Gamma_{\mathrm{i}})} = \left(\int_{\Gamma_{\mathrm{i}}} |\xi|^\alpha_{L_z^\beta}(x,y)\,\mathrm{d}\Gamma_{\mathrm{i}}\right)^{1/\alpha}.$$

Notice also that $|\xi|_{L^2_{\Gamma_{\mathrm{i}}}L^2_z} = |\xi|_{L^2(\mathcal{M})}$ and that, from the expression (2.40) of $w(\mathbf{v})$, and (2.39),

$$\left|w(\mathbf{v})\right|_{L_z^\infty} = \left|\int_z^0 \operatorname{div}\mathbf{v}\,\mathrm{d}z'\right|_{L_z^\infty} \leqslant \bar{h}^{1/2}|\operatorname{div}\mathbf{v}|_{L_z^2}.$$

Now we remember that (in space dimension two) there exists a constant $c = c(\Gamma_{\mathrm{i}})$ such that, for every function ζ in $H^1(\Gamma_{\mathrm{i}})$,

$$|\zeta|_{L^4(\Gamma_{\mathrm{i}})} \leqslant c|\zeta|^{1/2}_{L^2(\Gamma_{\mathrm{i}})}|\zeta|^{1/2}_{H^1(\Gamma_{\mathrm{i}})}, \tag{3.36}$$

from which we infer, for a function ξ as above (setting $\zeta = |\xi|_{L_z^2}$):

$$|\xi|_{L^4_{\Gamma_{\mathrm{i}}}L^2_z} \leqslant c|\xi|^{1/2}_{L^2_{\Gamma_{\mathrm{i}}}L^2_z}\left||\xi|_{L_z^2}\right|^2_{H^1(\Gamma_{\mathrm{i}})}. \tag{3.37}$$

As before $|\xi|_{L^2_{\Gamma_{\mathrm{i}}}L^2_z} = |\xi|_{L^2(\mathcal{M})}$, whereas

$$\left||\xi|_{L_z^2}\right|^2_{H^1(\Gamma_{\mathrm{i}})} = |\xi|^2_{L^2_{\Gamma_{\mathrm{i}}}L^2_z} + \left|\nabla|\xi|_{L_z^2}\right|^2_{L^2(\Gamma_{\mathrm{i}})} = |\xi|^2_{L^2(\mathcal{M})} + |\nabla\theta|^2_{L^2(\Gamma_{\mathrm{i}})}, \tag{3.38}$$

with

$$\theta = \theta(x, y) = |\xi|_{L^2_z}(x, y) = \left(\int_{-h}^{0} |\xi(x, y, z)|^2 \, dz \right)^{1/2}.$$

We intend to show that for ξ in $H^1(\mathcal{M})$ (scalar or vector ξ):

$$|\xi|_{L^4_{\Gamma_i} L^2_z} \leqslant c|\xi|^{1/2}_{L^2(\mathcal{M})} |\xi|^{1/2}_{H^1(\mathcal{M})} \tag{3.39}$$

for some suitable constant $c = c(\mathcal{M})$. For that purpose we note that

$$\nabla\theta = \left(\int_{-h}^{0} |\xi|^2 \, dz \right)^{1/2} \left\{ \int_{-h}^{0} \xi \nabla\xi \, dz + |\xi(-h)|^2 \nabla h \right\},$$

where $\xi(-h)$ and below, $\xi(z)$, are simplified notations for $\xi(x, y, -h(x, y))$ and $\xi(x, y, z)$ respectively. With the Schwarz inequality we find that pointwise a.e. (for a.e. $(x, y) \in \Gamma_i$).

$$|\nabla\theta| \leqslant |\nabla\xi|_{L^2_z} + c|\xi|^{-1}_{L^2_z} |\xi(-h)|^2. \tag{3.40}$$

We have classically:

$$|\xi(-h)|^2 = |\xi(z)|^2 - 2\int_{-h}^{z} \xi \frac{\partial\xi}{\partial z} \, dz \leqslant |\xi(z)|^2 + 2|\xi|_{L^2_z} \left| \frac{\partial\xi}{\partial z} \right|_{L^2_z},$$

and by integration in z from $-h(x, y)$ to 0:

$$h|\xi(-h)|^2 \leqslant |\xi|^2_{L^2_z} + 2\bar{h}|\xi|_{L^2_z} \left| \frac{\partial\xi}{\partial z} \right|_{L^2_z}.$$

Hence (3.40) yields pointwise a.e.

$$|\nabla\theta| \leqslant |\nabla\xi|_{L^2_z} + c|\xi|_{L^2_z} + c\left| \frac{\partial\xi}{\partial z} \right|_{L^2_z}.$$

By integration on Γ_i, we find

$$|\nabla\theta|_{L^2(\Gamma_i)} \leqslant c|\xi|_{H^1(\mathcal{M})},$$

and then (3.37) and (3.38) yield (3.39).

Having established (3.39), we return to (3.35): applying (3.39) with $\xi = \operatorname{div} \mathbf{v}$ and $\xi = \partial\tilde{\mathbf{v}}^\flat/\partial z$, we can bound the left-hand side of (3.35) by the second term on the right-hand side of (3.26), thus concluding the proof of Lemma 3.2. □

3.2. *Strong solutions of the two-dimensional primitive equations: Physical boundary conditions*

In this section, we are concerned with the global existence and the uniqueness of strong solutions of the two-dimensional primitive equations of the ocean. We will first derive the equations formally from the three-dimensional PEs under the assumption of invariance with respect to the y-variable, i.e., we will assume that the initial data, the forcing terms, as well as the depth function h are independent of the variable y. The uniqueness of weak solutions implies that the solution will be independent of y.

In Sections 3.2.1 and 3.2.2 we introduce the two-dimensional PEs and present their weak formulation. In Sections 3.2.2 and 3.2.3 we show that the strong solutions provided by an analog of Theorem 3.1 are in fact defined for all $t > 0$; this result is based on improved and more involved a priori estimates which are described hereafter.

3.2.1. *The two-dimensional primitive equations.* We assume that the domain occupied by the ocean is represented by

$$\{(x, y, z) \in \mathbb{R}^3, x \in (0, L), y \in \mathbb{R}, -h(x) < z < 0\},$$

and we denote by $\mathcal{M}$ its cross section:

$$\mathcal{M} = \{(x, z), x \in (0, L), -h(x) < z < 0\}. \tag{3.41}$$

Here L is a positive number and $h : [0, L] \to \mathbb{R}$ satisfies $h \in \mathcal{C}^3([0, L])$,

$$h(x) \geqslant \underline{h} > 0 \quad \text{for } x \in (0, L), h'(0) = h'(L) = 0. \tag{3.42}$$

By dropping all the terms containing a derivative with respect to y in the three-dimensional primitive equations (2.44)–(2.50), we obtain the following system:

$$\frac{\partial u}{\partial t} + u\frac{\partial u}{\partial x} + w\frac{\partial u}{\partial z} - \mu_{\mathrm{v}}\frac{\partial^2 u}{\partial x^2} - \nu_{\mathrm{v}}\frac{\partial^2 u}{\partial z^2} - fv + \frac{\partial p_{\mathrm{s}}}{\partial x} = g\int_z^0 \frac{\partial \rho}{\partial x}\,\mathrm{d}z' + F_u, \tag{3.43}$$

$$\frac{\partial v}{\partial t} + u\frac{\partial v}{\partial x} + w\frac{\partial v}{\partial z} - \mu_{\mathrm{v}}\frac{\partial^2 v}{\partial x^2} - \nu_{\mathrm{v}}\frac{\partial^2 v}{\partial z^2} + fu = F_v, \tag{3.44}$$

$$\frac{\partial T}{\partial t} + u\frac{\partial T}{\partial x} + w\frac{\partial T}{\partial z} - \mu_T\frac{\partial^2 T}{\partial x^2} - \nu_T\frac{\partial^2 T}{\partial z^2} = F_T, \tag{3.45}$$

$$\frac{\partial S}{\partial t} + u\frac{\partial S}{\partial x} + w\frac{\partial S}{\partial z} - \mu_S\frac{\partial^2 S}{\partial x^2} - \nu_S\frac{\partial^2 S}{\partial z^2} = F_S, \tag{3.46}$$

$$\frac{\partial u}{\partial x} + \frac{\partial w}{\partial z} = 0, \tag{3.47}$$

where

$$\rho = 1 - \beta_T(T - T_r) + \beta_S(S - S_r), \tag{3.48}$$

$$\int_{\mathcal{M}} S \, d\mathcal{M} = 0. \tag{3.49}$$

Here u and v are the two components of the horizontal velocity $\mathbf{v}$. Note that, despite y-invariance, v does not vanish in the problem of physical relevance (unlike the two-dimensional Navier–Stokes equations). The quantity p_s above is the same as p in (2.49), whereas the expression P in (2.49) has been replaced by $g \int_z^0 \rho \, dz'$, with ρ being a function of T and S through (3.48). Finally, as in the three-dimensional case, $F = (F_u, F_v, F_T, F_S)$ vanishes in the physical problem and it is added here for mathematical generality.

Boundary conditions. These equations are supplemented with the same set of boundary conditions and initial data as in Section 2. On the top boundary of $\mathcal{M}$, denoted Γ_i, $\Gamma_i = \{(x, y); x \in (0, L); z = 0\}$, we have (see also after (2.54)):

$$\begin{aligned} &\nu_v \frac{\partial u}{\partial z} + \alpha_v u = g_u, \qquad &&\nu_v \frac{\partial v}{\partial z} + \alpha_v v = g_v, \\ &\nu_T \frac{\partial T}{\partial z} + \alpha_T T = g_T, \qquad &&\frac{\partial S}{\partial z} = 0. \end{aligned} \tag{3.50}$$

On the remaining part of the boundary, we assume the Dirichlet boundary condition for the velocity and the Neumann condition for the temperature and the salinity. That is,

$$\begin{aligned} &(u, v, w) = (0, 0, 0) \quad \text{on } \Gamma_\ell \cup \Gamma_b, \\ &\frac{\partial T}{\partial n_T} = \frac{\partial S}{\partial n_S} = 0 \quad \text{on } \Gamma_\ell \cup \Gamma_b, \end{aligned} \tag{3.51}$$

where

$$\begin{aligned} &\Gamma_\ell = \{(x, y); x = 0 \text{ or } L, -h(x) < z < 0\}, \\ &\Gamma_b = \{(x, z); x \in (0, L), z = -h(x)\}. \end{aligned} \tag{3.52}$$

We also have the initial data given by

$$u|_{t=0} = u_0, \qquad v|_{t=0} = v_0, \qquad T|_{t=0} = T_0 \quad \text{and} \quad S|_{t=0} = S_0. \tag{3.53}$$

3.2.2. *Weak formulation. The main result.* We now proceed, as in Section 2, towards the weak and functional formulations of this problem (3.43)–(3.49) with some simplifications due to the invariance with respect to y, and some other aspects which are specific to dimension two.

We introduce, as in Section 2.2.1, the following spaces:

$$
\begin{aligned}
&V = V_1 \times V_2 \times V_3, \qquad H = H_1 \times H_2 \times H_3,\\
&V_1 = \left\{ \mathbf{v} = (u, v) \in H^1(\mathcal{M})^2, \frac{\partial}{\partial x}\int_{-h(x)}^{0} u(x,z)\,\mathrm{d}z = 0, \mathbf{v} = 0 \text{ on } \Gamma_\ell \cup \Gamma_\mathrm{b} \right\},\\
&V_2 = H^1(\mathcal{M}),\\
&V_3 = \dot{H}^1(\mathcal{M}) = \left\{ S \in H^1(\mathcal{M}), \int_{\mathcal{M}} S\,\mathrm{d}\mathcal{M} = 0 \right\},\\
&H_1 = \left\{ \mathbf{v} = (u, v) \in L^2(\mathcal{M})^2, \int_{-h(x)}^{0} u(x,z)\,\mathrm{d}z = 0, u = 0 \text{ on } \Gamma_\ell \right\},\\
&H_2 = L^2(\mathcal{M}),\\
&H_3 = \dot{L}^2(\mathcal{M}) = \left\{ S \in L^2(\mathcal{M}), \int_{\mathcal{M}} S\,\mathrm{d}\mathcal{M} = 0 \right\}.
\end{aligned}
\tag{3.54}
$$

The scalar products are defined exactly as in Section 2.2.1, ∇ being replaced by $\partial/\partial x$. The condition $\int_{-h(x)}^{0} u(x,z)\,\mathrm{d}z = 0$ comes from the fact that the derivative in x of this quantity vanishes (the two-dimensional analog of (2.48)) and that this quantity vanishes at $x = 0$ and L (see Section 2.2.1 for the three-dimensional analog).

We also introduce the spaces $\mathcal{V}_1, \mathcal{V}_2, \mathcal{V}_3$ and $\mathcal{V} = \mathcal{V}_1 \times \mathcal{V}_2 \times \mathcal{V}_3$ with a similar definition.

Similarly we consider the forms a, b, c, e, defined exactly as in dimension three, just deleting all quantities involving a y-derivative; the associated operators A, B, E are defined in the same way.

With these notations, the weak formulation is exactly as in dimension three (see (2.79) and (2.80) or, in operational form, (2.81) and (2.82)). There is no new difficulty in proving the analogue of Theorem 2.4 giving the existence, for all time, of weak solutions.

Similarly we can prove, exactly as in Section (3.1), an analogue of Theorem 3.1. Our aim in this section is to show that $t_* = t_1$ in space dimension two, for the t_* appearing in the statement of Theorem 3.1.

More precisely we will prove the following (cf. Theorems 3.1 and 2.2):

THEOREM 3.2. *We assume that $\mathcal{M}$ is as in* (3.41) *and that* (3.42) *is satisfied.*

We are given $t_1 > 0$, $U_0 \in V$, $F = (F_\mathbf{v}, F_T, F_S)$, and $g = (g_\mathbf{v}, g_T)$ such that F and $\mathrm{d}F/\mathrm{d}t$ are in $L^2(0, t_1; H)$ (or $L^2(0, t_1; L^2(\mathcal{M})^4)$ and g and $\mathrm{d}g/\mathrm{d}t$ are in $L^2(0, t_1; H_0^1(\Gamma_\mathrm{i})^3)$.

Then there exists a unique solution U of the primitive equations (2.79) *and* (2.80) *such that*

$$
U \in \mathcal{C}\big([0, t_1]; V\big) \cap L^2\big(0, t_1; H^2(\mathcal{M})^4\big). \tag{3.55}
$$

PROOF. The proof of uniqueness is easy and done as in Theorem 3.1 for dimension three. To prove the existence of solutions, we start from the strong solution given by the two-dimensional analogue of Theorem 3.1 and prove by contradiction that $t_* = t_1$. Indeed let us denote by $[0, t_0]$ the maximal interval of existence of a strong solution, that is,[9]

$$U \in L^\infty\big(0, t'; V\big) \tag{3.56}$$

for every $t' < t_0$ and (3.56) does not occur for $t' = t_0$, which means, in particular, that

$$\limsup_{t \to t_0 - 0} \big\|U(t)\big\| = +\infty. \tag{3.57}$$

We will show that (3.57) cannot occur: we will derive a finite bound for $\|U(t')\|$ on $[0, t_0]$, thus contradicting (3.57).

The bounds for $\|U(t)\|$ will be derived *sequentially*: we will show successively that u_z, u_x are in $L^\infty(0, t_0; L^2(\mathcal{M}))$ and $L^2(0, t_0; H^1(\mathcal{M}))$, where $\varphi_x = \partial\varphi/\partial x$ and $\varphi_z = \partial\varphi/\partial z$; then we will prove at once that $\mathbf{v}$, T and S are in $L^\infty(0, t_0; H^1)$ and $L^2(0, t_0; H^2)$. In fact, we will give the proofs for u_z, u_x, T; the other quantities being estimated in exactly the same way. For the sake of simplicity, we assume hereafter that $g = (g_{\mathbf{v}}, g_T) = 0$. When $g \neq 0$, we need to "homogenize" the boundary conditions by considering $U' = U - U^*$, with U^* defined exactly as in Section (3.1), then perform the following calculations for U'.[10]

Before we proceed, let us recall that we have already available the a priori estimates for U in $L^\infty(0, t_1; L^2)$ and $L^2(0, t_1; H^1)$ used to prove the analog of Theorem 2.2 (that is (2.93) in the discrete case). □

3.2.3. *Vertical averaging.* To derive the new a priori estimates, we need some operators related to vertical averaging that we now define.

For any function φ defined and integrable on $\mathcal{M}$, we set

$$\begin{aligned} \overline{P}\varphi(x) &= \int_{-h(x)}^0 \varphi(x, z)\, \mathrm{d}z, \\ P\varphi &= \frac{1}{h}\overline{P}\varphi, \qquad Q\varphi = \varphi - P\varphi. \end{aligned} \tag{3.58}$$

We now establish some useful properties of these operators, some simple, some more involved.

We first note that $PQ = 0$, so that,

$$\int_{-h(x)}^0 Q\varphi(x, z)\, \mathrm{d}z = 0 \quad \forall \varphi \in L^1(\mathcal{M}), \tag{3.59}$$

and

$$\int_{\mathcal{M}} (P\varphi)(x)(Q\psi)(x, z)\, \mathrm{d}x\, \mathrm{d}z = 0 \quad \forall \varphi, \psi \in L^2(\mathcal{M}). \tag{3.60}$$

[9] It is easy to see that, if $U \in L^\infty(0, t'; V)$, then $U \in L^2(0, t'; H^2(\mathcal{M})^4)$ as well.

[10] Note that, in (3.1), U^* was chosen so that the initial and boundary conditions for U' vanish. Here we do not need to homogenize the initial condition, but we can use the same U^*.

Also, for all φ sufficiently regular,

$$\begin{aligned}
\overline{P}\frac{\partial\varphi}{\partial x} &= \frac{\partial}{\partial x}\overline{P}\varphi - h'(x)\varphi\big(x,-h(x)\big),\\
\overline{P}\frac{\partial^2\varphi}{\partial x^2} &= \frac{\partial^2}{\partial x^2}\overline{P}\varphi - 2h'(x)\frac{\partial\varphi}{\partial x}\big(x,h(x)\big)\\
&\quad + h'(x)^2\frac{\partial\varphi}{\partial z}\big(x,-h(x)\big) - h''(x)\varphi\big(x,-h(x)\big).
\end{aligned}\tag{3.61}$$

Now, if φ vanishes on Γ_{b}, $\varphi(x,-h(x))=0$, $0<x<L$, then

$$\frac{\partial\varphi}{\partial x}\big(x,-h(x)\big) = h'(x)\frac{\partial\varphi}{\partial z}\big(x,-h(x)\big),\tag{3.62}$$

and hence

$$\begin{aligned}
\overline{P}\frac{\partial\varphi}{\partial x} &= \frac{\partial}{\partial x}\overline{P}\varphi,\\
\overline{P}\frac{\partial^2\varphi}{\partial x^2} &= \frac{\partial^2}{\partial x^2}\overline{P}\varphi - h'(x)\frac{\partial\varphi}{\partial x}\big(x,-h(x)\big),\\
P\frac{\partial^2\varphi}{\partial x^2}(x) &= -\frac{h'}{h}\frac{\partial\varphi}{\partial x}\big(x,-h(x)\big) = -\frac{h'^2}{h}\frac{\partial\varphi}{\partial z}\big(x,-h(x)\big).
\end{aligned}\tag{3.63}$$

Finally, the following lemma will be needed.

LEMMA 3.2. *For any $v \in H^2(\mathcal{M})$ such that*

$$\nu\frac{\partial v}{\partial z} + \alpha v = 0 \quad \text{on } \Gamma_{\mathrm{i}}, \qquad v = 0 \quad \text{on } \Gamma_\ell \cup \Gamma_{\mathrm{b}},$$

we have

$$\begin{aligned}
\int_{\mathcal{M}} \frac{\partial^2 v}{\partial x^2}\frac{\partial^2 v}{\partial z^2}\,\mathrm{d}z\,\mathrm{d}x &= \int_{\mathcal{M}} \left|\frac{\partial^2 v}{\partial x\partial z}\right|^2 \mathrm{d}z\,\mathrm{d}x + \alpha\int_0^L \left|\frac{\partial v}{\partial x}(x,0)\right|^2 \mathrm{d}x\\
&\quad - \frac{1}{2}\int_0^L h''(x)\left|\frac{\partial v}{\partial z}\big(x,-h(x)\big)\right|^2 \mathrm{d}x.
\end{aligned}\tag{3.64}$$

PROOF. We give the proof for v smooth, say $v \in \mathcal{C}^3(\mathcal{M})$; the result extends then to $v \in H^2(\mathcal{M})$ using a density argument (that we skip).

We write

$$\begin{aligned}
\frac{\partial^2 v}{\partial x^2}\frac{\partial^2 v}{\partial z^2} &= \frac{\partial}{\partial x}\left(\frac{\partial v}{\partial x}\frac{\partial^2 v}{\partial z^2}\right) - \frac{\partial v}{\partial x}\frac{\partial^3 v}{\partial x\,\partial z^2}\\
&= \frac{\partial}{\partial x}\left(\frac{\partial v}{\partial x}\frac{\partial^2 v}{\partial z^2}\right) - \frac{\partial}{\partial z}\left(\frac{\partial v}{\partial x}\frac{\partial^2 v}{\partial x\,\partial z}\right) + \left|\frac{\partial^2 v}{\partial x\,\partial z}\right|^2.
\end{aligned}\tag{3.65}$$

We first integrate in z and, taking into account (3.61) we obtain

$$\begin{aligned}
&\int_{-h}^{0} \frac{\partial^2 v}{\partial x^2}\frac{\partial^2 v}{\partial z^2}\,\mathrm{d}z \\
&\quad = I - h'(x)\frac{\partial v}{\partial x}\big(x,-h(x)\big)\frac{\partial^2 v}{\partial z^2}\big(x,-h(x)\big) - \frac{\partial v}{\partial x}(x,0)\frac{\partial^2 v}{\partial x\,\partial z}(x,0) \\
&\qquad + \frac{\partial v}{\partial x}\big(x,-h(x)\big)\frac{\partial^2 v}{\partial x\,\partial z}\big(x,-h(x)\big) + \int_{-h}^{0}\left|\frac{\partial^2 v}{\partial x\,\partial z}\right|^2 \mathrm{d}z, \qquad (3.66)\\
&I = \frac{\partial}{\partial x}\int_{-h}^{0}\frac{\partial v}{\partial x}\frac{\partial^2 v}{\partial z^2}\,\mathrm{d}z.
\end{aligned}$$

We now integrate in x. The integral of $I = I(x)$ vanishes because $v(0,z) = v(L,z) = 0$ for all z, so that

$$\left(\frac{\partial^2 v}{\partial z^2}\right)(0,z) = \left(\frac{\partial^2 v}{\partial z^2}\right)(L,z) = 0 \quad \forall z.$$

The third term on the right-hand side of (3.66) is equal to $\alpha \int_0^L |(\partial v/\partial x)(x,0)|^2\,\mathrm{d}x$. The sum of the second and fourth terms is equal to

$$\int_0^L \frac{\partial v}{\partial x}\big(x,-h(x)\big)\left(\frac{\partial^2 v}{\partial x\,\partial z} - h'\frac{\partial^2 v}{\partial z^2}\right)\big(x,-h(x)\big)\,\mathrm{d}x. \qquad (3.67)$$

Setting $\varphi(x) = (\partial v/\partial z)(x,-h(x))$, we see that

$$\varphi'(x) = \frac{\partial^2 v}{\partial x\,\partial z}\big(x,-h(x)\big) - h'(x)\frac{\partial^2 v}{\partial z^2}\big(x,-h(x)\big),$$

and since $v(x,-h(x)) = 0$, we have

$$\frac{\partial v}{\partial x}\big(x,-h(x)\big) - h'(x)\frac{\partial v}{\partial z}\big(x,-h(x)\big) = 0,$$

and the integral in (3.67) is equal to

$$\int_0^L h'(x)\varphi(x)\varphi'(x)\,\mathrm{d}x = -\frac{1}{2}\int_0^L h''(x)\varphi^2(x)\,\mathrm{d}x;$$

for the last relation we have used $\varphi(0) = \varphi(L) = 0$. The lemma is proved. □

3.2.4. *Estimates for u_z.* To show that $u_z \in L^\infty(0,t_0;L^2(\mathcal{M})) \cap L^2(0,t_0;H^1(\mathcal{M}))$, we multiply (3.43) by Qu_{zz}, integrate over $\mathcal{M}$ and integrate by parts, and remember that $Qu = u$. Thus, for each term successively, omitting the variable t, we find:

$$\begin{aligned}
\int_{\mathcal{M}} u_t Q(-u_{zz})\,\mathrm{d}\mathcal{M} &= -\int_{\mathcal{M}} u_t u_{zz}\,\mathrm{d}\mathcal{M}\\
&= -\int_0^L u_t u_z]_{-h}^0\,\mathrm{d}x + \int_{\mathcal{M}} u_{tz}u_z\,\mathrm{d}\mathcal{M}\\
&= \frac{\alpha_{\mathbf{v}}}{\nu_{\mathbf{v}}}\int_0^L u_t(x,0)u(x,0)\,\mathrm{d}x + \frac{1}{2}\frac{\mathrm{d}}{\mathrm{d}t}|u_z|_{L^2}^2\\
&= \frac{1}{2}\frac{\mathrm{d}}{\mathrm{d}t}\left(|u_z|_{L^2}^2 + \frac{\alpha_{\mathbf{v}}}{\nu_{\mathbf{v}}}|u(x,0)|_{L^2(\Gamma_{\mathrm{i}})}^2\right),
\end{aligned}$$

$$\begin{aligned}
\int_{\mathcal{M}} uu_x Q(-u_{zz})\,\mathrm{d}\mathcal{M} &= \int_{\mathcal{M}} uu_x u_{zz}\,\mathrm{d}\mathcal{M} + \int_{\mathcal{M}} uu_x Pu_{zz}\,\mathrm{d}\mathcal{M}\\
&= -\int_L^0 uu_x u_z]_{-h}^0\,\mathrm{d}x + \int_{\mathcal{M}} uu_{xz}u_z\,\mathrm{d}\mathcal{M}\\
&\quad + \int_{\mathcal{M}} u_x u_z^2\,\mathrm{d}\mathcal{M} + \int_{\mathcal{M}} uu_x Pu_{zz}\,\mathrm{d}\mathcal{M}\\
&= \frac{\alpha_{\mathbf{v}}}{\nu_{\mathbf{v}}}\int_0^L u^2(x,0)u_x(x,0)\,\mathrm{d}x + \frac{1}{2}\int_{\partial\mathcal{M}} un_x u_z^2\,\mathrm{d}(\partial\mathcal{M})\\
&\quad + \frac{1}{2}\int_{\mathcal{M}} u_x u_z^2\,\mathrm{d}\mathcal{M} + \int_{\mathcal{M}} uu_z Pu_{zz}\,\mathrm{d}\mathcal{M}\\
&= \frac{\alpha_{\mathbf{v}}}{3\nu_{\mathbf{v}}}u^3(x,0)]_0^L + \frac{1}{2}\int_{\mathcal{M}} u_x u_z^2\,\mathrm{d}\mathcal{M} + \int_{\mathcal{M}} uu_x Pu_{zz}\,\mathrm{d}\mathcal{M}\\
&= \frac{1}{2}\int_{\mathcal{M}} u_x u_z^2\,\mathrm{d}\mathcal{M} + \int_{\mathcal{M}} uu_x Pu_{zz}\,\mathrm{d}\mathcal{M}.
\end{aligned}$$

In the relations above, $n = (n_x, n_z)$ is the unit outward normal on $\partial\mathcal{M}$, and we used the fact that $un_x = 0$ on $\partial\mathcal{M}$, and that $u(0,0) = u(L,0) = 0$ because $u = 0$ on Γ_ℓ.

$$\begin{aligned}
\int_{\mathcal{M}} wu_z Q(-u_{zz})\,\mathrm{d}\mathcal{M} &= -\int_{\mathcal{M}} wu_z u_{zz}\,\mathrm{d}\mathcal{M} + \int_{\mathcal{M}} wu_z Pu_{zz}\,\mathrm{d}\mathcal{M}\\
&= -\frac{1}{2}\int_0^L wu_z^2]_{-h}^0\,\mathrm{d}x + \frac{1}{2}\int_{\mathcal{M}} w_z u_z^2\,\mathrm{d}\mathcal{M}\\
&\quad + \int_{\mathcal{M}} wu_z Pu_{zz}\,\mathrm{d}\mathcal{M}\\
&= (\text{since } w = 0 \text{ on } \Gamma_{\mathrm{i}} \text{ and } \Gamma_{\mathrm{b}} \text{ and } w_z = -u_z)\\
&= -\frac{1}{2}\int_{\mathcal{M}} u_x u_z^2\,\mathrm{d}\mathcal{M} + \int_{\mathcal{M}} wu_z Pu_{zz}\,\mathrm{d}\mathcal{M}.
\end{aligned}$$

Since Pu_{zz} is independent of z and w vanishes on Γ_{i} and Γ_{b}, we have

$$\int_{\mathcal{M}} wu_z Pw_{zz}\,\mathrm{d}\mathcal{M} = \int_0^L wu\big]_{-h}^0 Pu_{zz}\,\mathrm{d}x - \int_{\mathcal{M}} uw_z Pu_{zz}\,\mathrm{d}\mathcal{M}$$
$$= \int_{\mathcal{M}} uu_z Pu_{zz}\,\mathrm{d}\mathcal{M}.$$

Finally the last two terms add up in the following way:

$$\int_{\mathcal{M}} (uu_x + wu_z)Q(-u_{zz})\,\mathrm{d}\mathcal{M} = 2\int_{\mathcal{M}} uu_x Pu_{zz}\,\mathrm{d}\mathcal{M}$$
$$= 2\int_{\mathcal{M}} \frac{1}{h} uu_x\big[u_z(x,0) - u_z(x,h)\big]\,\mathrm{d}\mathcal{M},$$

$$\int_{\mathcal{M}} p_{sx} Q(-u_{zz})\,\mathrm{d}\mathcal{M} = 0 \quad (\text{since } p_{\mathrm{s}} \text{ is independent of } z),$$

$$-\nu_{\mathbf{v}} \int_{\mathcal{M}} u_{zz} Q(-u_{zz})\,\mathrm{d}\mathcal{M} = \nu_{\mathbf{v}}|Qu_{zz}|_{L^2}^2 = \nu_{\mathbf{v}}|u_{zz}|_{L^2}^2 - \nu_{\mathbf{v}}|Pu_{zz}|_{L^2}^2,$$

$$-\mu_{\mathbf{v}} \int_{\mathcal{M}} u_{xx} Q(-u_{zz})\,\mathrm{d}\mathcal{M} = \mu_{\mathbf{v}} \int_{\mathcal{M}} u_{xx}u_{zz}\,\mathrm{d}\mathcal{M} - \mu_{\mathbf{v}} \int_{\mathcal{M}} (Pu_{xx})u_{zz}\,\mathrm{d}\mathcal{M}.$$

Using (3.63) and Lemma 3.2, we see that this expression is equal to

$$\mu_{\mathbf{v}}|u_{xz}|_{L^2}^2 + \alpha_{\mathbf{v}}\mu_{\mathbf{v}} \int_0^L \big|u_x(x,0)\big|^2\,\mathrm{d}x$$
$$- \mu_{\mathbf{v}} \int_0^L \left(\frac{h'(x)^2}{h(x)} + \frac{1}{2}h''(x)\right)\big|u_z\big(x,-h(x)\big)\big|^2\,\mathrm{d}x$$
$$- \frac{\alpha_{\mathbf{v}}}{\mu_{\mathbf{v}}\nu_{\mathbf{v}}} \int_0^L u_z(x,-h)u(x,0)\,\mathrm{d}x.$$

The other terms are left unchanged, then gathering all these terms we find

$$\begin{aligned}
&\frac{1}{2}\frac{\mathrm{d}}{\mathrm{d}t}\left(|u_z|_{L^2(\mathcal{M})}^2 + \frac{\alpha_{\mathbf{v}}}{\nu_{\mathbf{v}}}\big|u(x,0)\big|_{L^2(\Gamma_{\mathrm{i}})}^2\right) + \mu_{\mathbf{v}}|u_{xz}|_{L^2(\mathcal{M})}^2 \\
&\quad + \alpha_{\mathbf{v}}\mu_{\mathbf{v}}\big|u_x(x,0)\big|_{L^2(\Gamma_{\mathrm{i}})}^2 + \nu_{\mathbf{v}}|u_{zz}|_{L^2(\mathcal{M})}^2 \\
&= \nu_{\mathbf{v}}|Pu_{zz}|_{L^2}^2 + \mu_{\mathbf{v}} \int_0^L \left(\frac{h'^2}{h} + \frac{1}{2}h''\right)\big|u_z\big(x,-h(x)\big)\big|^2\,\mathrm{d}x \\
&\quad + \frac{\alpha_{\mathbf{v}}}{\mu_{\mathbf{v}}\nu_{\mathbf{v}}} \int_0^L u_z(x,-h)u(x,0)\,\mathrm{d}x \\
&\quad + \int_{\mathcal{M}} fvQu_{zz}\,\mathrm{d}\mathcal{M} - g\int_{\mathcal{M}}\left(\int_z^0 \rho_x\,\mathrm{d}z'\right)u_{zz}\,\mathrm{d}\mathcal{M}.
\end{aligned} \tag{3.68}$$

We estimate the right-hand side of (3.68) as follows, c denoting a constant depending only on $\mathcal{M}$ and on the coefficients $\alpha_{\mathbf{v}}$, $\mu_{\mathbf{v}}$, $\nu_{\mathbf{v}}$,

$$\begin{aligned}
&Pu_{zz} = \frac{1}{h}\big(u_z(x,0) - u_z(x,-h)\big) = \frac{1}{h}\frac{\alpha_{\mathbf{v}}}{\nu_{\mathbf{v}}}u(x,0) - \frac{1}{h}u_z\big(x,-h(x)\big),\\
&|Pu_{zz}|_{L^2(\mathcal{M})} \leqslant c\big|u(x,0)\big|_{L^2(\Gamma_{\mathrm{i}})} + c\big|u_z\big(x,-h(x)\big)\big|_{L^2(\Gamma_{\mathrm{i}})}. \qquad (3.69)
\end{aligned}$$

The last two norms are bounded by the trace theorems:

$$\begin{aligned}
&|u|_{L^2(\Gamma_{\mathrm{i}})} \leqslant c|u|^{1/2}_{L^2(\mathcal{M})}\|u\|^{1/2}, \qquad (3.70)\\
&\big|u_z\big(x,-h(x)\big)\big|_{L^2(\Gamma_{\mathrm{i}})} \leqslant c|u_z|_{L^2(\Gamma_{\mathrm{b}})}\\
&\qquad \leqslant c|u_z|^{1/2}_{L^2(\mathcal{M})}|\nabla u_z|^{1/2}_{L^2(\mathcal{M})}\\
&\qquad \leqslant c\|u\|^{1/2}\big(|u_{zz}|^2_{L^2(\mathcal{M})} + |u_{zx}|^2_{L^2(\mathcal{M})}\big)^{1/4}, \qquad (3.71)\\
&|Pu_{zz}|^2_{L^2(\mathcal{M})} \leqslant c|u|_{L^2(\mathcal{M})}\|u\| + c\|u\|\,\|u_z\|.
\end{aligned}$$

We write also

$$\begin{aligned}
\left|\int_{\mathcal{M}} f v Q u_{zz}\,\mathrm{d}\mathcal{M}\right| &\leqslant c|v|_{L^2(\mathcal{M})}|u_{zz}|_{L^2(\mathcal{M})}\\
&\leqslant c|v|_{L^2(\mathcal{M})}\|u_z\|.
\end{aligned}$$

Finally using again (3.70) and (3.71) for the other terms on the right-hand side of (3.68), we obtain

$$\begin{aligned}
&\frac{1}{2}\frac{\mathrm{d}}{\mathrm{d}t}\left(|u_z|^2_{L^2(\mathcal{M})} + \frac{\alpha_{\mathbf{v}}}{\nu_{\mathbf{v}}}\big|u(x,0)\big|^2_{L^2(\Gamma_{\mathrm{i}})}\right) + \nu\|u_z\|^2 + \alpha_{\mathbf{v}}\mu_{\mathbf{v}}\big|u_x(x,0)\big|^2_{L^2(\Gamma_{\mathrm{i}})}\\
&\quad \leqslant c|u|_{L^2(\mathcal{M})}\|u\| + c\|u\|\,\|u_z\| + c|v|_{L^2}\|u_z\|\\
&\qquad + c\big(|T_x|_{L^2} + |S_x|_{L^2}\big)|u_{zz}|_{L^2} + |F_u|_{L^2}|u_{zz}|_{L^2}\\
&\quad \leqslant \frac{\nu}{2}\|u_z\|^2 + c\|u\|^2 + c|v|^2_{L^2} + c\|U\|^2 + c|F_u|^2_{L^2},
\end{aligned}$$

where $\nu = \min(\mu_{\mathbf{v}}, \nu_{\mathbf{v}})$. Hence

$$\frac{\mathrm{d}}{\mathrm{d}t}\left(|u_z|^2_{L^2(\mathcal{M})} + \frac{\alpha_{\mathbf{v}}}{\nu_{\mathbf{v}}}\big|u_x(x,0)\big|^2_{L^2(\Gamma_{\mathrm{i}})}\right) + \nu\|u_z\|^2 \leqslant c\|U\|^2 + c|F_u|^2_{L^2(\mathcal{M})}. \qquad (3.72)$$

Taking into account the earlier estimates of U in $L^\infty(0,t_1;H)$ and $L^2(0,t_1;V)$, we obtain an a priori bound of u_z in $L^\infty(0,t_0;L^2(\mathcal{M}))$, a first step in proving that t_0 can not be less than t_1.

REMARK 3.1. We recall that the estimates above were made under the simplifying assumption that $g=(g_{\mathbf{v}},g_T)=0$. When this is not the case, we explained that we ought to consider $U'=U-U^*$, U^* defined as in Section 3.1. Then the calculations above are made for the equation for u'. This equation will involve some additional terms such as $u^*\partial u'/\partial x$, $u'\partial u^*/\partial x$, etc.; these additional terms are estimated in a similar way, leading to the same conclusions. The same remark applies for the estimates below concerning u_x, v_z, etc.

3.2.5. *Estimates for u_x.* To show that u_x is bounded in $L^\infty(0,t_0;L^2(\mathcal{M}))\cap L^2(0,t_0;H^1(\mathcal{M}))$, we multiply (3.43) by $-u_{xx}$, integrate over $\mathcal{M}$ and integrate by parts. At any fixed time t, each term can be written as follows:

$$
\begin{aligned}
-\int_{\mathcal{M}} u_t u_{xx}\,\mathrm{d}\mathcal{M} &= -\int_{\partial\mathcal{M}} u_t u_x n_x\,\mathrm{d}(\partial\mathcal{M}) + \int_{\mathcal{M}} u_{tx}u_x\,\mathrm{d}\mathcal{M}\\
&= \frac{1}{2}\frac{\mathrm{d}}{\mathrm{d}t}\int_{\mathcal{M}} u_x^2\,\mathrm{d}\mathcal{M},\\
-\int_{\mathcal{M}} u u_x u_{xx}\,\mathrm{d}\mathcal{M} &= -\frac{1}{2}\int_{\partial\mathcal{M}} u u_x^2 n_x\,\mathrm{d}(\partial\mathcal{M}) + \frac{1}{2}\int_{\mathcal{M}} u_x^3\,\mathrm{d}\mathcal{M}\\
&= \frac{1}{2}\int_{\mathcal{M}} u_x^3\,\mathrm{d}\mathcal{M},\\
-\int_{\mathcal{M}} w u_z u_{xx}\,\mathrm{d}\mathcal{M} &= -\int_{\partial\mathcal{M}} w u_z u_x n_x\,\mathrm{d}(\partial\mathcal{M}) + \int_{\mathcal{M}} w_x u_z u_x\,\mathrm{d}\mathcal{M}\\
&\quad + \int_{\mathcal{M}} w u_{zx} u_x\,\mathrm{d}\mathcal{M}\\
&= \int_{\mathcal{M}} w_x u_z u_x\,\mathrm{d}\mathcal{M} + \frac{1}{2}\int_0^L w u_x^2\big]_{-h}^0\,\mathrm{d}x\\
&\quad - \frac{1}{2}\int_{\mathcal{M}} w_z u_x^2\,\mathrm{d}\mathcal{M}\\
&= \int_{\mathcal{M}} w_x u_z u_x\,\mathrm{d}\mathcal{M} + \frac{1}{2}\int_{\mathcal{M}} u_x^3\,\mathrm{d}\mathcal{M},\\
\nu_{\mathbf{v}}\int_{\mathcal{M}} u_{zz}u_{xx}\,\mathrm{d}\mathcal{M} &= \text{(thanks to Lemma 3.2)}\\
&= \nu_{\mathbf{v}}|u_{zx}|_{L^2}^2 + \alpha_{\mathbf{v}}\nu_{\mathbf{v}}\int_0^L \big|u_x(x,0)\big|^2\,\mathrm{d}x\\
&\quad - \frac{1}{2}\int_0^L h''(x)\big|u_z\big(x,-h(x)\big)\big|^2\,\mathrm{d}x.
\end{aligned}
$$

The other terms are left unchanged and, with $\nu = \min(\mu_{\mathbf{v}}, \nu_{\mathbf{v}})$, we arrive at

$$\begin{aligned}
&\frac{1}{2}\frac{\mathrm{d}}{\mathrm{d}t}|u_x|_{L^2}^2 + \nu\|u_x\|^2 + \alpha_{\mathbf{v}}\nu_{\mathbf{v}}\int_0^L \big|u_x(x,0)\big|^2\,\mathrm{d}x \\
&\quad = -\int_{\mathcal{M}} u_x^3\,\mathrm{d}\mathcal{M} - \int_{\mathcal{M}} w_x u_z u_x\,\mathrm{d}\mathcal{M} \\
&\qquad - \int_{\mathcal{M}} f v u_{xx}\,\mathrm{d}\mathcal{M} + \frac{1}{2}\int_0^L h''(x)\big|u_z\big(x,-h(x)\big)\big|^2\,\mathrm{d}x \\
&\qquad - g\int_{\mathcal{M}}\Big(\int_z^0 \rho_x\,\mathrm{d}z'\Big)u_{xx}\,\mathrm{d}\mathcal{M} + \int_{\mathcal{M}} F_u u_{xx}\,\mathrm{d}\mathcal{M}.
\end{aligned}$$

We write

$$\begin{aligned}
\left|\int_{\mathcal{M}} u_x^3\,\mathrm{d}\mathcal{M}\right| &= |u_x|_{L^3}^3 \\
&\leqslant \text{(by Sobolev embedding and interpolation)} \\
&\leqslant c|u_x|_{H^{1/3}}^3 \leqslant c|u_x|_{L^2}^2|u_x|_{H^1} \\
&\leqslant c\|u\|^2\|u_x\| \\
&\leqslant \frac{\nu}{10}\|u_x\|^2 + c\|u\|^4,
\end{aligned}$$

$$\begin{aligned}
\left|\int_{\mathcal{M}} w_x u_z u_x\,\mathrm{d}\mathcal{M}\right| &\leqslant |w_x|_{L^2}|u_z|_{L^4}|u_x|_{L^4} \\
&\leqslant \text{(by Sobolev embedding and interpolation)} \\
&\leqslant c|u_{xx}|_{L^2}|u_z|_{L^2}^{1/2}\|u_z\|^{1/2}|u_x|_{L^2}^{1/2}\|u_x\|^{1/2} \\
&\leqslant c|u_z|_{L^2}^{1/2}\|u_z\|^{1/2}|u_x|_{L^2}^{1/2}\|u_x\|^{3/2} \\
&\leqslant \frac{\nu}{10}\|u_x\|^2 + c|u_z|_{L^2}^2\|u_z\|^2|u_x|_{L^2}^2
\end{aligned}$$

and

$$\left|\int_{\mathcal{M}} f v u_{xx}\,\mathrm{d}\mathcal{M}\right| \leqslant |f v|_{L^2}|u_{xx}|_{L^2} \leqslant \frac{\nu_{\mathbf{v}}}{10}\|u_x\|^2 + c|v|_{L^2}^2.$$

The next terms are bounded as before and the last term is easy. Hence

$$\begin{aligned}
&\frac{\mathrm{d}}{\mathrm{d}t}|u_x|^2_{L^2(\mathcal{M})} + \nu\|u_x\|^2 + \alpha_{\mathbf{v}}\nu_{\mathbf{v}}\int_0^L |u_x(x,0)|^2\,\mathrm{d}x \\
&\quad \leqslant c\|u\|^4 + c|v|^2_{L^2} + c|u_z|^2_{L^2(\mathcal{M})}\|u_z\|^2|w_z|^2_{L^2(\mathcal{M})} \\
&\qquad + c\|u\|\|u_z\| + c\|U\|^2 + c|F_u|^2_{L^2(\mathcal{M})}.
\end{aligned} \tag{3.73}$$

Remembering that $\|u\|^2 = |u_x|^2_{L^2(\mathcal{M})} + |u_z|^2_{L^2(\mathcal{M})}$, we see that the right-hand side of (3.73) is of the form $\xi(t) + \eta(t)|u_x|^2_{L^2(\mathcal{M})}$, where ξ, η are in $L^1(0,t_1)$; for $\eta = c\|u\|^2 + c|u_z|^2_{L^2(\mathcal{M})}\|u_z\|^2$, this follows from the previous estimates on U and on u_z; similarly the contribution of $c\|u\|\|u_z\|$ to ξ is in $L^1(0,t_1)$ due to the previous results on u_z and U. Therefore the Gronwall lemma applied to (3.73) provides an a priori bound of u_x in $L^\infty(0,t_0;L^2(\mathcal{M}))$ and in $L^2(0,t_0;H^1(\mathcal{M}))$.

3.2.6. *Estimates for v, T and S.* We now prove at once that T is bounded in $L^\infty(0,t_0;H^1(\mathcal{M}))$ and $L^2(0,t_0;H^2(\mathcal{M}))$; the proof is similar for v and S, and this will thus conclude the proof of Theorem 3.2.

For this, we multiply each side of (3.45) by $A_2T = -\mu_T\,\partial^2T/\partial x^2 - \nu_T\,\partial^2T/\partial z^2$. We recall that $g_T = 0$ here; see the end of Section 3.2.2. We have

$$\begin{aligned}
-\int_{\mathcal{M}} T_t(\mu_T T_{xx} + \nu_T T_{zz})\,\mathrm{d}\mathcal{M} &= -\int_{\partial\mathcal{M}} T_t(\mu_T T_x n_x + \nu_T T_z n_z)\,\mathrm{d}(\partial\mathcal{M}) \\
&\quad + \frac{1}{2}\int_{\mathcal{M}} (\mu_T T_{tx}T_x + \nu_T T_{tz}T_z)\,\mathrm{d}\mathcal{M} \\
&= \big(\text{see the notations in (2.38) and after (2.54)}\big) \\
&= -\int_{\partial\mathcal{M}} T_t\frac{\partial T}{\partial \mathbf{n}_T}\,\mathrm{d}(\partial\mathcal{M}) + \frac{1}{2}\frac{\mathrm{d}}{\mathrm{d}t}a_2(T,T) \\
&= \big(\text{with (2.35), (2.55) and } g_T = 0\big) \\
&= \frac{1}{2}\frac{\mathrm{d}}{\mathrm{d}t}a_2(T,T) + \int_{\Gamma_{\mathrm{i}}} \alpha_T T T_t\,\mathrm{d}\Gamma_{\mathrm{i}} \\
&= \frac{1}{2}\frac{\mathrm{d}}{\mathrm{d}t}a_2(T,T) + \alpha_T|T(x,0)|^2_{L^2(\Gamma_{\mathrm{i}})}
\end{aligned}$$

Hence, we find

$$\begin{aligned}
&\frac{1}{2}\frac{\mathrm{d}}{\mathrm{d}t}\big(|A_2^{1/2}T|^2_{L^2} + \alpha_T|T(x,0)|^2_{L^2(\Gamma_{\mathrm{i}})}\big) + |A_2T|^2_{L^2} \\
&\quad = -\int_{\mathcal{M}} (uT_x + wT_z)A_2T\,\mathrm{d}\mathcal{M} + \int_{\mathcal{M}} F_T A_2T\,\mathrm{d}\mathcal{M}.
\end{aligned} \tag{3.74}$$

Each term on the right-hand side of (3.74) is bounded as follows:

$$\begin{aligned}\left|\int_{\mathcal{M}} uT_x A_2 T\,\mathrm{d}\mathcal{M}\right| &\leqslant |u|_{L^4}|T_x|_{L^4}|A_2T|_{L^2}\\ &\leqslant c|u|_{L^2}^{1/2}\|u\|^{1/2}|T_x|_{L^2}^{1/2}\|T_x\|^{1/2}|A_2T|_{L^2}\\ &\leqslant c|u|_{L^2}^{1/2}\|u\|^{1/2}|A_2^{1/2}T|_{L^2}^{1/2}|A_2T|_{L^2}^{3/2}\\ &\leqslant \frac{1}{6}|A_2T|^2 + c|u|_{L^2}^2\|u\|^2|A_2^{1/2}T|_{L^2}^2,\end{aligned}$$

$$\begin{aligned}\left|\int_{\mathcal{M}} wT_z A_2 T\,\mathrm{d}\mathcal{M}\right| &\leqslant |w|_{L^4}|T_z|_{L^4}|A_2T|_{L^2}\\ &\leqslant c|u_x|_{L^4}|T_z|_{L^4}|A_2T|_{L^2}\\ &\leqslant c\|u\|^{1/2}\|u_x\|^{1/2}|A_2^{1/2}T|_{L^2}^{1/2}|A_2T|^{3/2}\\ &\leqslant \frac{1}{6}|A_2T|^2 + c\|u\|^2\|u_x\|^2|A_2^{1/2}T|_{L^2}^2\end{aligned}$$

and

$$\begin{aligned}\left|\int_{\mathcal{M}} F_T A_2 T\,\mathrm{d}\mathcal{M}\right| &\leqslant |F_T|_{L^2}|A_2T|_{L^2}\\ &\leqslant \frac{1}{6}|A_2T|_{L^2}^2 + c|F_T|_{L^2}^2.\end{aligned}$$

Here we have used the fact (easy to prove) that $|A_2^{1/2}T|_{L^2}$ is a norm equivalent to $\|T\|$ in V_2, and the much more involved result, proved in Section 4.3, that $|A_2T|_{L^2(\mathcal{M})}$ is, on $D(A_2)$, a norm equivalent to $|T|_{H^2(\mathcal{M})}$; for the application of Theorem 4.3, we required (3.42) which is the one-dimensional analog of (4.54). Note that, as explained in Remark 4.1, we believe that this purely technical hypothesis can be removed.

With this we infer from (3.74) that

$$\begin{aligned}&\frac{\mathrm{d}}{\mathrm{d}t}\left(|A_2^{1/2}T|_{L^2(\mathcal{M})}^2 + \alpha_T|T(x,0)|_{L^2(\Gamma_\mathrm{i})}^2\right) + |A_2T|_{L^2(\mathcal{M})}^2\\ &\quad\leqslant \xi(t) + \eta(t)|A_2^{1/2}T|_{L^2(\mathcal{M})}^2,\end{aligned}\tag{3.75}$$

with $\xi = c|F_T|_{L^2(\mathcal{M})}^2$ and $\eta = c(|u|_{L^2(\mathcal{M})}^2 + \|u_x\|^2)\|u\|^2$. By assumption $\xi \in L^1(0,t_0)$ and the earlier estimates on U, u_x and u_z show that $\eta \in L^1(0,t_0)$. Then, Gronwall's lemma implies that $|A_2^{1/2}T|_{L^2(\mathcal{M})}$ is in $L^2(0,t_0)$, which means that T is in $L^\infty(0,t_0;H^1(\mathcal{M}))$ and $L^2(0,t_0;H^2(\mathcal{M}))$. This concludes the proof of Theorem 3.2.

3.3. *The space periodic case in dimension two: Higher regularities*[11]

Our aim in this section is to present some existence, uniqueness and regularity results for the PEs of the ocean in space dimension two with periodic boundary conditions. We prove the existence of weak solutions for the PEs, the existence and uniqueness of strong solutions and the existence of more regular solutions, up to $\mathcal{C}^\infty$ regularity.

For the sake of simplicity and to follow [29], we do not consider the salinity; introducing the salinity would not produce any additional technical difficulty. In this case ρ is a linear function of T,[12] and, in what follows, ρ is the prognostic variable instead of T.

Because of the hydrostatic equation it is not possible to produce a solution that is space periodic in all variables; for that reason ρ, p and T below represent the deviation from a stratified solution $\bar\rho$ for which $N^2=-(g/\rho_0)(\mathrm{d}\bar\rho/\mathrm{d}z)$ is a constant, and, as usual $\mathrm{d}\bar p/\mathrm{d}z=-g\bar\rho$ and $\bar\rho=\rho_0(1-\alpha(\overline{T}-T_0))$, ρ_0, T_0 being reference values of ρ and T (of the same order as $\bar\rho$ and $\overline{T}$). Furthermore the periodic (disturbance) solutions that we consider present certain symmetries that are described below (see (3.77)). We refer the reader to [29] for more details on the physical background. Unlike the preceding sections (but this is not important), we consider here the PEs written in nondimensional form, that is (see [29]),

$$\frac{\partial u}{\partial t}+u\frac{\partial u}{\partial x}+w\frac{\partial u}{\partial z}-\frac{1}{R_0}v+\frac{1}{R_0}\frac{\partial p}{\partial x}=\nu_{\mathrm{v}}\Delta u+F_u, \tag{3.76a}$$

$$\frac{\partial v}{\partial t}+u\frac{\partial v}{\partial x}+w\frac{\partial v}{\partial z}+\frac{1}{R_0}u=\nu_{\mathrm{v}}\Delta v+F_v, \tag{3.76b}$$

$$\frac{\partial p}{\partial z}=-\rho, \tag{3.76c}$$

$$\frac{\partial u}{\partial x}+\frac{\partial w}{\partial z}=0, \tag{3.76d}$$

$$\frac{\partial \rho}{\partial t}+u\frac{\partial \rho}{\partial x}+w\frac{\partial \rho}{\partial z}-\frac{N^2}{R_0}w=\nu_\rho\Delta\rho+F_\rho. \tag{3.76e}$$

Here (u,v,w) are the three components of the velocity vector and, as usual, we denote by p and ρ the pressure and density deviations, respectively, from the background state mentioned above. The (dimensionless) parameters are the Rossby number R_0, the Burgers number N, and the inverse (eddy) Reynolds numbers ν_{v} and ν_ρ.

[11]This section essentially reproduces the article [29], with the authorization of the publisher of the journal.

[12]In fact ρ is an affine function of T but the deviation from the density considered below is a linear function of the deviation from the temperature.

We notice easily that if u, v, ρ, w, p are solutions of (3.76) for $F = (F_u, F_v, F_\rho)$, then $\tilde{u}, \tilde{v}, \tilde{\rho}, \tilde{w}, \tilde{p}$ are solutions of (3.76) for $\widetilde{F}_u, \widetilde{F}_v, \widetilde{F}_\rho$, where

$$\begin{aligned}
&\tilde{u}(x,z,t) = u(x,-z,t),\\
&\tilde{v}(x,z,t) = v(x,-z,t),\\
&\tilde{w}(x,z,t) = -w(x,-z,t),\\
&\tilde{p}(x,z,t) = p(x,-z,t),\\
&\tilde{\rho}(x,z,t) = -\rho(x,-z,t),\\
&\widetilde{F}_u(x,z,t) = F_u(x,-z,t),\\
&\widetilde{F}_v(x,z,t) = F_v(x,-z,t),\\
&\widetilde{F}_\rho(x,z,t) = -F_\rho(x,-z,t).
\end{aligned} \tag{3.77}$$

Therefore if we assume that F_u, F_v, F_ρ are periodic and F_u, F_v are even in z and F_ρ is odd in z, then we can anticipate the existence of a solution of (3.76) such that

$$u, v, w, p, \rho \text{ are periodic in } x \text{ and } z \text{ with periods } L_1 \text{ and } L_3, \text{ and} \tag{3.78}$$

$$u, v \text{ and } p \text{ are even in } z;\ w \text{ and } \rho \text{ are odd in } z, \tag{3.79}$$

provided the initial conditions satisfy the same symmetry properties. Our aim is to solve the problem (3.76) with the periodicity and symmetry properties above and with initial data

$$u = u_0, \qquad v = v_0, \qquad \rho = \rho_0 \quad \text{at } t = 0.$$

One motivation for considering periodic boundary conditions is that they are needed in numerical studies of rotating stratified turbulence (see, e.g., [3]) and also for the study of the renormalized equations considered in [30].

The two spatial directions are $0x$ and $0z$, corresponding to the west–east and vertical directions in the so-called f-plane approximation for geophysical flows; $\Delta = \partial^2/\partial x^2 + \partial^2/\partial z^2$.

The rest of this section is organized as follows: We start in Section 3.3.1 by recalling the variational formulation of problem (3.76) under suitable assumptions and we say a few words about (the now standard) proof of existence of weak solutions for the PEs. We continue in Section 3.3.2 by proving the existence and uniqueness of strong solutions. Finally in Section 3.3.3 we prove the existence of more regular solutions, up to $\mathcal{C}^\infty$ regularity.

3.3.1. *Existence of the weak solutions for the PEs.* We work in the limited domain

$$\mathcal{M} = (0, L_1) \times (-L_3/2, L_3/2), \tag{3.80}$$

and, as mentioned, we assume space periodicity with period $\mathcal{M}$, that is, all functions are taken to satisfy $f(x + L_1, z, t) = f(x, z, t) = f(x, z + L_3, t)$ when extended to $\mathbb{R}^2$. Moreover, we assume that the symmetries (3.77) hold.

Our aim is to solve the problem (3.76) with initial data

$$u = u_0, \qquad v = v_0, \qquad \rho = \rho_0 \quad \text{at } t = 0. \tag{3.81}$$

Hence the natural function spaces for this problem are as follows:

$$V = \Big\{ (u, v, \rho) \in \big(\dot{H}^1_{\text{per}}(\mathcal{M})\big)^3,$$

$$u, v \text{ even in } z, \rho \text{ odd in } z, \int_{-L_3/2}^{L_3/2} u(x, z')\, \mathrm{d}z' = 0 \Big\}, \tag{3.82}$$

$$H = \text{closure of } V \text{ in } \big(\dot{L}^2(\mathcal{M})\big)^3. \tag{3.83}$$

Here the dot above $\dot{H}^1_{\text{per}}$ or $\dot{L}^2$ denotes the functions with average in $\mathcal{M}$ equal to zero. These spaces are endowed with Hilbert scalar products; in H the scalar product is

$$\big(U, \widetilde{U}\big)_H = \big(u, \tilde{u}\big)_{L^2} + \big(v, \tilde{v}\big)_{L^2} + \kappa\big(\rho, \tilde{\rho}\big)_{L^2}, \tag{3.84}$$

and in $\dot{H}^1_{\text{per}}$ and V the scalar product is (using the same notation when there is no ambiguity):

$$\big(\big(U, \widetilde{U}\big)\big) = \big(\big(u, \tilde{u}\big)\big) + \big(\big(v, \tilde{v}\big)\big) + \kappa\big(\big(\rho, \tilde{\rho}\big)\big), \tag{3.85}$$

where we have written $\mathrm{d}\mathcal{M}$ for $\mathrm{d}x\, \mathrm{d}z$, and

$$\big(\big(\phi, \tilde{\phi}\big)\big) = \int_{\mathcal{M}} \left(\frac{\partial \phi}{\partial x} \frac{\partial \tilde{\phi}}{\partial x} + \frac{\partial \phi}{\partial z} \frac{\partial \tilde{\phi}}{\partial z} \right) \mathrm{d}\mathcal{M}. \tag{3.86}$$

The positive constant κ is defined below. We have

$$|U|_H \leqslant c_0 \|U\| \quad \forall U \in V, \tag{3.87}$$

where $c_0 > 0$ is a positive constant related to κ and the Poincaré constant in $\dot{H}^1_{\text{per}}(\mathcal{M})$. More generally, the c_i, c_i', c_i'' will denote various positive constants. Inequality (3.87) implies that $\|U\| = ((U, U))^{1/2}$ is indeed a norm on V.

We first show how we can express the diagnostic variables w and p in terms of the prognostic variables u, v and ρ, the situation being slightly different here due to the boundary conditions. For each $U = (u, v, \rho) \in V$ we can determine uniquely $w = w(U)$ from (3.76d),

$$w(U) = w(x, z, t) = -\int_0^z u_x\big(x, z', t\big)\, \mathrm{d}z', \tag{3.88}$$

since $w(x,0)=0$, w being odd in z. Furthermore, writing that $w(x,-L_3/2,t)=w(x,L_3/2,t)$, we also have

$$\int_{-L_3/2}^{L_3/2} u_x(x,z',t)\,\mathrm{d}z'=0. \tag{3.89}$$

As for the pressure, we obtain from (3.76c),

$$p(x,z,t)=p_{\mathrm{s}}(x,t)-\int_0^z \rho(x,z',t)\,\mathrm{d}z', \tag{3.90}$$

where $p_{\mathrm{s}}=p(x,0,t)$ is the surface pressure. Thus, we can uniquely determine the pressure p in terms of ρ up to p_{s}.

It is appropriate to use Fourier series and we write, e.g., for u,

$$u(x,z,t)=\sum_{(k_1,k_3)\in\mathbb{Z}} u_{k_1,k_3}(t)\mathrm{e}^{\mathrm{i}(k_1'x+k_3'z)}, \tag{3.91}$$

where for notational conciseness we set $k_1'=2\pi k_1/L_1$ and $k_3'=2\pi k_3/L_3$. Since u is real and even in z, we have $u_{-k_1,-k_3}=\bar{u}_{k_1,k_3}=\bar{u}_{k_1,-k_3}$, where $\bar{u}$ denotes the complex conjugate of u. Regarding the pressure, we obtain from (3.76c)

$$\begin{aligned}
p(x,z,t)&=p(x,0,t)-\int_0^z \sum_{(k_1,k_3)} \rho_{k_1,k_3}\mathrm{e}^{\mathrm{i}(k_1'x+k_3'z')}\,\mathrm{d}z'\\
&=\sum_{k_1} p_{\mathrm{s}k_1}\mathrm{e}^{\mathrm{i}k_1'x}-\sum_{(k_1,k_3),k_3\neq 0}\frac{\rho_{k_1,k_3}}{\mathrm{i}k_3'}\mathrm{e}^{\mathrm{i}k_1'x}\left(\mathrm{e}^{\mathrm{i}k_3'z}-1\right)\\
&\qquad\text{[using the fact that } \rho_{k_1,0}=0,\ \rho \text{ being odd in } z]\\
&=\sum_{k_1}\left(p_{\mathrm{s}k_1}+\sum_{k_3\neq 0}\frac{\rho_{k_1,k_3}}{\mathrm{i}k_3'}\right)\mathrm{e}^{\mathrm{i}k_1'x}-\sum_{(k_1,k_3),k_3\neq 0}\frac{\rho_{k_1,k_3}}{\mathrm{i}k_3'}\mathrm{e}^{\mathrm{i}(k_1'x+k_3'z)}\\
&=\sum_{k_1} p_{\star k_1}\mathrm{e}^{\mathrm{i}k_1'x}-\sum_{(k_1,k_3),k_3\neq 0}\frac{\rho_{k_1,k_3}}{\mathrm{i}k_3'}\mathrm{e}^{\mathrm{i}(k_1'x+k_3'z)},
\end{aligned}$$

where we denoted by p_{s} the surface pressure and $p_\star=\sum_{k_1\in\mathbb{Z}} p_{\star k_1}\mathrm{e}^{\mathrm{i}k_1'x}$, which is the average of p in the vertical direction, is defined by

$$p_{\star,k_1}=p_{\mathrm{s}k_1}+\sum_{k_3\neq 0}\frac{\rho_{k_1,k_3}}{\mathrm{i}k_3'}.$$

Note that p is fully determined by ρ, up to one of the terms p_{s} or $p_\star$ which are connected by the relation above.

We now obtain the variational formulation of problem (3.76). For that purpose we consider a test function $\widetilde{U}=(\tilde{u},\tilde{v},\tilde{\rho})\in V$ and we multiply (3.76a), (3.76b) and (3.76e), respectively by $\tilde{u}$, $\tilde{v}$ and $\kappa\tilde{\rho}$, where the constant κ (which was already introduced in (3.84) and (3.85)) will be chosen later. We add the resulting equations and integrate over $\mathcal{M}$. We find

$$\frac{\mathrm{d}}{\mathrm{d}t}(U,\widetilde{U})_H + b(U,U,\widetilde{U}) + a(U,\widetilde{U}) + \frac{1}{R_0}e(U,\widetilde{U}) = (F,\widetilde{U})_H \quad \forall \widetilde{U}\in V. \tag{3.92}$$

Here we set

$$\begin{aligned}
a(U,\widetilde{U}) &= \nu_{\mathrm{v}}((u,\tilde{u})) + \nu_{\mathrm{v}}((v,\tilde{v})) + \kappa\nu_\rho((\rho,\tilde{\rho})),\\
e(U,\widetilde{U}) &= \int_{\mathcal{M}}(u\tilde{v} - v\tilde{u})\,\mathrm{d}\mathcal{M} + \int_{\mathcal{M}}(\rho\tilde{w} - \kappa N^2 w\tilde{\rho})\,\mathrm{d}\mathcal{M},\\
b(U,U^\sharp,\widetilde{U}) &= \int_{\mathcal{M}}\left(u\frac{\partial u^\sharp}{\partial x} + w(U)\frac{\partial u^\sharp}{\partial z}\right)\tilde{u}\,\mathrm{d}\mathcal{M} + \int_{\mathcal{M}}\left(u\frac{\partial v^\sharp}{\partial x} + w(U)\frac{\partial v^\sharp}{\partial z}\right)\tilde{v}\,\mathrm{d}\mathcal{M}\\
&\quad + \int_{\mathcal{M}}\left(u\frac{\partial \rho^\sharp}{\partial x} + w(U)\frac{\partial \rho^\sharp}{\partial z}\right)\tilde{\rho}\,\mathrm{d}\mathcal{M}.
\end{aligned}$$

We now choose $\kappa = 1/N^2$ and this way we find $e(U,U) = 0$. Also it can be easily seen that:

$$\begin{aligned}
&a: V\times V \to \mathbb{R} \text{ is bilinear, continuous, coercive, } a(U,U) \geqslant c_1\|U\|^2,\\
&e: V\times V \to \mathbb{R} \text{ is bilinear, continuous, } e(U,U) = 0,\\
&b \text{ is trilinear, continuous from } V\times V_2\times V \text{ into } \mathbb{R}, \text{ and}\\
&\quad \text{from } V\times V\times V_2 \text{ into } \mathbb{R},
\end{aligned}\tag{3.93}$$

where V_2 is the closure of $V\cap(H^2_{\mathrm{per}}(\mathcal{M}))^3$ in $(H^2_{\mathrm{per}}(\mathcal{M}))^3$. Furthermore,

$$\begin{aligned}
&b(U,\widetilde{U},U^\sharp) = -b(U,U^\sharp,\widetilde{U}),\\
&b(U,\widetilde{U},\widetilde{U}) = 0,
\end{aligned}\tag{3.94}$$

when $U,\widetilde{U},U^\sharp\in V$ with $\widetilde{U}$ or $U^\sharp$ in V_2. We also have the following:

LEMMA 3.3. *There exists a constant $c_2 > 0$ such that, for all $U\in V$, $\widetilde{U}\in V_2$ and $U^\sharp\in V$:*

$$\begin{aligned}
|b(U,U^\sharp,\widetilde{U})| &\leqslant c_2|U|_{L^2}^{1/2}\|U\|^{1/2}\|U^\sharp\||\widetilde{U}|_{L^2}^{1/2}\|\widetilde{U}\|^{1/2}\\
&\quad + c_2\|U\|\|U^\sharp\|^{1/2}|U^\sharp|_{V_2}^{1/2}|\widetilde{U}|_{L^2}^{1/2}\|\widetilde{U}\|^{1/2}.
\end{aligned}\tag{3.95}$$

PROOF. We only estimate two typical terms, the other terms are estimated exactly in the same way. Using the Hölder, Sobolev and interpolation inequalities, we write

$$
\begin{aligned}
\left|\int_{\mathcal{M}} u \frac{\partial u^{\sharp}}{\partial x} \tilde{u} \, \mathrm{d}\mathcal{M}\right| &\leqslant |u|_{L^4} \left|\frac{\partial u^{\sharp}}{\partial x}\right|_{L^2} |\tilde{u}|_{L^4} \\
&\leqslant c_1' |u|_{L^2}^{1/2} \|u\|^{1/2} \left|\frac{\partial u^{\sharp}}{\partial x}\right|_{L^2} |\tilde{u}|_{L^2}^{1/2} \|\tilde{u}\|^{1/2}, \\
\left|\int_{\mathcal{M}} w(U) \frac{\partial u^{\sharp}}{\partial z} \tilde{u} \, \mathrm{d}\mathcal{M}\right| &\leqslant |w(U)|_{L^2} \left|\frac{\partial u^{\sharp}}{\partial z}\right|_{L^4} |\tilde{u}|_{L^4} \\
&\leqslant c_2' \|u\| \left|\frac{\partial u^{\sharp}}{\partial z}\right|_{L^2}^{1/2} \left\|\frac{\partial u^{\sharp}}{\partial z}\right\|^{1/2} |\tilde{u}|^{1/2} \|\tilde{u}\|^{1/2};
\end{aligned}
$$

(3.95) follows from these estimates and the analogous estimates for the other terms. □

We now recall the result regarding the existence of weak solutions for the PEs of the ocean; the proof is exactly the same as that of Theorem 2.2 in space dimension three (see also Theorem 3.1 for the two-dimensional case with different boundary conditions).

THEOREM 3.3. *Given* $U_0 \in H$ *and* $F \in L^{\infty}(\mathbb{R}_+; H)$, *there exists at least one solution* U *of* (3.92), $U \in L^{\infty}(\mathbb{R}_+; H) \cap L^2(0, t_\star; V)$ $\forall t_\star > 0$, *with* $U(0) = U_0$.

As for Theorem 2.2, the proof of this theorem is based on the a priori estimates given below, which gives, as in [22], that $U \in L^{\infty}(0, t_\star; H)$ $\forall t_\star > 0$; however, as shown below, we have in fact,[13]

$$U \in L^{\infty}(\mathbb{R}_+; H).$$

Taking $\widetilde{U} = U$ in equation (3.92), after some simple computations and using (3.93), we obtain:

$$\frac{\mathrm{d}}{\mathrm{d}t}|U|_H^2 + c_0 c_1 |U|_H^2 \leqslant \frac{\mathrm{d}}{\mathrm{d}t}|U|_H^2 + c_1 \|U\|^2 \leqslant c_1' |F|_{\infty}^2, \tag{3.96}$$

where $|F|_{\infty}$ is the norm of F in $L^{\infty}(\mathbb{R}_+; H)$. Using the Gronwall inequality, we infer from (3.96) that:

$$|U(t)|_H^2 \leqslant |U(0)|_H^2 \mathrm{e}^{-c_1 c_0 t} + \frac{c_1'}{c_1 c_0}\left(1 - \mathrm{e}^{-c_1 c_0 t}\right)|F|_{\infty}^2 \quad \forall t > 0. \tag{3.97}$$

Hence

$$\limsup_{t \to \infty} |U(t)|_H^2 \leqslant \frac{c_1'}{c_1 c_0}|F|_{\infty}^2 =: r_0^2,$$

[13] The same holds in the previous cases in dimension 3 and 2, although the result was not stated in this form. At all orders, we present here results uniform in time, $t \in \mathbb{R}_+$.

and any ball $B(0, r_0')$ in H with $r_0' > r_0$ is an absorbing ball; that is, for all U_0, there exists $t_0 = t_0(|U_0|_H)$ depending increasingly on $|U_0|_H$ (and depending also on r_0', $|F|_\infty$ and other data), such that $|U(t)|_H \leqslant r_0'$ $\forall t \geqslant t_0(|U_0|_H)$. Furthermore, integrating (3.96) from t to $t + r$, with $r > 0$ arbitrarily chosen, we find

$$\int_t^{t+r} \|U(t')\|^2 \, dt' \leqslant K_1 \quad \text{for all } t \geqslant t_0(|U_0|_H), \tag{3.98}$$

where K_1 denotes a constant depending on the data but not on U_0. As mentioned before, (3.97) implies also that

$$U \in L^\infty(\mathbb{R}_+; H), \qquad |U(t)|_H \leqslant \max(|U_0|_H, r_0).$$

REMARK 3.2. We notice that, in the inviscid case ($\nu_{\mathrm{v}} = \nu_\rho = 0$ with $F = 0$), taking $\widetilde{U} = U$ in (3.92), we find, at least formally,

$$\frac{d}{dt}\left(|u|_{L^2}^2 + |v|_{L^2}^2 + \frac{1}{N^2}|\rho|_{L^2}^2\right) = 0. \tag{3.99}$$

The physical meaning of (3.99) is that the sum of the kinetic energy (given by $\frac{1}{2}(|u|_{L^2}^2 + |v|_{L^2}^2)$) and the available potential energy (given by $\frac{1}{2N^2}|\rho|_{L^2}^2$) is conserved in time. This is the physical justification of the introduction of the constant $\kappa = N^{-2}$ in (3.84).

3.3.2. *Existence and uniqueness of strong solutions for the PEs.* The solutions given by Theorem 3.3 are called weak solutions as usual. We are now interested in strong solutions (and even more regular solutions in Section 3.3.3). We use here the same terminology as before: weak solutions are those in $L^\infty(L^2) \cap L^2(H^1)$, strong solutions are those in $L^\infty(H^1) \cap L^2(H^2)$. We notice that as for Theorem 3.1, we cannot obtain directly the global existence of strong solutions for the PEs from a single a priori estimate. Instead, we will proceed as for Theorem 3.1 and derive the necessary a priori estimates by steps: we successively derive estimates in $L^\infty(L^2)$ and $L^2(H^1)$ for u_z, u_x, v_z, v_x, ρ_z and ρ_x (here the subscripts t, x, z denote differentiation). Notice that the order in which we obtain these estimates cannot be changed in the calculations below.[14]

Firstly, using (3.90) we rewrite (3.76a) as

$$\frac{\partial u}{\partial t} + u\frac{\partial u}{\partial x} + w\frac{\partial u}{\partial z} - \frac{1}{R_0}\frac{\partial p_s}{\partial x} + \frac{1}{R_0}\int_0^z \rho_x(x, z', t)\, dz' = \nu_{\mathrm{v}}\Delta u + F_u. \tag{3.100}$$

We differentiate (3.100) with respect to z and we find, with $w_z = -u_x$,

$$u_{tz} + uu_{xz} + wu_{zz} - \frac{1}{R_0}v_z - \frac{1}{R_0}\rho_x - \nu_{\mathrm{v}}u_{xxz} - \nu_{\mathrm{v}}u_{zzz} = F_{u,z},$$

[14]However, as for Theorem 3.2 we could, at once, obtain the estimates for v_x and v_z, and then for ρ_x and ρ_z.

where $F_{u,z} = \partial_z F_u = \partial F_u/\partial z$. After multiplying this equation by u_z and integrating over $\mathcal{M}$, we find:

$$\frac{1}{2}\frac{\mathrm{d}}{\mathrm{d}t}|u_z|_{L^2}^2 + \nu_{\mathrm{v}}\|u_z\|^2 + \int_{\mathcal{M}} u u_z u_{xz}\,\mathrm{d}\mathcal{M} + \int_{\mathcal{M}} w u_z u_{zz}\,\mathrm{d}\mathcal{M}$$
$$- \frac{1}{R_0}\int_{\mathcal{M}} v_z u_z\,\mathrm{d}\mathcal{M} - \frac{1}{R_0}\int_{\mathcal{M}} \rho_x u_z\,\mathrm{d}\mathcal{M} = \int_{\mathcal{M}} u_z F_{u,z}\,\mathrm{d}\mathcal{M}.$$

Integrating by parts and taking into account the periodicity and the conservation of mass equation (3.76d) we obtain:

$$\frac{1}{2}\frac{\mathrm{d}}{\mathrm{d}t}|u_z|_{L^2}^2 + \nu_{\mathrm{v}}\|u_z\|^2 - \frac{1}{R_0}\int_{\mathcal{M}} v_z u_z\,\mathrm{d}\mathcal{M} - \frac{1}{R_0}\int_{\mathcal{M}} \rho_x u_z\,\mathrm{d}\mathcal{M}$$
$$= \int_{\mathcal{M}} u_z F_{u,z}\,\mathrm{d}\mathcal{M}. \tag{3.101}$$

In all that follows $K, K', K'', \ldots$, denote constants depending on the data but not on U_0; we use the same symbol for different constants. We easily obtain the following estimates:

$$\frac{1}{R_0}\left|\int_{\mathcal{M}} v_z u_z\,\mathrm{d}\mathcal{M}\right| = \frac{1}{R_0}\left|\int_{\mathcal{M}} v u_{zz}\,\mathrm{d}\mathcal{M}\right| \leqslant K|v|_{L^2}^2 + \frac{\nu_{\mathrm{v}}}{6}\|u_z\|^2,$$
$$\frac{1}{R_0}\left|\int_{\mathcal{M}} \rho_x u_z\,\mathrm{d}\mathcal{M}\right| = \frac{1}{R_0}\left|\int_{\mathcal{M}} \rho u_{xz}\,\mathrm{d}\mathcal{M}\right| \leqslant \frac{\nu_{\mathrm{v}}}{6}\|u_z\|^2 + K|\rho|_{L^2}^2,$$
$$\left|\int_{\mathcal{M}} F_{u,z} u_z\,\mathrm{d}\mathcal{M}\right| = \left|\int_{\mathcal{M}} F_u u_{zz}\,\mathrm{d}\mathcal{M}\right| \leqslant \frac{\nu_{\mathrm{v}}}{6}\|u_z\|^2 + c_1'|F_u|_{L^2}^2.$$

Applied to (3.101), these give

$$\frac{\mathrm{d}}{\mathrm{d}t}|u_z|_{L^2}^2 + \nu_{\mathrm{v}}\|u_z\|^2 \leqslant K\left(|v|_{L^2}^2 + |\rho|_{L^2}^2\right) + c_1'|F_u|_{L^2}^2. \tag{3.102}$$

We apply the Poincaré inequality (3.87) and we find

$$\frac{\mathrm{d}}{\mathrm{d}t}|u_z|_{L^2}^2 + c_0\nu_{\mathrm{v}}|u_z|_{L^2}^2 \leqslant K\left(|v|_{L^2}^2 + |\rho|_{L^2}^2\right) + c_1'|F_u|_{L^2}^2. \tag{3.103}$$

Using Gronwall lemma, we infer from (3.103) that

$$\begin{aligned}
\left|u_z(t)\right|_{L^2}^2 &\leqslant \left|u_z(0)\right|_{L^2}^2 \mathrm{e}^{-c_0\nu_{\mathrm{v}}t} + K\mathrm{e}^{-c_0\nu_{\mathrm{v}}t}\int_0^t \left(\left|v(t')\right|_{L^2}^2 + \left|\rho(t')\right|_{L^2}^2\right)\mathrm{e}^{c_0\nu_{\mathrm{v}}t'}\,\mathrm{d}t' \\
&\quad + c_2'|F_u|_\infty^2 \\
&\leqslant \left|u_z(0)\right|_{L^2}^2 \mathrm{e}^{-c_0\nu_{\mathrm{v}}t} + K'\left(1 - \mathrm{e}^{-c_0\nu_{\mathrm{v}}t}\right)\left(|v|_\infty^2 + |\rho|_\infty^2\right) + c_2'|F_u|_\infty^2 \\
&\leqslant \left|u_z(0)\right|_{L^2}^2 \mathrm{e}^{-c_0\nu_{\mathrm{v}}t} + K'\left(|v|_\infty^2 + |\rho|_\infty^2\right) + c_2'|F_u|_\infty^2,
\end{aligned} \tag{3.104}$$

where $|v|_\infty = |v|_{L^\infty(\mathbb{R}_+;L^2(\mathcal{M}))}$, and similarly for ρ and F_u. We obtain an explicit bound for the norm of u_z in $L^\infty(\mathbb{R}_+; H)$,

$$\left|u_z(t)\right|_{L^2}^2 \leqslant \left|u_z(0)\right|_{L^2}^2 + K'\left(|v|_\infty^2 + |\rho|_\infty^2\right) + c_2'|F_u|_\infty^2. \tag{3.105}$$

For what follows, we recall here the uniform Gronwall lemma (see, e.g., [35]):

If ξ, η and y are three positive locally integrable functions on (t_1, ∞) such that y' is locally integrable on (t_1, ∞) and which satisfy

$$\begin{aligned}
&y' \leqslant \xi y + \eta, \\
&\int_t^{t+r} \xi(s)\,\mathrm{d}s \leqslant a_1, \qquad \int_t^{t+r} \eta(s)\,\mathrm{d}s \leqslant a_2, \\
&\int_t^{t+r} y(s)\,\mathrm{d}s \leqslant a_3 \quad \forall t \geqslant t_1,
\end{aligned} \tag{3.106}$$

where r, a_1, a_2 and a_3 are positive constants, then

$$y(t+r) \leqslant \left(\frac{a_3}{r} + a_2\right)\mathrm{e}^{a_1}, \quad t \geqslant t_1. \tag{3.107}$$

The bound (3.105) depends on the initial data U_0. In order to obtain a bound independent of U_0 we apply the uniform Gronwall lemma to the equation

$$\frac{\mathrm{d}}{\mathrm{d}t}|u_z|_{L^2}^2 \leqslant K\left(|v|_{L^2}^2 + |\rho|_{L^2}^2\right) + c_1'|F_u|_{L^2}^2, \tag{3.108}$$

to obtain

$$\left|u_z(t)\right| \leqslant K'\left(r, r_0'\right) \quad \forall t \geqslant t_1', \tag{3.109}$$

where $t_1' = t_0(|U_0|_{L^2}) + r$ and $r > 0$ is fixed. Integrating equation (3.102) from t to $t + r$ with $r > 0$ as before, we also find:

$$\int_t^{t+r} \left\|u_z(s)\right\|^2 \mathrm{d}s \leqslant K''\left(r, r_0'\right) \quad \forall t \geqslant t_1'. \tag{3.110}$$

We now derive the same kind of estimates for u_x: We differentiate (3.100) with respect to x and we obtain

$$\begin{aligned}
&u_{tx} + u_x^2 + uu_{xx} + wu_{xz} + w_x u_z \\
&\quad - \frac{1}{R_0}v_x + \frac{1}{R_0}p_{s,xx} + \int_z^0 \rho_{xx}\left(z'\right)\mathrm{d}z' - \nu_{\mathrm{v}}u_{xxx} - \nu_{\mathrm{v}}u_{zzx} = F_{u,x}.
\end{aligned} \tag{3.111}$$

Multiplying this equation by u_x and integrating over $\mathcal{M}$ we find, using (3.76d):

$$\frac{1}{2}\frac{\mathrm{d}}{\mathrm{d}t}|u_x|_{L^2}^2 + \int_{\mathcal{M}} u_x^3\,\mathrm{d}\mathcal{M} + \int_{\mathcal{M}} w_x u_z u_x\,\mathrm{d}\mathcal{M} - \frac{1}{R_0}\int_{\mathcal{M}} v_x u_x\,\mathrm{d}\mathcal{M} - \frac{1}{R_0}\int_{\mathcal{M}} p_{s,xx} u_x\,\mathrm{d}\mathcal{M} + \int_{\mathcal{M}}\left(\int_z^0 \rho_{xx}(z')\,\mathrm{d}z'\right)u_x\,\mathrm{d}\mathcal{M} + \nu_{\mathbf{v}}\|u_x\|^2 = \int_{\mathcal{M}} u_x F_{u,x}\,\mathrm{d}\mathcal{M}. \tag{3.112}$$

Based on the Hölder, Sobolev and interpolation inequalities, we derive the following estimates:

$$\left|\int_{\mathcal{M}} u_x^3\,\mathrm{d}\mathcal{M}\right| \leqslant |u_x|_{L^3(\mathcal{M})}^3 \leqslant c_4'|u_x|_{H^{1/3}(\mathcal{M})}^3 \leqslant c_5'|u_x|_{L^2}^2\|u_x\| \leqslant \frac{\nu_{\mathbf{v}}}{12}\|u_x\|^2 + c_6'|u_x|_{L^2}^4,$$

$$\begin{aligned}\left|\int_{\mathcal{M}} w_x u_z u_x\,\mathrm{d}\mathcal{M}\right| &\leqslant c_7'|w_x|_{L^2}|u_z|_{L^2}^{1/2}\|u_z\|^{1/2}|u_x|_{L^2}^{1/2}\|u_x\|^{1/2}\\ &\leqslant c_8'|u_{xx}|_{L^2}|u_z|_{L^2}^{1/2}\|u_z\|^{1/2}|u_x|_{L^2}^{1/2}\|u_x\|^{1/2}\\ &\leqslant \frac{\nu_{\mathbf{v}}}{12}\|u_x\|^2 + c_9'|u_z|_{L^2}^2\|u_z\|^2|u_x|_{L^2}^2.\end{aligned}$$

By the definition of V, and since p_{s} is independent of z, we find

$$\frac{1}{R_0}\left|\int_{\mathcal{M}} p_{\mathrm{s},xx} u_x\,\mathrm{d}\mathcal{M}\right| = \frac{1}{R_0}\left|\int_0^L p_{\mathrm{s},xx}\int_{-L_3/2}^{L_3/2} u_x\,\mathrm{d}z\,\mathrm{d}x\right| = 0.$$

We can also prove the following estimates:

$$\frac{1}{R_0}\left|\int_{\mathcal{M}} v_x u_x\,\mathrm{d}\mathcal{M}\right| \leqslant \frac{\nu_{\mathbf{v}}}{12}\|u_x\|^2 + K'|v|_{L^2}^2,$$

$$\begin{aligned}\frac{1}{R_0}\left|\int_{\mathcal{M}}\left(\int_z^0 \rho_{xx}(z')\,\mathrm{d}z'\right)u_x\,\mathrm{d}\mathcal{M}\right| &= \frac{1}{R_0}\left|\int_{\mathcal{M}}\left(\int_z^0 \rho_x(z')\,\mathrm{d}z'\right)u_{xx}\,\mathrm{d}\mathcal{M}\right|\\ &\leqslant \frac{\nu_{\mathbf{v}}}{12}\|u_x\|^2 + K''|\rho_x|_{L^2}^2,\end{aligned}$$

$$\left|\int_{\mathcal{M}} u_x F_{u,x}\,\mathrm{d}\mathcal{M}\right| \leqslant \frac{\nu_{\mathbf{v}}}{12}\|u_x\|^2 + c_{10}'|F_u|_\infty^2.$$

With these relations (3.112) implies

$$\frac{\mathrm{d}}{\mathrm{d}t}|u_x|_{L^2}^2 + \nu_{\mathrm{v}}\|u_x\|^2 \leqslant \xi|u_x|_{L^2}^2 + \eta, \tag{3.113}$$

where we denoted

$$\xi = \xi(t) = 2c_6'|u_x|_{L^2}^2 + 2c_9'|u_z|_{L^2}^2\|u_z\|^2,$$

and

$$\eta = \eta(t) = 2K'|v|_{L^2}^2 + 2K''|\rho_x|_{L^2}^2 + 2c_{10}'|F_u|_\infty^2.$$

We easily conclude from (3.113) that

$$u_x \in L^\infty\big(0, t_\star; L^2\big) \cap L^2\big(0, t_\star; H^1\big) \quad \forall t_\star > 0. \tag{3.114}$$

However, for later purposes, (3.114) is not sufficient, and we need estimates uniform in time.

We will apply the uniform Gronwall lemma to (3.113) with $t_1 = t_1'$ as in (3.109). Noting that

$$\begin{aligned}\int_t^{t+r} \xi\big(t'\big)\,\mathrm{d}t' &= \int_t^{t+r} \big[2c_6'|u_x|_{L^2}^2 + 2c_9'\big|u_z\big(t'\big)\big|_{L^2}^2\big\|u_z\big(t'\big)\big\|^2\big]\,\mathrm{d}t' \\ &\leqslant 2c_6' \int_t^{t+r} \big|u_x\big(t'\big)\big|_{L^2}^2\,\mathrm{d}t' + 2c_9'|u_z|_\infty^2 \int_t^{t+r} \big\|u_z\big(t'\big)\big\|^2\,\mathrm{d}t' \\ &\leqslant a_1 \qquad \forall t \geqslant t_1',\end{aligned} \tag{3.115}$$

$$\begin{aligned}\int_t^{t+r} \eta\big(t'\big)\,\mathrm{d}t' &= \int_t^{t+r} \big[2K'|v|_{L^2}^2 + 2K''|\rho_x|_{L^2}^2 + 2c_{10}'|F_u|_\infty^2\big]\,\mathrm{d}t' \\ &\leqslant K + 2c_{10}'r|F_u|_\infty^2 \\ &= a_2 \qquad \forall t \geqslant t_1',\end{aligned} \tag{3.116}$$

$$\int_t^{t+r} \big|u_x\big(t'\big)\big|_{L^2}^2\,\mathrm{d}t' \leqslant a_3 \quad \forall t \geqslant t_1', \tag{3.117}$$

(3.107) then yields

$$\big|u_x(t)\big|_{L^2}^2 \leqslant \left(\frac{a_3}{r} + a_2\right)\mathrm{e}^{a_1} \quad \forall t \geqslant t_1' + r, \tag{3.118}$$

and thus

$$|u_x|_{L^2} \in L^\infty(\mathbb{R}_+). \tag{3.119}$$

Note that in (3.115)–(3.117) we can use bounds on $|u_z|_\infty$ (and other similar terms) independent of U_0, since $t \geqslant t_0(|U_0|_{L^2}) + r$. Integrating (3.113) from 0 to $t_1' + r$ where $t_1' = t_1'(|U_0|_{L^2})$, we obtain a bound for u_x in $L^2(0, t_1' + r; H^1)$ which depends on $\|U_0\|$. A bound independent of U_0 is obtained if we work with $t \geqslant t_1' + r = t_1'' = t_1''(|U_0|_{L^2})$: Integrating (3.113) from t to $t + r$ with r as before, we find

$$\int_t^{t+r} \|u_x(s)\|^2 \, \mathrm{d}s \leqslant K \quad \forall t \geqslant t_1''. \tag{3.120}$$

We perform similar computations for v_z: We differentiate (3.76b) with respect to z, multiply the resulting equation by v_z and integrate over $\mathcal{M}$. Using again the conservation of mass relation, we arrive at

$$\begin{aligned} &\frac{1}{2}\frac{\mathrm{d}}{\mathrm{d}t}|v_z|_{L^2}^2 + \int_{\mathcal{M}} u_z v_x v_z \, \mathrm{d}\mathcal{M} + \int_{\mathcal{M}} w_z v_z^2 \, \mathrm{d}\mathcal{M} \\ &\quad + \frac{1}{R_0}\int_{\mathcal{M}} u_z v_z \, \mathrm{d}\mathcal{M} + \nu_{\mathrm{v}} \|v_z\|^2 = \int_{\mathcal{M}} v_z F_{u,z} \, \mathrm{d}\mathcal{M}. \end{aligned} \tag{3.121}$$

We notice the following estimate

$$\begin{aligned} \left| \int_{\mathcal{M}} u_z v_x v_z \, \mathrm{d}\mathcal{M} \right| &\leqslant c_{11}' |u_z|_{L^2}^{1/2} \|u_z\|^{1/2} |v_x|_{L^2} |v_z|_{L^2}^{1/2} \|v_z\|^{1/2} \\ &\leqslant \frac{\nu_{\mathrm{v}}}{8} \|v_z\|^2 + c_{12}' |u_z|_{L^2}^{2/3} \|u_z\|^{2/3} |v_x|_{L^2}^{4/3} |v_z|_{L^2}^{2/3} \\ &\leqslant \frac{\nu_{\mathrm{v}}}{8} \|v_z\|^2 + c_{12}' |u_z|_{L^2}^{2/3} \|u_z\|^{2/3} |v_x|_{L^2}^{4/3} \left(1 + |v_z|_{L^2}^2\right). \end{aligned}$$

We also see that

$$\begin{aligned} \left| \int_{\mathcal{M}} w_z v_z v_z \, \mathrm{d}\mathcal{M} \right| &= \left| \int_{\mathcal{M}} u_x v_z v_z \, \mathrm{d}\mathcal{M} \right| \\ &\leqslant c_{13}' |u_x|_{L^2}^{1/2} \|u_x\|^{1/2} |v_z|_{L^2}^{3/2} \|v_z\|^{1/2} \\ &\leqslant \frac{\nu_{\mathrm{v}}}{8} \|v_z\|^2 + c_{14}' |u_x|_{L^2}^{2/3} \|u_x\|^{2/3} |v_z|_{L^2}^2, \end{aligned}$$

$$\frac{1}{R_0}\left| \int_{\mathcal{M}} u_z v_z \, \mathrm{d}\mathcal{M} \right| = \frac{1}{R_0}\left| \int_{\mathcal{M}} u v_{zz} \, \mathrm{d}\mathcal{M} \right| \leqslant \frac{\nu_{\mathrm{v}}}{8} \|v_z\|^2 + K|u|_{L^2}^2,$$

$$\left| \int_{\mathcal{M}} F_{v,z} v_z \, \mathrm{d}\mathcal{M} \right| = \left| \int_{\mathcal{M}} F_v v_{zz} \, \mathrm{d}\mathcal{M} \right| \leqslant \frac{\nu_{\mathrm{v}}}{8} \|v_z\|^2 + c_{15}' |F_v|_\infty^2,$$

which gives

$$\frac{\mathrm{d}}{\mathrm{d}t}|v_z|_{L^2}^2 + \nu_{\mathbf{v}}\|v_z\|^2 \leqslant \xi|v_z|^2 + \eta, \tag{3.122}$$

where we denoted

$$\eta = \eta(t) = 2c_{12}'|u_z|_{L^2}^{2/3}\|u_z\|^{2/3}|v_x|_{L^2}^{4/3} + 2K|u|^2 + 2c_{15}'|F_v|_\infty^2$$

and

$$\xi = \xi(t) = 2c_{12}'|u_z|_{L^2}^{2/3}\|u_z\|^{2/3}|v_x|_{L^2}^{4/3} + 2c_{14}'|u_x|_{L^2}^{2/3}\|u_x\|^{2/3}.$$

From (3.122), using the estimates obtained before and applying the classical Gronwall lemma we obtain bounds depending on the initial data for v_z in $L^\infty_{\mathrm{loc}}(0, t_\star; L^2)$ and $L^2_{\mathrm{loc}}(0, t_\star; H^1)$, valid for any finite interval of time $(0, t_\star)$.

To obtain estimates valid for all time, we apply the uniform Gronwall lemma observing that

$$\begin{aligned}
&\int_t^{t+r} \eta(t')\,\mathrm{d}t' \\
&\quad\leqslant 2c_{12}'|u_z|_\infty^{2/3}\left(\int_t^{t+r}\|u_z(t')\|\,\mathrm{d}t'\right)^{1/3}\left(\int_t^{t+r}|v_x(t')|_{L^2}^2\,\mathrm{d}t'\right)^{2/3} \\
&\qquad + 2K|u|_\infty^2 r + 2c_{15}' r|F_v|_\infty^2 \\
&\quad\leqslant a_1 \quad \forall t \geqslant t_1'',
\end{aligned} \tag{3.123}$$

$$\begin{aligned}
&\int_t^{t+r} \xi(t')\,\mathrm{d}t' \\
&\quad\leqslant 2c_{12}'|u_z|_\infty^{2/3}\left(\int_t^{t+r}\|u_z(t')\|\,\mathrm{d}t'\right)^{1/3}\left(\int_t^{t+r}|v_x(t')|_{L^2}^2\,\mathrm{d}t'\right)^{2/3} \\
&\qquad + 2c_{14}'|u_x|_\infty^{2/3}\int_t^{t+r}\|u_x(t')\|^{2/3}\,\mathrm{d}t' \\
&\quad\leqslant a_2 \quad \forall t \geqslant t_1'',
\end{aligned} \tag{3.124}$$

$$\int_t^{t+r}|v_z(t')|^2\,\mathrm{d}t' \leqslant a_3 \quad \forall t \geqslant t_1''. \tag{3.125}$$

Then the uniform Gronwall lemma gives

$$|v_z(t)|_{L^2}^2 \leqslant \left(\frac{a_3}{r} + a_2\right)\mathrm{e}^{a_1} \quad \forall t \geqslant t_1'' + r, \tag{3.126}$$

with a_1, a_2, a_3 as in (3.123), (3.124) and (3.125). Integrating (3.122) from t to $t+r$ with $r>0$ as above and $t \geqslant t_1'' + r$, we find

$$\int_t^{t+r} \left\| v_z(s) \right\|^2 \mathrm{d}s \leqslant K \quad \forall t \geqslant t_1'' + r. \tag{3.127}$$

The same methods apply to v_x, ρ_z and ρ_x, noticing that at each step we precisely use the estimates from the previous steps, so the order cannot be changed in this calculations.

With these estimates, the Galerkin method as used for the proof of Theorem 2.2 gives the existence of strong solutions.

THEOREM 3.4. *Given $U_0 \in V$ and $F \in L^\infty(\mathbb{R}_+; H)$, there is a unique solution U of equation* (3.92) *with $U(0) = U_0$ such that*

$$U \in L^\infty(\mathbb{R}_+; V) \cap L^2\bigl(0, t_\star; \bigl(\dot{H}^2(\mathcal{M})\bigr)^3\bigr) \quad \forall t_\star > 0. \tag{3.128}$$

PROOF. As we said, the existence of strong solutions follows from the previous estimates. The proof for the uniqueness follows the same idea as of the Theorem 3.1 so we skip it. □

3.3.3. *More regular solutions for the PEs.* In this section we show how to obtain estimates on the higher-order derivatives from which one can derive the existence of solutions of the PEs in $(\dot{H}^m(\mathcal{M}))^3$ for all $m \in \mathbb{N}$, $m \geqslant 2$ (hence up to $\mathcal{C}^\infty$ regularity). In all that follows we work with U_0 in $(\dot{H}^m_{\text{per}}(\mathcal{M}))^3$.

We set $|U|_m = (\sum_{[\alpha]=m} |D^\alpha U|^2_{L^2})^{1/2}$. We fix $m \geqslant 2$ and, proceeding by induction, we assume that for all $0 \leqslant l \leqslant m-1$, we have shown that

$$U \in L^\infty\bigl(\mathbb{R}_+; \bigl(\dot{H}^l(\mathcal{M})\bigr)^3\bigr) \cap L^2\bigl(0, t_\star; \bigl(\dot{H}^{l+1}(\mathcal{M})\bigr)^3\bigr) \quad \forall t_\star > 0, \tag{3.129}$$

with

$$\int_t^{t+r} \left| U(t') \right|^2_{l+1} \mathrm{d}t' \leqslant a_l \quad \forall t \geqslant t_l(U_0), \tag{3.130}$$

where a_l is a constant depending on the data (and l) but not on U_0, and $r > 0$ is fixed (the same as before). We then want to establish the same results for $l = m$.

In equation (3.92) we take $\widetilde{U} = \Delta^m U(t)$ with $m \geqslant 2$ and t arbitrarily fixed, and we obtain:

$$\begin{aligned} &\left(\frac{\mathrm{d}U}{\mathrm{d}t}, \Delta^m U \right)_{L^2} + a\bigl(U, \Delta^m U\bigr) + b\bigl(U, U, \Delta^m U\bigr) + \frac{1}{R_0} e\bigl(U, \Delta^m U\bigr) \\ &\quad = \bigl(F, \Delta^m U\bigr)_{L^2}. \end{aligned} \tag{3.131}$$

Integrating by parts, using periodicity and the coercivity of a and the fact that $e(U, U) = 0$, we find

$$\frac{1}{2} \frac{\mathrm{d}}{\mathrm{d}t} \left| U(t) \right|^2_m + c_1 |U|^2_{m+1} \leqslant \left| b\bigl(U, U, \Delta^m U\bigr) \right| + \left| \bigl(F, \Delta^m U\bigr)_{L^2} \right|. \tag{3.132}$$

We need to estimate the terms on the right-hand side of (3.132). We first notice that

$$\left|\left(F, \Delta^m U\right)_L^2\right| \leqslant c|F|_{m-1}^2 + \frac{c_1}{2(m+3)}|U|_{m+1}^2, \tag{3.133}$$

and it remains to estimate $|b(U, U, \Delta^m U)|$.

By the definition of b we have

$$\begin{aligned} b\left(U, U, \Delta^m U\right) = &\int_{\mathcal{M}} \left(u u_x + w(U) u_z\right) \Delta^m u \,\mathrm{d}\mathcal{M} \\ &+ \int_{\mathcal{M}} \left(u v_x + w(U) v_z\right) \Delta^m v \,\mathrm{d}\mathcal{M} \\ &+ \int_{\mathcal{M}} \left(u \rho_x + w(U) \rho_z\right) \Delta^m \rho \,\mathrm{d}\mathcal{M}. \end{aligned} \tag{3.134}$$

The computations are similar for all the terms, and, for simplicity, we shall only estimate the first integral on the right-hand side of (3.134).

We notice that $b(U, U, \Delta^m U)$ is a sum of integrals of the type

$$\int_{\mathcal{M}} u \frac{\partial u}{\partial x} D_1^{2\alpha_1} D_3^{2\alpha_3} u \,\mathrm{d}\mathcal{M}, \qquad \int_{\mathcal{M}} w(U) \frac{\partial u}{\partial z} D_1^{2\alpha_1} D_3^{2\alpha_3} u \,\mathrm{d}\mathcal{M},$$

where $\alpha_i \in \mathbb{N}$ with $\alpha_1 + \alpha_3 = m$. By D_i we denoted the differential operator $\partial/\partial x_i$. Integrating by parts and using periodicity, the integrals take the form

$$\int_{\mathcal{M}} D^\alpha \left(u \frac{\partial u}{\partial x}\right) D^\alpha u \,\mathrm{d}\mathcal{M}, \qquad \int_{\mathcal{M}} D^\alpha \left(w(U) \frac{\partial u}{\partial z}\right) D^\alpha u \,\mathrm{d}\mathcal{M}, \tag{3.135}$$

where $D^\alpha = D_1^{\alpha_1} D_3^{\alpha_3}$. Using Leibniz formula, we see that the integrals are sums of integrals of the form

$$\int_{\mathcal{M}} u D^\alpha \frac{\partial u}{\partial x} D^\alpha u \,\mathrm{d}\mathcal{M}, \qquad \int_{\mathcal{M}} w(U) D^\alpha \frac{\partial u}{\partial z} D^\alpha u \,\mathrm{d}\mathcal{M}, \tag{3.136}$$

and of integrals of the form

$$\int_{\mathcal{M}} \delta^k u \delta^{m-k} \frac{\partial u}{\partial x} D^\alpha u \,\mathrm{d}\mathcal{M}, \qquad \int_{\mathcal{M}} \delta^k w(U) \delta^{m-k} \frac{\partial u}{\partial z} D^\alpha u \,\mathrm{d}\mathcal{M}, \tag{3.137}$$

with $k = 1, \ldots, m$, where δ^k is some differential operator D^α with $[\alpha] = \alpha_1 + \alpha_3 = k$. For each α, after integration by parts we see that the sum of the two integrals in (3.136) is zero because of the mass conservation equation (3.76d). It remains to estimate the integrals of type (3.137). We use here the Sobolev and interpolation inequalities. For the first term in (3.137) we write

$$
\begin{aligned}
&\left|\int_{\mathcal{M}} \delta^k u \delta^{m-k} \frac{\partial u}{\partial x} D^\alpha u \,\mathrm{d}\mathcal{M}\right| \\
&\quad \leqslant \left|\delta^k u\right|_{L^4} \left|\delta^{m-k} \frac{\partial u}{\partial x}\right|_{L^4} \left|D^\alpha u\right|_{L^2} \\
&\quad \leqslant c_1' \left|\delta^k u\right|_{L^2}^{1/2} \left|\delta^k u\right|_{H^1}^{1/2} \left|\delta^{m-k} \frac{\partial u}{\partial x}\right|_{L^2}^{1/2} \left|\delta^{m-k} \frac{\partial u}{\partial x}\right|_{H^1}^{1/2} \left|D^\alpha u\right|_{L^2} \\
&\quad \leqslant c_1' |U|_k^{1/2} |U|_{k+1}^{1/2} |U|_{m-k+1}^{1/2} |U|_{m-k+2}^{1/2} |U|_m,
\end{aligned}
\tag{3.138}
$$

where $k = 1, \ldots, m$.

The second term from (3.137) is estimated as follows

$$
\begin{aligned}
&\left|\int_{\mathcal{M}} \delta^k w(U) \delta^{m-k} \frac{\partial u}{\partial z} D^\alpha u \,\mathrm{d}\mathcal{M}\right| \\
&\quad \leqslant \left|\delta^k w(U)\right|_{L^2} \left|\delta^{m-k} \frac{\partial u}{\partial z}\right|_{L^4} \left|D^\alpha u\right|_{L^4} \\
&\quad \leqslant c_2' \left|\delta^k w(U)\right|_{L^2} \left|\delta^{m-k} \frac{\partial u}{\partial z}\right|_{L^2}^{1/2} \left|\delta^{m-k} \frac{\partial u}{\partial z}\right|_{H^1}^{1/2} \left|D^\alpha u\right|_{L^2}^{1/2} \left|D^\alpha u\right|_{H^1}^{1/2} \\
&\quad \leqslant c_3' |U|_{k+1} |U|_{m-k+1}^{1/2} |U|_{m-k+2}^{1/2} |U|_m^{1/2} |U|_{m+1}^{1/2},
\end{aligned}
\tag{3.139}
$$

where $k = 1, \ldots, m$.

From (3.138) and (3.139) we obtain that

$$
\begin{aligned}
&\left|b\left(U, U, \Delta^m U\right)\right| \\
&\quad \leqslant c_3 \sum_{k=1}^{m} |U|_k^{1/2} |U|_{k+1}^{1/2} |U|_{m-k+1}^{1/2} |U|_{m-k+2}^{1/2} |U|_m \\
&\qquad + c_3 \sum_{k=1}^{m} |U|_{k+1} |U|_{m-k+1}^{1/2} |U|_{m-k+2}^{1/2} |U|_m^{1/2} |U|_{m+1}^{1/2}.
\end{aligned}
\tag{3.140}
$$

We now need to bound the terms on the right-hand side of (3.140). The terms corresponding to $k = 2, \ldots, m-1$ in the first sum do not contain $|U|_{m+1}$ and we leave them as they are. For $k = 1$ and $k = m$, we apply Young's inequality and we obtain

$$
\begin{aligned}
&c_3 |U|_1^{1/2} |U|_2^{1/2} |U|_m^{3/2} |U|_{m+1}^{1/2} \\
&\quad \leqslant \frac{c_1}{2(m+3)} |U|_{m+1}^2 + c_4' |U|_1^{2/3} |U|_2^{2/3} |U|_m^2.
\end{aligned}
\tag{3.141}
$$

For the terms in the second sum in (3.140) we distinguish between $k = 1$, $k = m$ and $k = 2, \ldots, m-1$. The term corresponding to $k = 1$ is bounded by

$$c_3|U|_2|U|_m|U|_{m+1} \leqslant \frac{c_1}{2(m+3)}|U|_{m+1}^2 + c_5'|U|_2^2|U|_m^2. \tag{3.142}$$

For $k = m$ we find

$$c_3|U|_1^{1/2}|U|_2^{1/2}|U|_m^{1/2}|U|_{m+1}^{3/2} \leqslant \frac{c_1}{2(m+3)}|U|_{m+1}^2 + c_6'|U|_1^2|U|_2^2|U|_m^2. \tag{3.143}$$

For the terms corresponding to $k = 2, \ldots, m-1$, we apply Young's inequality in the following way:

$$\begin{aligned} &c_3|U|_{k+1}|U|_{m-k+1}^{1/2}|U|_{m-k+2}^{1/2}|U|_m^{1/2}|U|_{m+1}^{1/2} \\ &\quad\leqslant \frac{c_1}{2(m+3)}|U|_{m+1}^2 + c_7'|U|_{k+1}^{4/3}|U|_{m-k+1}^{2/3}|U|_{m-k+2}^{2/3}|U|_m^{2/3}. \end{aligned} \tag{3.144}$$

Gathering all the estimates above we find

$$\frac{\mathrm{d}}{\mathrm{d}t}|U|_m^2 + c_1|U|_{m+1}^2 \leqslant \xi + \eta|U|_m^2,$$

where the expressions of ξ and η are easily derived from (3.132) and (3.141)–(3.144). Using the Gronwall lemma and the induction hypotheses (3.129) and (3.130) we obtain a bound for U in $L^\infty(0, t_\star; H^m)$ and $L^2(0, t_\star; H^{m+1})$, for all fixed $t_\star > 0$, this bound depending also on $|U_0|_m$. We also see that, because of the induction hypotheses (3.129) and (3.130), we can apply the uniform Gronwall lemma and we obtain U bounded in $L^\infty(\mathbb{R}_+; H^m)$ with a bound independent of $|U_0|_m$ when $t \geqslant t_m(U_0)$; we also obtain an analogue of (3.130). The details regarding the way we apply the uniform Gronwall lemma and derive these bounds are similar to the developments in Section 3.3.2.

In summary we have proven the following result:

THEOREM 3.5. *Given $m \in \mathbb{N}$, $m \geqslant 1$, $U_0 \in V \cap (\dot{H}_{\mathrm{per}}^m(\mathcal{M}))^3$ and $F \in L^\infty(\mathbb{R}_+; H \cap (\dot{H}_{\mathrm{per}}^{m-1}(\mathcal{M}))^3)$, (3.92) has a unique solution U such that*

$$U \in L^\infty\big(\mathbb{R}_+; \big(\dot{H}_{\mathrm{per}}^m(\mathcal{M})\big)^3\big) \cap L^2\big(0, t_\star; \big(\dot{H}_{\mathrm{per}}^{m+1}(\mathcal{M})\big)^3\big) \quad \forall t_\star > 0. \tag{3.145}$$

REMARK 3.3. Since $\bigcap_{m\geqslant 0} \dot{H}_{\mathrm{per}}^m(\mathcal{M}) = \dot{\mathcal{C}}_{\mathrm{per}}^\infty(\mathcal{M})$, given $U_0 \in (\dot{\mathcal{C}}_{\mathrm{per}}^\infty(\mathcal{M}))^3$ and $F \in L^\infty(\mathbb{R}_+; (\dot{\mathcal{C}}_{\mathrm{per}}^\infty(\mathcal{M}))^3)$, (3.92) has a unique solution U belonging to $L^\infty(\mathbb{R}_+; (\dot{H}_{\mathrm{per}}^m(\mathcal{M}))^3)$ for all $m \in \mathbb{N}$; that is, U is in $L^\infty(\mathbb{R}_+; (\dot{\mathcal{C}}_{\mathrm{per}}^\infty(\mathcal{M}))^3)$. Regularity (differentiability) in time can be also derived if F is also $\mathcal{C}^\infty$ in time. However the arguments above do not provide the existence of an absorbing set in $(\dot{\mathcal{C}}_{\mathrm{per}}^\infty(\mathcal{M}))^3$.

4. Regularity for the elliptic linear problems in GFD

We have used many times, in particular in Section 3 the result of H^2 regularity of the solutions to certain linear elliptic problems. Following the general results of Agmon, Douglis and Nirenberg [1,2], we know that the solutions of second-order elliptic problems are in H^{m+2}, if the right-hand sides of the equations are in $H^m, m \geqslant 0$, and the other data are in suitable spaces; see also Lions and Magenes [20] for $m < 0$. Results of this type are proved in this section.

There are several specific aspects and several specific difficulties which justify the lengthy and technical developments of this section which do not allow us to directly refer to the general results of [1] and [2]:

For the whole atmosphere (not studied in detail here) and for the space periodic case studied in Section 3.3 the domains are smooth, making the results of this section easy.

(i) The (linear, stationary) GFD–Stokes problem (see Section 4.4.1), involves an integral equation (the second equation in (4.96)), which prevents from a purely local treatment, like for the classical Stokes problem of incompressible fluid mechanics.

(ii) The boundary conditions of the problem can be a combination of Dirichlet, Neumann and/or Robin boundary conditions.

(iii) The domains that we have considered and that we consider in this section are not smooth, they have angles in two dimensions and edges in three dimensions. This is automatically the case for the ocean and for regional atmosphere or ocean problems. For this reason, technique pertaining to the theory of elliptic problems in nonsmooth domains (see, e.g., [11,17,25]), are needed and used here.

(iv) Because the domain is not smooth, only the H^2 regularity is proved here, $m = 0$. The H^3 regularity, $m \geqslant 1$ is not expected in general.

(v) Another aspect of the study in this section concerns the shape of the ocean or the atmosphere (shallowness). A "small" parameter ε is introduced, the depth being called εh instead of h, $0 < \varepsilon < 1$, and we want to see how the regularity constants (which depend on the domain) depend on ε.

The small depth hypothesis was considered in [16] and is not considered in this chapter. Introducing the parameter ε makes the proofs of this section generally more involved than needed for this chapter. However, these results usefully complement the article [15] used in [16].

Many of the results presented in this section are new although some related results appeared in [44] and [45]. The results are fairly general, except for the orthogonality condition appearing in (4.54) (Γ_{b} orthogonal to Γ_ℓ). This condition is not physically desirable; and it is not either mathematically needed (most likely), as no such condition appears for the regularity theory of elliptic problems in angles or edges [11,25,17]. We believe that it can be removed, but this problem is open. Let us recall also that all the results in Section 3 are valid whenever the necessary H^2-regularity results can be proved.

For the notations, the basic domain under consideration is $\mathcal{M}_\varepsilon$:

$$\mathcal{M}_\varepsilon = \{(x, y, z), (x, y) \in \Gamma_{\mathrm{i}}, -\varepsilon h(x, y) < z < 0\}.$$

For $\varepsilon = 1$, we recover the domain $\mathcal{M}(\mathcal{M}_1 = \mathcal{M})$ used in Sections 2 and 3. Below, for technical reasons, we introduce auxiliary domains $\widehat{\mathcal{M}}_\varepsilon$ and Q_ε. A number of unspecified constants independent of ε are generically denoted by C_0.

Note also that this section is closer to PDE theory than to geoscience and therefore we stay closer to the PDE notations than to the geoscience notations. Hence the notations are not necessarily the same as in the rest of the chapter; in particular, we do not use bold faces for vectors and the current point of $\mathbb{R}^2$ or $\mathbb{R}^3$ is denoted $x = (x_1, x_2)$ or $x = (x_1, x_2, x_3)$ instead of (x, y) or (x, y, z).

4.1. *Regularity of solutions of elliptic boundary value problems in cylinder type domains*

We study in this section the H^2 regularity of solutions of elliptic problems of the second order in a cylinder type domain; the boundary condition is either of Dirichlet or of Neumann type on all the boundary.

Since the domain contains wedges, it is not smooth and we rely heavily on the results of [11] about regularity for elliptic problems in nonsmooth domains. However a convexity assumption of the domain is essential in [11], that we want to avoid: this section is mainly devoted to the implementation of a suitable technique, corresponding to a tubular (cylindrical) covering of the domain under consideration.

Let $\widehat{\mathcal{M}}_\varepsilon = \{(x_1, x_2, x_3) \in \mathbb{R}^3;\ (x_1, x_2) \in \Gamma_{\mathrm{i}}, -\varepsilon h(x_1, x_2) < x_3 < \varepsilon h(x_1, x_2)\}$, where Γ_{i} is a bounded open subset of $\mathbb{R}^2$, with $\partial\Gamma_{\mathrm{i}}$ a C^2-curve, and $h : \overline{\Gamma_{\mathrm{i}}} \to \mathbb{R}_+$ is a positive function, $h \in C^4(\overline{\Gamma_{\mathrm{i}}})$ and there exist two positive constants $\underline{h}$ and $\bar{h}$ such that $\underline{h} \leqslant h(x_1, x_2) \leqslant \bar{h}$ for all (x_1, x_2) in $\overline{\Gamma_{\mathrm{i}}}$. Define the elliptic operator A by

$$Au = -\sum_{k,\ell=1}^{3} \frac{\partial}{\partial x_k}\left(a^{k\ell}(x)\frac{\partial u}{\partial x_\ell}\right) + \sum_{k=1}^{3} b^k(x)\frac{\partial u}{\partial x_k} + c(x)u, \tag{4.1}$$

where the coefficients $a^{k,\ell}, k, \ell = 1, 2, 3$ are of class $C^2(\overline{\widehat{\mathcal{M}}_\varepsilon})$, $b^k, k = 1, 2, 3$, are of class $C^1(\overline{\widehat{\mathcal{M}}_\varepsilon})$ and c is of class $C^0(\overline{\widehat{\mathcal{M}}_\varepsilon})$. Furthermore, we assume that A is uniformly strongly elliptic, i.e., there exists a positive constant α independent of x and ε such that

$$\sum_{k,\ell=1}^{3} a^{k\ell}(x)\xi_k\xi_\ell \geqslant \alpha|\xi|^2 \quad \forall x \in \overline{\widehat{\mathcal{M}}_\varepsilon},\ \forall \xi \in \mathbb{R}^3. \tag{4.2}$$

We also assume that the functions $a^{k\ell}, b^k, c, k, \ell = 1, 2, 3$, are independent of ε. We aim to study the regularity and the dependence on ε of the solutions to the Dirichlet problem

$$\begin{cases} Au = f & \text{in } \widehat{\mathcal{M}}_\varepsilon, \\ u = 0 & \text{in } \partial\widehat{\mathcal{M}}_\varepsilon, \end{cases} \tag{4.3}$$

and the solutions of the Neumann problem,

$$\begin{cases} Au = f & \text{in } \widehat{\mathcal{M}}_\varepsilon, \\ \frac{\partial u}{\partial n_A} = 0 & \text{on } \partial\hat{\mathcal{M}}_\varepsilon, \end{cases} \tag{4.4}$$

where $\frac{\partial}{\partial n_A}$ denotes the co-normal boundary operator defined by

$$\frac{\partial u}{\partial n_A} = \sum_{k,\ell} a^{k\ell} \frac{\partial u}{\partial x_\ell} n_k, \tag{4.5}$$

and $n = (n_1, n_1, n_3)$ denotes the unit vector in the direction of the outward normal to $\partial\widehat{\mathcal{M}}_\varepsilon$. Our goal is to prove the H^2 regularity of solutions to the Dirichlet problem (4.3) or the Neumann problem (4.4), and to obtain the dependence on ε of the constant C_ε appearing in the inequality

$$\sum_{k,\ell} \left| \frac{\partial^2 u}{\partial x_k \, \partial x_\ell} \right|^2_{L^2(\widehat{\mathcal{M}}_\varepsilon)} \leqslant C_\varepsilon |Au|^2_{L^2(\widehat{\mathcal{M}}_\varepsilon)}. \tag{4.6}$$

In fact we will show that $C_\varepsilon = C_0$ is independent of ε and, more precisely, we will prove the following.

THEOREM 4.1. *Let* $\widehat{\mathcal{M}}_\varepsilon = \{(x_1, x_2, x_3) \in \mathbb{R}^3;\ (x_1, x_2) \in \Gamma_{\mathrm{i}},\ -\varepsilon h(x_1, x_2) < x_3 < \varepsilon h(x_1, x_2)\}$, *where* Γ_{i} *is a bounded subset of* $\mathbb{R}^2$, *with* $\partial\Gamma_{\mathrm{i}}$ *a* C^2-*curve, and* $h \in C^4(\Gamma_{\mathrm{i}})$, *and there exists two positive constants* $\underline{h}, \bar{h}$ *such that* $\underline{h} \leqslant h(x_1, x_2) \leqslant \bar{h}$ *for all* (x_1, x_2) *in* $\overline{\Gamma_{\mathrm{i}}}$. *Let* $f \in L^2(\widehat{\mathcal{M}}_\varepsilon)$ *and* $u \in H_0^1(\widehat{\mathcal{M}}_\varepsilon)$, *with* $\|u\|_{H^1(\widehat{\mathcal{M}}_\varepsilon)} \leqslant C_0 |f|_{L^2(\widehat{\mathcal{M}}_\varepsilon)}$ *with* C_0 *independent of* ε.

If u satisfies

$$Au = -\sum_{k,\ell=1}^{3} \frac{\partial}{\partial x_k}\left(a^{k\ell}(x) \frac{\partial u}{\partial x_\ell} \right) + \sum_{k=1}^{3} b^k(x) \frac{\partial u}{\partial x_k} + c(x)u = f,$$

where a^{ij}, b^i *are in* $C^2(\overline{\widehat{\mathcal{M}}_\varepsilon})$, c *in* $C^1(\overline{\widehat{\mathcal{M}}_\varepsilon})$ *and A is uniformly strongly elliptic in the sense of* (4.2), *then* $u \in H^2(\widehat{\mathcal{M}}_\varepsilon)$ *and there exists a constant* C_0 *independent of* ε *such that*

$$\sum_{i,j=1} \left| \frac{\partial^2 u}{\partial x_i \, \partial x_j} \right|^2_{L^2(\widehat{\mathcal{M}}_\varepsilon)} \leqslant C_0 |f|^2_{L^2(\widehat{\mathcal{M}}_\varepsilon)}. \tag{4.7}$$

PROOF. The proof of Theorem 4.1 is divided into four steps.

STEP 1 (Flattening the top and bottom boundaries). We straighten the bottom and top boundaries and transform the domain $\widehat{\mathcal{M}}_\varepsilon$ into the cylinder $Q_\varepsilon = \Gamma_{\mathrm{i}} \times (-\varepsilon, \varepsilon)$, and the operator A is transformed to an operator $\widetilde{A}$ of the same form and satisfying the same assumptions as A. In fact let

$$\begin{aligned}&\Psi:(x_1,x_2,x_3)\mapsto(y_1,y_2,y_3),\\&y_1=x_1,\quad y_2=x_2,\quad y_3=\frac{x_3}{h(x_1,x_2)}.\end{aligned}\tag{4.8}$$

Since $h\in C^4(\overline{\Gamma_{\rm i}})$, we have $\Psi\in C^4(\overline{\widehat{\mathcal{M}}_\varepsilon})$. Furthermore, we note that Au may be written as

$$Au=-\sum_{k,\ell=1}^{3}a^{k\ell}(x)\frac{\partial^2u}{\partial x_k\,\partial x_\ell}+\sum_{k=1}^{3}d^k(x)\frac{\partial u}{\partial x_k}+c(x)u,\tag{4.9}$$

where

$$d^k(x)=b^k(x)-\sum_{\ell=1}^{3}\frac{\partial a^{\ell k}(x)}{\partial x_\ell}\in C^1(\overline{\widehat{\mathcal{M}}_\varepsilon}).\tag{4.10}$$

Now let $\tilde u:Q_\varepsilon\to\mathbb{R}$, $\tilde u(y_1,y_2,y_3)=u(x_1,x_2,x_3)$. Since $\Psi\in C^4(\overline{\widehat{\mathcal{M}}_\varepsilon})$ and Ψ is independent of ε, the H^2-norm of $\tilde u$ is equivalent to the H^2-norm of u with constants independent of ε. More precisely, there exists a constant C_0 independent of ε such that

$$\begin{aligned}C_0^{-1}\sum_{k,\ell}\left|\frac{\partial^2\tilde u}{\partial y_k\,\partial y_\ell}\right|^2_{L^2(Q_\varepsilon)}&\leqslant\sum_{k,\ell}\left|\frac{\partial^2u}{\partial x_k\,\partial x_\ell}\right|^2_{L^2(\widehat{\mathcal{M}}_\varepsilon)}\\&\leqslant C_0\sum_{k,\ell}\left|\frac{\partial^2\tilde u}{\partial y_k\,\partial y_\ell}\right|^2_{L^2(Q_\varepsilon)}.\end{aligned}\tag{4.11}$$

Furthermore, we can easily check that

$$\widetilde{A}\tilde u(y)=-\sum_{k,\ell}\tilde a^{k\ell}(y)\frac{\partial^2\tilde u}{\partial y_k\,\partial y_\ell}+\sum_k\tilde d^k(y)\frac{\partial\tilde u}{\partial y_k}+\tilde c(y)\tilde u(y)=\tilde f(y),\tag{4.12}$$

where

$$\begin{cases}\tilde a^{k\ell}(y)=\sum_{r,s}\frac{\partial\Psi_k}{\partial x_r}\frac{\partial\Psi_\ell}{\partial x_s}a^{rs}(\Psi^{-1}(y)),\\ \tilde d^k(y)=-\sum_{r,s}\frac{\partial^2\Psi_k}{\partial x_r\partial x_s}a^{sr}(\Psi^{-1}(y))+\sum_r\frac{\partial\Psi_k}{\partial x_r}d^r(\Psi^{-1}(y)),\\ \tilde c(y)=c(\Psi^{-1}(y))\quad\text{and}\quad\tilde f(y)=f(\Psi^{-1}(y)).\end{cases}\tag{4.13}$$

It is clear, since $\Psi \in C^4(\widehat{\mathcal{M}_\varepsilon})$, that $\tilde{a}^{k,\ell} \in C^2(\overline{Q})$, $\tilde{d}^k \in C^1(\overline{Q})$ and $\tilde{c} \in C^0(\overline{Q}_\varepsilon)$. Finally, if $u = 0$ on $\partial\widehat{\mathcal{M}}_\varepsilon$ then $\tilde{u} = 0$ on ∂Q_ε and if $\partial u/\partial n_A = 0$ on $\partial\widehat{\mathcal{M}}_\varepsilon$ then $\partial\tilde{u}/\partial n_{\tilde{A}} = 0$ on ∂Q_ε, where, as in (4.5),

$$\frac{\partial \tilde{u}}{\partial n_{\tilde{A}}} = \sum_{k,\ell} \tilde{a}^{k\ell} \frac{\partial \tilde{u}}{\partial y_\ell} n_k. \tag{4.14}$$

This is classical, but we include the verification here at the bottom boundary $x_3 = -\varepsilon h$ or equivalently $y_3 = -\varepsilon$. By (4.5) we have

$$\begin{aligned} &\left(1+\varepsilon^2|\nabla h|^2\right)^{1/2} \frac{\partial u}{\partial n_A} \\ &\quad = \varepsilon \sum_{j=1}^{3} a^{1j} \frac{\partial u}{\partial x_j}\frac{\partial h}{\partial x_1} + \varepsilon \sum_{j=1}^{3} a^{2j} \frac{\partial u}{\partial x_j}\frac{\partial h}{\partial x_2} + \sum_{j=1}^{3} a^{3j} \frac{\partial u}{\partial x_j}, \end{aligned} \tag{4.15}$$

but $\frac{\partial u}{\partial x_j} = \sum_{r=1}^{3} \frac{\partial \tilde{u}}{\partial y_r}\frac{\partial \Psi_r}{\partial x_j}$, and thus

$$\begin{aligned} &\left(1+\varepsilon^2|\nabla h|^2\right)^{1/2} \frac{\partial u}{\partial n_A} \\ &\quad = \sum_{r=1}^{3}\sum_{j=1}^{3} \left(a^{1j}\left(\varepsilon \frac{\partial h}{\partial x_1}\right) + a^{2j}\left(\varepsilon \frac{\partial h}{\partial x_2}\right) + a^{3j}\right) \frac{\partial \Psi_r}{\partial x_j}\frac{\partial \tilde{u}}{\partial y_r}. \end{aligned} \tag{4.16}$$

On the other hand, by definition, we have

$$\frac{\partial \tilde{u}}{\partial n_{\tilde{A}}} = -\sum_{r=1}^{3} \tilde{a}^{3r}(-\varepsilon)\frac{\partial \tilde{u}}{\partial y_r} \quad \text{(the normal is in the direction of } y_3 < 0), \tag{4.17}$$

but

$$\tilde{a}^{3r} = \sum_{m,n} \frac{\partial \Psi_3}{\partial x_m}\frac{\partial \Psi_r}{\partial x_n} a^{mn} = \sum_{n=1}^{3} \frac{\partial \Psi_r}{\partial x_n}\left(\sum_{m=1}^{3} a^{mn}\frac{\partial \Psi_3}{\partial x_m}\right), \tag{4.18}$$

and since $\Psi_3(x_1, x_2, x_3) = x_3/h(x_1, x_2)$, we have

$$\tilde{a}^{3r}(-\varepsilon) = \frac{1}{h}\sum_{j=1}^{3}\left(a^{1j}\left(\varepsilon \frac{\partial h}{\partial x_1}\right) + a^{2j}\left(\varepsilon \frac{\partial h}{\partial x_2}\right) + a^{3j}\right)\frac{\partial \Psi_r}{\partial x_j}. \tag{4.19}$$

Hence

$$\frac{\partial \tilde{u}}{\partial n_{\tilde{A}}} = 0 \quad \text{at } y_3 = -\varepsilon. \tag{4.20}$$

A similar computation yields $\partial\tilde{u}/\partial n_{\widetilde{A}} = 0$ at $y_3 = \varepsilon$. Now we check the Neumann condition at the lateral boundary. First write

$$\frac{\partial u}{\partial n_A} = \sum_{k=1}^{2}\sum_{\ell=1}^{3} a^{k\ell}\frac{\partial u}{\partial x_\ell}n_k = \sum_{k=1}^{2}\sum_{\ell,r=1}^{3} a^{k\ell}\frac{\partial \tilde{u}}{\partial y_r}\frac{\partial \Psi_r}{\partial x_\ell}n_k \tag{4.21}$$

and

$$\frac{\partial \tilde{u}}{\partial n_{\widetilde{A}}} = \sum_{k=1}^{2}\sum_{\ell=1}^{3} \tilde{a}^{k\ell}\frac{\partial \tilde{u}}{\partial y_\ell}n_k = \sum_{k=1}^{2}\sum_{\ell,r,s=1}^{3} a^{sr}\frac{\partial \Psi_k}{\partial x_s}\frac{\partial \Psi_\ell}{\partial x_r}\frac{\partial \tilde{u}}{\partial y_\ell}n_k, \tag{4.22}$$

but since for $k = 1, 2$, $\Psi_k(x_1, x_2, x_3) = x_k$, we have $\partial\Psi_k/\partial x_s = \delta_{sk}$ (the Kronecker symbol). Hence

$$\frac{\partial \tilde{u}}{\partial n_{\widetilde{A}}} = \sum_{k,\ell,r=1}^{3} a^{kr}\frac{\partial \Psi_\ell}{\partial x_r}\frac{\partial \tilde{u}}{\partial y_\ell}n_k. \tag{4.23}$$

Interchanging ℓ and r, we obtain $\partial\tilde{u}/\partial n_{\widetilde{A}} = \partial u/\partial n_A = 0$ on the lateral boundary. From now on we concentrate on the Dirichlet boundary condition, the Neumann condition case follows in the same manner.

STEP 2 (Interior regularity). Let B_R be an open ball, with $B_R \subset\subset \Gamma_{\rm i}$; without loss of generality we assume that B_R is centered at 0. By Step 1, we may assume that $\widehat{\mathcal{M}}_\varepsilon$ is a right cylinder, i.e., $\widehat{\mathcal{M}}_\varepsilon = Q_\varepsilon = \Gamma_{\rm i} \times (-\varepsilon, \varepsilon)$. Now let $\theta \in C_0^\infty(B_R)$ (θ independent of x_3) with $\theta \equiv 1$ in $B_{R/2}$ and $0 \leqslant \theta \leqslant 1$. Then

$$\widetilde{A}(\theta\tilde{u}) = \theta\tilde{f} + E_\theta\tilde{u},$$

where E_θ is a first-order differential operator, which implies that $|E_\theta\tilde{u}|_{L^2(Q_\varepsilon)} \leqslant C_0|f|_\varepsilon$ where C_0 is independent on ε, and $|f|_\varepsilon$ is an alternate notation for $|f|_{L^2(\widehat{\mathcal{M}}_\varepsilon)}$. Hence

$$\widetilde{A}(\theta\tilde{u}) = f_{\rm loc}, \quad \text{with } |f_{\rm loc}|_{L^2(B_R\times(-\varepsilon,\varepsilon))} \leqslant C_0|f|_\varepsilon,$$

and, for either boundary condition (Dirichlet, Neumann), $\theta\tilde{u} = 0$ on $\partial(B_R) \times (-\varepsilon, \varepsilon)$ and, in the Dirichlet case, $\theta\tilde{u} = 0$ on $B_R \times (-\varepsilon, \varepsilon)$, and, in the Neumann case, $\partial(\theta\tilde{u})/\partial z = 0$ on $B_R \times (-\varepsilon, \varepsilon)$.

We quote the following theorems from [11], we start first with the case of the Dirichlet boundary condition.

THEOREM A (Grisvard [11], Theorem 3.2.1.2). *Let Ω be a convex, bounded and open subset of $\mathbb{R}^n$. Then for each $f \in L^2(\Omega)$, there exists a unique $u \in H^2(\Omega)$, the solution of $Au = f$ in $\Omega, u = 0$ on $\partial\Omega$.*

The proof of the theorem, given in [11], pp. 148–149, is based on a priori bounds for solutions in $H^2(\Omega)$. These bounds depend in the case of general domains on the curvature of $\partial\Omega$, however in the case of a convex domain the curvature is negative and the constants in the bounds are therefore independent on the domain.

Similarly, in the case of Neumann boundary condition, we have the following theorem.

THEOREM B (Grisvard [11], Theorem 3.2.1.3). *Let Ω be a convex, bounded and open subset of $\mathbb{R}^n$. Then for each $f \in L^2(\Omega)$ and for each $\lambda > 0$, there exists a unique $u \in H^2(\Omega)$, the solution of*

$$-\sum_{k,\ell=1}^{3} \int_\Omega a^{k\ell}(x) \frac{\partial u}{\partial x_\ell} \frac{\partial v}{\partial x_k} \,\mathrm{d}x + \lambda \int_\Omega uv \,\mathrm{d}x = \int_\Omega fv \,\mathrm{d}x \tag{4.24}$$

for all $v \in H^1(\Omega)$.

We note that (4.24) is the weak form of the Neumann problem for the equation

$$-\sum_{k,\ell=1}^{3} \frac{\partial}{\partial x_k}\left(a^{k\ell}(x)\frac{\partial u}{\partial x_\ell}\right) + \lambda u = f \quad \text{in } \Omega$$

together with the boundary condition $\frac{\partial u}{\partial n_A} = 0$ on $\partial\Omega$. Again, here, the convexity of the domain implies that the curvature of the boundary of the domain is negative, and therefore the constants in the bounds on the L^2 norm of the mixed second derivatives in terms of the L^2 norm of f are independent on the domain. For more details, see [11].

Now we use Theorem A above by first rewriting $\widetilde{A}$ in a divergence form and moving the extra terms to the right-hand side. As above, since $\theta\tilde{u} \in H^1(B_R \times (-\varepsilon, \varepsilon))$ the extra terms are in $L^2(B_R \times (-\varepsilon, \varepsilon))$ and

$$\sum_{i,j} \left| \frac{\partial^2(\theta\tilde{u})}{\partial y_i \,\partial y_j} \right|^2_{L^2(B_R\times(-\varepsilon,\varepsilon))} \leqslant C_0 |f|^2_{L^2(\widehat{\mathcal{M}}_\varepsilon)}.$$

Finally, since $\theta \equiv 1$ in $B_{R/2}$, we have $\tilde{u} \in H^2(B_{R/2} \times (-\varepsilon, \varepsilon))$ and

$$\sum_{i,j} \left| \frac{\partial^2\tilde{u}}{\partial y_i \,\partial y_j} \right|^2_{L^2(B_{R/2}\times(-\varepsilon,\varepsilon))} \leqslant C_0 |f|^2_{L^2(\widehat{\mathcal{M}}_\varepsilon)}.$$

STEP 3 (Boundary regularity). Let $\mathbb{R}^2_+ = \{x \in \mathbb{R}^2;\, x_2 > 0\}$ and let $B_r^+ = \{x \in \mathbb{R}^2_+;\, |x| < r\}$ be the open half-ball with center at the origin and radius r contained in $\mathbb{R}^2$. By the assumption on Γ_{i}, for all $x^0 \in \partial\Gamma_{\mathrm{i}}$, there exists a neighborhood V of x^0 in $\mathbb{R}^2$ and a diffeomorphism $\widetilde{\Psi}$ such that

$$\widetilde{\Psi}(V \cap \Gamma_{\mathrm{i}}) = B_r^+,$$
$$\widetilde{\Psi}(x^0) = 0.$$

Using the diffeomorphism $\widetilde{\Psi}$, we can construct a (tubular) diffeomorphism Ψ in $\mathbb{R}^3$ such that

$$\Psi\big(V \times (-\varepsilon, \varepsilon) \cap Q_\varepsilon\big) = B_r^+ \times (-\varepsilon, \varepsilon),$$

by setting $\Psi_i(y_1, y_2, y_3) = \widetilde{\Psi}_i(y_1, y_2), i = 1, 2$, and $\Psi_3(y_1, y_2, y_3) = y_3$. Following the same procedure as in Steps 1 and 2, let W be an open set of $\mathbb{R}^2$ containing x^0 such that $\overline{W} \subset V$ and let $\theta \in C_0^\infty(V)$ be such that $0 \leqslant \theta \leqslant 1$ and $\theta \equiv 1$ in W. Then

$$\widetilde{A}\big(\theta\tilde{u}\big) = f_{\mathrm{loc}} \quad \text{with } |f_{\mathrm{loc}}|_{L^2(V\times(-\varepsilon,\varepsilon)\cap Q_\varepsilon)} \leqslant C_0|f|_\varepsilon,$$

and, in the case of the Dirichlet boundary condition $\theta\tilde{u} = 0$ on $\partial(V \times (-\varepsilon, \varepsilon) \cap Q_\varepsilon)$. Next we use the transformation Ψ which is independent of ε and which transforms the domain $V \times (-\varepsilon, \varepsilon) \cap Q_\varepsilon$ into $B_r^+ \times (-\varepsilon, \varepsilon)$, $\theta\tilde{u}$ into u^*, and $\widetilde{A}$ into A^* with A^*u^* given as in Step 1. Now, $u^* = 0$ on $\partial(B_r^+ \times (-\varepsilon, \varepsilon))$ and $B_r^+ \times (-\varepsilon, \varepsilon)$ is convex; hence rewriting A^* in a divergence form we obtain using [11], $u^* \in H^2(B_r^+ \times (-\varepsilon, \varepsilon))$ and

$$\sum_{i,j} \left|\frac{\partial^2 u^*}{\partial z_i\, \partial z_j}\right|^2_{L^2(B_R^+\times(-\varepsilon,\varepsilon))} \leqslant C_0|f|_\varepsilon^2.$$

Going back to $V \times (-\varepsilon, \varepsilon) \cap Q_\varepsilon$ using Ψ^{-1}, we obtain

$$\sum_{i,j} \left|\frac{\partial^2 (\theta\tilde{u})}{\partial y_i\, \partial y_j}\right|^2_{L^2(V\times(-\varepsilon,\varepsilon)\cap Q_\varepsilon)} \leqslant C_0|f|_\varepsilon^2.$$

Hence

$$\sum_{i,j} \left|\frac{\partial^2 \tilde{u}}{\partial y_i\, \partial y_j}\right|^2_{L^2(V\times(-\varepsilon,\varepsilon)\cap Q_\varepsilon)} \leqslant C_0|f|_\varepsilon^2.$$

STEP 4 (Partition of unity and conclusion). Let $V_0, V_1, \ldots, V_N$ and $W_0, \ldots, W_N$ be two finite open coverings of $\overline{\Gamma_{\mathrm{i}}}$ satisfying $\overline{V_0} \subset \Gamma_{\mathrm{i}}$; $V_k, k \geqslant 1$ is contained in the domain of a local map $\widetilde{\Psi}_{(k)}$ such that

$$\widetilde{\Psi}_{(k)}(V_k \cap \Gamma_{\mathrm{i}}) = B_r^+,$$
$$W_0 = V_0,$$
$$\overline{W}_k \subset V_k \quad \text{for all } k \geqslant 1.$$

Finally let $\{\varphi_k\}_k$ be a partition of unity subordinated to the covering $\{W_k\}_k$ of Γ_{i}. Then $\tilde{u} = \sum_{k=1}^N \varphi_k\tilde{u}$ and by Steps 1–3, $\varphi_k\tilde{u} \in H^2(Q_\varepsilon)$ and

$$\sum_{k=0}^{N} \sum_{i,j=1}^{3} \left|\frac{\partial^2 (\varphi_k\tilde{u})}{\partial y_i\, \partial y_j}\right|^2_{L^2(Q_\varepsilon)} \leqslant C_0|f|_\varepsilon^2.$$

Therefore $\tilde{u} \in H^2(Q_\varepsilon)$ and

$$\sum_{i,j=1}^{3} \left| \frac{\partial^2 \tilde{u}}{\partial y_i \, \partial y_j} \right|_{L^2(Q_\varepsilon)}^2 \leqslant C_0 |f|_\varepsilon^2.$$

Finally, we go back to the domain $\widehat{\mathcal{M}}_\varepsilon$ and conclude that $u \in H^2(\widehat{\mathcal{M}}_\varepsilon)$ and

$$\sum_{i,j=1}^{3} \left| \frac{\partial^2 u}{\partial x_i \, \partial x_j} \right|_{L^2(\widehat{\mathcal{M}}_\varepsilon)}^2 \leqslant C_0 |f|_\varepsilon^2.$$

Theorem 4.1 is proved. □

4.2. *Regularity of solutions of a Dirichlet–Robin mixed boundary value problem*

We now want to derive a result similar to that of Section 4.1, for a boundary value problem with mixed Dirichlet–Robin boundary conditions, the elliptic operator being the same as in (4.1). The proof consists in reducing the boundary condition to a full Dirichlet boundary condition and then use Theorem 4.1.

From now on, we will consider the actual domain

$$\mathcal{M}_\varepsilon = \left\{ (x_1, x_2, x_3) \in \mathbb{R}^3;\ (x_1, x_2) \in \Gamma_\mathrm{i},\ -\varepsilon h(x_1, x_2) < x_3 < 0 \right\};$$

$|\cdot|_\varepsilon$ will denote the norm in $L^2(\mathcal{M}_\varepsilon)$ (or product of such spaces) and $|\cdot|_i$ will denote the norm in $L^2(\Gamma_\mathrm{i})$ (or product of such spaces), $\|g\|_i = |\nabla g|_i$.

We will prove the following theorem:

THEOREM 4.2. *Assume that Γ_i is an open bounded set of $\mathbb{R}^2$, with C^3-boundary $\partial\Gamma_\mathrm{i}$ and $h \in C^4(\Gamma_\mathrm{i})$. Then, for $f \in L^2(\mathcal{M}_\varepsilon)$ and $g \in H_0^1(\Gamma_\mathrm{i})$, there exists a unique $\Psi \in H^2(\mathcal{M}_\varepsilon)$ solution of*

$$\begin{cases} -\Delta_3 \Psi = f & \text{in } \mathcal{M}_\varepsilon, \\ \frac{\partial \Psi}{\partial x_3} + \alpha \Psi = g & \text{on } \Gamma_\mathrm{i}, \\ \Psi = 0 & \text{on } \Gamma_\mathrm{b} \cup \Gamma_\ell. \end{cases} \tag{4.25}$$

Furthermore, there exists a constant $C = C(h, \Gamma_\mathrm{i}, \alpha)$ independent on ε such that

$$\sum_{i,j=1}^{3} \left| \frac{\partial^2 \Psi}{\partial x_i \, \partial x_j} \right|_\varepsilon^2 \leqslant C(h, \Gamma_\mathrm{i}, \alpha) \left[|f|_\varepsilon^2 + \|g\|_{H^1(\Gamma_\mathrm{i})}^2 \right]. \tag{4.26}$$

PROOF. The proof is divided into several steps.

First we construct a function Ψ^* satisfying the boundary conditions in (4.25), and find the precise dependence on ε of the L^2-norm of the second-order derivatives of Ψ^* (see Lemma 4.1). Then we set

$$\widehat{\Phi} = \mathrm{e}^{\alpha x_3}\left(\Psi - \Psi^*\right) \tag{4.27}$$

and verify that $\widehat{\Phi}$ satisfies the homogeneous Neumann condition on Γ_{i} ($\partial\widehat{\Phi}/\partial x_3 = 0$ on Γ_{i}) and the homogeneous Dirichlet boundary condition on $\Gamma_\ell \cup \Gamma_{\mathrm{b}}$ ($\widehat{\Phi} = 0$ on $\Gamma_\ell \cup \Gamma_{\mathrm{b}}$). By a reflection argument, we extend f to $x_3 > 0$ to be an even function and consider a homogeneous Dirichlet problem on the extended domain $\widehat{\mathcal{M}_\varepsilon} = \{(x_1, x_2, x_3);\ (x_1, x_2) \in \Gamma_{\mathrm{i}},\ -\varepsilon h(x_1, x_2) < x_3 < \varepsilon h(x_1, x_2)\}$, the solution of which, $\widehat{W}$, coincides with $\widehat{\Phi}$ on $\mathcal{M}_\varepsilon$. Finally we invoke Theorem 4.1 to conclude the H^2 regularity of $\widehat{W}$ and thus of $\widehat{\Phi}$, along with an estimate of the type (4.26) for $\widehat{\Phi}$; we therefore obtain the H^2 regularity of Ψ and (4.26) by simply using (4.27).

Thus the whole proof of Theorem 4.2 hinges on the following lifting lemma.

LEMMA 4.1. *Let $h \in C^4(\overline{\Gamma_{\mathrm{i}}})$ and $g \in H_0^1(\Gamma_{\mathrm{i}})$. There exists $\Psi^* \in H^2(\mathcal{M}_\varepsilon)$ such that $\frac{\partial \Psi^*}{\partial x_3} + \alpha\Psi^* = g$ on Γ_{i}, $\Psi^* = 0$ on $\Gamma_\ell \cup \Gamma_{\mathrm{b}}$, and there exists a constant $C = C(h, \Gamma_{\mathrm{i}})$ independent on ε, such that for $0 < \varepsilon \leqslant 1$,*

$$\sum_{i,j} \left|\frac{\partial^2 \Psi^*}{\partial x_k\, \partial x_j}\right|_\varepsilon^2 \leqslant C(h, \Gamma_{\mathrm{i}}) \|g\|_{H^1(\Gamma_{\mathrm{i}})}^2. \tag{4.28}$$

PROOF. First we construct a function $\widetilde{\Psi}$ as a solution of the heat equation with $-x_3$ corresponding to time

$$\begin{cases} \dfrac{\partial\widetilde{\Psi}}{\partial x_3} = -\Delta\widetilde{\Psi} & \text{in } \Gamma_{\mathrm{i}} \times (-\infty, 0), \\ \widetilde{\Psi} = 0 & \text{on } \partial\Gamma_{\mathrm{i}} \times (-\infty, 0), \\ \widetilde{\Psi}(x_1, x_2, 0) = g(x_1, x_2) & \text{on } \Gamma_{\mathrm{i}}. \end{cases} \tag{4.29}$$

Here $\Delta = \Delta_2 = (\partial^2/\partial x_1^2 + \partial^2/\partial x_2^2)$ and, below, $\nabla = \nabla_2 = (\partial/\partial x_1, \partial/\partial x_2)$. The function Ψ^* is then constructed as

$$\Psi^*(x_1, x_2, x_3) = \mathrm{e}^{-\alpha x_3} \int_{-\varepsilon h(x_1,x_2)}^{x_3} \widetilde{\Psi}(x_1, x_2, z)\, \mathrm{d}z. \tag{4.30}$$

It is clear that $\Psi^* \equiv 0$ on $\Gamma_\ell \cup \Gamma_{\mathrm{b}}$, and $\partial\Psi^*/\partial x_3 + \alpha\Psi^* = \mathrm{e}^{-\alpha x_3}\widetilde{\Psi}(x_1, x_2, x_3)$ in $\Gamma_{\mathrm{i}} \times (-\infty, 0)$, which implies $\partial\Psi^*/\partial x_3 + \alpha\Psi^* = g$ on Γ_{i}. We only need to check that $\Psi^* \in H^2(\mathcal{M}_\varepsilon)$ and that the inequality (4.28) is valid. This will be done using the classical energy estimates on the solution of the heat equation which are recalled in Lemmas 4.2 and 4.3.

We note that for $k=1,2$,

$$e^{\alpha x_3}\frac{\partial\Psi^*}{\partial x_k}=\int_{-\varepsilon h(x_1,x_2)}^{x_3}\frac{\partial\widetilde{\Psi}}{\partial x_k}(x_1,x_2,z)\,dz+\varepsilon\frac{\partial h}{\partial x_k}\widetilde{\Psi}\big(x_1,x_2,-\varepsilon h(x_1,x_2)\big) \tag{4.31}$$

and, for $k, j=1,2$,

$$\begin{aligned}e^{\alpha x_3}\frac{\partial^2\Psi^*}{\partial x_k\,\partial x_j}&=\int_{-\varepsilon h(x_1,x_2)}^{x_3}\frac{\partial^2\widetilde{\Psi}}{\partial x_k\,\partial x_j}(x_1,x_2,z)\,dz\\&\quad+\varepsilon\frac{\partial h}{\partial x_j}\frac{\partial\widetilde{\Psi}}{\partial x_k}\big(x_1,x_2,-\varepsilon h(x_1,x_2)\big)\\&\quad+\varepsilon\frac{\partial h}{\partial x_k}\frac{\partial\widetilde{\Psi}}{\partial x_j}\big(x_1,x_2,-\varepsilon h(x_1,x_2)\big)\\&\quad+\varepsilon\frac{\partial^2 h}{\partial x_k\,\partial x_j}\widetilde{\Psi}\big(x_1,x_2,-\varepsilon h(x_1,x_2)\big)\\&\quad-\varepsilon^2\frac{\partial h}{\partial x_k}\frac{\partial h}{\partial x_j}\frac{\partial\widetilde{\Psi}}{\partial x_3}\big(x_1,x_2,-\varepsilon h(x_1,x_2)\big).\end{aligned} \tag{4.32}$$

Here we need bounds on the L^2-norm (on Γ_{i}) of $(\partial\widetilde{\Psi}/\partial x_k)(x_1,x_2,-\varepsilon h(x_1,x_2))$ and $(\partial\widetilde{\Psi}/\partial x_3)(x_1,x_2,-\varepsilon h(x_1,x_2))$ which are provided by Lemma 4.3. We have

$$\begin{aligned}&\int_{\Gamma_{\mathrm{i}}}\big|\widetilde{\Psi}\big(x_1,x_2,-\varepsilon h(x_1,x_2)\big)\big|^2\,dx_1\,dx_2\leqslant C_0\|g\|_i^2,\\&\int_{\Gamma_{\mathrm{i}}}\big|\nabla\widetilde{\Psi}\big(x_1,x_2,-\varepsilon h(x_1,x_2)\big)\big|^2\,dx_1\,dx_2\leqslant C_0\|g\|_i^2,\\&\int_{\Gamma_{\mathrm{i}}}\left|\frac{\partial\widetilde{\Psi}}{\partial x_3}\big(x_1,x_2,-\varepsilon h(x_1,x_2)\big)\right|^2\,dx_1\,dx_2\leqslant\frac{1}{\varepsilon}C_0\|g\|_i^2.\end{aligned} \tag{4.33}$$

Now, using (4.30)–(4.32) and (4.33), we obtain

$$\left|\frac{\partial^2\Psi^*}{\partial x_k\,\partial x_j}\right|_\varepsilon^2\leqslant C_0\|g\|_i^2,\quad k,j=1,2. \tag{4.34}$$

Similar relations hold for Ψ^* and $\nabla\Psi^*$, using (4.30) and (4.31). Furthermore, since $\partial\Psi^*/\partial x_3=-\alpha\Psi^*+e^{-\alpha x_3}\widetilde{\Psi}$, we have

$$\left|\frac{\partial\Psi^*}{\partial x_3}\right|_\varepsilon^2\leqslant 2\alpha^2|\Psi^*|_\varepsilon^2+2e^{2\alpha\bar{h}}|\widetilde{\Psi}|_\varepsilon^2\leqslant\varepsilon^2C(h)\|g\|_i^2, \tag{4.35}$$

$$\left|\frac{\partial}{\partial x_3}\nabla\Psi^*\right|_\varepsilon^2\leqslant 2\alpha^2|\nabla\Psi^*|_\varepsilon^2+2e^{2\alpha\bar{h}}|\nabla\widetilde{\Psi}|_\varepsilon^2\leqslant\varepsilon^2C(h)\|g\|_i^2. \tag{4.36}$$

Finally

$$\frac{\partial^2 \Psi^*}{\partial x_3^2} = -\alpha \frac{\partial \Psi^*}{\partial x_3} - \alpha \mathrm{e}^{-\alpha x_3} \widetilde{\Psi} + \mathrm{e}^{-\alpha x_3} \frac{\partial \widetilde{\Psi}}{\partial x_3}, \tag{4.37}$$

implies

$$\left| \frac{\partial^2 \Psi^*}{\partial x_3^2} \right|_{\varepsilon}^2 \leqslant C(h) \| g \|_i^2. \tag{4.38}$$

The proof of Lemma 4.1 is complete. □

The proof of Theorem 4.2 relied on estimates given by Lemmas 4.2 and 4.3 which we now state and prove

LEMMA 4.2 (Estimates on solutions of the heat equation). *Let Ψ be the solution of the heat equation*

$$\begin{aligned} &\frac{\partial \Psi}{\partial x_3} = -\Delta \Psi \quad \text{in } \Gamma_{\mathrm{i}} \times (-\infty, 0), \\ &\Psi(x_1, x_2, 0) = g(x_1, x_2) \quad \text{on } \Gamma_{\mathrm{i}}, \end{aligned} \tag{4.39}$$

with either Dirichlet or Neumann boundary condition, and

$$g \in H_0^1(\Gamma_{\mathrm{i}}) \quad \text{and} \quad \Psi = 0 \quad \text{on } \partial \Gamma_{\mathrm{i}} \times (-\infty, 0)$$

or

$$g \in H^1(\Gamma_{\mathrm{i}}) \quad \text{and} \quad \frac{\partial \Psi}{\partial n_{\Gamma_{\mathrm{i}}}} = 0 \quad \text{on } \partial \Gamma_{\mathrm{i}} \times (-\infty, 0).$$

Then

$$\begin{aligned} &\frac{1}{2} |x_3|^k \left| \frac{\partial^j \Psi}{\partial x_3^j} \right|_i^2 (x_3) + \int_{x_3}^0 |z|^k \left| \nabla \frac{\partial^j \Psi}{\partial x_3^j} \right|_i^2 (z) \, \mathrm{d}z \\ &\quad \leqslant \begin{cases} \frac{1}{2} |g|_i^2 & \text{for } k = j = 0, \\ C |x_3|^{k-2j+1} \| g \|_i^2 & \text{for } k \geqslant 2j - 1,\ j \geqslant 1, \end{cases} \end{aligned} \tag{4.40}$$

and

$$\begin{aligned} &\frac{1}{2} |x_3|^k \left| \nabla \frac{\partial^j \Psi}{\partial x_3^j} \right|_i^2 (x_3) + \int_{x_3}^0 |z|^k \left| \frac{\partial^{j+1} \Psi}{\partial x_3^{j+1}} \right|_i^2 (z) \, \mathrm{d}z \\ &\quad \leqslant \begin{cases} \frac{1}{2} \| g \|_i^2 & \text{for } k = 0,\ j = 0, \\ C |x_3|^{k-2j} \| g \|_i^2 & \text{for } k \geqslant 2j,\ j \geqslant 1. \end{cases} \end{aligned} \tag{4.41}$$

In (4.40) and (4.41), C is a constant depending on k, j and h. As before, $\nabla = \nabla_2 = (\partial/\partial x_1, \partial/\partial x_2)$.

Proof of Lemma 4.2. Denote by $e_{k,j}$ and $f_{k,j}$ the left-hand sides of (4.40) and (4.41).

We differentiate (4.39) j times in x_3, multiply the resulting equation by $x_3^k\, \partial^j\Psi/\partial x_3^j$ and integrate over $\Gamma_{\rm i}$. Using Stokes formula and observing that $\partial^j\Psi/\partial x_3^j$ satisfies the same boundary condition on $\partial\Gamma_{\rm i}$ as Ψ, we obtain after multiplication by $(-1)^k$:

$$e_{k,j} = \begin{cases} \frac{k}{2}\int_{x_3}^0 |z|^{k-1}\left|\frac{\partial^j\Psi}{\partial x_3^j}\right|_i^2(z)\,\mathrm{d}z & \text{for } k \geqslant 1,\\ \frac{1}{2}\left|\frac{\partial^j\Psi}{\partial x_3^j}\right|_i^2(0) = \frac{1}{2}\left|\Delta^j\Psi\right|_i^2(0) = \frac{1}{2}\left|\Delta^j g\right|_i^2 & \text{for } k = 0. \end{cases} \tag{4.42}$$

Similarly, if we differentiate (4.39) j times in x_3, multiply the resulting equation by $x_3^k\, \partial^{j+1}\Psi/\partial x_3^{j+1}$ and integrate over $\Gamma_{\rm i}$, we find

$$f_{k,j} = \begin{cases} \frac{k}{2}\int_{x_3}^0 |z|^{k-1}\left|\nabla\frac{\partial^j\Psi}{\partial x_3^j}\right|_i^2(z)\,\mathrm{d}z & \text{for } k \geqslant 1,\\ \frac{1}{2}\left|\nabla\frac{\partial^j\Psi}{\partial x_3^j}\right|_i^2(0) = \frac{1}{2}\left|\nabla\Delta^j g\right|_i^2 & \text{for } k = 0. \end{cases} \tag{4.43}$$

Using (4.42), (4.43) with $k = j = 0$, (4.42) with $k = j = 1$ and (4.43) with $k = 2, j = 1$, we find some of the relations (4.40), (4.41), namely

$$\begin{aligned} &\frac{1}{2}\left|\Psi(x_3)\right|_i^2 + \int_{x_3}^0 |\nabla\Psi|_i^2(z)\,\mathrm{d}z \leqslant \frac{1}{2}|g|_i^2,\\ &\frac{1}{2}\left|\nabla\Psi(x_3)\right|_i^2 + \int_{x_3}^0 \left|\frac{\partial\Psi}{\partial x_3}\right|_i^2(z)\,\mathrm{d}z \leqslant \frac{1}{2}\|g\|_i^2,\\ &\frac{1}{2}|x_3|\left|\frac{\partial\Psi}{\partial x_3}\right|_i^2(x_3) + \int_{x_3}^0 |z|\left|\nabla\frac{\partial\Psi}{\partial x_3}\right|_i^2(z)\,\mathrm{d}z \leqslant \frac{1}{4}\|g\|_i^2,\\ &\frac{1}{2}|x_3|^2\left|\nabla\frac{\partial\Psi}{\partial x_3}\right|_i^2(x_3) + \int_{x_3}^0 z^2\left|\frac{\partial^2\Psi}{\partial x_3^2}\right|_i^2(z)\,\mathrm{d}z \leqslant \frac{1}{4}\|g\|_i^2. \end{aligned} \tag{4.44}$$

To derive the other relations (4.40) and (4.41), we first integrate (4.42) from x_3 to 0, with $x_3 < 0, k \geqslant 1$ and $j \geqslant 0$; we obtain

$$\begin{aligned} \int_{x_3}^0 |z|^k\left|\frac{\partial^j\Psi}{\partial x_3^j}\right|_i^2(z)\,\mathrm{d}z &\leqslant k\int_{x_3}^0\int_t^0 |z|^{k-1}\left|\frac{\partial^j\Psi}{\partial x_3^j}\right|_i^2(z)\,\mathrm{d}z\,\mathrm{d}t\\ &\leqslant k\int_{x_3}^0 (z - x_3)|z|^{k-1}\left|\frac{\partial^j\Psi}{\partial x_3^j}\right|_i^2(z)\,\mathrm{d}z. \end{aligned}$$

Thus, for $k \geqslant 1$, $j \geqslant 0$,

$$\int_{x_3}^{0} |z|^k \left|\frac{\partial^j \Psi}{\partial x_3^j}\right|_i^2 (z)\,\mathrm{d}z \leqslant \frac{k}{k+1}|x_3| \int_{x_3}^{0} |z|^{k-1} \left|\frac{\partial^j \Psi}{\partial x_3^j}\right|_i^2 (z)\,\mathrm{d}z \tag{4.45}$$

and, by induction on k and using (4.41),

$$\begin{aligned} \int_{x_3}^{0} |z|^k \left|\frac{\partial^j \Psi}{\partial x_3^j}\right|_i^2 (z)\,\mathrm{d}z &\leqslant \frac{|x_3|^k}{k+1} \int_{x_3}^{0} \left|\frac{\partial^j \Psi}{\partial x_3^j}\right|_i^2 (z)\,\mathrm{d}z, \\ &\leqslant \begin{cases} \frac{|x_3|^{k+1}}{2(k+1)} |g|_i^2 & \text{for } j = 0, \\ \frac{|x_3|^k}{2(k+1)} \|g\|_i^2 & \text{for } j = 1 \ (k \geqslant 1). \end{cases} \end{aligned} \tag{4.46}$$

Now, combining (4.42) and (4.43), we find for $k \geqslant 2$, $j \geqslant 1$,

$$e_{k,j} \leqslant \frac{k(k-1)}{2^2} e_{k-2,j-1}.$$

Hence, by induction

$$e_{k,j} \leqslant \frac{k!}{2^{2r}(k-2r)!} e_{k-2r,j-r},$$

$$e_{k,j} \leqslant \frac{k!}{2^{2j-2}(k-2j+2)!} e_{s,1},$$

$s = k - 2j + 2$, for $k \geqslant 2j - 2$, $j \geqslant 1$. For $s \geqslant 1$, i.e., $k \geqslant 2j - 1$ and $j \geqslant 1$, we have, thanks to (4.42) and (4.46),

$$\begin{aligned} e_{s,1} &= \frac{1}{2} \int_{x_3}^{0} |z|^{s-1} \left|\frac{\partial \Psi}{\partial x_3}\right|_i^2 (z)\,\mathrm{d}z \\ &\leqslant \frac{1}{2} |x_3|^{s-1} \|g\|_i^2, \\ e_{k,j} &\leqslant \frac{k!}{2^{2j-1}(k-2j+2)!} |x_3|^{k-2j+1} \|g\|_i^2, \end{aligned} \tag{4.47}$$

for $k \geqslant 2j - 1$, $j \geqslant 1$. The relations (4.40) are proven, the relations (4.41) follow from (4.44) for $j = 0$ and from $f_{k,j} \leqslant (k/2)e_{k-1,j}$ for $j \geqslant 1$.

Lemma 4.2 is proved. □

From Lemma 4.2 we easily infer the following lemma:

LEMMA 4.3. *Under the hypotheses of Lemma* 4.2,

$$\int_{\Gamma_{\mathrm{i}}} \left|\nabla \frac{\partial^j \Psi}{\partial x_3^j}\right|^2 \bigl(x_1, x_2, -\varepsilon h(x_1, x_2)\bigr)\,\mathrm{d}x_1\,\mathrm{d}x_2 \leqslant C\varepsilon^{-2j}\|g\|^2_{H^1(\Gamma_{\mathrm{i}})} \quad \text{for } j \geqslant 0, \tag{4.48}$$

$$\int_{\Gamma_{\mathrm{i}}} \left|\frac{\partial^j \Psi}{\partial x_3^j}\right|^2 \bigl(x_1, x_2, -\varepsilon h(x_1, x_2)\bigr)\,\mathrm{d}x_1\,\mathrm{d}x_2 \leqslant \begin{cases} C\|g\|^2_{H^1(\Gamma_{\mathrm{i}})} & \text{for } j = 0,\\ C\varepsilon^{-2j+1}\|g\|^2_{H^1(\Gamma_{\mathrm{i}})} & \text{for } j \geqslant 1, \end{cases} \tag{4.49}$$

$$\int_{\Gamma_{\mathrm{i}}} |\nabla^j \Psi|^2 \bigl(x_1, x_2, -\varepsilon h(x_1, x_2)\bigr)\,\mathrm{d}x_1\,\mathrm{d}x_2 \leqslant C\varepsilon^{-j+1}\|g\|^2_{H^1(\Gamma_{\mathrm{i}})}, \quad j = 2, 3, \tag{4.50}$$

$$\int_{\Gamma_{\mathrm{i}}} \left|\frac{\partial}{\partial x_3}\nabla^2 \Psi\right|^2 \bigl(x_1, x_2, -\varepsilon h(x_1, x_2)\bigr)\,\mathrm{d}x_1\,\mathrm{d}x_2 \leqslant C\varepsilon^{-3}\|g\|^2_{H^1(\Gamma_{\mathrm{i}})}, \tag{4.51}$$

where $\|g\|^2_{H^1(\Gamma_{\mathrm{i}})} = |g|_i^2 + \|g\|_i^2$, *and* C *is a constant depending on* j *and* h *and independent of* ε.

PROOF. We write

$$\begin{aligned} &\left(\nabla \frac{\partial^j \Psi}{\partial x_3^j}\right)^2 \bigl(x_1, x_2, -\varepsilon h(x_1, x_2)\bigr)\\ &\quad = \left(\nabla \frac{\partial^j \Psi}{\partial x_3^j}\right)^2 (x_1, x_2, -\varepsilon \bar{h})\\ &\qquad + 2\int_{-\varepsilon \bar{h}}^{-\varepsilon h(x_1,x_2)} \left(\nabla \frac{\partial^j \Psi}{\partial x_3^j}\right)\left(\nabla \frac{\partial^{j+1} \Psi}{\partial x_3^{j+1}}\right)(x_1, x_2, z)\,\mathrm{d}z. \end{aligned}$$

Integrating in x_1, x_2 on Γ_{i} we find

$$\begin{aligned} &\int_{\Gamma_{\mathrm{i}}} \left(\nabla \frac{\partial^j \Psi}{\partial x_3^j}\right)^2 \bigl(x_1, x_2, -\varepsilon h(x_1, x_2)\bigr)\,\mathrm{d}x_1\,\mathrm{d}x_2\\ &\quad \leqslant \left|\nabla \frac{\partial^j \Psi}{\partial x_3^j}\right|_i^2 (-\varepsilon \bar{h})\\ &\qquad + 2\left(\int_{-\varepsilon \bar{h}}^{-\varepsilon \underline{h}} \left|\nabla \frac{\partial^j \Psi}{\partial x_3^j}\right|_i^2 (z)\,\mathrm{d}z\right)^{1/2} \left(\int_{-\varepsilon \bar{h}}^{-\varepsilon \underline{h}} \left|\nabla \frac{\partial^{j+1} \Psi}{\partial x_3^{j+1}}\right|_i^2 (z)\,\mathrm{d}z\right)^{1/2}. \end{aligned}$$

By (4.41),

$$\left|\nabla \frac{\partial^j \Psi}{\partial x_3^j}\right|_i^2 (-\varepsilon \bar{h}) \leqslant C(\varepsilon \bar{h})^{-2j} \|g\|_{H^1(\Gamma_{\mathrm{i}})}^2 \quad (j \geqslant 0)$$

and by integration of (4.41),

$$\begin{aligned}\int_{-\varepsilon \bar{h}}^{-\varepsilon \underline{h}} \left|\nabla \frac{\partial^j \Psi}{\partial x_3^j}\right|_i^2 (z)\,\mathrm{d}z &\leqslant C(\varepsilon \bar{h})^{-k} (\varepsilon \bar{h})^{k-2j+1} \|g\|_{H^1(\Gamma_{\mathrm{i}})}^2 \\ &\leqslant C\varepsilon^{-2j+1} \|g\|_{H^1(\Gamma_{\mathrm{i}})}^2 \quad (j \geqslant 0),\end{aligned}$$

(4.48) follows. The proof of (4.49) is similar.

For (4.50) and (4.51) we observe that, by the regularity property for the Neumann problem in Γ_{i},

$$\begin{aligned}\sum_{k,\ell=1}^{2} \left|\frac{\partial^2 \Psi}{\partial x_k\, \partial x_\ell}\right|_{L^2(\Gamma_{\mathrm{i}})}^2 (x_3) &\leqslant C\big(|\Psi|_{L^2(\Gamma_{\mathrm{i}})}^2 (x_3) + |\Delta \Psi|_{L^2(\Gamma_{\mathrm{i}})}^2 (x_3)\big) \\ &\leqslant C\left(|\Psi|_{L^2(\Gamma_{\mathrm{i}})}^2 (x_3) + \left|\frac{\partial \Psi}{\partial x_3}\right|_{L^2(\Gamma_{\mathrm{i}})}^2 (x_3)\right).\end{aligned} \tag{4.52}$$

By repeating the argument above, it appears that the bounds for the left-hand-sides of (4.50) and (4.51) are the same as those of (4.48) and (4.49), for $j = 1$ and 2 respectively; (4.50) and (4.51) are proved. □

4.3. *Regularity of solutions of a Neumann–Robin boundary value problem*

We now want to derive a result similar to that of Sections 4.1 and 4.2. for a mixed Neumann–Robin type boundary condition, that is for the problem (4.53) below. Our result is quite general except for the restrictions (4.54) below.

We will prove the following theorem:

THEOREM 4.3. *Assume Γ_{i} is an open bounded set of $\mathbb{R}^2$, with a C^3-boundary $\partial\Gamma_{\mathrm{i}}$ and $h \in C^4(\overline{\Gamma_{\mathrm{i}}})$. For $f \in L^2(\mathcal{M}_\varepsilon)$ and $g \in H^1(\Gamma_{\mathrm{i}})$, there exists a unique $\Psi \in H^2(\mathcal{M}_\varepsilon)$ solution of*

$$\begin{aligned}&-\Delta_3 \Psi = f, \\ &\frac{\partial \Psi}{\partial x_3} + \alpha \Psi = g \quad \text{on } \Gamma_{\mathrm{i}}, \\ &\frac{\partial \Psi}{\partial n} = 0 \quad \text{on } \Gamma_{\mathrm{b}} \cup \Gamma_\ell.\end{aligned} \tag{4.53}$$

Furthermore if

$$\nabla h \cdot n_{\Gamma_\mathrm{i}} = 0 \quad \text{on } \partial \Gamma_\mathrm{i}, \tag{4.54}$$

then there exists $C = C(h, \Gamma_\mathrm{i}, \alpha)$ *such that*

$$\sum_{k,\ell=1}^{3} \left| \frac{\partial^2 \Psi}{\partial x_k \, \partial x_\ell} \right|_\varepsilon^2 \leqslant C\big[|f|_\varepsilon^2 + \|g\|_{H^1}^2\big].$$

REMARK 4.1. Condition (4.54) means that Γ_b and Γ_ℓ meet at a wedge angle of $\pi/2$. This is a technical condition needed in the method of proof used below, this condition is not required by the theory of regularity of elliptic problems in nonsmooth three-dimensional problems [11]. It might be possible to remove this condition with a different proof. Note that this condition is needed for the dependence on ε *and not for the sole* H^2*-regularity*.

PROOF OF THEOREM 4.3. The proof is divided into several steps. First we reduce the problem to the case $\alpha = 0$, then we reduce the problem to the case where $g = 0$ and $\alpha = 0$. Thus we obtain a homogeneous Neumann problem. Then by a reflection around $x_3 = 0$, we can use Theorem 4.1 and conclude the proof.

STEP 1 (Reduction to the case $g = 0$). Our goal now is to find an explicit function $\overline{T}$ that satisfies the nonhomogeneous boundary conditions imposed on the temperature. Since we are interested in obtaining a sharp dependence on the thickness ε, we have to construct $\overline{T}$ explicitly instead of the classical method of lifting by localization and straightening the boundary which yields constants which do not have the right dependence on ε. We will carry the computations only in the case where $\nabla h \cdot n_{\Gamma_\mathrm{i}} = 0$ on $\partial \Gamma_\mathrm{i}$.

LEMMA 4.4. *Let* $h \in C^4(\overline{\Gamma_\mathrm{i}})$ *with* $\nabla h \cdot n_{\Gamma_\mathrm{i}} = 0$ *on* $\partial \Gamma_\mathrm{i}$ *and let* $g \in H^1(\Gamma_\mathrm{i})$. *There exists a function* $\overline{T} \in H^2(\mathcal{M}_\varepsilon)$ *such that*

$$\begin{aligned} &\frac{\partial \overline{T}}{\partial x_3} + \alpha \overline{T} = g \quad \text{on } \Gamma_\mathrm{i}, \\ &\frac{\partial \overline{T}}{\partial n} = 0 \quad \text{on } \Gamma_\ell \cup \Gamma_\mathrm{b}. \end{aligned} \tag{4.55}$$

Furthermore

$$\sum_{k,l=1}^{3} \left| \frac{\partial^2 \overline{T}}{\partial x_k \, \partial x_l} \right|_\varepsilon^2 \leqslant C \|g\|_{H^1(\Gamma_\mathrm{i})}^2, \tag{4.56}$$

where C *is a constant independent on* ε.

PROOF. Let $\overline{\Psi}$ be a solution of the heat equation, where $-x_3$ corresponds to time:

$$\begin{cases} \frac{\partial \overline{\Psi}}{\partial x_3} = -\Delta \overline{\Psi} & \text{in } \Gamma_{\mathrm{i}} \times (-\infty, 0), \\ \frac{\partial \overline{\Psi}}{\partial n_{\Gamma_{\mathrm{i}}}} = 0 & \text{on } \partial \Gamma_{\mathrm{i}} \times (-\infty, 0), \\ \overline{\Psi}(x_1, x_2, 0) = g(x_1, x_2), \end{cases} \tag{4.57}$$

and define $\overline{T}$ by

$$\begin{aligned} \overline{T}(x_1, x_2, x_3) = {} & \mathrm{e}^{-\alpha x_3} \int_{-\varepsilon h(x_1, x_2)}^{x_3} \overline{\Psi}(x_1, x_2, z)\, \mathrm{d}z \\ & - \left(x_3 - \frac{1}{\alpha} \right) \theta_1(x_1, x_2) + x_3^2 (x_3 + \varepsilon h) \theta_2(x_1, x_2), \end{aligned} \tag{4.58}$$

where

$$\theta_1(x_1, x_2) = \mathrm{e}^{\alpha \varepsilon h(x_1, x_2)} \overline{\Psi}\big(x_1, x_2, -\varepsilon h(x_1, x_2)\big)\big(1 + \varepsilon^2 |\nabla h|^2\big) \tag{4.59}$$

and

$$\theta_2(x_1, x_2) = \frac{-(\varepsilon h + \frac{1}{\alpha})}{\varepsilon h^2 (1 + \varepsilon^2 |\nabla h|^2)} \nabla \theta_1 \cdot \nabla h. \tag{4.60}$$

Then

$$\begin{aligned} \frac{\partial \overline{T}}{\partial x_3} = {} & -\alpha \mathrm{e}^{-\alpha x_3} \int_{-\varepsilon h}^{x_3} \overline{\Psi}(x_1, x_2, z)\, \mathrm{d}z + \mathrm{e}^{-\alpha x_3} \overline{\Psi}(x_1, x_2, x_3) \\ & - \theta_1(x_1, x_2) + 2 x_3 (x_3 + \varepsilon h) \theta_2(x_1, x_2) + x_3^2 \theta_2(x_1, x_2), \end{aligned} \tag{4.61}$$

$$\left. \frac{\partial \overline{T}}{\partial x_3} + \alpha \overline{T} \right|_{x_3 = 0} = \overline{\Psi}(x_1, x_2, 0) = g, \tag{4.62}$$

$$\begin{aligned} \nabla \overline{T} = {} & \mathrm{e}^{-\alpha x_3} \int_{-\varepsilon h}^{x_3} \nabla \overline{\Psi}(x_1, x_2, z)\, \mathrm{d}z + \varepsilon \mathrm{e}^{-\alpha x_3} \overline{\Psi}(x_1, x_2, -\varepsilon h) \nabla h \\ & - \left(x_3 - \frac{1}{\alpha} \right) \nabla \theta_1 + x_3^2 (x_3 + \varepsilon h) \nabla \theta_2 + \varepsilon x_3^2 \theta_2 \nabla h, \end{aligned} \tag{4.63}$$

$$\begin{aligned} \varepsilon \nabla \overline{T} \cdot \nabla h \big|_{x_3 = -\varepsilon h} = {} & \varepsilon^2 \mathrm{e}^{\alpha \varepsilon h} \overline{\Psi}(x_1, x_2, -\varepsilon h) |\nabla h|^2 \\ & + \varepsilon \left(\varepsilon h + \frac{1}{\alpha} \right) \nabla \theta_1 \cdot \nabla h + \varepsilon^4 h^2 \theta_2 |\nabla h|^2, \end{aligned} \tag{4.64}$$

$$\left. \frac{\partial \overline{T}}{\partial x_3} \right|_{x_3 = -\varepsilon h} = \mathrm{e}^{\alpha \varepsilon h} \overline{\Psi}(x_1, x_2, -\varepsilon h) - \theta_1(x_1, x_2) + \varepsilon^2 h^2 \theta_2$$

and

$$\frac{\partial \overline{T}}{\partial x_3} + \varepsilon \nabla \overline{T} \cdot \nabla h \bigg|_{x_3=-\varepsilon h} = \mathrm{e}^{\alpha \varepsilon h} \overline{\Psi}(x_1, x_2, -\varepsilon h)\big(1 + \varepsilon^2 |\nabla h|^2\big)$$

$$+ \varepsilon^2 h^2 \theta_2 \big(1 + \varepsilon^2 |\nabla h|^2\big) - \theta_1$$

$$+ \varepsilon \left(\varepsilon h + \frac{1}{\alpha}\right) \nabla \theta_1 \cdot \nabla h.$$

Hence, with θ_1, θ_2 as in (4.59) and (4.60), we have

$$\frac{\partial \overline{T}}{\partial x_3} + \varepsilon \nabla \overline{T} \cdot \nabla h = 0 \quad \text{on } \Gamma_{\mathrm{b}}.$$

Now, we use the assumption (4.54) and prove that

$$\nabla \theta_1 \cdot n_{\Gamma_{\mathrm{i}}} = 0, \nabla \theta_2 \cdot n_{\Gamma_{\mathrm{i}}} = 0 \qquad \text{on } \partial \Gamma_{\mathrm{i}}, \tag{4.65}$$

which implies that $\nabla \overline{T} \cdot n_{\Gamma_{\mathrm{i}}} = 0$ on $\partial \Gamma_{\mathrm{i}}$.

First, by working in local coordinates (s, t) where s is the coordinate in the normal direction of $\partial \Gamma_{\mathrm{i}}$ and t the coordinate in the tangential direction, the condition $\nabla h \cdot n_{\Gamma_{\mathrm{i}}} = 0$ on Γ_{i} implies (since $\partial \Gamma_{\mathrm{i}}$ is smooth)

$$\frac{\partial h}{\partial s} = 0 \quad \text{and} \quad \frac{\partial^2 h}{\partial s \, \partial t} = 0 \quad \text{on } \partial \Gamma_{\mathrm{i}}. \tag{4.66}$$

Therefore

$$\frac{\partial}{\partial s} |\nabla h|^2 = \frac{\partial}{\partial s} \left(\left| \frac{\partial h}{\partial s} \right|^2 + \left| \frac{\partial h}{\partial t} \right|^2 \right) = 2 \left(\frac{\partial h}{\partial s} \frac{\partial^2 h}{\partial s^2} + \frac{\partial h}{\partial t} \frac{\partial^2 h}{\partial s \, \partial t} \right) = 0. \tag{4.67}$$

Now

$$\begin{aligned} \nabla \theta_1 \cdot n_{\Gamma_{\mathrm{i}}} &= \alpha \varepsilon \mathrm{e}^{\alpha \varepsilon h} \overline{\Psi}(x_1, x_2, -\varepsilon h)\big(1 + \varepsilon^2 |\nabla h|^2\big) \nabla h \cdot n_{\Gamma_{\mathrm{i}}} \\ &\quad + \mathrm{e}^{\alpha \varepsilon h} \big(1 + \varepsilon^2 |\nabla h|^2\big) \nabla \overline{\Psi} \cdot n_{\Gamma_{\mathrm{i}}} \\ &\quad - \varepsilon \mathrm{e}^{\alpha \varepsilon h} \frac{\partial \overline{\Psi}}{\partial x_3}(x_1, x_2, -\varepsilon h)\big(1 + \varepsilon^2 |\nabla h|^2\big) \nabla h \cdot n_{\Gamma_{\mathrm{i}}} \\ &\quad + \mathrm{e}^{\alpha \varepsilon h} \overline{\Psi}(x_1, x_2, -\varepsilon h) \nabla\big(|\nabla h|^2\big) \cdot n_{\Gamma_{\mathrm{i}}} \end{aligned} \tag{4.68}$$

and since $\nabla h \cdot n_{\Gamma_{\mathrm{i}}} = 0$, $\nabla \overline{\Psi} \cdot n_{\Gamma_{\mathrm{i}}} = 0$ and $\nabla(|\nabla h|^2) \cdot n_{\Gamma_{\mathrm{i}}} = 0$ on $\partial \Gamma_{\mathrm{i}}$, we have $\nabla \theta_1 \cdot n_{\Gamma_{\mathrm{i}}} = 0$ on $\partial \Gamma_{\mathrm{i}}$.

Next we check that $\nabla\theta_2 \cdot n_{\Gamma_i} = 0$ on $\partial\Gamma_i$. Here we only need to show that $\nabla(\nabla\theta_1 \cdot \nabla h) \cdot n_{\Gamma_i} = 0$ on $\partial\Gamma_i$. Again, this can be done by working in local coordinates. We have

$$\nabla\theta_1 \cdot \nabla h = \frac{\partial\theta_1}{\partial s}\frac{\partial h}{\partial s} + \frac{\partial\theta_1}{\partial t}\frac{\partial h}{\partial t} \tag{4.69}$$

and therefore

$$\begin{aligned} \nabla(\nabla\theta_1 \cdot \nabla h) \cdot n_{\Gamma_i} &= \frac{\partial}{\partial n_{\Gamma_i}}(\nabla\theta_1 \cdot \nabla h) \\ &= \frac{\partial\theta_1}{\partial s}\frac{\partial^2 h}{\partial s^2} + \frac{\partial^2\theta_1}{\partial s^2}\frac{\partial h}{\partial s} + \frac{\partial^2\theta_1}{\partial t\,\partial s}\frac{\partial h}{\partial t} + \frac{\partial\theta_1}{\partial t}\frac{\partial^2 h}{\partial s\,\partial t} \end{aligned} \tag{4.70}$$

but since $\partial h/\partial s = 0$ and $\partial\theta_1/\partial s = 0$ we have $\partial^2 h/\partial s\,\partial t = 0$ and $\partial^2\theta_1/\partial s\,\partial t = 0$. Thus

$$\nabla(\nabla\theta_1 \cdot \nabla h) \cdot n_{\Gamma_i} = 0 \quad \text{on } \partial\Gamma_i. \tag{4.71}$$

Finally, since θ_2 is the product of functions each of which has normal derivative to $\partial\Gamma_i$ vanishing on $\partial\Gamma_i$, we have

$$\nabla\theta_2 \cdot n_{\Gamma_i} = 0 \quad \text{on } \partial\Gamma_i. \tag{4.72}$$

This concludes the verification of $\overline{T}$ satisfying the boundary conditions. We now use estimates on solutions of the heat equation and the explicit expression of $\overline{T}$ to establish the inequality (4.56) of the lemma. Here we fully rely on the estimates for $\overline{\Psi}$ provided by Lemmas 4.2 and 4.3.

With $\partial\overline{T}/\partial x_3$ and $\nabla\overline{T}$ given by (4.61) and (4.63), we write

$$\begin{aligned} \frac{\partial^2\overline{T}}{\partial x_3^2} &= \alpha^2 e^{-\alpha x_3}\int_{-\varepsilon h}^{x_3} \overline{\Psi}(x_1, x_2, z)\,dz \\ &\quad - 2\alpha e^{-\alpha x_3}\overline{\Psi}(x_1, x_2, x_3) + e^{-\alpha x_3}\frac{\partial\overline{\Psi}}{\partial x_3}(x_1, x_2, x_3) \\ &\quad + 2(x_3 + \varepsilon h)\theta_2(x_1, x_2) + 4x_3\theta_2(x_1, x_2), \end{aligned} \tag{4.73}$$

and, for $k = 1, 2$,

$$\begin{aligned} &\frac{\partial^2\overline{T}}{\partial x_k\,\partial x_3} \\ &= -\alpha e^{-\alpha x_3}\int_{-\varepsilon h}^{x_3} \frac{\partial\overline{\Psi}}{\partial x_k}(x_1, x_2, z)\,dz - \alpha\varepsilon e^{-\alpha x_3}\overline{\Psi}(x_1, x_2, -\varepsilon h)\frac{\partial h}{\partial x_k} \\ &\quad + e^{-\alpha x_3}\frac{\partial\overline{\Psi}}{\partial x_k} - \frac{\partial\theta_1}{\partial x_k} + 2\varepsilon x_3\theta_2\frac{\partial h}{\partial x_k} + 2x_3(x_3 + \varepsilon h)\frac{\partial\theta_2}{\partial x_k} + x_3^2\frac{\partial\theta_2}{\partial x_k}. \end{aligned} \tag{4.74}$$

Finally, for $k, \ell = 1, 2$,

$$
\begin{aligned}
\frac{\partial^2 \overline{T}}{\partial x_k\, \partial x_\ell} &= \mathrm{e}^{-\alpha x_3} \int_{-\varepsilon h(x_1,x_2)}^{x_3} \frac{\partial^2 \overline{\Psi}}{\partial x_k\, \partial x_\ell}(x_1, x_2, z)\, \mathrm{d}z \\
&= +\varepsilon \mathrm{e}^{-\alpha x_3} \frac{\partial \overline{\Psi}}{\partial x_\ell}\big(x_1, x_2, -\varepsilon h(x_1, x_2)\big) \frac{\partial h}{\partial x_k} \\
&\quad + \varepsilon \mathrm{e}^{-\alpha x_3} \frac{\partial \overline{\Psi}}{\partial x_k}\big(x_1, x_2, -\varepsilon h(x_1, x_2)\big) \frac{\partial h}{\partial x_\ell} \\
&\quad - \varepsilon^2 \mathrm{e}^{-\alpha x_3} \frac{\partial \overline{\Psi}}{\partial x_3}(x_1, x_2, -\varepsilon h) \frac{\partial h}{\partial x_k} \frac{\partial h}{\partial x_\ell} \\
&\quad + \varepsilon \mathrm{e}^{-\alpha x_3} \overline{\Psi}(x_1, x_2, -h) \frac{\partial^2 h}{\partial x_k\, \partial x_\ell} - \left(x_3 - \frac{1}{\alpha} \right) \frac{\partial^2 \theta_1}{\partial x_k\, \partial x_\ell} \\
&\quad + x_3^2 (x_3 + \varepsilon h) \frac{\partial^2 \theta_2}{\partial x_k\, \partial x_\ell} + \varepsilon x_3^2 \frac{\partial h}{\partial x_k} \frac{\partial \theta_2}{\partial x_\ell} + \varepsilon x_3^2 \theta_2 \frac{\partial^2 h}{\partial x_k\, \partial x_\ell}.
\end{aligned} \tag{4.75}
$$

To estimate the L^2-norms of the second derivatives of $\overline{T}$, we need to bound the L^2-norms of θ_1, θ_2 and their derivatives, which we do in Lemma 4.5.

Using Lemma 4.5, we estimate as follows, the norm, in $L^2(\mathcal{M}_\varepsilon)$, of $\partial^2 \overline{T}/\partial x_k\, \partial x_\ell$, $k, \ell = 1, 2$, as given by (4.75).

The first term in the right-hand side of (4.75) is bounded by a constant times the norm of $\partial^2 \overline{\Psi}/\partial x_k\, \partial x_\ell$ in $Q_\varepsilon = \Gamma_{\mathrm{i}} \times (-\varepsilon \bar{h}, 0)$; using (4.51) and (4.41), this term is bounded as in (4.56). We then use Lemma 4.3 to estimate the four subsequent terms, and the bounds are consistent with (4.56). The remaining terms involve θ_1, θ_2 and their derivatives; the integration over Γ_{i} of these functions provide the bounds given by Lemma 4.5, and, for each of these terms there is a factor of $C\varepsilon^m$, $m \geqslant 2$, which is due to the integration in x_3. The bound (4.56) follows.

We proceed similarly for $\partial^2 \overline{T}/\partial x_3^2$ given by (4.75) and for $\partial^2 \overline{T}/\partial x_k\, \partial x_3$, $k = 1, 2$, given by (4.74). Lemma 4.4 follows. □

We now conclude the proof of Lemma 4.4 by proving Lemma 4.5.

LEMMA 4.5. *The functions* θ_1 *and* θ_2 *introduced in* (4.59) *and* (4.60), *are bounded as follows*:

$$
|\theta_1|_{L^2(\Gamma_{\mathrm{i}})} + |\nabla \theta_1|_{L^2(\Gamma_{\mathrm{i}})} + \varepsilon^{1/2} \big|\nabla^2 \theta_1\big|_{L^2(\Gamma_{\mathrm{i}})} \leqslant C \|g\|^2_{H^1(\Gamma_{\mathrm{i}})}, \tag{4.76}
$$

$$
|\theta_2|_{L^2(\Gamma_{\mathrm{i}})} + \varepsilon^{1/2} \big|\nabla \theta_2\big|_{L^2(\Gamma_{\mathrm{i}})} + \varepsilon \big|\nabla^2 \theta_2\big|_{L^2(\Gamma_{\mathrm{i}})} \leqslant C \varepsilon^{-1} \|g\|^2_{H^1(\Gamma_{\mathrm{i}})}, \tag{4.77}
$$

where C *is a constant independent of* ε.

PROOF. The proof strongly relies on the definition (4.59) and (4.60) of θ_1 and θ_2, and on the estimates on $\overline{\Psi}$ given by Lemmas 4.2 and 4.3.

We write $\widetilde{\Psi}(x_1,x_2)=\overline{\Psi}(x_1,x_2,-\varepsilon h(x_1,x_2))$, and observe that, pointwise,

$$\begin{aligned}
\nabla\widetilde{\Psi} &= \nabla\overline{\Psi}+\varepsilon\xi\frac{\partial\overline{\Psi}}{\partial x_3},\\
\nabla^2\widetilde{\Psi} &= \nabla^2\overline{\Psi}+\varepsilon\xi\frac{\partial\nabla\overline{\Psi}}{\partial x_3}+\varepsilon\xi\frac{\partial\overline{\Psi}}{\partial x_3}+\varepsilon^2\xi\frac{\partial^2\overline{\Psi}}{\partial x_3^2},\\
\nabla^3\widetilde{\Psi} &= \nabla^3\overline{\Psi}+\varepsilon\xi\frac{\partial\nabla^2\overline{\Psi}}{\partial x_3}+\varepsilon\xi\frac{\partial\nabla\overline{\Psi}}{\partial x_3}+\varepsilon\xi\frac{\partial^2\overline{\Psi}}{\partial x_3^2}+\varepsilon\xi\frac{\partial\overline{\Psi}}{\partial x_3}\\
&\quad+\varepsilon^2\xi\frac{\partial^2\nabla\overline{\Psi}}{\partial x_3^2}+\varepsilon^3\xi\frac{\partial^2\overline{\Psi}}{\partial x_3^3}+\varepsilon\xi\frac{\partial\nabla\overline{\Psi}}{\partial x_3}+\varepsilon^2\xi\frac{\partial^2\overline{\Psi}}{\partial x_3^2}.
\end{aligned}\tag{4.78}$$

Here the ξ are (different) continuous (scalar, vector or tensor) functions bounded on Γ_{i} independently of ε ($\varepsilon\leqslant 1$), $\widetilde{\Psi}$ and its derivatives are evaluated at $(x_1,x_2)\in\Gamma_{\mathrm{i}}$, $\overline{\Psi}$ and its derivatives are evaluated at $(x_1,x_2,-\varepsilon h(x_1,x_2))$.

It follows then from Lemma 4.3 that

$$\begin{aligned}
&|\widetilde{\Psi}|_{L^2(\Gamma_{\mathrm{i}})}\leqslant C, && |\nabla\widetilde{\Psi}|_{L^2(\Gamma_{\mathrm{i}})}\leqslant C,\\
&|\nabla^2\widetilde{\Psi}|_{L^2(\Gamma_{\mathrm{i}})}\leqslant C\varepsilon^{1/2}, && |\nabla^3\widetilde{\Psi}|_{L^2(\Gamma_{\mathrm{i}})}\leqslant C\varepsilon^{-1}.
\end{aligned}\tag{4.79}$$

Now, similarly, θ_1 and its first, second and third derivatives are of the following form:

$$\begin{aligned}
&\theta_1=\xi\widetilde{\Psi},\qquad \nabla\theta_1=\nabla\xi\cdot\widetilde{\Psi}+\xi\nabla\widetilde{\Psi},\\
&\nabla^2\theta_1=\nabla^2\xi\cdot\widetilde{\Psi}+2\nabla\xi\otimes\nabla\widetilde{\Psi}+\xi\nabla^2\widetilde{\Psi},\\
&\nabla^3\theta_1=\nabla^3\xi\cdot\widetilde{\Psi}+3\nabla^2\xi\otimes\nabla\widetilde{\psi}+3\nabla\xi\otimes\nabla^2\widetilde{\Psi}+\xi\nabla^3\widetilde{\Psi},
\end{aligned}$$

where ξ and its first, second and third derivatives are uniformly bounded on Γ_{i} (for $\varepsilon\leqslant 1$); hence (4.76) using (4.78). To obtain (4.77), we observe that, with a different ξ, θ_2 is of the form $\varepsilon^{-1}\xi\cdot\nabla\theta_1$.

Lemma 4.5 is proved. □

STEP 2 (Reduction to the case $\alpha=0$ (and $g=0$)). Let $\widetilde{T}$ be the solution of

$$\begin{aligned}
&-\Delta_3\widetilde{T}=f_2 && \text{in } \mathcal{M}_\varepsilon,\\
&\frac{\partial\widetilde{T}}{\partial x_3}+\alpha\widetilde{T}=0 && \text{on } \Gamma_{\mathrm{i}},\\
&\frac{\partial\widetilde{T}}{\partial n}=0 && \text{on } \Gamma_{\mathrm{b}}\cup\Gamma_\ell.
\end{aligned}\tag{4.80}$$

We first note that

$$|\nabla\widetilde{T}|_\varepsilon^2 + \left|\frac{\partial\widetilde{T}}{\partial x_3}\right|_\varepsilon^2 + \alpha|\widetilde{T}|_i^2 \leqslant |f_2|_\varepsilon|\widetilde{T}|_\varepsilon. \tag{4.81}$$

Also by a density argument and, since

$$\begin{aligned}|\widetilde{T}(x_1,x_2,x_3)| &\leqslant |\widetilde{T}(x_1,x_2,0)| + \int_0^{x_3}\left|\frac{\partial\widetilde{T}}{\partial x_3}\right| \mathrm{d}x_3' \\ &\leqslant |\widetilde{T}(x_1,x_2,0)| + \sqrt{\varepsilon\bar{h}}\left(\int_{-\varepsilon h(x_1,x_2)}^0 \left|\frac{\partial\widetilde{T}}{\partial x_3}\right|^2 \mathrm{d}x_3\right)^{1/2},\end{aligned} \tag{4.82}$$

and

$$\int_{M_\varepsilon} |\widetilde{T}(x_1,x_2,x_3)|^2 \,\mathrm{d}x_1\,\mathrm{d}x_2\,\mathrm{d}x_3 \leqslant 2\varepsilon\bar{h}|\widetilde{T}|_i^2 + 2\varepsilon\bar{h}\left|\frac{\partial\widetilde{T}}{\partial x_3}\right|_\varepsilon^2. \tag{4.83}$$

We infer from (4.81) that

$$\begin{aligned}|\nabla\widetilde{T}|_\varepsilon^2 + \left|\frac{\partial\widetilde{T}}{\partial x_3}\right|_\varepsilon^2 + \alpha|\widetilde{T}|_i^2 &\leqslant \sqrt{2\varepsilon\bar{h}}|f_2|_\varepsilon|\widetilde{T}|_i + \sqrt{2\varepsilon\bar{h}}|f_2|_\varepsilon\left|\frac{\partial\widetilde{T}}{\partial x_3}\right|_\varepsilon \\ &\leqslant \frac{\alpha}{2}|\widetilde{T}|_i^2 + \frac{\varepsilon\bar{h}}{2\alpha}|f_2|_\varepsilon^2 + \frac{1}{2}\left|\frac{\partial\widetilde{T}}{\partial x_3}\right|_\varepsilon^2 + \varepsilon\bar{h}|f_2|_\varepsilon^2.\end{aligned} \tag{4.84}$$

Therefore

$$|\nabla\widetilde{T}|_\varepsilon^2 + \frac{1}{2}\left|\frac{\partial\widetilde{T}}{\partial x_3}\right|_\varepsilon^2 + \frac{\alpha}{2}|\widetilde{T}|_i^2 \leqslant \varepsilon\bar{h}|f_2|_\varepsilon^2\left(\frac{1}{2\alpha}+1\right). \tag{4.85}$$

Hence

$$|\widetilde{T}|_i^2 \leqslant C\varepsilon|f_2|_\varepsilon^2 \tag{4.86}$$

and

$$\left|\frac{\partial\widetilde{T}}{\partial x_3}\right|_\varepsilon^2 \leqslant C\varepsilon|f_2|_\varepsilon^2,$$

where C is a constant independent of ε. Then by (4.82),

$$|\widetilde{T}|_\varepsilon^2 \leqslant C\varepsilon^2|f_2|_\varepsilon^2. \tag{4.87}$$

Next, transform (4.79) into a homogeneous Neumann condition. Let $T^* = \eta\widetilde{T}$, where

$$\eta(x_1, x_2, x_3) = \exp\left[\alpha x_3 + \frac{\alpha x_3^2}{2\varepsilon h(x_1, x_2)} + \alpha x_3^2\big(x_3 + \varepsilon h(x_1, x_2)\big)\varphi(x_1, x_2)\right] \tag{4.88}$$

with

$$\varphi = \frac{|\nabla h|^2}{2h^2(1 + \varepsilon^2|\nabla h|^2)}. \tag{4.89}$$

Noting that

$$\frac{\partial T^*}{\partial x_3} = \frac{\partial \eta}{\partial x_3}\widetilde{T} + \eta\frac{\partial \widetilde{T}}{\partial x_3}, \qquad \frac{\partial T^*}{\partial n} = \frac{\partial \eta}{\partial n}\widetilde{T} + \eta\frac{\partial \widetilde{T}}{\partial n},$$
$$\frac{\partial \eta}{\partial x_3} = \eta\left[\alpha + \frac{\alpha x_3}{\varepsilon h} + 2\alpha x_3(x_3 + \varepsilon h)\varphi + \alpha x_3^2\varphi\right], \tag{4.90}$$

and at $x_3 = 0$, $\partial\eta/\partial x_3 = \alpha\eta$ and $\eta = 1$, which implies that

$$\frac{\partial T^*}{\partial x_3} = \eta\left(\frac{\partial \widetilde{T}}{\partial x_3} + \alpha\widetilde{T}\right) = 0.$$

Furthermore, at $x_3 = -\varepsilon h(x_1, x_2)$, we have

$$\frac{\partial \eta}{\partial x_3} = \alpha\eta\varepsilon^2 h^2(x_1, x_2)\varphi(x_1, x_2). \tag{4.91}$$

Now we compute $\nabla\eta$, where $\nabla = (\partial/\partial x_1, \partial/\partial x_2)$:

$$\nabla\eta = \eta\left[-\frac{\alpha x_3^2}{2\varepsilon h^2}\nabla h + \varepsilon\alpha x_3^2\varphi\nabla h + \alpha x_3^2(x_3 + \varepsilon h)\nabla\varphi\right]. \tag{4.92}$$

Hence at $x_3 = -\varepsilon h(x_1, x_2)$, $\nabla\eta = \eta[-\frac{1}{2}\alpha\varepsilon\xi\nabla h + \varepsilon^3\alpha h^2\xi\nabla h]$ and

$$\begin{aligned}\frac{\partial \eta}{\partial n} &= \frac{\partial \eta}{\partial x_3} + \varepsilon\nabla h\nabla\eta \\ &= \eta\left[\alpha\varepsilon^2 h^2\varphi - \frac{\varepsilon^2\alpha}{2}|\nabla h|^2 + \varepsilon^4\alpha h^2\varphi|\nabla h|^2\right] \\ &= \varepsilon^2\eta\alpha\left[h^2\varphi\big(1 + \varepsilon^2|\nabla h|^2\big) - \frac{1}{2}|\nabla h|^2\right] = 0.\end{aligned} \tag{4.93}$$

That is, $\partial\eta/\partial n|_{\Gamma_b} = 0$, and $\partial T^*/\partial n|_{\Gamma_b} = 0$. Assume now (4.54), that is,

$$\nabla h \cdot n_{\Gamma_i} = 0 \quad \text{on } \partial\Gamma_i. \tag{4.94}$$

One can easily check that $\nabla\varphi \cdot n_{\Gamma_{\mathrm{i}}} = 0$: using a system (s, t) of local coordinates on $\partial\Gamma_{\mathrm{i}}$, with s normal to $\partial\Gamma_{\mathrm{i}}$ and t tangential, and using (4.66) and (4.67). Hence

$$\nabla h \cdot n_{\Gamma_{\mathrm{i}}} = 0 \quad \text{and} \quad \nabla\varphi \cdot n_{\Gamma_{\mathrm{i}}} = 0 \quad \text{on } \partial\Gamma_{\mathrm{i}}, \tag{4.95}$$

and we have immediately

$$\nabla\eta \cdot n_{\Gamma_{\mathrm{i}}} = 0 \quad \text{on } \partial\Gamma_{\mathrm{i}},$$

and therefore

$$\frac{\partial T^*}{\partial n_{\Gamma_{\mathrm{i}}}} = 0 \quad \text{on } \partial\Gamma_{\mathrm{i}}.$$

The conclusion of these computations and of Step 2 is summarized by the following lemma.

LEMMA 4.6. *Assume that $\partial\Gamma_{\mathrm{i}}$ is of class C^2, $h : \overline{\Gamma_{\mathrm{i}}} \to \mathbb{R}_+$ belong to $C^4(\overline{\Gamma_{\mathrm{i}}})$ and*

$$\nabla h \cdot n_{\Gamma_{\mathrm{i}}} = 0 \quad \textit{on } \partial\Gamma_{\mathrm{i}}.$$

Let $\widetilde{T}$ be a solution of (4.79) *and $T^* = \eta\widetilde{T}$, with*

$$\eta = \exp\left[\alpha x_3 + \frac{\alpha x_3^2}{2\varepsilon h(x_1, x_2)} + \alpha x_3^2\big(x_3 + \varepsilon h(x_1, x_2)\big)\varphi(x_1, x_2)\right],$$

where

$$\varphi(x_1, x_2) = \frac{|\nabla h(x_1, x_2)|^2}{2h^2(x_1, x_2)(1 + \varepsilon^2|\nabla h(x_1, x_2)|^2)}.$$

Then

$$-\Delta T^* = f^*,$$

$$\frac{\partial T^*}{\partial x_3} = 0 \quad \textit{on } \Gamma_{\mathrm{i}},$$

$$\frac{\partial T^*}{\partial n} = 0 \quad \textit{on } \Gamma_{\mathrm{b}} \cup \Gamma_{\ell},$$

where $f^* = \eta f_2 - 2\nabla_3\eta \cdot \nabla_3\widetilde{T} - \widetilde{T}\Delta_3\eta$ and $|f^*|_\varepsilon \leqslant C_0|f_2|_\varepsilon$, with C_0 independent on ε.

PROOF. It remains only to estimate the L^2-norm of f^*. First note that there exists a constant C_0 independent of ε (depending on α and h) such that

$$\frac{1}{C_0} \leqslant \eta(x_1, x_2, x_3) \leqslant C_0 \qquad \text{for } (x_1, x_2, x_3) \in \mathcal{M}_\varepsilon,$$

and using (4.90) and (4.92), there exists another constant still denoted C_0 such that

$$\left|\frac{\partial\eta}{\partial x_3}\right|_{L^\infty(\mathcal{M}_\varepsilon)} \leqslant C_0 \quad \text{and} \quad |\nabla\eta|_{L^\infty(\mathcal{M}_\varepsilon)} \leqslant C_0.$$

Now we compute $\Delta\eta$ ($\Delta = \partial^2/\partial x_1^2 + \partial^2/\partial x_2^2$):

$$\begin{aligned}\Delta\eta &= \operatorname{div}(\nabla\eta)\\ &= \eta\left[\frac{\alpha x_3^2}{2\varepsilon}\Delta\left(\frac{1}{h}\right) + \varepsilon\alpha x_3^2 \operatorname{div}(\varphi\nabla h) + \alpha x_3^2 \operatorname{div}\big((x_3+\varepsilon h)\nabla\varphi\big)\right],\end{aligned}$$

and therefore, since $h \in C^4(\overline{\Gamma_{\mathrm{i}}})$, we have

$$|\Delta\eta|_{L^\infty(\mathcal{M}_\varepsilon)} \leqslant C_0, \quad \text{with } C_0 \text{ independent on } \varepsilon.$$

Finally, we compute $\partial^2\eta/\partial x_3^2$:

$$\begin{aligned}\frac{\partial^2\eta}{\partial x_3^2} &= \frac{\partial\eta}{\partial x_3}\left[\alpha + \frac{\alpha x_3}{\varepsilon h} + 2\alpha x_3(x_3+\varepsilon h)\varphi + \alpha x_3^2\varphi\right]\\ &\quad + \eta\left[\frac{\alpha}{\varepsilon h} + 2\alpha(x_3+\varepsilon h)\varphi + 6\alpha x_3\varphi\right].\end{aligned}$$

Therefore

$$\left|\frac{\partial^2\eta}{\partial x_3^2}\right|_{L^\infty(\mathcal{M}_\varepsilon)} \leqslant \frac{C}{\varepsilon},$$

where C is independent on ε. Now

$$\left|f^*\right|_\varepsilon \leqslant |\eta|_{L^\infty}|f_2|_\varepsilon + |\nabla_3\eta|_{L^\infty}\left|\nabla_3\widetilde{T}\right|_\varepsilon + |\Delta\eta|_{L^\infty}\left|\widetilde{T}\right|_\varepsilon + \left|\frac{\partial^2\eta}{\partial x_3^2}\right|_{L^\infty}\left|\widetilde{T}\right|_\varepsilon,$$

and since by (4.84) and (4.87),

$$\left|\nabla\widetilde{T}\right|_\varepsilon \leqslant C|f_2|_\varepsilon \quad \text{and} \quad \left|\widetilde{T}\right|_\varepsilon \leqslant C\varepsilon|f_2|_\varepsilon,$$

we have

$$\left|f^*\right|_\varepsilon \leqslant C|f_2|_\varepsilon,$$

where C is independent of ε. □

4.4. *Regularity of the velocity*

In this section, we study the H^2 regularity of the velocity, solution of the GFD–Stokes problem (we use either x_3 or z to denote the vertical variable):

$$\begin{cases} -\left(\Delta v + \frac{\partial^2 v}{\partial x_3^2}\right) + \nabla p = f_v & \text{in } \mathcal{M}_\varepsilon, \\ \operatorname{div} \int_{-\varepsilon h}^0 v \,\mathrm{d}z = 0 & \text{in } \Gamma_\mathrm{i}, \\ v = 0 & \text{on } \Gamma_\ell \cup \Gamma_\mathrm{b}, \\ \frac{\partial v}{\partial x_3} + \alpha_v v = g_v & \text{on } \Gamma_\mathrm{i}. \end{cases} \tag{4.96}$$

The H^2 regularity of problems similar to (4.96) are given in [44] where $\varepsilon = 1$ and $g_v = 0$, and in [15], in the case of constant depth function and under a convexity condition of $\mathcal{M}_\varepsilon$. We study here the H^2 regularity of solutions to (4.96), and give the dependence on ε of the constant appearing in the Cattabriga–Solonnikov type inequality associated to the H^2 regularity of solutions. By contrast with the articles quoted above, our analysis here will be carried out in the case where $\mathcal{M}_\varepsilon$ is not necessarily convex, and with a varying bottom topography. This regularity result is discussed in Section 4.4.2, and we start in Section 4.4.1 with a discussion of the weak formulation of the GFD–Stokes problem (4.96).

4.4.1. *Weak formulation of the GFD–Stokes problem.* In this section we drop the index ε which is irrelevant ($\varepsilon = 1$, $\mathcal{M}_\varepsilon = \mathcal{M}$). For the weak formulation of (4.96) we consider the space

$$V = \left\{ v \in H^1(\mathcal{M})^2, \operatorname{div} \int_{-h}^0 v \,\mathrm{d}z = 0 \text{ on } \Gamma_\mathrm{i}, v = 0 \text{ on } \Gamma_\ell \cup \Gamma_\mathrm{b} \right\};$$

thanks to the Poincaré inequality, this space is Hilbert for the scalar product

$$((v, \tilde{v})) = \sum_{i=1}^{2} \sum_{j=1}^{3} \int_{\mathcal{M}} \frac{\partial v_i}{\partial x_j} \frac{\partial \tilde{v}_i}{\partial x_j} \,\mathrm{d}\mathcal{M}.$$

To obtain the weak formulation, we multiply the first equation (4.96) by a test function $\tilde{v} \in V$ and integrate over $\mathcal{M}$; assuming regularity for v, p and $\tilde{v}$, the term involving p (independent of x_3) disappears:

$$\begin{aligned} &\int_{\mathcal{M}} \nabla p \tilde{v} \,\mathrm{d}\mathcal{M} \\ &\quad = \int_{\partial\mathcal{M}} p \tilde{v} \cdot n \,\mathrm{d}(\partial\mathcal{M}) - \int_{\mathcal{M}} p \operatorname{div} \tilde{v} \,\mathrm{d}\mathcal{M} \quad \text{(by Stokes formula)} \\ &\quad = - \int_{\mathcal{M}} p \operatorname{div} \tilde{v} \,\mathrm{d}\mathcal{M} \end{aligned}$$

$$= -\int_{\Gamma_\mathrm{i}} p\left(\int_{-h}^{0} \operatorname{div} \tilde{v}\, \mathrm{d}x_3\right) \mathrm{d}x_1\, \mathrm{d}x_2$$

$$= -\int_{\Gamma_\mathrm{i}} p\left(\operatorname{div} \int_{-h}^{0} \tilde{v}\, \mathrm{d}x_3\right) \mathrm{d}x_1\, \mathrm{d}x_2$$

$$= 0 \quad \text{(by the properties of } \tilde{v}\text{)}.$$

We also have, with Stokes formula, and since $\tilde{v}$ vanishes on $\Gamma_\mathrm{b} \cup \Gamma_\ell$,

$$-\int_{\mathcal{M}} \left(\Delta v + \frac{\partial^2 v}{\partial x_3^2}\right) \tilde{v}\, \mathrm{d}\mathcal{M} = -\int_{\Gamma_\mathrm{i}} \frac{\partial v}{\partial x_3} \tilde{v}\, \mathrm{d}\Gamma_\mathrm{i} + ((v, \tilde{v}))$$

$$= \alpha_v \int_{\Gamma_\mathrm{i}} v\tilde{v}\, \mathrm{d}\Gamma_\mathrm{i} - \int_{\Gamma_\mathrm{i}} g_v \tilde{v}\, \mathrm{d}\Gamma_\mathrm{i} + ((v, \tilde{v})).$$

Hence the weak formulation of (4.96):
To find $v \in V$ such that

$$a(v, \tilde{v}) = \ell(\tilde{v}) \quad \forall \tilde{v} \in V, \tag{4.97}$$

with

$$\begin{aligned} a(v, \tilde{v}) &= ((v, \tilde{v})) + \alpha_v \int_{\Gamma_\mathrm{i}} v\tilde{v}\, \mathrm{d}\Gamma_\mathrm{i}, \\ \ell(\tilde{v}) &= (f_v, \tilde{v})_H + \int_{\Gamma_\mathrm{i}} g_v \tilde{v}\, \mathrm{d}\Gamma_\mathrm{i}. \end{aligned} \tag{4.98}$$

Existence and uniqueness of a solution $v \in V$ of (4.97) follow promptly from the Lax–Milgram theorem. More delicate is the question of showing that, conversely, v is, in some sense, solution of (4.96). The second equation (4.96) and $v = 0$ on $\Gamma_\ell \cup \Gamma_\mathrm{b}$ follow from the fact that $v \in V$; hence we need to show that there exists a distribution p independent of x_3 such that

$$-\left(\Delta v + \frac{\partial^2 v}{\partial x_3^2}\right) + \nabla p = f_v \quad \text{in } \mathcal{M}, \tag{4.99}$$

and also that

$$\frac{\partial v}{\partial x_3} + \alpha_v v = g_v \quad \text{on } \Gamma_\mathrm{i}. \tag{4.100}$$

For (4.99), consider a test function $\varphi \in \mathcal{C}_0^\infty(\mathcal{M})$ ($\mathcal{C}^\infty$ with compact support in $\mathcal{M}$), and observe that $\tilde{v} = (\tilde{v}_1, 0)$, $\tilde{v}_1 = \partial\varphi/\partial x_3$, belongs to V. Writing (4.97) with this $\tilde{v}$, we conclude that

$$\frac{\partial}{\partial x_3}(\Delta_3 v_1 + f_{v1}) = 0.$$

In the same way we prove that

$$\frac{\partial}{\partial x_3}(\Delta_3 v_2 + f_{v2}) = 0, \tag{4.101}$$

showing that each component of $\Delta v + f_v$ is a distribution on $\mathcal{M}$ independent of x_3.

Distributions independent of x_3. Now we can identify a distribution G on $\mathcal{M}$ independent of x_3, with a distribution on Γ_{i} as follows: let θ be any $\mathcal{C}^\infty$ scalar function with compact support in $(-\underline{h}, 0)$, and such that

$$\int_{-\underline{h}}^{0} \theta(z)\,\mathrm{d}z = 1. \tag{4.102}$$

Then, if $\varphi \in \mathcal{C}_0^\infty(\Gamma_{\mathrm{i}})$ is a $\mathcal{C}^\infty$ scalar function with a compact support in Γ_{i}, $\varphi\theta \in \mathcal{C}_0^\infty(\mathcal{M}_\varepsilon)$, and we associate to G a distribution $\widetilde{G}$ on Γ_{i} by setting

$$\langle \widetilde{G}, \varphi \rangle_{\Gamma_{\mathrm{i}}} = \langle G, \varphi\theta \rangle_{\mathcal{M}}. \tag{4.103}$$

The right-hand side of (4.103) is independent of θ; indeed if θ_1 and θ_2 are two such functions then $\langle G, \varphi\theta_1 \rangle = \langle G, \varphi\theta_2 \rangle$ because

$$\int_{-\underline{h}}^{0} (\theta_1 - \theta_2)(z)\,\mathrm{d}z = 0,$$

so that $\theta_0(x_3) = \int_{x_3}^{0} (\theta_1 - \theta_2)(z)\,\mathrm{d}z$ is a $\mathcal{C}^\infty$ function with compact support in $(-\underline{h}, 0)$, and

$$\langle G, \varphi(\theta_1 - \theta_2) \rangle_{\mathcal{M}} = -\left\langle G, \frac{\partial}{\partial x_3}(\varphi\theta_0) \right\rangle_{\mathcal{M}} = \left\langle \frac{\partial G}{\partial x_3}, \varphi\theta_0 \right\rangle_{\mathcal{M}} = 0.$$

It is then easy to see that (4.103) defines $\widetilde{G}$ as a distribution on Γ_{i}.

Now, conversely, assume that $\widetilde{G}$ is a distribution on Γ_{i}, and let $\varphi \in \mathcal{C}_0^\infty(\mathcal{M})$. It is clear that $\tilde{\varphi} = \int_{-\underline{h}}^{0} \varphi\,\mathrm{d}x_3$ belongs to $\mathcal{C}_0^\infty(\Gamma_{\mathrm{i}})$ and we associate to $\widetilde{G}$ a distribution G on $\mathcal{M}$ by setting

$$\langle G, \varphi \rangle_{\mathcal{M}} = \langle \widetilde{G}, \tilde{\varphi} \rangle_{\Gamma_{\mathrm{i}}} \quad \forall \varphi \in \mathcal{C}_0^\infty(\mathcal{M}).$$

Introduction of p. Thanks to the previous discussion, we can now consider $\Delta v + f_v$ as a distribution on Γ_{i}. As in the theory of Navier–Stokes equations, consider now a vector function $v^* \in \mathcal{V}(\Gamma_{\mathrm{i}})$, that is, v^* is (two-dimensional) $\mathcal{C}^\infty$ with compact support in Γ_{i}, and $\operatorname{div} v^* = 0$. It is clear that $\tilde{v} = v^*\theta$ belongs to V, where θ is a function as above (see (4.102)). Writing (4.97) with this $\tilde{v}$, we obtain

$$\begin{aligned} &\big((v, v^*\theta)\big) = \big(f_v, v^*\theta\big)_H, \\ &\big\langle \Delta v + f_v, v^*\theta\big\rangle_{\mathcal{M}} = 0, \\ &\big\langle \Delta v + f_v, v^*\big\rangle_{\Gamma_\mathrm{i}} = 0 \quad \forall v^* \in \mathcal{V}(\Gamma_\mathrm{i}). \end{aligned} \tag{4.104}$$

The last equation, which is well known in the theory of Navier–Stokes equations (see, e.g., Lions [19], Temam [36]), implies that there exists a distribution p on Γ_i such that

$$\Delta v + f_v = \nabla p \quad \text{in } \Gamma_\mathrm{i} \text{ (or } \mathcal{M}),$$

and (4.99) is proven.

A trace theorem. The proof of (4.100) necessitates establishing first a trace theorem: we need to show that, for a function v in V, such that (4.99) holds, one can define the trace of $\partial v/\partial x_3$ on Γ_i, as an element of $(H_{00}^{1/2}(\Gamma_\mathrm{i}))'$, the dual of $H_{00}^{1/2}(\Gamma_\mathrm{i})$ (that is the 1/2 interpolate between $H_0^1(\Gamma_\mathrm{i})$ and $L^2(\Gamma_\mathrm{i})$).

Observe first that the trace on Γ_i of any function Φ in $H^1(\mathcal{M})$ which vanishes on Γ_ℓ belongs to $H_{00}^{1/2}(\Gamma_\mathrm{i})$. Indeed by odd symmetry and truncation one can extend such a Φ as a function Φ^* in $H_0^1(\Gamma_\mathrm{i} \times \mathbb{R})$, vanishing for $|x_3|$ sufficiently large, and the trace of such a function on any plane $x_3 = c_0$, belongs to $H_{00}^{1/2}(\Gamma_\mathrm{i})$. Conversely, if φ belongs to $H_{00}^{1/2}(\Gamma_\mathrm{i})$, there exists Φ^* in $H_0^1(\Gamma_\mathrm{i} \times \mathbb{R})$ such that the trace of Φ^* on Γ_i is φ, the mapping $\varphi \to \Phi^*$ being linear continuous (lifting operator). From the remark above we infer that the trace on Γ_i of a function in V belongs to $H_{00}^{1/2}(\Gamma_\mathrm{i})$.

We then show that the traces on Γ_i of the functions of V are all in $H_{00}^{1/2}(\Gamma_\mathrm{i})^2$. Indeed let $\varphi \in H_{00}^{1/2}(\Gamma_\mathrm{i})^2$. Using the previous lifting operator, there exists $\widetilde{\Phi} \in H_0^1(\Gamma_\mathrm{i} \times \mathbb{R})^2$ such that $\widetilde{\Phi}|_{\Gamma_\mathrm{i}} = \varphi$; by truncation we can assume that $\widetilde{\Phi} \in H^1(\Gamma_\mathrm{i} \times (-\underline{h}, 0))^2$ and $\widetilde{\Phi}$ vanishes on $\partial\Gamma_\mathrm{i}$ and at $x_3 = -\underline{h}$. Let

$$\xi = \operatorname{div} \int_{-\underline{h}}^{0} \widetilde{\Phi}\, \mathrm{d}x_3,$$

and observe that $\xi \in L^2(\Gamma_\mathrm{i})$ and

$$\int_{\Gamma_\mathrm{i}} \xi\, \mathrm{d}\Gamma_\mathrm{i} = \int_{Q_{\underline{h}}} \operatorname{div} \widetilde{\Phi}\, \mathrm{d}\mathcal{M} = \int_{\partial Q_\varepsilon} \widetilde{\Phi} \cdot n_h\, \mathrm{d}(\partial\mathcal{M}) = 0, \tag{4.105}$$

where n_h is the horizontal component of the unit outward normal n on $\partial\mathcal{M}$. Because of (4.105), we can solve in $Q_{\underline{h}} = \Gamma_{\mathrm{i}} \times (-\underline{h}, 0)$, the (usual) Stokes problem

$$\begin{cases} \Delta \Phi^* + \nabla \pi = 0 & \text{in } Q_{\underline{h}}, \\ \operatorname{div}_3 \Phi^* = \frac{1}{\underline{h}} \xi & \text{in } Q_{\underline{h}}, \\ \Phi^* = 0 & \text{on } \partial Q_{\underline{h}}, \end{cases} \tag{4.106}$$

and $\Phi^* \in H^1(Q_{\underline{h}})^3$, $\pi \in L^2(Q_{\underline{h}})$. Now,

$$\int_{-\underline{h}}^{0} \operatorname{div}_3 \Phi^* \, \mathrm{d}x_3 = \operatorname{div} \int_{-\underline{h}}^{0} (\phi_1^*, \phi_2^*) \, \mathrm{d}x_3 = \xi,$$

and it is easy to see that the function $\Phi = \widetilde{\Phi} - (\Phi_1^*, \Phi_2^*)$ extended by 0 in $\mathcal{M} \backslash Q_{\underline{h}}$ belongs to V, and its trace on Γ_{i} is precisely φ. We can furthermore observe that with the construction above, the mapping $\varphi \mapsto \Phi$ is linear continuous from $H_{00}^{1/2}(\Gamma_{\mathrm{i}})$ into V.

Finally, (4.100) follows promptly from (4.97), (4.99) and the following proposition.

PROPOSITION 4.1. *Let v be a function in $H^1(\mathcal{M})^2$ which vanishes on $\Gamma_{\mathrm{b}} \cup \Gamma_\ell$ and assume that $-\Delta v + \nabla p \in L^2(\mathcal{M})^2$, for some distribution p independent of x_3.*

Then there exists $\gamma_1 v \in (H_{00}^{1/2}(\Gamma_{\mathrm{i}}))^2$ such that

$$\gamma_1 v = \left. \frac{\partial v}{\partial x_3} \right|_{\Gamma_{\mathrm{i}}} \quad \text{if } v \in \mathcal{C}^2(\overline{\mathcal{M}})^2, \tag{4.107}$$

and $\gamma_1 v$ is defined by

$$\langle \gamma_1 v, \varphi \rangle = \big((v, \Phi)\big) - \int_{\mathcal{M}} (-\Delta v + \nabla p) \Phi \, \mathrm{d}\mathcal{M}, \tag{4.108}$$

where φ is arbitrary in $H_{00}^{1/2}(\Gamma_{\mathrm{i}})$ and Φ is any function of V such that $\Phi|_{\Gamma_{\mathrm{i}}} = \varphi$.

PROOF. We first show that the right-hand side $X(\Phi)$ of (4.108) depends on φ and not on Φ. Indeed, let Φ_1 and Φ_2 be two functions of V such that $\Phi_1|_{\Gamma_{\mathrm{i}}} = \Phi_2|_{\Gamma_{\mathrm{i}}} = \varphi$. Then $\Phi^* = \Phi_1 - \Phi_2$ belongs to $H_0^1(\mathcal{M})^2$ and $\operatorname{div} \int_{-\underline{h}}^0 \Phi^* \, \mathrm{d}x_3 = 0$. It was shown in [21] that Φ^* is limit in $H_0^1(\mathcal{M})^2$ of $\mathcal{C}^\infty$ functions Φ_n^* with compact support in $\mathcal{M}$ such that $\operatorname{div} \int_{-\underline{h}}^0 \Phi_n^* \, \mathrm{d}x_3 = 0$. It is easy to see that $X(\Phi_n^*) = 0$ and, by continuity, $X(\Phi^*) = 0$, i.e., $X(\Phi_1) = X(\Phi_2)$.

After this observation we choose Φ as constructed above, so that the mapping $\varphi \to \Phi$ is linear continuous from $H_{00}^{1/2}(\Gamma_{\mathrm{i}})^2$ into V. It then appears that the right-hand side of (4.108) is a linear form continuous on $H_{00}^{1/2}(\Gamma_{\mathrm{i}})^2$, and thus $\gamma_1 v$ is defined and belongs to $(H_{00}^{1/2}(\Gamma_{\mathrm{i}}))^2$. Finally, (4.107) follows from the fact that (4.108) is easy when v and Φ are smooth and $\gamma_1 v$ is replaced by $\partial v / \partial x_3|_{\Gamma_{\mathrm{i}}}$. □

REMARK 4.2. We have shown the complete equivalence of (4.96) with its variational formulation (4.97).

4.4.2. *H^2 regularity for the GFD–Stokes problem.* For convenience, we use hereafter the classical notation $\mathbb{L}^2$, $\mathbb{H}^1$, etc., for spaces of vector functions with components in L^2, H^1, etc.

The main result of this section is the following theorem:

THEOREM 4.4. *Assume that h is a positive function in $C^4(\overline{\Gamma_i})$, $h \geqslant \underline{h} > 0$ and that $f_v \in \mathbb{L}^2(\mathcal{M}_\varepsilon)$ and $g_v \in \mathbb{H}_0^1(\Gamma_{\mathrm{i}})$. Let $(v, p) \in \mathbb{H}^1(M_\varepsilon) \times L^2(\Gamma_{\mathrm{i}})$ be a weak solution of* (4.96). *Then*

$$(v, p) \in \mathbb{H}^2(\mathcal{M}_\varepsilon) \times H^1(\mathcal{M}_\varepsilon). \tag{4.109}$$

Moreover, the following inequality holds:

$$|v|^2_{\mathbb{H}^2(\mathcal{M}_\varepsilon)} + \varepsilon |p|^2_{H^1(\Gamma_{\mathrm{i}})} \leqslant C\big[|f_v|^2_\varepsilon + |g_v|^2_{L^2(\Gamma_{\mathrm{i}})} + \varepsilon |\nabla g_v|^2_{L^2(\Gamma_{\mathrm{i}})}\big]. \tag{4.110}$$

The approach to the proof of the H^2 regularity in Theorem 4.4 is the same as in the articles of Ziane [44,45] and Hu, Temam and Ziane [15], it is based on the following observation: the weak solution of (4.96) satisfies $p \in L^2(\Gamma_{\mathrm{i}})$; assume further that the solution v of (4.96) satisfies $\frac{\partial v}{\partial x_3}|_{\Gamma_{\mathrm{i}}} \in \mathbb{L}^2(\Gamma_{\mathrm{i}})$, $\frac{\partial v}{\partial x_3}|_{\Gamma_{\mathrm{b}}} \in \mathbb{L}^2(\Gamma_{\mathrm{b}})$ and $\frac{\partial v}{\partial x_k}|_{\Gamma_{\mathrm{b}}} \in \mathbb{L}^2(\Gamma_{\mathrm{b}})$, $k = 1, 2$, then an integration of the first equation in (4.96) with respect to x_3 over $(-\varepsilon h, 0)$ yields a two-dimensional Stokes problem on the smooth domain Γ_{i} with a homogeneous boundary condition. By the classical regularity theory of the two-dimensional-Stokes problem in smooth domains, see, for instance, Ghidaglia [9], Temam [36] and Constantin and Foias [7], p belongs to $H^1(\Gamma_{\mathrm{i}})$. Then, by moving the pressure term to the right-hand side, problem (4.96) reduces to an elliptic problem of the type studied in Section 4.2, and the H^2 regularity of v follows. The estimates on the L^2-norms of the second derivatives are then obtained using the trace theorem and the estimates in Section 4.2.

We start this proof by showing that $\frac{\partial v}{\partial x_3}|_{\Gamma_{\mathrm{i}}} \in \mathbb{L}^2(\Gamma_{\mathrm{i}})$, $\frac{\partial v}{\partial x_3}|_{\Gamma_{\mathrm{b}}} \in \mathbb{L}^2(\Gamma_{\mathrm{b}})$ and $\frac{\partial v}{\partial x_k}|_{\Gamma_{\mathrm{i}}} \in \mathbb{L}^2(\Gamma_{\mathrm{i}})$, $k = 1, 2$. The following lemma is just a rewriting of Theorem 4.2.

LEMMA 4.7. *Assume that $h \in C^2(\overline{\Gamma_{\mathrm{i}}})$. For $f \in L^2(\mathcal{M}_\varepsilon)$ and $g \in H_0^1(\Gamma_{\mathrm{i}})$, there exists a unique $\Psi \in H^2(\mathcal{M}_\varepsilon)$ solution of*

$$\begin{cases} -\Delta_3 \Psi = f & \text{in } \mathcal{M}_\varepsilon, \\ \frac{\partial \Psi}{\partial x_3} + \alpha \Psi = g & \text{on } \Gamma_{\mathrm{i}}, \\ \Psi = 0 & \text{on } \Gamma_{\mathrm{b}} \cup \Gamma_{\mathrm{l}}. \end{cases} \tag{4.111}$$

Furthermore, there exists a constant $C(h, \alpha)$ depending only on α and h (and Γ_{i}), such that

$$\sum_{k,j=1}^{3} \left| \frac{\partial^2 \Psi}{\partial x_k \, \partial x_j} \right|^2_\varepsilon \leqslant C(h, \alpha)\big[|f|^2_\varepsilon + |g|^2_i + |\nabla g|^2_i\big].$$

As we said, this lemma is just a rewriting of Theorem 4.2. We will also need the following intermediate result simply obtained by interpolation between H^1 and H^2.

LEMMA 4.8. *Under the assumptions of Lemma 4.7, and with $g=0$: if $f \in H^{-1/2+\delta}(\mathcal{M}_\varepsilon)$ with $-\frac{1}{2} < \delta < \frac{1}{2}, \delta \neq 0$, then $\Psi \in H^{3/2+\delta}(\mathcal{M}_\varepsilon)$.*

Before we start the proof of the main result of this section (Theorem 4.1), we first prove

LEMMA 4.9. *Assume that $h \in C^3(\overline{\Gamma_i})$. For $f \in L^2(\mathcal{M}_\varepsilon)$, $g \in H_0^1(\Gamma_\text{i})$, and $\psi_i \in H_0^{1+\gamma}(\Gamma_\text{b})$, $-\frac{1}{2} < \gamma < \frac{1}{2}, \gamma \neq 0$, there exists a unique $\Psi_i \in H^{3/2+\gamma}(M_\varepsilon)$ solution of*

$$\begin{cases} -\Delta_3 \Psi_i = f & \text{in } \mathcal{M}_\varepsilon, \\ \frac{\partial \Psi_i}{\partial x_3} + \alpha_v \Psi_i = g - \alpha_v \psi_i & \text{on } \Gamma_\text{i}, \\ \Psi_i = -\psi_i & \text{on } \Gamma_\text{b}, \\ \Psi_i = 0 & \text{on } \Gamma_\text{l}. \end{cases} \tag{4.112}$$

PROOF. Using Lemma 4.7, we reduce the problem to the case $f=0$ and $g=0$, by replacing Ψ_i with $\Psi_i - \Psi$, where Ψ is the function constructed in Lemma 4.7. Thus, without loss of generality, we will assume from now on that $f=0$ and $g=0$. Our next step is to construct a function $\tilde{v}_p$ which agrees with Ψ_i on $\partial\mathcal{M}_\varepsilon$. This will be done by first constructing an auxiliary function v_p on a straight cylinder and then the explicit expression of $\tilde{v}_p$ will be given.

Let Q_ε be the cylinder $Q_\varepsilon = \Gamma_\text{i} \times (-\varepsilon, 0)$, and let v_p be the unique solution of

$$\begin{cases} \Delta_3 v_p = 0 & \text{in } Q_\varepsilon, \\ v_p = 0 & \text{on } \partial\Gamma_\text{i} \times (-\varepsilon, 0), \\ v_p = -\psi_i & \text{on } \Gamma_\text{i} \times \{-\varepsilon\}, \\ v_p = \varepsilon h \alpha_v \psi_i & \text{on } \Gamma_\text{i} \times \{0\}. \end{cases} \tag{4.113}$$

We will show that $v_p \in H^{3/2+\gamma}(Q_\varepsilon)$ for all $-\frac{1}{2} < \gamma < \frac{1}{2}, \gamma \neq 0$. Then, setting

$$\begin{aligned} &\tilde{v}_p(x_1, x_2, x_3) \\ &\quad = -\frac{x_3}{\varepsilon h(x_1, x_2)} v_p\left(x_1, x_2, \frac{x_3}{h(x_1, x_2)}\right) \quad \text{for } (x_1, x_2, x_3) \in \mathcal{M}_\varepsilon, \end{aligned} \tag{4.114}$$

it is obvious that $\tilde{v}_p \in H^{3/2+\gamma}(\mathcal{M}_\varepsilon)$, $\tilde{v}_p(x_1, x_2, -\varepsilon h(x_1, x_2)) = -\psi_i(x_1, x_2)$, and $\frac{\partial \tilde{v}_p}{\partial x_3} + \alpha_v \tilde{v}_p = -\alpha_v \psi_i$ on Γ_i. Therefore setting $\widetilde{V} = \Psi_i - \tilde{v}_p$, we have

$$\begin{aligned} &\Delta_3 \widetilde{V} = -\Delta_3 \tilde{v}_p \in H^{-1/2+\gamma}(\mathcal{M}_\varepsilon), \\ &\widetilde{V} = 0 \quad \text{on } \Gamma_\text{l} \cup \Gamma_\text{b}, \\ &\frac{\partial \widetilde{V}}{\partial x_3} + \alpha_v \widetilde{V} = 0 \quad \text{on } \Gamma_\text{i}. \end{aligned} \tag{4.115}$$

Hence, thanks to Lemmas 4.7 and 4.8, we see that $\widetilde{V}$ and thus Ψ_i are in $H^{3/2+\gamma}(M_\varepsilon)$ for $-\frac{1}{2} < \gamma < \frac{1}{2}, \gamma \neq 0$.

To complete the proof of Lemma 4.9, it remains only to show that $v_p \in H^{3/2+\gamma}(Q_\varepsilon)$ for all $-\frac{1}{2} < \gamma < \frac{1}{2}$, $\gamma \neq 0$. To this end, let $\widehat{Q}_\varepsilon$ be any C^2-domain containing Q_ε such that $\Gamma_{\rm i} \times \{-\varepsilon, 0\} \subset \partial\widehat{Q}_\varepsilon$. Since ψ_i (resp. $h\alpha_v\psi_i$) is in $H_0^{1+\gamma}(\Gamma_{\rm i} \times \{-\varepsilon\})$ (resp $H_0^{1+\gamma}(\Gamma_{\rm i} \times \{0\})$), we can define a function $V_i \in H^1(\partial\widehat{Q}_\varepsilon)$ by setting $V_i = -\psi_i$ on $\Gamma_{\rm i} \times \{-\varepsilon\}$, $V_i = \varepsilon h\alpha_v\psi_i$ on $\Gamma_{\rm i} \times \{0\}$, and $V_i = 0$ on $\partial\widehat{Q}_\varepsilon \backslash \Gamma_{\rm i} \times \{-\varepsilon, 0\}$. Now let V_p be the unique solution of $\Delta_3 V_p = 0$ in $\widehat{Q}_\varepsilon$ and $V_p = V_i$ on $\partial\widehat{Q}_\varepsilon$. Since $\partial\widehat{Q}_\varepsilon$ is of class C^2, the classical regularity results for elliptic problems (see, e.g., Lions and Magenes [20]) yield $V_p \in H^{3/2+\gamma}(\widehat{Q}_\varepsilon)$ for $-\frac{1}{2} < \gamma < \frac{1}{2}$, $\gamma \neq 0$. Now let $\widetilde{V}_i$ be the trace of V_p on $\partial\Gamma_{\rm i} \times (-\varepsilon, 0)$. It is easy to see that $\widetilde{V}_i \in H_0^{1+\gamma}(\partial\Gamma_{\rm i} \times (-\varepsilon, 0))$. Let $\widetilde{V}_p = V_p - v_p$, we have

$$\begin{aligned} &\Delta_3 \widetilde{V}_p = 0 && \text{in } Q_\varepsilon, \\ &\widetilde{V}_p = 0 && \text{on } \Gamma_{\rm i} \times \{-\varepsilon, 0\}, \\ &\widetilde{V}_p = \widetilde{V}_i && \text{on } \partial\Gamma_{\rm i} \times (-\varepsilon, 0). \end{aligned} \tag{4.116}$$

Using a reflection argument around $x_3 = 0$ (resp. $x_3 = -\varepsilon$) by extending $\widetilde{V}_i$ in a "symmetrically" odd function defined on $\partial\Gamma_{\rm i} \times (-\varepsilon, \varepsilon)$ (resp. $\partial\Gamma_{\rm i} \times (-2\varepsilon, 0)$), and using the classical local regularity theory (see, e.g., Lions and Magenes [20]), we conclude that $\widetilde{V}_p \in H^{3/2+\gamma}(Q_\varepsilon)$ for $-\frac{1}{2} < \gamma < \frac{1}{2}$, $\gamma \neq 0$. Therefore, since $V_p \in H^{3/2+\gamma}(Q_\varepsilon)$, we have $v_p = V_p - \widetilde{V}_p \in H^{3/2+\gamma}(Q_\varepsilon)$. □

LEMMA 4.10. *Assume that $h \in C^3(\overline{\Gamma_{\rm i}})$, with $h \geqslant h_1 > 0$ on $\overline{\Gamma_{\rm i}}$. Let (v, p) be the weak solution of* (4.96), *then $v \in \mathbb{H}^{2-\delta}(\mathcal{M}_\varepsilon)$ for $0 < \delta < \frac{1}{2}$ and consequently,*

$$\left.\frac{\partial v}{\partial x_3}\right|_{\Gamma_{\rm i}} \in \mathbb{L}^2(\Gamma_{\rm i}), \quad \nabla v|_{\Gamma_{\rm b}} \text{ and } \left.\frac{\partial v}{\partial x_3}\right|_{\Gamma_{\rm b}} \in \mathbb{L}^2(\Gamma_{\rm b}). \tag{4.117}$$

PROOF. We saw in Section 4.1.1 that (4.97) has a unique solution $v \in H^1(\mathcal{M})^2$, and that there exists p such that (v, p) satisfy (4.96). By (4.96), ∇p belongs to $H^{-1}(\mathcal{M}_\varepsilon)$ and thus to $H^{-1}(\Gamma_{\rm i})$ since p is independent of x_3 (see Section 4.4.1). Let $v_i \in H_0^1(\Gamma_{\rm i})$ be the unique solution of the two-dimensional Dirichlet problem on $\Gamma_{\rm i}$:

$$\begin{cases} \Delta v_i = \nabla p & \text{in } \Gamma_{\rm i}, \\ v_i = 0 & \text{on } \partial\Gamma_{\rm i}. \end{cases} \tag{4.118}$$

Let $\tilde{v} = v - v_i$, then $\tilde{v}$ satisfies

$$\begin{cases} \Delta_3\tilde{v} = f_v & \text{in } \mathcal{M}_\varepsilon, \\ \tilde{v} = 0 & \text{on } \Gamma_\ell, \\ \tilde{v} = -v_i & \text{on } \Gamma_{\rm b}, \\ \frac{\partial\tilde{v}}{\partial x_3} + \alpha_v\tilde{v} = g_v - \alpha_v v_i & \text{on } \Gamma_{\rm i}. \end{cases} \tag{4.119}$$

Thanks to Lemma 4.7, with $g = g_v$, $\psi_i = v_i$ and $\gamma = -\delta$ for some $0 < \delta < \frac{1}{2}$, we have $\tilde{v} \in H^{3/2-\delta}(\mathcal{M}_\varepsilon)$. Hence,

$$g_i = -\frac{1}{\varepsilon h}\int_{-\varepsilon h}^{0} \operatorname{div}\tilde{v}\,\mathrm{d}x_3 \in H^{1/2-\delta}(\Gamma_{\mathrm{i}}). \tag{4.120}$$

Therefore, since $\operatorname{div} v_i = g_i$, we rewrite the equation for v_i in the form of a two-dimensional Stokes problem

$$\begin{cases} -\Delta v_i + \nabla p = 0 & \text{in } \Gamma_{\mathrm{i}},\\ \operatorname{div} v_i = g_i \in H^{1/2-\delta}(\Gamma_{\mathrm{i}}), & \\ v_i = 0 & \text{on } \partial\Gamma_{\mathrm{i}}, \end{cases} \tag{4.121}$$

and thanks to the classical regularity result for the nonhomogeneous Stokes problem on Γ_{i}, (see, e.g., Ghidaglia [9], Temam [36]), we have $v_i \in H^{3/2-\delta}(\Gamma_{\mathrm{i}}) \cap H_0^1(\Gamma_{\mathrm{i}}) = H_0^{3/2-\delta}(\Gamma_{\mathrm{i}})$. With this new information on the regularity of v_i, we return to problem (4.119) and using Lemma 4.9 with $\psi_i = v_i$ and $\gamma = \frac{1}{2} - \delta, 0 < \gamma < \frac{1}{2}$, we conclude that $\tilde{v} \in H^{2-\delta}(\mathcal{M}_\varepsilon)$. Therefore g_i, given by (4.120), belongs to $H^{1-\delta}(\Gamma_{\mathrm{i}})$. This in turn implies by the classical regularity of the two-dimensional Stokes problem, that the solution v_i of (4.121) is in $H^{2-\delta}(\Gamma_{\mathrm{i}})$. Therefore $v = \tilde{v} + v_i$ belongs to $H^{2-\delta}(\mathcal{M}_\varepsilon)$. Consequently the trace on Γ_{i} of the normal derivative $\partial v/\partial x_3|_{\Gamma_{\mathrm{i}}}$ belongs to $H^{1/2-\delta}(\Gamma_{\mathrm{i}})$, hence to $L^2(\Gamma_{\mathrm{i}})$, taking, e.g., $\delta = 1/4$. Similarly the traces on Γ_{b} of v and its normal derivative $\partial v/\partial n$ belong to $H^{3/2-\delta}(\Gamma_{\mathrm{b}})$ and $H^{1/2-\delta}(\Gamma_{\mathrm{b}})$ respectively, from which we infer that $\nabla v|_{\Gamma_{\mathrm{b}}}$ and $\partial v/\partial x_3|_{\Gamma_{\mathrm{b}}}$ are in $H^{1/2-\delta}(\Gamma_{\mathrm{b}})$ and therefore in $L^2(\Gamma_{\mathrm{b}})$. The proof of the lemma is now complete. □

PROOF OF THEOREM 4.4. The proof is divided into two steps. In Step 1, we prove the H^2 regularity of solutions, i.e., $v \in H^2(\mathcal{M}_\varepsilon)$ and $p \in H^1(\Gamma_{\mathrm{i}})$. Then, in Step 2, we establish the Cattabriga–Solonnikov type inequality on the solutions, i.e., establish the bounds (2.97) on the L^2-norms of the second derivatives of v and the H^1-norm on the pressure, in particular, we establish their (non)dependence on ε.

STEP 1 (The H^2 regularity of solutions). Let $\bar{v} = \int_{-\varepsilon h}^{0} v\,\mathrm{d}z$; we have

$$\frac{\partial^2}{\partial x_k^2}\bar{v}(x_1,x_2,x_3) = \int_{-\varepsilon h}^{0}\frac{\partial^2}{\partial x_k^2}v(x_1,x_2,z)\,\mathrm{d}z + I_k(v), \tag{4.122}$$

$$\begin{aligned} I_k(v) = {}& 2\varepsilon\frac{\partial h}{\partial x_k}\frac{\partial v}{\partial x_3}\big(x_1,x_2,-\varepsilon h(x_1,x_2)\big)\\ &- \varepsilon^2\left(\frac{\partial h}{\partial x_k}\right)^2 v\big(x_1,x_2,-\varepsilon h(x_1,x_2)\big), \quad k = 1,2. \end{aligned} \tag{4.123}$$

Therefore, by integrating the first equation in (4.141) with respect to x_3 we obtain the two-dimensional Stokes problem:

$$\begin{cases} -\Delta\bar{v} + \nabla(\varepsilon h p) = \bar{f} & \text{in } \Gamma_{\mathrm{i}}, \\ \operatorname{div}\bar{v} = 0 \quad \text{in } \Gamma_{\mathrm{i}}, & v = 0 \quad \text{on } \partial\Gamma_{\mathrm{i}}, \end{cases} \tag{4.124}$$

where

$$\bar{f} = \int_{-\varepsilon h}^{0} f_v \,\mathrm{d}z + \left.\frac{\partial v}{\partial x_3}\right|_{x_3=0} - \left.\frac{\partial v}{\partial x_3}\right|_{x_3=-\varepsilon h} + I_1(v) + I_2(v) + \varepsilon p \nabla h. \tag{4.125}$$

Thanks to Lemma 4.10, each term on the right-hand side of (4.125) is in $L^2(\Gamma_{\mathrm{i}})$, which implies $\bar{f} \in L^2(\Gamma_{\mathrm{i}})$: this is stated in (4.117) for $\partial v/\partial x_3|_{\Gamma_{\mathrm{i}}}$ and $\partial v/\partial x_3|_{\Gamma_{\mathrm{b}}}$; similarly each term in I_1 and I_2 belongs to $L^2(\Gamma_{\mathrm{b}})$ (and thus $L^2(\Gamma_{\mathrm{i}})$) because $v \in H^{2-\delta}(\mathcal{M}_\varepsilon)$, $0 < \delta < 1/2$; finally for p we recall from (4.118) that $\nabla p \in H^{-1}(\Gamma_{\mathrm{i}})$ and thus $p \in L^2(\Gamma_{\mathrm{i}})$. Therefore from the classical regularity theory of the two-dimensional Stokes problem, we conclude that $\nabla(hp) \in L^2(\Gamma_{\mathrm{i}})$, and then $\nabla p \in L^2(\Gamma_{\mathrm{i}})$. We return to problem (4.96), and move the gradient of the pressure to the right hand-side and obtain, thanks to Lemma 4.7, $v \in H^2(\mathcal{M}_\varepsilon)$ and

$$\sum_{k,j=1}^{3} \left|\frac{\partial^2 v}{\partial x_k\,\partial x_j}\right|_\varepsilon^2 \leqslant C(h,\alpha_v)\big[|f_v|_\varepsilon^2 + |g_v|_i^2 + |\nabla g_v|_i^2\big] + C(h,\alpha)\varepsilon|\nabla p|_i^2. \tag{4.126}$$

Note that we have the pressure term on the right-hand side of (4.126). Removing that term is done in the second step below.

STEP 2 (The Cattabriga–Solonnikov type inequality). Our aim is now to bound $|\nabla p|_i$ properly and to derive (4.110) from (4.126). First we homogenize the boundary condition in (4.96). Let $v_\ell = (\Psi_1, \Psi_2)$ where Ψ_1 and Ψ_2 are constructed using Lemma 4.7, i.e.,

$$\begin{cases} -\Delta_3 \Psi_k = f_{v,k} & \text{in } \mathcal{M}_\varepsilon,\ k = 1, 2, \\ \frac{\partial \Psi_k}{\partial x_3} + \alpha_v \Psi_k = g_{v,k} & \text{on } \Gamma_{\mathrm{i}},\ k = 1, 2, \\ \Psi_k = 0 & \text{on } \Gamma_{\mathrm{b}} \cup \Gamma_\ell,\ k = 1, 2, \end{cases}$$

where $f_v = (f_{v,1}, f_{v,2})$, $g_v = (g_{v,1}, g_{v,2})$. Thanks to Lemma 4.7,

$$\sum_{k,j=1}^{3} \left| \frac{\partial^2 v_l}{\partial x_k\, \partial x_j} \right|_\varepsilon^2 \leqslant C(h, \alpha_v)\big(|f_v|_\varepsilon^2 + |g_v|_i^2 + |\nabla g_v|_i^2\big). \tag{4.127}$$

Setting $v^* = v - v_\ell$, it suffices to establish (4.127) with v_l replaced by v^*. We have:

$$\begin{cases} -\big(\Delta v^* + \frac{\partial^2 v^*}{\partial x_3^2}\big) + \nabla p = 0 & \text{in } \mathcal{M}_\varepsilon, \\ \operatorname{div} \int_{-\varepsilon h}^{0} v^*\, \mathrm{d}z = g^* & \text{on } \Gamma_{\mathrm{i}}, \\ v^* = 0 & \text{on } \Gamma_\ell \cup \Gamma_{\mathrm{b}}, \\ \frac{\partial v^*}{\partial x_3} + \alpha_v v^* = 0 & \text{on } \Gamma_{\mathrm{i}}, \end{cases} \tag{4.128}$$

where

$$g^* = -\operatorname{div} \int_{-\varepsilon h}^{0} v_l\, \mathrm{d}x_3.$$

Note that inequality (4.127) together with the Cauchy–Schwarz inequality imply

$$\|g^*\|_{H^1(\Gamma_{\mathrm{i}})}^2 \leqslant C(h, \alpha_v)\varepsilon\big[|f_1|_\varepsilon^2 + |g_v|_i^2 + |\nabla g_v|_i^2\big]. \tag{4.129}$$

Define

$$V^* = \int_{-\varepsilon h}^{0} v^*\, \mathrm{d}x_3;$$

V^* is the solution of the two-dimensional Stokes problem

$$\begin{cases} -\Delta V^* + \nabla(\varepsilon p) = F^* & \text{in } \Gamma_{\mathrm{i}}, \\ \operatorname{div} V^* = g^*, & \\ V^* = 0 & \text{on } \partial\Gamma_{\mathrm{i}}, \end{cases} \tag{4.130}$$

where

$$F^* = \left.\frac{\partial v^*}{\partial x_3}\right|_{x_3=0} - \left.\frac{\partial v^*}{\partial x_3}\right|_{x_3=-\varepsilon h} + I_1(v^*) + I_2(v^*),$$

with I_1 and I_2 as in (4.123). Hence

$$|F^*|_{L^2(\Gamma_{\mathrm{i}})}^2 \leqslant C(h)\left[\left|\frac{\partial v^*}{\partial x_3}\right|_{L^2(\Gamma_{\mathrm{i}})}^2 + \left|\frac{\partial v^*}{\partial x_3}\right|_{L^2(\Gamma_{\mathrm{b}})}^2\right]. \tag{4.131}$$

Now, since $v^* = 0$ on Γ_{b}, we have $\frac{\partial v^*}{\partial x_k} = \varepsilon \frac{\partial h}{\partial x_k}\frac{\partial v^*}{\partial x_3}$ on Γ_{b} and, by the Poincaré inequality and the boundary condition satisfied by v^* on Γ_{i}, we have $|\frac{\partial v^*}{\partial x_3}|^2_{L^2(\Gamma_{\mathrm{i}})} \leqslant 2\alpha_v^2 \varepsilon \bar{h} |\frac{\partial v^*}{\partial x_3}|^2_\varepsilon$. Furthermore, we can write

$$
\begin{aligned}
\left|\frac{\partial v^*}{\partial x_3}\right|^2_{L^2(\Gamma_{\mathrm{b}})} &\leqslant \left|\frac{\partial v^*}{\partial x_3}\right|^2_{L^2(\Gamma_{\mathrm{i}})} + 2\left|\frac{\partial v^*}{\partial x_3}\right|_\varepsilon \left|\frac{\partial^2 v^*}{\partial x_3^2}\right|_\varepsilon \\
&\leqslant 2\alpha_v^2 \varepsilon \bar{h}\left|\frac{\partial v^*}{\partial x_3}\right|^2_\varepsilon + \theta\varepsilon\left|\frac{\partial^2 v^*}{\partial x_3^2}\right|^2_\varepsilon + \frac{C\varepsilon}{\theta}\left|\frac{\partial v^*}{\partial x_3}\right|^2_\varepsilon,
\end{aligned} \tag{4.132}
$$

where θ is a positive constant independent of ε, that will be chosen below.

Therefore

$$
|F^*|^2_{L^2(\Gamma_{\mathrm{i}})} \leqslant C\left|\frac{\partial v^*}{\partial x_k}\right|^2_\varepsilon + C(h)\theta\varepsilon\left|\frac{\partial^2 v^*}{\partial x_3^2}\right|^2_\varepsilon. \tag{4.133}
$$

We estimate the H^1-norm of v^*, using $v^* = v - v_l$ and the H^1-estimates of v and v_l. Therefore we can easily obtain

$$
|F^*|^2_{L^2(\Gamma_{\mathrm{i}})} \leqslant C(h)\varepsilon\big[|f_1|^2_\varepsilon + |g_v|^2_i + |\nabla g_v|^2_i\big] + C(h)\theta\varepsilon\left|\frac{\partial^2 v^*}{\partial x_3^2}\right|^2_\varepsilon. \tag{4.134}
$$

Now using the Cattabriga–Solonnikov inequality for the two-dimensional Stokes problem (4.130), there exists a constant C independent of ε such that

$$
|V^*|^2_{H^2(\Gamma_{\mathrm{i}})} + \varepsilon^2|\nabla(hp)|^2_{L^2(\Gamma_{\mathrm{i}})} \leqslant C|F^*|^2_{L^2(\Gamma_{\mathrm{i}})}. \tag{4.135}
$$

From this we obtain

$$
\begin{aligned}
&\varepsilon^2|\nabla(hp)|^2_{L^2(\Gamma_{\mathrm{i}})} \leqslant C(h,\theta)\varepsilon\big[|f_v|^2_\varepsilon + |g_v|^2_i + |\nabla g_v|^2_i\big] + C(h)\theta\varepsilon\left|\frac{\partial^2 v^*}{\partial x_3^2}\right|^2_\varepsilon, \\
&\varepsilon^2|\nabla p|^2_{L^2(\Gamma_{\mathrm{i}})} \leqslant C(h,\theta)\varepsilon^2|p|^2_{L^2(\Gamma_{\mathrm{i}})} \\
&\qquad + C(h,\theta)\varepsilon\big[|f_v|^2_\varepsilon + |g_v|^2_i + |\nabla g_v|^2_i\big] + C|h|\theta\varepsilon\left|\frac{\partial^2 v^*}{\partial x_3^2}\right|^2_\varepsilon.
\end{aligned} \tag{4.136}
$$

From (4.96) and the weak formulation (4.98) of (4.96), we see that

$$
\begin{aligned}
|p|_{L^2(\Gamma_{\mathrm{i}})/\mathbb{R}} &\leqslant C|\nabla p|_{H^{-1}(\Gamma_{\mathrm{i}})} \\
&\leqslant C\|v\|_{H^1(\mathcal{M}_\varepsilon)} \leqslant C|f_v|_\varepsilon,
\end{aligned}
$$

so that we actually have the same type of estimate (4.136) for ∇p as for $\nabla(hp)$. Finally, since $\Delta_3 v^* = \nabla p$, in $\mathcal{M}_\varepsilon$, $v^* = 0$ on $\Gamma_{\mathrm{b}} \cup \Gamma_\ell$ and $\frac{\partial v^*}{\partial x_3} + \alpha^* v = 0$ on Γ_{i}, we have thanks to

Lemma 4.7,

$$\sum_{k,j=1}^{3}\left|\frac{\partial^2 v^*}{\partial x_k\,\partial x_j}\right|_{\varepsilon}^{2} \leqslant C(h,\alpha_v)\varepsilon|\nabla p|_{L^2(\Gamma_{\rm i})}^{2}$$
$$\leqslant C(h,\alpha_v)\big[|f_1|_{\varepsilon}^{2}+|g_v|_{i}^{2}+|\nabla g_v|_{i}^{2}\big]+C(h,\alpha_v)\theta\left|\frac{\partial^2 v^*}{\partial x_3^2}\right|_{\varepsilon}^{2},$$

and therefore for θ small enough, so that $C(h,\alpha_v)\theta \leqslant \frac{1}{2}$, we conclude that

$$\sum_{k,j=1}^{3}\left|\frac{\partial^2 v^*}{\partial x_k\,\partial x_j}\right|_{\varepsilon}^{2} \leqslant C(h,\alpha)\varepsilon|\nabla p|_{L^2(\Gamma_{\rm i})}^{2} \leqslant C(h,\alpha)\big[|f_1|_{\varepsilon}^{2}+|g_v|_{i}^{2}+|\nabla g_v|_{i}^{2}\big].$$

The proof of Theorem 4.4 is now complete. □

By interpolation, it is easy to derive from Theorem 4.4 the following result:

THEOREM 4.5. *Assume that h is a positive function in $\mathcal{C}^3(\overline{\Gamma_i})$. Let $(v,p)\in\mathbb{H}^1(M_\varepsilon)\times L^2(\Gamma_{\rm i})$ be a weak solution of*

$$\begin{cases} -\big(\Delta v+\frac{\partial^2 v}{\partial x_3^2}\big)+\nabla p=f_v & \text{in } \mathcal{M}_\varepsilon,\\ \operatorname{div}\int_{-\varepsilon h}^{0} v\,{\rm d}z=0 & \text{on } \Gamma_{\rm i},\\ v=0 & \text{on } \Gamma_\ell\cup\Gamma_{\rm b},\\ \frac{\partial v}{\partial x_3}+\alpha_v v=g_v & \text{on } \Gamma_{\rm i}. \end{cases} \tag{4.137}$$

Then, if $f_v\in\mathbb{L}^2(\mathcal{M}_\varepsilon)$ and $g_v\in\mathbb{H}^s(\Gamma_{\rm i})$, $0\leqslant s\leqslant 1$,

$$(v,p)\in\mathbb{H}^{s+1}(\mathcal{M}_\varepsilon)\times H^s(\mathcal{M}_\varepsilon). \tag{4.138}$$

Moreover the following inequality holds:

$$|v|_{\mathbb{H}^{1+s}(\mathcal{M}_\varepsilon)}^{2}+\varepsilon|p|_{H^s(\Gamma_{\rm i})}^{2}\leqslant C_0\big[|f_v|_{\varepsilon}^{2}+\varepsilon^{1-s}\|g_v\|_{H^s(\Gamma_{\rm i})}^{2}\big], \tag{4.139}$$

where C_0 is a constant depending on the data but not on ε.

4.5. *Regularity of the coupled system*

In this section we prove the H^2 regularity of the solution of a coupled system of equations corresponding to the linear part of the primitive equations of the coupled atmosphere–ocean. We will concentrate on the velocity part; the temperature and salinity parts follow in the same manner.

The unknown is $v = (v^{\mathrm{a}}, v^{\mathrm{s}})$, with v^{a}, v^{s} corresponding to the horizontal velocities in the air and in the ocean.[15] These functions satisfy the following equations and boundary conditions:

$$\begin{cases} -\Delta v^{\mathrm{a}} - \frac{\partial^2 v^{\mathrm{a}}}{\partial x_3^2} + \nabla p^{\mathrm{a}} = f_v^{\mathrm{a}} & \text{in } \mathcal{M}_\varepsilon^{\mathrm{a}}, \\ \operatorname{div} \int_0^L v^{\mathrm{a}}(x_1, x_2, z)\,\mathrm{d}z = 0, & (x_1, x_2) \in \mathbb{R}^2, \\ v^{\mathrm{a}} = 0 & \text{on } \Gamma_{\mathrm{u}} \cup \Gamma_\ell^{\mathrm{a}}, \\ \frac{\partial v^{\mathrm{a}}}{\partial x_3} + \alpha_v \big(v^{\mathrm{a}} - v^{\mathrm{s}}\big) = g_v & \text{on } \Gamma_{\mathrm{i}}, \\ \frac{\partial v^{\mathrm{a}}}{\partial x_3} + \alpha_v v^{\mathrm{a}} = g_v & \text{on } \Gamma_{\mathrm{e}}, \end{cases} \tag{4.140}$$

and

$$\begin{cases} -\Delta v^{\mathrm{s}} - \frac{\partial^2 v^{\mathrm{s}}}{\partial x_3^2} + \nabla p^{\mathrm{s}} = f_v^{\mathrm{s}} & \text{in } \mathcal{M}_\varepsilon^{\mathrm{s}}, \\ \operatorname{div} \int_{-\varepsilon h}^0 v^{\mathrm{s}}\,\mathrm{d}z = 0 & \text{in } \Gamma_{\mathrm{i}}, \\ v^{\mathrm{s}} = 0 & \text{on } \Gamma_\ell^{\mathrm{s}} \cup \Gamma_{\mathrm{b}}, \\ -\frac{\partial v^{\mathrm{s}}}{\partial x_3} + \alpha_{v^{\mathrm{a}}} \big(v^{\mathrm{a}} - v^{\mathrm{s}}\big) = g_v & \text{on } \Gamma_{\mathrm{i}}. \end{cases} \tag{4.141}$$

The domain $\mathcal{M}_\varepsilon^{\mathrm{s}}$ is the domain occupied by the ocean while $\mathcal{M}_\varepsilon^{\mathrm{a}}$ is the domain occupied by the atmosphere and $\mathcal{M}_\varepsilon = \mathcal{M}_\varepsilon^{\mathrm{a}} \cup \mathcal{M}_\varepsilon^{\mathrm{s}}$:

$$\mathcal{M}_\varepsilon^{\mathrm{s}} = \big\{(x_1, x_2, x_3) \in \mathbb{R}^3;\ (x_1, x_2) \in \Gamma_{\mathrm{i}}, -\varepsilon h(x_1, x_2) < x_3 < 0\big\},$$
$$\mathcal{M}_\varepsilon^{\mathrm{a}} = \big\{(x_1, x_2, x_3) \in \mathbb{R}^3;\ (x_1, x_2) \in \Gamma, 0 < x_3 < \varepsilon L\big\}.$$

Here, Γ, which is a bounded domain in the plane $x_3 = 0$, is the lower boundary of the atmosphere; it consists of the interface Γ_{i} with the ocean and Γ_{e}, the interface with the Earth, $\Gamma = \Gamma_{\mathrm{i}} \cup \Gamma_{\mathrm{e}}$ ($\Gamma_{\mathrm{i}} \cap \Gamma_{\mathrm{e}} = \emptyset$) (see Section 2.5); furthermore, and as in Section 2.5,

$$\Gamma_{\mathrm{b}} = \big\{\big(x', x_3\big);\ x' \in \Gamma_{\mathrm{i}}, x_3 = -\varepsilon h\big(x'\big)\big\},$$
$$\Gamma_\ell^{\mathrm{a}} = \big\{\big(x', x_3\big),\ x' \in \partial\Gamma, 0 < x_3 < \varepsilon L\big\},$$
$$\Gamma_\ell^{\mathrm{s}} = \big\{\big(x', x_3\big);\ x' \in \partial\Gamma_{\mathrm{i}}, -\varepsilon h\big(x'\big) < x_3 < 0\big\},$$
$$\Gamma_{\mathrm{u}} = \big\{\big(x', x_3\big);\ x' \in \Gamma_{\mathrm{i}}, x_3 = \varepsilon L\big\},$$
$$\Gamma_{\mathrm{e}} = \Gamma \backslash \Gamma_{\mathrm{i}}, \Gamma \quad \text{and} \quad \Gamma_{\mathrm{i}} \text{ as above.}$$

The coefficient α_{v} is a positive number, and g_{v} is a function defined on Γ.

Problem (4.140)–(4.141) is the stationary linearized form of the primitive equations of the coupled system atmosphere–ocean. Besides its intrinsic interest, the study of this problem is needed for the study of the full nonlinear (stationary or time dependent) coupled atmosphere–ocean system.

[15] We recall that we use the superscript "s" as sea, instead of "o" as ocean which can be confused with a zero.

4.5.1. *Weak formulation of the coupled system.* As in Section 4.4.1 we start with the weak formulation of (4.140) and (4.141). In this section we drop the index ε which is irrelevant ($\varepsilon = 1$).

We are given f_{v} in $\mathbb{L}^2(\mathcal{M})$ and g_{v} in $\mathbb{H}^{1/2}(\Gamma)$.

For the weak formulation of (4.140) and (4.141) we consider the space

$$V = \Big\{ v = (v^{\mathrm{a}}, v^{\mathrm{s}}) \in \mathbb{H}^1(\mathcal{M}^{\mathrm{a}}) \times \mathbb{H}^1(\mathcal{M}^{\mathrm{s}}),\ \operatorname{div} \int_{-h}^{0} v^{\mathrm{s}}\, \mathrm{d}z = 0,$$

$$v^{\mathrm{a}} = 0 \text{ on } \Gamma_{\mathrm{u}} \cup \Gamma_{\ell}^{\mathrm{a}},\ v^{\mathrm{s}} = 0 \text{ on } \Gamma_{\mathrm{b}} \cup \Gamma_{\ell}^{\mathrm{s}} \Big\}.$$

Here $v^{\mathrm{a}} = v|_{\mathcal{M}^{\mathrm{a}}}$ and $v^{\mathrm{s}} = v|_{\mathcal{M}^{\mathrm{s}}}$; note that the traces of v^{a} and v^{s} on Γ_{i} are not necessarily equal, as explained in Remark 2.7(iii). We set, with obvious notations:

$$\begin{aligned} ((v, \tilde{v})) &= ((v^{\mathrm{a}}, \tilde{v}^{\mathrm{a}}))_{\mathrm{a}} + ((v^{\mathrm{s}}, \tilde{v}^{\mathrm{s}}))_{\mathrm{s}} \\ &= \sum_{i=1}^{2} \sum_{j=1}^{3} \int_{\mathcal{M}} \frac{\partial v_i}{\partial x_j} \frac{\partial \tilde{v}_i}{\partial x_j}\, \mathrm{d}\mathcal{M}; \end{aligned}$$

because of the Poincaré inequality, $\|v\| = ((v, v))^{1/2}$ is a Hilbert norm on V.

To obtain the weak formulation, we consider a test function $\tilde{v} = (\tilde{v}^{\mathrm{a}}, \tilde{v}^{\mathrm{s}}) \in V$; we multiply the first equation (4.140) by $\tilde{v}^{\mathrm{a}}$ and the first equation in (4.141) by $\tilde{v}^{\mathrm{s}}$. We integrate over $\mathcal{M}^{\mathrm{a}}$ and $\mathcal{M}^{\mathrm{s}}$ respectively and add the resulting equations; we proceed exactly as in Section 4.4.1, using the boundary condition in (4.140) and (4.141) and we arrive at the following: To find $v \in V$ such that

$$a(v, \tilde{v}) = \ell(\tilde{v}) \quad \forall \tilde{v} \in V, \tag{4.142}$$

with

$$\begin{aligned} a(v, \tilde{v}) &= ((v, \tilde{v})) + \int_{\Gamma_{\mathrm{i}}} \alpha_{\mathrm{v}} (v^{\mathrm{a}} - v^{\mathrm{s}})(\tilde{v}^{\mathrm{a}} - \tilde{v}^{\mathrm{s}})\, \mathrm{d}\Gamma_{\mathrm{i}} + \int_{\Gamma_{\mathrm{e}}} \alpha_v v^{\mathrm{a}} \tilde{v}^{\mathrm{a}}\, \mathrm{d}\Gamma_{\mathrm{e}}, \\ \ell(\tilde{v}) &= \int_{\mathcal{M}^{\mathrm{a}}} f_v^{\mathrm{a}} \tilde{v}^{\mathrm{a}}\, \mathrm{d}\mathcal{M}^{\mathrm{a}} + \int_{\mathcal{M}^{\mathrm{s}}} f_v^{\mathrm{a}} \tilde{v}^{\mathrm{s}}\, \mathrm{d}\mathcal{M}^{\mathrm{s}} \\ &\quad + \int_{\Gamma_{\mathrm{i}}} g_v (\tilde{v}^{\mathrm{a}} - \tilde{v}^{\mathrm{s}})\, \mathrm{d}\Gamma_{\mathrm{i}} + \int_{\Gamma_{\mathrm{e}}} g_v \tilde{v}^{\mathrm{a}}\, \mathrm{d}\Gamma_{\mathrm{e}}. \end{aligned} \tag{4.143}$$

The existence and uniqueness of a solution $v \in V$ of (4.142) is elementary, and follows from the Lax–Milgram theorem. The more delicate question of showing that $v = (v^{\mathrm{a}}, v^{\mathrm{s}})$ actually satisfies all the equations (4.140) and (4.141) is handled as follows: we find p^{a} and p^{s} such that the first equation (4.140) and (4.141) are valid exactly as we did in Section 4.4.1, for the ocean and the atmosphere. Using also Proposition 4.1 for the ocean and the atmosphere, we obtain the boundary conditions on Γ_{i} and Γ_{e}; the other equations and boundary conditions follow from $v \in V$.

REMARK 4.3. Setting $\tilde{v} = v$ in (4.142), we find

$$\begin{aligned} &\left|\nabla v^{\mathrm{a}}\right|^2_{L^2(\mathcal{M}^{\mathrm{a}}_\varepsilon)^4} + \left|\nabla v^{\mathrm{s}}\right|^2_{L^2(\mathcal{M}^{\mathrm{s}}_\varepsilon)^4} + \int_{\Gamma_{\mathrm{i}}} \alpha_v \left|v^{\mathrm{a}} - v^{\mathrm{s}}\right|^2 \mathrm{d}\Gamma_{\mathrm{i}} + \int_{\Gamma_{\mathrm{e}}} \alpha_v \left|v^{\mathrm{a}}\right|^2 \mathrm{d}\Gamma_{\mathrm{e}} \\ &\quad = \int_{\mathcal{M}^{\mathrm{a}}} f_v^{\mathrm{a}} v^{\mathrm{a}} \,\mathrm{d}\mathcal{M}^{\mathrm{a}}_\varepsilon + \int_{\mathcal{M}^{\mathrm{s}}} f_v^{\mathrm{s}} v^{\mathrm{s}} \,\mathrm{d}\mathcal{M}^{\mathrm{s}}_\varepsilon + \int_{\Gamma_{\mathrm{i}}} g_v \left(v^{\mathrm{a}} - v^{\mathrm{s}}\right) \mathrm{d}\Gamma_{\mathrm{i}} + \int_{\Gamma_{\mathrm{e}}} g_v v^{\mathrm{a}} \,\mathrm{d}\Gamma_{\mathrm{e}}. \end{aligned} \tag{4.144}$$

4.5.2. *H^2 regularity for the coupled system.* Having established the complete equivalence of (4.140)–(4.141) with (4.142), we now want to show that the solution of this system possesses the H^2 regularity, namely

$$\left(v^{\mathrm{a}}, p^{\mathrm{a}}\right) \in \mathbb{H}^2\left(\mathcal{M}^{\mathrm{a}}\right) \times H^1\left(\mathcal{M}^{\mathrm{a}}\right) \quad \text{and} \quad \left(v^{\mathrm{s}}, p^{\mathrm{s}}\right) \in \mathbb{H}^2\left(\mathcal{M}^{\mathrm{s}}\right) \times H^1\left(\mathcal{M}^{\mathrm{s}}\right), \tag{4.145}$$

whenever $f_v^{\mathrm{a}} \in \mathbb{L}^2(\mathcal{M}^{\mathrm{a}})$, $f_v^{\mathrm{s}} \in \mathbb{L}^2(\mathcal{M}^{\mathrm{s}})$ and $g_v \in \mathbb{H}^{1/2}(\Gamma)$.

More precisely, we will prove the following theorem:

THEOREM 4.6. *Assume that h is a positive function in $\mathcal{C}^3(\overline{\Gamma_i})$. Let $(v^{\mathrm{a}}, p^{\mathrm{a}}) \in \mathbb{H}^1(\mathcal{M}^{\mathrm{a}}_\varepsilon) \times L^2(\Gamma_{\mathrm{i}} \cup \Gamma_{\mathrm{e}})$, and $(v^{\mathrm{s}}, p^{\mathrm{s}}) \in \mathbb{H}^1(\mathcal{M}^{\mathrm{s}}_\varepsilon) \times L^2(\Gamma_{\mathrm{i}})$ be a weak solution of* (4.140) *and* (4.141) *(or* (4.142)*). If $f_v^{\mathrm{a}} \in \mathbb{L}^2(\mathcal{M}^{\mathrm{a}}_\varepsilon)$, $f_v^{\mathrm{s}} \in \mathbb{L}^2(\mathcal{M}^{\mathrm{s}}_\varepsilon)^2$ and $g_v \in \mathbb{H}^1(\Gamma)$, $g_v = 0$ on $\partial\Gamma_{\mathrm{e}}$, then*

$$\left(v^{\mathrm{a}}, p^{\mathrm{a}}\right) \in \mathbb{H}^2\left(\mathcal{M}^{\mathrm{a}}_\varepsilon\right) \times H^1(\Gamma_{\mathrm{i}} \cup \Gamma_{\mathrm{e}}) \quad \textit{and} \quad \left(v^{\mathrm{s}}, p^{\mathrm{s}}\right) \in \mathbb{H}^2\left(\mathcal{M}^{\mathrm{s}}_\varepsilon\right) \times H^1(\Gamma_{\mathrm{i}}). \tag{4.146}$$

Moreover, the following inequality holds:

$$\begin{aligned} &\left|v^{\mathrm{a}}\right|^2_{\mathbb{H}^2(\mathcal{M}^{\mathrm{a}}_\varepsilon)} + \left|v^{\mathrm{s}}\right|^2_{\mathbb{H}^2(\mathcal{M}^{\mathrm{s}}_\varepsilon)} + \varepsilon\left|p^{\mathrm{a}}\right|^2_{H^1(\Gamma_{\mathrm{i}})} + \varepsilon\left|p^{\mathrm{s}}\right|^2_{H^1(\Gamma_{\mathrm{i}})} \\ &\quad \leqslant C_0\left[\left|f_v^{\mathrm{a}}\right|^2_\varepsilon + \left|f_v^{\mathrm{s}}\right|^2_\varepsilon + \left|\nabla g_v\right|^2_{\mathbb{L}^2(\Gamma)}\right]. \end{aligned} \tag{4.147}$$

PROOF. Since $v^{\mathrm{a}} \in \mathbb{H}^1(\mathcal{M}^{\mathrm{a}}_\varepsilon)$ and $v^{\mathrm{s}} \in \mathbb{H}^1(\mathcal{M}^{\mathrm{s}}_\varepsilon)$, $v^{\mathrm{a}}|_{\Gamma_{\mathrm{i}}}$ and $v^{\mathrm{s}}|_{\Gamma_{\mathrm{i}}}$ belong to $\mathbb{H}^{1/2-\delta}(\Gamma_{\mathrm{i}})$ for all δ, $0 < \delta < 1/2$, and there exists a constant C_0 independent of ε such that

$$\left|v^{\mathrm{a}}\right|^2_{\mathbb{H}^{1/2-\delta}(\Gamma_{\mathrm{i}})} + \left|v^{\mathrm{s}}\right|^2_{\mathbb{H}^{1/2-\delta}(\Gamma_{\mathrm{i}})} \leqslant C_0\left[\left\|v^{\mathrm{a}}\right\|^2_{\mathbb{H}^1(\mathcal{M}^{\mathrm{a}}_\varepsilon)} + \left\|v^{\mathrm{s}}\right\|^2_{\mathbb{H}^1(\mathcal{M}^{\mathrm{s}}_\varepsilon)}\right]. \tag{4.148}$$

Furthermore, (4.144) implies that the right-hand side of (4.148) can be bounded by an expression identical to the right-hand side of (4.146).

The boundary conditions on Γ_{i} imply then that

$$\frac{\partial v^{\mathrm{a}}}{\partial x_3} + \alpha_v v^{\mathrm{a}} \quad \text{and} \quad -\frac{\partial v^{\mathrm{s}}}{\partial x_3} + \alpha_v v^{\mathrm{s}}$$

belong to $\mathbb{H}^{1/2-\delta}(\Gamma_{\mathrm{i}})$ and their norm in these spaces are bounded similarly.

Therefore, by Theorem 4.5 applied separately to $\mathcal{M}_\varepsilon^{\mathrm{a}}$ and $\mathcal{M}_\varepsilon^{\mathrm{s}}$, we conclude that

$$(v^{\mathrm{a}}, p^{\mathrm{a}}) \in \mathbb{H}^{3/2-\delta}(M_\varepsilon^{\mathrm{a}}) \times H^{1/2-\delta}(\Gamma),$$
$$(v^{\mathrm{s}}, p^{\mathrm{s}}) \in \mathbb{H}^{3/2-\delta}(M_\varepsilon^{\mathrm{s}}) \times H^{1/2-\delta}(\Gamma_{\mathrm{i}})$$

and

$$|v^{\mathrm{a}}|^2_{\mathbb{H}^{3/2-\delta}(\mathcal{M}_\varepsilon^{\mathrm{a}})} + |v^{\mathrm{s}}|^2_{\mathbb{H}^{3/2-\delta}(\mathcal{M}_\varepsilon^{\mathrm{s}})} + \varepsilon|p^{\mathrm{a}}|^2_{H^{1/2-\delta}(\Gamma_{\mathrm{i}})} + \varepsilon|p^{\mathrm{s}}|^2_{H^{1/2-\delta}(\Gamma_{\mathrm{i}})} \leqslant \tilde{\kappa}, \tag{4.149}$$

where $\tilde{\kappa}$ is the right-hand side of (4.139) with a possibly different constant C_0.

Using the trace theorem again, we see that

$$\frac{\partial v^{\mathrm{a}}}{\partial x_3} + \alpha_v v^{\mathrm{a}} \quad \text{and} \quad -\frac{\partial v^{\mathrm{s}}}{\partial x_3} + \alpha_v v^{\mathrm{s}} \quad \text{belong to } \mathbb{H}_0^{1-\delta}(\Gamma_{\mathrm{i}})^2\ \forall\delta,$$

and there exists a constant C_0 independent of ε such that

$$\|v^{\mathrm{a}}\|^2_{\mathbb{H}^{1-\delta}(\Gamma)} + \|v^{\mathrm{s}}\|^2_{\mathbb{H}^{1-\delta}(\Gamma_{\mathrm{i}})} \leqslant C_0\left(\|v^{\mathrm{a}}\|^2_{\mathbb{H}^{3/2-\delta}(\mathcal{M}_\varepsilon^{\mathrm{a}})} + \|v^{\mathrm{s}}\|^2_{\mathbb{H}^{3/2-\delta}(\mathcal{M}_\varepsilon^{\mathrm{s}})}\right). \tag{4.150}$$

Therefore, by Theorem 4.5, we conclude that

$$(v^{\mathrm{a}}, p^{\mathrm{a}}) \in \mathbb{H}^{2-\delta}(\mathcal{M}_\varepsilon^{\mathrm{a}}) \times H^{1-\delta}(\Gamma),$$
$$(v^{\mathrm{s}}, p^{\mathrm{s}}) \in \mathbb{H}^{2-\delta}(\mathcal{M}_\varepsilon^{\mathrm{s}}) \times H^{1-\delta}(\Gamma_{\mathrm{i}})$$

and

$$\|v^{\mathrm{a}}\|^2_{\mathbb{H}^{2-\delta}(\mathcal{M}_\varepsilon^{\mathrm{a}})} + \|v^{\mathrm{s}}\|^2_{\mathbb{H}^{2-\delta}(\mathcal{M}_\varepsilon^{\mathrm{s}})} + \varepsilon|p^{\mathrm{a}}|^2_{H^{1-\delta}(\Gamma)} + \varepsilon|p^{\mathrm{s}}|^2_{H^{1-\delta}(\Gamma_{\mathrm{i}})} \leqslant \tilde{\kappa},$$

$\tilde{\kappa}$ as above. A final application of the trace theorem and of Theorem 4.5 to $\mathcal{M}_\varepsilon^{\mathrm{a}}$ and $\mathcal{M}_\varepsilon^{\mathrm{s}}$ yields

$$(v^{\mathrm{a}}, p^{\mathrm{a}}) \in \mathbb{H}^{2-\delta}(\mathcal{M}_\varepsilon^{\mathrm{a}}) \times H^{1-\delta}(\Gamma),$$
$$(v^{\mathrm{s}}, p^{\mathrm{s}}) \in \mathbb{H}^{2-\delta}(\mathcal{M}_\varepsilon^{\mathrm{s}}) \times H^{1-\delta}(\Gamma_{\mathrm{i}})$$

and

$$|v^{\mathrm{a}}|^2_{\mathbb{H}^2(\mathcal{M}_\varepsilon^{\mathrm{a}})} + |v^{\mathrm{s}}|^2_{\mathbb{H}^2(\mathcal{M}_\varepsilon^{\mathrm{s}})} + \varepsilon|p^{\mathrm{a}}|^2_{H^1(\Gamma_{\mathrm{i}})} + \varepsilon|p^{\mathrm{s}}|^2_{H^1(\Gamma_{\mathrm{i}})} \leqslant \tilde{\kappa},$$

$\tilde{\kappa}$ as above. The proof is complete. □

Acknowledgments

This work was partially supported by the National Science Foundation under the grants NSF-DMS-0074334, NSF-DMS-0204863 and NSF-DMS-0305110 and by the Research Fund of Indiana University.

References

[1] S. Agmon, A. Douglis and L. Nirenberg, *Estimates near the boundary for solutions of elliptic partial differential equations satisfying general boundary conditions, I*, Comm. Pure Appl. Math. **12** (1959), 623–727.

[2] S. Agmon, A. Douglis and L. Nirenberg, *Estimates near the boundary for solutions of elliptic partial differential equations satisfying general boundary conditions, II*, Comm. Pure Appl. Math. **17** (1964), 35–92.

[3] P. Bartello, *Geostrophic adjustment and inverse cascades in rotating stratified turbulence*, J. Atmos. Sci. **52** (24) (1995), 4410–4428.

[4] T. Bewley, R. Temam and M. Ziane, *A general framework for robust control in fluid mechanics*, Phys. D **138** (2000), 360–392.

[5] T. Bewley, R. Temam and M. Ziane, *Existence and uniqueness of optimal control to the Navier–Stokes equations*, C. R. Acad. Sci. Paris Sér. I **330** (2000), 1007–1011.

[6] D. Bresch, A. Kazhikov and J. Lemoine, *On the two-dimensional hydrostatic Navier–Stokes equations*, to appear.

[7] P. Constantin and C. Foias, *Navier–Stokes Equations*, Univ. Chicago Press, Chicago, IL (1988).

[8] B. Ewald and R. Temam, *Maximum principles for the primitive equations of the atmosphere*, Discrete Contin. Dyn. Syst. **7** (2) (2001), 343–362.

[9] J.M. Ghidaglia, *Régularité des solutions de certains problèmes aux limites linéaires liés aux équations d'Euler* (in French) [Regularity of the solutions of certain linear boundary value problems associated with the Euler equations], Comm. Partial Differential Equations **9** (13) (1984), 1265–1298.

[10] A.F. Gill, *Atmosphere–Ocean Dynamics*, Academic Press, New York (1982).

[11] P. Grisvard, *Elliptic Problems in Non-Smooth Domains*, Monographs Stud. in Math., Vol. 24, Pitman, London (1985).

[12] F. Guillén-González, N. Masmoudi and M.A. Rodríguez-Bellido, *Anisotropic estimates and strong solutions of the Primitive Equations*, Differential Integral Equations **14** (11) (2001), 1381–1408.

[13] G. Haltiner and R. Williams, *Numerical Weather Prediction and Dynamic Meteorology*, Second edn, Wiley, New York (1980).

[14] R. Haney, *Surface thermal boundary condition for ocean circulation models*, J. Phys. Oceanogr. **1** (1971), 241–248.

[15] C. Hu, R. Temam and M. Ziane, *Regularity results for linear elliptic problems related to the primitive equations*, Chinese Ann. Math. Ser. B **23** (2) (2002), 277–292.

[16] C. Hu, R. Temam and M. Ziane, *The primitive equations of the large scale ocean under the small depth hypothesis*, Discrete Cont. Dyn. Syst. **9** (1) (2003), 97–131.

[17] V.A. Kozlov, V.G. Mazya and J. Rossmann, *Elliptic Boundary Value Problems in Domains with Point Singularities*, Math. Surveys Monogr., Vol. 52, Amer. Math. Soc., Providence, RI (1997).

[18] R. Laprise, *The Euler equation of motion with hydrostatic pressure as independent coordinate*, Mon. Wea. Rev. **120** (1) (1992), 197–207.

[19] J.L. Lions, *Quelques méthodes de résolution des problèmes aux limites non linéaires*, Dunod, Paris (1969); reedited (2002).

[20] J.L. Lions and E. Magenes, *Nonhomogeneous Boundary Value Problems and Applications*, Vol. 1, Springer-Verlag, New York (1972).

[21] J.L. Lions, R. Temam and S. Wang, *New formulations of the primitive equations of the atmosphere and applications*, Nonlinearity **5** (1992), 237–288.

[22] J.L. Lions, R. Temam and S. Wang, *On the equations of the large-scale ocean*, Nonlinearity **5** (1992), 1007–1053.

[23] J.L. Lions, R. Temam and S. Wang, *Models of the coupled atmosphere and ocean (CAO I)*, Comput. Mech. Adv. **1** (1993), 5–54. *Numerical analysis of the coupled models of atmosphere and ocean (CAO II)*, Comput. Mech. Adv. **1** (1993), 55–119.
[24] J.L. Lions, R. Temam and S. Wang, *Mathematical study of the coupled models of atmosphere and ocean (CAO III)*, J. Math. Pures Appl. **74** (1995), 105–163.
[25] V.G. Mazya and J. Rossmann, *On the behaviour of solutions to the Dirichlet problem for second order elliptic equations near edges and polyhedral vertices with critical angles*, Z. Anal. Anwendungen **13** (1) (1994), 19–47.
[26] I. Moise, E. Simonnet, R. Temam and M. Ziane, *Numerical simulation of differential systems displaying rapidly oscillating solutions*, J. Engrg. Math. **34** (1998), 201–214.
[27] I. Moise, R. Temam and M. Ziane, *Asymptotic analysis of the Navier–Stokes equations in thin domains*, Topol. Methods Nonlinear Anal. (volume dedicated to O.A. Ladyzhenskaya) **10** (2) (1997), 249–282.
[28] J. Pedlosky, *Geophysical Fluid Dynamics*, Second edn, Springer-Verlag, New York (1987).
[29] M. Petcu, R. Temam and D. Wirosoetisno, *Existence and regularity results for the primitive equations in two space dimensions*, Comm. Pure Appl. Anal. **3** (1) (2004), 115–131.
[30] M. Petcu, R. Temam and D. Wirosoetisno, *Renormalization group method applied to the primitive equation*, J. Differential Equations, to appear.
[31] P. Orenga, *Un théorème d'existence de solutions d'un problème de shallow water* (in French) [A theorem on the existence of solutions of a shallow-water problem], Arch. Ration. Mech. Anal. **130** (2) (1995), 183–204.
[32] H. Schlichting, *Boundary Layer Theory*, Seventh edn, McGraw-Hill, New York (1979).
[33] J. Smagorinsky, *General circulation experiments with the primitive equations, I. The basic experiment*, Mon. Wea. Rev. **91** (1963), 98–164.
[34] P.K. Smolarkiewicz, L.G. Margolin and A.A. Wyszogrodzki, *A class of nonhydrostatic global models*, J. Atmosph. Sci. **58** (4) (2001), 349–364.
[35] R. Temam, *Infinite-Dimensional Dynamical Systems in Mechanics and Physics*, Second edn, Springer-Verlag, New York (1997).
[36] R. Temam, *Navier–Stokes Equations*, Stud. Math. Appl., North-Holland, Amsterdam (1977); reedited, AMS-Chelsea Ser., Amer. Math. Soc., Providence, RI (2001).
[37] R. Temam and M. Ziane, *Navier–Stokes equations in three-dimensional thin domains with various boundary conditions*, Adv. Differential Equations **1** (4) (1996), 499–546.
[38] R. Temam and M. Ziane, *Navier–Stokes equations in thin spherical domains*, Contemp. Math. **209** (1997), 281–314.
[39] K.E. Trenberth, *Climate System Modeling*, Cambridge Univ. Press, Cambridge (1992).
[40] S. Wang, *On the 2-D model of large-scale atmospheric motion: Well-posedness and attractors*, Nonlinear Anal. **18** (1) (1992), 17–60.
[41] S. Wang, *Attractors for the 3-D baroclinic quasi-geostrophic equations of large-scale atmosphere*, J. Math. Anal. Appl. **165** (1) (1992), 266–283.
[42] W.M. Washington and C.L. Parkinson, *An Introduction to Three Dimensional Climate Modeling*, Oxford Univ. Press, Oxford (1986).
[43] Q.C. Zeng, *Mathematical and Physical Foundations of Numerical Weather Prediction*, Science Press, Beijing (1979) (in Chinese).
[44] M. Ziane, *Regularity results for Stokes type systems*, Appl. Anal. **58** (1995), 263–292.
[45] M. Ziane, *Regularity results for stationary primitive equations of atmosphere and the ocean*, Nonlinear Anal. **28** (1997), 289–313.
[46] M. Ziane, unpublished manuscript (2000).

Author Index

Roman numbers refer to pages on which the author (or his/her work) is mentioned. Italic numbers refer to reference pages. Numbers between brackets are the reference numbers. No distinction is made between first and co-author(s).

Subject Index